NINTH EDITION

Meteorology Today

AN INTRODUCTION TO WEATHER, CLIMATE, AND THE ENVIRONMENT

C. Donald Ahrens
Emeritus, Modesto Junior College

BROOKS/COLE
CENGAGE Learning™

Australia • Brazil • Japan • Korea • Mexico • Singapore • Spain • United Kingdom • United States

BROOKS/COLE
CENGAGE Learning™

Meteorology Today: An Introduction to Weather, Climate, and the Environment

Ninth Edition

C. Donald Ahrens

Development Editor: Jake Warde

Assistant Editor: Liana Monari

Editorial Assistant: Paige Leeds

Technology Project Manager: Alexandria Brady

Marketing Manager: Joe Rogove

Marketing Assistant: Elizabeth Eong

Marketing Communications Manager:
  Belinda Krohmer

Project Manager, Editorial Production:
  Hal Humphrey

Art Director: Vernon Boes

Print Buyer: Rebecca Cross

Permissions Editor:
  Margaret Chamberlain-Gaston

Production Service: Janet Bollow Associates

Text Designer: Janet Bollow

Art Editor: Janet Bollow

Copy Editor: Stuart Kenter

Illustrator: Charles Preppernau

Cover Designer: William Stanton

Cover Image: Copyright R. Hoelzl/
Peter Arnold, Inc.

Compositor: Graphic World, Inc.

For product information and technology assistance, contact us at
Cengage Learning Customer & Sales Support, 1-800-354-9706

For permission to use material from this text or product,
submit all requests online at **cengage.com/permissions**
Further permissions questions can be emailed to
permissionrequest@cengage.com

Library of Congress Control Number: 2008928602

ISBN-13: 978-0-495-55573-5

ISBN-10: 0-495-55573-8

Brooks/Cole
10 Davis Drive
Belmont, CA 94002
USA

Cengage Learning is a leading provider of customized learning solutions with office locations around the globe, including Singapore, the United Kingdom, Australia, Mexico, Brazil, and Japan. Locate your local office at:
international.cengage.com/region

Cengage Learning products are represented in Canada by Nelson Education, Ltd.

For your course and learning solutions, visit **academic.cengage.com**

Purchase any of our products at your local college store or at our preferred online store **www.ichapters.com**

Printed in China by China Translation & Printing Services Limited
2 3 4 5 6 7 11 10 09

BRIEF CONTENTS

CONTENTS

© Frans Lanting/Minden Pictures

© C. Donald Ahrens

© J. L. Medeiros

© C. Donald Ahrens

CHAPTER 8

Air Pressure and Winds   192

CHAPTER 9

Wind: Small Scale and Local Systems   222

CHAPTER 10

Wind: Global Systems 258

CHAPTER 11

Air Masses and Fronts 286

© C. Donald Ahrens

© C. Donald Ahrens

NASA

© C. Donald Ahrens

© J. L. Medeiros

CHAPTER 19

Light, Color, and Atmospheric Optics   528

The world is an ever-changing picture of naturally occurring events. From drought and famine to devastating floods, some of the greatest challenges we face come in the form of natural disasters created by weather. Yet, dealing with weather and climate is an inevitable part of our lives. Sometimes it is as small as deciding what to wear for the day or how to plan a vacation. But it can also have life-shattering consequences, especially for those who are victims to a hurricane or a tornado.

In recent years, weather and climate have become front page news from the record-setting hurricane year of 2005, when some of the strongest hurricanes ever observed moved over the Gulf of Mexico, to environmental issues such as global warming and ozone depletion. The dynamic nature of the atmosphere seems to demand our attention and understanding more these days than ever before. Almost daily, there are newspaper articles describing some weather event or impending climate change. For this reason, and the fact that weather influences our daily lives in so many ways, interest in meteorology (the study of the atmosphere) has been growing. This rapidly developing and popular science is giving us more information about the workings of the atmosphere than ever before. Although the atmosphere will always provide challenges for us, as research and technology advance, our ability to understand our atmosphere improves, as well. The information available to you in this book, therefore, is intended to aid in your own personal understanding and appreciation of our earth's dynamic atmosphere.

About This Book

Meteorology Today is written for college-level students taking an introductory course on the atmospheric environment. The main purpose of the text is to convey meteorological concepts in a visual and practical manner, while simultaneously providing students with a comprehensive background in basic meteorology. This ninth edition includes up-to-date information on important topics, such as global warming, ozone depletion, and El Niño. Also included are discussions of weather events, such as the devastating fires associated with strong Santa Ana winds that roared through areas of Southern California during October, 2007. As was the case in previous editions, no special prerequisites are necessary.

Written expressly for the student, this book emphasizes the understanding and application of meteorological principles. The text encourages watching the weather so that it becomes "alive," allowing readers to immediately apply textbook material to the world around them. To assist with this endeavor, a color Cloud Chart appears at the end of this text. The Cloud Chart can be separated from the book and used as a learning tool any place where one chooses to observe the sky. To strengthen points and clarify concepts, illustrations are rendered in full color throughout. Color photographs were carefully selected to illustrate features, stimulate interest, and show how exciting the study of weather can be.

This edition, organized into nineteen chapters, is designed to provide maximum flexibility to instructors of atmospheric science courses. Thus, chapters can be covered in any desired order. For example, the chapter on atmospheric optics, Chapter 19, is self-contained and can be covered before or after any chapter. Instructors, then, are able to tailor this text to their particular needs. This book basically follows a traditional approach. After an introductory chapter on the composition, origin, and structure of the atmosphere, it then covers energy, temperature, moisture, precipitation, and winds. Then come chapters that deal with air masses and middle-latitude cyclones. Weather prediction and severe storms are next. A chapter on hurricanes is followed by a chapter on climate change. A chapter on global climate is next. A chapter on air pollution precedes the final chapter on atmospheric optics.

Each chapter contains at least two Focus sections, which expand on material in the main text or explore a subject closely related to what is being discussed. Focus sections fall into one of four distinct categories: Observations, Special Topics, Environmental Issues, and Advanced Topics. Some include material that is not always found in introductory meteorology textbooks, subjects such as temperature extremes, cloud seeding, and the weather on other planets. Others help to bridge theory and practice. Focus sections new to this edition include "The Wavy Warm Front" in Chapter 11, and "Hurricanes in a Warmer World" in Chapter 15, and "When a Dry Spell Is Not a Drought, and a Drought Does Not Mean Dry" in Chapter 17. Quantitative discussions of important equations, such as the geostrophic wind equation and the hydrostatic equation, are found in Focus sections on advanced topics.

Set apart as "Weather Watch" features in each chapter is weather information that may not be commonly known, yet pertains to the topic under discussion. Designed to bring the reader into the text, most of these weather highlights relate to some interesting weather fact or astonishing event.

Each chapter incorporates other effective learning aids:

- A major topic outline begins each chapter.

- Interesting introductory pieces draw the reader naturally into the main text.

- Important terms are boldfaced, with their definitions appearing in the glossary or in the text.

- Key phrases are italicized.

- English equivalents of metric units in most cases are immediately provided in parentheses.

- A brief review of the main points is placed toward the middle of most chapters.

- Summaries at the end of each chapter review the chapter's main ideas.

- A list of key terms following each chapter allows students to review and reinforce their knowledge of the chief concepts they have encountered. Each key term is followed by the number of the page on which the term appears in the text.

- Questions for Review act to check how well students assimilate the material.

- Questions for Thought require students to synthesize learned concepts for deeper understanding.

- Problems and Exercises require mathematical calculations that provide a technical challenge to the student.

Questions for exploration, flashcards, and more can be found on the companion website. Animations, including ten new animations specifically designed for the ninth edition of *Meteorology Today* can be found in the Meteorology Resource Center, an online learning companion.

Ten appendices conclude the book. For easy access, the map of annual global precipitation is now Appendix G. In addition, at the end of the book, a compilation of supplementary material is presented, as is an extensive glossary.

On the endsheet at the back of the book is a new freature: A geophysical map of North America. The map serves as a quick reference for locating states, provinces, and geographical features, such as mountain ranges and large bodies of water.

Ninth Edition Changes

One exciting change to this ninth edition of *Meteorology Today* is the extensive expanded art program. To help the student visualize how exciting meteorology can be, more than 200 new and revised color illustrations and many new photographs have been added to this edition. Moreover, all satellite images have been rendered in full color. To complement the photographs and new art, the ninth edition of *Meteorology Today* has been extensively updated and revised to reflect the changing nature of the field.

Chapter 1, "The Earth and Its Atmosphere," still serves to present a broad overview of the atmosphere. To help with this endeavor, many new illustrations and photographs have been added.

Chapter 2, "Energy: Warming the Earth and the Atmosphere," contains the latest information on greenhouse warming along with additional information on heat transfer.

Many of the sections in Chapter 3, "Seasonal and Daily Temperatures," have been strengthened with new illustrations.

Chapter 4, "Atmospheric Humidity," has been reorganized so that the hydrologic cycle now appears at the beginning of the chapter. Many sections in this chapter have been rewritten and restructured so that the text flows more easily. Moreover, additional information on the role of the dew point in predicting the minimum temperature has been added.

Chapter 5, "Condensation: Dew, Fog, and Clouds," contains new material on the *TRMM Satellite* as well as additional information on radiation fog, advection fog, and mixing fog. The first part of Chapter 6, "Stability and Cloud Development," has been reorganized for clarity. In addition, many diagrams have been redrawn. Chapter 7, "Precipitation," contains a new section on measuring precipitation from space, as well as a rewritten section on the formation of hail. The beginning of the chapter on air pressure and winds (Chapter 8) has been rewritten with new art to complement this endeavor.

Chapter 9, "Wind: Small-Scale and Local Systems," contains new material on determining wind direction and speed, along with a new Focus section on observing winds from space, and the latest information on the Santa Ana wind-driven fires during October, 2007. The surface and upper air charts in Chapter 10, "Wind Global Systems," have all been redrawn for additional clarity, and the Ocean-Niño Index (ONI) has been added to the chapter. Also, the section on jet stream has been restructured for clarity.

The chapter on Air Masses and Fronts (Chapter 11) now contains information on the arctic front along with a new Focus section on a wavy warm front. Chapter 12, "Middle-Latitude Cyclones," has many new and redrawn illustrations to help with comprehension of concepts presented in this chapter. Chapter 13, "Weather Forecasting," has been restructured so that the section on forecasting tools now precedes the section on forecasting methods. The chapter now contains a section on TV weather forecasters, as well as new maps and charts throughout.

Chapter 14, "Thunderstorms and Tornadoes," has been extensively revised so that thunderstorms are now divided into three categories: ordinary cell, multicell, and supercell. This chapter also contains many new illustrations and photos. Chapter 15, "Hurricanes," has been reorganized so that both the section on naming hurricanes and the section on the Saffir-Simpson Scale are now closer to the beginning of the chapter. The section on hurricane formation and development has been rewritten to reflect the latest ideas in this area. In addition, this chapter contains a new section on hurricane forecasting, new information on the frequency of hurricanes, and a new Focus section on hurricanes and global warming.

Chapter 16, "The Earth's Changing Climate," has been revised to include the latest information on global warming from the 2007 Report of the Intergovernment Panel on Climate Change (IPCC). Many new diagrams appear in this chapter, as well as a rewritten section on feedback mechanisms, and a new section on curbing global warming. Chapter 17, "Global Climate," contains all new climographs and a new map of Köppen's climatic classification of the world. Moreover, the chapter contains new material on climate controls and a new Focus section on drought and the Palmer Drought Severity Index. The chapter on air pollution (Chapter 18) has been updated and revised with new information on air pollution trends. Chapter 19, "Light, Color, and Atmospheric Optics," contains many new illustrations and photos to graphically convey the excitement of the atmosphere.

Acknowledgments

A special thank you to the many people who have helped make this ninth edition of *Meteorology Today* a reality. My very special thanks to my wife, Lita, for her proofreading, indexing, and invaluable assistance. Thanks also goes to Charles Preppernau for his meticulous rendering of the art, to Jan Null for researching some of the photographs, and to Michelle Richey for her overall assistance.

A huge thank you goes to Janet Alleyn, who designed the book and, once again, transformed the manuscript into a beautiful product. Thank you to Stuart Kenter for his careful editing and many helpful comments. My thanks also go to the many people at Cengage Learning who worked on this project, especially Marcus Boggs, Jake Warde, and Hal Humphrey. Thank you to my friends and colleagues who provided comments, suggestions, and thoughtful input. I am indebted to those individuals who were kind enough to review all or part of this edition, including:

Richard R. Brandt
Salem State University

James Brothen
Inver Hills Community College

Jongnam Choice
Western Illinois University

Andrew Grundstein
University of Georgia

Peter S. Ray
Florida State University

Alan Robock
Rutgers University

Andy White
University of Oklahoma

To the Student

Learning about the atmosphere can be an enjoyable experience, especially if you become involved. This book is intended to give you some insight into the workings of the atmosphere, but for a real appreciation of your atmospheric environment, you must go outside and observe. Mountains take millions of years to form, while a cumulus cloud can develop into a raging thunderstorm in less than an hour. To help with your observations, a color Cloud Chart is at the back of the book for easy reference. Remove it and keep it with you. And remember, all of the information in this book is out there—please, take the time to look.

Donald Ahrens

Meteorology Today

Although these clouds may resemble scoops of ice cream heaped on a plate, in reality they are towering clouds that form as warm air rises, cools, and condenses into clouds over the central plains of North America.

© C. Donald Ahrens

CHAPTER 1

The Earth and Its Atmosphere

I well remember a brilliant red balloon which kept me completely happy for a whole afternoon, until, while I was playing, a clumsy movement allowed it to escape. Spellbound, I gazed after it as it drifted silently away, gently swaying, growing smaller and smaller until it was only a red point in a blue sky. At that moment I realized, for the first time, the vastness above us: a huge space without visible limits. It was an apparent void, full of secrets, exerting an inexplicable power over all the earth's inhabitants. I believe that many people, consciously or unconsciously, have been filled with awe by the immensity of the atmosphere. All our knowledge about the air, gathered over hundreds of years, has not diminished this feeling.

Theo Loebsack, *Our Atmosphere*

 CONTENTS

Our *atmosphere* is a delicate life-giving blanket of air that surrounds the fragile earth. In one way or another, it influences everything we see and hear—it is intimately connected to our lives. Air is with us from birth, and we cannot detach ourselves from its presence. In the open air, we can travel for many thousands of kilometers in any horizontal direction, but should we move a mere eight kilometers above the surface, we would suffocate. We may be able to survive without food for a few weeks, or without water for a few days, but, without our atmosphere, we would not survive more than a few minutes. Just as fish are confined to an environment of water, so we are confined to an ocean of air. Anywhere we go, it must go with us.

The earth without an atmosphere would have no lakes or oceans. There would be no sounds, no clouds, no red sunsets. The beautiful pageantry of the sky would be absent. It would be unimaginably cold at night and unbearably hot during the day. All things on the earth would be at the mercy of an intense sun beating down upon a planet utterly parched.

Living on the surface of the earth, we have adapted so completely to our environment of air that we sometimes forget how truly remarkable this substance is. Even though air is tasteless, odorless, and (most of the time) invisible, it protects us from the scorching rays of the sun and provides us with a mixture of gases that allows life to flourish. Because we cannot see, smell, or taste air, it may seem surprising that between your eyes and the pages of this book are trillions of air molecules. Some of these may have been in a cloud only yesterday, or over another continent last week, or perhaps part of the life-giving breath of a person who lived hundreds of years ago.

In this chapter, we will examine a number of important concepts and ideas about the earth's atmosphere, many of which will be expanded in subsequent chapters.

Overview of the Earth's Atmosphere

The universe contains billions of galaxies and each galaxy is made up of billions of stars. Stars are hot, glowing balls of gas that generate energy by converting hydrogen into helium near their centers. Our sun is an average size star situated near the edge of the Milky Way galaxy. Revolving around the sun are the earth and seven other planets (see ● Fig. 1.1).* These plan-

*Pluto was once classified as a true planet. But recently it has been reclassified as a planetary object called a *dwarf planet.*

ets, along with a host of other material (comets, asteroids, meteors, dwarf planets, etc.), comprise our solar system.

Warmth for the planets is provided primarily by the sun's energy. At an average distance from the sun of nearly 150 million kilometers (km) or 93 million miles (mi), the earth intercepts only a very small fraction of the sun's total energy output. However, it is this *radiant energy* (or *radiation*)* that drives the atmosphere into the patterns of everyday wind and weather and allows the earth to maintain an average surface temperature of about 15°C (59°F).† Although this temperature is mild, the earth experiences a wide range of temperatures, as readings can drop below −85°C (−121°F) during a frigid Antarctic night and climb, during the day, to above 50°C (122°F) on the oppressively hot subtropical desert.

The earth's **atmosphere** is a thin, gaseous envelope comprised mostly of nitrogen and oxygen, with small amounts of other gases, such as water vapor and carbon dioxide. Nestled in the atmosphere are clouds of liquid water and ice crystals. Although our atmosphere extends upward for many hundreds of kilometers, almost 99 percent of the atmosphere lies within a mere 30 km (19 mi) of the earth's surface (see ● Fig. 1.2). In fact, if the earth were to shrink to the size of a beach ball, its inhabitable atmosphere would be thinner than a piece of paper. This thin blanket of air constantly shields the surface and its inhabitants from the sun's dangerous ultraviolet radiant energy, as well as from the onslaught of material from interplanetary space. There is no definite upper limit to the atmosphere; rather, it becomes thinner and thinner, eventually merging with empty space, which surrounds all the planets.

COMPOSITION OF THE ATMOSPHERE ▼ Table 1.1 shows the various gases present in a volume of air near the earth's surface. Notice that **nitrogen** (N_2) occupies about 78 percent and **oxygen** (O_2) about 21 percent of the total volume of dry air. If all the other gases are removed, these percentages for nitrogen and oxygen hold fairly constant up to an elevation of about 80 km (50 mi). (For a closer look at the composition of a breath of air at the earth's surface, read the Focus section on p. 6.)

*Radiation is energy transferred in the form of waves that have electrical and magnetic properties. The light that we see is radiation, as is ultraviolet light. More on this important topic is given in Chapter 2.

†The abbreviation °C is used when measuring temperature in degrees Celsius, and °F is the abbreviation for degrees Fahrenheit. More information about temperature scales is given in Appendix B and in Chapter 2.

● FIGURE 1.1
The relative sizes and positions of the planets in our solar system. Pluto is included as an object called a dwarf planet. (Positions are not to scale.)

WEATHER WATCH

When it rains, it rains pennies from heaven — sometimes. On July 17, 1940, a tornado reportedly picked up a treasure of over 1000 sixteenth-century silver coins, carried them into a thunderstorm, then dropped them on the village of Merchery in the Gorki region of Russia.

At the surface, there is a balance between destruction (output) and production (input) of these gases. For example, nitrogen is removed from the atmosphere primarily by biological processes that involve soil bacteria. In addition, nitrogen is taken from the air by tiny ocean-dwelling plankton that convert it into nutrients that help fortify the ocean's food chain. It is returned to the atmosphere mainly through the decaying of plant and animal matter. Oxygen, on the other hand, is removed from the atmosphere when organic matter decays and when oxygen combines with other substances, producing oxides. It is also taken from the atmosphere during breathing, as the lungs take in oxygen and release carbon dioxide (CO_2). The addition of oxygen to the atmosphere occurs during photosynthesis, as plants, in the presence of sunlight, combine carbon dioxide and water to produce sugar and oxygen.

The concentration of the invisible gas **water vapor** (H_2O), however, varies greatly from place to place, and from time to time. Close to the surface in warm, steamy, tropical locations, water vapor may account for up to 4 percent of the atmospheric gases, whereas in colder arctic areas, its concentration may dwindle to a mere fraction of a percent (see Table 1.1). Water vapor molecules are, of course, invisible. They become visible only when they transform into larger liquid or solid particles, such as cloud droplets and ice crystals, which may grow in size and eventually fall to the earth as rain or snow. The changing of water vapor into liquid water is called *condensation*, whereas the process of liquid water becoming water vapor is called *evaporation*. The falling rain and snow is

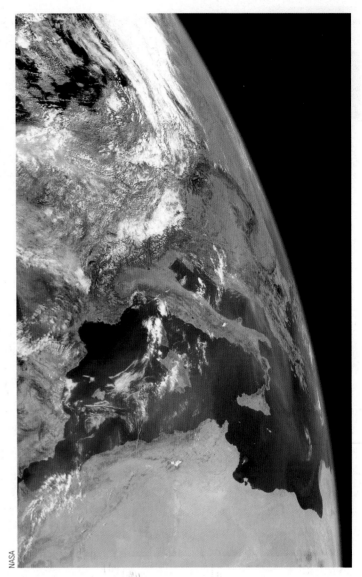

NASA

● FIGURE 1.2 The earth's atmosphere as viewed from space. The atmosphere is the thin blue region along the edge of the earth.

▼ TABLE 1.1 Composition of the Atmosphere near the Earth's Surface

PERMANENT GASES			VARIABLE GASES			
Gas	Symbol	Percent (by Volume) Dry Air	Gas (and Particles)	Symbol	Percent (by Volume)	Parts per Million (ppm)*
Nitrogen	N_2	78.08	Water vapor	H_2O	0 to 4	
Oxygen	O_2	20.95	Carbon dioxide	CO_2	0.038	385*
Argon	Ar	0.93	Methane	CH_4	0.00017	1.7
Neon	Ne	0.0018	Nitrous oxide	N_2O	0.00003	0.3
Helium	He	0.0005	Ozone	O_3	0.000004	0.04†
Hydrogen	H_2	0.00006	Particles (dust, soot, etc.)		0.000001	0.01−0.15
Xenon	Xe	0.000009	Chlorofluorocarbons (CFCs)		0.00000002	0.0002

*For CO_2, 385 parts per million means that out of every million air molecules, 385 are CO_2 molecules.

†Stratospheric values at altitudes between 11 km and 50 km are about 5 to 12 ppm.

FOCUS ON A SPECIAL TOPIC

A Breath of Fresh Air

If we could examine a breath of air, we would see that air (like everything else in the universe) is composed of incredibly tiny particles called *atoms.* We cannot see atoms individually. Yet, if we could see one, we would find electrons whirling at fantastic speeds about an extremely dense center, somewhat like hummingbirds darting and circling about a flower. At this center, or nucleus, are the protons and neutrons. Almost all of the atom's mass is concentrated here, in a trillionth of the atom's entire volume. In the nucleus, the proton carries a positive charge, whereas the neutron is electrically neutral. The circling electron carries a negative charge. As long as the total number of protons in the nucleus equals the

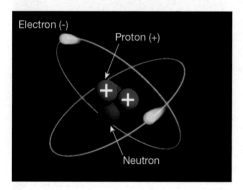

● FIGURE 1
An atom has neutrons and protons at its center with electrons orbiting this center (or nucleus). Molecules are combinations of two or more atoms. The air we breathe is mainly molecular nitrogen (N_2) and molecular oxygen (O_2).

number of orbiting electrons, the atom as a whole is electrically neutral (see Fig. 1).

Most of the air particles are *molecules,* combinations of two or more atoms (such as nitrogen, N_2, and oxygen, O_2), and most of the molecules are electrically neutral. A few, however, are electrically charged, having lost or gained electrons. These charged atoms and molecules are called *ions.*

An average breath of fresh air contains a tremendous number of molecules. With every deep breath, trillions of molecules from the atmosphere enter your body. Some of these inhaled gases become a part of you, and others are exhaled.

The volume of an average size breath of air is about a liter.* Near sea level, there are roughly ten thousand million million million (10^{22})† air molecules in a liter. So,

1 breath of air $\approx 10^{22}$ molecules.

We can appreciate how large this number is when we compare it to the number of stars in the universe. Astronomers have estimated that there are about 100 billion (10^{11}) stars in an average size galaxy and that there may be as many as 10^{11} galaxies in the universe. To deter-

_____
*One cubic centimeter is about the size of a sugar cube, and there are a thousand cubic centimeters in a liter.

†The notation 10^{22} means the number one followed by twenty-two zeros. For a further explanation of this system of notation see Appendix A.

mine the total number of stars in the universe, we multiply the number of stars in a galaxy by the total number of galaxies and obtain

$$10^{11} \times 10^{11} = 10^{22} \text{ stars in the universe.}$$

Therefore, each breath of air contains about as many molecules as there are stars in the known universe.

In the entire atmosphere, there are nearly 10^{44} molecules. The number 10^{44} is 10^{22} squared; consequently

$$10^{22} \times 10^{22} = 10^{44} \text{ molecules}$$
$$\text{in the atmosphere.}$$

We thus conclude that there are about 10^{22} breaths of air in the entire atmosphere. In other words, there are as many molecules in a single breath as there are breaths in the atmosphere.

Each time we breathe, the molecules we exhale enter the turbulent atmosphere. If we wait a long time, those molecules will eventually become thoroughly mixed with all of the other air molecules. If none of the molecules were consumed in other processes, eventually there would be a molecule from that single breath in every breath that is out there. So, considering the many breaths people exhale in their lifetimes, it is possible that in our lungs are molecules that were once in the lungs of people who lived hundreds or even thousands of years ago. In a very real way then, we all share the same atmosphere.

called *precipitation.* In the lower atmosphere, water is everywhere. It is the only substance that exists as a gas, a liquid, and a solid at those temperatures and pressures normally found near the earth's surface (see ● Fig. 1.3).

Water vapor is an *extremely* important gas in our atmosphere. Not only does it form into both liquid and solid cloud particles that grow in size and fall to earth as precipitation, but it also releases large amounts of heat—called *latent heat*—when it changes from vapor into liquid water or ice. Latent heat is an important source of atmospheric energy, especially for storms, such as thunderstorms and hurricanes. Moreover, water vapor is a potent *greenhouse gas* because it strongly absorbs a portion of the earth's outgoing radiant energy (somewhat like the glass of a greenhouse prevents the

heat inside from escaping and mixing with the outside air). Thus, water vapor plays a significant role in the earth's heat-energy balance.

Carbon dioxide (CO_2), a natural component of the atmosphere, occupies a small (but important) percent of a volume of air, about 0.038 percent. Carbon dioxide enters the atmosphere mainly from the decay of vegetation, but it also comes from volcanic eruptions, the exhalations of animal life, from the burning of fossil fuels (such as coal, oil, and natural gas), and from deforestation. The removal of CO_2 from the atmosphere takes place during *photosynthesis,* as plants consume CO_2 to produce green matter. The CO_2 is then stored in roots, branches, and leaves. The oceans act as a huge reservoir for CO_2, as phytoplankton (tiny drifting plants) in surface

● FIGURE 1.3 The earth's atmosphere is a rich mixture of many gases, with clouds of condensed water vapor and ice crystals. Here, water evaporates from the ocean's surface. Rising air currents then transform the invisible water vapor into many billions of tiny liquid droplets that appear as puffy cumulus clouds. If the rising air in the cloud should extend to greater heights, where air temperatures are quite low, some of the liquid droplets would freeze into minute ice crystals.

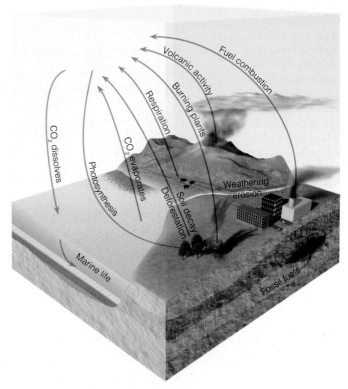

● FIGURE 1.4 The main components of the atmospheric carbon dioxide cycle. The gray lines show processes that put carbon dioxide into the atmosphere, whereas the red lines show processes that remove carbon dioxide from the atmosphere.

water fix CO_2 into organic tissues. Carbon dioxide that dissolves directly into surface water mixes downward and circulates through greater depths. Estimates are that the oceans hold more than 50 times the total atmospheric CO_2 content. ● Figure 1.4 illustrates important ways carbon dioxide enters and leaves the atmosphere.

● Figure 1.5 reveals that the atmospheric concentration of CO_2 has risen more than 20 percent since 1958, when it was first measured at Mauna Loa Observatory in Hawaii. This increase means that CO_2 is entering the atmosphere at a greater rate than it is being removed. The increase appears to be due mainly to the burning of fossil fuels; however, deforestation also plays a role as cut timber, burned or left to rot, releases CO_2 directly into the air, perhaps accounting for about 20 percent of the observed increase. Measurements of CO_2 also come from ice cores. In Greenland and Antarctica, for example, tiny bubbles of air trapped within the ice sheets reveal that before the industrial revolution, CO_2 levels were stable at about 280 parts per million (ppm). (See ● Fig. 1.6.) Since the early 1800s, however, CO_2 levels have increased more than 37 percent. With CO_2 levels presently increasing by about 0.4 percent annually (1.9 ppm/year), scientists now estimate that the concentration of CO_2 will likely rise from its current value of about 385 ppm to a value near 500 ppm toward the end of this century.

Carbon dioxide is another important greenhouse gas because, like water vapor, it traps a portion of the earth's outgoing energy. Consequently, with everything else being equal, as the atmospheric concentration of CO_2 increases, so should the average global surface air temperature. In fact, over the last 100 years or so, the earth's average surface temperature has warmed by more than 0.8°C. Mathematical climate models that predict future atmospheric conditions estimate that if increasing levels of CO_2 (and other greenhouse gases) continue at their present rates, the earth's surface could warm by an additional 3°C (5.4°F) by the end of this century. As we shall see in Chapter 16, the negative consequences of *global warming,* such as rising sea levels and the rapid melting of polar ice, will be felt worldwide.

Carbon dioxide and water vapor are not the only greenhouse gases. Recently, others have been gaining notoriety, primarily because they, too, are becoming more concentrated. Such gases include *methane* (CH_4), *nitrous oxide* (N_2O), and *chlorofluorocarbons* (CFCs).*

Levels of methane, for example, have been rising over the past century, increasing recently by about one-half of one percent per year. Most methane appears to derive from the

---

*Because these gases (including CO_2) occupy only a small fraction of a percent in a volume of air near the surface, they are referred to collectively as *trace gases.*

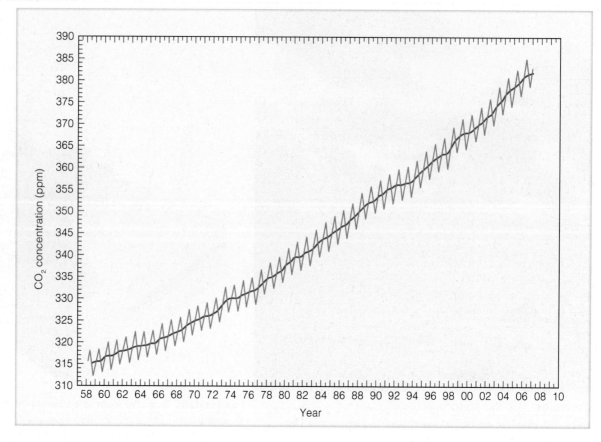

● FIGURE 1.5
Measurements of CO_2 in parts per million (ppm) at Mauna Loa Observatory, Hawaii. Higher readings occur in winter when plants die and release CO_2 to the atmosphere. Lower readings occur in summer when more abundant vegetation absorbs CO_2 from the atmosphere. The solid line is the average yearly value. Notice that the concentration of CO_2 has increased by more than 20 percent since 1958.

breakdown of plant material by certain bacteria in rice paddies, wet oxygen-poor soil, the biological activity of termites, and biochemical reactions in the stomachs of cows. Just why methane should be increasing so rapidly is currently under study. Levels of nitrous oxide—commonly known as laugh-

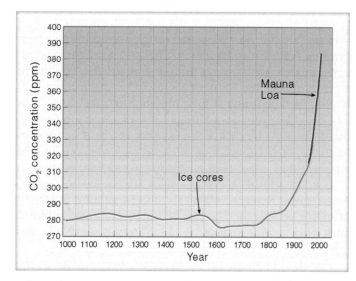

● FIGURE 1.6 Carbon dioxide values in parts per million during the past 1000 years from ice cores in Antarctica (blue line) and from Mauna Loa Observatory in Hawaii (red line). (Data courtesy of Carbon Dioxide Information Analysis Center, Oak Ridge National Laboratory.)

ing gas—have been rising annually at the rate of about one-quarter of a percent. Nitrous oxide forms in the soil through a chemical process involving bacteria and certain microbes. Ultraviolet light from the sun destroys it.

Chlorofluorocarbons (CFCs) represent a group of greenhouse gases that, up until recently, had been increasing in concentration. At one time, they were the most widely used propellants in spray cans. Today, however, they are mainly used as refrigerants, as propellants for the blowing of plastic-foam insulation, and as solvents for cleaning electronic microcircuits. Although their average concentration in a volume of air is quite small (see Table 1.1, p. 5), they have an important effect on our atmosphere as they not only have the potential for raising global temperatures, they also play a part in destroying the gas ozone in the stratosphere, a region in the atmosphere located between about 11 km and 50 km above the earth's surface.

At the surface, **ozone** (O_3) is the primary ingredient of *photochemical smog,** which irritates the eyes and throat and damages vegetation. But the majority of atmospheric ozone (about 97 percent) is found in the upper atmosphere—in the stratosphere—where it is formed naturally, as oxygen atoms combine with oxygen molecules. Here, the concentration of ozone averages less than 0.002 percent by volume. This small

*Originally the word *smog* meant the combining of smoke and fog. Today, however, the word usually refers to the type of smog that forms in large cities, such as Los Angeles, California. Because this type of smog forms when chemical reactions take place in the presence of sunlight, it is termed *photochemical smog.*

The Earth and Its Atmosphere 9

quantity is important, however, because it shields plants, animals, and humans from the sun's harmful ultraviolet rays. It is ironic that ozone, which damages plant life in a polluted environment, provides a natural protective shield in the upper atmosphere so that plants on the surface may survive.

When CFCs enter the stratosphere, ultraviolet rays break them apart, and the CFCs release ozone-destroying chlorine. Because of this effect, ozone concentration in the stratosphere has been decreasing over parts of the Northern and Southern Hemispheres. The reduction in stratospheric ozone levels over springtime Antarctica has plummeted at such an alarming rate that during September and October, there is an **ozone hole** over the region. ● Figure 1.7 illustrates the extent of the ozone hole above Antarctica during September, 2004.

Impurities from both natural and human sources are also present in the atmosphere: Wind picks up dust and soil from the earth's surface and carries it aloft; small saltwater drops from ocean waves are swept into the air (upon evaporating, these drops leave microscopic salt particles suspended in the atmosphere); smoke from forest fires is often carried high above the earth; and volcanoes spew many tons of fine ash particles and gases into the air (see ● Fig. 1.8). Collectively, these tiny solid or liquid suspended particles of various composition are called **aerosols.**

Some natural impurities found in the atmosphere are quite beneficial. Small, floating particles, for instance, act as surfaces on which water vapor condenses to form clouds. However, most human-made impurities (and some natural ones) are a nuisance, as well as a health hazard. These we call **pollutants.** For example, automobile engines emit copious amounts of *nitrogen dioxide* (NO_2), *carbon monoxide* (CO), and *hydrocarbons.* In sunlight, nitrogen dioxide reacts with hydrocarbons and other gases to produce ozone. Carbon monoxide is a major pollutant of city air. Colorless and odor-

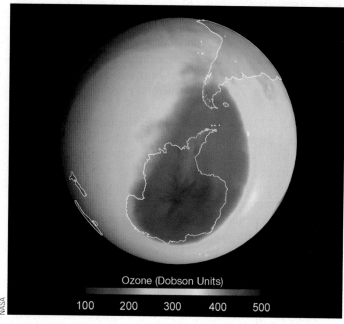

Ozone (Dobson Units)

100   200   300   400   500

NASA

● FIGURE 1.7 The darkest color represents the area of lowest ozone concentration, or ozone hole, over the Southern Hemisphere on September 22, 2004. Notice that the hole is larger than the continent of Antarctica. A Dobson unit (DU) is the physical thickness of the ozone layer if it were brought to the earth's surface, where 500 DU equals 5 millimeters.

less, this poisonous gas forms during the incomplete combustion of carbon-containing fuel. Hence, over 75 percent of carbon monoxide in urban areas comes from road vehicles.

The burning of sulfur-containing fuels (such as coal and oil) releases the colorless gas *sulfur dioxide* (SO_2) into the air. When the atmosphere is sufficiently moist, the SO_2 may transform into tiny dilute drops of sulfuric acid. Rain con-

● FIGURE 1.8 Erupting volcanoes can send tons of particles into the atmosphere, along with vast amounts of water vapor, carbon dioxide, and sulfur dioxide.

© David Weintraub/Photo Researchers

taining sulfuric acid corrodes metals and painted surfaces, and turns freshwater lakes acidic. **Acid rain** is a major environmental problem, especially downwind from major industrial areas. In addition, high concentrations of SO_2 produce serious respiratory problems in humans, such as bronchitis and emphysema, and have an adverse effect on plant life.

THE EARLY ATMOSPHERE   The atmosphere that originally surrounded the earth was probably much different from the air we breathe today. The earth's first atmosphere (some 4.6 billion years ago) was most likely *hydrogen* and *helium*—the two most abundant gases found in the universe—as well as hydrogen compounds, such as methane (CH_4) and ammonia (NH_3). Most scientists feel that this early atmosphere escaped into space from the earth's hot surface.

A second, more dense atmosphere, however, gradually enveloped the earth as gases from molten rock within its hot interior escaped through volcanoes and steam vents. We assume that volcanoes spewed out the same gases then as they do today: mostly water vapor (about 80 percent), carbon dioxide (about 10 percent), and up to a few percent nitrogen. These gases (mostly water vapor and carbon dioxide) probably created the earth's second atmosphere.

As millions of years passed, the constant outpouring of gases from the hot interior—known as **outgassing**—provided a rich supply of water vapor, which formed into clouds.* Rain fell upon the earth for many thousands of years, forming the rivers, lakes, and oceans of the world. During this time, large amounts of CO_2 were dissolved in the oceans. Through chemical and biological processes, much of the CO_2 became locked up in carbonate sedimentary rocks, such as limestone. With much of the water vapor already condensed and the concentration of CO_2 dwindling, the atmosphere gradually became rich in nitrogen (N_2), which is usually not chemically active.

It appears that oxygen (O_2), the second most abundant gas in today's atmosphere, probably began an extremely slow increase in concentration as energetic rays from the sun split water vapor (H_2O) into hydrogen and oxygen during a process called *photodissociation*. The hydrogen, being lighter, probably rose and escaped into space, while the oxygen remained in the atmosphere.

This slow increase in oxygen may have provided enough of this gas for primitive plants to evolve, perhaps 2 to 3 billion years ago. Or the plants may have evolved in an almost oxygen-free (anaerobic) environment. At any rate, plant growth greatly enriched our atmosphere with oxygen. The reason for this enrichment is that, during the process of photosynthesis, plants, in the presence of sunlight, combine carbon dioxide and water to produce oxygen. Hence, after plants evolved, the atmospheric oxygen content increased more rapidly, probably reaching its present composition about several hundred million years ago.

*It is now believed that some of the earth's water may have originated from numerous collisions with small meteors and disintegrating comets when the earth was very young.

BRIEF REVIEW

Before going on to the next several sections, here is a review of some of the important concepts presented so far:

- The earth's atmosphere is a mixture of many gases. In a volume of dry air near the surface, nitrogen (N_2) occupies about 78 percent and oxygen (O_2) about 21 percent.
- Water vapor, which normally occupies less than 4 percent in a volume of air near the surface, can condense into liquid cloud droplets or transform into delicate ice crystals. Water is the only substance in our atmosphere that is found naturally as a gas (water vapor), as a liquid (water), and as a solid (ice).
- Both water vapor and carbon dioxide (CO_2) are important greenhouse gases.
- Ozone (O_3) in the stratosphere protects life from harmful ultraviolet (UV) radiation. At the surface, ozone is the main ingredient of photochemical smog.
- The majority of water on our planet is believed to have come from its hot interior through outgassing.

Vertical Structure of the Atmosphere

A vertical profile of the atmosphere reveals that it can be divided into a series of layers. Each layer may be defined in a number of ways: by the manner in which the air temperature varies through it, by the gases that comprise it, or even by its electrical properties. At any rate, before we examine these various atmospheric layers, we need to look at the vertical profile of two important variables: air pressure and air density.

A BRIEF LOOK AT AIR PRESSURE AND AIR DENSITY   Earlier in this chapter we learned that most of our atmosphere is crowded close to the earth's surface. The reason for this fact is that air molecules (as well as everything else) are held near the earth by *gravity*. This strong invisible force pulling down on the air above squeezes (compresses) air molecules closer together, which causes their number in a given volume to increase. The more air above a level, the greater the squeezing effect or compression.

Gravity also has an effect on the weight of objects, including air. In fact, *weight* is the force acting on an object due to gravity. Weight is defined as the mass of an object times the acceleration of gravity; thus

$$\text{Weight} = \text{mass} \times \text{gravity}.$$

An object's *mass* is the quantity of matter in the object. Consequently, the mass of air in a rigid container is the same everywhere in the universe. However, if you were to instantly travel to the moon, where the acceleration of gravity is much less than that of earth, the mass of air in the container would be the same, but its weight would decrease.

When mass is given in grams (g) or kilograms (kg), volume is given in cubic centimeters (cm³) or cubic meters (m³). Near sea level, air density is about 1.2 kilograms per cubic meter (nearly 1.2 ounces per cubic foot).

The **density** of air (or any substance) is determined by the masses of atoms and molecules and the amount of space between them. In other words, density tells us how much matter is in a given space (that is, volume). We can express density in a variety of ways. The molecular density of air is the number of molecules in a given volume. Most commonly, however, density is given as the mass of air in a given volume; thus

$$\text{Density} = \frac{\text{mass}}{\text{volume}}.$$

Because there are appreciably more molecules within the same size volume of air near the earth's surface than at higher levels, air density is greatest at the surface and decreases as we move up into the atmosphere. Notice in ● Fig. 1.9 that, because air near the surface is compressed, air density normally decreases rapidly at first, then more slowly as we move farther away from the surface.

Air molecules are in constant motion. On a mild spring day near the surface, an air molecule will collide about 10 billion times each second with other air molecules. It will also bump against objects around it—houses, trees, flowers, the ground, and even people. Each time an air molecule bounces against a person, it gives a tiny push. This small force (push) divided by the area on which it pushes is called **pressure**; thus

$$\text{Pressure} = \frac{\text{force}}{\text{area}}.$$

If we weigh a column of air 1 square inch in cross section, extending from the average height of the ocean surface (sea level) to the "top" of the atmosphere, it would weigh nearly 14.7 pounds (see Fig. 1.9). Thus, normal atmospheric pressure near sea level is close to 14.7 pounds per square inch. If more molecules are packed into the column, it becomes more dense, the air weighs more, and the surface pressure goes up. On the other hand, when fewer molecules are in the column, the air weighs less, and the surface pressure goes down. So, the surface air pressure can be changed by changing the mass of air above the surface.

Pounds per square inch is, of course, just one way to express air pressure. Presently, the most common unit found on surface weather maps is the *millibar** (mb) although the *hectopascal* (hPa) is gradually replacing the millibar as the preferred unit of pressure on surface charts. Another unit of

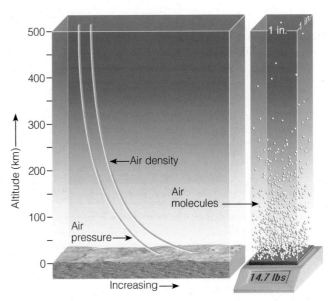

● FIGURE 1.9 Both air pressure and air density decrease with increasing altitude. The weight of all the air molecules above the earth's surface produces an average pressure near 14.7 lbs/in.²

pressure is *inches of mercury* (Hg), which is commonly used in the field of aviation and on television and radio weather broadcasts. At sea level, the *standard value* for atmospheric pressure is

1013.25 mb = 1013.25 hPa = 29.92 in. Hg.

Billions of air molecules push constantly on the human body. This force is exerted equally in all directions. We are not crushed by it because billions of molecules inside the body push outward just as hard. Even though we do not actually feel the constant bombardment of air, we can detect quick changes in it. For example, if we climb rapidly in elevation, our ears may "pop." This experience happens because air collisions outside the eardrum lessen. The popping comes about as air collisions between the inside and outside of the ear equalize. The drop in the number of collisions informs us that the pressure exerted by the air molecules decreases with height above the earth. A similar type of ear-popping occurs as we drop in elevation, and the air collisions outside the eardrum increase.

Air molecules not only take up space (freely darting, twisting, spinning, and colliding with everything around

---

*By definition, a *bar* is a force of 100,000 newtons (N) acting on a surface area of 1 square meter (m²). A *newton* is the amount of force required to move an object with a mass of 1 kilogram (kg) so that it increases its speed at a rate of 1 meter per second (m/sec) each second. Because the bar is a relatively large unit, and because surface pressure changes are usually small, the unit of pressure most commonly found on surface weather maps is the *millibar*, where 1 bar = 1000 mb. The unit of pressure designed by the International System (SI) of measurement is the *pascal* (Pa), where 1 pascal is the force of 1 newton acting on a surface of 1 square meter. A more common unit is the *hectopascal* (hPa), as 1 hectopascal equals 1 millibar.

WEATHER WATCH

The air density in the mile-high city of Denver, Colorado, is normally about 15 percent less than the air density at sea level. As the air density decreases, the drag force on a baseball in flight also decreases. Because of this fact, a baseball hit at Denver's Coors Field will travel farther than one hit at sea level. Hence, on a warm, calm day, a baseball hit for a 340-foot home run down the left field line at Coors Field would simply be a 300-foot out if hit at Camden Yards Stadium in Baltimore, Maryland.

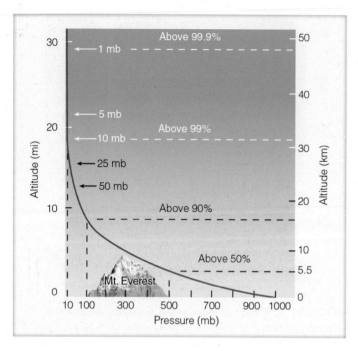

● FIGURE 1.10 Atmospheric pressure decreases rapidly with height. Climbing to an altitude of only 5.5 km, where the pressure is 500 mb, would put you above one-half of the atmosphere's molecules.

them), but—as we have seen—these same molecules have weight. In fact, air is surprisingly heavy. The weight of all the air around the earth is a staggering 5600 trillion tons, or about 5.136×10^{18} kg. The weight of the air molecules acts as a force upon the earth. The amount of force exerted over an area of surface is called *atmospheric pressure* or, simply, **air pressure.*** The pressure at any level in the atmosphere may be measured in terms of the total mass of air above any point. As we climb in elevation, fewer air molecules are above us; hence, *atmospheric pressure always decreases with increasing height.* Like air density, air pressure decreases rapidly at first, then more slowly at higher levels, as illustrated in Fig. 1.9.

● Figure 1.10 also illustrates how rapidly air pressure decreases with height. Near sea level, atmospheric pressure is usually close to 1000 mb. Normally, just above sea level, atmospheric pressure decreases by about 10 mb for every 100 meters (m) increase in elevation—about 1 inch of mercury for every 1000 feet (ft) of rise. At higher levels, air pressure decreases much more slowly with height. With a sea-level pressure near 1000 mb, we can see in Fig. 1.10 that, at an altitude of only 5.5 km (3.5 mi), the air pressure is about 500 mb, or half of the sea-level pressure. This situation means that, if you were at a mere 5.5 km (about 18,000 ft) above the earth's surface, you would be above one-half of all the molecules in the atmosphere.

At an elevation approaching the summit of Mt. Everest (about 9 km, or 29,000 ft—the highest mountain peak on

earth), the air pressure would be about 300 mb. The summit is above nearly 70 percent of all the air molecules in the atmosphere. At an altitude approaching 50 km, the air pressure is about 1 mb, which means that 99.9 percent of all the air molecules are below this level. Yet the atmosphere extends upwards for many hundreds of kilometers, gradually becoming thinner and thinner until it ultimately merges with outer space. (Up to now, we have concentrated on the earth's atmosphere. For a brief look at the atmospheres of the other planets, read the Focus section on pp. 14–15.)

LAYERS OF THE ATMOSPHERE   We have seen that both air pressure and density decrease with height above the earth—rapidly at first, then more slowly. *Air temperature,* however, has a more complicated vertical profile.*

Look closely at ● Fig. 1.11 and notice that air temperature normally decreases from the earth's surface up to an altitude of about 11 km, which is nearly 36,000 ft, or 7 mi. This decrease in air temperature with increasing height is due primarily to the fact (investigated further in Chapter 2) that sunlight warms the earth's surface, and the surface, in turn, warms the air above it. The rate at which the air temperature decreases with height is called the temperature **lapse rate.** The *average* (or *standard*) *lapse rate* in this region of the lower atmosphere is about 6.5°C for every 1000 m or about 3.6°F for every 1000 ft rise in elevation. Keep in mind that these values are only averages. On some days, the air becomes colder more quickly as we move upward. This would increase or steepen the lapse rate. On other days, the air temperature would decrease more slowly with height, and the lapse rate would be less. Occasionally, the air temperature may actually *increase* with height, producing a condition known as a **temperature inversion.** So the lapse rate fluctuates, varying from day to day and season to season.

The region of the atmosphere from the surface up to about 11 km contains all of the weather we are familiar with on earth. Also, this region is kept well stirred by rising and descending air currents. Here, it is common for air molecules to circulate through a depth of more than 10 km in just a few days. This region of circulating air extending upward from the earth's surface to where the air stops becoming colder with height is called the **troposphere**—from the Greek *tropein,* meaning to turn or change.

Notice in Fig. 1.11 that just above 11 km the air temperature normally stops decreasing with height. Here, the lapse rate is zero. This region, where, on average, the air temperature remains constant with height, is referred to as an *isothermal* (equal temperature) zone.† The bottom of this zone marks the top of the troposphere and the beginning of another layer, the **stratosphere.** The boundary separating the

---

*Because air pressure is measured with an instrument called a *barometer,* atmospheric pressure is often referred to as *barometric pressure.*

Air temperature is the degree of hotness or coldness of the air and, as we will see in Chapter 2, it is also a measure of the average speed of the air molecules.

†In many instances, the isothermal layer is not present, and the air temperature begins to increase with increasing height.

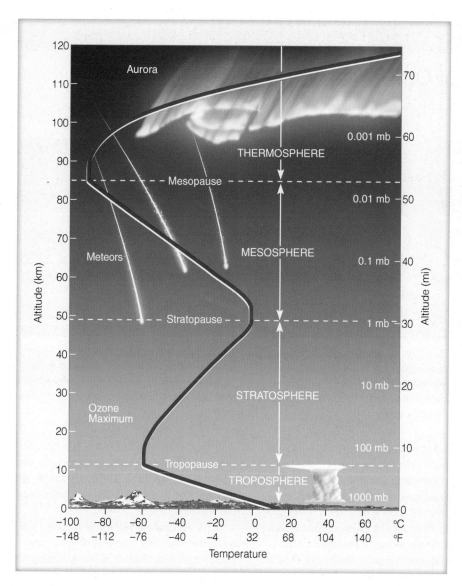

● FIGURE 1.11 Layers of the atmosphere as related to the average profile of air temperature above the earth's surface. The heavy line illustrates how the average temperature varies in each layer.

troposphere from the stratosphere is called the **tropopause.** The height of the tropopause varies. It is normally found at higher elevations over equatorial regions, and it decreases in elevation as we travel poleward. Generally, the tropopause is higher in summer and lower in winter at all latitudes. In some regions, the tropopause "breaks" and is difficult to locate and, here, scientists have observed tropospheric air mixing with stratospheric air and vice versa. These breaks also mark the position of *jet streams*—high winds that meander in a narrow channel, like an old river, often at speeds exceeding 100 knots.*

From Fig. 1.11 we can see that, in the stratosphere, the air temperature begins to increase with height, producing a *temperature inversion.* The inversion region, along with the lower isothermal layer, tends to keep the vertical currents of the troposphere from spreading into the stratosphere. The inversion also tends to reduce the amount of vertical motion in the stratosphere itself; hence, it is a stratified layer.

---

*A knot is a nautical mile per hour. One knot is equal to 1.15 miles per hour (mi/hr), or 1.9 kilometers per hour (km/hr).

Even though the air temperature is increasing with height, the air at an altitude of 30 km is extremely cold, averaging less than −46°C. At this level above polar latitudes, air temperatures can change dramatically from one week to the next, as a *sudden warming* can raise the temperature in one week by more than 50°C. Such a rapid warming, although not well understood, is probably due to sinking air associated with circulation changes that occur in late winter or early spring as well as with the poleward displacement of strong jet stream winds in the lower stratosphere. (The instrument that measures the vertical profile of air temperature in the atmosphere

WEATHER WATCH

If you are flying in a jet aircraft at 30,000 feet above the earth, the air temperature outside your window would typically be about −60°F. Due to the fact that air temperature normally decreases with increasing height, the air temperature outside your window may be more than 110°F colder than the air at the surface directly below you.

FOCUS ON A SPECIAL TOPIC

The Atmospheres of Other Planets

Earth is unique. Not only does it lie at just the right distance from the sun so that life may flourish, it also provides its inhabitants with an atmosphere rich in nitrogen and oxygen — two gases that are not abundant in the atmospheres of either Venus or Mars, our closest planetary neighbors.

The Venusian atmosphere is mainly carbon dioxide (95 percent) with minor amounts of water vapor and nitrogen. An opaque acid-cloud deck encircles the planet, hiding its surface. The atmosphere is quite turbulent, as instruments reveal twisting eddies and fierce winds in excess of 125 mi/hr. This thick dense atmosphere produces a surface air pressure of about 90,000 mb, which is 90 times greater than that on earth. To experience such a pressure on earth, one would have to descend in the ocean to a depth of about 900 m (2950 ft). Moreover, this thick atmosphere of CO_2 produces a strong greenhouse effect, with a scorching hot surface temperature of 480°C (900°F).

The atmosphere of Mars, like that of Venus, is mostly carbon dioxide, with only small amounts of other gases. Unlike Venus, the Martian atmosphere is very thin, and heat escapes from the surface rapidly. Thus, surface temperatures on Mars are much lower, averaging around −60°C (−76°F). Because of its

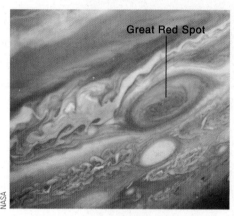

● FIGURE 2 A portion of Jupiter extending from the equator to the southern polar latitudes. The Great Red Spot, as well as the smaller ones, are spinning eddies similar to storms that exist in the earth's atmosphere.

thin cold atmosphere, there is no liquid water on Mars and virtually no cloud cover — only a barren desertlike landscape. In addition, this thin atmosphere produces an average surface air pressure of about 7 mb, which is less than one-hundredth of that experienced at the surface of the earth. Such a pressure on earth would be observed above the surface at an altitude near 35 km (22 mi).

Occasionally, huge dust storms develop near the Martian surface. Such storms may be

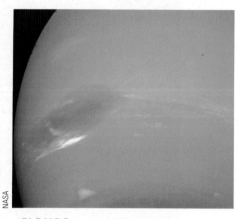

● FIGURE 3 The Great Dark Spot on Neptune. The white wispy clouds are similar to the high wispy cirrus clouds on earth. However, on Neptune, they are probably composed of methane ice crystals.

accompanied by winds of several hundreds of kilometers per hour. These winds carry fine dust around the entire planet. The dust gradually settles out, coating the landscape with a thin reddish veneer.

The atmosphere of the largest planet, Jupiter, is much different from that of Venus and Mars. Jupiter's atmosphere is mainly hydrogen (H_2) and helium (He), with minor amounts of methane (CH_4) and ammonia (NH_3). A prominent feature on Jupiter is the Great Red

---

up to an elevation sometimes exceeding 30 km [100,000 ft] is the **radiosonde.** More information on this instrument is given in the Focus section on p. 16.)

The reason for the inversion in the stratosphere is that the gas ozone plays a major part in heating the air at this altitude. Recall that ozone is important because it absorbs energetic ultraviolet (UV) solar energy. Some of this absorbed energy warms the stratosphere, which explains why there is an inversion. If ozone were not present, the air probably would become colder with height, as it does in the troposphere.

Notice in Fig. 1.11 that the level of maximum ozone concentration is observed near 25 km (at middle latitudes), yet the stratospheric air temperature reaches a maximum near 50 km. The reason for this phenomenon is that the air at 50 km is less dense than at 25 km, and so the absorption of intense solar energy at 50 km raises the temperature of fewer

molecules to a much greater degree. Moreover, much of the solar energy responsible for the heating is absorbed in the upper part of the stratosphere and, therefore, does not reach down to the level of ozone maximum. And due to the low air density, the transfer of energy downward from the upper stratosphere is quite slow.

Above the stratosphere is the **mesosphere** (middle sphere). The boundary near 50 km, which separates these layers, is called the *stratopause.* The air at this level is extremely thin and the atmospheric pressure is quite low, averaging about 1 mb, which means that only one-thousandth of all the atmosphere's molecules are above this level and 99.9 percent of the atmosphere's mass is located below it.

The percentage of nitrogen and oxygen in the mesosphere is about the same as at sea level. Given the air's low density in this region, however, we would not survive very long breathing

Spot—a huge atmospheric storm about three times larger than earth—that spins counter-clockwise in Jupiter's southern hemisphere (see Fig. 2 p. 14). Large white ovals near the Great Red Spot are similar but smaller storm systems. Unlike the earth's weather machine, which is driven by the sun, Jupiter's massive swirling clouds appear to be driven by a collapsing core of hot hydrogen. Energy from this lower region rises toward the surface; then it (along with Jupiter's rapid rotation) stirs the cloud layer into more or less horizontal bands of various colors.

Swirling storms exist on other planets, too, such as on Saturn and Neptune. In fact, the large dark oval on Neptune (Fig. 3) appears to be a storm similar to Jupiter's Great Red Spot. The white wispy clouds in the photograph are probably composed of methane ice crystals. Studying the atmospheric behavior of other planets may give us added insight into the workings of our own atmosphere. (Additional information about size, surface temperature, and atmospheric composition of planets is given in Table 1.)

▼ TABLE 1   Data on Planets and the Sun

	DIAMETER	AVERAGE DISTANCE FROM SUN	AVERAGE SURFACE TEMPERATURE		MAIN ATMOSPHERIC COMPONENTS
	Kilometers	Millions of Kilometers	°C	°F	
Sun	$1,392 \times 10^3$		5,800	10,500	—
Mercury	4,880	58	260*	500	—
Venus	12,112	108	480	900	CO_2
Earth	12,742	150	15	59	N_2, O_2
Mars	6,800	228	−60	−76	CO_2
Jupiter	143,000	778	−110	−166	H_2, He
Saturn	121,000	1,427	−190	−310	H_2, He
Uranus	51,800	2,869	−215	−355	H_2, CH_4
Neptune	49,000	4,498	−225	−373	N_2, CH_4
Pluto	3,100	5,900	−235	−391	CH_4

*Sunlit side.

here, as each breath would contain far fewer oxygen molecules than it would at sea level. Consequently, without proper breathing equipment, the brain would soon become oxygen-starved—a condition known as *hypoxia*. Pilots who fly above 3 km (10,000 ft) for too long without oxygen-breathing apparatus may experience this. With the first symptoms of hypoxia, there is usually no pain involved, just a feeling of exhaustion. Soon, visual impairment sets in and routine tasks become difficult to perform. Some people drift into an incoherent state, neither realizing nor caring what is happening to them. Of course, if this oxygen deficiency persists, a person will lapse into unconsciousness, and death may result. In fact, in the mesosphere, we would suffocate in a matter of minutes.

There are other effects besides suffocating that could be experienced in the mesosphere. Exposure to ultraviolet solar energy, for example, could cause severe burns on exposed parts of the body. Also, given the low air pressure, the blood in one's veins would begin to boil at normal body temperatures.

The air temperature in the mesosphere decreases with height, a phenomenon due, in part, to the fact that there is little ozone in the air to absorb solar radiation. Consequently, the molecules (especially those near the top of the mesosphere) are able to lose more energy than they absorb, which results in an energy deficit and cooling. So we find air in the mesosphere becoming colder with height up to an elevation near 85 km. At this altitude, the temperature of the atmosphere reaches its lowest average value, −90°C (−130°F).

The "hot layer" above the mesosphere is the **thermosphere.** The boundary that separates the lower, colder mesosphere from the warmer thermosphere is the *mesopause.* In the thermosphere, oxygen molecules (O_2) absorb energetic solar rays, warming the air. Because there are relatively few

FOCUS ON AN OBSERVATION

The Radiosonde

The vertical distribution of temperature, pressure, and humidity up to an altitude of about 30 km can be obtained with an instrument called a *radiosonde*.* The radiosonde is a small, lightweight box equipped with weather instruments and a radio transmitter. It is attached to a cord that has a parachute and a gas-filled balloon tied tightly at the end (see Fig. 4). As the balloon rises, the attached radiosonde measures air temperature with a small electrical thermometer — a thermistor — located just outside the box. The radiosonde measures humidity electrically by sending an electric current across a carbon-coated plate. Air pressure is obtained by a small barometer located inside the box. All of this information is transmitted to the surface by radio. Here, a computer rapidly reconverts the various frequencies into values of temperature, pressure, and moisture. Special tracking equipment at the surface may also be used to pro-

vide a vertical profile of winds.* (When winds are added, the observation is called a *rawinsonde*.) When plotted on a graph, the vertical distribution of temperature, humidity, and wind is called a *sounding*. Eventually, the balloon bursts and the radiosonde returns to earth, its descent being slowed by its parachute.

At most sites, radiosondes are released twice a day, usually at the time that corresponds to midnight and noon in Greenwich, England. Releasing radiosondes is an expensive operation because many of the instruments are never retrieved, and many of those that are retrieved are often in poor working condition. To complement the radiosonde, modern satellites (using instruments that measure radiant energy) are providing scientists with vertical temperature profiles in inaccessible regions.

*A radiosonde that is dropped by parachute from an aircraft is called a *dropsonde*.

*A modern development in the radiosonde is the use of satellite Global Positioning System (GPS) equipment. Radiosondes can be equipped with a GPS device that provides more accurate position data back to the computer for wind computations.

© C. Donald Ahrens

● FIGURE 4 The radiosonde with parachute and balloon.

atoms and molecules in the thermosphere, the absorption of a small amount of energetic solar energy can cause a large increase in air temperature. Furthermore, because the amount of solar energy affecting this region depends strongly on solar activity, temperatures in the thermosphere vary from day to day (see ● Fig. 1.12). The low density of the thermosphere also means that an air molecule will move an average distance (called *mean free path*) of over one kilometer before colliding with another molecule. A similar air molecule at the earth's surface will move an average distance of less than one millionth of a centimeter before it collides with another molecule. Moreover, it is in the thermosphere where charged particles from the sun interact with air molecules to produce dazzling aurora displays. (We will look at the aurora in more detail in Chapter 2.)

Because the air density in the upper thermosphere is so low, air temperatures there are not measured directly. They can, however, be determined by observing the orbital change of satellites caused by the drag of the atmosphere. Even though the air is extremely tenuous, enough air molecules strike a satellite to slow it down, making it drop into a slightly

lower orbit. (For this reason, the spacecraft *Solar Max* fell to earth in December, 1989, as did the Russian space station, *Mir*, in March, 2001.) The amount of drag is related to the density of the air, and the density is related to the temperature. Therefore, by determining air density, scientists are able to construct a vertical profile of air temperature.

At the top of the thermosphere, about 500 km (300 mi) above the earth's surface, molecules can move distances of 10 km before they collide with other molecules. Here, many of the lighter, faster-moving molecules traveling in the right direction actually escape the earth's gravitational pull. The region where atoms and molecules shoot off into space is sometimes referred to as the **exosphere,** which represents the upper limit of our atmosphere.

Up to this point, we have examined the atmospheric layers based on the vertical profile of temperature. The atmosphere, however, may also be divided into layers based on its composition. For example, the composition of the atmosphere begins to slowly change in the lower part of the thermosphere. Below the thermosphere, the composition of air remains fairly uniform (78 percent nitrogen, 21 percent oxygen) by turbulent mixing. This lower, well-mixed region is known as the **homosphere** (see Fig. 1.12). In the thermosphere, collisions between atoms and molecules are infrequent, and the air is unable to keep itself stirred. As a result, diffusion takes over as heavier atoms and molecules (such as oxygen and nitrogen) tend to settle to the bottom of the layer, while lighter gases (such as hydrogen and helium) float to the top. The region from about the base of the thermosphere to the top of the atmosphere is often called the **heterosphere.**

THE IONOSPHERE The **ionosphere** is not really a layer, but rather an electrified region within the upper atmosphere where fairly large concentrations of ions and free electrons exist. *Ions* are atoms and molecules that have lost (or gained) one or more electrons. Atoms lose electrons and become positively charged when they cannot absorb all of the energy transferred to them by a colliding energetic particle or the sun's energy.

The lower region of the ionosphere is usually about 60 km above the earth's surface. From here (60 km), the ionosphere extends upward to the top of the atmosphere. Hence, the bulk of the ionosphere is in the thermosphere, as illustrated in Fig. 1.12.

The ionosphere plays a major role in AM radio communications. The lower part (called the *D* region) reflects standard AM radio waves back to earth, but at the same time it seriously weakens them through absorption. At night, though, the *D* region gradually disappears and AM radio waves are able to penetrate higher into the ionosphere (into the *E* and *F* regions—see • Fig. 1.13), where the waves are reflected back to earth. Because there is, at night, little absorption of radio waves

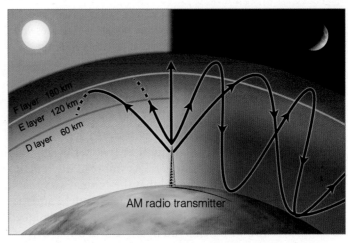

• FIGURE 1.13 At night, the higher region of the ionosphere (*F* region) strongly reflects AM radio waves, allowing them to be sent over great distances. During the day, the lower *D* region strongly absorbs and weakens AM radio waves, preventing them from being picked up by distant receivers.

in the higher reaches of the ionosphere, such waves bounce repeatedly from the ionosphere to the earth's surface and back to the ionosphere again. In this way, standard AM radio waves are able to travel for many hundreds of kilometers at night.

Around sunrise and sunset, AM radio stations usually make "necessary technical adjustments" to compensate for the changing electrical characteristics of the *D* region. Because they can broadcast over a greater distance at night, most AM stations reduce their output near sunset. This reduction prevents two stations—both transmitting at the same frequency but hundreds of kilometers apart—from interfering with each other's radio programs. At sunrise, as the *D* region intensifies, the power supplied to AM radio transmitters is normally increased. FM stations do not need to make these adjustments because FM radio waves are shorter than AM waves, and are able to penetrate through the ionosphere without being reflected.

BRIEF REVIEW

We have, in the last several sections, been examining our atmosphere from a vertical perspective. A few of the main points are:

- Atmospheric pressure at any level represents the total mass of air above that level, and atmospheric pressure always decreases with increasing height above the surface.

- The rate at which the air temperature decreases with height is called the *lapse rate*. A measured increase in air temperature with height is called an *inversion*.

- The atmosphere may be divided into layers (or regions) according to its vertical profile of temperature, its gaseous composition, or its electrical properties.

- The warmest atmospheric layer is the thermosphere; the coldest is the mesosphere. Most of the gas ozone is found in the stratosphere.

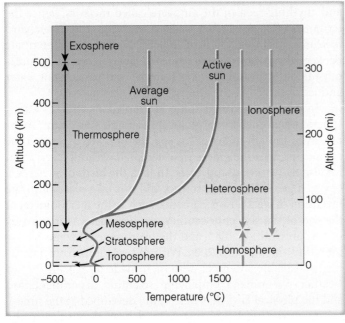

• FIGURE 1.12 Layers of the atmosphere based on temperature (red line), composition (green line), and electrical properties (dark blue line). (An active sun is associated with large numbers of solar eruptions.)

- We live at the bottom of the troposphere, which is an atmospheric layer where the air temperature normally decreases with height. The troposphere is a region that contains all of the weather we are familiar with.
- The ionosphere is an electrified region of the upper atmosphere that normally extends from about 60 km to the top of the atmosphere.

We will now turn our attention to weather events that take place in the lower atmosphere. As you read the remainder of this chapter, keep in mind that the content serves as a broad overview of material to come in later chapters, and that many of the concepts and ideas you encounter are designed to familiarize you with items you might read about in a newspaper or magazine, or see on television.

Weather and Climate

When we talk about the **weather,** we are talking about the condition of the atmosphere at any particular time and place. Weather—which is always changing—is comprised of the elements of:

1. *air temperature*—the degree of hotness or coldness of the air
2. *air pressure*—the force of the air above an area
3. *humidity*—a measure of the amount of water vapor in the air
4. *clouds*—a visible mass of tiny water droplets and/or ice crystals that are above the earth's surface
5. *precipitation*—any form of water, either liquid or solid (rain or snow), that falls from clouds and reaches the ground
6. *visibility*—the greatest distance one can see
7. *wind*—the horizontal movement of air

If we measure and observe these *weather elements* over a specified interval of time, say, for many years, we would obtain the "average weather" or the **climate** of a particular region. Climate, therefore, represents the accumulation of daily and seasonal weather events (the average range of weather) over a long period of time. The concept of climate is much more than this, for it also includes the extremes of weather—the heat waves of summer and the cold spells of winter—that occur in a particular region. The *frequency* of these extremes is what helps us distinguish among climates that have similar averages.

If we were able to watch the earth for many thousands of years, even the climate would change. We would see rivers of ice moving down stream-cut valleys and huge glaciers—sheets of moving snow and ice—spreading their icy fingers over large portions of North America. Advancing slowly from Canada, a single glacier might extend as far south as Kansas and Illinois, with ice several thousands of meters thick covering the region now occupied by Chicago. Over an interval of 2 million years or so, we would see the ice advance and retreat several times. Of course, for this phenomenon to happen, the average temperature of North America would have to decrease and then rise in a cyclic manner.

Suppose we could photograph the earth once every thousand years for many hundreds of millions of years. In time-lapse film sequence, these photos would show that not only is the climate altering, but the whole earth itself is changing as well: Mountains would rise up only to be torn down by erosion; isolated puffs of smoke and steam would appear as volcanoes spew hot gases and fine dust into the atmosphere; and the entire surface of the earth would undergo a gradual transformation as some ocean basins widen and others shrink.*

In summary, the earth and its atmosphere are dynamic systems that are constantly changing. While major transformations of the earth's surface are completed only after long spans of time, the state of the atmosphere can change in a matter of minutes. Hence, a watchful eye turned skyward will be able to observe many of these changes.

Up to this point, we have looked at the concepts of weather and climate without discussing the word *meteorology*. What does this term actually mean, and where did it originate?

METEOROLOGY—A BRIEF HISTORY **Meteorology** is the study of the atmosphere and its phenomena. The term itself goes back to the Greek philosopher Aristotle who, about 340 B.C., wrote a book on natural philosophy entitled *Meteorologica*. This work represented the sum of knowledge on weather and climate at that time, as well as material on astronomy, geography, and chemistry. Some of the topics covered included clouds, rain, snow, wind, hail, thunder, and hurricanes. In those days, all substances that fell from the sky, and anything seen in the air, were called meteors, hence the term *meteorology*, which actually comes from the Greek word *meteoros*, meaning "high in the air." Today, we differentiate between those meteors that come from extraterrestrial sources outside our atmosphere (meteoroids) and particles of water and ice observed in the atmosphere (hydrometeors).

In *Meteorologica*, Aristotle attempted to explain atmospheric phenomena in a philosophical and speculative manner. Even though many of his speculations were found to be erroneous, Aristotle's ideas were accepted without reservation for almost two thousand years. In fact, the birth of meteorology as a genuine natural science did not take place until the invention of weather instruments, such as the thermometer at the end of the sixteenth century, the barometer (for measuring air pressure) in 1643, and the hygrometer (for measuring humidity) in the late 1700s. With observations from instruments available, attempts were then made to explain certain weather phenomena employing scientific experimentation and the physical laws that were being developed at the time.

As more and better instruments were developed in the 1800s, the science of meteorology progressed. The invention

---

*The movement of the ocean floor and continents is explained in the widely acclaimed theory of *plate tectonics*.

of the telegraph in 1843 allowed for the transmission of routine weather observations. The understanding of the concepts of wind flow and storm movement became clearer, and in 1869 crude weather maps with *isobars* (lines of equal pressure) were drawn. Around 1920, the concepts of air masses and weather fronts were formulated in Norway. By the 1940s, daily upper-air balloon observations of temperature, humidity, and pressure gave a three-dimensional view of the atmosphere, and high-flying military aircraft discovered the existence of jet streams.

Meteorology took another step forward in the 1950s, when high-speed computers were developed to solve the mathematical equations that describe the behavior of the atmosphere. At the same time, a group of scientists in Princeton, New Jersey, developed numerical means for predicting the weather. Today, computers plot the observations, draw the lines on the map, and forecast the state of the atmosphere at some desired time in the future.

After World War II, surplus military radars became available, and many were transformed into precipitation-measuring tools. In the mid-1990s, these conventional radars were replaced by the more sophisticated *Doppler radars,* which have the ability to peer into a severe thunderstorm and unveil its winds and weather, as illustrated in ● Fig. 1.14.

In 1960, the first weather satellite, *Tiros I,* was launched, ushering in space-age meteorology. Subsequent satellites provided a wide range of useful information, ranging from day and night time-lapse images of clouds and storms to images that depict swirling ribbons of water vapor flowing around the globe. Throughout the 1990s, and into the twenty-first century, even more sophisticated satellites were developed to supply computers with a far greater network of data so that more accurate forecasts—perhaps up to two weeks or more—will be available in the future.

With this brief history of meterology we are now ready to observe weather events that occur at the earth's surface.

A SATELLITE'S VIEW OF THE WEATHER   A good view of the weather can be seen from a weather satellite. ● Figure 1.15 is a satellite image showing a portion of the Pacific Ocean and the North American continent. The image was obtained from a *geostationary satellite* situated about 36,000 km (22,300 mi) above the earth. At this elevation, the satellite travels at the same rate as the earth spins, which allows it to remain positioned above the same spot so it can continuously monitor what is taking place beneath it.

The solid black lines running from north to south on the satellite image are called *meridians,* or lines of longitude. Since the zero meridian (or prime meridian) runs through Greenwich, England, the *longitude* of any place on earth is simply how far east or west, in degrees, it is from the prime meridian. North America is west of Great Britain and most of the United States lies between 75°W and 125°W longitude.

The solid black lines that parallel the equator are called *parallels of latitude.* The latitude of any place is how far north or south, in degrees, it is from the equator. The latitude of the

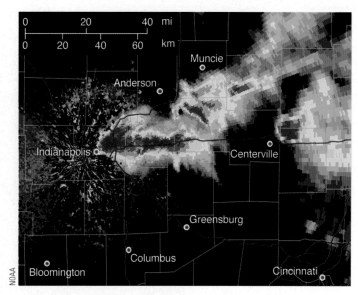

● **FIGURE 1.14** Doppler radar image showing the heavy rain and hail of a severe thunderstorm (dark red area) over Indianapolis, Indiana, on April 14, 2006.

equator is 0°, whereas the latitude of the North Pole is 90°N and that of the South Pole is 90°S. Most of the United States is located between latitude 30°N and 50°N, a region commonly referred to as the **middle latitudes.**

Storms of All Sizes   Probably the most dramatic spectacle in Fig. 1.15 is the whirling cloud masses of all shapes and sizes. The clouds appear white because sunlight is reflected back to space from their tops. The largest of the organized cloud masses are the sprawling storms. One such storm shows as an extensive band of clouds, over 2000 km long, west of the Great Lakes. Superimposed on the satellite image is the storm's center (indicated by the large red L) and its adjoining weather fronts in red, blue, and purple. This **middle-latitude cyclonic storm** system (or *extratropical cyclone*) forms outside the tropics and, in the Northern Hemisphere, has winds spinning counterclockwise about its center, which is presently over Minnesota.

A slightly smaller but more vigorous storm is located over the Pacific Ocean near latitude 12°N and longitude 116°W. This tropical storm system, with its swirling band of rotating clouds and surface winds in excess of 64 knots* (74 mi/hr), is known as a **hurricane.** The diameter of the hurricane is about 800 km (500 mi). The tiny dot at its center is called the *eye.* Near the surface, in the eye, winds are light, skies are generally clear, and the atmospheric pressure is lowest. Around the eye, however, is an extensive region where heavy rain and high surface winds are reaching peak gusts of 100 knots.

Smaller storms are seen as white spots over the Gulf of Mexico. These spots represent clusters of towering *cumulus* clouds that have grown into **thunderstorms,** that is, tall churning clouds accompanied by lightning, thunder, strong

---

*Recall from p. 13 that 1 knot equals 1.15 miles per hour.

• FIGURE 1.15
This satellite image (taken in visible reflected light) shows a variety of cloud patterns and storms in the earth's atmosphere.

gusty winds, and heavy rain. If you look closely at Fig. 1.15, you will see similar cloud forms in many regions. There were probably thousands of thunderstorms occurring throughout the world at that very moment. Although they cannot be seen individually, there are even some thunderstorms embedded in the cloud mass west of the Great Lakes. Later in the day on which this image was taken, a few of these storms spawned the most violent disturbance in the atmosphere — the **tornado.**

A tornado is an intense rotating column of air that extends downward from the base of a thunderstorm. Sometimes called *twisters,* or *cyclones,* they may appear as ropes or as a large circular cylinder. The majority are less than a kilometer wide and many are smaller than a football field. Tornado winds may exceed 200 knots but most probably peak at less than 125 knots. The rotation of some tornadoes never reaches the ground, and the rapidly rotating funnel appears to hang from the base of its parent cloud. Often they dip down, then rise up before disappearing.

A Look at a Weather Map We can obtain a better picture of the middle-latitude storm system by examining a simplified surface weather map for the same day that the satellite image was taken. The weight of the air above different regions varies and, hence, so does the atmospheric pressure. In • Fig. 1.16, the red letter L on the map indicates a region of low atmospheric pressure, often called a *low,* which marks the center of the middle-latitude storm. (Compare the center of the storm in Fig. 1.16 with that in Fig. 1.15.) The two blue letters H on the map represent regions of high atmospheric pressure, called *highs,* or *anticyclones.* The circles on the map represent either individual weather stations or cities where observations are taken. The **wind** is the horizontal movement of air. The **wind direction** — the direction *from which* the wind is blowing* — is given by lines that parallel the wind and extend outward from the center of the station. The *wind*

*If you are facing north and the wind is blowing in your face, the wind would be called a "north wind."

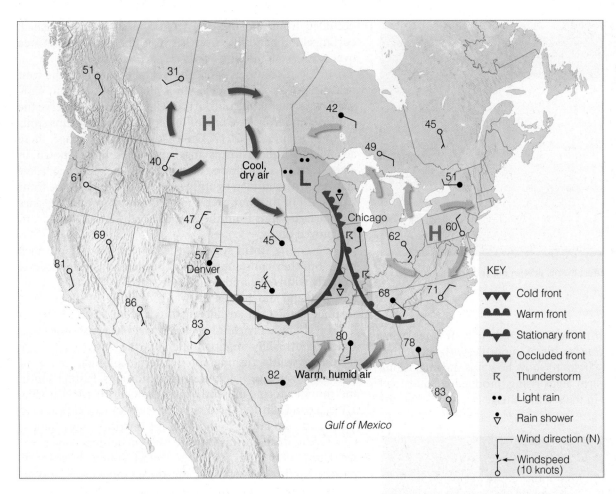

● FIGURE 1.16 Simplified surface weather map that correlates with the satellite image shown in Fig. 1.15. The shaded green area represents precipitation. The numbers on the map represent air temperatures in °F.

speed—the rate at which the air is moving past a stationary observer—is indicated by barbs.

Notice how the wind blows around the highs and the lows. The horizontal pressure differences create a force that starts the air moving from higher pressure toward lower pressure. Because of the earth's rotation, the winds are deflected from their path toward the right in the Northern Hemisphere.* This deflection causes the winds to blow *clockwise* and *outward* from the center of the highs, and *counterclockwise* and *inward* toward the center of the low.

As the surface air spins into the low, it flows together and rises, much like toothpaste does when its open tube is squeezed. The rising air cools, and the water vapor in the air condenses into clouds. Notice in Fig. 1.16 that the area of precipitation (the shaded green area) in the vicinity of the low corresponds to an extensive cloudy region in the satellite image (Fig. 1.15).

Also notice by comparing Figs. 1.15 and 1.16 that, in the regions of high pressure, skies are generally clear. As the surface air flows outward away from the center of a high, air

sinking from above must replace the laterally spreading surface air. Since sinking air does not usually produce clouds, we find generally clear skies and fair weather associated with the regions of high atmospheric pressure.

The swirling air around areas of high and low pressure are the major weather producers for the middle latitudes. Look at the middle-latitude storm and the surface temperatures in Fig. 1.16 and notice that, to the southeast of the storm, southerly winds from the Gulf of Mexico are bringing warm, humid air northward over much of the southeastern portion of the nation. On the storm's western side, cool dry northerly breezes combine with sinking air to create generally clear weather over the Rocky Mountains. The boundary that separates the warm and cool air appears as a heavy, colored lines on the map—a **front,** across which there is a sharp change in temperature, humidity, and wind direction.

Where the cool air from Canada replaces the warmer air from the Gulf of Mexico, a *cold front* is drawn in blue, with arrowheads showing the front's general direction of movement. Where the warm Gulf air is replacing cooler air to the north, a *warm front* is drawn in red, with half circles showing its general direction of movement. Where the cold front has

---

*This deflecting force, known as the *Coriolis force,* is discussed more completely in Chapter 8, as are the winds.

C. Donald Ahrens

● FIGURE I.17 Thunderstorms developing and advancing along an approaching cold front.

caught up to the warm front and cold air is now replacing cool air, an *occluded front* is drawn in purple, with alternating arrowheads and half circles to show how it is moving. Along each of the fronts, warm air is rising, producing clouds and precipitation. Notice in the satellite image (Fig. 1.15) that the

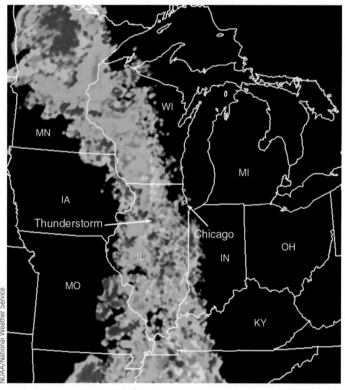

NOAA/National Weather Service

● FIGURE I.18 Doppler radar has the capacity of estimating rainfall intensity. In this composite image, the areas shaded green and blue indicate where light-to-moderate rain is falling. Yellow indicates heavier rainfall. The red-shaded area represents the heaviest rainfall and the possibility of intense thunderstorms. (Notice that a thunderstorm is approaching Chicago from the west.)

occluded front and the cold front appear as an elongated, curling cloud band that stretches from the low-pressure area over Minnesota into the northern part of Texas.

In Fig. 1.16 observe that the weather front is to the west of Chicago. As the westerly winds aloft push the front eastward, a person on the outskirts of Chicago might observe the approaching front as a line of towering thunderstorms similar to those in ● Fig. 1.17. On a Doppler radar image, these advancing thunderstorms may appear as those shown in ● Fig. 1.18. In a few hours, Chicago should experience heavy showers with thunder, lightning, and gusty winds as the front passes. All of this, however, should give way to clearing skies and surface winds from the west or northwest after the front has moved on by.

Observing storm systems, we see that not only do they move but they constantly change. Steered by the upper-level westerly winds, the middle-latitude storm in Fig. 1.16 gradually weakens and moves eastward, carrying its clouds and weather with it. In advance of this system, a sunny day in Ohio will gradually cloud over and yield heavy showers and thunderstorms by nightfall. Behind the storm, cool dry northerly winds rushing into eastern Colorado cause an overcast sky to give way to clearing conditions. Farther south, the thunderstorms presently over the Gulf of Mexico (Fig. 1.15) expand a little, then dissipate as new storms appear over water and land areas. To the west, the hurricane over the Pacific Ocean drifts northwestward and encounters cooler water. Here, away from its warm energy source, it loses its punch; winds taper off, and the storm soon turns into an unorganized mass of clouds and tropical moisture.

WEATHER AND CLIMATE IN OUR LIVES Weather and climate play a major role in our lives. Weather, for example, often dictates the type of clothing we wear, while climate influences the type of clothing we buy. Climate determines when to plant crops as well as what type of crops can be planted. Weather determines if these same crops will grow to maturity. Although weather and climate affect our lives in many ways, perhaps their most immediate effect is on our comfort. In order to survive the cold of winter and heat of summer, we build homes, heat them, air condition them, insulate them—only to find that when we leave our shelter, we are at the mercy of the weather elements.

Even when we are dressed for the weather properly, wind, humidity, and precipitation can change our perception of how cold or warm it feels. On a cold, windy day the effects of *wind chill* tell us that it feels much colder than it really is, and, if not properly dressed, we run the risk of *frostbite* or even *hypothermia* (the rapid, progressive mental and physical collapse that accompanies the lowering of human body temperature). On a hot, humid day we normally feel uncomfortably warm and blame it on the humidity. If we become too warm, our bodies overheat and *heat exhaustion* or *heat stroke* may result. Those most likely to suffer these maladies are the elderly with impaired circulatory systems and infants, whose heat regulatory mechanisms are not yet fully developed.

Weather affects how we feel in other ways, too. Arthritic pain is most likely to occur when rising humidity is accompanied by falling pressures. In ways not well understood, weather does seem to affect our health. The incidence of heart attacks shows a statistical peak after the passage of warm fronts, when rain and wind are common, and after the passage of cold fronts, when an abrupt change takes place as showery precipitation is accompanied by cold gusty winds. Headaches are common on days when we are forced to squint, often due to hazy skies or a thin, bright overcast layer of high clouds.

For some people, a warm dry wind blowing down-slope (a *chinook wind*) adversely affects their behavior (they often become irritable and depressed). Just how and why these winds impact humans physiologically is not well understood. We will, however, take up the question of why these winds are warm and dry in Chapter 9.

When the weather turns colder or warmer than normal, it impacts directly on the lives and pocketbooks of many people. For example, the exceptionally warm January of 2006 over the United States saved people millions of dollars in heating costs. On the other side of the coin, the colder than normal winter of 2000–2001 over much of North America sent heating costs soaring as demand for heating fuel escalated.

Major cold spells accompanied by heavy snow and ice can play havoc by snarling commuter traffic, curtailing airport services, closing schools, and downing power lines, thereby cutting off electricity to thousands of customers (see • Fig. 1.19). For example, a huge ice storm during January, 1998, in northern New England and Canada left millions of people without power and caused over a billion dollars in damages, and a devastating snow storm during March, 1993, buried parts of the East Coast with 14-foot snow drifts and left Syracuse, New York, paralyzed with a snow depth of 36 inches. When the frigid air settles into the Deep South, many millions of dollars worth of temperature-sensitive fruits and vegetables may be ruined, the eventual consequence being higher produce prices in the supermarket.

Prolonged dry spells, especially when accompanied by high temperatures, can lead to a shortage of food and, in some places, widespread starvation. Parts of Africa, for example, have periodically suffered through major droughts and famine. During the summer of 2007, the southeastern section of the United States experienced a terrible drought as searing summer temperatures wilted crops, causing losses in excess of a billion dollars. When the climate turns hot and dry, animals suffer too. In 1986, over 500,000 chickens perished in Georgia during a two-day period at the peak of a summer heat wave. Severe drought also has an effect on water reserves, often forcing communities to ration water and restrict its use. During periods of extended drought, vegetation often becomes tinder-dry and, sparked by lightning or a careless human, such a dried-up region can quickly become a raging inferno. During the winter of 2005–2006, hundreds of thousands of acres in drought-stricken Oklahoma and northern Texas were ravaged by wildfires.

• FIGURE 1.19 Ice storm near Oswego, New York, caused utility polls and power lines to be weighed down, forcing road closure.

Every summer, scorching *heat waves* take many lives. During the past 20 years, an annual average of more than 300 deaths in the United States were attributed to excessive heat exposure. In one particularly devastating heat wave that hit Chicago, Illinois, during July, 1995, high temperatures coupled with high humidity claimed the lives of more than 500 people. And Europe suffered through a devastating heat wave during the summer of 2003 when many people died, including 14,000 in France alone. In California during July, 2006, more than 100 people died as air temperatures climbed to over 46°C (115°F).

Every year, the violent side of weather influences the lives of millions. It is amazing how many people whose family roots are in the Midwest know the story of someone who was severely injured or killed by a tornado. Tornadoes have not only taken many lives, but annually they cause damage to buildings and property totaling in the hundreds of millions of dollars, as a single large tornado can level an entire section of a town (see • Fig. 1.20).

Although the gentle rains of a typical summer thunderstorm are welcome over much of North America, the heavy downpours, high winds, and hail of the *severe thunderstorms* are not. Cloudbursts from slowly moving, intense thunderstorms can provide too much rain too quickly, creating *flash floods* as small streams become raging rivers composed of

WEATHER WATCH

During September, 2005, Hurricane Katrina slammed into Mississippi and Louisiana. In the city of New Orleans several levees (that protected the city from flooding) broke, and flood waters over 20 feet deep inundated parts of the city, killing over 1200 people.

• FIGURE I.20 A tornado and a rainbow form over south-central Kansas during June, 2004. White streaks in the sky are descending hailstones.

• FIGURE I.21 Flooding during April, 1997, inundates Grand Forks, North Dakota, as flood waters of the Red River extend over much of the city.

• FIGURE I.22 Estimates are that lightning strikes the earth about 100 times every second. About 25 million lightning strikes hit the United States each year. Consequently, lightning is a very common, and sometimes deadly, weather phenomenon.

mud and sand entangled with uprooted plants and trees (see • Fig. 1.21). On the average, more people die in the United States from floods and flash floods than from any other natural disaster. Strong downdrafts originating inside an intense thunderstorm (a *downburst*) create turbulent winds that are capable of destroying crops and inflicting damage upon surface structures. Several airline crashes have been attributed to the turbulent *wind shear* zone within the downburst. Annually, hail damages crops worth millions of dollars, and lightning takes the lives of about eighty people in the United States and starts fires that destroy many thousands of acres of valuable timber (see • Fig. 1.22).

Even the quiet side of weather has its influence. When winds die down and humid air becomes more tranquil, fog may form. Dense fog can restrict visibility at airports, causing flight delays and cancellations. Every winter, deadly fog-related auto accidents occur along our busy highways and turnpikes. But fog has a positive side, too, especially during a dry spell, as fog moisture collects on tree branches and drips to the ground, where it provides water for the tree's root system.

Weather and climate have become so much a part of our lives that the first thing many of us do in the morning is to listen to the local weather forecast. For this reason, many radio and television newscasts have their own "weatherperson" to present weather information and give daily forecasts. More and more of these people are professionally trained in meteorology, and many stations require that the weathercaster obtain a seal of approval from the American Meteorological Society (AMS), or a certificate from the National Weather Association (NWA). To make their weather presentation as up-to-the-minute as possible, an increasing number of stations are taking advantage of the information provided by the National Weather Service (NWS), such as computerized weather forecasts, time-lapse satellite images, and color Doppler radar

FOCUS ON A SPECIAL TOPIC

What Is a Meteorologist?

Most people associate the term "meteorologist" with the weatherperson they see on television or hear on the radio. Many television and radio weathercasters are in fact professional meterologists, but some are not. A professional meterologist is usually considered to be a person who has completed the requirements for a college degree in meteorology or atmospheric science. This individual has strong, fundamental knowledge concerning how the atmosphere behaves, along with a substantial background of coursework in mathematics, physics, and chemistry.

A meterologist uses scientific principles to explain and to forecast atmospheric phenomena. About half of the approximately 9000 meteorologists and atmospheric scientists in the United States work doing weather forecasting for the National Weather Service, the military, or for a television or radio station. The other half work mainly in research, teach atmospheric science courses in colleges and universities, or do meteorological consulting work.

Scientists who do atmospheric research may be investigating how the climate is changing, how snowflakes form, or how pollution impacts temperature patterns. Aided by supercomputers, much of the work of a research meteorologist involves simulating the atmosphere to see how it behaves (see Fig. 5). Researchers often work closely with scientists from other fields, such as chemists, physicists, oceanographers, mathematicians, and environmental scientists to determine

how the atmosphere interacts with the entire ecosystem. Scientists doing work in physical meteorology may well study how radiant energy warms the atmosphere; those at work in the field of dynamic meteorology might be using the mathematical equations that describe air flow to learn more about jet streams. Scientists working in operational meteorology might be preparing a weather forecast by analyzing upper-air information over North America. A climatologist, or climate scientist, might be studying the interaction of the atmosphere and ocean to see what influence such interchange might have on planet Earth many years from now.

Meteorologists also provide a variety of services not only to the general public in the form of weather forecasts but also to city planners, contractors, farmers, and large corporations. Meteorologists working for private weather firms create the forecasts and graphics that are found in newspapers, on television, and on the Internet. Overall, there are many exciting jobs that fall under the heading of "meteorologist" — too many to mention here. However, for more information on this topic, visit this Web site: http://www.ametsoc.org/ and click on "Students."

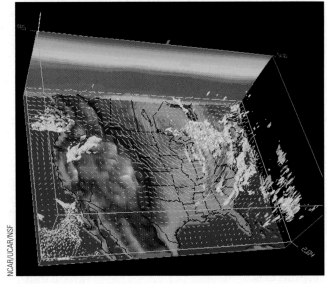

● FIGURE 5
A model that simulates a 3-dimensional view of the atmosphere. This computer model predicts how winds and clouds over the United States will change with time.

NCAR/UCAR/NSF

---

displays. (At this point it's interesting to note that many viewers believe the weather person they see on TV is a meteorologist and that all meteorologists forecast the weather. If you are interested in learning what a meteorologist or atmospheric scientist is and what he or she might do for a living (other than forecast the weather) read the Focus section above.)

For many years now, a staff of trained professionals at "The Weather Channel" have provided weather information twenty-four hours a day on cable television. And finally, the National Oceanic and Atmospheric Administration (NOAA), in cooperation with the National Weather Service, sponsors weather radio broadcasts at selected locations across the United States. Known as *NOAA weather radio* (and transmitted at VHF—FM frequencies), this service provides continuous weather information and regional forecasts (as well as special weather advisories, including watches and warnings) for over 90 percent of the United States.

SUMMARY

This chapter provides an overview of the earth's atmosphere. Our atmosphere is one rich in nitrogen and oxygen as well as smaller amounts of other gases, such as water vapor, carbon dioxide, and other greenhouse gases whose increasing levels are resulting in global warming. We examined the earth's early atmosphere and found it to be much different from the air we breathe today.

We investigated the various layers of the atmosphere: the troposphere (the lowest layer), where almost all weather events occur, and the stratosphere, where ozone protects us from a portion of the sun's harmful rays. In the stratosphere, ozone appears to be decreasing in concentration over parts of the Northern and Southern Hemispheres. Above the stratosphere lies the mesosphere, where the air temperature drops dramatically with height. Above the mesosphere lies the warmest part of the atmosphere, the thermosphere. At the top of the thermosphere is the exosphere, where collisions between gas molecules and atoms are so infrequent that fast-moving lighter molecules can actually escape the earth's gravitational pull and shoot off into space. The ionosphere represents that portion of the upper atmosphere where large numbers of ions and free electrons exist.

We looked briefly at the weather map and a satellite image and observed that dispersed throughout the atmosphere are storms and clouds of all sizes and shapes. The movement, intensification, and weakening of these systems, as well as the dynamic nature of air itself, produce a variety of weather events that we described in terms of weather elements. The sum total of weather and its extremes over a long period of time is what we call climate. Although sudden changes in weather may occur in a moment, climatic change takes place gradually over many years. The study of the atmosphere and all of its related phenomena is called *meteorology,* a term whose origin dates back to the days of Aristotle. Finally, we discussed some of the many ways weather and climate influence our lives.

KEY TERMS

The following terms are listed (with page number) in the order they appear in the text. Define each. Doing so will aid you in reviewing the material covered in this chapter.

atmosphere, 4
nitrogen, 4
oxygen, 4
water vapor, 5
carbon dioxide, 6
ozone, 8
ozone hole, 9
aerosol, 9
pollutant, 9
acid rain, 10
outgassing, 10
density, 11
pressure, 11
air pressure, 12
lapse rate, 12
temperature inversion, 12
radiosonde, 12
stratosphere, 12
tropopause, 13
troposphere, 14
mesosphere, 14
thermosphere, 15
exosphere, 16
homosphere, 17
heterosphere, 17
ionosphere, 17
weather, 18
climate, 18
meteorology, 18
middle latitudes, 19
middle-latitude cyclonic storm, 19
hurricane, 19
thunderstorm, 19
tornado, 20
wind, 20
wind direction, 20
front, 21

QUESTIONS FOR REVIEW

1. What is the primary source of energy for the earth's atmosphere?
2. List the four most abundant gases in today's atmosphere.
3. Of the four most abundant gases in our atmosphere, which one shows the greatest variation at the earth's surface?
4. What are some of the important roles that water plays in our atmosphere?
5. Briefly explain the production and natural destruction of carbon dioxide near the earth's surface. Give two reasons for the increase of carbon dioxide over the past 100 years.
6. List the two most abundant greenhouse gases in the earth's atmosphere. What makes them greenhouse gases?
7. Explain how the atmosphere "protects" inhabitants at the earth's surface.
8. What are some of the aerosols in our atmosphere?
9. How has the composition of the earth's atmosphere changed over time? Briefly outline the evolution of the earth's atmosphere.
10. (a) Explain the concept of air pressure in terms of mass of air above some level.
    (b) Why does air pressure always decrease with increasing height above the surface?
11. What is standard atmospheric pressure at sea level in
    (a) inches of mercury
    (b) millibars, and
    (c) hectopascals?
12. What is the average or standard temperature lapse rate in the troposphere?
13. Briefly describe how the air temperature changes from the earth's surface to the lower thermosphere.
14. On the basis of temperature, list the layers of the atmosphere from the lowest layer to the highest.
15. What atmospheric layer contains all of our weather?
16. (a) In what atmospheric layer do we find the lowest average air temperature?
    (b) The highest average temperature?
    (c) The highest concentration of ozone?

17. Above what region of the world would you find the ozone hole?
18. How does the ionosphere affect AM radio transmission during the day versus during the night?
19. Even though the actual concentration of oxygen is close to 21 percent (by volume) in the upper stratosphere, explain why, without proper breathing apparatus, you would not be able to survive there.
20. Define *meteorology* and discuss the origin of this word.
21. When someone says that "the wind direction today is south," does this mean that the wind is blowing *toward the south* or *from the south*?
22. Describe some of the features observed on a surface weather map.
23. Explain how wind blows around low- and high-pressure areas in the Northern Hemisphere.
24. How are fronts defined?
25. Rank the following storms in size from largest to smallest: hurricane, tornado, middle-latitude cyclonic storm, thunderstorm.
26. Weather in the middle latitudes tends to move in what general direction?
27. How does weather differ from climate?
28. Describe some of the ways weather and climate influence the lives of people.

QUESTIONS FOR THOUGHT

1. Which of the following statements relate more to weather and which relate more to climate?
   (a) The summers here are warm and humid.
   (b) Cumulus clouds presently cover the entire sky.
   (c) Our lowest temperature last winter was −29°C (−18°F).
   (d) The air temperature outside is 22°C (72°F).
   (e) December is our foggiest month.
   (f) The highest temperature ever recorded in Phoenixville, Pennsylvania, was 44°C (111°F) on July 10, 1936.
   (g) Snow is falling at the rate of 5 cm (2 in.) per hour.
   (h) The average temperature for the month of January in Chicago, Illinois, is −3°C (26°F).
2. A standard pressure of 1013.25 millibars is also known as one atmosphere (1 ATM).

(a) Look at Fig. 1.10 and determine at approximately what levels you would record a pressure of 0.5 ATM and 0.1 ATM.
(b) The surface air pressure on the planet Mars is about 0.007 ATM. If you were standing on Mars, the surface air pressure would be equivalent to a pressure observed at approximately what elevation in the earth's atmosphere?
3. If you were suddenly placed at an altitude of 100 km (62 mi) above the earth, would you expect your stomach to expand or contract? Explain.

PROBLEMS AND EXERCISES

1. Keep track of the weather. On an outline map of North America, mark the daily position of fronts and pressure systems for a period of several weeks or more. (This information can be obtained from newspapers, the TV news, or from the Internet.) Plot the general upper-level flow pattern on the map. Observe how the surface systems move. Relate this information to the material on wind, fronts, and cyclones covered in later chapters.
2. Compose a one-week journal, including daily newspaper weather maps and weather forecasts from the newspaper or from the Internet. Provide a commentary for each day regarding the coincidence of actual and predicted weather.
3. Formulate a short-term climatology for your city for one month by recording maximum and minimum temperatures and precipitation amounts every day. You can get this information from television, newspapers, the Internet, or from your own measurements. Compare this data to the actual climatology for that month. How can you explain any large differences between the two?

The aurora borealis, which forms as energetic particles from the sun interact with the earth's atmosphere, is seen here over Edmonton, Alberta, Canada.

CHAPTER 2

Energy: Warming the Earth and the Atmosphere

At high latitudes after darkness has fallen, a faint, white glow may appear in the sky. Lasting from a few minutes to a few hours, the light may move across the sky as a yellow green arc much wider than a rainbow; or, it may faintly decorate the sky with flickering draperies of blue, green, and purple light that constantly change in form and location, as if blown by a gentle breeze.

For centuries curiosity and superstition have surrounded these eerie lights. Eskimo legend says they are the lights from demons' lanterns as they search the heavens for lost souls. Nordic sagas called them a reflection of fire that surrounds the seas of the north. Even today there are those who proclaim that the lights are reflected sunlight from polar ice fields. Actually, this light show in the Northern Hemisphere is the aurora borealis—the northern lights—which is caused by invisible energetic particles bombarding our upper atmosphere. Anyone who witnesses this, one of nature's spectacular color displays, will never forget it.

❂ CONTENTS

Energy is everywhere. It is the basis for life. It comes in various forms: It can warm a house, melt ice, and drive the atmosphere, producing our everyday weather events. When the sun's energy interacts with our upper atmosphere we see energy at work in yet another form, a shimmering display of light from the sky—the aurora. What, precisely, is this common, yet mysterious, quantity we call "energy"? What is its primary source? How does it warm our earth and provide the driving force for our atmosphere? And in what form does it reach our atmosphere to produce a dazzling display like the aurora?

To answer these questions, we must first begin with the concept of energy itself. Then we will examine energy in its various forms and how energy is transferred from one form to another in our atmosphere. Finally, we will look more closely at the sun's energy and its influence on our atmosphere.

Energy, Temperature, and Heat

By definition, **energy** is the ability or capacity to do work on some form of matter. (Matter is anything that has mass and occupies space.) Work is done on matter when matter is either pushed, pulled, or lifted over some distance. When we lift a brick, for example, we exert a force against the pull of gravity—we "do work" on the brick. The higher we lift the brick, the more work we do. So, by doing work on something, we give it "energy," which it can, in turn, use to do work on other things. The brick that we lifted, for instance, can now do work on your toe—by falling on it.

The total amount of energy stored in any object (internal energy) determines how much work that object is capable of doing. A lake behind a dam contains energy by virtue of its position. This is called *gravitational potential energy* or simply **potential energy** because it represents the potential to do work—a great deal of destructive work if the dam were to break. The potential energy (PE) of any object is given as

$$PE = mgh,$$

where m is the object's mass, g is the acceleration of gravity, and h is the object's height above the ground.

A volume of air aloft has more potential energy than the same size volume of air just above the surface. This fact is so because the air aloft has the potential to sink and warm through a greater depth of atmosphere. A substance also possesses potential energy if it can do work when a chemical change takes place. Thus, coal, natural gas, and food all contain chemical potential energy.

Any moving substance possesses energy of motion, or **kinetic energy.** The kinetic energy (KE) of an object is equal to half its mass multiplied by its velocity squared; thus

$$KE = \tfrac{1}{2}\, mv^2.$$

Consequently, the faster something moves, the greater its kinetic energy; hence, a strong wind possesses more kinetic energy than a light breeze. Since kinetic energy also depends on the object's mass, a volume of water and an equal volume of air may be moving at the same speed, but, because the water has greater mass, it has more kinetic energy. The atoms and molecules that comprise all matter have kinetic energy due to their motion. This form of kinetic energy is often referred to as *heat energy*. Probably the most important form of energy in terms of weather and climate is the energy we receive from the sun—*radiant energy*.

Energy, therefore, takes on many forms, and it can change from one form into another. But the total amount of energy in the universe remains constant. *Energy cannot be created nor can it be destroyed.* It merely changes from one form to another in any ordinary physical or chemical process. In other words, the energy lost during one process must equal the energy gained during another. This is what we mean when we say that energy is conserved. This statement is known as the *law of conservation of energy,* and is also called the *first law of thermodynamics.*

We know that air is a mixture of countless billions of atoms and molecules. If they could be seen, they would appear to be moving about in all directions, freely darting, twisting, spinning, and colliding with one another like an angry swarm of bees. Close to the earth's surface, each individual molecule will travel only about a thousand times its diameter before colliding with another molecule. Moreover, we would see that all the atoms and molecules are not moving at the same speed, as some are moving faster than others. The temperature of the air (or any substance) is a measure of its average kinetic energy. Simply stated, **temperature** *is a measure of the average speed of the atoms and molecules,* where higher temperatures correspond to faster average speeds.

Suppose we examine a volume of surface air about the size of a large flexible balloon, as shown in • Fig. 2.1a. If we warm the air inside, the molecules would move faster, but they also would move slightly farther apart—the air becomes less dense, as illustrated in Fig. 2.1b. Conversely, if we cool the air back to its original temperature, the molecules would slow down, crowd closer together, and the air would become more dense. This molecular behavior is why, in many places throughout the book, we refer to surface air as either *warm, less-dense air* or as *cold, more-dense air.*

The atmosphere and oceans contain *internal energy,* which is the total energy (potential and kinetic) stored in their molecules. As we have just seen, the temperature of air and water is determined only by the *average* kinetic energy (average speed) of *all* their molecules. Since temperature only indicates how "hot" or "cold" something is relative to some set standard value, it does not always tell us how much internal energy that something possesses. For example, two identical mugs, each half-filled with water and each with the same temperature, contain the same internal energy. If the water from one mug is poured into the other, the total internal energy of the filled mug has doubled because its mass has doubled. Its temperature, however, has not changed, since the average speed of all of the molecules is still the same.

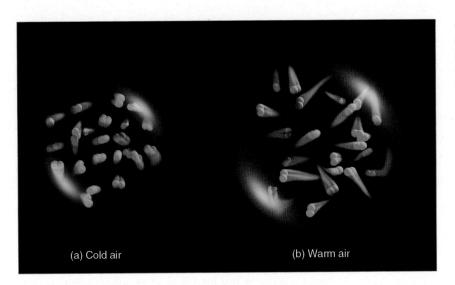

(a) Cold air   (b) Warm air

● FIGURE 2.1 Air temperature is a measure of the average speed of the molecules. In the cold volume of air, the molecules move more slowly and crowd closer together. In the warm volume, they move faster and farther apart.

Now, imagine that you are sipping a hot cup of tea on a small raft in the middle of a lake. The tea has a much higher temperature than the lake, yet the lake contains more internal energy because it is composed of many more molecules. If the cup of tea is allowed to float on top of the water, the tea would cool rapidly. The energy that would be transferred from the hot tea to the cool water (because of their temperature difference) is called *heat.*

In essence, **heat** *is energy in the process of being transferred from one object to another because of the temperature difference between them.* After heat is transferred, it is stored as internal energy. How is this energy transfer process accomplished? In the atmosphere, heat is transferred by *conduction, convection,* and *radiation.* We will examine these mechanisms of energy transfer after we look at temperature scales and at the important concepts of *specific heat* and *latent heat.*

> View this concept in action on the Meteorology Resource Center at academic.cengage.com/login

TEMPERATURE SCALES Suppose we take a small volume of air (like the one shown in Fig. 2.1a) and allow it to cool. As the air slowly cools, its atoms and molecules would move slower and slower until the air reaches a temperature of $-273°C$ ($-459°F$), which is the lowest temperature possible. At this temperature, called **absolute zero,** the atoms and molecules would possess a minimum amount of energy and theoretically no thermal motion. At absolute zero, we can begin a temperature scale called the *absolute scale,* or **Kelvin scale** after Lord Kelvin (1824–1907), a famous British scientist who first introduced it. Since the Kelvin scale begins at absolute zero, it contains no negative numbers and is, therefore, quite convenient for scientific calculations.

Two other temperature scales commonly used today are the Fahrenheit and Celsius (formerly centigrade). The **Fahrenheit scale** was developed in the early 1700s by the physicist G. Daniel Fahrenheit, who assigned the number 32 to the temperature at which water freezes, and the number 212 to the temperature at which water boils. The zero point was simply the lowest temperature that he obtained with a mixture of ice, water, and salt. Between the freezing and boiling points are 180 equal divisions, each of which is called a degree. A thermometer calibrated with this scale is referred to as a Fahrenheit thermometer, for it measures an object's temperature in degrees Fahrenheit (°F).

The **Celsius scale** was introduced later in the eighteenth century. The number 0 (zero) on this scale is assigned to the temperature at which pure water freezes, and the number 100 to the temperature at which pure water boils at sea level. The space between freezing and boiling is divided into 100 equal degrees. Therefore, each Celsius degree is 180/100 or 1.8 times larger than a Fahrenheit degree. Put another way, an increase in temperature of 1°C equals an increase of 1.8°F. A formula for converting °F to °C is

$$°C = ⁵⁄₉ (°F - 32).$$

On the Kelvin scale, degrees Kelvin are called *Kelvins* (abbreviated K). Each degree on the Kelvin scale is exactly the same size as a degree Celsius, and a temperature of 0 K is equal to $-273°C$. Converting from °C to K can be made by simply adding 273 to the Celsius temperature, as

$$K = °C + 273.$$

● Figure 2.2 compares the Kelvin, Celsius, and Fahrenheit scales. Converting a temperature from one scale to another can be done by simply reading the corresponding temperature from the adjacent scale. Thus, 303 on the Kelvin scale is the equivalent of 30°C and 86°F.*

In most of the world, temperature readings are taken in °C. In the United States, however, temperatures above the surface are taken in °C, while temperatures at the surface are typically read in °F. Currently, then, temperatures on upper-level maps are plotted in °C, while, on surface weather maps,

---

*A more complete table of conversions is given in Appendix A.

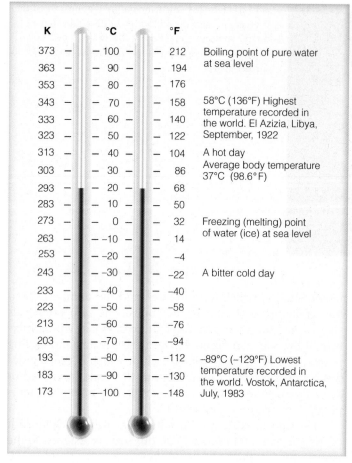

amount of heat energy absorbed by that substance to its corresponding temperature rise. The heat capacity of a substance per unit mass is called **specific heat.** In other words, specific heat is the amount of heat needed to raise the temperature of one gram (g) of a substance one degree Celsius.

If we heat 1 g of liquid water on a stove, it would take about 1 calorie (cal)* to raise its temperature by 1°C. So water has a specific heat of 1. If, however, we put the same amount (that is, same mass) of compact dry soil on the flame, we would see that it would take about one-fifth the heat (about 0.2 cal) to raise its temperature by 1°C. The specific heat of water is therefore 5 times greater than that of soil. In other words, water must absorb 5 times as much heat as the same quantity of soil in order to raise its temperature by the same amount. The specific heat of various substances is given in ▼ Table 2.1.

Not only does water heat slowly, it cools slowly as well. It has a much higher capacity for storing energy than other common substances, such as soil and air. A given volume of water can store a large amount of energy while undergoing only a small temperature change. Because of this attribute, water has a strong modifying effect on weather and climate. Near large bodies of water, for example, winters usually remain warmer and summers cooler than nearby inland regions—a fact well known to people who live adjacent to oceans or large lakes.

LATENT HEAT—THE HIDDEN WARMTH We know from Chapter 1 that water vapor is an invisible gas that becomes visible when it changes into larger liquid or solid (ice) particles. This process of transformation is known as a *change of state* or, simply, a *phase change*. The heat energy required to change a substance, such as water, from one state to another is called **latent heat.** But why is this heat referred to as "latent"? To answer this question, we will begin with something familiar to most of us—the cooling produced by evaporating water.

Suppose we microscopically examine a small drop of pure water. At the drop's surface, molecules are constantly escaping (evaporating). Because the more energetic, faster-moving molecules escape most easily, the average motion of all the molecules left behind decreases as each additional molecule evaporates. Since temperature is a measure of average molecular motion, the slower motion suggests a lower water temperature. *Evaporation is, therefore, a cooling process.* Stated another way, evaporation is a cooling process because the energy needed to evaporate the water—that is, to change its phase from a liquid to a gas—may come from the water or other sources, including the air.

In the everyday world, we experience evaporational cooling as we step out of a shower or swimming pool into a dry area. Because some of the energy used to evaporate the water

• FIGURE 2.2 Comparison of the Kelvin, Celsius, and Fahrenheit scales.

they are in °F. Since both scales are in use, temperature readings in this book will, in most cases, be given in °C followed by their equivalent in °F.

SPECIFIC HEAT A watched pot never boils, or so it seems. The reason for this is that water requires a relatively large amount of heat energy to bring about a small temperature change. The **heat capacity** of a substance is the ratio of the

▼ TABLE 2.1 Specific Heat of Various Substances

SUBSTANCE	SPECIFIC HEAT (Cal/g × °C)	J/(kg × °C)
Water (pure)	1.00	4186
Wet mud	0.60	2512
Ice (0°C)	0.50	2093
Sandy clay	0.33	1381
Dry air (sea level)	0.24	1005
Quartz sand	0.19	795
Granite	0.19	794

*By definition, a calorie is the amount of heat required to raise the temperature of 1 g of water from 14.5°C to 15.5°C. The kilocalorie is 1000 calories and is the heat required to raise 1 kg of water 1°C. In the International System (SI), the unit of energy is the joule (J), where 1 calorie = 4.186 J. (For pronunciation: joule rhymes with pool.)

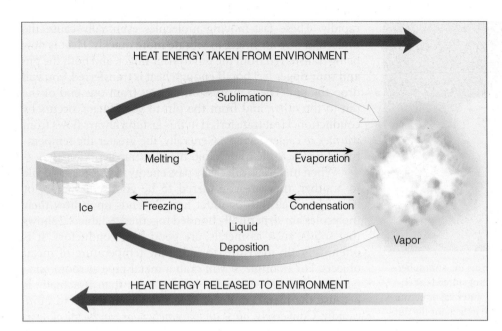

comes from our skin, we may experience a rapid drop in skin temperature, even to the point where goose bumps form. In fact, on a hot, dry, windy day in Tucson, Arizona, cooling may be so rapid that we begin to shiver even though the air temperature is hovering around 38°C (100°F).

The energy lost by liquid water during evaporation can be thought of as carried away by, and "locked up" within, the water vapor molecule. The energy is thus in a "stored" or "hidden" condition and is, therefore, called *latent heat.* It is latent (hidden) in that the temperature of the substance changing from liquid to vapor is still the same. However, the heat energy will reappear as **sensible heat** (the heat we can feel, "sense," and measure with a thermometer) when the vapor condenses back into liquid water. Therefore, *condensation (the opposite of evaporation) is a warming process.*

The heat energy released when water vapor condenses to form liquid droplets is called *latent heat of condensation.* Conversely, the heat energy used to change liquid into vapor at the same temperature is called *latent heat of evaporation* (vaporization). Nearly 600 cal (2500 J) are required to evaporate a single gram of water at room temperature. With many hundreds of grams of water evaporating from the body, it is no wonder that after a shower we feel cold before drying off.

In a way, latent heat is responsible for keeping a cold drink with ice colder than one without ice. As ice melts, its temperature does not change. The reason for this fact is that the heat added to the ice only breaks down the rigid crystal pattern, changing the ice to a liquid without changing its temperature. The energy used in this process is called *latent heat of fusion* (melting). Roughly 80 cal (335 J) are required to melt a single gram of ice. Consequently, heat added to a cold drink with ice primarily melts the ice, while heat added to a cold drink without ice warms the beverage. If a gram of water at 0°C changes back into ice at 0°C, this same amount of heat (80 cal) would be released as sensible heat to the en-

vironment. Therefore, when ice melts, heat is taken in; when water freezes, heat is liberated.

The heat energy required to change ice into vapor (a process called *sublimation*) is referred to as *latent heat of sublimation.* For a single gram of ice to transform completely into vapor at 0°C requires nearly 680 cal—80 cal for the latent heat of fusion plus 600 cal for the latent heat of evaporation. If this same vapor transformed back into ice (a process called *deposition*), approximately 680 cal (2850 J) would be released.

● Figure 2.3 summarizes the concepts examined so far. When the change of state is from left to right, heat is absorbed by the substance and taken away from the environment. The processes of melting, evaporation, and sublimation all cool the environment. When the change of state is from right to left, heat energy is given up by the substance and added to the environment. The process of freezing, condensation, and deposition all warm their surroundings.

Latent heat is an important source of atmospheric energy. Once vapor molecules become separated from the earth's surface, they are swept away by the wind, like dust before a broom. Rising to high altitudes where the air is cold, the vapor changes into liquid and ice cloud particles. During these processes, a tremendous amount of heat energy is released into the environment. This heat provides energy for storms, such as hurricanes, middle latitude cyclones, and thunderstorms (see ● Fig. 2.4).

Water vapor evaporated from warm, tropical water can be carried into polar regions, where it condenses and gives up its heat energy. Thus, as we will see, evaporation–transportation–condensation is an extremely important mechanism for the relocation of heat energy (as well as water) in the atmosphere. (Before going on to the next section, you may wish to read the Focus section on p. 35, which summarizes some of the concepts considered thus far.)

● FIGURE 2.4 Every time a cloud forms, it warms the atmosphere. Inside this developing thunderstorm a vast amount of stored heat energy (latent heat) is given up to the air, as invisible water vapor becomes countless billions of water droplets and ice crystals. In fact, for the duration of this storm alone, more heat energy is released inside this cloud than is unleashed by a small nuclear bomb.

Heat Transfer in the Atmosphere

CONDUCTION   The transfer of heat from molecule to molecule within a substance is called **conduction.** Hold one end of a metal straight pin between your fingers and place a flaming candle under the other end (see ● Fig. 2.5). Because of the energy they absorb from the flame, the molecules in the pin vibrate faster. The faster-vibrating molecules cause adjoining molecules to vibrate faster. These, in turn, pass vibrational energy on to their neighboring molecules, and so on, until the molecules at the finger-held end of the pin begin to vibrate

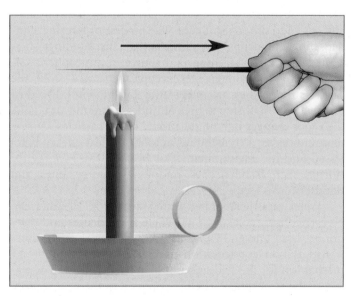

● FIGURE 2.5 The transfer of heat from the hot end of the metal pin to the cool end by molecular contact is called *conduction.*

rapidly. These fast-moving molecules eventually cause the molecules of your finger to vibrate more quickly. Heat is now being transferred from the pin to your finger, and both the pin and your finger feel hot. If enough heat is transferred, you will drop the pin. The transmission of heat from one end of the pin to the other, and from the pin to your finger, occurs by conduction. Heat transferred in this fashion always flows from *warmer to colder* regions. Generally, the greater the temperature difference, the more rapid the heat transfer.

When materials can easily pass energy from one molecule to another, they are considered to be good conductors of heat. How well they conduct heat depends upon how their molecules are structurally bonded together. ▼ Table 2.2 shows that solids, such as metals, are good heat conductors. It is often difficult, therefore, to judge the temperature of metal objects. For example, if you grab a metal pipe at room temperature, it will seem to be much colder than it actually is because the metal conducts heat away from the hand quite rapidly. Conversely, *air is an extremely poor conductor of heat,* which is why most insulating materials have a large number of air spaces trapped within them. Air is such a poor heat conductor that, in calm weather, the hot ground only warms a shallow layer of air a few centimeters thick by conduction. Yet, air can carry this energy rapidly from one region to another. How then does this phenomenon happen?

CONVECTION   The transfer of heat by the mass movement of a fluid (such as water and air) is called **convection.** This type of heat transfer takes place in liquids and gases because

▼ TABLE 2.2   Heat Conductivity* of Various Substances

SUBSTANCE	HEAT CONDUCTIVITY (Watts† per meter per °C)
Still air	0.023 (at 20°C)
Wood	0.08
Dry soil	0.25
Water	0.60 (at 20°C)
Snow	0.63
Wet soil	2.1
Ice	2.1
Sandstone	2.6
Granite	2.7
Iron	80
Silver	427

*Heat (thermal) conductivity describes a substance's ability to conduct heat as a consequence of molecular motion.

†A watt (W) is a unit of power where one watt equals one joule (J) per second (J/s). One joule equals 0.24 calories.

FOCUS ON A SPECIAL TOPIC

The Fate of a Sunbeam

Consider sunlight in the form of radiant energy striking a large lake. (See Fig. 1.) Part of the incoming energy heats the water, causing greater molecular motion and, hence, an increase in the water's kinetic energy. This greater kinetic energy allows more water molecules to evaporate from the surface. As each molecule escapes, work is done to break it away from the remaining water molecules. This energy becomes the latent heat energy that is carried with the water vapor.

Above the lake, a large bubble* of warm, moist air rises and expands. In order for this expansion to take place, the gas molecules inside the bubble must use some of their kinetic energy to do work against the bubble's sides. This results in a slower molecular speed and a lower temperature. Well above the surface, the water vapor in the rising, cooling bubble of moist air condenses into clouds. The condensation of water vapor releases latent heat energy into the atmosphere, warming the air. The tiny suspended cloud droplets possess potential energy, which becomes kinetic energy when these droplets grow into raindrops that fall earthward.

---

*A bubble of rising (or sinking) air about the size of a large balloon is often called *a parcel of air*.

• FIGURE 1 Solar energy striking a large body of water goes through many transformations.

When the drops reach the surface, their kinetic energy erodes the land. As rain-swollen streams flow into a lake behind a dam, there is a buildup of potential energy, which can be transformed into kinetic energy as water is harnessed to flow down a chute. If the moving water drives a generator, kinetic energy is converted into electrical energy, which is sent to cities. There, it heats, cools, and lights the buildings in which people work and live. Meanwhile, some of the water in the lake behind the dam evaporates and is free to repeat the cycle. Hence, the energy from the sunlight on a lake can undergo many transformations and help provide the moving force for many natural and human-made processes.

---

they can move freely, and it is possible to set up currents within them.

Convection happens naturally in the atmosphere. On a warm, sunny day, certain areas of the earth's surface absorb more heat from the sun than others; as a result, the air near the earth's surface is heated somewhat unevenly. Air molecules adjacent to these hot surfaces bounce against them, thereby gaining some extra energy by conduction. The heated air expands and becomes less dense than the surrounding cooler air. The expanded warm air is buoyed upward and rises. In this manner, large bubbles of warm air rise and transfer heat energy upward. Cooler, heavier air flows toward the surface to replace the rising air. This cooler air becomes heated in turn, rises, and the cycle is repeated. In meteorology, this vertical exchange of heat is called *convection,* and the rising air bubbles are known as **thermals** (see • Fig. 2.6).

The rising air expands and gradually spreads outward. It then slowly begins to sink. Near the surface, it moves back

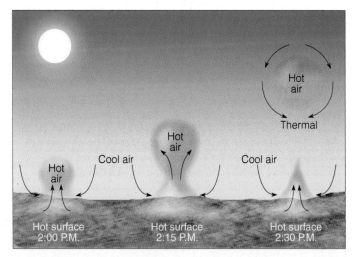

• FIGURE 2.6 The development of a thermal. A thermal is a rising bubble of air that carries heat energy upward by *convection.*

FOCUS ON A SPECIAL TOPIC

Rising Air Cools and Sinking Air Warms

To understand why rising air cools and sinking air warms we need to examine some air. Suppose we place air in an imaginary thin, elastic wrap about the size of a large balloon (see Fig. 2). This invisible balloonlike "blob" is called an *air parcel*. The air parcel can expand and contract freely, but neither external air nor heat is able to mix with the air inside. By the same token, as the parcel moves, it does not break apart, but remains as a single unit.

At the earth's surface, the parcel has the same temperature and pressure as the air surrounding it. Suppose we lift the parcel. Recall from Chapter I that air pressure always decreases as we move up into the atmosphere. Consequently, as the parcel rises, it enters a region where the surrounding air pressure is lower. To equalize the pressure, the parcel molecules inside push the parcel walls outward, expanding it. Because there is no other energy source, the air molecules inside use some of their own energy to expand the parcel. This energy loss shows up as slower molecular speeds, which represent a lower parcel temperature. Hence, *any air that rises always expands and cools.*

If the parcel is lowered to the earth (as shown in Fig. 2), it returns to a region where the air pressure is higher. The higher outside pressure squeezes (compresses) the parcel back to its original (smaller) size. Because air molecules have a faster rebound velocity after striking the sides of a collapsing parcel, the average speed of the molecules inside goes up. (A Ping-Pong ball moves faster after striking a paddle that is moving toward it.) This increase in molecular speed represents a warmer parcel temperature. Therefore, *any air that sinks (subsides), warms by compression.*

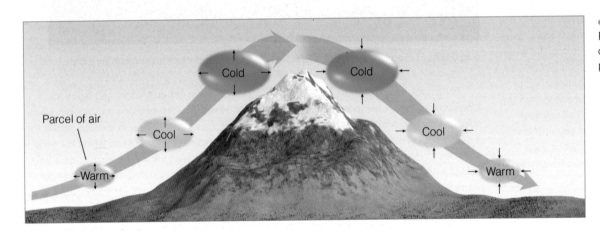

● FIGURE 2
Rising air expands and cools; sinking air is compressed and warms.

into the heated region, replacing the rising air. In this way, a *convective circulation,* or thermal "cell," is produced in the atmosphere. In a convective circulation, the warm, rising air cools. In our atmosphere, *any air that rises will expand and cool,* and *any air that sinks is compressed and warms.* This important concept is detailed in the Focus section above.

Although the entire process of heated air rising, spreading out, sinking, and finally flowing back toward its original location is known as a convective circulation, meteorologists usually restrict the term *convection* to the process of the rising and sinking part of the circulation.

The horizontally moving part of the circulation (called *wind*) carries properties of the air in that particular area with it. The transfer of these properties by horizontally moving air is called **advection.** For example, wind blowing across a body of water will "pick up" water vapor from the evaporating surface and transport it elsewhere in the atmosphere. If the air cools, the water vapor may condense into cloud droplets and release latent heat. In a sense, then, heat is advected (carried) by the water vapor as it is swept along with the wind. Earlier we saw that this is an important way to redistribute heat energy in the atmosphere.

WEATHER WATCH

Although we can't see air, there are signs that tell us where the air is rising. One example: On a calm day you can watch a hawk circle and climb high above level ground while its wings remain motionless. A rising thermal carries the hawk upward as it scans the terrain for prey. Another example: If the water vapor of a rising thermal condenses into liquid cloud droplets, the thermal becomes visible to us as a puffy cumulus cloud. Flying in a light aircraft beneath these clouds usually produces a bumpy ride, as passengers are jostled around by the rising and sinking air associated with convection.

BRIEF REVIEW

Before moving on to the next section, here is a summary of some of the important concepts and facts we have covered:

- The temperature of a substance is a measure of the average kinetic energy (average speed) of its atoms and molecules.
- Evaporation (the transformation of liquid into vapor) is a cooling process that can cool the air, whereas condensation (the transformation of vapor into liquid) is a warming process that can warm the air.
- Heat is energy in the process of being transferred from one object to another because of the temperature difference between them.
- In conduction, which is the transfer of heat by molecule-to-molecule contact, heat always flows from warmer to colder regions.
- Air is a poor conductor of heat.
- Convection is an important mechanism of heat transfer, as it represents the vertical movement of warmer air upward and cooler air downward.

There is yet another mechanism for the transfer of energy—radiation, or *radiant energy,* which is what we receive from the sun. In this method, energy may be transferred from one object to another without the space between them necessarily being heated.

Radiation

On a summer day, you may have noticed how warm and flushed your face feels as you stand facing the sun. Sunlight travels through the surrounding air with little effect upon the air itself. Your face, however, absorbs this energy and converts it to thermal energy. Thus, sunlight warms your face without actually warming the air. The energy transferred from the sun to your face is called **radiant energy,** or **radiation.** It travels in the form of waves that release energy when they are absorbed by an object. Because these waves have magnetic and electrical properties, we call them **electromagnetic waves.** Electromagnetic waves do not need molecules to propagate them. In a vacuum, they travel at a constant speed of nearly 300,000 km (186,000 mi) per second—the speed of light.

• Figure 2.7 shows some of the different wavelengths of radiation. Notice that the **wavelength** (which is usually expressed by the Greek letter lambda, λ) is the distance measured along a wave from one crest to another. Also notice that some of the waves have exceedingly short lengths. For example, radiation that we can see (visible light) has an average wavelength of less than one-millionth of a meter—a distance nearly one-hundredth the diameter of a human hair. To measure these short lengths, we introduce a new unit of measurement called a **micrometer** (represented by the symbol μm), which is equal to one-millionth of a meter (m); thus

$$1 \text{ micrometer } (\mu m) = 0.000001 \text{ m} = 10^{-6} \text{ m}.$$

In Fig. 2.7, we can see that the average wavelength of visible light is about 0.0000005 m, which is the same as 0.5 μm. To give you a common object for comparison, the average height of a letter on this page is about 2000 μm, or 2 millimeters (2 mm), whereas the thickness of this page is about 100 μm.

We can also see in Fig. 2.7 that the longer waves carry less energy than do the shorter waves. When comparing the energy carried by various waves, it is useful to give electromagnetic radiation characteristics of particles in order to explain some of the waves' behavior. We can actually think of radiation as streams of particles or **photons** that are discrete packets of energy.*

---

*Packets of photons make up waves, and groups of waves make up a beam of radiation.

• FIGURE 2.7
Radiation characterized according to wavelength. As the wavelength decreases, the energy carried per wave increases.

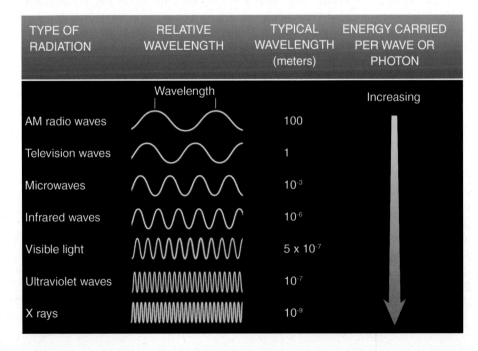

TYPE OF RADIATION	RELATIVE WAVELENGTH	TYPICAL WAVELENGTH (meters)	ENERGY CARRIED PER WAVE OR PHOTON
AM radio waves		100	Increasing
Television waves		1	
Microwaves		10^{-3}	
Infrared waves		10^{-6}	
Visible light		5×10^{-7}	
Ultraviolet waves		10^{-7}	
X rays		10^{-9}	

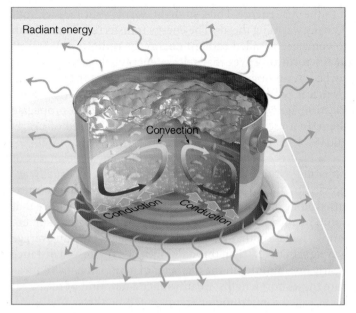

Radiant energy

Convection

Conduction   Conduction

● FIGURE 2.8 The hot burner warms the bottom of the pot by conduction. The warm pot, in turn, warms the water in contact with it. The warm water rises, settings up convection currents. The pot, water, burner, and everything else constantly emit radiant energy (orange arrows) in all directions.

An ultraviolet photon carries more energy than a photon of visible light. In fact, certain ultraviolet photons have enough energy to produce sunburns and penetrate skin tissue, sometimes causing skin cancer. As we discussed in Chapter 1, it is ozone in the stratosphere that protects us from the vast majority of these harmful rays. ● Figure 2.8 illustrates the concept of radiation along with the other forms of heat transfer—conduction and convection.

RADIATION AND TEMPERATURE   *All things (whose temperature is above absolute zero), no matter how big or small, emit radiation.* This book, your body, flowers, trees, air, the earth, the stars are all radiating a wide range of electromagnetic waves. The energy originates from rapidly vibrating electrons, billions of which exist in every object.

The wavelengths that each object emits depend primarily on the object's temperature. The higher the temperature, the faster the electrons vibrate, and the shorter are the wavelengths of the emitted radiation. This can be visualized by attaching one end of a rope to a post and holding the other end. If the rope is shaken rapidly (high temperature), numerous short waves travel along the rope; if the rope is shaken slowly (lower temperature), longer waves appear on the rope. Although objects at a temperature of about 500°C radiate waves with many lengths, some of them are short enough to stimulate the sensation of vision. We actually see these objects glow red. Objects cooler than this radiate at wavelengths that are too long for us to see. The page of this book, for example, is radiating electromagnetic waves. But because its temperature is only about 20°C (68°F), the waves emitted are much too long to stimulate vision. We are able to see the

page, however, because light waves from other sources (such as light bulbs or the sun) are being *reflected* (bounced) off the paper. If this book were carried into a completely dark room, it would continue to radiate, but the pages would appear black because there are no visible light waves in the room to reflect off the pages.

Objects that have a very high temperature emit energy at a greater rate or intensity than objects at a lower temperature. Thus, *as the temperature of an object increases, more total radiation is emitted each second.* This can be expressed mathematically as

$$E = \sigma T^4 \text{ (Stefan-Boltzmann law)},$$

where E is the maximum rate of radiation emitted by each square meter of surface area of the object, σ (the Greek letter sigma) is the Stefan-Boltzmann constant,* and T is the object's surface temperature in degrees Kelvin. This relationship, called the **Stefan-Boltzmann law** after Josef Stefan (1835–1893) and Ludwig Boltzmann (1844–1906), who derived it, states that all objects with temperatures above absolute zero (0 K or −273°C) emit radiation at a rate proportional to the fourth power of their absolute temperature. Consequently, a small increase in temperature results in a large increase in the amount of radiation emitted because doubling the absolute temperature of an object increases the maximum energy output by a factor of 16, which is 2^4.

RADIATION OF THE SUN AND EARTH   Most of the sun's energy is emitted from its surface, where the temperature is nearly 6000 K (10,500°F). The earth, on the other hand, has an average surface temperature of 288 K (15°C, 59°F). The sun, therefore, radiates a great deal more energy than does the earth (see ● Fig. 2.9). At what wavelengths do the sun and the earth radiate most of their energy? Fortunately, the sun and the earth both have characteristics (discussed in a later section) that enable us to use the following relationship called **Wien's law** (or *Wien's displacement law*) after the German physicist Wilhelm Wien (pronounced Ween, 1864–1928), who discovered it:

$$\lambda_{max} = \frac{constant}{T} \text{ (Wien's law)}$$

where λ_{max} is the wavelength in micrometers at which maximum radiation emission occurs, T is the object's temperature in Kelvins, and the constant is 2897 μm K. To make the numbers easy to deal with, we will round off the constant to the number 3000.

For the sun, with a surface temperature of about 6000 K, the equation becomes

$$\lambda_{max} = \frac{300 \, \mu m \, K}{6000 \, K} = 0.5 \, \mu m.$$

---

*The Stefan-Boltzmann constant σ in SI units is 5.67×10^{-8} W/m²k⁴. A watt (W) is a unit of power where one watt equals one joule (J) per second (J/s). One joule is equal to 0.24 cal. More conversions are given in Appendix A.

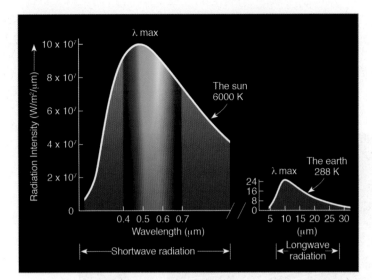

● FIGURE 2.9 The hotter sun not only radiates more energy than that of the cooler earth (the area under the curve), but it also radiates the majority of its energy at much shorter wavelengths. (The area under the curves is equal to the total energy emitted, and the scales for the two curves differ by a factor of 100,000.)

Thus, the sun emits a maximum amount of radiation at wavelengths near 0.5 μm. The cooler earth, with an average surface temperature of 288 K (rounded to 300 K), emits maximum radiation near wavelengths of 10 μm, since

$$\lambda_{max} = \frac{3000 \ \mu m \ K}{300 \ K} = 10 \ \mu m.$$

Thus, the earth emits most of its radiation at longer wavelengths between about 5 and 25 μm, while the sun emits the majority of its radiation at wavelengths less than 2 μm. For this reason, the earth's radiation *(terrestrial radiation)* is often called **longwave radiation,** whereas the sun's energy *(solar radiation)* is referred to as **shortwave radiation.**

Wien's law demonstrates that, as the temperature of an object increases, the wavelength at which maximum emission occurs is shifted toward shorter values. For example, if the sun's surface temperature were to double to 12,000 K, its

wavelength of maximum emission would be halved to about 0.25 μm. If, on the other hand, the sun's surface cooled to 3000 K, it would emit its maximum amount of radiation near 1.0 μm.

Even though the sun radiates at a maximum rate at a particular wavelength, it nonetheless emits some radiation at almost all other wavelengths. If we look at the amount of radiation given off by the sun at each wavelength, we obtain the sun's *electromagnetic spectrum.* A portion of this spectrum is shown in ● Fig. 2.10.

Since our eyes are sensitive to radiation between 0.4 and 0.7 μm, these waves reach the eye and stimulate the sensation of color. This portion of the spectrum is referred to as the **visible region,** and the radiant energy that reaches our eye is called *visible light.* The sun emits nearly 44 percent of its radiation in this zone, with the peak of energy output found at the wavelength corresponding to the color blue-green. The color violet is the shortest wavelength of visible light. Wavelengths shorter than violet (0.4 μm) are **ultraviolet (UV).** X-rays and gamma rays with exceedingly short wavelengths also fall into this category. The sun emits only about 7 percent of its total energy at ultraviolet wavelengths.

The longest wavelengths of visible light correspond to the color red. Wavelengths longer than red (0.7 μm) are **infrared (IR).** These waves cannot be seen by humans. Nearly 37 percent of the sun's energy is radiated between 0.7 μm and 1.5 μm, with only 12 percent radiated at wavelengths longer than 1.5 μm.

Whereas the hot sun emits only a part of its energy in the infrared portion of the spectrum, the relatively cool earth emits almost all of its energy at infrared wavelengths. Although we cannot see infrared radiation, there are instru-

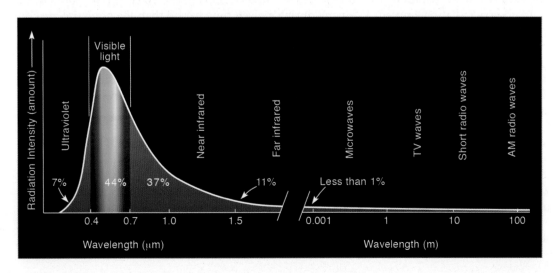

● FIGURE 2.10 The sun's electromagnetic spectrum and some of the descriptive names of each region. The numbers underneath the curve approximate the percent of energy the sun radiates in various regions.

FOCUS ON AN ENVIRONMENTAL ISSUE

Wave Energy, Sun Burning, and UV Rays

Standing close to a fire makes us feel warmer than we do when we stand at a distance from it. Does this mean that, as we move away from a hot object, the waves carry less energy and are, therefore, weaker? Not really. The intensity of radiation decreases as we move away from a hot object because radiant energy spreads outward in all directions. Figure 3 illustrates that, as the distance from a radiating object increases, a given amount of energy is distributed over a larger area, so that the energy received over a given area and over a given time decreases. In fact, at twice the distance from the source, the radiation is spread over four times the area.

Another interesting fact about radiation that we learned earlier in this chapter is that shorter waves carry much more energy than do longer waves. Hence, a photon of ultraviolet light carries more energy than a photon of visible light. In fact, ultraviolet (UV) wavelengths in the range of 0.20 and 0.29 μm (known as *UV–C radiation*) are harmful to living things, as certain waves can cause chromosome mutations, kill single-celled organisms, and damage the cornea of the eye. Fortunately, virtually all the ultraviolet radiation at wavelengths in the UV–C range is absorbed by ozone in the stratosphere.

Ultraviolet wavelengths between about 0.29 and 0.32 μm (known as *UV–B radiation*)

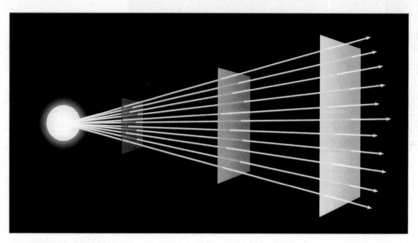

● FIGURE 3 The intensity, or amount, of radiant energy transported by electromagnetic waves decreases as we move away from a radiating object because the same amount of energy is spread over a larger area.

reach the earth in small amounts. Photons in this wavelength range have enough energy to produce sunburns and penetrate skin tissues, sometimes causing skin cancer. About 90 percent of all skin cancers are linked to sun exposure and UV–B radiation. Oddly enough, these same wavelengths activate provitamin D in the skin and convert it into vitamin D, which is essential to health.

Longer ultraviolet waves with lengths of about 0.32 to 0.40 μm (called *UV–A radiation*) are less energetic, but can still tan the skin. Al-

though UV–B is mainly responsible for burning the skin, UV–A can cause skin redness. It can also interfere with the skin's immune system and cause long-term skin damage that shows up years later as accelerated aging and skin wrinkling. Moreover, recent studies indicate that longer UV–A exposures needed to create a tan pose about the same cancer risk as a UV–B tanning dose.

Upon striking the human body, ultraviolet radiation is absorbed beneath the outer layer of skin. To protect the skin from these harmful

---

ments called *infrared sensors* that can. Weather satellites that orbit the globe use these sensors to observe radiation emitted by the earth, the clouds, and the atmosphere. Since objects of different temperatures radiate their maximum energy at different wavelengths, infrared photographs can distinguish among objects of different temperatures. Clouds always radiate infrared energy; thus, cloud images using infrared sensors can be taken during both day and night.

In summary, both the sun and earth emit radiation. The *hot sun* (6000 K) radiates nearly 88 percent of its energy at wavelengths less than 1.5 μm, with maximum emission in the *visible region* near 0.5 μm. The *cooler earth* (288 K) radiates nearly all its energy between 5 and 25 μm with a peak intensity in the *infrared region* near 10 μm (look back at Fig. 2.9). The sun's surface is nearly 20 times hotter than the earth's surface. From the Stefan-Boltzmann relationship, this fact means that

a unit area on the sun emits nearly 160,000 (20^4) times more energy during a given time period than the same size area on the earth. And since the sun has such a huge surface area from which to radiate, the total energy emitted by the sun each minute amounts to a staggering 6 billion, billion, billion calories! (Additional information on radiation intensity and its effect on humans is given in the Focus section above.)

Balancing Act—Absorption, Emission, and Equilibrium

If the earth and all things on it are continually radiating energy, why doesn't everything get progressively colder? The answer is that all objects not only radiate energy, they absorb it as well. If an object radiates more energy than it absorbs, it gets colder; if

rays, the body's defense mechanism kicks in. Certain cells (when exposed to UV radiation) produce a dark pigment *(melanin)* that begins to absorb some of the UV radiation. (It is the production of melanin that produces a tan.) Consequently, a body that produces little melanin—one with pale skin—has little natural protection from UV–B.

Additional protection can come from a sunscreen. Unlike the old lotions that simply moisturized the skin before it baked in the sun, sunscreens today block UV rays from ever reaching the skin. Some contain chemicals (such as zinc oxide) that reflect UV radiation. (These are the white pastes once seen on the noses of lifeguards.) Others consist of a mixture of chemicals (such as benzophenone and paraaminobenzoic acid, PABA) that actually absorb ultraviolet radiation, usually UV–B, although new products with UV–A-absorbing qualities are now on the market. The *Sun Protection Factor* (SPF) number on every container of sunscreen dictates how effective the product is in protecting from UV–B—the higher the number, the better the protection.

Protecting oneself from excessive exposure to the sun's energetic UV rays is certainly wise. Estimates are that, in a single year, over 30,000 Americans will be diagnosed with malignant melanoma, the most deadly form of skin cancer. And if the protective ozone shield should diminish even more over certain areas of the world, there is an ever-increasing risk of problems associated with UV–B. Using a good sunscreen and proper clothing can certainly help. The best way to protect yourself from too much sun, however, is to limit your time in direct sunlight, especially between the hours of 11 A.M. and 3 P.M. when the sun is highest in the sky and its rays are most direct.

Presently, the National Weather Service makes a daily prediction of UV radiation levels for selected cities throughout the United States.

EXPOSURE CATEGORY	UV INDEX	PROTECTIVE MEASURES
Minimal	0–2	Apply SPF 15 sunscreen
Low	3–4	Wear a hat and apply SPF 15 sunscreen
Moderate	5–6	Wear a hat, protective clothing, and sunglasses with UV-A and UV-B protection; apply SPF 15+ sunscreen
High	7–9	Wear a hat, protective clothing, and sunglasses; stay in shady areas; apply SPF 15+ sunscreen
Very high	10+	Wear a hat, protective clothing, and sunglasses; use SPF 15+ sunscreen; avoid being in sun between 10 A.M. and 4 P.M.

● FIGURE 4   The UV Index.

The forecast, known as the UV Index, gives the UV level at its peak, around noon standard time or 1 P.M. daylight savings time. The 15-point index corresponds to five exposure categories set by the Environmental Protection Agency (EPA). An index value of between 0 and 2 is considered "minimal," whereas a value of 10 or greater is deemed "very high" (see Fig. 4). Depending on skin type, a UV index of 10 means that in direct sunlight, (without sunscreen protection) a person's skin will likely begin to burn in about 6 to 30 minutes.

it absorbs more energy than it emits, it gets warmer. On a sunny day, the earth's surface warms by absorbing more energy from the sun and the atmosphere than it radiates, while at night the earth cools by radiating more energy than it absorbs from its surroundings. When an object emits and absorbs energy at equal rates, its temperature remains constant.

The rate at which something radiates and absorbs energy depends strongly on its surface characteristics, such as color, texture, and moisture, as well as temperature. For example, a black object in direct sunlight is a good absorber of visible radiation. It converts energy from the sun into internal energy, and its temperature ordinarily increases. You need only walk barefoot on a black asphalt road on a summer afternoon to experience this. At night, the blacktop road will cool quickly by emitting infrared radiation and, by early morning, it may be cooler than surrounding surfaces.

Any object that is a perfect absorber (that is, absorbs all the radiation that strikes it) and a perfect emitter (emits the maximum radiation possible at its given temperature) is called a **blackbody.** Blackbodies do not have to be colored black; they simply must absorb and emit all possible radiation. Since the earth's surface and the sun absorb and radiate with nearly 100 percent efficiency for their respective temperatures, they both behave as blackbodies. This is the reason we were able to use Wien's law and the Stefan-Boltzmann law to determine the characteristics of radiation emitted from the sun and the earth.

When we look at the earth from space, we see that half of it is in sunlight, the other half is in darkness. The outpouring of solar energy constantly bathes the earth with radiation, while the earth, in turn, constantly emits infrared radiation. If we assume that there is no other method of transferring heat, then, when the rate of absorption of solar radiation equals the rate of emission of infrared earth radiation, a state of *radiative equilibrium* is achieved. The average temperature at which this occurs is called the **radiative equilibrium temperature.** At this temperature, the earth (behaving as a blackbody) is absorbing solar radiation and emitting infrared radiation at equal rates, and its average temperature does not change. Because the earth is about 150 million km (93 million mi) from the sun, the earth's *radiative equilibrium temperature* is about 255 K ($-18°C$, $0°F$). But this temperature is *much* lower than the earth's observed average surface temperature of 288 K ($15°C$, $59°F$). Why is there such a large difference?

The answer lies in the fact that *the earth's atmosphere absorbs and emits infrared radiation.* Unlike the earth, the atmosphere does *not* behave like a blackbody, as it absorbs some wavelengths of radiation and is transparent to others. Objects that selectively absorb and emit radiation, such as gases in our atmosphere, are known as **selective absorbers.** Let's examine this concept more closely.

SELECTIVE ABSORBERS AND THE ATMOSPHERIC GREEN-HOUSE EFFECT

Just as some people are selective eaters of certain foods, most substances in our environment are selective absorbers; that is, they absorb only certain wavelengths of radiation. Glass is a good example of a selective absorber in that it absorbs some of the infrared and ultraviolet radiation it receives, but not the visible radiation that is transmitted through the glass. As a result, it is difficult to get a sunburn through the windshield of your car, although you can see through it.

Objects that selectively absorb radiation also selectively emit radiation at the same wavelength. This phenomenon is called **Kirchhoff's law.** This law states that *good absorbers are good emitters at a particular wavelength, and poor absorbers are poor emitters at the same wavelength.*[*]

Snow is a good absorber as well as a good emitter of infrared energy (white snow actually behaves as a blackbody in the infrared wavelengths). The bark of a tree absorbs sunlight and emits infrared energy, which the snow around it absorbs. During the absorption process, the infrared radiation is converted into internal energy, and the snow melts outward away from the tree trunk, producing a small depression that encircles the tree, like the ones shown in • Fig. 2.11.

• Figure 2.12 shows some of the most important selectively absorbing gases in our atmosphere. The shaded area represents the absorption characteristics of each gas at various wavelengths. Notice that both water vapor (H_2O) and carbon dioxide (CO_2) are strong absorbers of infrared radiation and poor absorbers of visible solar radiation. Other, less important, selective absorbers include nitrous oxide (N_2O), methane (CH_4), and ozone (O_3), which is most abundant in the stratosphere. As these gases absorb infrared radiation emitted from the earth's surface, they gain kinetic energy

● FIGURE 2.11 The melting of snow outward from the trees causes small depressions to form. The melting is caused mainly by the snow's absorption of the infrared energy being emitted from the warmer tree and its branches. The trees are warmer because they are better absorbers of sunlight than is the snow.

(energy of motion). The gas molecules share this energy by colliding with neighboring air molecules, such as oxygen and nitrogen (both of which are poor absorbers of infrared energy). These collisions increase the average kinetic energy of the air, which results in an increase in air temperature. Thus, most of the infrared energy emitted from the earth's surface keeps the lower atmosphere warm.

Besides being selective absorbers, water vapor and CO_2 selectively emit radiation at infrared wavelengths.[*] This radiation travels away from these gases in all directions. A portion of this energy is radiated toward the earth's surface and absorbed, thus heating the ground. The earth, in turn, constantly radiates infrared energy upward, where it is absorbed and warms the lower atmosphere. In this way, water vapor and CO_2 absorb and radiate infrared energy and act as an insulating layer around the earth, keeping part of the earth's infrared radiation from escaping rapidly into space. Consequently, the earth's surface and the lower atmosphere are much warmer than they would be if these selectively absorbing gases were not present. In fact, as we saw earlier, the earth's mean radiative equilibrium temperature without CO_2 and water vapor would be around $-18°C$ ($0°F$), or about $33°C$ ($59°F$) lower than at present.

The absorption characteristics of water vapor, CO_2, and other gases such as methane and nitrous oxide (depicted in Fig. 2.12) were, at one time, thought to be similar to the glass of a florist's greenhouse. In a greenhouse, the glass allows visible radiation to come in, but inhibits to some degree the passage of outgoing infrared radiation. For this reason, the absorption of infrared radiation from the earth by water vapor and CO_2 is popularly called the **greenhouse effect.** However, studies have shown that the warm air inside a

---

[*]Strictly speaking, this law only applies to gases.

[*]Nitrous oxide, methane, and ozone also emit infrared radiation, but their concentration in the atmosphere is much smaller than water vapor and carbon dioxide (see Table 1.1, p. 5.)

greenhouse is probably caused more by the air's inability to circulate and mix with the cooler outside air, rather than by the entrapment of infrared energy. Because of these findings, some scientists suggest that the greenhouse effect should be called the *atmosphere effect.* To accommodate everyone, we will usually use the term *atmospheric greenhouse effect* when describing the role that water vapor, CO_2, and other greenhouse gases* play in keeping the earth's mean surface temperature higher than it otherwise would be.

Look again at Fig. 2.12 and observe that, in the bottom diagram, there is a region between about 8 and 11 μm where neither water vapor nor CO_2 readily absorb infrared radiation. Because these wavelengths of emitted energy pass upward through the atmosphere and out into space, the wavelength range (between 8 and 11 μm) is known as the **atmospheric window.** Clouds can enhance the atmospheric greenhouse effect. Tiny liquid cloud droplets are selective absorbers in that they are good absorbers of infrared radiation but poor absorbers of visible solar radiation. Clouds even absorb the wavelengths between 8 and 11 μm, which are otherwise "passed up" by water vapor and CO_2. Thus, they have the effect of enhancing the atmospheric greenhouse effect by closing the atmospheric window.

Clouds—especially low, thick ones—are excellent emitters of infrared radiation. Their tops radiate infrared energy upward and their bases radiate energy back to the earth's surface where it is absorbed and, in a sense, radiated back to the clouds. This process keeps calm, cloudy nights warmer than calm, clear ones. If the clouds remain into the next day, they prevent much of the sunlight from reaching the ground by reflecting it back to space. Since the ground does not heat up as much as it would in full sunshine, cloudy, calm days are normally cooler than clear, calm days. Hence, the presence of clouds tends to keep nighttime temperatures higher and daytime temperatures lower.

In summary, the atmospheric greenhouse effect occurs because water vapor, CO_2, and other greenhouse gases are selective absorbers. They allow most of the sun's visible radiation to reach the surface, but they absorb a good portion of the earth's outgoing infrared radiation, preventing it from escaping into space (see ● Fig. 2.13). It is the atmospheric

---

*The term "greenhouse gases" derives from the standard use of "greenhouse effect." Greenhouse gases include, among others, water vapor, carbon dioxide, methane, nitrous oxide, and ozone.

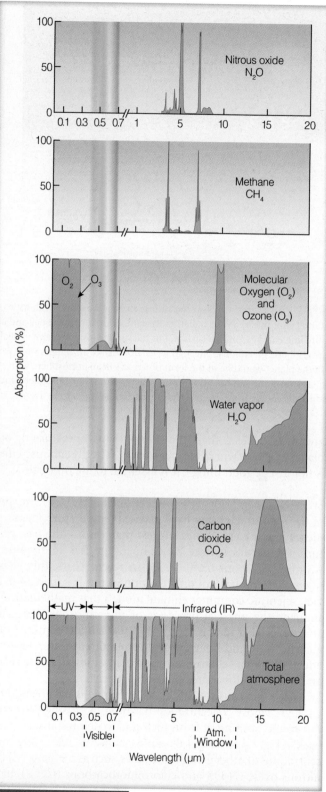

ACTIVE FIGURE 2.12 Absorption of radiation by gases in the atmosphere. The shaded area represents the percent of radiation absorbed by each gas. The strongest absorbers of infrared radiation are water vapor and carbon dioxide. The bottom figure represents the percent of radiation absorbed by all of the atmospheric gases. Visit the Meteorology Resource Center to view this and other active figures at academic.cengage.com/login

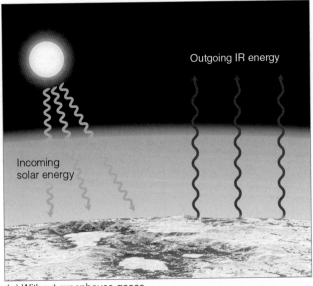

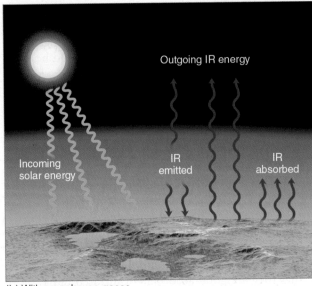

(a) Without greenhouse gases

(b) With greenhouse gases

● FIGURE 2.13 (a) Near the surface in an atmosphere with little or no greenhouse gases, the earth's surface would constantly emit infrared (IR) radiation upward, both during the day and at night. Incoming energy from the sun would equal outgoing energy from the surface, but the surface would receive virtually no IR radiation from its lower atmosphere. (No atmospheric greenhouse effect.) The earth's surface air temperature would be quite low, and small amounts of water found on the planet would be in the form of ice. (b) In an atmosphere with greenhouse gases, the earth's surface not only receives energy from the sun but also infrared energy from the atmosphere. Incoming energy still equals outgoing energy, but the added IR energy from the greenhouse gases raises the earth's average surface temperature to a more habitable level.

greenhouse effect, then, that keeps the temperature of our planet at a level where life can survive. The greenhouse effect is not just a "good thing"; it is essential to life on earth.

ENHANCEMENT OF THE GREENHOUSE EFFECT In spite of the inaccuracies that have plagued temperature measurements in the past, studies suggest that, during the past century, the earth's surface air temperature has undergone a warming of about 0.6°C (1°F). In recent years, this *global warming* trend has not only continued, but has increased. In fact, scientific computer climate models that mathematically simulate the physical processes of the atmosphere, oceans, and ice, predict that, if such a warming should continue unabated, we would be irrevocably committed to the negative effects of climate change, such as a continuing rise in sea level and a shift in global precipitation patterns.

The main cause of this global warming is the greenhouse gas CO_2, whose concentration has been increasing primarily due to the burning of fossil fuels and to deforestation. (Look back at Fig. 1.5 and Fig. 1.6 on p. 8). However, increasing concentrations of other greenhouse gases, such as methane (CH_4), nitrous oxide (N_2O), and chlorofluorocarbons (CFCs), have collectively been shown to have an effect almost equal to that of CO_2. Look at Fig. 2.12 and notice that both CH_4 and N_2O absorb strongly at infrared wavelengths. Moreover, a particular CFC (CFC-12) absorbs in the region of the atmospheric window between 8 and 11 μm. Thus, in terms of its absorption impact on infrared radiation, the addition of a single CFC-12 molecule to the atmosphere is the equivalent of adding 10,000 molecules of CO_2. Overall, water vapor accounts for about 60 percent of the atmospheric greenhouse effect, CO_2 accounts for about 26 percent, and the remaining greenhouse gases contribute about 14 percent.

Presently, the concentration of CO_2 in a volume of air near the surface is about 0.038 percent. Climate models predict that a continuing increase of CO_2 and other greenhouse gases will cause the earth's current average surface temperature to possibly rise an additional 3°C (5.4°F) by the end of the twenty-first century. How can increasing such a small quantity of CO_2 and adding miniscule amounts of other greenhouse gases bring about such a large temperature increase?

Mathematical climate models predict that rising ocean temperatures will cause an increase in evaporation rates. The added *water vapor*—the primary greenhouse gas—will enhance the atmospheric greenhouse effect and double the temperature rise in what is known as a *positive feedback*. But there are other feedbacks to consider.*

The two potentially largest and least understood feedbacks in the climate system are the clouds and the oceans. Clouds can change area, depth, and radiation properties simultaneously with climatic changes. The net effect of all these changes is not totally clear at this time. Oceans, on the other hand, cover 70 percent of the planet. The response of ocean circulations, ocean temperatures, and sea ice to global

---

*A feedback is a process whereby an initial change in a process will tend to either reinforce the process (positive feedback) or weaken the process (negative feedback). The *water vapor–greenhouse* feedback is a positive feedback because the initial increase in temperature is reinforced by the addition of more water vapor, which absorbs more of the earth's infrared energy, thus strengthening the greenhouse effect and enhancing the warming.

warming will determine the global pattern and speed of climate change. Unfortunately, it is not now known how quickly each of these feedbacks will respond.

Satellite data from the *Earth Radiation Budget Experiment* (ERBE) suggest that clouds overall appear to *cool* the earth's climate, as they reflect and radiate away more energy than they retain. (The earth would be warmer if clouds were not present.) So an increase in global cloudiness (if it were to occur) might offset some of the global warming brought on by an enhanced atmospheric greenhouse effect. Therefore, if clouds were to act on the climate system in this manner, they would provide a *negative feedback* on climate change.*

Uncertainties unquestionably exist about the impact that increasing levels of CO_2 and other greenhouse gases will have on enhancing the atmospheric greenhouse effect. Nonetheless, the most recent studies on climate change say that climate change is presently occurring worldwide due primarily to increasing levels of greenhouse gases. The evidence for this conclusion comes from increases in global average air and ocean temperatures, as well as from the widespread melting of snow and ice, and rising sea levels. (We will examine the topic of climate change in more detail in Chapter 16.)

BRIEF REVIEW

In the last several sections, we have explored examples of some of the ways radiation is absorbed and emitted by various objects. Before reading the next several sections, let's review a few important facts and principles:

- *All* objects with a temperature above absolute zero emit radiation.
- The higher an object's temperature, the greater the amount of radiation emitted per unit surface area and the shorter the wavelength of maximum emission.
- The earth absorbs solar radiation only during the daylight hours; however, it emits infrared radiation continuously, both during the day and at night.
- The earth's surface behaves as a blackbody, making it a much better absorber and emitter of radiation than the atmosphere.
- Water vapor and carbon dioxide are important atmospheric greenhouse gases that selectively absorb and emit infrared radiation, thereby keeping the earth's average surface temperature warmer than it otherwise would be.
- Cloudy, calm nights are often warmer than clear, calm nights because clouds strongly emit infrared radiation back to the earth's surface.
- It is *not* the greenhouse effect itself that is of concern, but the *enhancement* of it due to increasing levels of greenhouse gases.
- As greenhouse gases continue to increase in concentration, the average surface air temperature is projected to rise substantially by the end of this century.

---

*Overall, the most recent climate models tend to show that changes in clouds would provide a small positive feedback on climate change.

With these concepts in mind, we will first examine how the air near the ground warms; then we will consider how the earth and its atmosphere maintain a yearly energy balance.

WARMING THE AIR FROM BELOW   If you look back at Fig. 2.12 (p. 43), you'll notice that the atmosphere does not readily absorb radiation with wavelengths between 0.3 μm and 1.0 μm, the region where the sun emits most of its energy. Consequently, on a clear day, solar energy passes through the lower atmosphere with little effect upon the air. Ultimately it reaches the surface, warming it (see ● Fig. 2.14). Air molecules in contact with the heated surface bounce against it, gain energy by *conduction,* then shoot upward like freshly popped kernels of corn, carrying their energy with them. Because the air near the ground is very dense, these molecules only travel a short distance (about 10^{-7} m) before they collide with other molecules. During the collision, these more rapidly moving molecules share their energy with less energetic molecules, raising the average temperature of the air. But air is such a poor heat conductor that this process is only important within a few centimeters of the ground.

As the surface air warms, it actually becomes less dense than the air directly above it. The warmer air rises and the cooler air sinks, setting up thermals, or *free convection cells,* that transfer heat upward and distribute it through a deeper layer of air. The rising air expands and cools, and, if sufficiently moist, the water vapor condenses into cloud droplets, releasing latent heat that warms the air. Meanwhile, the earth constantly emits infrared energy. Some of this energy is absorbed by greenhouse gases (such as water vapor and carbon dioxide) that emit infrared energy upward and downward, back to the surface. Since the concentration of water vapor decreases rapidly above the earth, most of the absorption occurs in a layer near the surface. Hence, the lower atmosphere is mainly heated from the ground upward.

Incoming Solar Energy

As the sun's radiant energy travels through space, essentially nothing interferes with it until it reaches the atmosphere. At the top of the atmosphere, solar energy received on a surface perpendicular to the sun's rays appears to remain fairly constant at nearly two calories on each square centimeter each minute or 1367 W/m^2—a value called the **solar constant.***

SCATTERED AND REFLECTED LIGHT   When solar radiation enters the atmosphere, a number of interactions take place. For example, some of the energy is absorbed by gases, such as ozone, in the upper atmosphere. Moreover, when sunlight

---

*By definition, the solar constant (which, in actuality, is *not* "constant") is the rate at which radiant energy from the sun is received on a surface at the outer edge of the atmosphere perpendicular to the sun's rays when the earth is at an average distance from the sun. Satellite measurements from the *Earth Radiation Budget Satellite* suggest the solar constant varies slightly as the sun's radiant output varies. The average is about 1.96 $cal/cm^2/min$, or between 1365 W/m^2 and 1372 W/m^2 in the SI system of measurement.

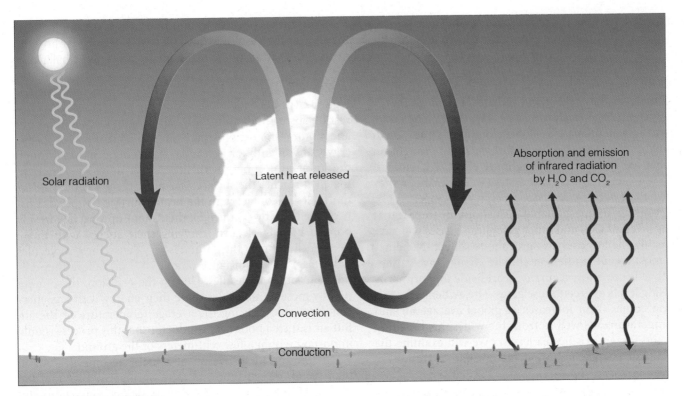

ACTIVE FIGURE 2.14 Air in the lower atmosphere is heated from the ground upward. Sunlight warms the ground, and the air above is warmed by conduction, convection, and infrared radiation. Further warming occurs during condensation as latent heat is given up to the air inside the cloud. Visit the Meteorology Resource Center to view this and other active figures at academic.cengage.com/login

WEATHER WATCH

Talk about an enhanced greenhouse effect! The atmosphere of Venus, which is mostly carbon dioxide, is considerably more dense than that of Earth. Consequently, the greenhouse effect on Venus is exceptionally strong, producing a surface air temperature of about 500°C, or nearly 950°F.

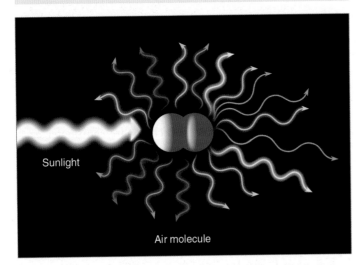

• FIGURE 2.15 The scattering of light by air molecules. Air molecules tend to selectively scatter the shorter (violet, green, and blue) wavelengths of visible white light more effectively than the longer (orange, yellow, and red) wavelengths.

strikes very small objects, such as air molecules and dust particles, the light itself is deflected in all directions—forward, sideways, and backwards (see • Fig. 2.15). The distribution of light in this manner is called **scattering.** (Scattered light is also called *diffuse light.*) Because air molecules are much smaller than the wavelengths of visible light, they are more effective scatterers of the shorter (blue) wavelengths than the longer (red) wavelengths. Hence, when we look away from the direct beam of sunlight, blue light strikes our eyes from all directions, turning the daytime sky blue. (More information on the effect of scattered light and what we see is given in the Focus section on p. 47.)

Sunlight can be **reflected** from objects. Generally, reflection differs from scattering in that during the process of reflection more light is sent *backwards*. **Albedo** is the percent of radiation returning from a given surface compared to the amount of radiation initially striking that surface. Albedo, then, represents the *reflectivity* of the surface. In ▼ Table 2.3, notice that thick clouds have a higher albedo than thin clouds. On the average, the albedo of clouds is near 60 percent. When solar energy strikes a surface covered with snow, up to 95 percent of the sunlight may be reflected. Most of this energy is in the visible and ultraviolet wavelengths. Consequently, reflected radiation, coupled with direct sunlight, can produce severe sunburns on the exposed skin of unwary snow skiers, and unprotected eyes can suffer the agony of snow blindness.

Water surfaces, on the other hand, reflect only a small amount of solar energy. For an entire day, a smooth water

Blue Skies, Red Suns, and White Clouds

We know that the sky is blue because air molecules selectively scatter the shorter wavelengths of visible light—green, violet, and blue waves—more effectively than the longer wavelengths of red, orange, and yellow (see Fig. 2.14). When these shorter waves reach our eyes, the brain processes them as the color "blue." Therefore, on a clear day when we look up, blue light strikes our eyes from all directions, making the sky appear blue.

At noon, the sun is perceived as white because all the waves of visible sunlight strike our eyes (see Fig. 5). At sunrise and sunset, the white light from the sun must pass through a thick portion of the atmosphere. Scattering of light by air molecules (and particles) removes the shorter waves (blue light) from the beam, leaving the longer waves of red, orange, and yellow to pass on through. This situation often creates the image of a ruddy sun at sunrise and sunset. An observer at sunrise or sunset in Fig. 5 might see a sun similar to the one shown in Fig. 6.

The sky is blue, but why are clouds white? Cloud droplets are much larger than air molecules and do not selectively scatter sunlight. Instead, these larger droplets scatter all wavelengths of visible light more or less equally (see Fig. 7). Hence, clouds appear white because millions of cloud droplets scatter all wavelengths of visible light about equally in all directions.

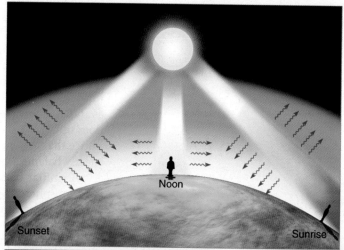

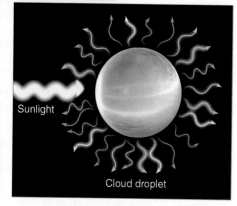

© C. Donald Ahrens

● FIGURE 5 At noon, the sun usually appears a bright white. At sunrise and at sunset, sunlight must pass through a thick portion of the atmosphere. Much of the blue light is scattered out of the beam, causing the sun to appear more red.

● FIGURE 6 A red sunset produced by the process of scattering.

● FIGURE 7 Cloud droplets scatter all wavelengths of visible white light about equally. This type of scattering by millions of tiny cloud droplets makes clouds appear white.

surface will have an average albedo of about 10 percent. Water has the highest albedo (and can therefore reflect sunlight best) when the sun is low on the horizon and the water is a little choppy. This may explain why people who wear brimmed hats while fishing from a boat in choppy water on a sunny day can still get sunburned during midmorning or midafternoon. Averaged for an entire year, the earth and its atmosphere (including its clouds) will redirect about 30 percent of the sun's incoming radiation back to space, which gives the earth and its atmosphere a combined albedo of 30 percent (see ● Fig. 2.16).

THE EARTH'S ANNUAL ENERGY BALANCE Although the average temperature at any one place may vary considerably

from year to year, the earth's overall average equilibrium temperature changes only slightly from one year to the next. This fact indicates that, each year, the earth and its atmosphere combined must send off into space just as much energy as they receive from the sun. The same type of energy balance must exist between the earth's surface and the atmosphere. That is, each year, the earth's surface must return to the atmosphere the same amount of energy that it absorbs. If this did not occur, the earth's average surface temperature would change. How do the earth and its atmosphere maintain this yearly energy balance?

Suppose 100 units of solar energy reach the top of the earth's atmosphere. We can see in Fig. 2.16 that, on the average, clouds, the earth, and the atmosphere reflect and scatter

▼ TABLE 2.3 Typical Albedo of Various Surfaces

SURFACE	ALBEDO (PERCENT)
Fresh snow	75 to 95
Clouds (thick)	60 to 90
Clouds (thin)	30 to 50
Venus	78
Ice	30 to 40
Sand	15 to 45
Earth and atmosphere	30
Mars	17
Grassy field	10 to 30
Dry, plowed field	5 to 20
Water	10*
Forest	3 to 10
Moon	7

*Daily average.

30 units back to space, and that the atmosphere and clouds together absorb 19 units, which leaves 51 units of direct and indirect solar radiation to be absorbed at the earth's surface. ● Figure 2.17 shows approximately what happens to the solar radiation that is absorbed by the surface and the atmosphere.

Out of 51 units reaching the surface, a large amount (23 units) is used to evaporate water, and about 7 units are lost through conduction and convection, which leaves 21 units to be radiated away as infrared energy. Look closely at Fig. 2.17 and notice that the earth's surface actually radiates upward a whopping 117 units. It does so because, although it receives solar radiation only during the day, it constantly emits infrared energy both during the day and at night. Additionally, the atmosphere above only allows a small fraction of this energy (6 units) to pass through into space. The majority of it (111 units) is absorbed mainly by the greenhouse gases water vapor and CO_2, and by clouds. Much of this energy (96 units) is radiated back to earth, producing the atmospheric greenhouse effect. Hence, the earth's surface receives nearly twice as much longwave infrared energy from its atmosphere as it does shortwave radiation from the sun. In all these exchanges, notice that the energy lost at the earth's surface (147 units) is exactly balanced by the energy gained there (147 units).

A similar balance exists between the earth's surface and its atmosphere. Again in Fig. 2.17 observe that the energy gained by the atmosphere (160 units) balances the energy lost. Moreover, averaged for an entire year, the solar energy received at the earth's surface (51 units) and that absorbed by the earth's atmosphere (19 units) balances the infrared energy lost to space by the earth's surface (6 units) and its atmosphere (64 units).

We can see the effect that conduction, convection, and latent heat play in the warming of the atmosphere if we look at the energy balance only in radiative terms. The earth's surface receives 147 units of radiant energy from the sun and its

● FIGURE 2.16 On the average, of all the solar energy that reaches the earth's atmosphere annually, about 30 percent ($^{30}/_{100}$) is reflected and scattered back to space, giving the earth and its atmosphere an albedo of 30 percent. Of the remaining solar energy, about 19 percent is absorbed by the atmosphere and clouds, and 51 percent is absorbed at the surface.

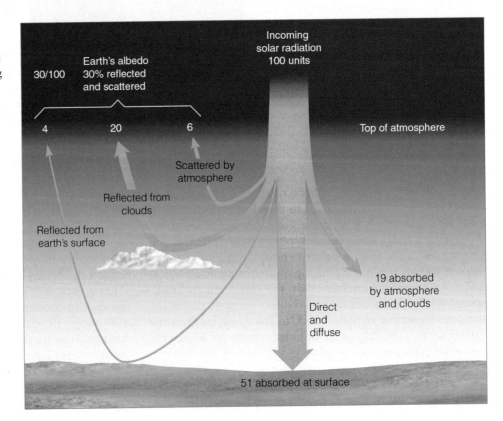

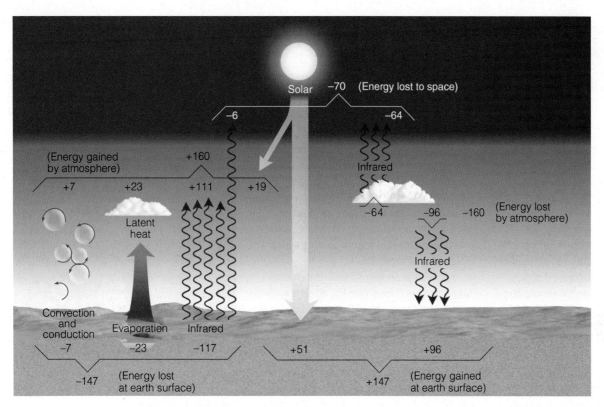

●FIGURE 2.17
The earth-atmosphere energy balance. Numbers represent approximations based on surface observations and satellite data. While the actual value of each process may vary by several percent, it is the relative size of the numbers that is important.

own atmosphere, while it radiates away 117 units, producing a *surplus* of 30 units. The atmosphere, on the other hand, receives 130 units (19 units from the sun and 111 from the earth), while it loses 160 units, producing a *deficit* of 30 units. The balance (30 units) is the warming of the atmosphere produced by the heat transfer processes of conduction and convection (7 units) and by the release of latent heat (23 units).

And so, the earth and the atmosphere absorb energy from the sun, as well as from each other. In all of the energy exchanges, a delicate balance is maintained. Essentially, there is no yearly gain or loss of total energy, and the average temperature of the earth and the atmosphere remains fairly constant from one year to the next. This equilibrium does not imply that the earth's average temperature does not change, but that the changes are small from year to year (usually less than one-tenth of a degree Celsius) and become significant only when measured over many years.

Even though the earth and the atmosphere together maintain an annual energy balance, such a balance is not maintained at each latitude. High latitudes tend to lose more energy to space each year than they receive from the sun, while low latitudes tend to gain more energy during the course of a year than they lose. From ● Fig. 2.18 we can see that only at middle latitudes near 38° does the amount of energy received each year balance the amount lost. From this situation, we might conclude that polar regions are growing colder each year, while tropical regions are becoming warmer. But this does not happen. To compensate for these gains and losses of energy, winds in the atmosphere and currents in the oceans circulate warm air and water toward the poles, and

cold air and water toward the equator. Thus, the transfer of heat energy by atmospheric and oceanic circulations prevents low latitudes from steadily becoming warmer and high latitudes from steadily growing colder. These circulations are extremely important to weather and climate, and will be treated more completely in Chapter 10.

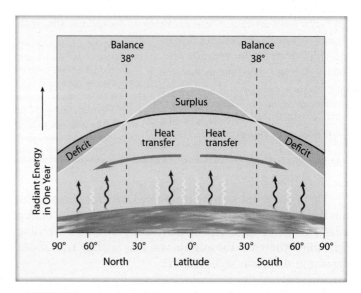

●FIGURE 2.18 The average annual incoming solar radiation (yellow lines) absorbed by the earth and the atmosphere along with the average annual infrared radiation (red lines) emitted by the earth and the atmosphere.

FOCUS ON A SPECIAL TOPIC

Characteristics of the Sun

The sun is our nearest star. It is some 150 million km (93 million mi) from earth. The next star, Alpha Centauri, is more than 250,000 times farther away. Even though the earth only receives about one two-billionths of the sun's total energy output, it is this energy that allows life to flourish. Sunlight determines the rate of photosynthesis in plants and strongly regulates the amount of evaporation from the oceans. It warms this planet and drives the atmosphere into the dynamic patterns we experience as everyday wind and weather. Without the sun's radiant energy, the earth would gradually cool, in time becoming encased in a layer of ice! Evidence of life on the cold, dark, and barren surface would be found only in fossils. Fortunately, the sun has been shining for billions of years, and it is likely to shine for at least several billion more.

The sun is a giant celestial furnace. Its core is extremely hot, with a temperature estimated to be near 15 million degrees Celsius. In the core, hydrogen nuclei (protons) collide at such fantastically high speeds that they fuse together to form helium nuclei. This thermonuclear process generates an enormous amount of energy, which gradually works its way to the sun's outer luminous surface—the *photosphere* ("sphere of light"). Temperatures here are much cooler than in the interior, generally near 6000°C. We have noted already that a body with this surface temperature emits radiation at a maximum rate in the visible region of the spectrum. The sun is, therefore, a shining example of such an object.

Dark blemishes on the photosphere called *sunspots* are huge, cooler regions that typically average more than five times the diameter of the earth. Although sunspots are not well understood, they are known to be regions of strong magnetic fields. They are cyclic, with the maximum number of spots occurring approximately every eleven years.

Above the photosphere are the *chromosphere* and the *corona* (see Fig. 8). The chromosphere ("color sphere") acts as a boundary between the relatively cool (6000°C) photosphere and the much hotter (2,000,000°C) corona, the outermost envelope of the solar atmosphere. During a solar eclipse, the corona is visible. It appears as a pale, milky cloud encircling the sun. Although much hotter than the photosphere, the corona radiates much less energy because its density is extremely low. This very thin solar atmosphere extends into space for many millions of kilometers.*

Violent solar activity occasionally occurs in the regions of sunspots. The most dramatic of these events are *prominences* and *flares*. Prominences are huge cloudlike jets of gas that often shoot up into the corona in the form of an arch. Solar flares are tremendous, but brief, eruptions. They emit large quantities of high-

---

*During a solar eclipse or at any other time, you should not look at the sun's corona either with sunglasses or through exposed negatives. Take this warning seriously. Viewing just a small area of the sun directly permits large amounts of UV radiation to enter the eye, causing serious and permanent damage to the retina. View the sun by projecting its image onto a sheet of paper, using a telescope or pinhole camera.

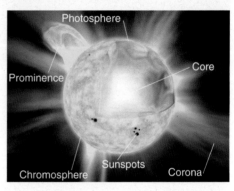

● FIGURE 8 Various regions of the sun.

energy ultraviolet radiation, as well as energized charged particles, mainly protons and electrons, which stream outward away from the sun at extremely high speeds.

An intense solar flare can disturb the earth's magnetic field, producing a so-called *magnetic storm.* Because these storms can intensify the electrical properties of the upper atmosphere, they are often responsible for interruptions in radio and satellite communications. One such storm knocked out electricity throughout the province of Quebec, Canada, during March, 1989. And in May, 1998, after a period of intense solar activity, a communications satellite failed, causing 45 million pagers to suddenly go dead.

More recently, a sudden burst of radio waves from an energetic flare overwhelmed dozens of radio receivers linked to the Global Positioning System (GPS) satellites, causing a widespread loss of GPS signals in New Mexico and Colorado.

Up to this point we have considered radiant energy of the sun and earth. Before we turn our attention to how incoming solar energy, in the form of particles, produces a dazzling light show known as the aurora, you may wish to read about the sun in the Focus section above.

SOLAR PARTICLES AND THE AURORA From the sun and its tenuous atmosphere comes a continuous discharge of particles. This discharge happens because, at extremely high temperatures, gases become stripped of electrons by violent

collisions and acquire enough speed to escape the gravitational pull of the sun.

As these charged particles (ions and electrons) travel through space, they are known as *plasma*, or **solar wind.** When the solar wind moves close enough to the earth, it interacts with the earth's magnetic field.

The magnetic field that surrounds the earth is much like the field around an ordinary bar magnet (see ● Fig. 2.19). Both have north and south magnetic poles, and both have invisible lines of force (field lines) that link the poles. On the

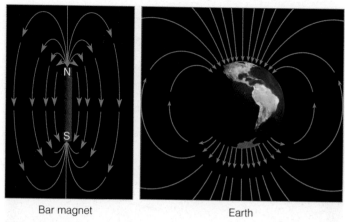

Bar magnet

Earth

●FIGURE 2.19 A magnetic field surrounds the earth just as it does a bar magnet.

NCAR/UCAR/NSF

●FIGURE 2.20 The stream of charged particles from the sun—called the *solar wind*—distorts the earth's magnetic field into a teardrop shape known as the *magnetosphere.*

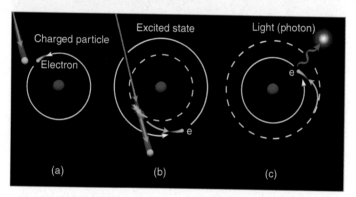

●FIGURE 2.21 When an excited atom, ion, or molecule de-excites, it can emit visible light. (a) The electron in its normal orbit becomes excited by a charged particle and (b) jumps into a higher energy level. When the electron returns to its normal orbit, it (c) emits a photon of light.

earth, these field lines form closed loops as they enter near the magnetic North pole and leave near the magnetic South pole. Most scientists believe that an electric current coupled with fluid motions deep in the earth's hot molten core is responsible for its magnetic field. This field protects the earth, to some degree, from the onslaught of the solar wind.

Observe in ● Fig. 2.20 that, when the solar wind encounters the earth's magnetic field, it severely deforms it into a teardrop-shaped cavity known as the *magnetosphere.* On the side facing the sun, the pressure of the solar wind compresses the field lines. On the opposite side, the magnetosphere stretches out into a long tail—the *magnetotail*—which reaches far beyond the moon's orbit. In a way, the magnetosphere acts as an obstacle to the solar wind by causing some of its particles to flow around the earth.

Inside the earth's magnetosphere are ionized gases. Some of these gases are solar wind particles, while others are ions from the earth's upper atmosphere that have moved upward along electric field lines into the magnetosphere.

Normally, the solar wind approaches the earth at an average speed of 400 km/sec. However, during periods of high solar activity (many sunspots and flares), the solar wind is more dense, travels much faster, and carries more energy. When these energized solar particles reach the earth, they cause a variety of effects, such as changing the shape of the magnetosphere and producing auroral displays.

The aurora is not reflected light from the polar ice fields, nor is it light from demons' lanterns as they search for lost souls. The *aurora* is produced by the solar wind disturbing the magnetosphere. The disturbance involves high-energy particles within the magnetosphere being ejected into the earth's upper atmosphere, where they excite atoms and molecules. The excited atmospheric gases emit visible radiation, which causes the sky to glow like a neon light. Let's examine this process more closely.

A high-energy particle from the magnetosphere will, upon colliding with an air molecule (or atom), transfer some of its energy to the molecule. The molecule then becomes

excited (see ● Fig. 2.21). Just as excited football fans leap up when their favorite team scores the winning touchdown, electrons in an excited molecule jump into a higher energy level as they orbit its center. As the fans sit down after all the excitement is over, so electrons quickly return to their lower level. When molecules de-excite, they release the energy originally received from the energetic particle, either all at once (one big jump), or in steps (several smaller jumps). This emitted energy is given up as radiation. If its wavelength is in the visible range, we see it as visible light. In the Northern Hemisphere, we call this light show the **aurora borealis,** or *northern lights;* its counterpart in the Southern Hemisphere is the **aurora australis,** or *southern lights.*

Since each atmospheric gas has its own set of energy levels, each gas has its own characteristic color. For example, the de-excitation of atomic oxygen can emit green or red light. Molecular nitrogen gives off red and violet light. The shades

In the figure 2.21 labels: Charged particle, Electron, Excited state, Light (photon), e, (a), (b), (c)

● FIGURE 2.22 The aurora borealis is a phenomenon that forms as energetic particles from the sun interact with the earth's atmosphere.

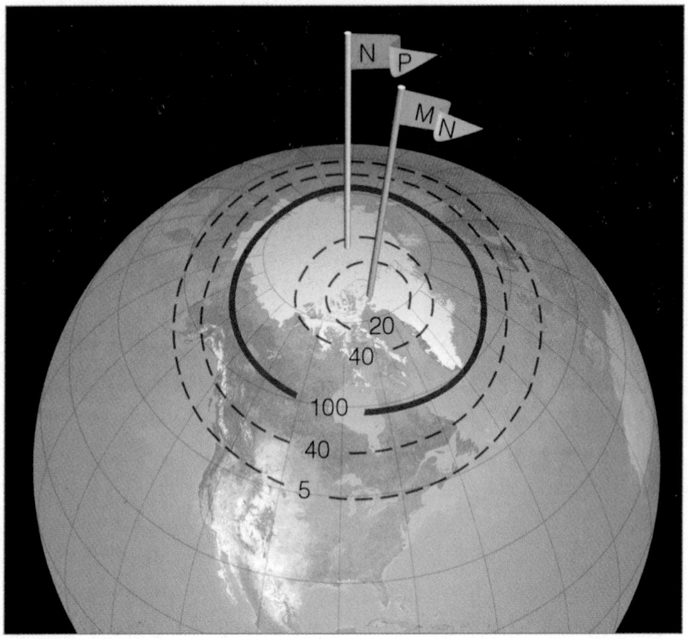

● FIGURE 2.23 The aurora belt (solid red line) represents the region where you would most likely observe the aurora on a clear night. (The numbers represent the average number of nights per year on which you might see an aurora if the sky were clear.) The flag MN denotes the magnetic North Pole, where the earth's magnetic field lines emerge from the earth. The flag NP denotes the geographic North Pole, about which the earth rotates.

of these colors can be spectacular as they brighten and fade, sometimes in the form of waving draperies, sometimes as unmoving, yet flickering, arcs and soft coronas. On a clear, quiet night the aurora is an eerie yet beautiful spectacle. (See ● Fig. 2.22 and the chapter-opening photograph on p. 28.)

The aurora is most frequently seen in polar latitudes. Energetic particles trapped in the magnetosphere move along the earth's magnetic field lines. Because these lines emerge from the earth near the magnetic poles, it is here that the particles interact with atmospheric gases to produce an aurora. Notice in ● Fig. 2.23 that the zone of most frequent auroral sightings (aurora belt) is not at the magnetic pole (marked by the flag MN), but equatorward of it, where the field lines emerge from the earth's surface. At lower latitudes, where the field lines are oriented almost horizontal to the earth's surface, the chances of seeing an aurora diminish rapidly.

On rare occasions, however, the aurora is seen in the southern United States. Such sightings happen only when the sun is very active—as giant flares hurl electrons and protons earthward at a fantastic rate. These particles move so fast that some of them penetrate unusually deep into the earth's magnetic field before they are trapped by it. In a process not fully understood, particles from the magnetosphere are acceler-ated toward the earth along electrical field lines that parallel the magnetic field lines. The acceleration of these particles gives them sufficient energy so that when they enter the upper atmosphere they are capable of producing an auroral display much farther south than usual.

How high above the earth is the aurora? The exact height appears to vary, but it is almost always observed within the thermosphere. The base of an aurora is rarely lower than 80 km, and it averages about 105 km. Since the light of an aurora gradually fades, it is difficult to define an exact upper limit. Most auroras, however, are observed below 200 km (124 mi).

In summary, energy for the aurora comes from the solar wind, which disturbs the earth's magnetosphere. This disturbance causes energetic particles to enter the upper atmosphere, where they collide with atoms and molecules. The atmospheric gases become excited and emit energy in the form of visible light.

But there is other light coming from the atmosphere—a faint glow at night much weaker than the aurora. This feeble luminescence, called **airglow,** is detected at all latitudes and shows no correlation with solar wind activity. Apparently, this light comes from ionized oxygen and nitrogen and other gases that have been excited by solar radiation.

SUMMARY

In this chapter, we have seen how the concepts of heat and temperature differ and how heat is transferred in our environment. We learned that latent heat is an important source of atmospheric heat energy. We also learned that conduction, the transfer of heat by molecular collisions, is most effective in solids. Because air is a poor heat conductor, conduction in the atmosphere is only important in the shallow layer of air in contact with the earth's surface. A more important process of atmospheric heat transfer is convection, which involves the mass movement of air (or any fluid) with its energy from one region to another. Another significant heat transfer process is radiation—the transfer of energy by means of electromagnetic waves.

The hot sun emits most of its radiation as shortwave radiation. A portion of this energy heats the earth, and the earth, in turn, warms the air above. The cool earth emits most of its radiation as longwave infrared radiation. Selective absorbers in the atmosphere, such as water vapor and carbon dioxide, absorb some of the earth's infrared radiation and radiate a portion of it back to the surface, where it warms the surface, producing the atmospheric greenhouse effect. Because clouds are both good absorbers and good emitters of infrared radiation, they keep calm, cloudy nights warmer than calm, clear nights. The average equilibrium temperature of the earth and the atmosphere remains fairly constant from one year to the next because the amount of energy they absorb each year is equal to the amount of energy they lose.

Finally, we examined how the sun's energy in the form of solar wind particles interacts with our atmosphere to produce auroral displays.

KEY TERMS

The following terms are listed (with page numbers) in the order they appear in the text. Define each. Doing so will aid you in reviewing the material covered in this chapter.

energy, 30
potential energy, 30
kinetic energy, 30
temperature, 30
heat, 31
absolute zero, 31
Kelvin scale, 31
Fahrenheit scale, 31
Celsius scale, 31
heat capacity, 32
specific heat, 32
latent heat, 32
sensible heat, 33
conduction, 34

convection, 34
thermals, 35
advection, 36
radiant energy (radiation), 37
electromagnetic waves, 37
radiant energy (radiation), 37
electromagnetic waves, 37
wavelength, 37
micrometer, 37
photon, 37
Stefan-Boltzmann law, 38
Wien's law, 38
longwave radiation, 39
shortwave radiation, 39

visible region, 39
ultraviolet (UV) radiation, 39
infrared (IR) radiation, 39
blackbody, 41
radiative equilibrium temperature, 42
selective absorbers, 42
Kirchhoff's law, 42
greenhouse effect, 42

atmospheric window, 43
solar constant, 45
scattering, 46
reflected (light), 46
albedo, 46
solar wind, 50
aurora borealis, 51
aurora australis, 51
airglow, 52

QUESTIONS FOR REVIEW

1. How does the average speed of air molecules relate to the air temperature?
2. Distinguish between temperature and heat.
3. (a) How does the Kelvin temperature scale differ from the Celsius scale?
   (b) Why is the Kelvin scale often used in scientific calculations?
   (c) Based on your experience, would a temperature of 250 K be considered warm or cold? Explain.
4. Explain how in winter heat is transferred by:
   (a) conduction;
   (b) convection;
   (c) radiation.
5. How is latent heat an important source of atmospheric energy?
6. In the atmosphere, how does advection differ from convection?
7. How does the temperature of an object influence the radiation that it emits?
8. How does the amount of radiation emitted by the earth differ from that emitted by the sun?
9. How do the wavelengths of most of the radiation emitted by the sun differ from those emitted by the surface of the earth?
10. Which photon carries the most energy—infrared, visible, or ultraviolet?
11. When a body reaches a radiative equilibrium temperature, what is taking place?
12. If the earth's surface continually radiates energy, why doesn't it become colder and colder?
13. Why are carbon dioxide and water vapor called selective absorbers?
14. Explain how the earth's atmospheric greenhouse effect works.
15. What gases appear to be responsible for the enhancement of the earth's greenhouse effect?

16. Why do most climate models predict that the earth's average surface temperature will increase by an additional 3.0°C (5.4°F) by the end of this century?

17. What processes contribute to the earth's albedo being 30 percent?

18. Explain how the atmosphere near the earth's surface is warmed from below.

19. If a blackbody is a theoretical object, why can both the sun and earth be treated as blackbodies?

20. What is the solar wind?

21. Explain how the aurora is produced.

QUESTIONS FOR THOUGHT

1. Explain why the bridge in the diagram is the first to become icy.

•FIGURE 2.24

2. Explain why the first snowfall of the winter usually "sticks" better to tree branches than to bare ground.

3. At night, why do materials that are poor heat conductors cool to temperatures less than the surrounding air?

4. Explain how, in winter, ice can form on puddles (in shaded areas) when the temperature above and below the puddle is slightly above freezing.

5. In northern latitudes, the oceans are warmer in summer than they are in winter. In which season do the oceans lose heat most rapidly to the atmosphere by conduction? Explain.

6. How is heat transferred away from the surface of the moon? (Hint: The moon has no atmosphere.)

7. Why is ultraviolet radiation more successful in dislodging electrons from air atoms and molecules than is visible radiation?

8. Why must you stand closer to a small fire to experience the same warmth you get when standing farther away from a large fire?

9. If water vapor were no longer present in the atmosphere, how would the earth's energy budget be affected?

10. Which will show the greatest increase in temperature when illuminated with direct sunlight: a plowed field or a blanket of snow? Explain.

11. Why does the surface temperature often increase on a clear, calm night as a low cloud moves overhead?

12. Which would have the greatest effect on the earth's greenhouse effect: removing all of the CO_2 from the atmosphere or removing all of the water vapor? Explain why you chose your answer.

13. Explain why an increase in cloud cover surrounding the earth would increase the earth's albedo, yet not necessarily lead to a lower earth surface temperature.

14. Could a liquid thermometer register a temperature of −273°C when the air temperature is actually 1000°C? Where would this happen in the atmosphere, and why?

15. Why is it that auroral displays above Colorado can be forecast several days in advance?

16. Why does the aurora usually occur more frequently above Maine than above Washington State?

PROBLEMS AND EXERCISES

1. Suppose that 500 g of water vapor condense to make a cloud about the size of an average room. If we assume that the latent heat of condensation is 600 cal/g, how much heat would be released to the air? If the total mass of air before condensation is 100 kg, how much warmer would the air be after condensation? Assume that the air is not undergoing any pressure changes. (Hint: Use the specific heat of air in Table 2.1, p. 32.)

2. Suppose planet A is exactly twice the size (in surface area) of planet B. If both planets have the same exact surface temperature (1500 K), which planet would be emitting the most radiation? Determine the wavelength of maximum energy emission of both planets, using Wien's law.

3. Suppose, in question 2, the temperature of planet B doubles.

   (a) What would be its wavelength of maximum energy emission?

(b) In what region of the electromagnetic spectrum would this wavelength be found?

(c) If the temperature of planet A remained the same, determine which planet (A or B) would now be emitting the most radiation (use the Stefan-Boltzmann relationship). Explain your answer.

4. Suppose your surface body temperature averages 90°F. How much radiant energy in W/m^2 would be emitted from your body?

A warm fall day in Denali National Park, Alaska. Here air temperatures may climb well above freezing during the day and drop to well below freezing at night.
© Pat Kennedy

CHAPTER 3

Seasonal and Daily Temperatures

The sun doesn't rise or fall: it doesn't move, it just sits there, and we rotate in front of it. Dawn means that we are rotating around into sight of it, while dusk means we have turned another 180 degrees and are being carried into the shadow zone. The sun never "goes away from the sky." It's still there sharing the same sky with us; it's simply that there is a chunk of opaque earth between us and the sun which prevents our seeing it. Everyone knows that, but I really see it now. No longer do I drive down a highway and wish the blinding sun would set; instead I wish we could speed up our rotation a bit and swing around into the shadows more quickly.

Michael Collins, *Carrying the Fire*

CONTENTS

As you sit quietly reading this book, you are part of a moving experience. The earth is speeding around the sun at thousands of kilometers per hour while, at the same time, it is spinning on its axis. When we look down upon the North Pole, we see that the direction of spin is counterclockwise, meaning that we are moving toward the east at hundreds of kilometers per hour. We normally don't think of it in that way, but, of course, this is what causes the sun, moon, and stars to rise in the east and set in the west. It is these motions coupled with the fact that the earth is tilted on its axis that causes our seasons. Therefore, we will begin this chapter by examining how the earth's motions and the sun's energy work together to produce temperature variations on a seasonal basis. Later, we will examine temperature variations on a daily basis.

Why the Earth Has Seasons

The earth revolves completely around the sun in an elliptical path (not quite a circle) in slightly longer than 365 days (one year). As the earth revolves around the sun, it spins on its own axis, completing one spin in 24 hours (one day). The average distance from the earth to the sun is 150 million km (93 million mi). Because the earth's orbit is an ellipse instead of a circle, the actual distance from the earth to the sun varies during the year. The earth comes closer to the sun in January (147 million km) than it does in July (152 million km)* (see ● Fig. 3.1). From this we might conclude that our warmest weather should occur in January and our coldest weather in July. But, in the Northern Hemisphere, we normally experience cold weather in January when we are closer to the sun and warm weather in July when we are farther away. If nearness to the sun were the primary cause of the seasons then, indeed, January would be warmer than July. However, nearness to the sun is only a small part of the story.

Our seasons are regulated by the amount of solar energy received at the earth's surface. This amount is determined primarily by the angle at which sunlight strikes the surface,

---

*The time around January 3rd, when the earth is closest to the sun, is called *perihelion* (from the Greek *peri*, meaning "near" and *helios*, meaning "sun"). The time when the earth is farthest from the sun (around July 4th) is called *aphelion* (from the Greek *ap*, "away from").

● **FIGURE 3.1** The elliptical path (highly exaggerated) of the earth about the sun brings the earth slightly closer to the sun in January than in July.

(Figure 3.1 labels: January, July, 147 million km, 152 million km)

and by how long the sun shines on any latitude (daylight hours). Let's look more closely at these factors.

Solar energy that strikes the earth's surface perpendicularly (directly) is much more intense than solar energy that strikes the same surface at an angle. Think of shining a flashlight straight at a wall—you get a small, circular spot of light (see ● Fig. 3.2). Now, tip the flashlight and notice how the spot of light spreads over a larger area. The same principle holds for sunlight. Sunlight striking the earth at an angle spreads out and must heat a larger region than sunlight impinging directly on the earth. Everything else being equal, an area experiencing more direct solar rays will receive more heat than the same size area being struck by sunlight at an angle. In addition, the more the sun's rays are slanted from the perpendicular, the more atmosphere they must penetrate. And the more atmosphere they penetrate, the more they can be scattered and absorbed (attenuated). As a consequence, when the sun is high in the sky, it can heat the ground to a much higher temperature than when it is low on the horizon.

The second important factor determining how warm the earth's surface becomes is the length of time the sun shines each day. Longer daylight hours, of course, mean that more energy is available from sunlight. In a given location, more solar energy reaches the earth's surface on a clear, long day than on a day that is clear but much shorter. Hence, more surface heating takes place.

From a casual observation, we know that summer days have more daylight hours than winter days. Also, the noontime summer sun is higher in the sky than is the noontime winter

ACTIVE FIGURE 3.2 Sunlight that strikes a surface at an angle is spread over a larger area than sunlight that strikes the surface directly. Oblique sun rays deliver less energy (are less intense) to a surface than direct sun rays. Visit the Meterology Resource Center to view this and other active figures at academic.cengage.com/login

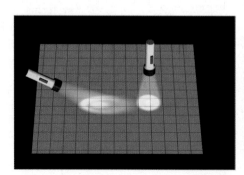

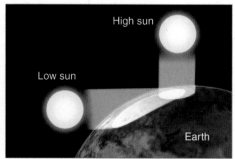

(Figure 3.2 labels: High sun, Low sun, Earth)

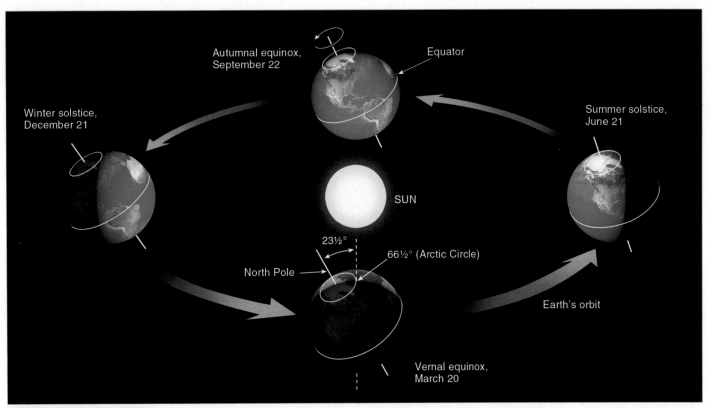

ACTIVE FIGURE 3.3 As the earth revolves about the sun, it is tilted on its axis by an angle of $23\frac{1}{2}°$. The earth's axis always points to the same area in space (as viewed from a distant star). Thus, in June, when the Northern Hemisphere is tipped toward the sun, more direct sunlight and long hours of daylight cause warmer weather than in December, when the Northern Hemisphere is tipped away from the sun. (Diagram, of course, is not to scale.) Visit the Meterology Resource Center to view this and other active figures at academic.cengage.com/login

sun. Both of these events occur because our spinning planet is inclined on its axis (tilted) as· it revolves around the sun. As ● Fig. 3.3 illustrates, the angle of tilt is $23\frac{1}{2}°$ from the perpendicular drawn to the plane of the earth's orbit. The earth's axis points to the same direction in space all year long; thus, the Northern Hemisphere is tilted toward the sun in summer (June), and away from the sun in winter (December).

> View this concept in action on the Meteorology Resource Center at academic.cengage.com/login

SEASONS IN THE NORTHERN HEMISPHERE Let's first discuss the *warm summer* season. Note in Fig. 3.3 that, on June 21, the northern half of the world is directed toward the sun. At noon on this day, solar rays beat down upon the Northern Hemisphere more directly than during any other time of year. The sun is at its highest position in the noonday sky, directly above $23\frac{1}{2}°$ north (N) latitude (Tropic of Cancer). If you were standing at this latitude on June 21, the sun at noon would be directly overhead. This day, called the **summer solstice,** is the astronomical first day of summer in the Northern Hemisphere.*

*As we will see later in this chapter, the seasons are reversed in the Southern Hemisphere. Hence, in the Southern Hemisphere, this same day is the winter solstice, or the astronomical first day of winter.

Study Fig. 3.3 closely and notice that, as the earth spins on its axis, the side facing the sun is in sunshine and the other side is in darkness. Thus, half of the globe is always illuminated. If the earth's axis were not tilted, the noonday sun would always be directly overhead at the equator, and there would be 12 hours of daylight and 12 hours of darkness at each latitude every day of the year. However, the earth is tilted. Since the Northern Hemisphere faces toward the sun on June 21, each latitude in the Northern Hemisphere will have more than 12 hours of daylight. The farther north we go, the longer are the daylight hours. When we reach the Arctic Circle ($66\frac{1}{2}°$N), daylight lasts for 24 hours. Notice in Fig. 3.3 how the region above $66\frac{1}{2}°$N never gets into the "shadow" zone as the earth spins. At the North Pole, the sun actually rises above the horizon on March 20 and has six months until it sets on September 22. No wonder this region is called the "Land of the Midnight Sun"! (See ● Fig. 3.4.)

Do longer days near polar latitudes mean that the highest daytime summer temperatures are experienced there? Not really. Nearly everyone knows that New York City (41°N) "enjoys" much hotter summer weather than Barrow, Alaska (71°N). The days in Barrow are much longer, so why isn't Barrow warmer? To figure this out, we must examine the *incoming solar radiation* (called *insolation*) on June 21. ● Figure 3.5 shows two curves: The upper curve represents the amount

● FIGURE 3.4 Land of the Midnight Sun. A series of exposures of the sun taken before, during, and after midnight in northern Alaska during July.

of insolation at the top of the earth's atmosphere on June 21, while the bottom curve represents the amount of radiation that eventually reaches the earth's surface on the same day.

The upper curve increases from the equator to the pole. This increase indicates that, during the entire day of June 21, more solar radiation reaches the top of the earth's atmosphere above the poles than above the equator. True, the sun shines on these polar latitudes at a relatively low angle, but it does so for 24 hours, causing the maximum to occur there. The lower curve shows that the amount of solar radiation eventually reaching the earth's surface on June 21 is maximum near 30°N. From there, the amount of insolation reaching the ground decreases as we move poleward.

The reason the two curves are different is that once sunlight enters the atmosphere, fine dust and air molecules scat-

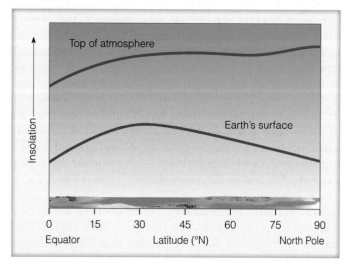

● FIGURE 3.5 The relative amount of radiant energy received at the top of the earth's atmosphere and at the earth's surface on June 21 — the summer solstice.

ter it, clouds reflect it, and some of it is absorbed by atmospheric gases. What remains reaches the surface. Generally, the greater the thickness of atmosphere that sunlight must penetrate, the greater are the chances that it will be either scattered, reflected, or absorbed by the atmosphere. During the summer in far northern latitudes, the sun is never very high above the horizon, so its radiant energy must pass through a thick portion of atmosphere before it reaches the earth's surface (see ● Fig. 3.6). And because of the increased cloud cover during the arctic summer, much of the sunlight is reflected before it reaches the ground.

Solar energy that eventually reaches the surface in the far north does not heat the surface effectively. A portion of the sun's energy is reflected by ice and snow, while some of it melts frozen soil. The amount actually absorbed is spread over a large area. So, even though northern cities, such as Barrow, experience 24 hours of continuous sunlight on June 21, they are not warmer than cities farther south. Overall, they receive less radiation at the surface, and what radiation they do receive does not effectively heat the surface.

In our discussion of Fig. 3.5, we saw that, on June 21, solar energy incident on the earth's surface is maximum near latitude 30°N. On this day, the sun is shining directly above latitude 23½°N. Why, then, isn't the most sunlight received here? A quick look at a world map shows that the major deserts of the world are centered near 30°N. Cloudless skies and drier air predominate near this latitude. At latitude 23½°N, the climate is more moist and cloudy, causing more sunlight to be scattered and reflected before reaching the surface. In addition, day length is longer at 30°N than at 23½°N on June 21. For these reasons, more radiation falls on 30°N latitude than at the Tropic of Cancer (23½°N).

Each day past June 21, the noon sun is slightly lower in the sky. Summer days in the Northern Hemisphere begin to shorten. June eventually gives way to September, and fall begins.

Look at Fig. 3.3 (p. 59) again and notice that, by September 22, the earth will have moved so that the sun is directly above the equator. Except at the poles, the days and nights throughout the world are of equal length. This day is called the **autumnal** (fall) **equinox,** and it marks the astronomical beginning of fall in the Northern Hemisphere. At the North Pole, the sun appears on the horizon for 24 hours, due to the bending of light by the atmosphere. The following day (or at least within several days), the sun disappears from view, not to rise again for a long, cold six months. Throughout the northern half of the world on each successive day, there are fewer hours of daylight, and the noon sun is slightly lower in the sky. Less direct sunlight and shorter hours of daylight spell cooler weather for the Northern Hemisphere. Reduced radiation, lower air temperatures, and cooling breezes stimulate the beautiful pageantry of fall colors (see • Fig. 3.7).

In some years around the middle of autumn, there is an unseasonably warm spell, especially in the eastern two-thirds of the United States. This warm period, referred to as **Indian**

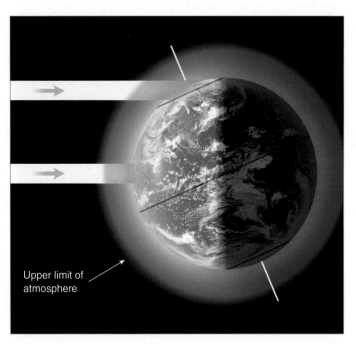

Upper limit of atmosphere

• FIGURE 3.6 During the Northern Hemisphere summer, sunlight that reaches the earth's surface in far northern latitudes has passed through a thicker layer of absorbing, scattering, and reflecting atmosphere than sunlight that reaches the earth's surface farther south. Sunlight is lost through both the thickness of the pure atmosphere and by impurities in the atmosphere. As the sun's rays become more oblique, these effects become more pronounced.

© Larry Ulrich/Stone

• FIGURE 3.7 The pageantry of fall colors in New England. The weather most suitable for an impressive display of fall colors is warm, sunny days followed by clear, cool nights with temperatures dropping below 7°C (45°F), but remaining above freezing.

▼ TABLE 3.1   Length of Time from Sunrise to Sunset for Various Latitudes on Different Dates in the Northern Hemisphere

LATITUDE	MARCH 20	JUNE 21	SEPT. 22	DEC. 21
0°	12 hr	12.0 hr	12 hr	12.0 hr
10°	12 hr	12.6 hr	12 hr	11.4 hr
20°	12 hr	13.2 hr	12 hr	10.8 hr
30°	12 hr	13.9 hr	12 hr	10.1 hr
40°	12 hr	14.9 hr	12 hr	9.1 hr
50°	12 hr	16.3 hr	12 hr	7.7 hr
60°	12 hr	18.4 hr	12 hr	5.6 hr
70°	12 hr	2 months	12 hr	0 hr
80°	12 hr	4 months	12 hr	0 hr
90°	12 hr	6 months	12 hr	0 hr

summer,[*] may last from several days up to a week or more. It usually occurs when a large high-pressure area stalls near the southeast coast. The clockwise flow of air around this system moves warm air from the Gulf of Mexico into the central or eastern half of the nation. The warm, gentle breezes and smoke from a variety of sources respectively make for mild, hazy days. The warm weather ends abruptly when an outbreak of polar air reminds us that winter is not far away.

On December 21 (three months after the autumnal equinox), the Northern Hemisphere is tilted as far away from the sun as it will be all year (see Fig. 3.3, p. 59). Nights are long and days are short. Notice in ▼ Table 3.1 that daylight decreases from 12 hours at the equator to 0 (zero) at latitudes above 66½°N. This is the shortest day of the year, called the **winter solstice,** the astronomical beginning of winter in the northern world. On this day, the sun shines directly above latitude 23½°S (Tropic of Capricorn). In the northern half of the world, the sun is at its lowest position in the noon sky. Its rays pass through a thick section of atmosphere and spread over a large area on the surface.

With so little incident sunlight, the earth's surface cools quickly. A blanket of clean snow covering the ground aids in the cooling. The snow reflects much of the sunlight that reaches the surface and continually radiates away infrared energy during the long nights. In northern Canada and Alaska, the arctic air rapidly becomes extremely cold as it lies poised, ready to do battle with the milder air to the south. Periodically, this cold arctic air pushes down into the northern United States, producing a rapid drop in temperature called a *cold wave,* which occasionally reaches far into the south during the winter. Sometimes, these cold spells arrive

[*]The origin of the term is uncertain, as it dates back to the eighteenth century. It may have originally referred to the good weather that allowed the Indians time to harvest their crops. Normally, a period of cool autumn weather must precede the warm weather period to be called Indian summer.

well before the winter solstice—the "official" first day of winter—bringing with them heavy snow and blustery winds. (More information on this "official" first day of winter is given in the Focus section on p. 64.)

On each winter day after December 21, the sun climbs a bit higher in the midday sky. The periods of daylight grow longer until days and nights are of equal length, and we have another equinox.

The date of March 20, which marks the astronomical arrival of spring, is called the **vernal** (spring) **equinox.** At this equinox, the noonday sun is shining directly on the equator, while, at the North Pole, the sun (after hiding for six months) peeks above the horizon. Longer days and more direct solar radiation spell warmer weather for the northern world.

Three months after the vernal equinox, it is June again. The Northern Hemisphere is tilted toward the sun, which shines high in the noonday sky. The days have grown longer and warmer, and another summer season has begun.

Up to now, we have seen that the seasons are controlled by solar energy striking our tilted planet, as it makes its annual voyage around the sun. This tilt of the earth causes a seasonal variation in both the length of daylight and the intensity of sunlight that reaches the surface. These facts are summarized in ● Fig. 3.8, which shows how the sun would appear in the sky to an observer at various latitudes at different times of the year. Earlier we learned that at the North Pole the sun rises above the horizon in March and stays above the horizon for six months until September. Notice in Fig. 3.8a that at the North Pole even when the sun is at its highest point in June, it is low in the sky—only 23½° above the horizon. Farther south, at the Arctic circle (Fig. 3.8b), the sun is always fairly low in the sky, even in June, when the sun stays above the horizon for 24 hours.

In the middle latitudes (Fig. 3.8c), notice that in December the sun rises in the southeast, reaches its highest point at noon (only about 26° above the southern horizon), and sets in the southwest. This apparent path produces little intense sunlight and short daylight hours. On the other hand, in June, the sun rises in the northeast, reaches a much higher position in the sky at noon (about 74° above the southern horizon) and sets in the northwest. This apparent path across the sky produces more intense solar heating, longer daylight hours, and, of course, warmer weather. Figure 3.8d illustrates how the tilt of the earth influences the sun's apparent path

WEATHER WATCH

The Land of Total Darkness. Does darkness (constant night) really occur at the Arctic Circle (66½°N) on the winter solstice? The answer is no. Due to the bending and scattering of sunlight by the atmosphere, the sky is not totally dark at the Arctic Circle on December 21. In fact, on this date, total darkness only happens north of about 82° latitude. Even at the North Pole, total darkness does not occur from September 22 through March 20, but rather from about November 5 through February 5.

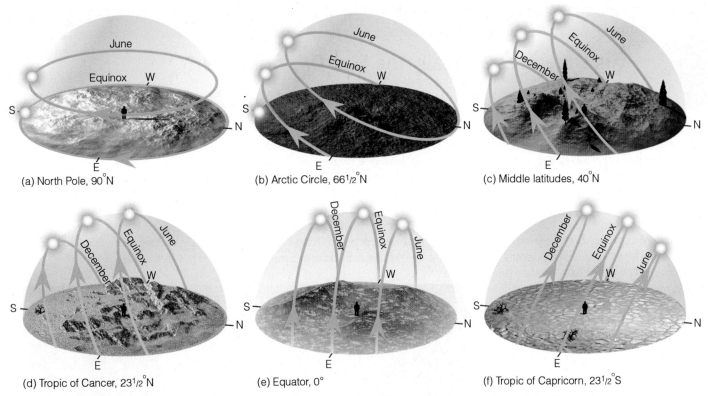

(a) North Pole, 90°N   (b) Arctic Circle, 66½°N   (c) Middle latitudes, 40°N

(d) Tropic of Cancer, 23½°N   (e) Equator, 0°   (f) Tropic of Capricorn, 23½°S

● FIGURE 3.8 The apparent path of the sun across the sky as observed at different latitudes on the June solstice (June 21), the December solstice (December 21), and the equinox (March 20 and September 22).

across the sky at the Tropic of Cancer (23½°). Figure 3.8e gives the same information for an observer at the equator.

At this point it is interesting to note that although sunlight is most intense in the Northern Hemisphere on June 21, the warmest weather in middle latitudes normally occurs weeks later, usually in July or August. This situation (called the *lag in seasonal temperature*) arises because although incoming energy from the sun is greatest in June, it still exceeds outgoing energy from the earth for a period of at least several weeks. When incoming solar energy and outgoing earth energy are in balance, the highest average temperature is attained. When outgoing energy exceeds incoming energy, the average temperature drops. Because outgoing earth energy exceeds incoming solar energy well past the winter solstice (December 21), we normally find our coldest weather occurring in January or February. As we will see later in this chapter, there is a similar lag in daily temperature between the time of most intense sunlight and the time of highest air temperature for the day.

View this concept in action on the Meteorology Resource Center at academic.cengage.com/login

SEASONS IN THE SOUTHERN HEMISPHERE   On June 21, the Southern Hemisphere is adjusting to an entirely different season. Again, look back at Fig. 3.3, (p. 59), and notice that this part of the world is now tilted away from the sun. Nights are long, days are short, and solar rays come in at an angle (see Fig. 3.8f). All of these factors keep air temperatures fairly low. The June solstice marks the astronomical beginning of winter in the Southern Hemisphere. In this part of the world, summer will not "officially" begin until the sun is over the Tropic of Capricorn (23½°S) — remember that this occurs on December 21. So, when it is winter and June in the Southern Hemisphere, it is summer and June in the Northern Hemisphere. Conversely, when it is summer and December in the Southern Hemisphere, it is winter and December in the Northern Hemisphere. So, if you are tired of the cold, December weather in your Northern Hemisphere city, travel to the summer half of the world and enjoy the warmer weather. The tilt of the earth as it revolves around the sun makes all this possible.

We know the earth comes nearer to the sun in January than in July. Even though this difference in distance amounts to only about 3 percent, the energy that strikes the top of the earth's atmosphere is almost 7 percent greater on January 3 than on July 4. These statistics might lead us to believe that summer should be warmer in the Southern Hemisphere than in the Northern Hemisphere, which, however, is not the case. A close examination of the Southern Hemisphere reveals that nearly 81 percent of the surface is water compared to 61 percent in the Northern Hemisphere. The added solar energy due to the closeness of the sun is absorbed by large bodies of water, becoming well mixed and circulated within them. This process keeps the average summer (January) temperatures in

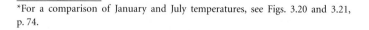

FOCUS ON A SPECIAL TOPIC

Is December 21 Really the First Day of Winter?

On December 21 (or 22, depending on the year) after nearly a month of cold weather, and perhaps a snowstorm or two (see Fig. 1), someone on the radio or television has the audacity to proclaim that "today is the first official day of winter." If during the last several weeks it was not winter, then what season was it?

Actually, December 21 marks the *astronomical* first day of winter in the Northern Hemisphere (NH), just as June 21 marks the *astronomical* first day of summer (NH). The earth is tilted on its axis by $23\frac{1}{2}°$ as it revolves around the sun. This fact causes the sun (as we view it from earth) to move in the sky from a point where it is directly above $23\frac{1}{2}°$ South latitude on December 21, to a point where it is directly above $23\frac{1}{2}°$ North latitude on June 21. The astronomical first day of spring (NH) occurs around March 20 as the sun crosses the equator moving northward and, likewise, the astronomical first day of autumn (NH) occurs around September 22 as the sun crosses the equator moving southward. Therefore the "official" beginning of any season is simply the day on which the sun passes over a particular latitude, and has nothing to do with how cold or warm the following day will be.

In the middle latitudes, summer is defined as the warmest season and winter the coldest season. If the year is divided into four seasons with each season consisting of three months, then the meteorological definition of summer over much of the Northern Hemisphere would be the three warmest months of June, July, and August. Winter would be the three coldest months of December, January, and February. Autumn would be September, October, and November—the transition between summer and winter. And spring would be March, April, and May—the transition between winter and summer.

So, the next time you hear someone remark on December 21 that "winter officially begins today," remember that this is the astronomical definition of the first day of winter. According to the meteorological definition, winter has been around for several weeks.

● FIGURE 1 A heavy snowfall covers New York City in early December. Since the snowstorm occurred before the winter solstice, is this a late fall storm or an early winter storm?

the Southern Hemisphere cooler than average summer (July) temperatures in the Northern Hemisphere. Because of water's large heat capacity, it also tends to keep winters in the Southern Hemisphere warmer than we might expect.*

Another difference between the seasons of the two hemispheres concerns their length. Because the earth describes an ellipse as it journeys around the sun, the total number of days from the vernal (March 20) to the autumnal (September 22) equinox is about 7 days longer than from the autumnal to vernal equinox (see ● Fig. 3.9). This means that spring and summer in the Northern Hemisphere not only last about a week longer than northern fall and winter, but also about a week longer than spring and summer in the Southern Hemisphere. Hence, the shorter spring and summer of the South-

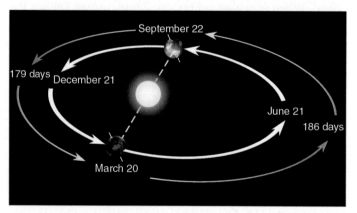

● FIGURE 3.9 Because the earth travels more slowly when it is farther from the sun, it takes the earth a little more than 7 days longer to travel from March 20 to September 22 than from September 22 to March 20.

*For a comparison of January and July temperatures, see Figs. 3.20 and 3.21, p. 74.

ern Hemisphere somewhat offset the extra insolation received due to a closer proximity to the sun.

Up to now, we have considered the seasons on a global scale. We will now shift to more local considerations.

Local Seasonal Variations

Look back at Fig. 3.8c, (p. 63), and observe that in the middle latitudes of the Northern Hemisphere, objects facing south will receive more sunlight during a year than those facing north. This fact becomes strikingly apparent in hilly or mountainous country.

Hills that face south receive more sunshine and, hence, become warmer than the partially shielded north-facing hills. Higher temperatures usually mean greater rates of evaporation and slightly drier soil conditions. Thus, south-facing hillsides are usually warmer and drier as compared to north-facing slopes at the same elevation. In many areas of the far west, only sparse vegetation grows on south-facing slopes, while, on the same hill, dense vegetation grows on the cool, moist hills that face north (see ● Fig. 3.10).

In northern latitudes, hillsides that face south usually have a longer growing season. Winemakers in western New York State do not plant grapes on the north side of hills. Grapes from vines grown on the warmer south side make better wine. Moreover, because air temperatures normally decrease with increasing height, trees found on the cooler north-facing side of mountains are often those that usually grow at higher elevations, while the warmer south-facing side of the mountain often supports trees usually found at lower elevations.

In the mountains, snow usually lingers on the ground for a longer time on north slopes than on the warmer south slopes. For this reason, ski runs are built facing north wherever possible. Also, homes and cabins built on the north side of a hill usually have a steep pitched roof as well as a reinforced deck to withstand the added weight of snow from successive winter storms.

The seasonal change in the sun's position during the year can have an effect on the vegetation around the home. In winter, a large two-story home can shade its own north side, keeping it much cooler than its south side. Trees that require warm, sunny weather should be planted on the south side, where sunlight reflected from the house can even add to the warmth.

● FIGURE 3.10 In areas where small temperature changes can cause major changes in soil moisture, sparse vegetation on the south-facing slopes will often contrast with lush vegetation on the north-facing slopes.

The design of a home can be important in reducing heating and cooling costs. Large windows should face south, allowing sunshine to penetrate the home in winter. To block out excess sunlight during the summer, a small eave or overhang should be built. A kitchen with windows facing east will let in enough warm morning sunlight to help heat this area. Because the west side warms rapidly in the afternoon, rooms having small windows (such as garages) should be placed here to act as a thermal buffer. Deciduous trees planted on the west or south side of a home provide shade in the summer. In winter, they drop their leaves, allowing the winter sunshine to warm the house. If you like the bedroom slightly cooler than the rest of the home, face it toward the north. Let nature help with the heating and air conditioning. Proper house design, orientation, and landscaping can help cut the demand for electricity, as well as for natural gas and fossil fuels, which are rapidly being depleted.

From our reading of the last several sections, it should be apparent that, when solar heating a home, proper roof angle is important in capturing much of the winter sun's energy. (The information needed to determine the angle at which sunlight will strike a roof is given in the Focus section on p. 66.)

Daily Temperature Variations

In a way, each sunny day is like a tiny season as the air goes through a daily cycle of warming and cooling. The air warms during the morning hours, as the sun gradually rises higher in the sky, spreading a blanket of heat energy over the ground. The sun reaches its highest point around noon, after which it begins its slow journey toward the western horizon. It is

Solar Heating and the Noonday Sun

The amount of solar energy that falls on a typical American home each summer day is many times the energy needed to heat the inside for a year. Thus, some people are turning to the sun as a clean, safe, and virtually inexhaustible source of energy. If solar collectors are used to heat a home, they should be placed on south-facing roofs to take maximum advantage of the energy provided. The roof itself should be constructed as nearly perpendicular to winter sun rays as possible. To determine the proper roof angle at any latitude, we need to know how high the sun will be above the southern horizon at noon.

The noon angle of the sun can be calculated in the following manner:

1. Determine the number of degrees between your latitude and the latitude where the sun is currently directly overhead.

2. Subtract the number you calculated in step 1 from 90°. This will give you the sun's elevation above the southern horizon at noon at your latitude.

For example, suppose you live in Denver, Colorado (latitude 39½°N), and the date is December 21. The difference between your latitude and where the sun is currently overhead is 63° (39½°N to 23½°S), so the sun is 27° (90° − 63°) above the southern horizon at noon. On March 20 in Denver, the angle of the sun is 50½° (90° − 39½°). To determine a reasonable roof angle, we must consider the average altitude of the midwinter sun (about 39° for Denver), building costs, and snow

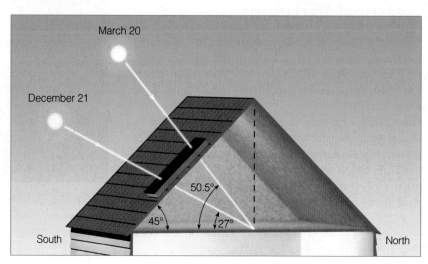

• FIGURE 2   The roof of a solar-heated home constructed in Denver, Colorado, at an angle of 45° absorbs the sun's energy in midwinter at nearly right angles.

loads. Figure 2 illustrates that a roof constructed in Denver, Colorado, at an angle of 45° will be nearly perpendicular to much of the winter sun's energy. Hence, the roofs of solar-heated homes in middle latitudes are generally built at an angle between 45° and 50°.

---

around noon when the earth's surface receives the most intense solar rays. However, somewhat surprisingly, noontime is usually not the warmest part of the day. Rather, the air continues to be heated, often reaching a maximum temperature later in the afternoon. To find out why this *lag in temperature* occurs, we need to examine a shallow layer of air in contact with the ground.

DAYTIME WARMING   As the sun rises in the morning, sunlight warms the ground, and the ground warms the air in contact with it by conduction. However, air is such a poor heat conductor that this process only takes place within a few centimeters of the ground. As the sun rises higher in the sky, the air in contact with the ground becomes even warmer, and there exists a thermal boundary separating the hot surface air from the slightly cooler air above. Given their random motion, some air molecules will cross this boundary: The "hot" molecules below bring greater kinetic energy to the cooler air; the "cool" molecules above bring a deficit of energy to the hot, surface air. However, on a windless day, this form of heat exchange is slow, and a substantial temperature difference

usually exists just above the ground (see • Fig. 3.11). This explains why joggers on a clear, windless, summer afternoon may experience air temperatures of over 50°C (122°F) at their feet and only 32°C (90°F) at their waist.

Near the surface, convection begins, and rising air bubbles (thermals) help to redistribute heat. In calm weather, these thermals are small and do not effectively mix the air near the surface. Thus, large vertical temperature gradients are able to exist. On windy days, however, turbulent eddies are able to mix hot surface air with the cooler air above. This form of mechanical stirring, sometimes called *forced convection,* helps the thermals to transfer heat away from the surface more efficiently. Therefore, on sunny, windy days the molecules near the surface are more quickly carried away than on sunny, calm days. • Figure 3.12 shows a typical vertical profile of air temperature on windy days and on calm days in summer.

We can now see why the warmest part of the day is usually in the afternoon. Around noon, the sun's rays are most intense. However, even though incoming solar radiation decreases in intensity after noon, it still exceeds outgoing heat energy from the surface for a time. This situation yields an

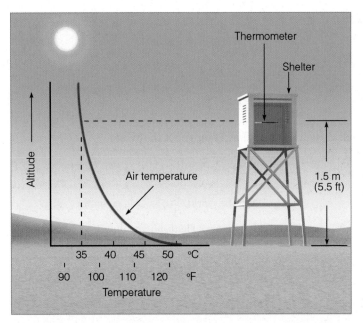

● FIGURE 3.11 On a sunny, calm day, the air near the surface can be substantially warmer than the air a meter or so above the surface.

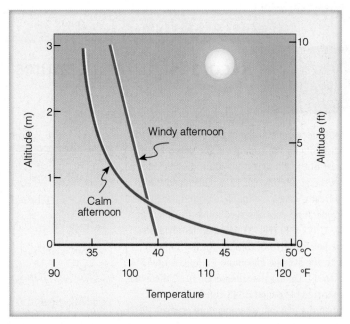

● FIGURE 3.12 Vertical temperature profiles above an asphalt surface for a windy and a calm summer afternoon.

energy surplus for two to four hours after noon and substantially contributes to a lag between the time of maximum solar heating and the time of maximum air temperature several meters above the surface (see ● Fig. 3.13).

The exact time of the highest temperature reading varies somewhat. Where the summer sky remains cloud-free all afternoon, the maximum temperature may occur sometime between 3:00 and 5:00 P.M. Where there is afternoon cloudiness or haze, the temperature maximum usually occurs an hour or two earlier. In Denver, afternoon clouds, which build over the mountains, drift eastward early in the afternoon. These clouds reflect sunlight, sometimes causing the maximum temperature to occur as early as noon. If clouds persist throughout the day, the overall daytime temperatures are usually lower.

Adjacent to large bodies of water, cool air moving inland may modify the rhythm of temperature change such that the warmest part of the day occurs at noon or before. In winter, atmospheric storms circulating warm air northward can even cause the highest temperature to occur at night.

Just how warm the air becomes depends on such factors as the type of soil, its moisture content, and vegetation cover. When the soil is a poor heat conductor (as loosely packed sand is), heat energy does not readily transfer into the ground. This fact allows the surface layer to reach a higher temperature, availing more energy to warm the air above. On the other hand, if the soil is moist or covered with vegetation, much of the available energy evaporates water, leaving less to heat the air. As you might expect, the highest summer temperatures usually occur over desert regions, where clear skies coupled with low humidities and meager vegetation permit the surface and the air above to warm up rapidly.

Where the air is humid, haze and cloudiness lower the maximum temperature by preventing some of the sun's rays from reaching the ground. In humid Atlanta, Georgia, the average maximum temperature for July is 30.5°C (87°F). In contrast, Phoenix, Arizona—in the desert southwest at the same latitude as Atlanta—experiences an average July maximum of 40.5°C (105°F). (Additional information on high daytime temperatures is given in the Focus section on p. 68.)

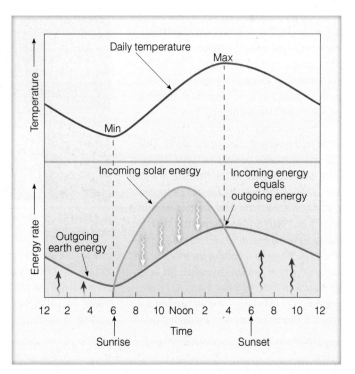

● FIGURE 3.13 The daily variation in air temperature is controlled by incoming energy (primarily from the sun) and outgoing energy from the earth's surface. Where incoming energy exceeds outgoing energy (orange shade), the air temperature rises. Where outgoing energy exceeds incoming energy (blue shade), the air temperature falls.

FOCUS ON A SPECIAL TOPIC

Record High Temperatures

Most people are aware of the extreme heat that exists during the summer in the desert southwest of the United States. But how hot does it get there? On July 10, 1913, Greenland Ranch in Death Valley, California, reported the highest temperature ever observed in North America: 57°C (134°F) (Fig. 3). Here, air temperatures are persistently hot throughout the summer, with the average maximum for July being 47°C (116°F). During the summer of 1917, there was an incredible period of 43 consecutive days when the maximum temperature reached 120°F or higher.

Probably the hottest urban area in the United States is Yuma, Arizona. Located along the California–Arizona border, Yuma's high temperature during July averages 42°C (108°F). In 1937, the high reached 100°F or more for 101 consecutive days.

In a more humid climate, the maximum temperature rarely climbs above 41°C (106°F). However, during the record heat wave of 1936, the air temperature reached 121°F near Alton, Kansas. And during the heat wave of 1983, which destroyed about $7 billion in crops and increased the nation's air-conditioning bill by an estimated $1 billion, Fayetteville reported North Carolina's all-time record high temperature when the mercury hit 110°F.

These readings, however, do not hold a candle to the hottest place in the world. That distinction probably belongs to Dallol, Ethiopia. Dallol is located south of the Red Sea, near latitude 12°N, in the hot, dry Danakil Depression. A prospecting company kept weather records at Dallol from 1960 to 1966. During this time, the average daily maximum temperature exceeded 38°C (100°F) every month of the year, except during December and January, when the average maximum lowered to 98°F and 97°F, respectively. On many days, the air temperature exceeded 120°F. The average annual tempera-

▼ TABLE 1   Some Record High Temperatures Throughout the World

LOCATION (LATITUDE)	RECORD HIGH TEMPERATURE (°C)	(°F)	RECORD FOR:	DATE
El Azizia, Libya (32°N)	58	136	The world	September 13, 1922
Death Valley, Calif. (36°N)	57	134	Western Hemisphere	July 10, 1913
Tirat Tsvi, Israel (32°N)	54	129	Middle East	June 21, 1942
Cloncurry, Queensland (21°S)	53	128	Australia	January 16, 1889
Seville, Spain (37°N)	50	122	Europe	August 4, 1881
Rivadavia, Argentina (35°S)	49	120	South America	December 11, 1905
Midale, Saskatchewan (49°N)	45	113	Canada	July 5, 1937
Fort Yukon, Alaska (66°N)	38	100	Alaska	June 27, 1915
Pahala, Hawaii (19°N)	38	100	Hawaii	April 27, 1931
Esparanza, Antarctica (63°S)	14	58	Antarctica	October 20, 1956

● FIGURE 3 The hottest place in North America, Death Valley, California, where the air temperature reached 57°C (134°F).

ture for the six years at Dallol was 34°C (94°F). In comparison, the average annual temperature in Yuma is 23°C (74°F) and at Death Valley, 24°C (76°F). The highest temperature reading on earth (under standard conditions)

occurred northeast of Dallol at El Azizia, Libya (32°N), when, on September 13, 1922, the temperature reached a scorching 58°C (136°F). Table 1 gives record high temperatures throughout the world.

NIGHTTIME COOLING   As the sun lowers, its energy is spread over a larger area, which reduces the heat available to warm the ground. Observe in Fig. 3.13 that sometime in late afternoon or early evening, the earth's surface and air above

begin to lose more energy than they receive; hence, they start to cool.

Both the ground and air above cool by radiating infrared energy, a process called **radiational cooling.** The ground, being a much better radiator than air, is able to cool more quickly. Consequently, shortly after sunset, the earth's surface is slightly cooler than the air directly above it. The surface air transfers some energy to the ground by conduction, which the ground, in turn, quickly radiates away.

As the night progresses, the ground and the air in contact with it continue to cool more rapidly than the air a few meters higher. The warmer upper air does transfer *some* heat downward, a process that is slow due to the air's poor thermal conductivity. Therefore, by late night or early morning, the coldest air is found next to the ground, with slightly warmer air above (see ● Fig. 3.14).

This measured increase in air temperature just above the ground is known as a **radiation inversion** because it forms mainly through radiational cooling of the surface. Because radiation inversions occur on most clear, calm nights, they are also called **nocturnal inversions.**

Radiation Inversions   A strong radiation inversion occurs when the air near the ground is much colder than the air higher up. Ideal conditions for a strong inversion (and, hence, very low nighttime temperatures) exist when the air is calm, the night is long, and the air is fairly dry and cloud-free. Let's examine these ingredients one by one.

A windless night is essential for a strong radiation inversion because a stiff breeze tends to mix the colder air at the surface with the warmer air above. This mixing, along with the cooling of the warmer air as it comes in contact with the cold ground, causes a vertical temperature profile that is almost isothermal (constant temperature) in a layer several meters thick. In the absence of wind, the cooler, more dense surface air does not readily mix with the warmer, less dense air above, and the inversion is more strongly developed, as illustrated in ● Fig. 3.15.

A long night also contributes to a strong inversion. Generally, the longer the night, the longer the time of radiational cooling and the better are the chances that the air near the ground will be much colder than the air above. Consequently, winter nights provide the best conditions for a strong radiation inversion, other factors being equal.

Finally, radiation inversions are more likely with a clear sky and dry air. Under these conditions, the ground is able to radiate its energy to outer space and thereby cool rapidly. However, with cloudy weather and moist air, much of the outgoing infrared energy is absorbed and radiated to the surface, retarding the rate of cooling. Also, on humid nights, condensation in the form of fog or dew will release latent

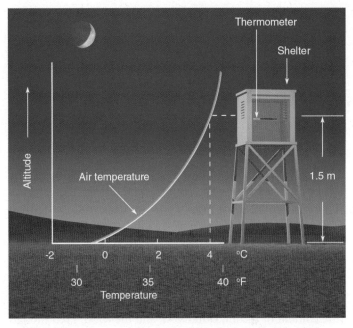

● FIGURE 3.14 On a clear, calm night, the air near the surface can be much colder than the air above. The increase in air temperature with increasing height above the surface is called a radiation temperature inversion.

heat, which warms the air. So, radiation inversions may occur on any night. But, during long winter nights, when the air is still, cloud-free, and relatively dry, these inversions can become strong and deep.

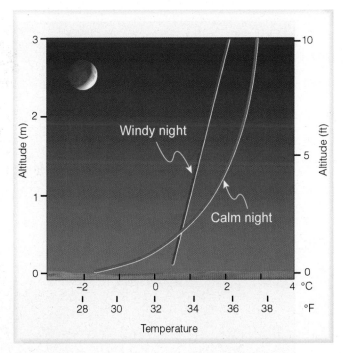

● FIGURE 3.15 Vertical temperature profiles just above the ground on a windy night and on a calm night. Notice that the radiation inversion develops better on the calm night.

On winter nights in middle latitudes, it is common to experience below-freezing temperatures near the ground and air 5°C (9°F) warmer at your waist. In middle latitudes, the top of the inversion—the region where the air temperature stops increasing with height—is usually not more than 100 m (330 ft) above the ground. In dry, polar regions, where winter nights are measured in months, the top of the inversion is often 1000 m (about 3300 ft) above the surface. It may, however, extend to as high as 3000 m (about 10,000 ft).

It should now be apparent that how cold the night air becomes depends primarily on the length of the night, the moisture content of the air, cloudiness, and the wind. Even though wind may initially bring cold air into a region, the coldest nights usually occur when the air is clear and relatively calm.

There are, however, other factors that determine how cold the night air becomes. For example, a surface that is wet or covered with vegetation can add water vapor to the air, retarding nighttime cooling. Likewise, if the soil is a good heat conductor, heat ascending toward the surface during the night adds warmth to the air, which restricts cooling. On the other hand, snow covering the ground acts as an insulating blanket that prevents heat stored in the soil from reaching the air. Snow, a good emitter of infrared energy, radiates away energy rapidly at night, which helps keep the air temperature above a snow surface quite low. (Up to this point we've been looking at low-nighttime temperatures. Additional information on this topic is given in the Focus section on p. 71.)

Look back at Fig. 3.13, (p. 67), and observe that the lowest temperature on any given day is usually observed around sunrise. However, the cooling of the ground and surface air may even continue beyond sunrise for a half hour or so, as outgoing energy can exceed incoming energy. This situation happens because light from the early morning sun passes through a thick section of atmosphere and strikes the ground at a low angle. Consequently, the sun's energy does not effectively heat the surface. Surface heating may be reduced further when the ground is moist and available energy is used for evaporation. (Any duck hunter lying flat in a marsh knows the sudden cooling that occurs as evaporation chills the air just after sunrise.) Hence, the lowest temperature may occur shortly after the sun has risen.

Cold, heavy surface air slowly drains downhill during the night and eventually settles in low-lying basins and valleys. Valley bottoms are thus colder than the surrounding hillsides (see Fig. 3.16). In middle latitudes, these warmer hillsides, called **thermal belts,** are less likely to experience freezing temperatures than the valley below. This encourages farmers to plant on hillsides those trees unable to survive the valley's low temperature.

On the valley floor, the cold, dense air is unable to rise. Smoke and other pollutants trapped in this heavy air restrict visibility. Therefore, valley bottoms are not only colder, but are also more frequently polluted than nearby hillsides. Even when the land is only gently sloped, cold air settles into lower-

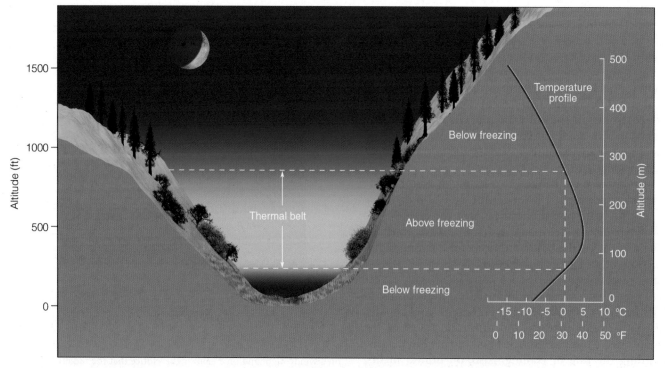

● FIGURE 3.16 On cold, clear nights, the settling of cold air into valleys makes them colder than surrounding hillsides. The region along the side of the hill where the air temperature is above freezing is known as a *thermal belt.*

FOCUS ON A SPECIAL TOPIC

Record Low Temperatures

One city in the United States that experiences very low temperatures is International Falls, Minnesota, where the average temperature for January is −16°C (3°F). Located several hundred miles to the south, Minneapolis–St. Paul, with an average temperature of −9°C (16°F) for the three winter months, is the coldest major urban area in the nation. For duration of extreme cold, Minneapolis reported 186 consecutive hours of temperatures below 0°F during the winter of 1911–1912. Within the forty-eight adjacent states, however, the record for the longest duration of severe cold belongs to Langdon, North Dakota, where the thermometer remained below 0°F for 41 consecutive days during the winter of 1936. The official record for the lowest temperature in the forty-eight adjacent states belongs to Rogers Pass, Montana, where on the morning of January 20, 1954, the mercury dropped to −57°C (−70°F). The lowest official temperature for Alaska, −62°C (−80°F), occurred at Prospect Creek on January 23, 1971.

The coldest areas in North America are found in the Yukon and Northwest Territories of Canada. Resolute, Canada (latitude 75°N), has an average temperature of −32°C (−26°F) for the month of January.

The lowest temperatures and coldest winters in the Northern Hemisphere are found in the interior of Siberia and Greenland. For example, the average January temperature in Yakutsk, Siberia (latitude 62°N), is −43°C (−46°F).

▼ TABLE 2   Some Record Low Temperatures Throughout the World

LOCATION (LATITUDE)	RECORD LOW TEMPERATURE (°C)	(°F)	RECORD FOR:	DATE
Vostok, Antarctica (78°S)	−89	−129	The world	July 21, 1983
Verkhoyansk, Russia (67°N)	−68	−90	Northern Hemisphere	February 7, 1892
Northice, Greenland (72°N)	−66	−87	Greenland	January 9, 1954
Snag, Yukon (62°N)	−63	−81	North America	February 3, 1947
Prospect Creek, Alaska (66°N)	−62	−80	Alaska	January 23, 1971
Rogers Pass, Montana (47°N)	−57	−70	U.S. (excluding Alaska)	January 20, 1954
Sarmiento, Argentina (34°S)	−33	−27	South America	June 1, 1907
Ifrane, Morocco (33°N)	−24	−11	Africa	February 11, 1935
Charlotte Pass, Australia (36°S)	−22	−8	Australia	July 22, 1949
Mt. Haleakala, Hawaii (20°N)	−10	14	Hawaii	January 2, 1961

There, the mean temperature for the entire year is a bitter cold −11°C (12°F). At Eismitte, Greenland, the average temperature for February (the coldest month) is −47°C (−53°F), with the mean annual temperature being a frigid −30°C (−22°F). Even though these temperatures are extremely low, they do not come close to the coldest area of the world: the Antarctic.

At the geographical South Pole, over nine thousand feet above sea level, where the Amundsen-Scott scientific station has been keeping records for more than forty years, the average temperature for the month of July (winter) is −59°C (−74°F) and the mean annual temperature is −49°C (−57°F). The lowest temperature ever recorded there (−83°C or −117°F) occurred under clear skies with a light wind on the morning of June 23, 1983. Cold as it was, it was not the record low for the world. That belongs to the Russian station at Vostok, Antarctica (latitude 78°S), where the temperature plummeted to −89°C (−129°F) on July 21, 1983. (See Table 2 for record low temperatures throughout the world.)

---

lying areas, such as river basins and floodplains. Because the flat floodplains are agriculturally rich areas, cold air drainage often forces farmers to seek protection for their crops.

So far, we have looked at how and why the air temperature near the ground changes during the course of a 24-hour day. We saw that during the day the air near the earth's surface can become quite warm, whereas at night it can cool off dramatically. ● Figure 3.17 summarizes these observations by illustrating how the average air temperature above the ground can change over a span of 24 hours. Notice in the figure that although the air several feet above the surface both cools and

View this concept in action on the Meteorology Resource Center at academic.cengage.com/login

warms, it does so at a slower rate than air at the surface. Also observe that the warmest part of the day several feet above the surface occurs at 3 P.M. (local time), while the surface reaches its maximum temperature at noon when the sun's energy is most intense.

Protecting Crops from the Cold   On cold nights, many plants may be damaged by low temperatures. To protect small plants or shrubs, cover them with straw, cloth, or plastic sheeting. This prevents ground heat from being radiated away to the colder surroundings. If you are a household gardener concerned about outside flowers and plants during cold weather, simply wrap them in plastic or cover each with a paper cup.

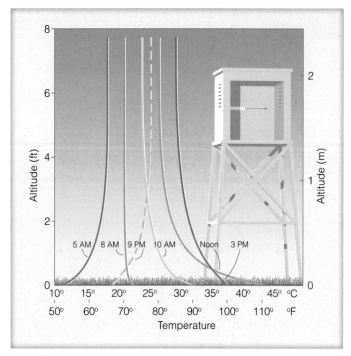

• FIGURE 3.17 An idealized distribution of air temperature above the ground during a 24-hour day. The temperature curves represent the variations in average air temperature above a grassy surface for a mid-latitude city during the summer under clear, calm conditions.

Fruit trees are particularly vulnerable to cold weather in the spring when they are blossoming. The protection of such trees presents a serious problem to the farmer. Since the lowest temperatures on a clear, still night occur near the surface, the lower branches of a tree are the most susceptible to damage.

Therefore, increasing the air temperature close to the ground may prevent damage. One way this increase can be achieved is to use **orchard heaters,** which warm the air around them by setting up convection currents close to the ground. In addition, heat energy radiated from oil or gas-fired orchard heaters is intercepted by the buds of the trees, which raises their temperature. Early forms of these heaters were called *smudge pots* because they produced large amounts of dense black smoke that caused severe pollution. People tolerated this condition only because they believed that the smoke acted like a blanket, trapping some of the earth's heat. Studies have shown this concept to be not as significant as previously thought. Orchard heaters are now designed to produce as little smoke as possible (see • Fig. 3.18).

Another way to protect trees is to mix the cold air at the ground with the warmer air above, thus raising the temperature of the air next to the ground. Such mixing can be accomplished by using **wind machines** (see • Fig. 3.19), which are power-driven fans that resemble airplane propellers. One significant benefit of wind machines is that they can be thermostatically controlled to turn off and on at prescribed temperatures. Farmers without their own wind machines can rent air mixers in the form of helicopters. Although helicopters are effective in mixing the air, they are expensive to operate.

If sufficient water is available, trees can be protected by irrigation. On potentially cold nights, farmers might flood the orchard. Because water has a high heat capacity, it cools more slowly than dry soil. Consequently, the surface does not become as cold as it would if it were dry. Furthermore, wet soil has a higher thermal conductivity than dry soil. Hence, in wet soil, heat is conducted upward from subsurface soil more rapidly, which helps to keep the surface warmer.

So far, we have discussed protecting trees against the cold air near the ground during a radiation inversion. Farmers often face another nighttime cooling problem. For instance, when subfreezing air blows into a region, the coldest air is not necessarily found at the surface; the air may actually become colder with height. This condition is known as a **freeze.** * A

• FIGURE 3.18 Orchard heaters circulate the air by setting up convection currents.

• FIGURE 3.19 Wind machines mix cooler surface air with warmer air above.

single freeze in California or Florida can cause several million dollars damage to citrus crops. As a case in point, several freezes during the spring of 2001 caused millions of dollars in damage to California's north coast vineyards, which resulted in higher wine prices.

Protecting an orchard from the damaging cold air blown by the wind can be a problem. Wind machines will not help because they would only mix cold air at the surface with the colder air above. Orchard heaters and irrigation are of little value as they would only protect the branches just above the ground. However, there is one form of protection that does work: An orchard's sprinkling system may be turned on so that it emits a fine spray of water. In the cold air, the water freezes around the branches and buds, coating them with a thin veneer of ice. As long as the spraying continues, the latent heat — given off as the water changes into ice — keeps the ice temperature at 0°C (32°F). The ice acts as a protective coating against the sub-freezing air by keeping the buds (or fruit) at a temperature higher than their damaging point. Care must be taken since too much ice can cause the branches to break. The fruit may be saved from the cold air, while the tree itself may be damaged by too much protection. Sprinklers work well when the air is fairly humid. They do not work well when the air is dry, as a good deal of the water may be lost through evaporation.

BRIEF REVIEW

Up to this point we have examined temperature variations on a seasonal and daily basis. Before going on, here is a review of some of the important concepts and facts we have covered:

- The seasons are caused by the earth being tilted on its axis as it revolves around the sun. The tilt causes annual variations in the amount of sunlight that strikes the surface as well as variations in the length of time the sun shines at each latitude.
- During the day, the earth's surface and air above will continue to warm as long as incoming energy (mainly sunlight) exceeds outgoing energy from the surface.
- At night, the earth's surface cools, mainly by giving up more infrared radiation than it receives — a process called radiational cooling.
- The coldest nights of winter normally occur when the air is calm, fairly dry (low water-vapor content), and cloud-free.
- The highest temperatures during the day and the lowest temperatures at night are normally observed at the earth's surface.
- Radiation inversions exist usually at night when the air near the ground is colder than the air above.

*A freeze occurs over a widespread area when the surface air temperature remains below freezing for a long enough time to damage certain agricultural crops. The terms *frost* and *freeze* are often used interchangeably by various segments of society. However, to the grower of perennial crops (such as apples and citrus) who have to protect the crop against damaging low temperatures, it makes no difference if visible "frost" is present or not. The concern is whether or not the plant tissue has been exposed to temperatures equal to or below 32°F. The actual freezing point of the plant, however, can vary because perennial plants can develop hardiness in the fall that usually lasts through the winter, then wears off gradually in the spring.

The Controls of Temperature

The main factors that cause variations in temperature from one place to another are called the **controls of temperature.** Earlier we saw that the greatest factor in determining temperature is the amount of solar radiation that reaches the surface. This, of course, is determined by the length of daylight hours and the intensity of incoming solar radiation. Both of these factors are a function of latitude; hence, latitude is considered an important control of temperature. The main controls are:

1. latitude
2. land and water distribution
3. ocean currents
4. elevation

We can obtain a better picture of these controls by examining ● Fig. 3.20 and ● Fig. 3.21, which show the average monthly temperatures throughout the world for January and July. The lines on the map are **isotherms** — lines connecting places that have the same temperature. Because air temperature normally decreases with height, cities at very high elevations are much colder than their sea level counterparts. Consequently, the isotherms in Figs. 3.20 and 3.21 are corrected to read at the same horizontal level (sea level) by adding to each station above sea level an amount of temperature that would correspond to an average temperature change with height.*

Figures 3.20 and 3.21 show the importance of latitude on temperature. Note that, on the average, temperatures decrease poleward from the tropics and subtropics in both January and July. However, because there is a greater variation in solar radiation between low and high latitudes in winter than in summer, the isotherms in January are closer together (a tighter gradient)† than they are in July. This fact means that if you travel from New Orleans to Detroit in January, you are more likely to experience greater temperature variations than if you make the same trip in July. Notice also in Fig. 3.20 and Fig. 3.21 that the isotherms do not run horizontally; rather, in many places they bend, especially where they approach an ocean-continent boundary.

On the January map, the temperatures are much lower in the middle of continents than they are at the same latitude near the oceans; on the July map, the reverse is true. The reason for these temperature variations can be attributed to the unequal heating and cooling properties of land and water. For one thing, solar energy reaching land is absorbed in a thin layer of soil; reaching water, it penetrates deeply. Because

*The amount of change is usually less than the standard temperature lapse rate of 6.5°C per 1000 m (3.6°F per 1000 ft). The reason is that the standard lapse rate is computed for altitudes above the earth's surface in the "free" atmosphere. In the less-dense air at high elevations, the absorption of solar radiation by the ground causes an overall slightly higher temperature than that of the free atmosphere at the same level.

†Gradient represents the rate of change of some quantity (in this case, temperature) over a given distance.

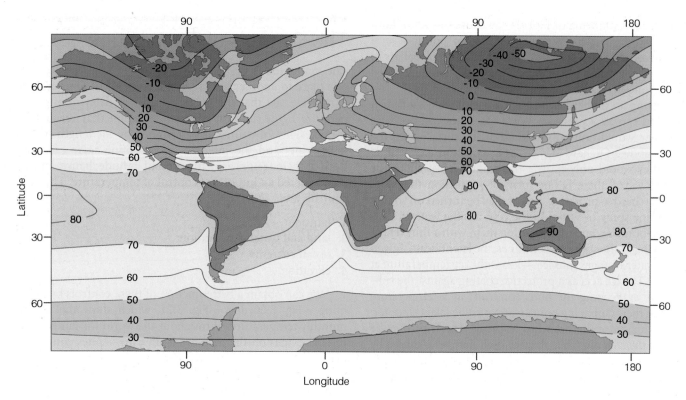

● FIGURE 3.20 Average air temperature near sea level in January (°F).

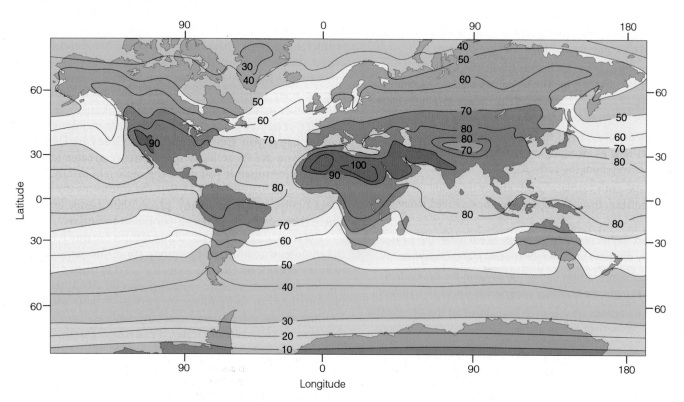

● FIGURE 3.21 Average air temperature near sea level in July (°F).

water is able to circulate, it distributes its heat through a much deeper layer. Also, some of the solar energy striking the water is used to evaporate it rather than heat it.

Another important reason for the temperature contrasts is that water has a high *specific heat.* As we saw in Chapter 2, it takes a great deal more heat to raise the temperature of 1 gram of water 1°C than it does to raise the temperature of 1 gram of soil or rock by 1°C. Water not only heats more slowly than land, it cools more slowly as well, and so the oceans act like huge heat reservoirs. Thus, mid-ocean surface temperatures change relatively little from summer to winter compared to the much larger annual temperature changes over the middle of continents.

Along the margin of continents, ocean currents often influence air temperatures. For example, along the eastern margins, warm ocean currents transport warm water poleward, while, along the western margins, they transport cold water equatorward. As we will see in Chapter 10, some coastal areas also experience upwelling, which brings cold water from below to the surface.

Even large lakes can modify the temperature around them. In summer, the Great Lakes remain cooler than the land. As a result, refreshing breezes blow inland, bringing relief from the sometimes sweltering heat. As winter approaches, the water cools more slowly than the land. The first blast of cold air from Canada is modified as it crosses the lakes, and so the first freeze is delayed on the eastern shores of Lake Michigan.

Air Temperature Data

The careful recording and application of temperature data are tremendously important to us all. Without accurate information of this type, the work of farmers, power company engineers, weather analysts, and many others would be a great deal more difficult. In these next sections, we will study the ways temperature data are organized and used. We will also examine the significance of daily, monthly, and yearly temperature ranges and averages in terms of practical application to everyday living.

DAILY, MONTHLY, AND YEARLY TEMPERATURES The greatest variation in daily temperature occurs right at the earth's surface. In fact, the difference between the daily maximum and minimum temperature—called the **daily** (or **diurnal**) **range of temperature**—is greatest next to the ground and becomes progressively smaller as we move away from the surface (see Fig. 3.22). This daily variation in temperature is also much larger on clear days than on cloudy ones.

The largest diurnal range of temperature occurs on high deserts, where the air is often cloud-free, and there is less CO_2 and water vapor above to radiate much infrared energy back to the surface. By day, clear summer skies allow the sun's en-

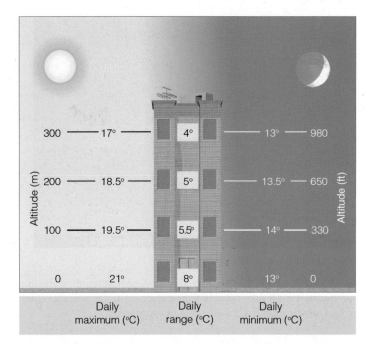

● FIGURE 3.22 The daily range of temperature decreases as we climb away from the earth's surface. Hence, there is less day-to-night variation in air temperature near the top of a high-rise apartment complex than at the ground level.

ergy to quickly warm the ground which, in turn, warms the air above to a temperature sometimes exceeding 35°C (95°F). At night, the ground cools rapidly by radiating infrared energy to space, and the minimum temperature in these regions occasionally dips below 5°C (41°F), thus giving a daily temperature range of 30°C (54°F).

A good example of a city with a large diurnal temperature range is Reno, Nevada, which is located on a plateau at an elevation of 1350 m (4400 ft) above sea level. Here, in the dry, thin summer air, the average daily maximum temperature for July is 33°C (92°F)—short-sleeve weather, indeed. But don't lose your shirt in Reno, for you will need it at night, as the average daily minimum temperature for July is 8°C (47°F). Reno has a daily range of 25°C (45°F)!

Clouds can have a large affect on the daily range in temperature. As we saw in Chapter 2, clouds (especially low, thick ones) are good reflectors of incoming solar radiation, and so they prevent much of the sun's energy from reaching the surface. This effect tends to lower daytime temperaures (see ● Fig 3.23a). If the clouds persist into the night, they tend to keep nighttime temperatures higher, as clouds are excellent absorbers and emitters of infrared radiation—the clouds actually emit a great deal of infrared energy back to the surface. Clouds, therefore, have the effect of lowering the daily range of temperature. In clear weather (Fig 3.23b), daytime air temperatures tend to be higher as the sun's rays impinge directly upon the surface, while nighttime temperatures are usually lower due to rapid radiational cooling. Therefore,

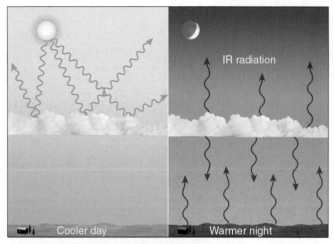
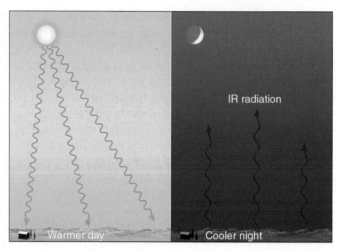

(a) Small daily temperature range

(b) Large daily temperature range

● FIGURE 3.23 (a) Clouds tend to keep daytime temperatures lower and nighttime temperatures higher, producing a small daily range in temperature. (b) In the absence of clouds, days tend to be warmer and nights cooler, producing a larger daily range in temperature.

clear days and clear nights combine to promote a large daily range in temperature.

Humidity can also have an effect on diurnal temperature ranges. For example, in humid regions, the diurnal temperature range is usually small. Here, haze and clouds lower the maximum temperature by preventing some of the sun's energy from reaching the surface. At night, the moist air keeps the minimum temperature high by absorbing the earth's infrared radiation and radiating a portion of it to the ground. An example of a humid city with a small summer diurnal temperature range is Charleston, South Carolina, where the average July maximum

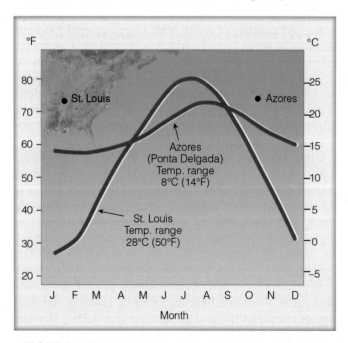

● FIGURE 3.24 Monthly temperature data and annual temperature range for St. Louis, Missouri, a city located near the middle of a continent and Ponta Delgada, a city located in the Azores in the Atlantic Ocean.

temperature is 32°C (90°F), the average minimum is 22°C (72°F), and the diurnal range is only 10°C (18°F).

Cities near large bodies of water typically have smaller diurnal temperature ranges than cities farther inland. This phenomenon is caused in part by the additional water vapor in the air and by the fact that water warms and cools much more slowly than land. Moreover, cities whose temperature readings are obtained at airports often have larger diurnal temperature ranges than those whose readings are obtained in downtown areas. The reason for this fact is that nighttime temperatures in cities tend to be warmer than those in outlying rural areas. This nighttime city warmth—called the *urban heat island*—is due to industrial and urban development.

The average of the highest and lowest temperature for a 24-hour period is known as the **mean (average) daily temperature.** Most newspapers list the mean daily temperature along with the highest and lowest temperatures for the preceding day. The average of the mean daily temperatures for a particular date averaged for a 30-year period gives the average (or "*normal*") temperatures for that date. The average temperature for each month is the average of the daily mean temperatures for that month. (Additional information on the concept of "normal" temperature is given in the Focus section on p. 77.)

At any location, the difference between the average temperature of the warmest and coldest months is called the **annual range of temperature.** Usually the largest annual ranges occur over land, the smallest over water (see ● Fig. 3.24). Moreover, inland cities have larger annual ranges than coastal cities. Near the equator (because daylight length varies little and the sun is always high in the noon sky), annual temperature ranges are small, usually less than 3°C (5°F). Quito, Ecuador—on the equator at an elevation of 2850 m (9350 ft)—experiences an annual range of less than 1°C. In middle and high latitudes, large seasonal variations in the amount of sunlight reaching the surface produce large tem-

FOCUS ON A SPECIAL TOPIC

When It Comes to Temperature, What's Normal?

When the weathercaster reports that "the normal high temperature for today is 68°F" does this mean that the high temperature on this day is usually 68°F? Or does it mean that we should expect a high temperature near 68°F? Actually, we should expect neither one.

Remember that the word *normal*, or *norm*, refers to weather data averaged over a period of 30 years. For example, Fig. 4 shows the high temperature measured for 30 years in a southwestern city on March 15. The average (mean) high temperature for this period is 68°F; hence, the normal high temperature for this date is 68°F (dashed line). Notice, however, that only on one day during this 30-year period did the high temperature actually measure 68°F (large red dot). In fact, the most common high temperature (called the *mode*) was 60°F, and occurred on 4 days (blue dots).

So what would be considered a typical high temperature for this date? Actually, any high temperature that lies between about 47°F and 89°F (two standard deviations* on either side of 68°F) would be considered typical for

this day. While a high temperature of 80°F may be quite warm and a high temperature of 47°F may be quite cool, they are both no more uncommon (unusual) than a high temperature of 68°F, which is the *normal* (average) high tem-

perature for the 30-year period. This same type of reasoning applies to *normal rainfall*, as the actual amount of precipitation will likely be greater or less than the 30-year average.

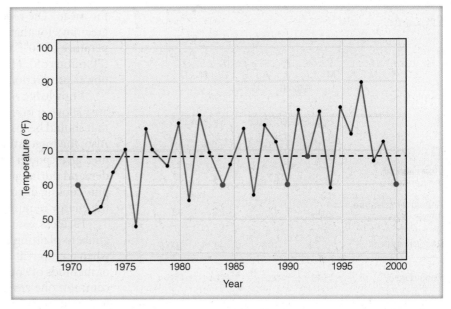

● FIGURE 4 The high temperature measured (for 30 years) on March 15 in a city located in the southwestern United States. The dashed line represents the *normal* temperature for the 30-year period.

*A standard deviation is a statistical measure of the spread of the data. Two standard deviations for this set of data mean that 95 percent of the time the high temperature occurs between 47°F and 89°F.

---

perature contrasts between winter and summer. Here, annual ranges are large, especially in the middle of a continent. Yakutsk, in northeastern Siberia near the Arctic Circle, has an extremely large annual temperature range of 62°C (112°F).

The average temperature of any station for the entire year is the **mean (average) annual temperature,** which represents the average of the twelve monthly average temperatures.* When two cities have the same mean annual temperature, it might first seem that their temperatures throughout the year are quite similar. However, often this is not the case. For example, San Francisco, California, and Richmond, Virginia, are at the same latitude (37°N). Both have similar hours of daylight during the year; both have the same mean annual temperature—14°C (57°F). Here, the similarities end. The

*The mean annual temperature may be obtained by taking the sum of the 12 monthly means and dividing that total by 12, or by obtaining the sum of the daily means and dividing that total by 365.

temperature differences between the two cities are apparent to anyone who has traveled to San Francisco during the summer with a suitcase full of clothes suitable for summer weather in Richmond.

● Figure 3.25 summarizes the average temperatures for San Francisco and Richmond. Notice that the coldest month for both cities is January. Even though January in Richmond averages only 8°C (14°F) colder than January in San Francisco, people in Richmond awaken to an average January

WEATHER WATCH

One of the greatest temperature ranges ever recorded in the Northern Hemisphere (56°C or 100°F) occurred at Browning, Montana, on January 23, 1916, when the air temperature plummeted from 7°C (44°F) to −49°C (−56°F) in less than 24 hours.

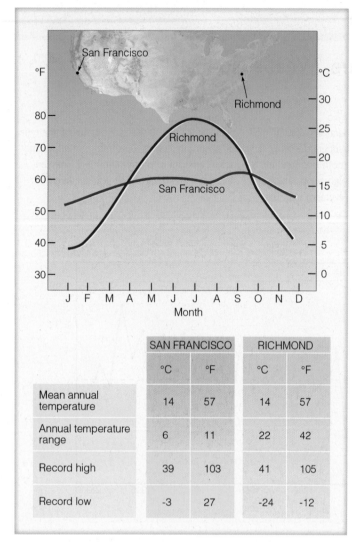

● FIGURE 3.25 Temperature data for San Francisco, California (37°N), and Richmond, Virginia (37°N)—two cities with the same mean annual temperature.

	SAN FRANCISCO		RICHMOND	
	°C	°F	°C	°F
Mean annual temperature	14	57	14	57
Annual temperature range	6	11	22	42
Record high	39	103	41	105
Record low	-3	27	-24	-12

minimum temperature of −3°C (27°F), which is the lowest temperature ever recorded in San Francisco. Trees that thrive in San Francisco's weather would find it difficult surviving a winter in Richmond. So, even though San Francisco and Richmond have the same mean annual temperature, the behavior and range of their temperatures differ greatly.

THE USE OF TEMPERATURE DATA An application of daily temperature developed by heating engineers in estimating energy needs is the **heating degree-day.** The heating degree-day is based on the assumption that people will begin to use their furnaces when the mean daily temperature drops below 65°F. Therefore, heating degree-days are determined by subtracting the mean temperature for the day from 65°F (18°C). Thus, if the mean temperature for a day is 64°F, there would be 1 heating degree-day on this day.*

_____
*In the United States, the National Weather Service and the Department of Agriculture use degrees Fahrenheit in their computations.

On days when the mean temperature is above 65°F, there are no heating degree-days. Hence, the lower the average daily temperature, the more heating degree-days and the greater the predicted consumption of fuel. When the number of heating degree-days for a whole year is calculated, the heating fuel requirements for any location can be estimated. ● Figure 3.26 shows the yearly average number of heating degree-days in various locations throughout the United States.

As the mean daily temperature climbs above 65°F, people begin to cool their indoor environment. Consequently, an index, called the **cooling degree-day,** is used during warm weather to estimate the energy needed to cool indoor air to a comfortable level. The forecast of mean daily temperature is converted to cooling degree-days by subtracting 65°F from the mean. The remaining value is the number of cooling degree-days for that day. For example, a day with a mean temperature of 70°F would correspond to 5 cooling degree-days (70 minus 65). High values indicate warm weather and high power production for cooling (see ● Fig. 3.27).

Knowledge of the number of cooling degree-days in an area allows a builder to plan the size and type of equipment that should be installed to provide adequate air conditioning. Also, the forecasting of cooling degree-days during the summer gives power companies a way of predicting the energy demand during peak energy periods. A composite of heating plus cooling degree-days would give a practical indication of the energy requirements over the year.

Farmers use an index called **growing degree-days** as a guide to planting and for determining the approximate dates when a crop will be ready for harvesting. There are a variety of methods of computing growing degree-days, but the most common one employs the mean daily temperature, since air temperature is the main factor that determines the physiological development of plants. Normally, a growing degree-day for a particular day is defined as a day on which the mean daily temperature is one degree above the *base temperature* (also known as *zero temperature*)—the minimum temperature required for growth of that crop. For sweet corn, the base temperature is 50°F and, for peas, it is 40°F.

On a summer day in Iowa, the mean temperature might be 80°F. From ▼ Table 3.2, we can see that, on this day, sweet corn would accumulate (80 − 50), or 30 growing degree-days. Theoretically, sweet corn can be harvested when it accumulates a total of 2200 growing degree-days. So, if sweet corn is planted in early April and each day thereafter averages about 20 growing degree-days, the corn would be ready for harvest about 110 days later, or around the middle of July.*

At one time, corn varieties were rated in terms of "days to maturity." This rating system was unsuccessful because, in actual practice, corn took considerably longer in some areas than in others. This discrepancy was the reason for defining "growing degree-days." Hence, in humid Iowa, where sum-

_____
*As a point of interest, in the corn belt when the air temperature climbs above 86°F, the hot air puts added stress on the growth of the corn. Consequently, the corn grows more slowly. Because of this fact, any maximum temperature over 86°F is reduced to 86°F when computing the mean air temperature.

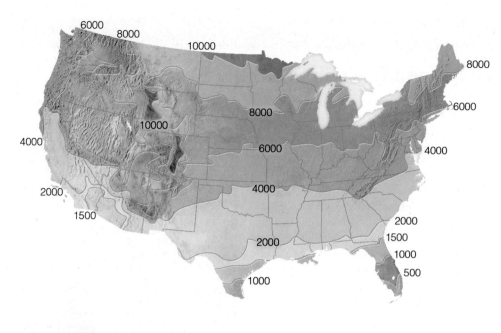

● FIGURE 3.26
Mean annual total heating degree-days across the United States (base 65°F).

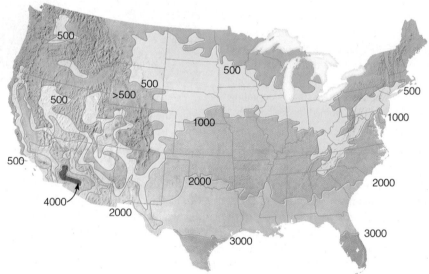

● FIGURE 3.27
Mean annual total cooling degree-days across the United States (base 65°F).

▼ TABLE 3.2   Estimated Growing Degree-Days for Certain Naturally Grown Agricultural Crops to Reach Maturity

CROP (VARIETY, LOCATION)	BASE TEMPERATURE (°F)	GROWING DEGREE-DAYS TO MATURITY
Beans (Snap/South Carolina)	50	1200–1300
Corn (Sweet/Indiana)	50	2200–2800
Cotton (Delta Smooth Leaf/Arkansas)	60	1900–2500
Peas (Early/Indiana)	40	1100–1200
Rice (Vegold/Arkansas)	60	1700–2100
Wheat (Indiana)	40	2100–2400

mer nighttime temperatures are high, growing degree-days accumulate much faster. Consequently, the corn matures in considerably fewer days than in the drier west, where summer nighttime temperatures are lower, and each day accumulates fewer growing degree-days. Although moisture and other conditions are not taken into account, growing degree-days nevertheless serve as a useful guide in forecasting approximate dates of crop maturity.

Air Temperature and Human Comfort

Probably everyone realizes that the same air temperature can feel differently on different occasions. For example, a temperature of 20°C (68°F) on a clear windless March afternoon in New York City can almost feel balmy after a long hard winter. Yet, this same temperature may feel uncomfortably cool

FOCUS ON AN OBSERVATION

A Thousand Degrees and Freezing to Death

Is there somewhere in our atmosphere where the air temperature can be exceedingly high (say above 1000°C or 1800°F) yet a person might feel extremely cold? There is a region, but it's not at the earth's surface.

You may recall from Chapter 1 (Fig. 1.10, p. 12) that in the upper reaches of our atmosphere (in the middle and upper thermosphere), air temperatures may exceed 1000°C. However, a thermometer shielded from the sun in this region of the atmosphere would indicate an extremely low temperature. This apparent discrepancy lies in the meaning of air temperature and how we measure it.

In Chapter 2, we learned that the air temperature is directly related to the average speed at which the air molecules are moving—faster speeds correspond to higher temperatures. In the middle and upper thermosphere (at altitudes approaching 300 km, or 200 mi) air molecules are zipping about at speeds correspond-

ing to extremely high temperatures. However, in order to transfer enough energy to heat something up by conduction (exposed skin or a thermometer bulb), an extremely large number of molecules must collide with the object. In the "thin" air of the upper atmosphere, air molecules are moving extraordinarily fast, but there are simply not enough of them bouncing against the thermometer bulb for it to register a high temperature. In fact, when properly shielded from the sun, the thermometer bulb loses far more energy than it receives and indicates a temperature near absolute zero. This explains why an astronaut, when space walking, will not only survive temperatures exceeding 1000°C, but will also feel a profound coldness when shielded from the sun's radiant energy. At these high altitudes, the traditional meaning of air temperature (that is, regarding how "hot" or "cold" something feels) is no longer applicable.

● **FIGURE 5** How can an astronaut survive when the "air" temperature is 1000°C?

NASA

---

on a summer afternoon in a stiff breeze. The human body's perception of temperature obviously changes with varying atmospheric conditions. The reason for these changes is related to how we exchange heat energy with our environment.

The body stabilizes its temperature primarily by converting food into heat (*metabolism*). To maintain a constant temperature, the heat produced and absorbed by the body must be equal to the heat it loses to its surroundings. There is, therefore, a constant exchange of heat—especially at the surface of the skin—between the body and the environment.

One way the body loses heat is by emitting infrared energy. But we not only emit radiant energy, we absorb it as well. Another way the body loses and gains heat is by conduction and convection, which transfer heat to and from the body by air motions. On a cold day, a thin layer of warm air molecules forms close to the skin, protecting it from the surrounding cooler air and from the rapid transfer of heat. Thus, in cold weather, when the air is calm, the temperature we perceive—called the **sensible temperature**—is often higher than a thermometer might indicate. (Could the opposite effect occur where the air temperature is very high and a person might feel exceptionally cold? If you are unsure, read the Focus section above.)

Once the wind starts to blow, the insulating layer of warm air is swept away, and heat is rapidly removed from the

skin by the constant bombardment of cold air. When all other factors are the same, the faster the wind blows, the greater the heat loss, and the colder we feel. How cold the wind makes us feel is usually expressed as a **wind-chill index (WCI)**.

The modern wind-chill index (see ▼ Table 3.3 and ▼ Table 3.4) was formulated in 2001 by a joint action group of the National Weather Service and other agencies. The new index takes into account the wind speed at about 1.5 m (5 ft) above the ground instead of the 10 m (33 ft) where "official" readings are usually taken. In addition, it translates the ability of the air to take heat away from a person's face (the air's cooling power) into a wind-chill equivalent temperature.* For example, notice in Table 3.3 that an air temperature of 10°F with a wind speed of 10 mi/hr produces a wind-chill equivalent temperature of −4°F. Under these conditions, the skin of a person's exposed face would lose as much heat in one minute in air with a temperature of 10°F and a wind speed of 10 mi/hr as it would in calm air with a temperature of −4°F. Of course, how cold we feel actually depends on a number of factors, including the fit and type of clothing we

---

*The wind-chill equivalent temperature formulas are as follows: Wind chill (°F) = $35.74 + 0.6215T − 35.75 (V^{0.16}) + 0.4275T (V^{0.16})$, where T is the air temperature in °F and V is the wind speed in mi/hr. Wind chill (°C) = $13.12 + 0.6215T − 11.37 (V^{0.16}) + 0.3965T (V^{0.16})$, where T is the air temperature in °C, and V is the wind speed in km/hr.

▼ TABLE 3.3 Wind-Chill Equivalent Temperature (°F). A 20-mi/hr Wind Combined with an Air Temperature of 20°F Produces a Wind-Chill Equivalent Temperature of 4°F.*

AIR TEMPERATURE (°F)																	
Calm	40	35	30	25	20	15	10	5	0	−5	−10	−15	−20	−25	−30	−35	−40
5	36	31	25	19	13	7	1	−5	−11	−16	−22	−28	−34	−40	−46	−52	−57
10	34	27	21	15	9	3	−4	−10	−16	−22	−28	−35	−41	−47	−53	−59	−66
15	32	25	19	13	6	0	−7	−13	−19	−26	−32	−39	−45	−51	−58	−64	−71
20	30	24	17	11	4	−2	−9	−15	−22	−29	−35	−42	−48	−55	−61	−68	−74
25	29	23	16	9	3	−4	−11	−17	−24	−31	−37	−44	−51	−58	−64	−71	−78
30	28	22	15	8	1	−5	−12	−19	−26	−33	−39	−46	−53	−60	−67	−73	−80
35	28	21	14	7	0	−7	−14	−21	−27	−34	−41	−48	−55	−62	−69	−76	−82
40	27	20	13	6	−1	−8	−15	−22	−29	−36	−43	−50	−57	−64	−71	−78	−84
45	26	19	12	5	−2	−9	−16	−23	−30	−37	−44	−51	−58	−65	−72	−79	−86
50	26	19	12	4	−3	−10	−17	−24	−31	−38	−45	−52	−60	−67	−74	−81	−88
55	25	18	11	4	−3	−11	−18	−25	−32	−39	−46	−54	−61	−68	−75	−82	−89
60	25	17	10	3	−4	−11	−19	−26	−33	−40	−48	−55	−62	−69	−76	−84	−91

*Dark blue shaded areas represent conditions where frostbite occurs in 30 minutes or less.

▼ TABLE 3.4 Wind-Chill Equivalent Temperature (°C)*

WIND SPEED (KM/HR)	AIR TEMPERATURE (°C)												
Calm	10	5	0	−5	−10	−15	−20	−25	−30	−35	−40	−45	−50
10	8.6	2.7	−3.3	−9.3	−15.3	−21.1	−27.2	−33.2	−39.2	−45.1	−51.1	−57.1	−63.0
15	7.9	1.7	−4.4	−10.6	−16.7	−22.9	−29.1	−35.2	−41.4	−47.6	−51.6	−59.9	−66.1
20	7.4	1.1	−5.2	−11.6	−17.9	−24.2	−30.5	−36.8	−43.1	−49.4	−55.7	−62.0	−68.3
25	6.9	0.5	−5.9	−12.3	−18.8	−25.2	−31.6	−38.0	−44.5	−50.9	−57.3	−63.7	−70.2
30	6.6	0.1	−6.5	−13.0	−19.5	−26.0	−32.6	−39.1	−45.6	−52.1	−58.7	−65.2	−71.7
35	6.3	−0.4	−7.0	−13.6	−20.2	−26.8	−33.4	−40.0	−46.6	−53.2	−59.8	−66.4	−73.1
40	6.0	−0.7	−7.4	−14.1	−20.8	−27.4	−34.1	−40.8	−47.5	−54.2	−60.9	−67.6	−74.2
45	5.7	−1.0	−7.8	−14.5	−21.3	−28.0	−34.8	−41.5	−48.3	−55.1	−61.8	−68.6	−75.3
50	5.5	−1.3	−8.1	−15.0	−21.8	−28.6	−35.4	−42.2	−49.0	−55.8	−62.7	−69.5	−76.3
55	5.3	−1.6	−8.5	−15.3	−22.2	−29.1	−36.0	−42.8	−49.7	−56.6	−63.4	−70.3	−77.2
60	5.1	−1.8	−8.8	−15.7	−22.6	−29.5	−36.5	−43.4	−50.3	−57.2	−64.2	−71.1	−78.0

*Dark blue shaded areas represent conditions where frostbite occurs in 30 minutes or less.

wear, the amount of sunshine striking the body, and the actual amount of exposed skin.

High winds, in below-freezing air, can remove heat from exposed skin so quickly that the skin may actually freeze and discolor. The freezing of skin, called **frostbite,** usually occurs on the body extremities first because they are the greatest distance from the source of body heat.

In cold weather, wet skin can be a factor in how cold we feel. A cold rainy day (drizzly, or even foggy) often feels colder than a "dry" one because water on exposed skin conducts heat away from the body better than air does. In fact, in cold, wet, and windy weather a person may actually lose body heat faster than the body can produce it. This may even occur in relatively mild weather with air temperatures as high as 10°C (50°F). The rapid loss of body heat may lower the body temperature below its normal level and bring on a condition known as **hypothermia**—the rapid, progressive mental and physical collapse that accompanies the lowering of human body temperature.

The first symptom of hypothermia is exhaustion. If exposure continues, judgment and reasoning power begin to disappear. Prolonged exposure, especially at temperatures near or below freezing, produces stupor, collapse, and death when the internal body temperature drops to 26°C (79°F). Most cases of hypothermia occur when the air temperature is between 0°C and 10°C (between 32°F and 50°F). This may be because many people apparently do not realize that wet clothing in windy weather greatly enhances the loss of body heat, even when the temperature is well above freezing.

In cold weather, heat is more easily dissipated through the skin. To counteract this rapid heat loss, the peripheral blood vessels of the body constrict, cutting off the flow of blood to the outer layers of the skin. In hot weather, the blood vessels enlarge, allowing a greater loss of heat energy to the surroundings. In addition to this, we perspire. As evaporation occurs, the skin cools because it supplies the large latent heat of vaporization (about 560 cal/g). When the air contains a great deal of water vapor and it is close to being saturated, perspiration does not readily evaporate from the skin. Less evaporational cooling causes most people to feel hotter than it really is, and a number of people start to complain about the "heat and humidity." (A closer look at how we feel in hot, humid weather will be given in Chapter 4 after we have examined the concepts of relative humidity and wet-bulb temperature.)

Measuring Air Temperature

Thermometers were developed to measure air temperature. Each thermometer has a definite scale and is calibrated so that a thermometer reading of 0°C in Vermont will indicate the same temperature as a thermometer with the same reading in North Dakota. If a particular reading were to represent different degrees of hot or cold, depending on location, thermometers would be useless.

Liquid-in-glass thermometers are often used for measuring surface air temperature because they are easy to read and

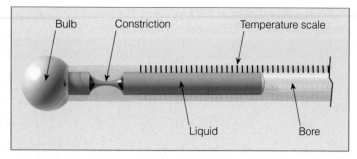

● FIGURE 3.28 A section of a maximum thermometer.

inexpensive to construct. These thermometers have a glass bulb attached to a sealed, graduated tube about 25 cm (10 in.) long. A very small opening, or bore, extends from the bulb to the end of the tube. A liquid in the bulb (usually mercury or red-colored alcohol) is free to move from the bulb up through the bore and into the tube. When the air temperature increases, the liquid in the bulb expands, and rises up the tube. When the air temperature decreases, the liquid contracts, and moves down the tube. Hence, the length of the liquid in the tube represents the air temperature. Because the bore is very narrow, a small temperature change will show up as a relatively large change in the length of the liquid column.

Maximum and minimum thermometers are liquid-in-glass thermometers used for determining daily maximum and minimum temperatures. The **maximum thermometer** looks like any other liquid-in-glass thermometer with one exception: It has a small constriction within the bore just above the bulb (see ● Fig. 3.28). As the air temperature increases, the mercury expands and freely moves past the constriction up the tube, until the maximum temperature occurs. However, as the air temperature begins to drop, the small constriction prevents the mercury from flowing back into the bulb. Thus, the end of the stationary mercury column indicates the maximum temperature for the day. The mercury will stay at this position until either the air warms to a higher reading or the thermometer is reset by whirling it on a special holder and pivot. Usually, the whirling is sufficient to push the mercury back into the bulb past the constriction until the end of the column indicates the present air temperature.*

A **minimum thermometer** measures the lowest temperature reached during a given period. Most minimum thermometers use alcohol as a liquid, since it freezes at a temperature of −130°C compared to −39°C for mercury. The minimum thermometer is similar to other liquid-in-glass thermometers except that it contains a small barbell-shaped index marker in the bore (see ● Fig. 3.29). The small index marker is free to slide back and forth within the liquid. It cannot move out of the liquid because the surface tension at the end of the liquid column (the *meniscus*) holds it in.

---

*Liquid-in-glass thermometers that measure body temperature are maximum thermometers, which is why they are shaken both before and after you take your temperature.

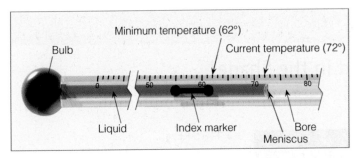

Minimum temperature (62°)

Current temperature (72°)

Bulb

Liquid

Index marker

Bore

Meniscus

● FIGURE 3.29 A section of a minimum thermometer showing both the current air temperature and the minimum temperature in °F.

A minimum thermometer is mounted horizontally. As the air temperature drops, the contracting liquid moves back into the bulb and brings the index marker down the bore with it. When the air temperature stops decreasing, the liquid and the index marker stop moving down the bore. As the air warms, the alcohol expands and moves freely up the tube past the stationary index marker. Because the index marker does not move as the air warms, the minimum temperature is read by observing the upper end of the marker.

To reset a minimum thermometer, simply tip it upside down. This allows the index marker to slide to the upper end of the alcohol column, which is indicating the current air temperature. The thermometer is then remounted horizontally, so that the marker will move toward the bulb as the air temperature decreases.

Highly accurate temperature measurements may be made with **electrical thermometers.** One type of electrical thermometer is the *electrical resistance thermometer,* which does not actually measure air temperature but rather the resistance of a wire, usually platinum or nickel, whose resistance increases as the temperature increases. An electrical meter measures the resistance, and is calibrated to represent air temperature.

Electrical resistance thermometers are the type of thermometers used in the measurement of air temperature at the over 900 fully automated surface weather stations (known as *ASOS* for *Automated Surface Observing System*) that exist at airports and military facilities throughout the United States (see ● Fig. 3.30). Hence, many of the liquid-in-glass thermometers have been replaced with electrical thermometers.

At this point it should be noted that the replacement of liquid-in-glass thermometers with electrical thermometers has raised concern among climatologists. For one thing, the response of the electrical thermometers to temperature change is faster. Thus, electrical thermometers may reach a brief extreme reading, which could have been missed by the slower-responding liquid-in-glass thermometer. In addition, many temperature readings, which were taken at airport weather offices, are now taken at ASOS locations that sit near or between runways at the airport. This change in instrumentation and relocation of the measurement site can sometimes introduce a small, but significant, temperature change at the reporting station.

Thermistors are another type of electrical thermometer. They are made of ceramic material whose resistance increases as the temperature decreases. A thermistor is the temperature-measuring device of the radiosonde—the instrument that measures air temperature from the surface up to an altitude near 30 kilometers.

Another electrical thermometer is the *thermocouple.* This device operates on the principle that the temperature difference between the junction of two dissimilar metals sets up a weak electrical current. When one end of the junction is maintained at a temperature different from that of the other end, an electrical current will flow in the circuit. This current is proportional to the temperature difference between the junctions.

Air temperature may also be obtained with instruments called *infrared sensors,* or **radiometers.** Radiometers do not measure temperature directly; rather, they measure emitted radiation (usually infrared). By measuring both the intensity of radiant energy and the wavelength of maximum emission of a particular gas, radiometers in orbiting satellites are now able to provide temperature readings at selected levels in the atmosphere.

A **bimetallic thermometer** consists of two different pieces of metal (usually brass and iron) welded together to form a single strip. As the temperature changes, the brass expands more than the iron, causing the strip to bend. The small amount of bending is amplified through a system of levers to a pointer on a calibrated scale. The bimetallic thermometer is usually the temperature-sensing part of the **thermograph,** an instrument that measures and records temperature (see ● Fig. 3.31).

● FIGURE 3.30 The instruments that comprise the ASOS system. The max-min temperature shelter is the middle box.

FOCUS ON AN OBSERVATION

Should Thermometers Be Read in the Shade?

When we measure air temperature with a common liquid thermometer, an incredible number of air molecules bombard the bulb, transferring energy either to or away from it. When the air is warmer than the thermometer, the liquid gains energy, expands, and rises up the tube; the opposite will happen when the air is colder than the thermometer. The liquid stops rising (or falling) when equilibrium between incoming and outgoing energy is established. At this point, we can read the temperature by observing the height of the liquid in the tube.

It is *impossible* to measure *air temperature* accurately in direct sunlight because the thermometer absorbs radiant energy from the sun in addition to energy from the air molecules. The thermometer gains energy at a much faster rate than it can radiate it away, and the liquid keeps expanding and rising until there is equilibrium between incoming and outgoing energy. Because of the direct absorption of solar

© Ross DePaola

● FIGURE 6
Instrument shelters such as the one shown here serve as a shady place for thermometers. Thermometers inside shelters measure the temperature of the air; whereas thermometers held in direct sunlight do not.

energy, the level of the liquid in the thermometer indicates a temperature *much* higher than the actual air temperature, and so a statement that says "today the air temperature measured 100 degrees in the sun," has no meaning. Hence, a thermometer must be kept in a shady place to measure the temperature of the air accurately.

Thermographs are gradually being replaced with *data loggers*. These small instruments have a thermistor connected to a circuit board inside the logger. A computer programs the interval at which readings are taken. The loggers are not only more responsive to air temperature than are thermographs, they are less expensive.

Chances are, you may have heard someone exclaim something like, "Today the thermometer measured 90 degrees in the shade!" Does this mean that the air temperature is sometimes measured in the sun? If you are unsure of the answer, read the Focus section above before reading the next section on instrument shelters.

Thermometers and other instruments are usually housed in an **instrument shelter.** The shelter completely encloses the instruments, protecting them from rain, snow, and the sun's direct rays. It is painted white to reflect sunlight, faces north to avoid direct exposure to sunlight, and has louvered sides, so that air is free to flow through it. This construction helps to keep the air inside the shelter at the same temperature as the air outside.

The thermometers inside a standard shelter are mounted about 1.5 to 2 m (5 to 6 ft) above the ground. As we saw in an earlier section, on a clear, calm night the air at ground level may be much colder than the air at the level of the shelter. As a result, on clear winter mornings it is possible to see ice or frost on the ground even though the minimum thermometer in the shelter did not reach the freezing point.

The older instrument shelters (such as the one shown in Focus Fig. 6, above) are gradually being replaced by the *Max-Min Temperature Shelter* of the ASOS system (the middle white box in Fig. 3.30, p. 83). The shelter is mounted on a pipe, and wires from the electrical temperature sensor inside are run to a building. A readout inside the building displays

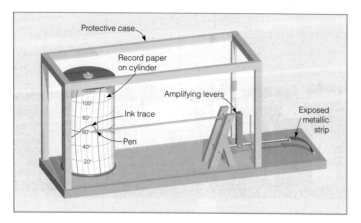

● FIGURE 3.31 The thermograph with a bimetallic thermometer.

the current air temperature and stores the maximum and minimum temperatures for later retrieval.

Because air temperatures vary considerably above different types of surfaces, where possible, shelters are placed over grass to ensure that the air temperature is measured at the same elevation over the same type of surface. Unfortunately, some shelters are placed on asphalt, others sit on concrete, while others are located on the tops of tall buildings, making it difficult to compare air temperature measurements from different locations. In fact, if either the maximum or minimum air temperature in your area seems suspiciously different from those of nearby towns, find out where the instrument shelter is situated.

SUMMARY

The earth has seasons because the earth is tilted on its axis as it revolves around the sun. The tilt of the earth causes a seasonal variation in both the length of daylight and the intensity of sunlight that reaches the surface. When the Northern Hemisphere is tilted toward the sun, the Southern Hemisphere is tilted away from the sun. Longer hours of daylight and more intense sunlight produce summer in the Northern Hemisphere, while, in the Southern Hemisphere, shorter daylight hours and less intense sunlight produce winter. On a more local setting, the earth's inclination influences the amount of solar energy received on the north and south side of a hill, as well as around a home.

The daily variation in air temperature near the earth's surface is controlled mainly by the input of energy from the sun and the output of energy from the surface. On a clear, calm day, the surface air warms, as long as heat input (mainly sunlight) exceeds heat output (mainly convection and radiated infrared energy). The surface air cools at night, as long as heat output exceeds input. Because the ground at night cools more quickly than the air above, the coldest air is normally found at the surface where a radiation inversion usually forms. When the air temperature in agricultural areas drops to dangerously low readings, fruit trees and grape vineyards can be protected from the cold by a variety of means, from mixing the air to spraying the trees and vines with water.

The greatest daily variation in air temperature occurs at the earth's surface. Both the diurnal and annual ranges of temperature are greater in dry climates than in humid ones. Even though two cities may have similar average annual temperatures, the range and extreme of their temperatures can differ greatly. Temperature information impacts our lives in many ways, from influencing decisions on what clothes to take on a trip to providing critical information for energy-use predictions and agricultural planning. We reviewed some of the many types of thermometers in use. Those designed to measure air temperatures near the surface are housed in instrument shelters to protect them from direct sunlight and precipitation.

KEY TERMS

The following terms are listed (with page number) in the order they appear in the text. Define each. Doing so will aid you in reviewing the material covered in this chapter.

summer solstice, 59
autumnal equinox, 61
Indian summer, 61
winter solstice, 62
vernal equinox, 62
radiational cooling, 68
radiation inversion, 69
nocturnal inversion, 69
thermal belts, 70
orchard heaters, 72
wind machines, 72
freeze, 73
controls of temperature, 73
isotherms, 73
daily (diurnal) range of
   temperature, 75
mean (average) daily
   temperature, 76
annual range of
   temperature, 76

mean (average) annual
   temperature, 77
heating degree-day, 78
cooling degree-day, 78
growing degree-days, 78
sensible temperature, 80
wind-chill index (WCI), 80
frostbite, 81
hypothermia, 82
liquid-in-glass
   thermometers, 82
maximum thermometer, 82
minimum thermometer, 82
electrical thermometers, 83
radiometers, 83
bimetallic thermometer, 83
thermograph, 83
instrument shelter, 84

QUESTIONS FOR REVIEW

1. In the Northern Hemisphere, why are summers warmer than winters, even though the earth is actually closer to the sun in January?
2. What are the main factors that determine seasonal temperature variations?
3. During the Northern Hemisphere's summer, the daylight hours in northern latitudes are longer than in middle latitudes. Explain why northern latitudes are not warmer.
4. If it is winter and January in New York City, what is the season in Sydney, Australia?
5. Explain why Southern Hemisphere summers are not warmer than Northern Hemisphere summers.
6. Explain why the vegetation on the north-facing side of a hill is frequently different from the vegetation on the south-facing side of the same hill.
7. Look at Figures 3.12 and 3.15, which show vertical profiles of air temperature during different times of the day. Explain why the temperature curves are different.
8. What are some of the factors that determine the daily fluctuation of air temperature just above the ground?
9. Explain how incoming energy and outgoing energy regulate the daily variation in air temperature.

10. On a calm, sunny day, why is the air next to the ground normally much warmer than the air just above?

11. Explain why the warmest time of the day is usually in the afternoon, even though the sun's rays are most direct at noon.

12. Explain how radiational cooling at night produces a radiation temperature inversion.

13. What weather conditions are best suited for the formation of a cold night and a strong radiation inversion?

14. Explain why thermal belts are found along hillsides at night.

15. List some of the measures farmers use to protect their crops against the cold. Explain the physical principle behind each method.

16. Why are the lower tree branches most susceptible to damage from low temperatures?

17. Describe each of the controls of temperature.

18. Look at Fig. 3.20 (temperature map for January) and explain why the isotherms dip southward (equatorward) over the Northern Hemisphere continents.

19. Explain why the daily range of temperature is normally greater
    (a) in dry regions than in humid regions and
    (b) on clear days than on cloudy days.

20. Why is the largest annual range of temperatures normally observed over continents away from large bodies of water?

21. Two cities have the same mean annual temperature. Explain why this fact does not mean that their temperatures throughout the year are similar.

22. During a cold, calm, sunny day, why do we usually feel warmer than a thermometer indicates?

23. What atmospheric conditions can bring on hypothermia?

24. During the winter, white frost can form on the ground when the minimum thermometer indicates a low temperature above freezing. Explain.

25. Why do daily temperature ranges decrease as you increase in altitude?

26. Why do the first freeze in autumn and the last freeze in spring occur in low-lying areas?

27. Someone says, "The air temperature today measured 99°F in the sun." Why does this statement have no meaning?

28. Briefly describe how the following thermometers measure air temperature:
    (a) liquid-in-glass
    (b) bimetallic
    (c) electrical
    (d) radiometer

QUESTIONS FOR THOUGHT

1. Explain (with the aid of a diagram) why the morning sun shines brightly through a south-facing bedroom window in December, but not in June.

2. Consider these two scenarios:
   (a) The tilt of the earth decreased to 10°.
   (b) The tilt of the earth increased to 40°.
   How would this change the summer and winter temperatures in your area? Explain, using a diagram.

3. At the top of the earth's atmosphere during the early summer (Northern Hemisphere), above what latitude would you expect to receive the most solar radiation in one day? During the same time of year, where would you expect to receive the most solar radiation at the surface? Explain why the two locations are different. (If you are having difficulty with this question, refer to Fig. 3.5, p. 60.)

4. If a construction company were to build a solar-heated home in middle latitudes in the Southern Hemisphere, in which direction should the solar panels on the roof be directed for maximum daytime heating?

5. Aside from the aesthetic appeal (or lack of such), explain why painting the outside north-facing wall of a middle latitude house one color and the south-facing wall another color is not a bad idea.

6. How would the lag in daily temperature experienced over land compare to the daily temperature lag over water?

7. Where would you expect to experience the smallest variation in temperature from year to year and from month to month? Why?

8. The average temperature in San Francisco, California, for December, January, and February is 11°C (52°F). During the same three-month period the average temperature in Richmond, Virginia, is 4°C (39°F). Yet, San Francisco and Richmond have nearly the same yearly total of heating-degree-days. Explain why. (Hint: See Fig. 3.25, p. 78.)

9. On a warm summer day, one city experienced a daily range of 22°C (40°F), while another had a daily range of 10°C (18°F). One of these cities is located in New Jersey and the other in New Mexico. Which location most likely had the highest daily range, and which one had the smallest? Explain.

10. Minimum thermometers are usually read during the morning, yet they are reset in the afternoon. Explain why.

11. If clouds arrive at 2 A.M. in the middle of a calm, clear night it is quite common to see temperatures rise after 2 A.M. How does this happen?

12. In the Northern Hemisphere, south-facing mountain slopes normally have a greater diurnal range in temperature than north-facing slopes. Why?

13. If the poles have 24 hours of sunlight during the summer, why is the average summer temperature still below 0°F?

14. In Pennsylvania and New York, wine grapes are planted on the side of hills rather than in valleys. Explain why this practice is so common in these areas.

PROBLEMS AND EXERCISES

1. Draw a graph similar to Fig. 3.5 (p. 60). Include in it the amount of solar radiation reaching the earth's surface in the Northern Hemisphere on the equinox.

2. Each day past the winter solstice the noon sun is a little higher above the southern horizon.
   (a) Determine how much change takes place each day at your latitude.
   (b) Does the same amount of change take place at each latitude in the Northern Hemisphere? Explain.

3. On approximately what dates will the sun be overhead at noon at latitudes:
   (a) 10°N?
   (b) 15°S?

4. Design a solar-heated home that sits on the north side of an east-west running street. If the home is located at 40°N, draw a proper roof angle for maximum solar heating. Design windows, doors, overhangs, and rooms with the intent of reducing heating and cooling costs. Place trees around the home that will block out excess summer sunlight and yet let winter sunlight inside. Choose a paint color for the house that will add to the home's energy efficiency.

5. Suppose peas are planted in Indiana on May 1. If the peas need 1200 growing degree-days before they can be picked, and if the average maximum temperature for May and June is 80°F and the average minimum is 60°F, on about what date will the peas be ready to pick? (Assume a base temperature of 55°F.)

6. What is the wind-chill equivalent temperature when the air temperature is 5°F and the wind speed is 35 mi/hr? (Use Table 3.3, p. 81.)

As the sun disappears behind an approaching deck of clouds, the air above the snow-covered landscape slowly cools. As the air temperature lowers, the relative humidity increases, and the air gradually approaches saturation.

© Brad Perks

CHAPTER 4

Atmospheric Humidity

Sometimes it rains and still fails to moisten the desert—the falling water evaporates halfway down between cloud and earth. Then you see curtains of blue rain dangling out of reach in the sky while the living things wither below for want of water. Torture by tantalizing, hope without fulfillment. And the clouds disperse and dissipate into nothingness. . . . The sun climbed noon-high, the heat grew thick and heavy on our brains, the dust clouded our eyes and mixed with our sweat. My canteen is nearly empty and I'm afraid to drink what little water is left—there may never be any more. I'd like to cave in for a while, crawl under yonder cottonwood and die peacefully in the shade, drinking dust.

Edward Abbey, *Desert Solitaire—A Season in the Wilderness*

 CONTENTS

We know from Chapter 1 that, in our atmosphere, the concentration of the invisible gas water vapor is normally less than a few percent of all the atmospheric molecules. Yet water vapor is exceedingly important, for it transforms into cloud particles—particles that grow in size and fall to the earth as precipitation. The term *humidity* can describe the amount of water vapor in the air. To most of us, a moist day suggests high humidity. However, there is usually more water vapor in the hot, "dry" air of the Sahara Desert than in the cold, "damp" polar air in New England, which raises an interesting question: Does the desert air have a higher humidity? As we will see later in this chapter, the answer to this question is both yes and no, depending on the type of humidity we mean.

So that we may better understand the concept of humidity, we will begin this chapter by examining the circulation of water in the atmosphere. Then, we will look at different ways to express humidity. At the end of the chapter, we will investigate various ways to measure humidity.

Circulation of Water in the Atmosphere

Within the atmosphere, there is an unending circulation. Since the oceans occupy over 70 percent of the earth's surface, we can think of this circulation as beginning over the ocean. Here, the sun's energy transforms enormous quantities of liquid water into water vapor in a process called **evaporation.** Winds then transport the moist air to other regions, where the water vapor changes back into liquid, forming clouds, in a process called **condensation.** Under certain conditions, the liquid (or solid) cloud particles may grow in size and fall to the surface as **precipitation**—rain, snow, or hail.* If the precipitation falls into an ocean, the water is ready to begin its cycle again. If, on the other hand, the precipitation falls on a continent, a great deal of the water returns to the ocean in a complex journey. This cycle of moving and transforming water molecules from liquid to vapor and back to liquid again is called the **hydrologic** (water) **cycle.** In the form with which we are most concerned, water molecules travel from ocean to atmosphere to land and then back to the ocean.

● Figure 4.1 illustrates the complexities of the hydrologic cycle. For example, before falling rain ever reaches the ground, a portion of it evaporates back into the air. Some of the precipitation may be intercepted by vegetation, where it evaporates or drips to the ground long after a storm has ended. Once on the surface, a portion of the water soaks into the ground by percolating downward through small openings in the soil and rock, forming groundwater that can be tapped by wells. What does not soak in collects in puddles of standing water or runs off into streams and rivers, which find their way back to the ocean. Even the underground water moves slowly and eventually surfaces, only to evaporate or be carried seaward by rivers.

---

*Precipitation is any form of water that falls from a cloud and reaches the ground.

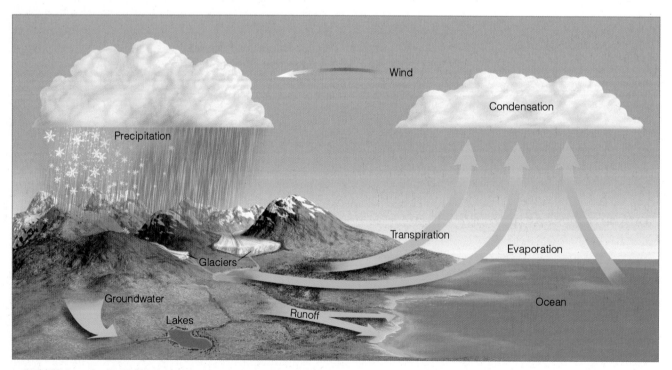

● FIGURE 4.1   The hydrologic cycle.

Over land, a considerable amount of water vapor is added to the atmosphere through evaporation from the soil, lakes, and streams. Even plants give up moisture by a process called *transpiration*. The water absorbed by a plant's root system moves upward through the stem and emerges from the plant through numerous small openings on the underside of the leaf. In all, evaporation and transpiration from continental areas amount to only about 15 percent of the nearly 1.5 billion billion gallons of water vapor that annually evaporate into the atmosphere; the remaining 85 percent evaporates from the oceans. If all of this water vapor were to suddenly condense and fall as rain, it would be enough to cover the entire globe with 2.5 centimeters, (or 1 inch) of water.* The total mass of water vapor stored in the atmosphere at any moment adds up to only a little over a week's supply of the world's precipitation. Since this amount varies only slightly from day to day, the hydrologic cycle is exceedingly efficient in circulating water in the atmosphere.

The Many Phases of Water

If we could see individual water molecules, we would find that, in the lower atmosphere, water is everywhere. If we could observe just one single water molecule by magnifying it billions of times, we would see an H_2O molecule in the shape of a tiny head that somewhat resembles Mickey Mouse (see ● Fig. 4.2). The bulk of the "head" of the molecule is the oxygen atom. The "mouth" is a region of excess negative charge. The "ears" are partially exposed protons of the hydrogen atom, which are regions of excess positive charge.

When we look at many H_2O molecules , we see that, as a gas, water vapor molecules move about quite freely, mixing well with neighboring atoms and molecules (see ● Fig. 4.3). As we learned in Chapter 2, the higher the temperature of the gas, the faster the molecules move. In the liquid state, the water molecules are closer together, constantly jostling and bumping into one another. If we lower the temperature of the liquid, water molecules would move slower and slower until, when cold enough, they arrange themselves into an orderly pattern with each molecule more or less locked into a rigid position, able to vibrate but not able to move about freely. In this solid state called *ice*, the shape and charge of the water molecule helps arrange the molecules into six-sided (hexagonal) crystals.

As we observe the ice crystal in freezing air, we see an occasional molecule gain enough energy to break away from its neighbors and enter into the air above. The molecule changes from an ice molecule directly into a vapor molecule without passing through the liquid state. This ice-to-vapor phase change is called **sublimation**. If a water vapor molecule should attach itself to the ice crystal, the vapor-to-ice phase change is called **deposition**. If we apply warmth to the ice

*If the water vapor in a column of air condenses and falls to the earth as rain, the depth of the rain on the surface is called *precipitable water*.

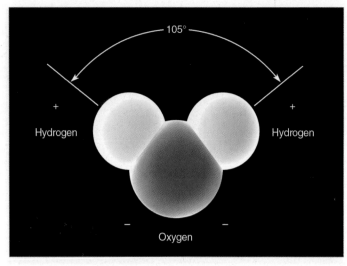

● FIGURE 4.2 The water molecule.

crystal, its molecules would vibrate faster. In fact, some of the molecules would actually vibrate out of their rigid crystal pattern into a disorderly condition—that is, the ice melts.

And so water vapor is a gas that becomes visible to us only when millions of molecules join together to form tiny cloud droplets or ice crystals. In this process—known as a *change of state* or, simply, *phase change*—water only changes its disguise, not its identity.

Evaporation, Condensation, and Saturation

Suppose we were able to observe individual water molecules in a beaker, as illustrated in ● Fig. 4.4a. What we would see are water molecules jiggling, bouncing, and moving about. However, we would also see that the molecules are not all moving

Gas (water vapor)   Liquid water   Ice

● FIGURE 4.3 The three states of matter. Water as a gas, as a liquid, and as a solid.

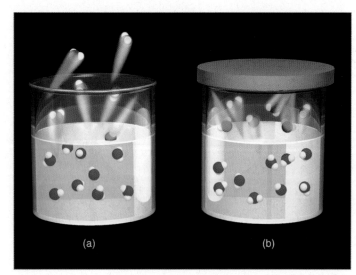

(a)   (b)

● FIGURE 4.4 (a) Water molecules at the surface of the water are evaporating (changing from liquid into vapor) and condensing (changing from vapor into liquid). Since more molecules are evaporating than condensing, net evaporation is occurring. (b) When the number of water molecules escaping from the liquid (evaporating) balances those returning (condensing), the air above the liquid is saturated with water vapor. (For clarity, only water molecules are illustrated.)

at the same speed — some are moving much faster than others. At the surface, molecules with enough speed (and traveling in the right direction) would occasionally break away from the liquid surface and enter into the air above. These molecules, changing from the *liquid state into the vapor state* are *evaporating*. While some water molecules are leaving the liquid, others are returning. Those returning are *condensing* as they are changing from a *vapor state to a liquid state*.

When a cover is placed over the beaker (see Fig. 4.4b), after a while the total number of molecules escaping from the liquid (evaporating) would be balanced by the number returning (condensing). When this condition exists, the air is said to be **saturated** with water vapor. For every molecule that evaporates, one must condense, and no net loss of liquid or vapor molecules results.

If we remove the cover and blow across the top of the water, some of the vapor molecules already in the air above would be blown away, creating a difference between the actual number of vapor molecules and the total number required for *saturation*. This would help prevent saturation from occurring and would allow for a greater amount of evaporation. Wind, therefore, enhances evaporation.

The temperature of the water also influences evaporation. All else being equal, warm water will evaporate more readily than cool water. The reason for this phenomenon is that, when heated, the water molecules will speed up. At higher temperatures, a greater fraction of the molecules have sufficient speed to break through the surface tension of the water and zip off into the air above. Consequently, the warmer the water, the greater the rate of evaporation.

If we could examine the air above the water in Fig. 4.4b, we would observe the water vapor molecules freely darting about and bumping into each other as well as neighboring molecules of oxygen and nitrogen. When these gas molecules collide, they tend to bounce off one another, constantly changing in speed and direction. However, the speed lost by one molecule is gained by another, and so the average speed of all the molecules does not change. Consequently, the temperature of the air does not change. Mixed in with all of the air molecules are microscopic bits of dust, smoke, salt, and other particles called *condensation nuclei* (so-called because water vapor condenses on them). In the warm air above the water, fast-moving vapor molecules strike the nuclei with such impact that they simply bounce away (see ● Figure 4.5a). However, if the air is chilled (Fig. 4.5b), the molecules move more slowly and are more apt to stick and condense to the nuclei. When many billions of these vapor molecules condense onto the nuclei, tiny liquid cloud droplets form.

We can see then that condensation is more likely to happen as the air cools and the speed of the vapor molecules decreases. As the air temperature increases, condensation is less likely because most of the molecules have sufficient speed (sufficient energy) to remain as a vapor. As we will see in this and other chapters, *condensation occurs primarily when the air is cooled.*[*]

Even though condensation is more likely to occur when the air cools, it is important to note that no matter how cold the air becomes, there will always be a few molecules with sufficient speed (sufficient energy) to remain as a vapor. It should be apparent, then, that with the same number of water vapor molecules in the air, saturation is more likely to occur in cool air than in warm air. This idea often leads to the statement that "warm air can hold more water vapor molecules before becoming saturated than can cold air" or, simply,

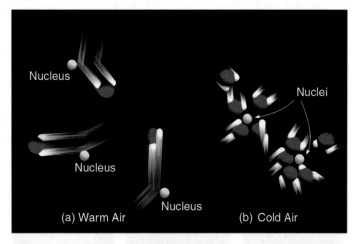

Nucleus

Nuclei

Nucleus

Nucleus

(a) Warm Air   (b) Cold Air

● FIGURE 4.5 Condensation is more likely to occur as the air cools. (a) In the warm air, fast-moving H$_2$O vapor molecules tend to bounce away after colliding with nuclei. (b) In the cool air, slow-moving vapor molecules are more likely to join together on nuclei. The condensing of many billions of water molecules produces tiny liquid water droplets.

----

*As we will see later, another way of explaining why cooling produces condensation is that the saturation vapor pressure decreases with lower temperatures.

"warm air has a greater capacity for water vapor than does cold air." At this point, it is important to realize that although these statements are correct, the use of such words as "hold" and "capacity" are misleading when describing water vapor content, as air does not really "hold" water vapor in the sense of making "room" for it.

Humidity

We are now ready to look more closely at the concept of **humidity,** which may refer to any one of a number of ways of specifying the amount of water vapor in the air. Since there are several ways to express atmospheric water vapor content, there are several meanings for the concept of humidity. The first type of humidity we'll take a look at is *absolute humidity.*

ABSOLUTE HUMIDITY   Suppose we enclose a volume of air in an imaginary thin elastic container—a *parcel*—about the size of a large balloon, as illustrated in ● Fig. 4.6. With a chemical drying agent, we can extract the water vapor from the air, weigh it, and obtain its mass. If we then compare the vapor's mass with the volume of air in the parcel, we would have determined the **absolute humidity** of the air—that is, the mass of water vapor in a given volume of air, which can be expressed as

$$\text{Absolute humidity} = \frac{\text{mass of water vapor}}{\text{volume of air}}.$$

Absolute humidity represents the *water vapor density* (mass/volume) in the parcel and, normally, is expressed as grams of

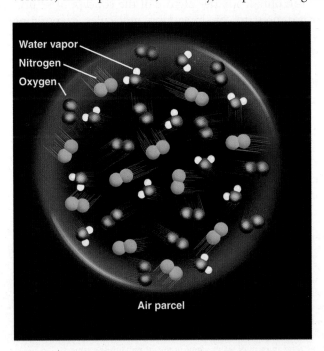

● FIGURE 4.6 The water vapor content (humidity) inside this air parcel can be expressed in a number of ways.

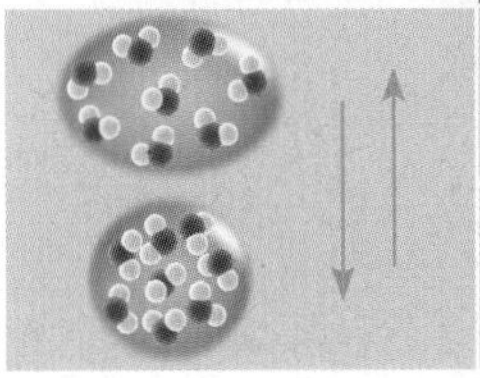

Parcel Size	Mass of H$_2$O Vapor	Absolute Humidity
2 m³	10 g	5 g/m³
1 m³	10 g	10 g/m³

● FIGURE 4.7 With the same amount of water vapor in a parcel of air, an increase in volume decreases absolute humidity, whereas a decrease in volume increases absolute humidity.

water vapor in a cubic meter of air. For example, if the water vapor in 1 cubic meter of air weighs 25 grams, the absolute humidity of the air is 25 grams per cubic meter (25 g/m³).

We learned in Chapter 2 that a rising or descending parcel of air will experience a change in its volume because of the changes in surrounding air pressure. Consequently, when a volume of air fluctuates, the absolute humidity changes—even though the air's vapor content has remained constant (see ● Fig. 4.7). For this reason, the absolute humidity is not commonly used in atmospheric studies.

SPECIFIC HUMIDITY AND MIXING RATIO   Humidity, however, can be expressed in ways that are not influenced by changes in air volume. When the mass of the water vapor in the air parcel in Fig. 4.6 is compared with the mass of all the air in the parcel (including vapor), the result is called the **specific humidity;** thus

$$\text{Specific humidity} = \frac{\text{mass of water vapor}}{\text{total mass of air}}.$$

Another convenient way to express humidity is to compare the mass of the water vapor in the parcel to the mass of the remaining dry air. Humidity expressed in this manner is called the **mixing ratio;** thus

$$\text{Mixing ratio} = \frac{\text{mass of water vapor}}{\text{mass of dry air}}.$$

Both specific humidity and mixing ratio are expressed as grams of water vapor per kilogram of air (g/kg).

The specific humidity and mixing ratio of an air parcel remain constant *as long as water vapor is not added to or removed from the parcel.* This happens because the total number of molecules (and, hence, the mass of the parcel) remains constant, even as the parcel expands or contracts (see ● Fig. 4.8). Since changes in parcel size do not affect specific humidity and mixing ratio, these two concepts are used extensively in the study of the atmosphere.

● Figure 4.9 shows how specific humidity varies with latitude. The average specific humidity is highest in the warm, muggy tropics. As we move away from the tropics, it decreases, reaching its lowest average value in the polar latitudes. Although the major deserts of the world are located

	Mass of Parcel	Mass of H_2O Vapor	Specific Humidity
	1 kg	1 g	1 g/kg
	1 kg	1 g	1 g/kg

● FIGURE 4.8 The specific humidity does not change as air rises and descends.

near latitude 30°, Fig. 4.9 shows that, at this latitude, the average air contains nearly twice the water vapor than does the air at latitude 50°N. Hence, the air of a desert is certainly not "dry," nor is the water vapor content extremely low. Since the hot, desert air of the Sahara often contains more water vapor than the cold, polar air farther north, we can say that *summertime Sahara air has a higher specific humidity*. (We will see later in what sense we consider desert air to be "dry.")

VAPOR PRESSURE   The air's moisture content may also be described by measuring the pressure exerted by the water vapor in the air. Suppose the air parcel in Fig. 4.6, (p. 93), is near sea level. The total pressure inside the parcel is due to the collision of all the molecules against the inside surface of the parcel. In other words, the total pressure inside the parcel is equal to the sum of the pressures of the individual gases. (This phenomenon is known as *Dalton's law of partial pressure*.) If

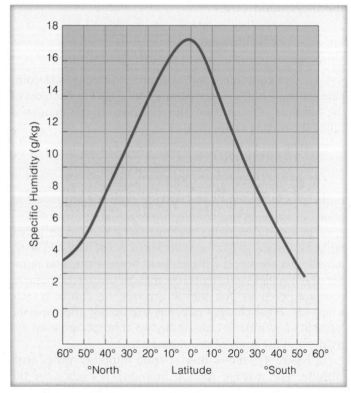

● FIGURE 4.9 The average specific humidity for each latitude. The highest average values are observed in the tropics and the lowest values in polar regions.

the total pressure inside the parcel is 1000 millibars (mb),* and the gases inside include nitrogen (78 percent), oxygen (21 percent), and water vapor (1 percent), then the partial pressure exerted by nitrogen would be 780 mb and by oxygen, 210 mb. The partial pressure of water vapor, called the **actual vapor pressure,** would be only 10 mb (1 percent of 1000).† It is evident, then, that because the number of water vapor molecules in any volume of air is small compared to the total number of air molecules in the volume, the actual vapor pressure is normally a small fraction of the total air pressure.

Everything else being equal, the more air molecules in a parcel, the greater the total air pressure. When you blow up a balloon, you increase its pressure by putting in more air. Similarly, an increase in the number of water vapor molecules will increase the total vapor pressure. Hence, the actual vapor pressure is a fairly good measure of the total amount of water vapor in the air: *High actual vapor pressure indicates large numbers of water vapor molecules, whereas low actual vapor pressure indicates comparatively small numbers of vapor molecules.*‡

In summer across North America, the highest vapor pressures are observed along the humid Gulf Coast, whereas the lowest values are experienced over the drier Great Basin, especially Nevada. In winter, the highest average vapor pressures are again observed along the Gulf Coast with lowest values over the northern Great Plains into Canada.

Actual vapor pressure indicates the air's total water vapor content, whereas **saturation vapor pressure** describes how much water vapor is necessary to make the air saturated at any given temperature. Put another way, *saturation vapor pressure is the pressure that the water vapor molecules would exert if the air were saturated with vapor at a given temperature.*§

We can obtain a better picture of the concept of saturation vapor pressure by imagining molecules evaporating from a water surface. Look back at Fig. 4.4b, (p. 92) and recall that when the air is saturated, the number of molecules escaping from the water's surface equals the number returning. Since the number of "fast-moving" molecules increases as the temperature increases, the number of water molecules escaping per second increases also. In order to maintain equilibrium, this situation causes an increase in the number of water vapor molecules in the air above the liquid. Consequently, at higher air temperatures, it takes more water vapor to saturate the air. And more vapor molecules exert a greater pressure. *Saturation vapor pressure, then, depends primarily on the air temperature.* From the graph in ● Fig. 4.10, we can see that at

---

*You may recall from Chapter 1 that the millibar is the unit of pressure most commonly found on surface weather maps, and that it expresses atmospheric pressure as a force over a given area.

†When we use the percentages of various gases in a volume of air, Dalton's law only gives us an approximation of the actual vapor pressure. The point here is that, near the earth's surface, the actual vapor pressure is often close to 10 mb.

‡Remember that actual vapor pressure is only an approximation of the total vapor content. A change in total air pressure will affect the actual vapor pressure even though the total amount of water vapor in the air remains the same.

§When the air is saturated, the amount of water vapor is the maximum possible at the existing temperature and pressure.

10°C, the saturation vapor pressure is about 12 mb, whereas at 30°C it is about 42 mb.

The insert in Fig. 4.10 shows that, when both water and ice exist at the same temperature below freezing, *the saturation vapor pressure just above the water is greater than the saturation vapor pressure over the ice.* In other words, at any temperature below freezing, it takes more vapor molecules to saturate air directly above water than it does to saturate air directly above ice. This situation occurs because it is harder for molecules to escape an ice surface than a water surface. Consequently, fewer molecules escape the ice surface at a given temperature, requiring fewer in the vapor phase to maintain equilibrium. Likewise, salts in solution bind water molecules, reducing the number escaping. These concepts are important and (as we will see in Chapter 7) play a role in the process of rain formation.

So far, we've described the amount of moisture actually in the air. If we want to report the moisture content of the air around us, we have several options:

1. *Absolute humidity* tells us the *mass* of water vapor in a fixed volume of air, or the *water vapor density.*

2. *Specific humidity* measures the *mass* of water vapor in a fixed *total mass* of air, and the *mixing ratio* describes the mass of water vapor in a fixed mass of the remaining dry air.

3. The *actual vapor pressure* of air expresses the amount of water vapor in terms of the amount of *pressure* that the water vapor molecules exert.

4. The *saturation vapor pressure* is the pressure that the water vapor molecules would exert if the air were saturated with vapor at a given temperature.

Each of these measures has its uses but, as we will see, the concepts of vapor pressure and saturation vapor pressure are critical to an understanding of the sections that follow. (Before looking at the most commonly used moisture variable—relative humidity—you may wish to read the Focus section on vapor pressure and boiling, p. 96.)

RELATIVE HUMIDITY   While relative humidity is the most common way of describing atmospheric moisture, it is also, unfortunately, the most misunderstood. The concept of relative humidity may at first seem confusing because it does not indicate the actual amount of water vapor in the air. Instead, it tells us how close the air is to being saturated. The **relative humidity** (RH) *is the ratio of the amount of water vapor actually in the air to the maximum amount of water vapor required for saturation at that particular temperature (and pressure).* It is the *ratio* of the air's water vapor *content* to its *capacity;* thus

$$RH = \frac{\text{water vapor content}}{\text{water vapor capacity}}.$$

We can think of the actual vapor pressure as a measure of the air's actual water vapor content, and the saturation vapor

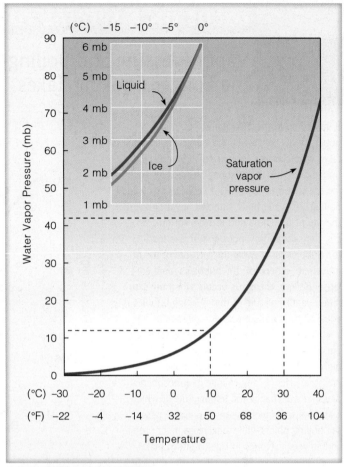

ACTIVE FIGURE 4.10 Saturation vapor pressure increases with increasing temperature. At a temperature of 10°C, the saturation vapor pressure is about 12 mb, whereas at 30°C it is about 42 mb. The insert illustrates that the saturation vapor pressure over water is greater than the saturation vapor pressure over ice.
Visit the Meteorology Resource Center to view this and other active figures at academic.cengage.com/login

pressure as a measure of air's total capacity for water vapor. Hence, the relative humidity can be expressed as

$$RH = \frac{\text{actual vapor pressure}}{\text{saturation vapor pressure}} \times 100 \text{ percent.}^*$$

Relative humidity is given as a percent. Air with a 50 percent relative humidity actually contains one-half the amount required for saturation. Air with a 100 percent relative humidity is said to be *saturated* because it is filled to capacity with water vapor. Air with a relative humidity greater than 100 percent is said to be **supersaturated.** Since relative humidity is used so much in the everyday world, let's examine it more closely.

---

*Relative humidity may also be expressed as

$$RH = \frac{\text{actual mixing ratio}}{\text{saturation mixing ratio}} \times 100 \text{ percent,}$$

where the actual mixing ratio is the mixing ratio of the air, and the saturation mixing ratio is the mixing ratio of saturated air at that particular temperature.

FOCUS ON A SPECIAL TOPIC

Vapor Pressure and Boiling—The Higher You Go, the Longer Cooking Takes

If you camp in the mountains, you may have noticed that, the higher you camp, the longer it takes vegetables to cook in boiling water. To understand this observation, we need to examine the relationship between vapor pressure and boiling. As water boils, bubbles of vapor rise to the top of the liquid and escape. For this to occur, the saturation vapor pressure exerted by the bubbles must equal the pressure of the atmosphere; otherwise, the bubbles would collapse. Boiling, therefore, occurs when the saturation vapor pressure of the escaping bubbles is equal to the total atmospheric pressure.

Because the saturation vapor pressure is directly related to the temperature of the liquid, higher water temperatures produce higher vapor pressures. Hence, any change in atmospheric pressure will change the temperature at which water boils: An increase in air pressure raises the boiling point, while a decrease in air pressure lowers it. Notice in Fig. 1 that, to make pure water boil at sea level, the water must be heated to a temperature of 100°C (212°F). At Denver, Colorado, which is situated about 1500 m (5000 ft) above sea level, the air pressure is near 850 millibars, and water boils at 95°C (203°F).

Once water starts to boil, its temperature remains constant, even if you continue to heat it. This happens because energy supplied to the

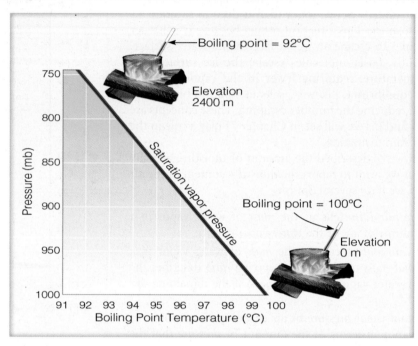

● **FIGURE 1** The lower the air pressure, the lower the saturation vapor pressure and, hence, the lower the boiling point temperature.

water is used to convert the liquid to a gas (steam). Now we can see why vegetables take longer to cook in the mountains. To be thoroughly cooked, they must boil for a longer time because the boiling water is cooler than at lower levels. In New York City, which is near

sea level, it takes about five minutes to hard boil an egg. An egg boiled for five minutes in the "mile high city" of Denver, Colorado, turns out to be runny.

A change in relative humidity can be brought about in two primary ways:

1. by changing the air's water vapor content
2. by changing the air temperature

In ● Fig. 4.11a, we can see that an increase in the water vapor content of the air (with no change in air temperature) increases the air's relative humidity. The reason for this increase resides in the fact that, as more water vapor molecules are added to the air, there is a greater likelihood that some of the vapor molecules will stick together and condense. Condensation takes place in saturated air. Therefore, as more and more water vapor molecules are added to the air, the air gradually approaches saturation, and the relative humidity of the air increases.* Conversely, removing water vapor from the air decreases the likelihood of saturation, which lowers the

air's relative humidity. In summary, with no change in air temperature, adding water vapor to the air increases the relative humidity; removing water vapor from the air lowers the relative humidity.

Figure 4.11b illustrates that, as the air temperature increases (with no change in water vapor content), the relative humidity decreases. This decrease in relative humidity occurs because in the warmer air the water vapor molecules are zipping about at such high speeds they are unlikely to join together and condense. The higher the temperature, the faster the molecular speed, the less saturation will occur, and

---

*We can also see in Fig. 4.11a that as the total number of vapor molecules increases (at a constant temperature), the actual vapor pressure increases and approaches the saturation vapor pressure at 20°C. As the actual vapor pressure approaches the saturation vapor pressure, the air approaches saturation and the relative humidity rises.

the lower the relative humidity.* As the air temperature lowers, the vapor molecules move more slowly, condensation becomes more likely as the air approaches saturation, and the relative humidity increases. In summary, with no change in water vapor content, an increase in air temperature lowers the relative humidity, while a decrease in air temperature raises the relative humidity.

In many places, the air's total vapor content varies only slightly during an entire day, and so it is the changing air temperature that primarily regulates the daily variation in relative humidity (see ● Fig. 4.12). As the air cools during the night, the relative humidity increases. Normally, the highest relative humidity occurs in the early morning, during the coolest part of the day. As the air warms during the day, the relative humidity decreases, with the lowest values usually occurring during the warmest part of the afternoon.

These changes in relative humidity are important in determining the amount of evaporation from vegetation and wet surfaces. If you water your lawn on a hot afternoon, when the relative humidity is low, much of the water will evaporate quickly from the lawn, instead of soaking into the ground. Watering the same lawn in the evening or during the early morning, when the relative humidity is higher, will cut down the evaporation and increase the effectiveness of the watering.

RELATIVE HUMIDITY AND DEW POINT Suppose it is early morning and the outside air is saturated. The air temperature is 10°C (50°F) and the relative humidity is 100 percent. We know from the previous section that relative humidity can be expressed as

$$RH = \frac{actual\ vapor\ pressure}{saturation\ vapor\ pressure} \times 100\ percent.$$

Looking back at Fig. 4.10 (p. 95), we can see that air with a temperature of 10°C has a saturation vapor pressure of 12 mb. Since the air is saturated and the relative humidity is 100 percent, the actual vapor pressure *must* be the same as the saturation vapor pressure (12 mb), since

$$RH = \frac{12\ mb}{12\ mb} \times 100\% = 100\ percent.$$

Suppose during the day the air warms to 30°C (86°F), with no change in water vapor content (or air pressure). Because there is no change in water vapor content, the actual vapor pressure must be the same (12 mb) as it was in the early morning when the air was saturated. The saturation vapor pressure, however, has increased because the air temperature has increased. From Fig. 4.10, note that air with a temperature of 30°C has a saturation vapor pressure of 42 mb. The

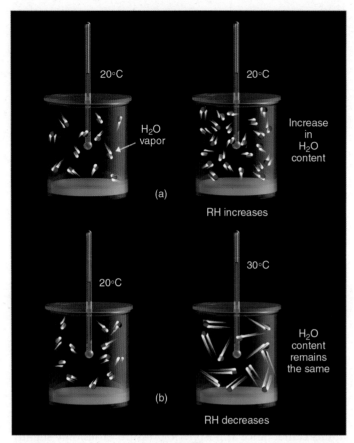

● FIGURE 4.11 (a) At the same air temperature, an increase in the water vapor content of the air increases the relative humidity as the air approaches saturation. (b) With the same water vapor content, an increase in air temperature causes a decrease in relative humidity as the air moves farther away from being saturated.

relative humidity of this unsaturated, warmer air is now much lower, as

$$RH = \frac{12\ mb}{42\ mb} \times 100\% = 29\ percent.$$

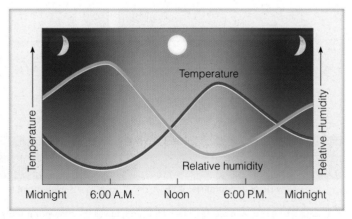

● FIGURE 4.12 When the air is cool (morning), the relative humidity is high. When the air is warm (afternoon), the relative humidity is low. These conditions exist in clear weather when the air is calm or of constant wind speed.

*Another way to look at this concept is to realize that, as the air temperature increases, the air's saturation vapor pressure also increases. As the saturation vapor pressure increases, with no change in water vapor content, the air moves farther away from saturation, and the relative humidity decreases.

To what temperature must the outside air, with a temperature of 30°C, be cooled so that it is once again saturated? The answer, of course, is 10°C. For this amount of water vapor in the air, 10°C is called the **dew-point temperature** or, simply, the **dew point.** It represents *the temperature to which air would have to be cooled (with no change in air pressure or moisture content) for saturation to occur.* The dew point is determined with respect to a flat surface of water. When the dew point is determined with respect to a flat surface of ice, it is called the **frost point.**

The dew point is an important measurement used in predicting the formation of dew, frost, fog, and even the minimum temperature (see ● Fig. 4.13). When used with an empirical formula (as illustrated in Chapter 6 on p. 155), the dew point can help determine the height of the base of a cumulus cloud. Since atmospheric pressure varies only slightly at the earth's surface, *the dew point is a good indicator of the air's actual water vapor content. High dew points indicate high water vapor content; low dew points, low water vapor content.* Addition of water vapor to the air increases the dew point; removing water vapor lowers it.

● Figure 4.14a shows the average dew-point temperatures across the United States and southern Canada for January. Notice that the dew points are highest (the greatest amount of water vapor in the air) over the Gulf Coast states and lowest over the interior. Compare New Orleans with Fargo. Cold,

dry winds from northern Canada flow relentlessly into the Center Plains during the winter, keeping this area dry. But warm, moist air from the Gulf of Mexico helps maintain a higher dew-point temperature in the southern states.

Figure 4.14b is a similar diagram showing the average dew-point temperatures for July. Again, the highest dew points are observed along the Gulf Coast, with some areas experiencing average dew-point temperatures near 75°F. Note, too, that the dew points over the eastern and central portion of the United States are much higher in July, meaning that the July air contains between 3 and 6 times more water vapor than the January air. The reason for the high dew points is that this region is almost constantly receiving humid air from the warm Gulf of Mexico. The lowest dew point, and hence the driest air, is found in the West, with Nevada experiencing the lowest values—a region surrounded by mountains that effectively shields it from significant amounts of moisture moving in from the southwest and northwest.

The difference between air temperature and dew point can indicate whether the relative humidity is low or high. When the air temperature and dew point are far apart, the relative humidity is low; when they are close to the same value, the relative humidity is high. When the air temperature and dew point are equal, the air is *saturated* and the relative humidity is 100 percent. Even though the relative humidity may be 100 percent, the air, under certain conditions, may be considered "dry."

Observe, for example, in ● Fig. 4.15a that, because the air temperature and dew point are the same in the polar air, the air is saturated and the relative humidity is 100 percent. On the other hand, the desert air (Fig. 4.15b), with a large separation between air temperature and dew point, has a much lower relative humidity—21 percent.* However, since dew

*The relative humidity can be computed from Fig. 4.10 (p. 95). The desert air with an air temperature of 35°C has a saturation vapor pressure of about 56 mb. A dew-point temperature of 10°C gives the desert air an actual vapor pressure of about 12 mb. These values produce a relative humidity of 12/56 × 100, or 21 percent.

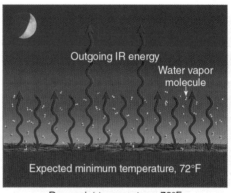
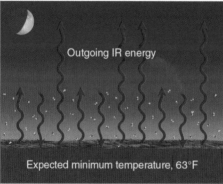
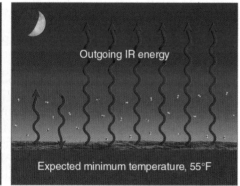

Outgoing IR energy
Water vapor molecule

Expected minimum temperature, 72°F

Dew-point temperature, 70°F

Outgoing IR energy

Expected minimum temperature, 63°F

Dew-point temperature, 60°F

Outgoing IR energy

Expected minimum temperature, 55°F

Dew-point temperature, 50°F

● FIGURE 4.13 On a calm, clear night, the lower the dew-point temperature, the lower the expected minimum temperature. With the same initial evening air temperature (80°F) and with no change in weather conditions during the night, as the dew point lowers, the expected minimum temperature lowers. This situation occurs because a lower dew point means that there is less water vapor in the air to absorb and radiate infrared energy back to the surface. More infrared energy from the surface is able to escape into space, producing more rapid radiational cooling at the surface. (Dots in each diagram represent the amount of water vapor in the air. Red wavy arrows represent infrared (IR) radiation.)

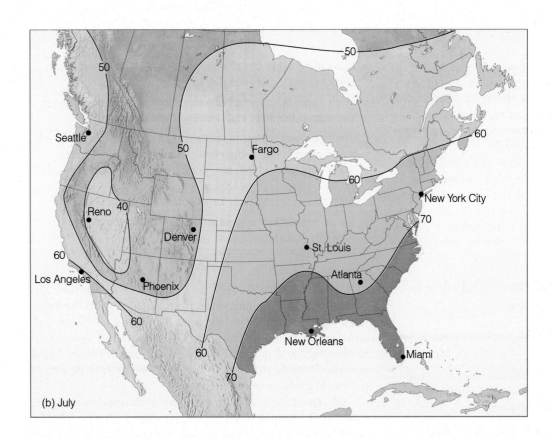

(a) POLAR AIR: Air temperature –2°C (28°F)
                Dew point –2°C (28°F)
                Relative humidity 100 percent

(b) DESERT AIR: Air temperature 35°C (95°F)
                 Dew point 10°C (50°F)
                 Relative humidity 21 percent

• FIGURE 4.15 The polar air has the higher relative humidity, whereas the desert air, with the higher dew point, contains more water vapor.

point is a measure of the amount of water vapor in the air, the desert air (with a higher dew point) must contain *more* water vapor. So even though the polar air has a higher relative humidity, the desert air that contains more water vapor has a higher water vapor density, or *absolute humidity,* and a higher specific humidity and mixing ratio as well.

Now we can see why polar air is often described as being "dry" when the relative humidity is high (often close to 100 percent). In cold, polar air, the dew point and air temperature are normally close together. But the low dew-point temperature means that there is little water vapor in the air. Consequently, the air is said to be "dry" even though the relative humidity is quite high.

View this concept in action on the Meteorology Resource Center at academic.cengage.com/login

BRIEF REVIEW

Up to this point we have looked at the different ways of describing humidity. Before going on, here is a review of some of the important concepts and facts we have covered:

- Relative humidity does not tell us how much water vapor is actually in the air; rather, it tells us how close the air is to being saturated.

- Relative humidity can change when the air's water-vapor content changes, or when the air temperature changes.

- With a constant amount of water vapor, cooling the air raises the relative humidity and warming the air lowers it.

- The dew-point temperature is a good indicator of the air's water-vapor content: High dew points indicate high water-vapor content; and low dew points, low water-vapor content.

- Dry air can have a high relative humidity. In polar air, when the dew-point temperature is low, the air is considered dry. But if the air temperature is close to the dew point, the relative humidity is high.

COMPARING HUMIDITIES • Figure 4.16 shows how the average relative humidity varies from the equator to the poles. High relative humidities are normally found in the tropics and near the poles, where there is little separation between air temperature and dew point. The average relative humidity is low near latitude 30°—a latitude where we find the deserts of the world girdling the globe.

Of course, not all locations near 30°N are deserts. Take, for example, humid New Orleans, Louisiana. During July, the air in New Orleans with an average dew-point temperature of 22°C (72°F) contains a great deal of water vapor—nearly 50 percent more than does the air along the southern California coast. Since both locations are adjacent to large bodies of water, why is New Orleans more humid?

● Figure 4.17 shows a summertime situation where air from the Pacific Ocean is moving into southern California and air from the Gulf of Mexico is moving into the southeastern states. Notice that the Pacific water is much cooler than the Gulf water. Westerly winds, blowing across the Pacific, cool to just about the same temperature as the water. Likewise, air over the warmer Gulf reaches a temperature near that of the water below it. Over the water, at both locations, the air is nearly saturated with water vapor. This means that the dew-point temperature of the air over the cooler Pacific Ocean is much lower than the dew-point temperature over the warmer Gulf. Consequently, the air from the Gulf of Mexico contains a great deal more water vapor than the Pacific air.

As the air moves inland, away from the source of moisture, the air temperature in both cases increases. But the amount of water vapor in the air (and, hence, the dew-point temperature) hardly changes. Therefore, as the humid air moves into the southeastern states, high air temperatures along with high dew-point temperatures produce high relative humidities, often greater than 75 percent during the hottest part of the day. On the other hand, over the southwestern part of the nation, high air temperatures and low dew-point temperatures produce low relative humidities, often less than 25 percent during the hottest part of the afternoon. Much of this inland area over the southwest is a desert. However, keep

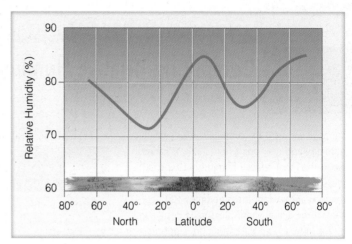

● **FIGURE 4.16** Relative humidity averaged for latitudes north and south of the equator.

in mind that although considered "dry," this area, with a dew-point temperature above freezing, still contains more water vapor than does the cold, arctic air in polar regions. (For more information on the computation of relative humidity and dew point, read the Focus section on p. 102.)

RELATIVE HUMIDITY IN THE HOME Question: How does the relative humidity of the winter air in your home compare with that in the Sahara Desert? Some homes actually have a

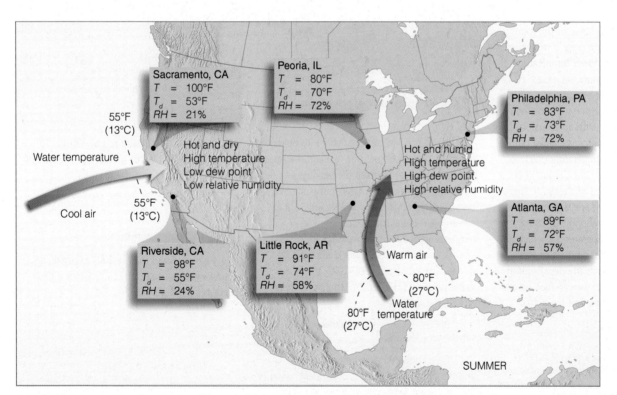

● **FIGURE 4.17** Air from the Pacific Ocean is hot and dry over land, whereas air from the Gulf of Mexico is hot and muggy over land. For each city, T represents the air temperature, T_d the dew point, and RH the relative humidity. (All data represent conditions during a July afternoon at 3 P.M. local time.)

FOCUS ON A SPECIAL TOPIC

Computing Relative Humidity and Dew Point

Suppose we want to compute the air's relative humidity and dew point from Table 1. Earlier, we learned that relative humidity may be expressed as the actual vapor pressure divided by the saturation vapor pressure times 100 percent. If the actual vapor pressure is designated by the letter e, and the saturation vapor pressure by e_s, then the expression for relative humidity becomes

$$RH = \frac{e}{e_s} \times 100\%.^*$$

Let's look at a practical example of using vapor pressure to measure relative humidity and obtain dew point. Suppose the air temperature in a room is 27°C (80°F). Because the saturation vapor pressure (e_s) is dependent on the temperature of the air, to obtain e_s from Table 1 we simply read the value adjacent to the air temperature much like we did in Fig. 4.10. Hence, air with a temperature of 27°C has a saturation vapor pressure of 35 mb.

Now, suppose that the air in the room is cooled suddenly with no change in moisture content. At successively lower temperatures, the saturation vapor pressure decreases. As the lowering saturation vapor pressure (e_s) approaches the actual vapor pressure (e), the relative humidity increases. With an actual vapor pressure of 25 mb, 100 percent relative humidity will be reached at a temperature of 21°C (70°F). This temperature (21°C) must then be the *dew-point temperature* of the air. If, then, we know the actual vapor pressure in a room, we can determine the dew point by using Table 1 to locate the temperature at which air will be saturated with that amount of vapor. Similarly, if we are told that the dew point in the room has some value, we can look up that temperature in Table 1 and find the actual vapor pressure.

*Relative humidity may also be expressed as RH = $w/w_s \times 100\%$, where w is the actual mixing ratio and w_s is the saturation mixing ratio. Relative humidity computations using mixing ratio and adiabatic charts are given in Chapter 6.

▼ TABLE 1 Saturation Vapor Pressure Over Water for Various Air Temperatures*

AIR TEMPERATURE (°C)	(°F)	SATURATION VAPOR PRESSURE (MB)	AIR TEMPERATURE (°C)	(°F)	SATURATION VAPOR PRESSURE (MB)
−18	(0)	1.5	18	(65)	21.0
−15	(5)	1.9	21	(70)	25.0
−12	(10)	2.4	24	(75)	29.6
−9	(15)	3.0	27	(80)	35.0
−7	(20)	3.7	29	(85)	41.0
−4	(25)	4.6	32	(90)	48.1
−1	(30)	5.6	35	(95)	56.2
2	(35)	6.9	38	(100)	65.6
4	(40)	8.4	41	(105)	76.2
7	(45)	10.2	43	(110)	87.8
10	(50)	12.3	46	(115)	101.4
13	(55)	14.8	49	(120)	116.8
16	(60)	17.7	52	(125)	134.2

*The data in this table can be obtained in Fig. 4.10 on p. 95 by reading where the air temperature intersects the saturation vapor pressure curve.

In essence, we can use Table 1 to obtain the saturation vapor pressure (e_s) and the actual vapor pressure (e) if the air temperature and dew point of the air are known. With this information we can calculate relative humidity. For example, what is the relative humidity of air with a temperature of 29°C and a dew point of 18°C?

Answer: At 29°C, Table 1 shows e_s = 41 mb. For a dew point of 18°C, the actual vapor pressure (e) is 21 mb; therefore, the relative humidity is

$$RH = \frac{e}{e_s} = \frac{21}{41} \times 100\% = 51\%.$$

If we know the air temperature is 27°C and the relative humidity is 60 percent, what is the dew-point temperature of the air? From Table 1, an air temperature of 27°C produces a saturation vapor pressure (e_s) of 35 mb. To obtain the actual vapor pressure (e), we simply plug the numbers into the formula

$$RH = \frac{e}{e_s} \times 100\%; \quad 60\% = \frac{e}{35}$$

$$e = 21 mb.$$

As we saw in the previous example, an actual vapor pressure of 21 mb yields a dew-point temperature of 18°C.

lower relative humidity than the desert, and the inhabitants are usually unaware of it. Remember that cold polar air contains only a little water vapor. Even when saturated, air with a temperature and dew point of −15°C (5°F) has an actual vapor pressure of only 1.9 mb. When this air is brought indoors and heated to 20°C (68°F), its saturation vapor pressure increases to 23.4 mb—about 12 times what it was outside. Notice in • Fig. 4.18 that the relative humidity of the heated air inside the house drops to 8 percent.* This relative humidity is lower than what you would normally experience in a desert during the hottest time of the day!

Very low relative humidities in a house can have an adverse effect on things living inside. For example, house plants have a difficult time surviving because the moisture from their leaves and the soil evaporates rapidly. Hence, house plants usually need watering more frequently in winter than in summer. People suffer, too, when the relative humidity is quite low. The rapid evaporation of moisture from exposed flesh causes skin to crack, dry, flake, or itch. These low humidities also irritate the mucous membranes in the nose and throat, producing an "itchy" throat. Similarly, dry nasal passages permit inhaled bacteria to incubate, causing persistent infections. The remedy for most of these problems is simply to increase the relative humidity. But how?

The relative humidity in a home can be increased just by heating water and allowing it to evaporate into the air. The added water vapor raises the relative humidity to a more comfortable level. In modern homes, a humidifier, installed near the furnace, adds moisture to the air at a rate of about one gallon per room per day. The air, with its increased water vapor, is circulated throughout the home by a forced air heating system. In this way, all rooms get their fair share of moisture—not just the room where the vapor is added.

To lower the air's moisture content, as well as the air temperature, many homes are air conditioned. Outside air cools as it passes through a system of cold coils located in the air conditioning unit. The cooling increases the air's relative humidity, and the air reaches saturation. The water vapor condenses into liquid water, which is carried away. The cooler, dehumidified air is now forced into the home.

In hot regions, where the relative humidity is low, *evaporative cooling systems* can be used to cool the air. These systems operate by having a fan blow hot, dry outside air across pads that are saturated with water. Evaporation cools the air, which is forced into the home, bringing some relief from the hot weather.

Evaporative coolers, also known as "swamp coolers," work best when the relative humidity is low and the air is warm. They do not work well in hot, muggy weather because a high relative humidity greatly reduces the rate of evaporation. Besides, swamp coolers add water vapor to the air—something that is not needed when the air is already

• FIGURE 4.18 When outside air with an air temperature and a dew point of −15°C (5°F) is brought indoors and heated to a temperature of 20°C (68°F) (without adding water vapor to the air), the relative humidity drops to 8 percent, placing adverse stress on plants, animals, and humans living inside. (*T* represents temperature; T_d, dew point; and *RH*, relative humidity.)

uncomfortably humid. That is why swamp coolers may be found on homes in Arizona, but not on homes in Alabama.

RELATIVE HUMIDITY AND HUMAN DISCOMFORT On a hot, muggy day when the relative humidity is high, it is common to hear someone exclaim (often in exasperation), "It's not so much the heat, it's the humidity." Actually, this statement has validity. In warm weather, the main source of body cooling is through evaporation of perspiration. Recall from Chapter 2 that evaporation is a cooling process, so when the air temperature is high and the relative humidity low, perspiration on the skin evaporates quickly, often making us feel that the air temperature is lower than it really is. However, when both the air temperature and relative humidity are high and the air is nearly saturated with water vapor, body moisture does not readily evaporate; instead, it collects on the skin as beads of perspiration. Less evaporation means less cooling, and so we usually feel warmer than we did with a similar air temperature, but a lower relative humidity.

A good measure of how cool the skin can become is the **wet-bulb temperature**—*the lowest temperature that can be reached by evaporating water into the air.** On a hot day when the wet-bulb temperature is low, rapid evaporation (and, hence, cooling) takes place at the skin's surface. As the wet-bulb

---

*RH = $\dfrac{1.9 \text{ mb}}{23.4 \text{ mb}} \times 100 = 8\%$.

---

*Notice that the wet-bulb temperature and the dew-point temperature are different. The wet-bulb temperature is attained by *evaporating water* into the air, whereas the dew-point temperature is reached by *cooling* the air.

temperature approaches the air temperature, less cooling occurs, and the skin temperature may begin to rise. When the wet-bulb temperature exceeds the skin's temperature, no net evaporation occurs, and the body temperature can rise quite rapidly. Fortunately, most of the time, the wet-bulb temperature is considerably below the temperature of the skin.

When the weather is hot and muggy, a number of heat-related problems may occur. For example, in hot weather when the human body temperature rises, the *hypothalamus* gland (a gland in the brain that regulates body temperature) activates the body's heat-regulating mechanism, and over ten million sweat glands wet the body with as much as two liters of liquid per hour. As this perspiration evaporates, rapid loss of water and salt can result in a chemical imbalance that may lead to painful *heat cramps*. Excessive water loss through perspiring coupled with an increasing body temperature may result in *heat exhaustion*—fatigue, headache, nausea, and even fainting. If one's body temperature rises above about 41°C (106°F), **heatstroke** can occur, resulting in complete failure of the circulatory functions. If the body temperature continues to rise, death may result. In fact, each year across North America, hundreds of people die from heat-related maladies. Even strong, healthy individuals can succumb to heatstroke, as did the Minnesota Vikings' all-pro offensive lineman, Korey Stringer, who collapsed after practice on July 31, 2001, and died 15 hours later. Before Korey fainted, temperatures on the practice field were in the 90s (°F) with the relative humidity above 55 percent.

In an effort to draw attention to this serious weather-related health hazard, an index called the **heat index (HI)** is used by the National Weather Service. The index combines air temperature with relative humidity to determine an **apparent temperature**—what the air temperature "feels like" to the average person for various combinations of air temperature and relative humidity. For example, in ● Fig. 4.19 an air temperature of 100°F and a relative humidity of 60 percent produce an apparent temperature of 132°F. As we can see in ▼ Table 4.1, heatstroke or sunstroke is imminent when the index reaches this level. However, as we saw in the preceding paragraph, heatstroke related deaths can occur when the heat index value is considerably lower than 130°F.

Tragically, many hundreds of people died of heat-related maladies during the great Chicago heat wave of July, 1995. On July 13, the afternoon air temperature reached 104°F. With a dew-point temperature of 76°F and a relative humidity near 40 percent, the apparent temperature soared to 119°F (see Table 4.1). In a van, with the windows rolled up, two small toddlers fell asleep and an hour later were found dead of heat exhaustion. Estimates are that, on a day like this one, temperatures inside a closed vehicle could approach 190°F within an hour.

At this point it is important to dispel a common myth that seems to circulate in hot, humid weather. After being outside for awhile, people will say that the air temperature today is 90 degrees and the relative humidity is 90 percent. We see in Fig. 4.19 that this weather condition would produce

● **FIGURE 4.19**
Air temperature (°F) and relative humidity are combined to determine an apparent temperature or heat index (HI). An air temperature of 95°F with a relative humidity of 55 percent produces an apparent temperature (HI) of 110°F.

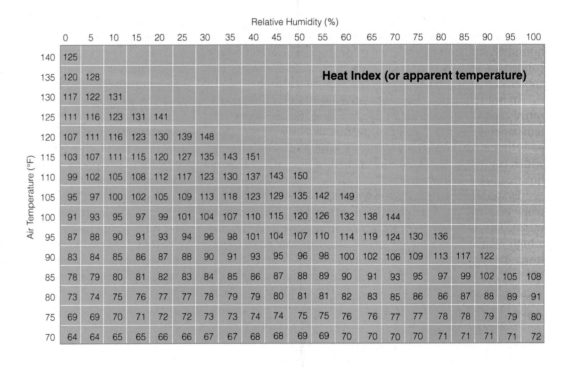

Relative Humidity (%)

Heat Index (or apparent temperature)

Air Temp (°F)	0	5	10	15	20	25	30	35	40	45	50	55	60	65	70	75	80	85	90	95	100
140	125																				
135	120	128																			
130	117	122	131																		
125	111	116	123	131	141																
120	107	111	116	123	130	139	148														
115	103	107	111	115	120	127	135	143	151												
110	99	102	105	108	112	117	123	130	137	143	150										
105	95	97	100	102	105	109	113	118	123	129	135	142	149								
100	91	93	95	97	99	101	104	107	110	115	120	126	132	138	144						
95	87	88	90	91	93	94	96	98	101	104	107	110	114	119	124	130	136				
90	83	84	85	86	87	88	90	91	93	95	96	98	100	102	106	109	113	117	122		
85	78	79	80	81	82	83	84	85	86	87	88	89	90	91	93	95	97	99	102	105	108
80	73	74	75	76	77	77	78	79	79	80	81	81	82	83	85	86	86	87	88	89	91
75	69	69	70	71	72	72	73	73	74	74	75	75	76	76	77	77	78	78	79	79	80
70	64	64	65	65	66	66	67	67	68	68	69	69	70	70	70	70	71	71	71	71	72

Air Temperature (°F)

▼ TABLE 4.1   The Heat Index and Related Syndrome

CATEGORY	APPARENT TEMPERATURE (°F)	HEAT SYNDROME
I	130° or higher	Heatstroke or sunstroke *imminent*
II	105°–130°	Sunstroke, heat cramps, or heat exhaustion *likely*, heatstroke *possible* with prolonged exposure and physical activity
III	90°–105°	Sunstroke, heat cramps, and heat exhaustion *possible* with prolonged exposure and physical activity
IV	80°–90°	Fatigue *possible* with prolonged exposure and physical activity

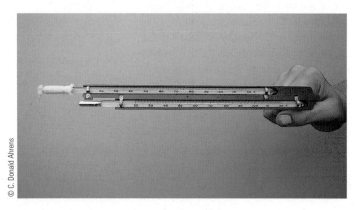

● FIGURE 4.20 The sling psychrometer.

a heat index of 122°F. Although this weather situation is remotely possible, it is *highly unlikely*, as a temperature of 90°F and a relative humidity of 90 percent can occur only if the dew-point temperature is incredibly high (nearly 87°F), and a dew-point temperature this high rarely, if ever, occurs in the United States, even on the muggiest of days.

During hot, humid weather some people remark about how "heavy" or how dense the air feels. Is hot, humid air really more dense than hot, dry air? If you are interested in the answer, read the Focus section on p. 106.

MEASURING HUMIDITY   One common instrument used to obtain dew point and relative humidity is a **psychrometer,** which consists of two liquid-in-glass thermometers mounted side by side and attached to a piece of metal that has either a handle or chain at one end (see ● Fig. 4.20). The thermometers are exactly alike except that one has a piece of cloth (wick) covering the bulb. The wick-covered thermometer—called the *wet bulb*—is dipped in clean (usually distilled) water, while the other thermometer is kept dry. Both thermometers are ventilated for a few minutes, either by whirling the instrument (*sling psychrometer),* or by drawing air past it with an electric fan (*aspirated psychrometer).* Water evaporates from the wick and the thermometer cools. The drier the air, the greater the amount of evaporation and cooling. After a few minutes, the wick-covered thermometer will cool to the lowest value possible. Recall from an earlier section that this is the *wet-bulb temperature*—the lowest temperature that can be attained by evaporating water into the air.

The dry thermometer (commonly called the *dry bulb*) gives the current air temperature, or *dry-bulb temperature.* The temperature difference between the dry bulb and the wet bulb is known as the *wet-bulb depression.* A large depression indicates that a great deal of water can evaporate into the air and that the relative humidity is low. A small depression indicates that little evaporation of water vapor is possible, so the air is close to saturation and the relative humidity is high. If there is no depression, the dry bulb, the wet bulb, and the dew point are the same; the air is saturated and the relative humidity is 100 percent. (Tables used to compute relative humidity and dew point are given at the back of the book in Appendix D.)

Instruments that measure humidity are commonly called **hygrometers.** One type—called the **hair hygrometer**—is constructed on the principle that the length of human hair increases by 2.5 percent as the relative humidity increases from 0 to 100 percent. This instrument uses human (or horse) hair to measure relative humidity. A number of strands of hair (with oils removed) are attached to a system of levers. A small change in hair length is magnified by a linkage system and transmitted to a dial (see ● Fig. 4.21) calibrated to show relative humidity, which can then be read directly or recorded on a chart. (Often, the chart is attached to a clock-driven rotating drum that gives a continuous record of relative humidity.) Because the hair hygrometer is not as accurate as the psychrometer (especially at very high and very low relative humidities and very low temperatures), it requires frequent calibration, principally in areas that experience large daily variations in relative humidity.

The *electrical hygrometer* is another instrument that measures humidity. It consists of a flat plate coated with a film of carbon. An electric current is sent across the plate. As water vapor is absorbed, the electrical resistance of the carbon coating changes. These changes are translated into relative humidity. This instrument is commonly used in the radiosonde, which gathers atmospheric data at various levels above the earth. Still

FOCUS ON A SPECIAL TOPIC

Is Humid Air "Heavier" Than Dry Air?

Does a volume of hot, humid air weigh more than a similar size volume of hot, dry air? The answer is no! At the same temperature and at the same level, humid air weighs *less* than dry air. (Keep in mind that we are referring strictly to water vapor—a gas—and not suspended liquid droplets.) To understand why, we must first see what determines the weight of atoms and molecules.

Almost all of the weight of an atom is concentrated in its nucleus, where the protons and neutrons are found. Neutrons weigh nearly the same as protons. To get some idea of how heavy an atom is, we simply add up the number of protons and neutrons in the nucleus. (Electrons are so light that we ignore them in comparing weights.) The larger this total, the heavier the atom. Now, we can compare one atom's weight with another's. For example, hydrogen, the lightest known atom, has only 1 proton in its center (no neutrons). Thus, it has an *atomic weight* of 1. Nitrogen, with 7 protons and 7 neutrons in its nucleus, has an atomic weight of 14. Oxygen, with 8 protons and 8 neutrons, weighs in at 16.

A molecule's weight is the sum of the atomic weights of its atoms. For example, molecular oxygen, with two oxygen atoms (O_2), has a molecular weight of 32. The most abundant atmospheric gas, molecular nitrogen (N_2), has a molecular weight of 28.

When we determine the weight of air, we are dealing with the weight of a mixture. As you might expect, a mixture's weight is a little more complex. We cannot just add the weights of all its atoms and molecules because the mixture might contain more of one kind than another. Air, for example, has far more nitrogen (78 percent) than oxygen (21 percent). We allow for this by multiplying the molecule's weight by its share in the mixture. Since dry air is essentially composed of N_2 and O_2 (99 percent), we ignore the other parts of air for the rough average shown in Table 2.

The symbol ≈ means "is approximately equal to." Therefore, dry air has a molecular weight of about 29. How does this compare

▼ TABLE 2

GAS	WEIGHT		NUMBER OF ATOMS		MOLECULAR WEIGHT		PERCENT BY VOLUME
Oxygen	16	×	2	=	32	×	21% ≈ 7
Nitrogen	14	×	2	=	28	×	78% ≈ 22

Molecular weight of dry air ≈ 29

with humid air?

Water vapor is composed of two atoms of hydrogen and one atom of oxygen (H_2O). It is an invisible gas, just as oxygen and nitrogen are invisible. It has a molecular weight; its two atoms of hydrogen (each with atomic weight of 1) and one atom of oxygen (atomic weight 16) give water vapor a molecular weight of 18. Obviously, air, at nearly 29, weighs appreciably more than water vapor.

Suppose we take a given volume of completely dry air and weigh it, then take exactly the same amount of water vapor at the same temperature and weigh it. We will find that the dry air weighs slightly more. If we replace dry air molecules one for one with water vapor molecules, the total number of molecules remains the same, but the total weight of the drier air decreases. Since density is mass per unit volume, *hot, humid air at the surface is less dense (lighter) than hot dry air.*

This fact can have an important influence on our weather. The lighter the air becomes, the more likely it is to rise. All other factors being equal, hot, humid (less-dense) air will rise more readily than hot, dry (more-dense) air. It is of course the water vapor in the rising air that changes into liquid cloud droplets and ice crystals, which, in turn, grow large enough to fall to the earth as precipitation (see ● Fig.2).

Of lesser importance to weather but of greater importance to sports is the fact that a baseball will "carry" farther in less-dense air. Consequently, without the influence of wind, a ball will travel slightly farther on a hot, humid day than it will on a hot, dry day. So when the sports announcer proclaims "the air today is heavy because of the high humidity" remember that this statement is not true and, in fact, a 404-foot home run on this humid day might simply be a 400-foot out on a very dry day.

● FIGURE 2
On this summer afternoon in Maryland, lighter (less-dense) hot, humid air rises and condenses into towering cumulus clouds.

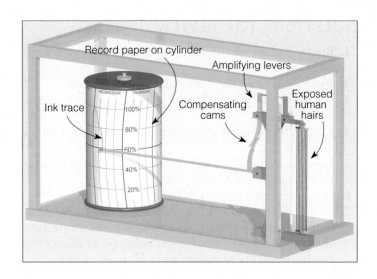

● FIGURE 4.21 The hair hygrometer measures relative humidity by amplifying and measuring changes in the length of human (or horse) hair.

another instrument—the *infrared hygrometer*—measures atmospheric humidity by measuring the amount of infrared energy absorbed by water vapor in a sample of air. The *dew-point hygrometer* measures the dew-point temperature by cooling the surface of a mirror until condensation (dew) forms. This sensor is the type that measures dew-point temperature in the hundreds of fully automated weather stations—*Automated Surface Observing System* (ASOS)—that exist throughout the United States. Finally, the *dew cell* determines the amount of water vapor in the air by measuring the air's actual vapor pressure.

SUMMARY

This chapter examines the concept of atmospheric humidity. The chapter begins by looking at the hydrologic cycle and the circulation of water in our atmosphere. It then looks at the different phases of water, showing how evaporation, condensation, and saturation occur at the molecular level. The next several sections look at the many ways of describing the amount of water vapor in the air. Here we learn that there are many ways of describing humidity. The absolute humidity represents the density of water vapor in a given volume of air. Specific humidity measures the mass of water vapor in a fixed mass of air, while the mixing ratio expresses humidity as the mass of water vapor in the fixed mass of remaining dry air. The actual vapor pressure indicates the air's total water vapor content by expressing the amount of water vapor in terms of the amount of pressure that the water vapor molecules exert. The saturation vapor pressure describes how much water vapor the air could hold at any given temperature in terms of how much pressure the water vapor molecules would exert if the air were saturated at that temperature. A good indicator of the air's actual water vapor content is the dew point—the temperature to which air would have to be cooled (at constant pressure) for saturation to occur.

Relative humidity is a measure of how close the air is to being saturated. Air with a high relative humidity does not necessarily contain a great deal of water vapor; it is simply close to being saturated. With a constant water-vapor content, cooling the air causes the relative humidity to increase, while warming the air causes the relative humidity to decrease. When the air temperature and dew point are close together, the relative humidity is high, and, when they are far apart, the relative humidity is low. High relative humidity in hot weather makes us feel hotter than it really is by retarding the evaporation of perspiration. The heat index is a measure of how hot it feels to an average person for various combinations of air temperature and relative humidity. Although relative humidity can be confusing (because it can change with either air temperature or moisture content), it is nevertheless the most widely used way of describing the air's moisture content.

The chapter concludes by examining the various instruments that measure humidity, such as the psychrometer and hair hygrometer.

KEY TERMS

The following terms are listed (with page number) in the order they appear in the text. Define each. Doing so will aid you in reviewing the material covered in this chapter.

evaporation, 90
condensation, 90
precipitation, 90
hydrologic cycle, 90
sublimation, 91
deposition, 91
saturated (air), 92
humidity, 93
absolute humidity, 93
specific humidity, 93
mixing ratio, 93
actual vapor pressure, 94
saturation vapor
   pressure, 94
relative humidity, 95

supersaturation, 95
dew-point temperature
   (dew point), 98
frost point, 98
wet-bulb temperature, 103
heatstroke, 104
heat index (HI), 104
apparent temperature, 104
psychrometer, 105
hygrometer, 105
hair hygrometer, 105

QUESTIONS FOR REVIEW

1. Briefly explain the movement of water in the hydrologic cycle.
2. Basically, how do the three states of water differ?
3. What are the primary factors that influence evaporation?
4. Explain why condensation occurs primarily when the air is cooled.
5. How are evaporation and condensation related to saturated air above a flat water surface?
6. How does condensation differ from precipitation?
7. Why are specific humidity and mixing ratio more commonly used in representing atmospheric moisture than absolute humidity? What is the only way to change the specific humidity or mixing ratio of an air parcel?
8. In a volume of air, how does the actual vapor pressure differ from the saturation vapor pressure? When are they the same?
9. What does saturation vapor pressure primarily depend upon?
10. Explain why it takes longer to cook vegetables in the mountains than at sea level.
11. (a) What does the relative humidity represent?
    (b) When the relative humidity is given, why is it also important to know the air temperature?
    (c) Explain two ways the relative humidity may be changed.
12. Explain why, during a summer day, the relative humidity will change as shown in Fig. 4.12, (p. 97).
13. Why do hot and humid summer days usually feel hotter than hot and dry summer days?
14. Why is the wet-bulb temperature a good measure of how cool human skin can become?
15. Explain why the air on a hot humid day is less dense than on a hot dry day.
16. (a) What is the dew-point temperature?
    (b) How is the difference between dew point and air temperature related to the relative humidity?
17. Why is cold polar air described as "dry" when the relative humidity of that air is very high?
18. How can a region have a high specific humidity and a low relative humidity? Give an example.
19. Why is the air from the Gulf of Mexico so much more humid than air from the Pacific Ocean at the same latitude?
20. How are the dew-point temperature and wet-bulb temperature different? Can they ever read the same? Explain.
21. When outside air is brought indoors on a cold winter day, the relative humidity of the heated air inside often drops below 25 percent. Explain why this situation occurs.
22. Describe how a sling psychrometer works. What does it measure? Does it give you dew point and relative humidity? Explain.
23. Why are human hairs often used in a hair hygrometer?

QUESTIONS FOR THOUGHT

1. Would you expect water in a glass to evaporate more quickly on a windy, warm, dry summer day or on a calm, cold, dry winter day? Explain.
2. How can frozen clothes "dry" outside in subfreezing weather? What exactly is taking place?
3. Explain how and why each of the following will change as a parcel of air with an unchanging amount of water vapor rises, expands, and cools:
   (a) absolute humidity;
   (b) relative humidity;
   (c) actual vapor pressure; and
   (d) saturation vapor pressure.
4. Where in the United States would you go to experience the *least* variation in dew point (actual moisture content) from January to July?
5. After completing a grueling semester of meteorological course work, you call your travel agent to arrange a much-needed summer vacation. When your agent suggests a trip to the desert, you decline because of a concern that the dry air will make your skin feel uncomfortable. The travel agent assures you that almost daily "desert relative humidities are above 90 percent." Could the agent be correct? Explain.
6. On a clear, calm morning, water condenses on the ground in a thick layer of dew. As the water slowly evaporates into the air, you measure a slow increase in dew point. Explain why.
7. Two cities have exactly the same amount of water vapor in the air. The 6:00 A.M. relative humidity in one city is 93 percent, while the 3:00 P.M. relative humidity in the other city is 28 percent. Explain how this can come about.
8. Suppose the dew point of cold outside air is the same as the dew point of warm air indoors. If the door is opened, and cold air replaces some of the warm inside air, would the new relative humidity indoors be (a) lower than before, (b) higher than before, or (c) the same as before? Explain your answer.
9. On a warm, muggy day, the air is described as "close." What are several plausible explanations for this expression?
10. Outside, on a very warm day, you swing a sling psychrometer for about a minute and read a dry-bulb temperature of 38°C and a wet-bulb temperature of 24°C. After swinging the instrument again, the dry bulb is still 38°C, but the wet bulb is now 26°C. Explain how this could happen.
11. Why are evaporative coolers used in Arizona, Nevada, and California but not in Florida, Georgia, or Indiana?
12. Devise a way of determining elevation above sea level if all you have is a thermometer and a pot of water.
13. A large family lives in northern Minnesota. This family gets together for a huge dinner three times a year: on Thanksgiving, on Christmas, and on the March solstice. The Thanksgiving and Christmas dinners consist of turkey, ham, mashed potatoes, and lots of boiled vegetables.

The solstice dinner is pizza. The air temperature inside the home is about the same for all three meals (70°F), yet everyone remarks about how "warm, cozy, and comfortable" the air feels during the Thanksgiving and Christmas dinners, and how "cool" the inside air feels during the solstice meal. Explain to the family members why they might feel "warmer" inside the house during Thanksgiving and Christmas, and "cooler" during the March solstice. (The answer has nothing to do with the amount or type of food consumed.)

PROBLEMS AND EXERCISES

1. On a bitter cold, snowy morning, the air temperature and dew point of the outside air are both −7°C. If this air is brought indoors and warmed to 21°C, with no change in vapor content, what is the relative humidity of the air inside the home? (Hint: See Table 1, p. 102.)

2. (a) With the aid of Fig. 4.14b (p. 99), determine the average July dew points in St. Louis, Missouri; New Orleans, Louisiana; and Los Angeles, California.
   (b) If the high temperature on a particular summer day in all three cities is 32°C (90°F), then calculate the afternoon relative humidity at each of the three cities. (Hint: Either Fig. 4.10, p. 95, or Table 1, p. 102, will be helpful.)

3. Suppose with the aid of a sling psychrometer you obtain an air temperature of 30°C and a wet-bulb temperature of 25°C. What is
   (a) the wet-bulb depression,
   (b) the dew point, and
   (c) the relative humidity of the air?
   (Use the tables in Appendix D at the back of the book.)

4. If the air temperature is 35°C and the dew point is 21°C, determine the relative humidity using
   (a) Table 1, p. 102;
   (b) Fig. 4.10, p. 95; and
   (c) Tables D.1 and D.2 in Appendix D.

5. Suppose the average vapor pressure in Nevada is about 8 mb.
   (a) Use Table 1 (p. 102), to determine the average dew point of this air.
   (b) Much of the state is above an elevation of 1500 m (5000 ft). At 1500 m, the normal pressure is about 12.5 percent less than at sea level. If the air over Nevada were brought down to sea level, without any change in vapor content, what would be the new vapor pressure of the air?

6. In Yellowstone National Park, there are numerous ponds of boiling water. If Yellowstone is about 2200 m (7200 ft) above sea level (where the air pressure is normally about 775 mb), what is the normal boiling point of water in Yellowstone? (Hint: See Fig. 1, p. 96.)

7. Three cities have the following temperature (T) and dew point (T_d) during a July afternoon:
   Atlanta, Georgia, $T = 90°F$; $T_d = 75°F$
   Baltimore, Maryland, $T = 80°F$; $T_d = 70°F$
   Norman, Oklahoma, $T = 70°F$; $T_d = 65°F$
   (a) Which city appears to have the highest relative humidity?
   (b) Which city appears to have the lowest relative humidity?
   (c) Which city has the *most* water vapor in the air?
   (d) Which city has the *least* water vapor in the air?
   (e) For each city use Table 1 on p. 102 and the information on the same page to calculate the relative humidity for each city.
   (f) Using both the relative humidity calculated in (e) and the air temperature, determine the heat index for each city using Fig. 4.19 (p. 104).

Visit the
Meteorology Resource Center
at
academic.cengage.com/login
for more assets, including questions for exploration, animations, videos, and more.

A lighthouse keeps a constant vigil as clouds and fog approach the coast of southern Australia.

CHAPTER 5

Condensation: Dew, Fog, and Clouds

The weather is an ever-playing drama before which we are a captive audience. With the lower atmosphere as the stage, air and water as the principal characters, and clouds for costumes, the weather's acts are presented continuously somewhere about the globe. The script is written by the sun; the production is directed by the earth's rotation; and, just as no theater scene is staged exactly the same way twice, each weather episode is played a little differently, each is marked with a bit of individuality.

Clyde Orr, Jr., *Between Earth and Space*

  CONTENTS

Have you walked barefoot across a lawn on a summer morning and felt the wet grass under your feet? Did you ever wonder how those glistening droplets of dew could form on a clear summer night? Or why they formed on grass but not on bushes several meters above the ground? In this chapter, we will investigate first the formation of dew and frost. Then we will examine the different types of fog. The chapter concludes with the identification and observation of clouds.

The Formation of Dew and Frost

On clear, calm nights, objects near the earth's surface cool rapidly by emitting infrared radiation. The ground and objects on it often become much colder than the surrounding air. Air that comes in contact with these cold surfaces cools by conduction. Eventually, the air cools to the *dew point*—the temperature at which saturation occurs. As surfaces such as twigs, leaves, and blades of grass cool below this temperature, water vapor begins to condense upon them, forming tiny visible specks of water called **dew** (see ● Fig. 5.1). If the air temperature should drop to freezing or below, the dew will freeze, becoming tiny beads of ice called **frozen dew.** Because the coolest air is usually at ground level, dew is more likely to form on blades of grass than on objects several meters above the surface. This thin coating of dew not only dampens bare feet, but is also a valuable source of moisture for many plants during periods of low rainfall. Averaged for an entire year in middle latitudes, dew yields a blanket of water between 12 and 50 mm (0.5 and 2 in.) thick.

Dew is more likely to form on nights that are clear and calm than on nights that are cloudy and windy. Clear nights allow objects near the ground to cool rapidly by emitting infrared radiation, and calm winds mean that the coldest air will be located at ground level. These atmospheric conditions are usually associated with large fair-weather, high-pressure sys-

tems. On the other hand, the cloudy, windy weather that inhibits rapid cooling near the ground and the forming of dew often signifies the approach of a rain-producing storm system. These observations inspired the following folk rhyme:

> When the dew is on the grass,
> rain will never come to pass.
> When grass is dry at morning light,
> look for rain before the night!

Visible white frost forms on cold, clear, calm mornings when the dew-point temperature is at or below freezing. When the air temperature cools to the dew point (now called the *frost point*) and further cooling occurs, water vapor can change directly to ice without becoming a liquid first—a process called *deposition.** The delicate, white crystals of ice that form in this manner are called *hoarfrost, white frost,* or simply **frost.** Frost has a treelike branching pattern that easily distinguishes it from the nearly spherical beads of frozen dew.

On cold winter mornings, frost may form on the inside of a windowpane in much the same way as it does outside, except that the cold glass chills the indoor air adjacent to it. When the temperature of the inside of the window drops below freezing, water vapor in the room forms a light, feathery deposit of frost (see ● Fig. 5.2).

In very dry weather, the air temperature may become quite cold and drop below freezing without ever reaching the frost point, and no visible frost forms. *Freeze* and *black frost* are words denoting this situation. These conditions can severely damage crops (see Chapter 3, pp. 71–73).

So, dew, frozen dew, and frost form in the rather shallow layer of air near the ground on clear, calm nights. But what happens to air as a deeper layer adjacent to the ground is cooled? We've seen in Chapter 4 that if air cools without any change in water-vapor content, the relative humidity in-

---

*Recall that when the ice changes back into vapor without melting, the process is called *sublimation.*

● FIGURE 5.1 Dew forms on clear nights when objects on the surface cool to a temperature below the dew point. If these beads of water should freeze, they would become frozen dew.

● FIGURE 5.2 These are the delicate ice-crystal patterns that frost exhibits on a window during a cold winter morning.

creases. When air cools to the dew point, the relative humidity becomes 100 percent and the air is saturated. Continued cooling condenses some of the vapor into tiny cloud droplets.

Condensation Nuclei

Actually, the condensation process that produces clouds is not quite so simple. Just as dew and frost need a surface to form on, there must be airborne particles on which water vapor can condense to produce cloud droplets.

Although the air may look clean, it never really is. On an ordinary day, a volume of air about the size of your index finger contains between 1000 and 150,000 particles. Since many of these serve as surfaces on which water vapor can condense, they are called **condensation nuclei.** Without them, relative humidities of several hundred percent would be required before condensation could begin.

Some condensation nuclei are quite small and have a radius less than 0.2 μm; these are referred to as *Aitken nuclei,* after the British physicist who discovered that water vapor condenses on nuclei. Particles ranging in size from 0.2 to 1 μm are called *large nuclei,* while others, called *giant nuclei,* are much larger and have radii exceeding 1 μm (see ▼ Table 5.1). The condensation nuclei most favorable for producing clouds (called *cloud condensation nuclei*) have radii of 0.1 μm or more. Usually, between 100 and 1000 nuclei of this size exist in a cubic centimeter of air. These particles enter the atmosphere in a variety of ways: dust, volcanoes, factory smoke, forest fires, salt from ocean spray, and even sulfate particles emitted by phytoplankton in the oceans. In fact, studies show that sulfates provide the major source of cloud condensation nuclei in the marine atmosphere. Because most particles are released into the atmosphere near the ground, the largest concentrations of nuclei are observed in the lower atmosphere near the earth's surface.

Condensation nuclei are extremely light (many have a mass less than one-trillionth of a gram), so they can remain suspended in the air for many days. They are most abundant over industrial cities, where highly polluted air may contain nearly 1 million particles per cubic centimeter. They decrease

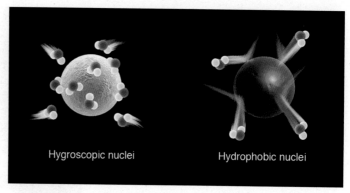

● **FIGURE 5.3** Hygroscopic nuclei are "water-seeking," and water vapor rapidly condenses on their surfaces. Hydrophobic nuclei are "water-repelling" and resist condensation.

in cleaner "country" air and over the oceans, where concentrations may dwindle to only a few nuclei per cubic centimeter.

Some particles are **hygroscopic** ("water-seeking"), and water vapor condenses upon these surfaces when the relative humidity is considerably lower than 100 percent. Ocean salt is hygroscopic, as is common table salt. In humid weather, it is difficult to pour salt from a shaker because water vapor condenses onto the salt crystals, sticking them together. Moreover, on a humid day, salty potato chips left outside in an uncovered bowl turn soggy. Other hygroscopic nuclei include sulfuric and nitric acid particles. Not all particles serve as good condensation nuclei. Some are **hydrophobic*** ("water-repelling")—such as oils, gasoline, and paraffin waxes—and resist condensation even when the relative humidity is above 100 percent (see ● Fig. 5.3). As we can see, condensation may begin on some particles when the relative humidity is well below 100 percent and on others only when the relative humidity is much higher than 100 percent. However, at any given time there are usually many nuclei present, so that haze, fog, and clouds will form at relative humidities near or below 100 percent.

Haze

Suppose you visit an area that has a layer of **haze** (that is, a layer of dust or salt particles) suspended above the region. There, you may notice that distant objects are usually more visible in the afternoon than in the morning, even when the concentration of particles in the air has not changed. Why? During the warm afternoon, the relative humidity of the air is often below the point where water vapor begins to condense, even on active hygroscopic nuclei. Therefore, the floating particles remain small—usually no larger than about one-tenth of a micrometer. These tiny *dry haze* particles selectively scatter some rays of sunlight, while allowing others to penetrate the air. The scattering effect of dry haze produces a bluish

▼ **TABLE 5.1**   Characteristic Sizes and Concentration of Condensation Nuclei and Cloud Droplets

TYPE OF PARTICLE	APPROXIMATE RADIUS (MICROMETERS)	NO. OF PARTICLES (PER CM³)	
		Range	Typical
Small (Aitken) condensation nuclei	<0.2	1000 to 10,000	1000
Large condensation nuclei	0.2 to 1.0	1 to 1000	100
Giant condensation nuclei	>1.0	<1 to 10	1
Fog and cloud droplets	>10	10 to 1000	300

*A synthetic hydrophobic is PTFE, or Teflon—the material used in rain-repellent fabric.

● FIGURE 5.4 The high relative humidity of the cold air above the lake is causing a layer of haze to form on a still winter morning.

© C. Donald Ahrens

color when viewed against a dark background and a yellowish tint when viewed against a light-colored background.

As the air cools during the night, the relative humidity increases. When the relative humidity reaches about 75 percent, condensation may begin on the most active hygroscopic nuclei, producing a *wet haze*. As water collects on the nuclei, their size increases and the particles, although still small, become large enough to scatter light much more efficiently. In fact, as the relative humidity increases from about 60 percent to 80 percent, the scattering effect increases by a factor of nearly 3. Since relative humidities are normally high during cool mornings, much of the light from distant objects is scattered away by the wet haze particles before reaching you; hence, it is difficult to see these distant objects.

Not only does wet haze restrict visibility more than dry haze, it also appears dull gray or white (see ● Fig. 5.4). Near seashores and in clean air over the open ocean, large salt particles suspended in air with a high relative humidity often produce a thin white veil across the horizon.

Fog

By now, it should be apparent that condensation is a continuous process beginning when water vapor condenses onto hygroscopic nuclei at relative humidities as low as 75 percent. As the relative humidity of the air increases, the visibility decreases, and the landscape becomes masked with a grayish tint. As the relative humidity gradually approaches 100 percent, the haze particles grow larger, and condensation begins on the less-active nuclei. Now a large fraction of the available nuclei have water condensing onto them, causing the droplets to grow even bigger, until eventually they become visible to the naked eye. The increasing size and concentration of droplets further restrict visibility. When the visibility lowers to less than 1 km (0.62 mi), and the air is wet with countless millions of tiny floating water droplets, the wet haze becomes a cloud resting near the ground, which we call **fog**.*

With the same water content, fog that forms in dirty city air often is thicker than fog that forms over the ocean. Normally, the smaller number of condensation nuclei over the middle of the ocean produce fewer, but larger, fog droplets. City air with its abundant nuclei produces many tiny fog droplets, which greatly increase the thickness (or opaqueness) of the fog and reduce visibility. A dramatic example of a thick fog forming in air with abundant nuclei occurred in London, England, during the early 1950s. The fog became so thick, and the air so laden with smoke particles, that sunlight could not penetrate the smoggy air, requiring that street lights be left on at midday. Moreover, fog that forms in polluted air can turn acidic as the tiny liquid droplets combine with gaseous impurities, such as oxides of sulfur and nitrogen. **Acid fog** poses a threat to human health, especially to people with preexisting respiratory problems. We'll examine in more detail the health problems associated with acid fog and other forms of pollution in Chapter 18.

As tiny fog droplets grow larger, they become heavier and tend to fall toward the earth. A fog droplet with a diameter of 25 μm settles toward the ground at about 5 cm (2 in.) each second. At this rate, most of the droplets in a fog layer 180 m (about 600 ft) thick would reach the ground in less than one hour. Therefore, two questions arise: How does fog form? How is fog maintained once it does form?

Fog, like any cloud, usually forms in one of two ways:

1. by cooling—air is cooled below its saturation point (dew point).

---

*This is the official international definition of fog. The United States Weather Service reports fog as a restriction to visibility when fog restricts the visibility to 6 miles or less and the spread between the air temperature and dew point is 5°F or less. When the visibility is less than one-quarter of a mile, the fog is considered dense.

2. by evaporation and mixing—water vapor is added to the air by evaporation, and the moist air mixes with relatively dry air.

Once fog forms it is maintained by new fog droplets, which constantly form on available nuclei. In other words, the air must maintain its degree of saturation either by continual cooling or by evaporation and mixing of vapor into the air. Let's examine both processes.

View this concept in action on the Meteorology Resource Center at academic.cengage.com/login

RADIATION FOG   How can the air cool so that a cloud will form near the surface? Radiation and conduction are the primary means for cooling nighttime air near the ground. Fog produced by the earth's radiational cooling is called **radiation fog,** or *ground fog.* It forms best on clear nights when a shallow layer of moist air near the ground is overlain by drier air. Since the moist layer is shallow, it does not absorb much of the earth's outgoing infrared radiation. The ground, therefore, cools rapidly and so does the air directly above it, and a surface inversion forms, with cooler air at the surface and warmer air above. The moist lower layer (chilled rapidly by the cold ground) quickly becomes saturated, and fog forms. The longer the night, the longer the time of cooling and the greater the likelihood of fog. Hence, radiation fogs are most common over land in late fall and winter.

Another factor promoting the formation of radiation fog is a light breeze of less than 5 knots. Although radiation fog may form in calm air, slight air movement brings more of the moist air in direct contact with the cold ground, and the transfer of heat occurs more rapidly. A strong breeze tends to prevent radiation fog from forming by mixing the air near the surface with the drier air above. The ingredients of clear skies and light winds are associated with large high-pressure areas (anticyclones). Consequently, during the winter, when a

high becomes stagnant over an area, radiation fog may form on many consecutive days.

Because cold, heavy air drains downhill and collects in valley bottoms, we normally see radiation fog forming in low-lying areas. Hence, radiation fog is frequently called *valley fog.* The cold air and high moisture content in river valleys make them susceptible to radiation fog. Since radiation fog normally forms in lowlands, hills may be clear all day long, while adjacent valleys are fogged in (see ● Fig. 5.5).

Radiation fogs form upward from the ground as the night progresses and are usually deepest around sunrise. However, fog may occasionally form after sunrise, especially when evaporation and mixing take place near the surface. This usually occurs at the end of a clear, calm night as radiational cooling brings the air temperature close to the dew point in a rather shallow layer above the ground. At the surface, the air becomes saturated, forming a thick blanket of dew on the grass. At daybreak, the sun's rays evaporate the dew, adding water vapor to the air. A light breeze then stirs the moist air with the drier air above, causing saturation (and, hence, fog) to form in a shallow layer near the ground.

Often a shallow fog layer will dissipate or *burn off* by the afternoon. Of course, the fog does not "burn"; rather, sunlight penetrates the fog and warms the ground, causing the air temperature in contact with the ground to increase. The warm air rises and mixes with the foggy air above, which increases the temperature of the foggy air. In the slightly warmer air, some of the fog droplets evaporate, allowing more sunlight to reach the ground, which produces more heating, and soon the fog completely evaporates and disappears.

Satellite images show that a blanket of radiation fog tends to evaporate ("burn off") first around its periphery, where the fog is usually thinnest. Sunlight rapidly warms this region, causing the fog to dissipate as the warmer air mixes in toward the denser foggy area.

If the fog is thick, with little sunlight penetrating it, and there is little mixing along the outside edges, the fog may not

● FIGURE 5.5 Radiation fog nestled in a valley.

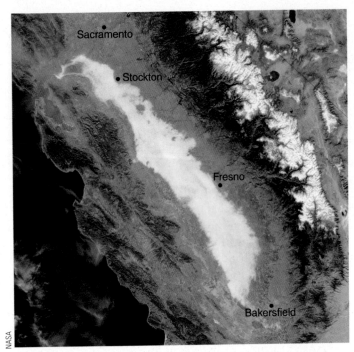

NASA

• FIGURE 5.6 Visible satellite image of dense radiation fog in the southern half of California's Central Valley on the morning of November 20, 2002. The white region to the east (right) of the fog is the snow-capped Sierra Nevada range. During the late fall and winter, the fog, nestled between two mountain ranges, can last for many days without dissipating. The fog on this day was responsible for several auto accidents, including a 14-car pileup near Fresno.

dissipate. This is often the case in the Central Valley area of California during the late fall and winter. A fog layer over 500 m (1700 ft) thick settles between two mountain ranges, while a strong inversion normally keeps the warmest air above the top of the fog. During the day, much of the light from the low winter sun reflects off the top of the fog, allowing only a

small amount of sunlight to penetrate the fog and warm the ground. As the air warms from below, the fog dissipates upward from the surface in a rather shallow layer less than 150 m (500 ft), creating the illusion that the fog is lifting. Since the fog no longer touches the ground, and a strong inversion exists above it, the fog is called a *high inversion fog*. (The low cloud above the ground is also called *stratus*, or, simply, *high fog*.) As soon as the sun sets, radiational cooling lowers the air temperature, and the fog once again forms on the ground. This daily lifting and lowering of the fog without the sun ever breaking through it may last for many days or even weeks during winter in California's Central Valley (see • Fig. 5.6).

ADVECTION FOG   Cooling surface air to its saturation point may be accomplished by warm moist air moving over a cold surface. The surface must be sufficiently cooler than the air above so that the transfer of heat from air to surface will cool the air to its dew point and produce fog. Fog that forms in this manner is called **advection fog.**

A good example of advection fog may be observed along the Pacific Coast during summer. The main reason fog forms in this region is that the surface water near the coast is much colder than the surface water farther offshore. Warm moist air from the Pacific Ocean is carried (advected) by westerly winds over the cold coastal waters. Chilled from below, the air temperature drops to the dew point, and fog is produced. Advection fog, unlike radiation fog, always involves the movement of air, so when there is a stiff summer breeze in San Francisco, it's common to watch advection fog roll in past the Golden Gate Bridge (see • Fig. 5.7). It is also more common to see advection fog forming at headlands that protrude seaward than in the mouths of bays. If you are curious as to why, read the Focus section on p. 117.

As summer winds carry the fog inland over the warmer land, the fog near the ground dissipates, leaving a sheet of low-lying gray clouds that block out the sun. Farther inland,

• FIGURE 5.7 Advection fog forms as the wind moves moist air over a cooler surface. Here advection fog, having formed over the cold, coastal water of the Pacific Ocean, is rolling inland past the Golden Gate Bridge in San Francisco. As fog moves inland, the air warms and the fog lifts above the surface. Eventually, the air becomes warm enough to totally evaporate the fog.

© H. Spichtinger/Zefa/CORBIS

Why Are Headlands Usually Foggier Than Beaches?

If you drive along a highway that parallels an irregular coastline, you may have observed that advection fog is more likely to form in certain regions. For example, headlands that protrude seaward usually experience more fog than do beaches that are nestled in the mouths of bays. Why?

As air moves onshore, it crosses the coastline at nearly a right angle. This causes the air to flow together or converge in the vicinity of the headlands (see Fig. 1). This area of weak convergence causes the surface air to rise and cool just a little. If the rising air is close to being saturated, it will cool to its dew point, and fog will form.

Meanwhile, near the beach area, the surface air spreads apart or diverges as it crosses the coastline. This area of weak divergence creates sinking and slightly warmer air. Because the sinking of air increases the separation between air temperature and dew point, fog is less likely to form in this region. Hence, the headlands can be shrouded in fog while the beaches are basking in sunshine.

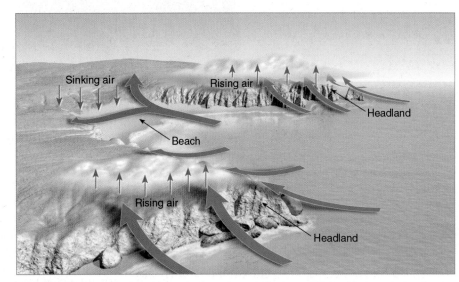

● FIGURE 1 Along an irregular coastline, advection fog is more likely to form at the headland (the region of land extending seaward) where moist surface air converges and rises than at the beach where air diverges and sinks.

the air is sufficiently warm so that even these low clouds evaporate and disappear. Since the fog is more likely to burn off during the warmer part of the day, a typical summertime weather forecast for coastal areas would read, "Fog and low cloudiness along the coast extending locally inland both nights and mornings with sunny afternoons."

Because they provide moisture to the coastal redwood trees, advection fogs are important to the scenic beauty of the Pacific Coast. Much of the fog moisture collected by the needles and branches of the redwoods drips to the ground (*fog drip*), where it is utilized by the tree's shallow root system (see ● Fig. 5.8). Without the summer fog, the coast's redwood trees would have trouble surviving the dry California summers. Hence, we find them nestled in the fog belt along the coast.

Advection fogs also prevail where two ocean currents with different temperatures flow next to one another. Such is the case in the Atlantic Ocean off the coast of Newfoundland, where the cold southward-flowing Labrador Current lies almost parallel to the warm northward-flowing Gulf Stream. Warm southerly air moving over the cold water produces fog in that region—so frequently that fog occurs on about two out of three days during summer.

Advection fog also forms over land. In winter, warm moist air from the Gulf of Mexico moves northward over progressively colder and slightly elevated land. As the air cools to its saturation point, a fog forms in the southern or

● FIGURE 5.8 Tiny drops, each one made from many fog droplets, drip from the needles of this tree and provide a valuable source of moisture during the otherwise dry summer along the coast of California.

● FIGURE 5.9 (a) Radiation fog tends to form on clear, relatively calm nights when cool, moist surface air is overlain by drier air and rapid radiational cooling occurs. (b) Advection fog forms when the wind moves moist air over a cold surface and the moist air cools to its dew point.

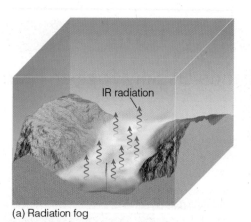

(a) Radiation fog

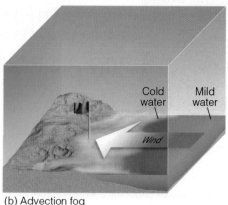

(b) Advection fog

central United States. Because the cold ground is often the result of radiational cooling, fog that forms in this manner is sometimes called **advection-radiation fog.** During this same time of year, air moving across the warm Gulf Stream encounters the colder land of the British Isles and produces the thick fogs of England. Similarly, fog forms as marine air moves over an ice or snow surface. In extremely cold arctic air, ice crystals form instead of water droplets, producing an *ice fog.* ● Figure 5.9 summarizes the ideas behind the formation of both advection and radiation fog.

UPSLOPE FOG Fog that forms as moist air flows up along an elevated plain, hill, or mountain is called **upslope fog.** Typically, upslope fog forms during the winter and spring on the eastern side of the Rockies, where the eastward-sloping plains are nearly a kilometer higher than the land farther east. Occasionally, cold air moves from the lower eastern plains

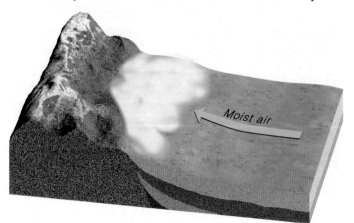

● FIGURE 5.10 Upslope fog forms as moist air slowly rises, cools, and condenses over elevated terrain.

westward. The air gradually rises, expands, becomes cooler, and—if sufficiently moist—a fog forms (see ● Fig. 5.10). Upslope fogs that form over an extensive area may last for many days.

Up to now, we have seen how the cooling of air produces fog. But remember that fog may also form by the mixing of two unsaturated masses of air. Fog that forms in this manner is usually called *evaporation fog* because evaporation initially enriches the air with water vapor. Probably, a more appropriate name for the fog is **evaporation (mixing) fog.** (For a better understanding of how mixing can produce fog, read the Focus section on p. 120.)

EVAPORATION (MIXING) FOG On a cold day, you may have unknowingly produced evaporation fog. When moist air from your mouth or nose meets the cold air and mixes with it, the air becomes saturated, and a tiny cloud forms with each exhaled breath.

A common form of evaporation-mixing fog is **steam fog,** which forms when cold air moves over warm water. This type of fog forms above a heated outside swimming pool in winter. As long as the water is warmer than the unsaturated air above, water will evaporate from the pool into the air. The increase in water vapor raises the dew point, and, if mixing is sufficient, the air above becomes saturated. The colder air directly above the water is heated from below and becomes warmer than the air directly above it. This warmer air rises and, from a distance, the rising condensing vapor appears as "steam."

It is common to see steam fog forming over lakes on autumn mornings, as cold air settles over water still warm from the long summer. On occasion, over the Great Lakes, columns of condensed vapor rise from the fog layer, forming whirling *steam devils,* which appear similar to the dust devils on land. If you travel to Yellowstone National Park, you will see steam fog forming above thermal ponds all year long (see ● Fig. 5.11). Over the ocean in polar regions, steam fog is referred to as *arctic sea smoke.*

Steam fog may form above a wet surface on a sunny day. This type of fog is commonly observed after a rainshower as sunlight shines on a wet road, heats the asphalt, and quickly evaporates the water. This added vapor mixes with the air above, producing steam fog. Fog that forms in this manner is short-lived and disappears as the road surface dries.

A warm rain falling through a layer of cold moist air can produce fog. Remember from Chapter 4 that the saturation vapor pressure depends on temperature: Higher temperatures correspond to higher saturation vapor pressures. When a warm raindrop falls into a cold layer of air, the saturation vapor pressure over the raindrop is greater than that of the air. This vapor-pressure difference causes water to evaporate from the raindrop into the air. This process may saturate the air and, if mixing occurs, fog forms. Fog of this type is often associated with warm air riding up and over a mass of colder surface air. The fog usually develops in the shallow layer of cold air just ahead of an approaching warm front or behind a cold front, which is why this type of evaporation fog is also known as *precipitation fog,* or **frontal fog.** Snow covering the ground is an especially favorable condition for the formation of frontal fog. The melting snow extracts heat from the environment, thereby cooling the already rain-saturated air.

● FIGURE 5.11 Even in summer, warm air rising above thermal pools in Yellowstone National Park condenses into a type of steam fog.

Foggy Weather

The foggiest regions in the United States are shown in ● Fig. 5.12. Notice that dense fog is more prevalent in coastal margins (especially those regions lapped by cold ocean currents) than in the center of the continent. In fact, the foggiest spot near sea level in the United States is Cape Disappointment, Washington. Located at the mouth of the Columbia River, it averages 2556 hours (or the equivalent of 106.5 twenty-four-hour days) of dense fog each year. Anyone who travels to this spot hoping to enjoy the sun during August and September would find its name appropriate indeed.

Although fog is basically a nuisance, it has many positive aspects. For example, the California Central Valley fog that many people scorn is extremely important to the economy of that area.* Fruit and nut trees that have finished growing dur-

---
*For reference, look back at Fig. 5.6 (p. 116), and see how the fog can cover a vast region. Also, note that the fog can last for many days on end.

ing the summer and fall require **winter chilling**—a large number of hours with the air temperature below 7°C (45°F) before trees will begin to grow again. The winter fog blocks out the sun and helps keep daytime temperatures quite cool, while keeping nighttime temperatures above freezing: The more continuous the fog, the more effective the chilling. Consequently, the agricultural economy of the region depends heavily on the fog, for without it and the winter chill it stimulates, many of the fruit and nut trees would not grow well. During the spring, when trees are in bloom, fog prevents nighttime air temperatures from dipping to dangerously low readings by trapping infrared energy that is radiated by the earth and releasing latent heat to the air as fog droplets form.

Unfortunately, fog also has many negative aspects. Along a gently sloping highway, the elevated sections may have excellent visibility, while in lower regions—only a few kilometers away—fog may cause poor visibility. Driving from the clear area into the fog on a major freeway can be extremely

● FIGURE 5.12
Average annual number of days with dense fog (visibility less than 0.25 miles) through the United States. (NOAA)

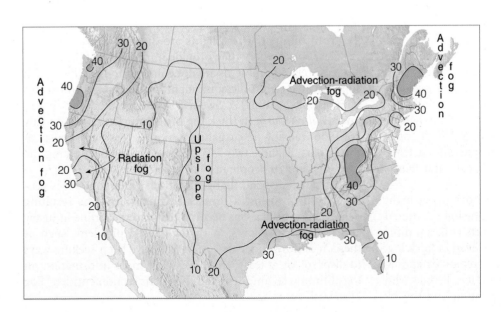

FOCUS ON A SPECIAL TOPIC

Fog That Forms by Mixing

How can unsaturated bodies of air mix together to produce fog (or a cloud)? To answer this question, let's first examine two unsaturated air parcels. (Later, we will look at the parcels mixed together.) The two parcels in Fig. 2 are essentially the same size and have a mass of 1 kg. Yet each has a different temperature and a different relative humidity. (We will assume that the parcels are near sea level where the atmospheric pressure is close to 1000 mb.)

Parcel A has an air temperature (T) of 20°C and a dew-point temperature (T_d) of 15°C. Remember from Chapter 4 that the air's relative humidity (RH) can be expressed as

$$RH = \frac{\text{actual mixing ratio } (w)}{\text{saturation mixing ratio } (w_s)} \times 100 \text{ percent.*}$$

To obtain the *saturation mixing ratio*, we look at Table 1 and read the value that corresponds to the parcel's air temperature. For a temperature of 20°C, the saturation mixing ratio is 15.0 g/kg. Likewise, the *actual mixing ratio* is obtained by reading the value in Table 1

---

*The actual mixing ratio (w) is the mass of water vapor per kilogram (kg) of dry air, usually expressed as grams per kilogram (g/kg). The saturation mixing ratio (w_s) is the mixing ratio of saturated air, also expressed as g/kg.

● FIGURE 2 The mixing of two unsaturated air parcels can produce fog. Notice in the saturated mixed parcel that the actual mixing ratio (w) is too high. As the mixed parcel cools below its saturation point, water vapor will condense onto nuclei, producing liquid droplets. This would keep the actual mixing ratio close to the saturation mixing ratio, and the relative humidity of the mixed parcel would remain close to 100 percent.

that corresponds to the parcel's dew-point temperature. For a dew-point temperature of 15°C, the actual mixing ratio is 10.8 g/kg. Hence, the relative humidity of the air in parcel A is

$$RH = \frac{w}{w_s} = \frac{10.8}{15.0} \times 100 \text{ percent}$$

$$RH = 72 \text{ percent.}$$

Air parcel B in Fig. 2 is considerably colder than parcel A with a temperature of −10°C, and considerably drier with a dew-point temperature of −15°C.

These temperatures yield a relative humidity of −15°C.

$$RH = \frac{w}{w_s} = \frac{1.2}{1.8} \times 100 \text{ percent}$$

$$RH = 67 \text{ percent.}$$

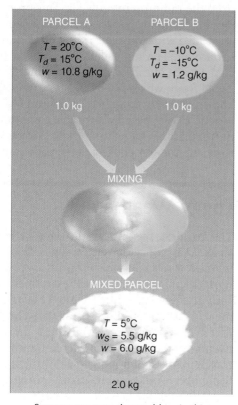

PARCEL A

$T = 20°C$
$T_d = 15°C$
$w = 10.8$ g/kg

1.0 kg

PARCEL B

$T = -10°C$
$T_d = -15°C$
$w = 1.2$ g/kg

1.0 kg

MIXING

MIXED PARCEL

$T = 5°C$
$w_S = 5.5$ g/kg
$w = 6.0$ g/kg

2.0 kg

Suppose we now thoroughly mix the two parcels in Fig 2. After mixing, the new parcel's temperature will be close to the average of parcel A and parcel B, or about 5°C. The total wa-

---

dangerous. In fact, every winter many people are involved in fog-related auto accidents. These usually occur when a car enters the fog and, because of the reduced visibility, the driver puts on the brakes to slow down. The car behind then slams into the slowed vehicle, causing a chain-reaction accident with many cars involved. One such accident actually occurred near Fresno, California, in February, 2002, when 87 vehicles smashed into each other along a stretch of foggy Highway 99. The accident left dozens of people injured, three people dead, and a landscape strewn with cars and trucks twisted into heaps of jagged steel.

Extremely limited visibility exists while driving at night in thick fog with the high-beam lights on. The light scattered back to the driver's eyes from the fog droplets makes it difficult to see very far down the road. However, even in thick fog, there is usually a drier and therefore clearer region extending about 35 cm (14 in.) above the road surface. People who drive a great deal in foggy weather take advantage of this by

installing extra head lamps—called *fog lamps*—just above the front bumper. These lights are directed downward into the clear space where they provide improved visibility.

Fog-related problems are not confined to land. Even with sophisticated electronic equipment, dense fog in the open sea hampers navigation. A Swedish liner rammed the luxury liner *Andrea Doria* in thick fog off Nantucket Island on July 25, 1956, causing 52 casualties. On a fog-covered runway in the Canary Islands, two 747 jet airliners collided, taking the lives of over 570 people in March, 1977.

Airports suspend flight operations when fog causes visibility to drop below a prescribed minimum. The resulting delays and cancellations become costly to the airline industry and irritate passengers. With fog-caused problems such as these, it is no wonder that scientists have been seeking ways to disperse, or at least "thin," fog. (For more information on fog-thinning techniques, read the Focus section entitled "Fog Dispersal" on p. 122.)

▼ TABLE 1   Saturation Mixing Ratios of Water Vapor for Various Air Temperatures (Air Pressure Is 1000 mb)

AIR TEMPERATURE (°C)	SATURATION MIXING RATIO (g/kg)
20	15.0
15	10.8
10	7.8
5	5.5
0	3.8
−5	2.6
−10	1.8
−15	1.2
−20	0.8

ter vapor content (the actual mixing ratio) of the mixed parcel will be the sum of the mixing ratios of parcel A and parcel B, or

$$\frac{10.8\text{ g}}{\text{kg}} + \frac{1.2\text{ g}}{\text{kg}} = \frac{12.0\text{ g}}{2\text{ kg}} = \frac{6.0\text{ g}}{\text{kg}}.$$

Look at Table 1 and observe that the saturation mixing ratio for a saturated parcel at 5°C

is only 5.5 g/kg. This means that the water vapor content of the mixed parcel, at 6.0 g/kg, is *above* that required for saturation and that the parcel is *supersaturated* with a relative humidity of 109 percent. Of course, such a high relative humidity is almost impossible to obtain, as water vapor would certainly condense on condensation nuclei, producing liquid water droplets as the two parcels mix together and the relative humidity approaches 100 percent. Hence, mixing two initially unsaturated masses of air can produce fog or a cloud.

Another way to look at this mixing process is to place the two unsaturated air parcels into Fig. 3, which is a graphic representation of Table 1. The solid blue line in Fig. 3 represents the saturation mixing ratio. Any air parcel with an air temperature and actual mixing ratio that falls on the blue line is saturated with a relative humidity of 100 percent. If an air parcel is located to the right of the blue line, the air parcel is unsaturated. If an air parcel lies to be left of the line, the parcel is supersaturated, and condensation will occur.

Notice that when parcel A and parcel B from Fig. 2 are plotted in Fig. 3, both unsaturated air parcels fall to the right of the blue line. However, when parcel A and parcel B are mixed, the final mixed air parcel (with an air

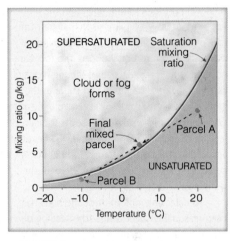

● FIGURE 3 The blue line is the saturation mixing ratio. The mixing of two unsaturated air parcels (A and B) can produce a saturated air parcel and fog.

temperature of 5°C and an actual mixing ratio of 6.0 g/kg) falls to the left of the blue line, indicating that water vapor inside the mixed parcel will condense either into fog or a cloud. Fog that forms in this manner is listed under the heading Evaporation (Mixing) Fog (p. 118). As you read that section, keep in mind that although evaporation has a part in fog formation, mixing plays the dominant role.

Up to this point, we have looked at the different forms of condensation that occur on or near the earth's surface. In particular, we learned that fog is simply many millions of tiny liquid droplets (or ice crystals) that form near the ground. In the following sections, we will see how these same particles, forming well above the ground, are classified and identified as clouds.

BRIEF REVIEW

However, before going on to the section on clouds, here is a brief review of some of the facts and concepts we covered so far:

- Dew, frost, and frozen dew generally form on clear nights when the temperature of objects on the surface cools below the air's dew-point temperature.

- Visible white frost forms in saturated air when the air temperature is at or below freezing. Under these conditions,

water vapor can change directly to ice, in a process called *deposition*.

- Condensation nuclei act as surfaces on which water vapor condenses. Those nuclei that have an affinity for water vapor are called *hygroscopic*.

- Fog is a cloud resting on the ground. It can be composed of water droplets, ice crystals, or a combination of both.

- Radiation fog, advection fog, and upslope fog all form as the air cools. The cooling for radiation fog is mainly radiational cooling at the earth's surface; for advection fog, the cooling is mainly warmer air moving over a colder surface; for upslope fog, the cooling occurs as moist air gradually rises and expands along sloping terrain.

- Evaporation (mixing) fog, such as steam fog and frontal fog, forms as water evaporates and mixes with drier air.

FOCUS ON AN ENVIRONMENTAL ISSUE

Fog Dispersal

In any airport fog-clearing operation the problem is to improve visibility so that aircraft can take off and land. Experts have tried various methods, which can be grouped into four categories: (1) increase the size of the fog droplets, so that they become heavy and settle to the ground as a light drizzle; (2) seed cold fog with dry ice (solid carbon dioxide), so that fog droplets are converted into ice crystals; (3) heat the air, so that the fog evaporates; and (4) mix the cooler saturated air near the surface with the warmer unsaturated air above.

To date, only one of these methods has been reasonably successful — the seeding of cold fog. *Cold fog* forms when the air temperature is below freezing, and most of the fog droplets remain as liquid water. (Liquid fog in below-freezing air is also called *supercooled fog*.) The fog can be cleared by injecting several hundred pounds of dry ice into it. As the tiny pieces of cold ($-78°C$) dry ice descend, they freeze some of the supercooled fog droplets in their path, producing ice crystals. As we will

see in Chapter 7, these crystals then grow larger at the expense of the remaining liquid fog droplets. Hence, the fog droplets evaporate and the larger ice crystals fall to the ground, which leaves a "hole" in the fog for aircraft takeoffs and landings.

Unfortunately, most of the fogs that close airports in the United States are *warm fogs* that form when the air temperature is above freezing. Since dry ice seeding does not work in warm fog, other techniques must be tried.

One method involves injecting hygroscopic particles into the fog. Large salt particles and other chemicals absorb the tiny fog droplets and form into larger drops. More large drops and fewer small drops improve the visibility; plus, the larger drops are more likely to fall as a light drizzle. Since the chemicals are expensive and the fog clears for only a short time, this method of fog dispersal is not economically feasible.

Another technique for fog dispersal is to warm the air enough so that the fog droplets

evaporate and visibility improves. Tested at Los Angeles International Airport in the early 1950s, this technique was abandoned because it was smoky, expensive, and not very effective. In fact, the burning of hundreds of dollars worth of fuel only cleared the runway for a short time. And the smoke particles, released during the burning of the fuel, provided abundant nuclei for the fog to recondense upon.

A final method of warm fog dispersal uses helicopters to mix the air. The chopper flies across the fog layer, and the turbulent downwash created by the rotor blades brings drier air above the fog into contact with the moist fog layer (see Fig. 4). The aim, of course, is to evaporate the fog. Experiments show that this method works well, as long as the fog is a shallow radiation fog with a relatively low-liquid water content. But many fogs are thick, have a high liquid water content, and form by other means. An inexpensive and practical method of dispersing warm fog has yet to be discovered.

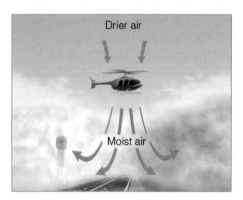

● FIGURE 4 Helicopters hovering above an area of shallow fog (left) can produce a clear area (right) by mixing the drier air into the foggy air below.

Clouds

Clouds are aesthetically appealing and add excitement to the atmosphere. Without them, there would be no rain or snow, thunder or lightning, rainbows or halos. How monotonous if one had only a clear blue sky to look at. A *cloud* is a visible aggregate of tiny water droplets or ice crystals suspended in the air. Some are found only at high elevations, whereas others nearly touch the ground. Clouds can be thick or thin, big

or little — they exist in a seemingly endless variety of forms. To impose order on this variety, we divide clouds into ten basic types. With a careful and practiced eye, you can become reasonably proficient in correctly identifying them.

CLASSIFICATION OF CLOUDS   Although ancient astronomers named the major stellar constellations about 2000 years ago, clouds were not formally identified and classified until the early nineteenth century. The French naturalist Lamarck (1744–1829) proposed the first system for classifying clouds in

1802; however, his work did not receive wide acclaim. One year later, Luke Howard, an English naturalist, developed a cloud classification system that found general acceptance. In essence, Howard's innovative system employed Latin words to describe clouds as they appear to a ground observer. He named a sheet-like cloud *stratus* (Latin for "layer"); a puffy cloud *cumulus* ("heap"); a wispy cloud *cirrus* ("curl of hair"); and a rain cloud *nimbus* ("violent rain"). In Howard's system, these were the four basic cloud forms. Other clouds could be described by combining the basic types. For example, nimbostratus is a rain cloud that shows layering, whereas cumulonimbus is a rain cloud having pronounced vertical development.

In 1887, Abercromby and Hildebrandsson expanded Howard's original system and published a classification system that, with only slight modification, is still in use today. Ten principal cloud forms are divided into four primary cloud groups. Each group is identified by the height of the cloud's base above the surface: high clouds, middle clouds, and low clouds. The fourth group contains clouds showing more vertical than horizontal development. Within each group, cloud types are identified by their appearance. ▼ Table 5.2 lists these four groups and their cloud types.

The approximate base height of each cloud group is given in ▼ Table 5.3. Note that the altitude separating the high and middle cloud groups overlaps and varies with latitude. Large temperature changes cause most of this latitudinal variation. For example, high cirriform clouds are composed almost entirely of ice crystals. In tropical regions, air temperatures low enough to freeze all liquid water usually occur only above 6000 m (about 20,000 ft). In polar regions, however, these same temperatures may be found at altitudes as low as 3000 m (about 10,000 ft). Hence, while you may observe cirrus clouds at 3600 m (about 12,000 ft) over northern Alaska, you will not see them at that elevation above southern Florida.

Clouds cannot be accurately identified strictly on the basis of elevation. Other visual clues are necessary. Some of these are explained in the following section.

CLOUD IDENTIFICATION

High Clouds High clouds in middle and low latitudes generally form above 6000 m (20,000 ft). Because the air at these elevations is quite cold and "dry," high clouds are composed almost exclusively of ice crystals and are also rather

▼ TABLE 5.2 The Four Major Cloud Groups and Their Types

1. High clouds	3. Low clouds
Cirrus (Ci)	Stratus (St)
Cirrostratus (Cs)	Stratocumulus (Sc)
Cirrocumulus (Cc)	Nimbostratus (Ns)
2. Middle clouds	**4. Clouds with vertical development**
Altostratus (As)	Cumulus (Cu)
Altocumulus (Ac)	Cumulonimbus (Cb)

thin.* High clouds usually appear white, except near sunrise and sunset, when the unscattered (red, orange, and yellow) components of sunlight are reflected from the underside of the clouds.

The most common high clouds are the **cirrus** (Ci), which are thin, wispy clouds blown by high winds into long streamers called *mares' tails*. Notice in ● Fig. 5.13 that they can look like a white, feathery patch with a faint wisp of a tail at one end. Cirrus clouds usually move across the sky from west to east, indicating the prevailing winds at their elevation, and they generally point to fair, pleasant weather.

Cirrocumulus (Cc) clouds, seen less frequently than cirrus, appear as small, rounded, white puffs that may occur individually or in long rows (see ● Fig. 5.14). When in rows, the cirrocumulus cloud has a rippling appearance that distinguishes it from the silky look of the cirrus and the sheetlike cirrostratus. Cirrocumulus seldom cover more than a small portion of the sky. The dappled cloud elements that reflect the red or yellow light of a setting sun make this one of the most beautiful of all clouds. The small ripples in the cirrocumulus strongly resemble the scales of a fish; hence, the expression *"mackerel sky"* commonly describes a sky full of cirrocumulus clouds.

The thin, sheetlike, high clouds that often cover the entire sky are **cirrostratus** (Cs) (see ● Fig. 5.15), which are so thin that the sun and moon can be clearly seen through them. The ice crystals in these clouds bend the light passing through them and will often produce a *halo*—a ring of light that en-

---

*Small quantities of liquid water in cirrus clouds at temperatures as low as $-36°C$ ($-33°F$) were discovered during research conducted above Boulder, Colorado.

▼ TABLE 5.3 Approximate Height of Cloud Bases Above the Surface for Various Locations

CLOUD GROUP	TROPICAL REGION	MIDDLE LATITUDE REGION	POLAR REGION
High	20,000 to 60,000 ft	16,000 to 43,000 ft	10,000 to 26,000 ft
Ci, Cs, Cc	(6,000 to 18,000 m)	(5000 to 13,000 m)	(3000 to 8000 m)
Middle	6500 to 26,000 ft	6500 to 23,000 ft	6500 to 13,000 ft
As, Ac	(2000 to 8000 m)	(2000 to 7000 m)	(2000 to 4000 m)
Low	surface to 6500 ft	surface to 6500 ft	surface to 6500 ft
St, Sc, Ns	(0 to 2000 m)	(0 to 2000 m)	(0 to 2000 m)

© C. Donald Ahrens

● FIGURE 5.13 Cirrus clouds.

circles the sun or moon. In fact, the veil of cirrostratus may be so thin that a halo is the only clue to its presence. Thick cirrostratus clouds give the sky a glary white appearance and frequently form ahead of an advancing storm; hence, they can be used to predict rain or snow within 12 to 24 hours, especially if they are followed by middle-type clouds.

© C. Donald Ahrens

● FIGURE 5.14 Cirrocumulus clouds.

© C. Donald Ahrens

● FIGURE 5.15 Cirrostratus clouds with a faint halo encircling the sun. The sun is the bright white area in the center of the circle.

Middle Clouds The middle clouds have bases between 2000 and 7000 m (6500 to 23,000 ft) in the middle latitudes. These clouds are composed of water droplets and—when the temperature becomes low enough—some ice crystals.

Altocumulus (Ac) clouds are middle clouds that are composed mostly of water droplets and are rarely more than 1 km thick. They appear as gray, puffy masses, sometimes rolled out in parallel waves or bands (see ● Fig. 5.16). Usually, one part of the cloud is darker than another, which helps to separate it from the higher cirrocumulus. Also, the individual puffs of the altocumulus appear larger than those of the cirrocumulus. A layer of altocumulus may sometimes be confused with altostratus; in case of doubt, clouds are called altocumulus if there are rounded masses or rolls present. Altocumulus clouds that look like "little castles" *(castellanus)* in the sky indicate the presence of rising air at cloud level. The appearance of these clouds on a warm, humid summer morning often portends thunderstorms by late afternoon.

The **altostratus** (As) is a gray or blue-gray cloud composed of ice crystals and water droplets. Altostratus clouds often cover the entire sky across an area that extends over many hundreds of square kilometers. In the thinner section of the cloud, the sun (or moon) may be *dimly visible* as a round disk, as if the sun were shining through ground glass. This appearance is sometimes referred to as a "watery sun" (see ● Fig. 5.17). Thick cirrostratus clouds are occasionally confused with thin altostratus clouds. The gray color, height, and dimness of the sun are good clues to identifying an altostratus. The fact that halos only occur with cirriform clouds also helps one distinguish them. Another way to separate the two is to look at the ground for shadows. If there are none, it is a good bet that the cloud is altostratus because cirrostratus are usually transparent enough to produce them. Altostratus clouds often form ahead of storms having widespread and relatively continuous precipitation. If precipitation falls from an alto-

● FIGURE 5.16 Altocumulus clouds.

● FIGURE 5.17 Altostratus clouds. The appearance of a dimly visible "watery sun" through a deck of gray clouds is usually a good indication that the clouds are altostratus.

● FIGURE 5.18 The nimbostratus is the sheetlike cloud from which light rain is falling. The ragged-appearing clouds beneath the nimbostratus is stratus fractus, or scud.

● FIGURE 5.19 Stratocumulus clouds forming along the south coast of Florida. Notice that the rounded masses are larger than those of the altocumulus.

stratus, its base usually lowers. If the precipitation reaches the ground, the cloud is then classified as *nimbostratus.*

Low Clouds   Low clouds, with their bases lying below 2000 m (6500 ft), are almost always composed of water droplets; however, in cold weather, they may contain ice particles and snow.

The **nimbostratus** (Ns) is a dark gray, "wet"-looking cloudy layer associated with more or less continuously falling rain or snow (see ● Fig. 5.18). The intensity of this precipitation is usually light or moderate—it is never of the heavy, showery variety, unless well-developed cumulus clouds are embedded within the nimbostratus cloud. The base of the nimbostratus cloud is normally impossible to identify clearly and its top may

be over 3 km (10,000 ft) higher. Nimbostratus is easily confused with the altostratus. Thin nimbostratus is usually darker gray than thick altostratus, and you normally cannot see the sun or moon through a layer of nimbostratus. Visibility below a nimbostratus cloud deck is usually quite poor because rain will evaporate and mix with the air in this region. If this air becomes saturated, a lower layer of clouds or fog may form beneath the original cloud base. Since these lower clouds drift rapidly with the wind, they form irregular shreds with a ragged appearance that are called *stratus fractus,* or *scud.*

Stratocumulus (Sc) are low lumpy clouds that appear in rows, in patches, or as rounded masses with blue sky visible between the individual cloud elements (see ● Fig. 5.19). Often

● FIGURE 5.20 A layer of low-lying stratus clouds hides these mountains in Iceland.

● FIGURE 5.21 Cumulus clouds. Small cumulus clouds such as these are sometimes called *fair weather cumulus*, or *cumulus humilis*.

they appear near sunset as the spreading remains of a much larger cumulus cloud. Occasionally, the sun will shine through the cloud breaks producing bands of light (called *crepuscular rays*) that appear to reach down to the ground. The color of stratocumulus ranges from light to dark gray. It differs from altocumulus in that it has a lower base and larger individual cloud elements. (Compare Fig. 5.16 with Fig. 5.19.) To distinguish between the two, hold your hand at arm's length and point toward the cloud. Altocumulus cloud elements will generally be about the size of your thumbnail; stratocumulus cloud elements will usually be about the size of your fist. Although precipitation rarely falls from stratocumulus, precipitation in the form of showers may occur in winter if the cloud elements develop vertically into much larger clouds and their tops grow colder than about −5°C (23°F).

    Stratus (St) is a uniform grayish cloud that often covers the entire sky. It resembles a fog that does not reach the ground (see ● Fig. 5.20). Actually, when a thick fog "lifts," the resulting cloud is a deck of low stratus. Normally, no pre-

cipitation falls from the stratus, but sometimes it is accompanied by a light mist or drizzle. This cloud commonly occurs over Pacific and Atlantic coastal waters in summer. A thick layer of stratus might be confused with nimbostratus, but the distinction between them can be made by observing the low base of the stratus cloud and remembering that light-to-moderate precipitation occurs with nimbostratus. Moreover, stratus often has a more uniform base than does nimbostratus. Also, a deck of stratus may be confused with a layer of altostratus. However, if you remember that stratus are lower and darker gray and that the sun normally appears "watery" through altostratus, the distinction can be made.

Clouds with Vertical Development Familiar to almost everyone, the puffy **cumulus** (Cu) cloud takes on a variety of shapes, but most often it looks like a piece of floating cotton with sharp outlines and a flat base (see ● Fig. 5.21). The base appears white to light gray, and, on a humid day, may be only 1000 m (3300 ft) above the ground and a kilometer or so

© C. Donald Ahrens

● FIGURE 5.22 Cumulus congestus. This line of cumulus congestus clouds is building along Maryland's eastern shore.

© T. Ansel Toney

● FIGURE 5.23 A cumulonimbus cloud (thunderstorm). Strong upper-level winds blowing from right to left produce a well-defined anvil. Sunlight scattered by falling ice crystals produces the white (bright) area beneath the anvil. Notice the heavy rain shower falling from the base of the cloud.

wide. The top of the cloud—often in the form of rounded towers—denotes the limit of rising air and is usually not very high. These clouds can be distinguished from stratocumulus by the fact that cumulus clouds are detached (usually a great deal of blue sky between each cloud) while stratocumulus usually occur in groups or patches. Also, the cumulus has a dome- or tower-shaped top as opposed to the generally flat tops of the stratocumulus. Cumulus clouds that show only slight vertical growth are called *cumulus humilis* and are associated with fair weather; therefore, we call these clouds *"fair weather cumulus."* Ragged-edge cumulus clouds that are smaller than cumulus humilis and scattered across the sky are called *cumulus fractus.*

Harmless-looking cumulus often develop on warm summer mornings and, by afternoon, become much larger and more vertically developed. When the growing cumulus resembles a head of cauliflower, it becomes a *cumulus congestus,* or *towering cumulus* (Tcu). Most often, it is a single large cloud, but, occasionally, several grow into each other, forming

a line of towering clouds, as shown in ● Fig. 5.22. Precipitation that falls from a cumulus congestus is always showery.

If a cumulus congestus continues to grow vertically, it develops into a giant **cumulonimbus** (Cb)—a thunderstorm cloud (see ● Fig. 5.23). While its dark base may be no more than 600 m (2000 ft) above the earth's surface, its top may

WEATHER WATCH

The updrafts and downdrafts inside a cumulonimbus cloud can exceed 70 knots. On July 26, 1959, Colonel William A. Rankin took a wild ride inside one of these clouds. Bailing out of his disabled military aircraft inside a thunderstorm at 14.5 km (about 47,500 ft), Rankin free-fell for about 3 km (10,000 ft). When his parachute opened, surging updrafts carried him higher into the cloud, where he was pelted by heavy rain and hail, and nearly struck by lightning. After being carried up and down violently several times, he landed on the ground, tattered and torn but thankful to be alive.

extend upward to the tropopause, over 12,000 m (39,000 ft) higher. A cumulonimbus can occur as an isolated cloud or as part of a line or "wall" of clouds.

Tremendous amounts of energy released by the condensation of water vapor within a cumulonimbus result in the development of violent up- and downdrafts, which may exceed 70 knots. The lower (warmer) part of the cloud is usually composed of only water droplets. Higher up in the cloud, water droplets and ice crystals both abound, while, toward the cold top, there are only ice crystals. Swift winds at these higher altitudes can reshape the top of the cloud into a huge flattened anvil*(cumulonimbus incus). These great thunderheads may contain all forms of precipitation—large raindrops, snowflakes, snow pellets, and sometimes hailstones—all of which can fall to earth in the form of heavy showers. Lightning, thunder, and even tornadoes are associated with the cumulonimbus. (More information on the violent nature of thunderstorms and tornadoes is given in Chapter 14.)

Cumulus congestus and cumulonimbus frequently look alike, making it difficult to distinguish between them. However, you can usually distinguish them by looking at the top of the cloud. If the sprouting upper part of the cloud is sharply defined and not fibrous, it is usually a cumulus congestus; conversely, if the top of the cloud loses its sharpness and becomes fibrous in texture, it is usually a cumulonimbus.

*An anvil is a heavy block of iron or steel with a smooth, flat top on which metals are shaped by hammering.

Compare Fig. 5.22 with Fig. 5.23. The weather associated with these clouds also differs: lightning, thunder, and large hail typically occur with cumulonimbus.

So far, we have discussed the ten primary cloud forms, summarized pictorially in ● Fig. 5.24. This figure, along with the cloud photographs and descriptions (and the cloud chart at the back of the book), should help you identify the more common cloud forms. Don't worry if you find it hard to estimate cloud heights. This is a difficult procedure, requiring much practice. You can use local objects (hills, mountains, tall buildings) of known height as references on which to base your height estimates.

To better describe a cloud's shape and form, a number of descriptive words may be used in conjunction with its name. We mentioned a few in the previous section; for example, a stratus cloud with a ragged appearance is a stratus fractus, and a cumulus cloud with marked vertical growth is a cumulus congestus. ▼ Table 5.4 lists some of the more common terms that are used in cloud identification.

SOME UNUSUAL CLOUDS Although the ten basic cloud forms are the most frequently seen, there are some unusual clouds that deserve mentioning. For example, moist air crossing a mountain barrier often forms into waves. The clouds that form in the wave crest usually have a lens shape and are, therefore, called **lenticular clouds** (see ● Fig. 5.25). Frequently, they form one above the other like a stack of pan-

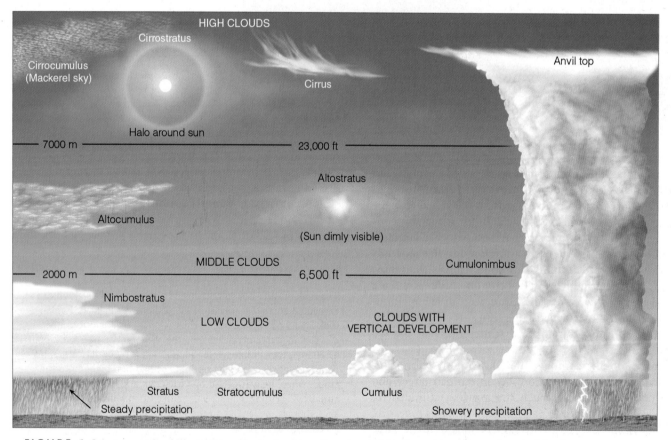

● FIGURE 5.24 A generalized illustration of basic cloud types based on height above the surface and vertical development.

▼ TABLE 5.4   Common Terms Used in Identifying Clouds

TERM	LATIN ROOT AND MEANING	DESCRIPTION
Lenticularis	(*lens, lenticula,* lentil)	Clouds having the shape of a lens or an almond, often elongated and usually with well-defined outlines. This term applies mainly to cirrocumulus, altocumulus, and stratocumulus
Fractus	(*frangere,* to break or fracture)	Clouds that have a ragged or torn appearance; applies only to stratus and cumulus
Humilis	(*humilis,* of small size)	Cumulus clouds with generally flattened bases and slight vertical growth
Congestus	(*congerere,* to bring together; to pile up)	Cumulus clouds of great vertical extent that from a distance may resemble a head of cauliflower
Calvus	(*calvus,* bald)	Cumulonimbus in which at least some of the upper part is beginning to lose its cumuliform outline
Capillatus	(*capillus,* hair; having hair)	Cumulonimbus characterized by the presence in the upper part of cirriform clouds with fibrous or striated structure
Undulatus	(*unda,* wave; having waves)	Clouds in patches, sheets, or layers showing undulations
Translucidus	(*translucere,* to shine through; transparent)	Clouds that cover a large part of the sky and are sufficiently translucent to reveal the position of the sun or moon
Incus	(*incus,* anvil)	The smooth cirriform mass of cloud in the upper part of a cumulonimbus that is anvil-shaped
Mammatus	(*mamma,* mammary)	Baglike clouds that hang like a cow's udder on the underside of a cloud; may occur with cirrus, altocumulus, altostratus, stratocumulus, and cumulonimbus
Pileus	(*pileus,* cap)	A cloud in the form of a cap or hood above or attached to the upper part of a cumuliform cloud, particularly during its developing stage
Castellanus	(*castellum,* a castle)	Clouds that show vertical development and produce towerlike extensions, often in the shape of small castles

cakes, and at a distance they may resemble a hovering spacecraft. Hence, it is no wonder a large number of UFO sightings take place when lenticular clouds are present. When a cloud forms over and extends downwind of an isolated mountain peak, as shown in ● Fig. 5.26, it is called a **banner cloud.**

Similar to the lenticular is the *cap cloud,* or **pileus,** that usually resembles a silken scarf capping the top of a sprouting cumulus cloud (see ● Fig. 5.27). Pileus clouds form when moist winds are deflected up and over the top of a building cumulus congestus or cumulonimbus. If the air flowing over the top of the cloud condenses, a pileus often forms.

Most clouds form in rising air, but the mammatus forms in sinking air. **Mammatus clouds** derive their name from their appearance—baglike sacs that hang beneath the cloud and resemble a cow's udder (see ● Fig. 5.28). Although mammatus most frequently form on the underside of cumulonimbus, they may develop beneath cirrocumulus, altostratus, altocumulus, and stratocumulus. For mammatus to form, the sinking air must be cooler than the air around it and have a high liquid water or ice content. As saturated air sinks, it

© Dick Hilton

● FIGURE 5.25 Lenticular clouds forming on the leeward side of the Sierra Nevada near Verdi, Nevada.

● FIGURE 5.26 The cloud forming over and downwind of Mt. Rainier is called a banner cloud.

● FIGURE 5.27 A pileus cloud forming above a developing cumulus cloud.

● FIGURE 5.28 Mammatus clouds forming beneath a thunderstorm.

warms, but the warming is retarded because of the heat taken from the air to evaporate the liquid or melt ice particles. If the sinking air remains saturated and cooler than the air around it, the sinking air can extend below the cloud base appearing as rounded masses we call mammatus clouds.

● FIGURE 5.29 A contrail forming behind a jet aircraft.

Jet aircraft flying at high altitudes often produce a cirrus-like trail of condensed vapor called a *condensation trail* or **contrail** (see ● Fig. 5.29). The condensation may come directly from the water vapor added to the air from engine exhaust. In this case, there must be sufficient mixing of the hot exhaust gases with the cold air to produce saturation. The release of particles in the exhaust may even provide nuclei on which ice crystals form. Contrails evaporate rapidly when the relative humidity of the surrounding air is low. If the relative humidity is high, however, contrails may persist for many hours. Contrails may also form by a cooling process. The reduced pressure produced by air flowing over the wing causes the air to cool. This cooling may supersaturate the air, producing an *aerodynamic contrail*. This type of trail usually disappears quickly in the turbulent wake of the aircraft.

Aside from the cumulonimbus cloud that sometimes penetrates into the stratosphere, all of the clouds described so far are observed in the lower atmosphere—in the troposphere. Occasionally, however, clouds may be seen above the troposphere. For example, soft pearly looking clouds called **nacreous clouds,** or *mother-of-pearl clouds*, form in the stratosphere at altitudes above 30 km (see ● Fig. 5.30). They are best viewed in polar latitudes during the winter months when the sun, being just below the horizon, is able to illuminate them because of their high altitude. Their exact composition is not known, although they appear to be composed of water in either solid or liquid (supercooled) form.

Wavy bluish-white clouds, so thin that stars shine brightly through them, may sometimes be seen in the upper mesosphere, at altitudes above 75 km (46 mi). The best place to view these clouds is in polar regions at twilight. At this time, because of their altitude, the clouds are still in sunshine. To a ground observer, they appear bright against a dark background and, for this reason, they are called **noctilucent clouds,** meaning "luminous night clouds" (see ● Fig. 5.31). Studies reveal that these clouds are composed of tiny ice crystals. The water to make the ice may originate in meteoroids that disintegrate when entering the upper atmosphere or from the chemical breakdown of methane gas at high levels in the atmosphere.

© Pekka Parviainen

● FIGURE 5.30 The clouds in this photograph are nacreous clouds. They form in the stratosphere and are most easily seen at high latitudes.

© Pekka Parviainen

● FIGURE 5.31 The wavy clouds in this photograph are noctilucent clouds. They are usually observed at high latitudes, at altitudes between 75 and 90 km above the earth's surface.

CLOUD OBSERVATIONS

Determining Sky Conditions Often, a daily weather forecast will include a phrase such as, "overcast skies with clouds becoming scattered by evening." To the average person, this means that the cloudiness will diminish, but to the meteorologist, the terms *overcast* and *scattered* have a more specific meaning. In meteorology, descriptions of sky conditions are defined by the fraction of sky covered by clouds. A *clear* sky, for example, is one where no clouds are present.* When there are between one-eighth and two-eighths clouds covering the sky, there are a *few* clouds present. When cloudiness increases to between three-eighths and four-eighths, the sky is described as being *scattered* with clouds. "Partly cloudy" also

*In automated (ASOS) station usage, the phrase "clear sky" means that no clouds are reported whose bases are at or below 12,000 ft.

describes these sky conditions. Clouds covering between five-eighths and seven-eighths of the sky denote a sky with *broken* clouds ("mostly cloudy"), and *overcast* conditions exist when the sky is covered (eight-eighths) with clouds. ▼ Table 5.5 presents a summary of sky cover conditions.

Observing sky conditions far away can sometimes fool even the trained observer. A broken cloud deck near the ho-

FOCUS ON AN OBSERVATION

Measuring Cloud Ceilings

In addition to knowing about sky conditions, it is usually important to have a good estimate of the height of cloud bases. Aircraft could not operate safely without accurate cloud height information, particularly at lower elevations.

The term *ceiling* is defined as the height of the lowest layer of clouds above the surface that are either broken or overcast, but not thin. Direct information on cloud height can be obtained from pilots who report the altitude at which they encounter the ceiling. Less directly, *ceiling balloons* can measure the height of clouds. A small balloon filled with a known amount of hydrogen or helium rises at a fairly constant and known rate. The ceiling is determined by measuring the time required for the balloon to enter the lowest cloud layer.* Ceiling balloon observations can be made at night simply by attaching a small battery-operated light to the balloon.

For many years, the *rotating-beam ceilometer* provided information on cloud ceiling, especially at airports. This instrument consists of a ground-based projector that rotates vertically from horizon to horizon. As it rotates, it sends out a powerful light beam that moves along the

*For example, if the balloon rises 125 m (about 400 ft) each minute, and it takes three minutes to enter a broken layer of stratocumulus, the ceiling would be 375 m (about 1200 ft).

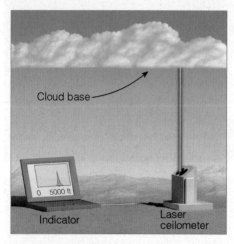

● FIGURE 5 The laser-beam ceilometer sends pulses of infrared radiation up to the cloud. Part of this beam is reflected back to the ceilometer. The interval of time between pulse transmission and return is a measure of cloud height, as displayed on the indicator screen.

base of the cloud. A light-sensitive detector, some known distance from the projector, points upward and picks up the light from the cloud base. By knowing the projector angle and its distance from the detector, the cloud height is determined mathematically.

Most of the rotating-beam ceilometers have been phased out and replaced with *laser-*

beam ceilometers. The laser ceilometer is a fixed-beam type whose transmitter and receiver point straight up at the cloud base (see Fig. 5). Short, intense pulses of infrared radiation from the transmitter strike the cloud base, and a portion of this radiation is reflected back to the receiver. The time interval between pulse transmission and its return from the cloud determines the cloud-base height.

The Automated Surface Observing System (ASOS) uses a laser beam ceilometer to measure cloud height. The ceilometer measures the cloud height and then infers the amount of cloud cover by averaging the amount of clouds that have passed over the sensor during a duration of 30 minutes. The ASOS laser ceilometer is unable to measure clouds that are not above the sensor. To help remedy this situation, a second laser ceilometer may be located nearby. Another limitation of the ASOS ceilometer is that it does not report clouds above 12,000 ft.* A new laser ceilometer that will provide cloud height information up to 25,000 ft is being developed.

*The latest geostationary satellites above North America are equipped to measure cloud heights above 12,000 ft over ASOS stations.

▼ TABLE 5.5   Description of Sky Conditions

DESCRIPTION	OBSERVATION		MEANING
	ASOS*	HUMAN	
Clear (CLR or SKC)	0 to 5%	0	No clouds
Few	>5 to ≤25%	0 to $2/8$	Few clouds visible
Scattered (SCT)	>25 to ≤50%	$3/8$ to $4/8$	Partly cloudy
Broken (BKN)	>50 to ≤87%	$5/8$ to $7/8$	Mostly cloudy
Overcast (OVC)	>87 to 100%	$8/8$	Sky is covered by clouds
Sky obscured	—	—	Sky is hidden by surface-based phenomena, such as fog, blowing snow, smoke, and so forth, rather than by cloud cover

*Automated Surface Observing System. Symbol > means greater than; < means less than; ≥ means equal to or greater than.

rizon usually appears as overcast because the open spaces between the clouds are less visible at a distance. Therefore, cloudiness is usually overestimated when clouds are near the horizon. Viewed from afar, clouds not normally associated with precipitation may appear darker and thicker than they actually are. The reason for this observation is that light from a distant cloud travels through more atmosphere and is more attenuated than the light from the same type of cloud closer to the observer. (Information on measuring the height of cloud bases is given in the Focus section on p. 132.)

Up to this point, we have seen how clouds look from the ground. We will now look at clouds from a different vantage point—the satellite view.

Satellite Observations   The weather satellite is a cloud-observing platform in earth's orbit. It provides extremely valuable cloud photographs of areas where there are no ground-based observations. Because water covers over 70 percent of the earth's surface, there are vast regions where few (if any) surface cloud observations are made. Before weather satellites were available, tropical storms, such as hurricanes and typhoons, often went undetected until they moved dangerously near inhabited areas. Residents of the regions affected had little advance warning. Today, satellites spot these storms while they are still far out in the ocean and track them accurately.

There are two primary types of weather satellites in use for viewing clouds. The first are called **geostationary satellites** (or *geosynchronous satellites*) because they orbit the equator at the same rate the earth spins and, hence, remain at nearly 36,000 km (22,300 mi) above a fixed spot on the earth's surface (see • Fig. 5.32). This positioning allows continuous monitoring of a specific region.

Geostationary satellites are also important because they use a "real time" data system, meaning that the satellites transmit images to the receiving system on the ground as soon as the camera takes the picture. Successive cloud images from these satellites can be put into a time-lapse movie sequence to show the cloud movement, dissipation, or development associated with weather fronts and storms. This information is a great help in forecasting the progress of large weather systems. Wind directions and speeds at various levels may also be approximated by monitoring cloud movement with the geostationary satellite.

To complement the geostationary satellites, there are **polar-orbiting satellites,** which closely parallel the earth's meridian lines. These satellites pass over the north and south polar regions on each revolution. As the earth rotates to the east beneath the satellite, each pass monitors an area to the west of the previous pass (see • Fig. 5.33). Eventually, the satellite covers the entire earth.

Polar-orbiting satellites have the advantage of photographing clouds directly beneath them. Thus, they provide sharp pictures in polar regions, where photographs from a geostationary satellite are distorted because of the low angle at which the satellite "sees" this region. Polar orbiters also

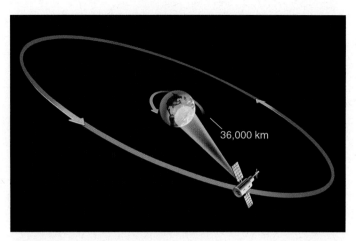

• FIGURE 5.32 The geostationary satellite moves through space at the same rate that the earth rotates, so it remains above a fixed spot on the equator and monitors one area constantly.

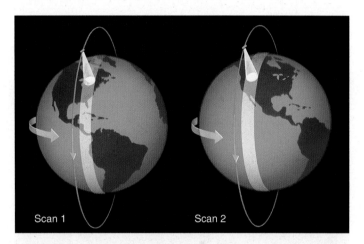

• FIGURE 5.33 Polar-orbiting satellites scan from north to south, and on each successive orbit the satellite scans an area farther to the west.

circle the earth at a much lower altitude (about 850 km) than geostationary satellites and provide detailed photographic information about objects, such as violent storms and cloud systems.

Continuously improved detection devices make weather observation by satellites more versatile than ever. Early satellites, such as *TIROS I*, launched on April 1, 1960, used television cameras to photograph clouds. Contemporary satellites use radiometers, which can observe clouds during both day and night by detecting radiation that emanates from the top of the clouds. Additionally, satellites have the capacity to obtain vertical profiles of atmospheric temperature and moisture by detecting emitted radiation from atmospheric gases, such as water vapor. In modern satellites, a special type of advanced radiometer (called an *imager*) provides satellite pictures with much better resolution than did previous imagers. Moreover, another type of special radiometer (called a *sounder*) gives a more accurate profile of temperature and moisture at different levels in the atmosphere than did earlier

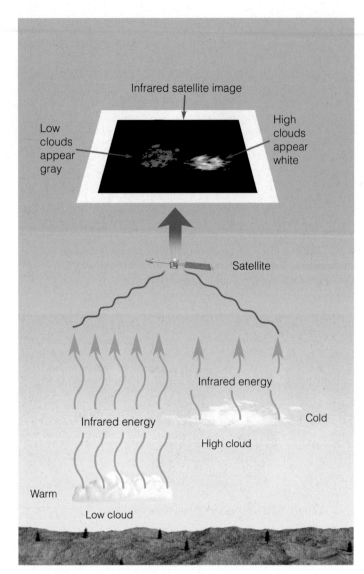

● FIGURE 5.34 Generally, the lower the cloud, the warmer its top. Warm objects emit more infrared energy than do cold objects. Thus, an infrared satellite picture can distinguish warm, low (gray) clouds from cold, high (white) clouds.

instruments. In the latest Geostationary Operational Environment Satellite (GOES) series, the imager and sounder are able to operate independent of each other.

Information on cloud thickness and height can be deduced from satellite images. Visible images show the sunlight reflected from a cloud's upper surface. Because thick clouds have a higher albedo (reflectivity) than thin clouds, they appear brighter on a visible satellite image. However, high, middle, and low clouds have just about the same albedo, so it is difficult to distinguish among them simply by using visible light photographs. To make this distinction, *infrared cloud images* are used. Such pictures produce a better image of the actual radiating surface because they do not show the strong visible reflected light. Since warm objects radiate more energy than cold objects, high temperature regions can be artificially made to appear darker on an infrared photograph.

Because the tops of low clouds are warmer than those of high clouds, cloud observations made in the infrared can distinguish between warm low clouds (dark) and cold high clouds (light) (see ● Fig. 5.34). Moreover, cloud temperatures can be converted by a computer into a three-dimensional image of the cloud. These are the 3-D cloud photos presented on television by many weathercasters.

● Figure 5.35a shows a visible satellite image (from a geostationary satellite) of a storm system in the eastern Pacific. Notice that all of the clouds in the image appear white. However, in the infrared image (see ● Fig. 5.35b), taken on the same day (and just about the same time), the clouds appear to have many shades of gray. In the visible image, the clouds covering part of Oregon and northern California appear relatively thin compared to the thicker, bright clouds to the west. Furthermore, these thin clouds must be high because they also appear bright in the infrared image. The elongated band

NOAA

ACTIVE FIGURE 5.35 (a) A visible image of the eastern Pacific taken at just about the same time on the same day as the image in Fig. 5.35 (b). Notice that the clouds in the visible image appear white. Visit the Meteorology Resource Center to view this and other active figures at academic.cengage.com/login

of clouds off the coast marks the position of an approaching weather front. Here, the clouds appear white and bright in both pictures, indicating a zone of thick, heavy clouds. Behind the front, the lumpy clouds are probably cumulus because they appear gray in the infrared image, indicating that their tops are low and relatively warm.

When temperature differences are small, it is difficult to directly identify significant cloud and surface features on an infrared image. Some way must be found to increase the contrast between features and their backgrounds. This can be done by a process called *computer enhancement*. Certain temperature ranges in the infrared image are assigned specific shades of gray—grading from black to white. Normally, clouds with cold tops, and those tops near freezing are assigned the darkest gray color.

● Figure 5.36 is an infrared-enhanced image for the same day as shown in Fig. 5.35. Often in this type of image, dark blue or red is assigned to clouds with the coldest (highest) tops. Hence, the dark red areas embedded along the front in Fig. 5.36 represent the region where the coldest and, therefore, highest and thickest clouds are found. It is here where the stormiest weather is probably occurring. Also notice that, near the southern tip of the image, the dark red blotches surrounded by areas of white are thunderstorms that have developed over warm tropical waters. They show up clearly as thick white clouds in both the visible and infrared images. By examining the movement of these clouds on successive satellite images, forecasters can predict the arrival of clouds and storms, as well as the passage of weather fronts.

In regions where there are no clouds, it is difficult to observe the movement of the air. To help with this situation, geostationary satellites are equipped with water-vapor sensors that can profile the distribution of atmospheric water vapor in the middle and upper troposphere (see ● Fig. 5.37).

ACTIVE FIGURE 5.35 (b) Infrared image of the eastern Pacific taken at just about the same time on the same day as the image in Fig. 5.35 (a). Notice that the low clouds in the infrared image appear in various shades of gray. Visit the Meteorology Resource Center to view this and other active figures at academic.cengage.com/login

● FIGURE 5.36 An enhanced infrared image of the eastern Pacific taken on the same day as the images shown in Fig. 5.35(a) and (b).

FOCUS ON A SPECIAL TOPIC

Satellites Do More Than Observe Clouds

The use of satellites to monitor weather is not restricted to observing clouds. For example, there are satellites that relay data communications and television signals, and provide military surveillance. Moreover, satellites measure radiation from the earth's surface and atmosphere, giving us information about the earth-atmosphere energy balance, discussed in Chapter 2. The infrared radiation measurements, obtained by an atmospheric *sounder,* are transformed into vertical profiles of temperature and moisture, which are fed into National Weather Service computer forecast models.

Radiation intensities from the ocean surface are translated into sea-surface temperature readings (see Fig. 6). This information is valuable to the fishing industry, as well as to the meteorologist. In fact, the *Tropical Rainfall Measuring Mission (TRMM)* satellite obtains sea-surface temperatures with a microwave scanner, even through clouds and atmospheric particles.

Satellites also monitor the amount of snow cover in winter, the extent of ice fields in the Arctic and Antarctic, the movement of large icebergs that drift into shipping lanes, and the height of the ocean's surface. One polar-orbiting satellite actually carries equipment that can detect faint distress signals anywhere on the globe, and relay them to rescue forces on the ground.

Infrared sensors on polar-orbiting satellites are able to assess conditions of crops, areas of deforestation, and regions of extensive drought. Satellites are also able to detect volcanic eruptions and follow the movement of ash clouds. During the winter, *GOES* satellites are able to monitor the southward progress of freezing air in Florida and Texas, allowing forecasters to warn growers of impending low temperatures so that they can take necessary measures to protect sensitive crops.

The *Global Positioning System (GPS)* consists of 24 polar-orbiting satellites that transmit

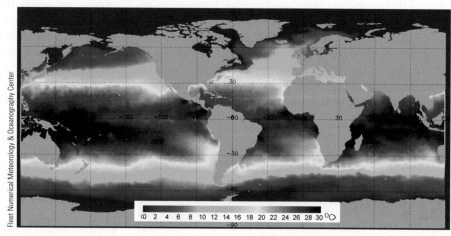

● FIGURE 6   Sea-surface temperatures for February 19, 2004. Temperatures are derived mainly from satellites, but temperature information also comes from buoys and ships.

radio signals to ground receivers, which then use the signals for navigation and relative positioning on earth. The signal the satellites send to earth is slowed by the amount of water vapor in the air. Because of this effect, the GPS can estimate the atmosphere's *precipitable water vapor* (the total atmospheric water vapor contained in a vertical column of air).

Geostationary satellites, such as *GOES,* are equipped with systems that receive environmental information from remote data-collection platforms on the surface. These platforms include instrumented buoys, river gauges, automatic weather stations, seismic and tsunami ("tidal" wave) stations, and ships. This information is transmitted to the satellite, which relays it to a central receiving station.

Normally, a network of five geostationary satellites positioned over the equator gives nearly complete global coverage from about latitude 60°N to 60°S. Along with monitoring clouds and the atmosphere, the latest *GOES* series provides forecasters and researchers with data from Doppler radars and the network of automated surface-observing stations. Geostationary satellites detect pollution and haze, and

provide accurate cloud-height measurements during the day. They even have the capacity to monitor the seasonal and daily trend in atmospheric ozone.

Satellites specifically designed to monitor the natural resources of the earth *(LandSat)* circle the earth 14 times a day in a near polar circular orbit. Photographs taken in several wavelength bands provide valuable information about this planet's geology, hydrology, oceanography, and ecology. *LandSat* also collects data transmitted from remote ground stations in North America. These stations monitor a variety of environmental data, with water quality, rainfall amount, and snow depth of particular interest to the meteorologist and hydrologist.

Satellite information is not confined to the lower atmosphere. There are satellites that monitor the concentrations of ozone, air temperature, and winds in the upper atmosphere. And both geostationary and polar-orbiting satellites carry instruments that monitor solar activity. Even with all this information available, there are more sophisticated satellites on the drawing board that will provide more and improved data in the future.

● FIGURE 5.37 Infrared water-vapor image. The darker areas represent dry air aloft; the brighter the gray, the more moist the air in the middle or upper troposphere. Bright white areas represent dense cirrus clouds or the tops of thunderstorms. The area in color represents the coldest cloud tops.

In time-lapse films, the swirling patterns of moisture clearly show wet regions and dry regions, as well as middle tropospheric swirling wind patterns and jet streams.

The *TRMM* (*Tropical Rainfall Measuring Mission*) satellite provides information on clouds and precipitation from about 35°N to 35°S. A joint venture of NASA and the National Space Agency of Japan, this satellite orbits the earth at an altitude of about 400 km (250 mi). From this vantage point the satellite, when looking straight down, can pick out individual cloud features as small as 2.4 km (1.5 mi) in diameter. Some of the instruments onboard the *TRMM* satellite include a visible and infrared scanner, a microwave imager, and precipitation radar. These instruments help provide three-dimensional images of clouds and storms, along with the intensity and distribution of precipitation (see ● Fig 5.38). Additional onboard instruments send back information concerning the earth's energy budget and lightning discharges in storms.

At this point, it should be apparent that today's satellites do a great deal more than simply observe clouds. More information on satellites and the information they provide is given in the Focus section on p. 136.

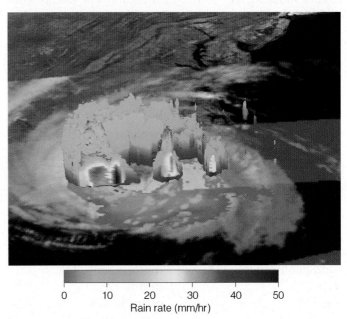

0 10 20 30 40 50
Rain rate (mm/hr)

● FIGURE 5.38 A three-dimensional *TRMM* satellite image of Hurricane Ophelia along the North Carolina coast on September 14, 2005. The light green areas in the cut-a-away view represent the region of lightest rainfall, whereas dark red and orange indicate regions of heavy rainfall.

SUMMARY

In this chapter, we examined the different forms of condensation. We saw that dew forms when the air temperature cools to the dew point in a shallow layer of air near the surface. If the dew should freeze, it produces tiny beads of ice called frozen dew. Frost forms when the air cools to a dew point that is at freezing or below.

As the air cools in a deeper layer near the surface, the relative humidity increases and water vapor begins to condense on hygroscopic condensation nuclei, forming wet haze. As the relative humidity approaches 100 percent, condensation occurs on most nuclei, and the air becomes filled with tiny liquid droplets (or ice crystals) called fog.

Fog forms in two primary ways: cooling of air and evaporating and mixing water vapor into the air. Radiation fog, advection fog, and upslope fog form by the cooling of air, while steam fog and frontal fog are two forms of evaporation (mixing) fog. Although fog has some beneficial effects—providing winter chilling for fruit trees and water for thirsty redwoods—in many places it is a nuisance, for it disrupts air traffic and it is the primary cause of a number of auto accidents.

Condensation above the earth's surface produces clouds. When clouds are classified according to their height and physical appearance, they are divided into four main groups: high, middle, low, and clouds with vertical development. Since each cloud has physical characteristics that distinguish it from all the others, careful cloud observations normally lead to correct identification.

Satellites enable scientists to obtain a bird's-eye view of clouds on a global scale. Polar-orbiting satellites obtain data covering the earth from pole to pole, while geostationary satellites located above the equator continuously monitor a desired portion of the earth. Both types of satellites use radiometers (imagers) that detect emitted radiation. As a consequence, clouds can be observed both day and night.

Visible satellite images, which show sunlight reflected from a cloud's upper surface, can distinguish thick clouds from thin clouds. Infrared images show an image of the cloud's radiating top and can distinguish low clouds from high clouds. To increase the contrast between cloud features, infrared photographs are enhanced.

Satellites do a great deal more than simply photograph clouds. They provide us with a wealth of physical information about the earth and the atmosphere.

KEY TERMS

The following terms are listed (with page number) in the order they appear in the text. Define each. Doing so will aid you in reviewing the material covered in this chapter.

dew, 112
frozen dew, 112
frost, 112
condensation nuclei, 113
hygroscopic nuclei, 113
hydrophobic nuclei, 113
haze, 113
fog, 114
acid fog, 114
radiation (ground) fog, 115
advection fog, 116
advection-radiation fog, 118
upslope fog, 118
evaporation (mixing) fog, 118
steam fog, 118
frontal fog, 119
winter chilling, 119
cirrus, 123

cirrocumulus, 123
cirrostratus, 123
altocumulus, 124
altostratus, 124
nimbostratus, 125
stratocumulus, 125
stratus, 126
cumulus, 126
cumulonimbus, 127
lenticular clouds, 128
banner cloud, 129
pileus, 129
mammatus clouds, 129
contrail, 130
nacreous clouds, 130
noctilucent clouds, 130
geostationary satellites, 133
polar-orbiting satellites, 133

QUESTIONS FOR REVIEW

1. Explain how dew, frozen dew, and visible frost each form.
2. Distinguish among dry haze, wet haze, and fog.
3. Why is fog that forms in industrial areas normally thick?
4. How can fog form when the air's relative humidity is less than 100 percent?
5. Name and describe four types of fog. What conditions are necessary for the formation of radiation fog?
6. Why do ground fogs usually "burn off" by early afternoon?
7. List as many positive consequences of fog as you can.
8. List and describe three methods of fog dispersal.
9. How does radiation fog normally form?
10. What atmospheric conditions are necessary for the development of advection fog?
11. How does evaporation (mixing) fog form?
12. Clouds are most generally classified by height above the earth's surface. List the major height categories and the cloud types associated with each.
13. List at least two distinguishable characteristics of each of the ten basic clouds.
14. Why are high clouds normally thin? Why are they composed almost entirely of ice crystals?
15. How can you distinguish altostratus from cirrostratus?
16. Which clouds are associated with each of the following characteristics:
    (a) lightning;
    (b) heavy rain showers;
    (c) mackerel sky;
    (d) mares' tails;
    (e) halos;
    (f) light continuous rain or snow;
    (g) hailstones;
    (h) anvil top.
17. Why does a broken layer of clouds near the horizon often appear as overcast?
18. How do geostationary satellites differ from polar-orbiting satellites?
19. Explain why visible and infrared images can be used to distinguish: (a) high clouds from low clouds; (b) thick clouds from thin clouds.
20. Why are infrared images enhanced?
21. Name two clouds that form above the troposphere.
22. List and explain the various types of environmental information obtained from satellites.

QUESTIONS FOR THOUGHT

1. Explain the reasoning behind the wintertime expression, "Clear moon, frost soon."
2. Explain why icebergs are frequently surrounded by fog.
3. During a summer visit to New Orleans, you stay in an air-conditioned motel. One afternoon, you put on your sunglasses, step outside, and within no time your glasses are "fogged up." Explain what has apparently caused this.
4. While driving from cold air (well below freezing) into much warmer air (well above freezing), frost forms on the windshield of the car. Does the frost form on the inside or outside of the windshield? How can the frost form when the air is so warm?
5. Why are really clean atmospheres and really dirty atmospheres undesirable?
6. Why do relative humidities seldom reach 100 percent in polluted air?

7. Why are advection fogs rare over tropical water?

8. A January snowfall covers central Arkansas with 5 inches of snow. The following day, a south wind brings heavy fog to this region. Explain what has apparently happened.

9. If all fog droplets gradually settle earthward, explain how fog can last (without disappearing) for many days at a time.

10. Near the shore of an extremely large lake, explain why steam fog is more likely to form during the autumn and advection fog in early spring.

11. The air temperature during the night cools to the dew point in a deep layer, producing fog. Before the fog formed, the air temperature cooled each hour about 2°C. After the fog formed, the air temperature cooled by only 0.5°C each hour. Give *two* reasons why the air cooled more slowly after the fog formed.

12. On a winter night, the air temperature cooled to the dew point and fog formed. Before the formation of fog, the dew point remained almost constant. After the fog formed, the dew point began to decrease. Explain why.

13. Why can you see your breath on a cold morning? Does the air temperature have to be below freezing for this to occur?

14. Explain why altocumulus clouds might be observed at 6400 m (21,000 ft) above the surface in Mexico City, Mexico, but never at that altitude above Fairbanks, Alaska.

15. The sky is overcast and it is raining. Explain how you could tell if the cloud above you is a nimbostratus or a cumulonimbus.

16. Suppose it is raining lightly from a deck of nimbostratus clouds. Beneath the clouds are small, ragged, puffy clouds that are moving rapidly with the wind. What would you call these clouds? How did they probably form?

17. You are sitting inside your house on a sunny afternoon. The shades are drawn and you look at the window and notice the sun disappears for about 10 seconds. The alternate light and dark period lasts for nearly 30 minutes. Are the clouds passing in front of the sun cirrocumulus, altocumulus, stratocumulus, or cumulus? Give a reasonable explanation for your answer.

PROBLEMS AND EXERCISES

1. The data in ▼ Table 5.6 below represent the dew-point temperature and expected minimum temperature near the ground for various clear winter mornings in a south-eastern city. Assume that the dew point remains constant throughout the night. Answer the following questions about the data.
   (a) On which morning would there be the greatest likelihood of observing visible frost? Explain why.
   (b) On which morning would frozen dew most likely form? Explain why.
   (c) On which morning would there be black frost with no sign of visible frost, dew, or frozen dew? Explain.
   (d) On which morning would you probably only observe dew on the ground? Explain why.

2. If a ceiling balloon rises at 120 m (about 400 ft) each minute, what is the ceiling of an overcast deck of stratus clouds 1500 m (about 5000 ft) thick if the balloon disappears into the clouds in 5 minutes?

3. Compare the visible satellite image (Fig. 5.35a) with the infrared image (Fig. 5.35b). With the aid of the infrared image, label on the visible image the regions of middle, high, and low clouds. On the enhanced infrared image (Fig. 5.36), label where the highest and thickest clouds appear to be located.

Visit the
Meteorology Resource Center
at
academic.cengage.com/login
for more assets, including questions for exploration, animations, videos, and more.

▼ TABLE 5.6

	MORNING 1	MORNING 2	MORNING 3	MORNING 4	MORNING 5
Dew-point temperature	2°C (35°F)	−7°C (20°F)	1°C (34°F)	−4°C (25°F)	3°C (38°F)
Expected minimum temperature	4°C (40°F)	−3°C (27°F)	0°C (32°F)	−4.5°C (24°F)	2°C (35°F)

A mass of moist, stable air gliding up and over Mt. Rainier condenses into a spectacular lenticular cloud.

© Jeffrey A. Schmidt

CHAPTER 6

Stability and Cloud Development

In July and August on the high desert the thunderstorms come. Mornings begin clear and dazzling bright, the sky as blue as the Virgin's cloak, unflawed by a trace of cloud in all that emptiness. . . . By noon, however, clouds begin to form over the mountains, coming it seems out of nowhere, out of nothing, a special creation.

The clouds multiply and merge, cumulonimbus piling up like whipped cream, like mashed potatoes, like sea foam, building upon one another into a second mountain range greater in magnitude than the terrestrial range below.

The massive forms jostle and grate, ions collide, and the sound of thunder is heard over the sun-drenched land. More clouds emerge from empty sky, anvil-headed giants with glints of lightning in their depths. An armada assembles and advances, floating on a plane of air that makes it appear, from below, as a fleet of ships must look to the fish in the sea.

Edward Abbey, *Desert Solitaire — A Season in the Wilderness*

CONTENTS

Clouds, spectacular features in the sky, add beauty and color to the natural landscape. Yet, clouds are important for nonaesthetic reasons, too. As they form, vast quantities of heat are released into the atmosphere. Clouds help regulate the earth's energy balance by reflecting and scattering solar radiation and by absorbing the earth's infrared energy. And, of course, without clouds there would be no precipitation. But clouds are also significant because they visually indicate the physical processes taking place in the atmosphere; to a trained observer, they are signposts in the sky. This chapter examines the atmospheric processes these signposts point to, the first of which is atmospheric stability.

Atmospheric Stability

Most clouds form as air rises and cools. Why does air rise on some occasions and not on others? And why do the size and shape of clouds vary so much when the air does rise? Let's see how knowing about the air's stability will help us to answer these questions.

When we speak of atmospheric stability, we are referring to a condition of equilibrium. For example, rock A resting in the depression in ● Fig. 6.1 is in *stable* equilibrium. If the rock is pushed up along either side of the hill and then let go, it will quickly return to its original position. On the other hand, rock B, resting on the top of the hill, is in a state of *unstable* equilibrium, as a slight push will set it moving away from its original position. Applying these concepts to the atmosphere, we can see that air is in stable equilibrium when, after being lifted or lowered, it tends to return to its original position—it resists upward and downward air motions. Air that is in unstable equilibrium will, when given a little push, move farther away from its original position—it favors vertical air currents.

To explore the behavior of rising and sinking air, we must first put some air in an imaginary thin elastic wrap. This small volume of air is referred to as a **parcel of air.*** Although the air parcel can expand and contract freely, it does not break apart,

---

*An air parcel is an imaginary body of air about the size of a large basketball. The concept of an air parcel is illustrated several places in the text, including Fig. 4.6, p. 93.

but remains as a single unit. At the same time, neither external air nor heat can mix with the air inside the parcel. The space occupied by the air molecules within the parcel defines the air density. The average speed of the molecules is directly related to the air temperature, and the molecules colliding against the parcel walls determine the air pressure inside.

At the earth's surface, the parcel has the same temperature and pressure as the air surrounding it. Suppose we lift the air parcel up into the atmosphere. We know from Chapter 1 that air pressure decreases with height. Consequently, the air pressure surrounding the parcel lowers. The lower pressure outside allows the air molecules inside to push the parcel walls outward, expanding the parcel. Because there is no other energy source, the air molecules inside must use some of their own energy to expand the parcel. This shows up as slower average molecular speeds, which result in a lower parcel temperature. If the parcel is lowered to the surface, it returns to a region where the surrounding air pressure is higher. The higher pressure squeezes (compresses) the parcel back into its original (smaller) volume. This squeezing increases the average speed of the air molecules and the parcel temperature rises. Hence, *a rising parcel of air expands and cools, while a sinking parcel is compressed and warms.*

If a parcel of air expands and cools, or compresses and warms, with no interchange of heat with its surroundings, this situation is called an **adiabatic process.** As long as the air in the parcel is unsaturated (the relative humidity is less than 100 percent), the rate of adiabatic cooling or warming remains constant. This rate of heating or cooling is about 10°C for every 1000 m of change in elevation (5.5°F per 1000 ft) and applies only to unsaturated air. For this reason, it is called the **dry adiabatic rate*** (see ● Fig. 6.2).

As the rising air cools, its relative humidity increases as the air temperature approaches the dew-point temperature. If the rising air cools to its dew-point temperature, the relative humidity becomes 100 percent. Further lifting results in condensation, a cloud forms, and latent heat is released inside the rising air parcel. Because the heat added during condensation offsets some of the cooling due to expansion, the air no longer cools at the dry adiabatic rate but at a lesser rate called the **moist adiabatic rate.** If a saturated parcel containing water droplets were to sink, it would compress and warm at the moist adiabatic rate because evaporation of the liquid droplets would offset the rate of compressional warming. Hence, the rate at which rising or sinking saturated air changes temperature—the moist adiabatic rate—is less than the dry adiabatic rate.†

---

*For aviation purposes, the dry adiabatic rate is sometimes expressed as 3°C per 1000 ft.

†Consider an air parcel initially at rest. Suppose the air parcel rises and cools, and a cloud forms. Further suppose that no precipitation (rain or snow) falls from the cloud (leaves the parcel). If the parcel should descend to its original level the latent heat released inside the parcel during condensation will be the same amount that is absorbed as the cloud evaporates. This process is called a *reversible adiabatic process.* If, on the other hand, rain or snow falls from the cloud during uplift and leaves the parcel, the sinking parcel will not recover during evaporation the same amount of latent heat released during condensation because the parcel's water content is lower. This process is known as an *irreversible pseudoadiabatic process.*

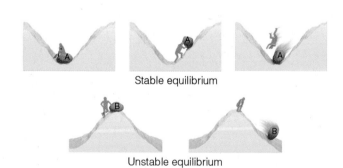

Stable equilibrium

Unstable equilibrium

● FIGURE 6.1 When rock A is disturbed, it will return to its original position; rock B, however, will accelerate away from its original position.

Unlike the dry adiabatic rate, the moist adiabatic rate is not constant, but varies greatly with temperature and, hence, with moisture content—as warm saturated air produces more liquid water than cold saturated air. The added condensation in warm, saturated air liberates more latent heat. Consequently, the moist adiabatic rate is much less than the dry adiabatic rate when the rising air is warm; however, the two rates are nearly the same when the rising air is very cold (see ▼ Table 6.1). Although the moist adiabatic rate does vary, to make the numbers easy to deal with, we will use an average of 6°C per 1000 m (3.3°F per 1000 ft) in most of our examples and calculations.

Determining Stability

We determine the stability of the air by comparing the temperature of a rising parcel to that of its surroundings. If the rising air is colder than its environment, it will be more dense* (heavier) and tend to sink back to its original level. In this case, the air is *stable* because it resists upward movement. If the rising air is warmer and, therefore, less dense (lighter) than the surrounding air, it will continue to rise until it reaches the same temperature as its environment. This is an example of *unstable* air. To figure out the air's stability, we need to measure the temperature both of the rising air and of its environment at various levels above the earth.

A STABLE ATMOSPHERE Suppose we release a balloon-borne instrument—a radiosonde (see Fig. 4, p. 16)—and it sends back temperature data as shown in ● Fig. 6.3. (Such a vertical profile of temperature is called a *sounding*.) We measure the air temperature in the vertical and find that it decreases by 4°C for every 1000 m (2°F per 1000 ft). Remember from Chapter 1 that the rate at which the air temperature changes with elevation is called the *lapse rate*. Because this is the rate at which the air temperature surrounding us will be changing if we were to climb upward into the atmosphere, we will refer to it as the **environmental lapse rate.** Now suppose in Fig. 6.3a that a parcel of unsaturated air with a temperature of 30°C is lifted from the surface. As it rises, it cools at the dry adiabatic rate (10°C per 1000 m), and the temperature inside the parcel at 1000 meters would be 20°C, or 6°C lower than the air surrounding it. Look at Fig. 6.3a closely and notice that, as the air parcel rises higher, the temperature difference between it and the surrounding air becomes even greater. Even if the parcel is initially saturated (see Fig. 6.3b), it will cool at the moist rate—6°C per 1000 m—and will be colder than its environment at all levels. In both cases, the rising air is colder and heavier than the air surrounding it. In this example, the atmosphere is **absolutely stable.** *The atmo-*

---

*When, at the same level in the atmosphere, we compare parcels of air that are equal in size but vary in temperature, we find that cold air parcels are more dense than warm air parcels; that is, in the cold parcel, there are more molecules that are crowded closer together.

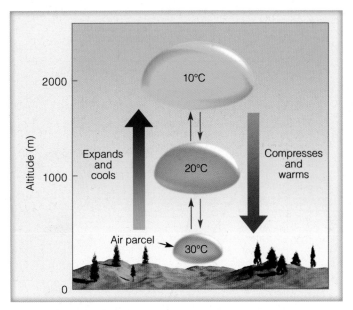

● FIGURE 6.2 The dry adiabatic rate. As long as the air parcel remains unsaturated, it expands and cools by 10°C per 1000 m; the sinking parcel compresses and warms by 10°C per 1000 m.

▼ TABLE 6.1 The Moist Adiabatic Rate for Different Temperatures and Pressures in °C/1000 m and °F/1000 ft

Pressure (mb)	TEMPERATURE (°C)					TEMPERATURE (°F)				
	−40	−20	0	20	40	−40	−5	30	65	100
1000	9.5	8.6	6.4	4.3	3.0	5.2	4.7	3.5	2.4	1.6
800	9.4	8.3	6.0	3.9		5.2	4.6	3.3	2.2	
600	9.3	7.9	5.4			5.1	4.4	3.0		
400	9.1	7.3				5.0	4.0			
200	8.6					4.7				

sphere is always absolutely stable when the environmental lapse rate is less than the moist adiabatic rate.

Since air in an absolutely stable atmosphere strongly resists upward vertical motion, it will, *if forced to rise,* tend to spread out horizontally. If clouds form in this rising air, they, too, will spread horizontally in relatively thin layers and usually have flat tops and bases. We might expect to see clouds—such as cirrostratus, altostratus, nimbostratus, or stratus—forming in stable air.

What conditions are necessary to bring about a stable atmosphere? As we have just seen, the atmosphere is stable when the environmental lapse rate is small; that is, when the difference in temperature between the surface air and the air aloft is relatively small. Consequently, the atmosphere tends to become more stable—that is, it stabilizes—as the air aloft warms or the surface air cools. If the air aloft is being replaced by warmer air (warm advection), and the surface air is not changing appreciably, the environmental lapse rate decreases and the atmosphere becomes more stable. Similarly, the envi-

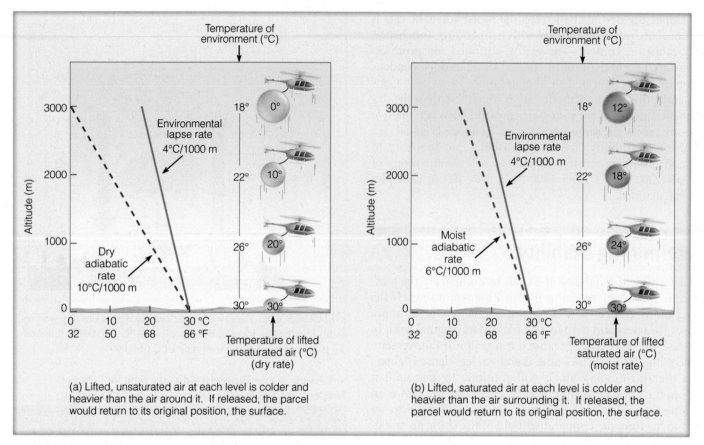

• FIGURE 6.3 An absolutely stable atmosphere occurs when the environmental lapse rate is less than the moist adiabatic rate. In a stable atmosphere, a rising air parcel is colder and more dense than the air surrounding it, and, if given the chance, it will return to its original position.

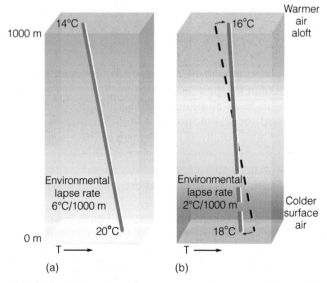

• FIGURE 6.4 The initial environmental lapse rate in diagram (a) will become more stable (stabilize) as the air aloft warms and the surface air cools, as illustrated in diagram (b).

ronmental lapse rate decreases and the atmosphere becomes more stable when the lower layer cools (see • Fig. 6.4). The *cooling* of the *surface air* may be due to:

1. nighttime radiational cooling of the surface
2. an influx of cold surface air brought in by the wind (cold advection)
3. air moving over a cold surface

Consequently, on any given day, the atmosphere is most stable in the early morning around sunrise, when the lowest surface air temperature is recorded. If the surface air becomes saturated in a stable atmosphere, a persistent layer of haze or fog may form (see • Fig. 6.5).

Another way the atmosphere becomes more stable is when an entire layer of air sinks. For example, if a layer of unsaturated air over 1000 m thick and covering a large area subsides, the entire layer will warm by adiabatic compression. As the layer subsides, it becomes compressed by the weight of the atmosphere and shrinks vertically. The upper part of the layer sinks farther, and, hence, warms more than the bottom part. This phenomenon is illustrated in • Fig. 6.6. After subsiding, the top of the layer is actually warmer than the bottom, and an inversion* is formed. Inversions that form as air

---

*Recall from Chapter 3 that an inversion represents an atmospheric condition where the air becomes warmer with height.

© J. L. Medeiros

● FIGURE 6.5 Cold surface air, on this morning, produces a stable atmosphere that inhibits vertical air motions and allows the fog and haze to linger close to the ground.

WEATHER WATCH

If you take a walk on a bitter cold, yet clear, winter morning, when the air is calm and a strong subsidence inversion exists, the air aloft—thousands of meters above you—may be more than 17°C (30°F) warmer than the air at the surface.

slowly sinks over a large area are called **subsidence inversions.** They sometimes occur at the surface, but more frequently, they are observed aloft and are often associated with large high-pressure areas because of the sinking air motions associated with these systems.

An inversion represents an atmosphere that is absolutely stable. Why? Within the inversion, warm air overlies cold air, and, if air rises into the inversion, it is becoming colder, while the air around it is getting warmer. Obviously, the colder air would tend to sink. Inversions, therefore, act as lids on vertical air motion. When an inversion exists near the ground, stratus clouds, fog, haze, and pollutants are all kept close to

the surface. In fact, as we will see in Chapter 18, most air pollution episodes occur with subsidence inversions. (For additional information on subsidence inversions, read the Focus section on p. 150.)

Before we turn our attention to unstable air, let's first examine a condition known as **neutral stability.** If the lapse rate is exactly equal to the dry adiabatic rate, rising or sinking unsaturated air will cool or warm at the same rate as the air around it. At each level, it would have the same temperature and density as the surrounding air. Because this air tends neither to continue rising nor sinking, the atmosphere is said to be neutrally stable. For saturated air, *neutral stability* exists when the environmental lapse rate is equal to the moist adiabatic rate.

AN UNSTABLE ATMOSPHERE   Suppose a radiosonde sends back the temperatures above the earth as plotted in ● Fig. 6.7a. Once again, we determine the atmosphere's stability by comparing the environmental lapse rate to the moist and dry adiabatic rates. In this case, the environmental lapse rate is

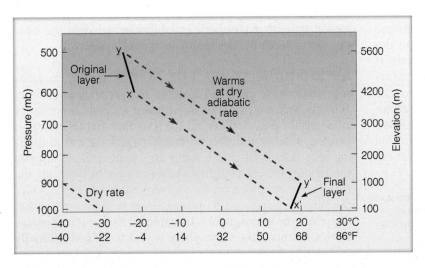

● FIGURE 6.6 The layer x–y is initially 1400 m thick. If the entire layer slowly subsides, it shrinks in the more-dense air near the surface. As a result of the shrinking, the top of the layer warms more than the bottom, and the entire layer (x′–y′) becomes more stable, and in this example forms an inversion.

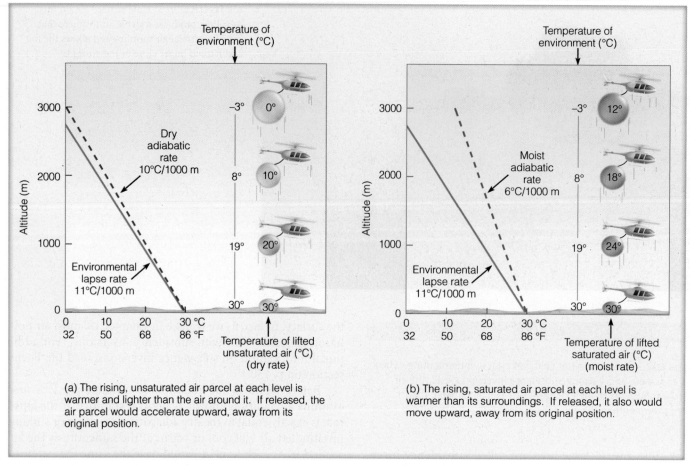

• FIGURE 6.7 An absolutely unstable atmosphere occurs when the environmental lapse rate is greater than the dry adiabatic rate. In an unstable atmosphere, a rising air parcel will continue to rise because it is warmer and less dense than the air surrounding it.

11°C per 1000 m (6°F per 1000 ft). A rising parcel of unsaturated surface air will cool at the dry adiabatic rate. Because the dry adiabatic rate is less than the environmental lapse rate, the parcel will be warmer than the surrounding air and will continue to rise, constantly moving upward, away from its original position. The atmosphere is unstable. Of course, a parcel of saturated air cooling at the lower moist adiabatic rate will be even warmer than the air around it (see • Fig. 6.7b). In both cases, the air parcels, once they start upward, will continue to rise on their own because the rising air parcels are warmer and less dense than the air around them. The atmosphere in this example is said to be **absolutely unstable.*** *Absolute instability results when the environmental lapse rate is greater than the dry adiabatic rate.*

It should be noted, however, that deep layers in the atmosphere are seldom, if ever, absolutely unstable. Absolute instability is usually limited to a very shallow layer near the ground on hot, sunny days. Here the environmental lapse rate can

exceed the dry adiabatic rate, and the lapse rate is called *superadiabatic.* On rare occasions when the environmental lapse rate exceeds about 3.4°C per 100 m (the *autoconvective lapse rate*), convection becomes spontaneous, resulting in the automatic overturning of the air.

So far, we have seen that the atmosphere is absolutely stable when the environmental lapse rate is less than the moist adiabatic rate and absolutely unstable when the environmental lapse rate is greater than the dry adiabatic rate. However, a typical type of atmospheric instability exists when the lapse rate lies between the moist and dry adiabatic rates.

A CONDITIONALLY UNSTABLE ATMOSPHERE   The environmental lapse rate in • Fig. 6.8 is 7°C per 1000 m (4°F per 1000 ft). When a parcel of unsaturated air rises, it cools dry adiabatically and is colder at each level than the air around it (see Fig. 6.8a). It will, therefore, tend to sink back to its original level because it is in a stable atmosphere. Now, suppose the rising parcel is saturated. As we can see in Fig. 6.8b, the rising air is warmer than its environment at each level. Once the parcel is given a push upward, it will tend to move in that direction; the atmosphere is unstable for the saturated parcel. In this example, the atmosphere is said to be **conditionally**

---

*When an air parcel is warmer (less dense) than the air surrounding it, there is an upward-directed force (called *buoyant force*) acting on it. The warmer the air parcel compared to its surroundings, the greater the buoyant force and the more rapidly the air rises.

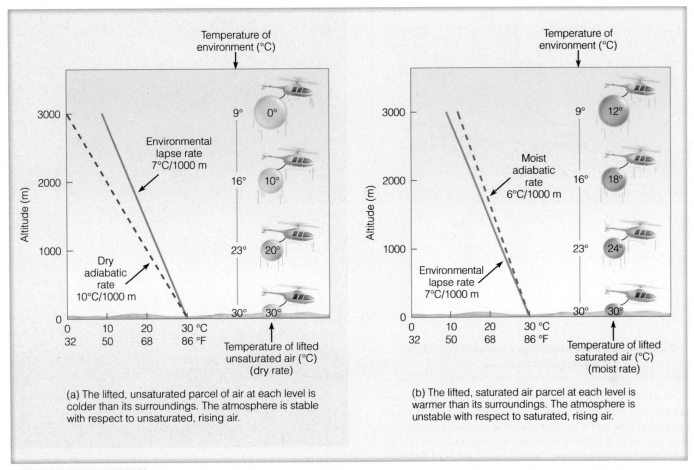

(a) The lifted, unsaturated parcel of air at each level is colder than its surroundings. The atmosphere is stable with respect to unsaturated, rising air.

(b) The lifted, saturated air parcel at each level is warmer than its surroundings. The atmosphere is unstable with respect to saturated, rising air.

ACTIVE FIGURE 6.8 Conditionally unstable atmosphere. The atmosphere is *stable* if the rising air is *unsaturated* (a), but *unstable* if the rising air is *saturated* (b). A conditionally unstable atmosphere occurs when the environmental lapse rate is between the moist adiabatic rate and the dry adiabatic rate. Visit the Meteorology Resource Center to view this and other active figures at academic.cengage.com/login

unstable. This type of stability depends upon whether or not the rising air is saturated. When the rising parcel of air is unsaturated, the atmosphere is stable; when the parcel of air is saturated, the atmosphere is unstable. Conditional instability means that, if unsaturated air could be lifted to a level where it becomes saturated, instability would result.

Conditional instability occurs whenever the environmental lapse rate is between the moist adiabatic rate and the dry adiabatic rate. Recall from Chapter 1 that the average lapse rate in the troposphere is about 6.5°C per 1000 m (3.6°F per 1000 ft). Since this value lies between the dry adiabatic rate and the average moist rate, *the atmosphere is ordinarily in a state of conditional instability.* (● Figure 6.9 summarizes the concept of unstable, conditionally unstable, and stable atmospheres.)

> View this concept in action on the Meteorology Resource Center at academic.cengage.com/login

CAUSES OF INSTABILITY What causes the atmosphere to become more unstable? The atmosphere becomes more unstable as the environmental lapse rate steepens; that is, as the air temperature drops rapidly with increasing height. This circumstance may be brought on by either air aloft becoming colder or the surface air becoming warmer (see ● Fig. 6.10).

The *cooling of the air aloft* may be due to:

1. winds bringing in colder air (cold advection)
2. clouds (or the air) emitting infrared radiation to space (radiational cooling)

The *warming of the surface air* may be due to:

1. daytime solar heating of the surface
2. an influx of warm air brought in by the wind (warm advection)
3. air moving over a warm surface

The combination of cold air aloft and warm surface air can produce a steep lapse rate and atmospheric instability (see ● Fig. 6.11).

At this point, we can see that the stability of the atmosphere changes during the course of a day. In clear, calm weather around sunrise, surface air is normally colder than the air above it, a radiation inversion exists, and the atmosphere is quite stable as indicated by smoke or haze lingering close to the ground. As the day progresses, sunlight warms the

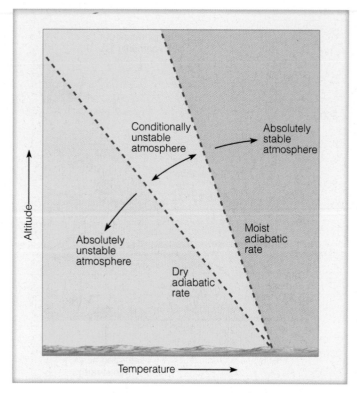

● FIGURE 6.9 When the environmental lapse rate is greater than the dry adiabatic rate, the atmosphere is absolutely unstable. When the environmental lapse rate is less than the moist adiabatic rate, the atmosphere is absolutely stable. And when the environmental lapse rate lies *between* the dry adiabatic rate and the moist adiabatic rate (shaded green area), the atmosphere is conditionally unstable.

● FIGURE 6.11 The warmth from this forest fire in the northern Sierra Nevada foothills heats the air, causing instability near the surface. Warm, less-dense air (and smoke) bubbles upward, expanding and cooling as it rises. Eventually the rising air cools to its dew point, condensation begins, and a cumulus cloud forms.

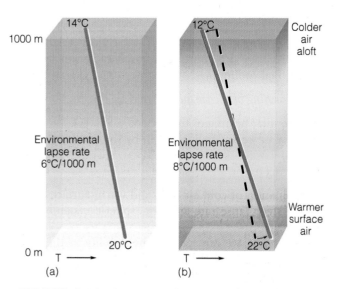

● FIGURE 6.10 The initial environmental lapse rate in diagram (a) will become more unstable (that is, destabilize) as the air aloft cools and the surface air warms, as illustrated in diagram (b).

surface and the surface warms the air above. As the air temperature near the ground increases, the lower atmosphere gradually becomes more unstable—that is, it *destabilizes*—with maximum instability usually occurring during the hottest part of the day.

Up to now, we have seen that a layer of air may become more unstable by either cooling the air aloft or warming the air at the surface. A layer of air may also be made more unstable by either mixing or lifting. Let's look at mixing first. In ● Fig. 6.12, the environmental lapse rate before mixing is less than the moist rate, and the layer is stable (*A*). Now, suppose the air in the layer is mixed either by convection or by wind-induced turbulent eddies. Air is cooled adiabatically as it is brought up from below and heated adiabatically as it is mixed downward. The up and down motion in the layer redistributes the air in such a way that the temperature at the top of the layer decreases, while, at the base, it increases. This steepens the environmental lapse rate and makes the layer more unstable. If this mixing continues for some time, and the air remains unsaturated, the vertical temperature distribution will eventually be equal to the dry adiabatic rate (*B*).

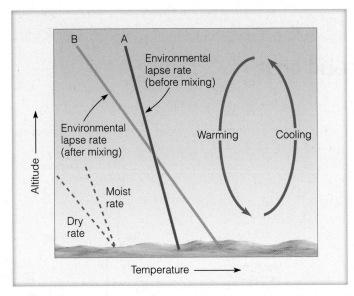

● FIGURE 6.12 Mixing tends to steepen the lapse rate. Rising, cooling air lowers the temperature toward the top of the layer, while sinking, warming air increases the temperature near the bottom.

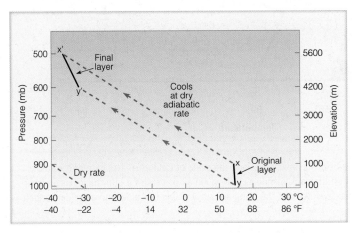

● FIGURE 6.13 The lifting of an entire layer of air tends to increase the instability of the layer. The initial stable layer (x–y) after lifting is now a conditionally unstable layer (x′–y′).

Just as lowering an entire layer of air makes it more stable, the lifting of a layer makes it more unstable. In ● Fig. 6.13, the air lying between 1000 and 900 mb is initially absolutely stable since the environmental lapse rate of layer *x–y* is less than the moist adiabatic rate. The layer is lifted, and, as it rises, the rapid decrease in air density aloft causes the layer to stretch out vertically. If the layer remains unsaturated, the entire layer cools at the dry adiabatic rate. Due to the stretching effect, however, the top of the layer cools more than the bottom. This steepens the environmental lapse rate. Note that the absolutely stable layer *x–y*, after rising, has become conditionally unstable between 500 and 600 mb (layer *x′–y′*).

A very stable air layer may be converted into an absolutely unstable layer when the lower portion of a layer is

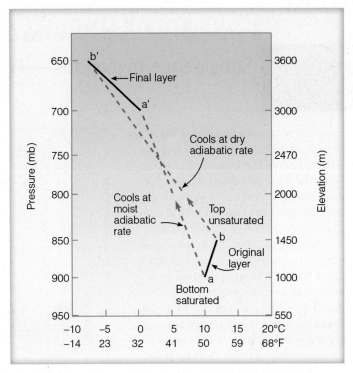

● FIGURE 6.14 Convective instability. The layer a–b is initially absolutely stable. The lower part of the layer is saturated, and the upper part is "dry." After lifting, the entire layer *(a′–b′)* becomes absolutely unstable.

moist and the upper portion is quite dry. In ● Fig. 6.14, the inversion layer between 900 and 850 mb is absolutely stable. Suppose the bottom of the layer is saturated while the air at the top is unsaturated. If the layer is forced to rise, even a little, the upper portion of the layer cools at the dry adiabatic rate and grows cold quite rapidly, while the air near the bottom cools more slowly at the moist adiabatic rate. It does not take much lifting before the upper part of the layer is much colder than the bottom part; the environmental lapse rate steepens and the entire layer becomes absolutely unstable (layer *a′–b′*). The potential instability, brought about by the lifting of a stable layer whose surface is humid and whose top is "dry," is called *convective instability*. Convective instability is associated with the development of severe storms, such as thunderstorms and tornadoes, which are investigated more thoroughly in Chapter 14.

WEATHER WATCH

Ever hear of a pyrocumulus? No, it's not a fiery-red cumulus cloud, but a cloud that often forms above a forest fire. Forest fires generate atmospheric instability by heating the air near the surface. The hot, rising air above the fire contains tons of tiny smoke particles that act as cloud condensation nuclei. As the air rises and cools, water vapor will often condense onto the nuclei, producing a cumuliform cloud directly above the fire called a *pyrocumulus*.

FOCUS ON A SPECIAL TOPIC

Subsidence Inversions—Put a Lid on It

Figure 1 shows a typical summertime vertical profile of air temperature and dew point measured with a radiosonde near the coast of California. Notice that the air temperature decreases from the surface up to an altitude of about 300 m (1000 ft). Notice also that, where the air temperature reaches the dew point, a cloud forms.

Above about 300 m, the air temperature increases rapidly up to an altitude near 900 m (about 3000 ft). This region of increasing air temperature with increasing height marks the region of the subsidence inversion. Within the inversion, air from aloft warms by compression. The sinking air at the top of the inversion is not only warm (about 24°C or 75°F) but also dry with a low relative humidity, as indicated by the large spread between air temperature and dew point. The subsiding air, which does not reach the surface, is associated with a large high-pressure area, located to the west of California.

Immediately below the base of the inversion lies cool, moist air. The cool air is unable to penetrate the inversion because a lifted parcel of cool, marine air within the inversion would be much colder and heavier than the air surrounding it. Since the colder air parcel would fall back to its original position, the atmosphere is absolutely stable within the inversion. The subsidence inversion, therefore, acts as a lid on the air below, preventing the air from mixing vertically into the inversion. And so the marine air with its pollution and clouds is confined to a relatively shallow region near the earth's sur-

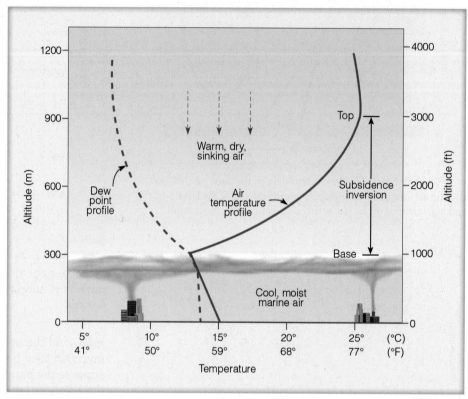

● FIGURE 1 A strong subsidence inversion along the coast of California. The base of the stable inversion acts as a cap or lid on the cool, marine air below. An air parcel rising into the inversion layer would sink back to its original level because the rising air parcel would be colder and more dense than the air surrounding it.

face. It is this trapping of air near the surface, associated with a strong subsidence inversion, that helps to make West Coast cities such as Los Angeles very polluted.

BRIEF REVIEW

Up to now, we have looked briefly at stability as it relates to cloud development. The next section describes how atmospheric stability influences the physical mechanisms responsible for the development of individual cloud types. However, before going on, here is a brief review of some of the facts and concepts concerning stability.

● The air temperature in a rising parcel of *unsaturated* air decreases at the dry adiabatic rate, whereas the air temperature in a rising parcel of *saturated* air decreases at the moist adiabatic rate.

● The dry adiabatic rate and moist adiabatic rate of cooling are different due to the fact that latent heat is released in a rising parcel of saturated air.

● In a *stable atmosphere*, a lifted parcel of air will be colder (heavier) than the air surrounding it. Because of this fact, the lifted parcel will tend to sink back to its original position.

● In an *unstable atmosphere*, a lifted parcel of air will be warmer (lighter) than the air surrounding it, and thus will continue to rise upward, away from its original position.

● The atmosphere becomes more stable (stabilizes) as the surface air cools, the air aloft warms, or a layer of air sinks (subsides) over a vast area.

● The atmosphere becomes more unstable (destabilizes) as the surface air warms, the air aloft cools, or a layer of air is either mixed or lifted.

- A conditionally unstable atmosphere exists when the environmental lapse rate is between the moist adiabatic rate and the dry adiabatic rate.
- The atmosphere is normally most stable in the early morning and most unstable in the afternoon.
- Layered clouds tend to form in a stable atmosphere, whereas cumuliform clouds tend to form in a conditionally unstable atmosphere.

Cloud Development

We know that most clouds form as air rises, cools, and condenses. Since air normally needs a "trigger" to start it moving upward, what is it that causes the air to rise so that clouds are able to form? Basically, the following mechanisms are responsible for the development of the majority of clouds we observe:

1. surface heating and free convection
2. uplift along topography
3. widespread ascent due to convergence of surface air
4. uplift along weather fronts (see ● Fig. 6.15)

The first mechanism that can cause the air to rise is convection. Although we briefly looked at convection in Chapter 2 when we examined rising thermals and how they transfer heat upward into the atmosphere, we will now look at convection from a slightly different perspective—how rising thermals are able to form into cumulus clouds.

CONVECTION AND CLOUDS Some areas of the earth's surface are better absorbers of sunlight than others and, therefore, heat up more quickly. The air in contact with these "hot spots" becomes warmer than its surroundings. A hot "bubble" of air—a *thermal*—breaks away from the warm surface and rises, expanding and cooling as it ascends. As the thermal rises, it mixes with the cooler, drier air around it and gradually loses its identity. Its upward movement now slows. Frequently, be-

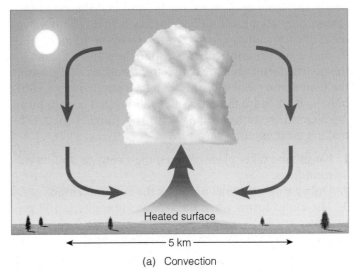

(a) Convection

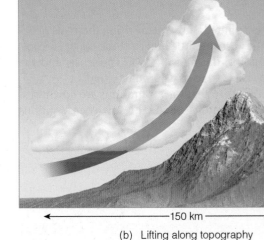

(b) Lifting along topography

(c) Convergence of air

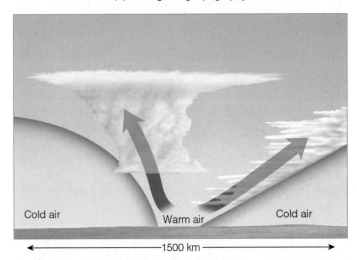

(d) Lifting along weather fronts

● FIGURE 6.15 The primary ways clouds form: (a) surface heating and convection; (b) forced lifting along topographic barriers; (c) convergence of surface air; (d) forced lifting along weather fronts.

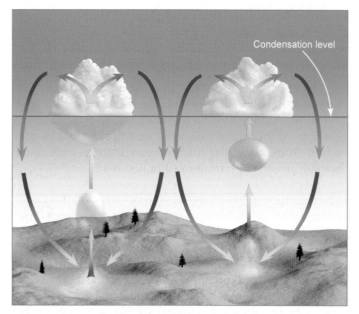

● FIGURE 6.16 Cumulus clouds form as hot, invisible air bubbles detach themselves from the surface, then rise and cool to the condensation level. Below and within the cumulus clouds, the air is rising. Around the cloud, the air is sinking.

fore it is completely diluted, subsequent rising thermals penetrate it and help the air rise a little higher. If the rising air cools to its saturation point, the moisture will condense, and the thermal becomes visible to us as a cumulus cloud.

Observe in ● Fig. 6.16 that the air motions are downward on the outside of the cumulus cloud. The downward motions

● FIGURE 6.17 Cumulus clouds building on a warm summer afternoon. Each cloud represents a region where thermals are rising from the surface. The clear areas between the clouds are regions where the air is sinking.

are caused in part by evaporation around the outer edge of the cloud, which cools the air, making it heavy. Another reason for the downward motion is the completion of the convection current started by the thermal. Cool air slowly descends to replace the rising warm air. Therefore, we have rising air in the cloud and sinking air around it. Since subsiding air greatly inhibits the growth of thermals beneath it, small cumulus clouds usually have a great deal of blue sky between them (see ● Fig. 6.17).

As the cumulus clouds grow, they shade the ground from the sun. This, of course, cuts off surface heating and upward convection. Without the continual supply of rising air, the cloud begins to erode as its droplets evaporate. Unlike the sharp outline of a growing cumulus, the cloud now has indistinct edges, with cloud fragments extending from its sides. As the cloud dissipates (or moves along with the wind), surface heating begins again and regenerates another thermal, which becomes a new cumulus. This is why you often see cumulus clouds form, gradually disappear, then reform in the same spot.

Suppose that it is a warm, humid summer afternoon and the sky is full of cumulus clouds. The cloud bases are all at nearly the same level above the ground and the cloud tops extend only about a thousand meters higher. The development of these clouds depends primarily upon the air's stability and moisture content. To illustrate how these factors influence the formation of a convective cloud, we will examine the temperature and moisture characteristics within a rising bubble of air. Since the actual air motions that go into forming a cloud are rather complex, we will simplify matters by making these assumptions:

1. No mixing takes place between the rising air and its surroundings.
2. Only a single thermal produces the cumulus cloud.
3. The cloud forms when the relative humidity is 100 percent.
4. The rising air in the cloud remains saturated.

The environmental lapse rate on this particular day is plotted in ● Fig. 6.18 and is represented as a dark gray line on the far left of the illustration. The changing environmental air temperature indicates changes in the atmosphere's stability. The environmental lapse rate in layer A is greater than the dry adiabatic rate, so the layer is absolutely unstable. The air layers above it—layer B and layer C—are both absolutely stable since the environmental lapse rate in each layer is less than the moist adiabatic rate. However, the overall environmental lapse rate from the surface up to the base of the inversion (2000 m) is 7.5°C per 1000 m (4.1°F per 1000 ft), which indicates a conditionally unstable atmosphere.

Now, suppose that a warm bubble of air with an air temperature and dew-point temperature of 35°C and 27°C (95°F and 80.5°F), respectively, breaks away from the surface and begins to rise (which is illustrated in the middle of Fig. 6.18). Notice that, a short distance above the ground, the air inside the bubble is warmer than the air around it, so it is buoyant and rises freely. This level in the atmosphere where the rising

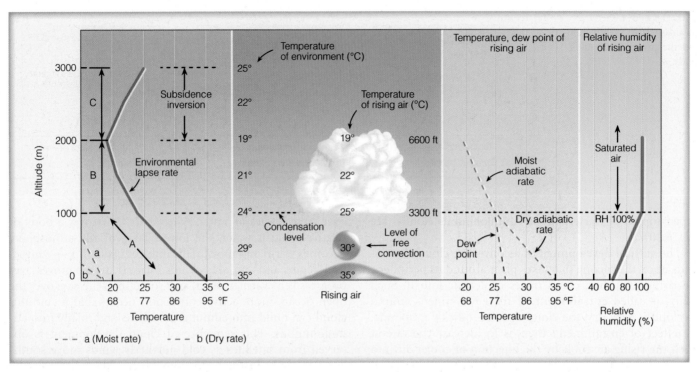

● FIGURE 6.18 The development of a cumulus cloud.

air becomes warmer than the surrounding air is called the *level of free convection.* The rising bubble will continue to rise as long as it is warmer than the air surrounding it.

The rising air cools at the dry adiabatic rate and the dew point falls, but not as rapidly.* The rate at which the dew point drops varies with the moisture content of the rising air, but an approximation of 2°C per 1000 m (1°F per 1000 ft) is commonly used. So, as unsaturated rising air cools, the air temperature and dew point approach each other at the rate of 8°C per 1000 m (4.5°F per 1000 ft). This process causes an increase in the air's relative humidity (illustrated in the far right-hand side of Fig. 6.18, by the dark green line).

At an elevation of 1000 m (3300 ft), the air has cooled to the dew point, the relative humidity is 100 percent, condensation begins, and a cloud forms. The elevation where the cloud forms is called the **condensation level.** Above the condensation level the rising air is saturated and cools at the moist adiabatic rate. Condensation continues to occur, and since water vapor is transforming into liquid cloud droplets, the dew point within the cloud now drops more rapidly with increasing height than before. The air remains saturated as both the air temperature and dew point decrease at the moist adiabatic rate (illustrated in the area of Fig. 6.18 shaded tan).

Notice that inside the cloud the rising air remains warmer than the environment and continues its spontaneous rise upward through layer B. The top of the bulging cloud at 2000 m (about 6600 ft) represents the top of the rising air, which has now cooled to a temperature equal to its surroundings. The air would have a difficult time rising much above this level because of the stable subsidence inversion directly above it. The subsidence inversion, associated with the downward air motions of a high-pressure system, prevents the clouds from building very high above their bases. Hence, an afternoon sky full of flat-base cumuli with little vertical growth indicates fair weather. (Recall from Chapter 5 that the proper name of these fair-weather cumulus clouds is *cumulus humilis.*)

As we can see, the stability of the air above the condensation level plays a major role in determining the vertical growth of a cumulus cloud. Notice in ● Fig. 6.19 that, when a deep stable layer begins a short distance above the cloud base, only cumulus humilis are able to form. If a deep conditionally unstable layer exists above the cloud base, cumulus congestus are likely to grow, with billowing cauliflowerlike tops. When the conditionally unstable layer is extremely deep—usually greater than 4 km (2.5 mi)—the cumulus congestus may even develop into a cumulonimbus.

Seldom do cumulonimbus clouds extend very far above the tropopause. The stratosphere is quite stable, so once a cloud penetrates the tropopause, it usually stops growing vertically and spreads horizontally. The low temperature at this altitude produces ice crystals in the upper section of the cloud. In the middle latitudes, high winds near the tropopause blow the ice crystals laterally, producing the flat anvil-

---

*The decrease in dew-point temperature is caused by the rapid decrease in air pressure within the rising air. Since the dew point is directly related to the actual vapor pressure of the rising air, a decrease in total air pressure causes a corresponding decrease in vapor pressure and, hence, a lowering of the dew-point temperature.

● FIGURE 6.19
The air's stability greatly influences the growth of cumulus clouds.

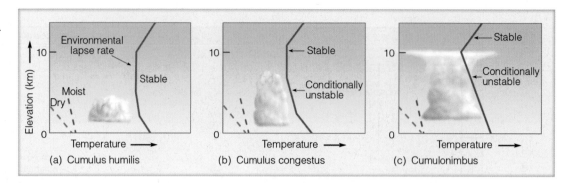

(a) Cumulus humilis   (b) Cumulus congestus   (c) Cumulonimbus

shaped top so characteristic of cumulonimbus clouds (see ● Fig. 6.20).

The vertical development of a convective cloud also depends upon the mixing that takes place around its periphery. The rising, churning cloud mixes cooler air into it. Such mixing is called **entrainment.** If the environment around the cloud is very dry, the cloud droplets quickly evaporate. The effect of entrainment, then, is to increase the rate at which the rising air cools by the injection of cooler air into the cloud and the subsequent evaporation of the cloud droplets. If the rate of cooling approaches the dry adiabatic rate, the air stops rising and the cloud no longer builds, even though the lapse rate may indicate a conditionally unstable atmosphere.

Up to now, we have looked at convection over land. Convection and the development of cumulus clouds also occur over large bodies of water. As cool air flows over a body of relatively warm water, the lowest layer of the atmosphere becomes warm and moist. This induces instability—convection begins and cumulus clouds form. If the air moves over progressively warmer water, as is sometimes the case over the open ocean, more active convection occurs and a cumulus cloud can build into cumulus congestus and finally into cumulonimbus. This sequence of cloud development is observed from satellites as cold northerly winds move southward over the northern portions of the Atlantic and Pacific oceans (see ● Fig. 6.21).

Once a convective cloud forms, stability, humidity, and entrainment all play a part in its vertical development. The level at which the cloud initially forms, however, is determined primarily by the surface temperature and moisture content of the original thermals. (The Focus section on p. 155

● FIGURE 6.20 Cumulus clouds developing into thunderstorms in a conditionally unstable atmosphere over the Great Plains. Notice that, in the distance, the cumulonimbus with the anvil top has reached the stable part of the atmosphere.

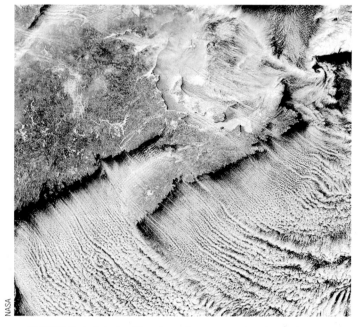

● FIGURE 6.21 Satellite view of stratocumulus clouds forming in rows over the Atlantic Ocean as cold, dry arctic air sweeps over Canada, then out over warmer water. Notice that the clouds are absent over the landmass and directly along the coast, but form and gradually thicken as the surface air warms and destabilizes farther offshore.

FOCUS ON AN OBSERVATION

Determining Convective Cloud Bases

The bases of cumulus clouds that form by convection on warm, sunny afternoons can be estimated quite easily when the surface air temperature and dew point are known. If the air is not too windy, we can assume that entrainment of air will not change the characteristics of a rising thermal. Since the rising air cools at the dry adiabatic rate of about 10°C per 1000 m, and the dew point drops at about 2°C per 1000 m, the air temperature and dew point approach each other at the rate of 8°C for every 1000 m of rise. Rising surface air with an air temperature and dew point spread of 8°C would produce saturation and a cloud at an elevation of 1000 m. Put another way, a 1°C difference between the surface air temperature and the dew point produces a cloud base at 125 m. Therefore, by finding the difference between surface air temperature (T) and dew point (T_d), and multiplying this value by 125, we can estimate the base of the convective cloud forming overhead, as

$$H_{meter} = 125 \ (T - T_d),* \qquad (1)$$

where H is the height of the base of the cumulus cloud in meters above the surface, with both T and T_d measured in degrees Celsius. If T and T_d are in °F, H can be calculated with the formula

$$H_{feet} = 228 \ (T - T_d). \qquad (2)$$

To illustrate the use of formula (1), let's determine the base of the cumulus cloud in Fig. 6.18. Recall that the surface air temperature and dew point were 35°C and 27°C, respectively. The difference, $T - T_d$, is 8°C. This value multiplied by 125 gives us a cumulus cloud with a base at 1000 m above the ground. This agrees with the condensation level we originally calculated.

*The formula works best when the air is well mixed from the surface up to the cloud base, such as in the afternoon on a sunny day. The formula does not work well at night or in the early morning.

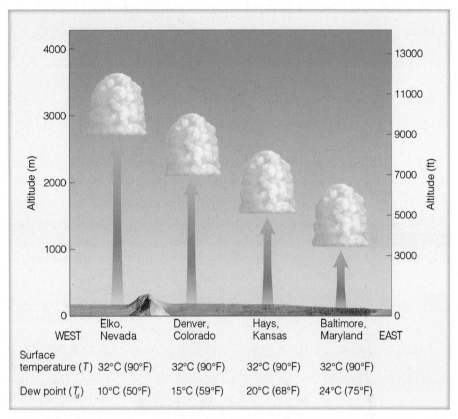

● FIGURE 2 During the summer, cumulus cloud bases typically increase in elevation above the ground as one moves westward into the drier air of the Central Plains.

	Elko, Nevada	Denver, Colorado	Hays, Kansas	Baltimore, Maryland
Surface temperature (T)	32°C (90°F)	32°C (90°F)	32°C (90°F)	32°C (90°F)
Dew point (T_d)	10°C (50°F)	15°C (59°F)	20°C (68°F)	24°C (75°F)

Along the East Coast in summer, when the air is warm and muggy, the separation between air temperature and dew point may be smaller than 9°C (16°F). The bases of afternoon cumulus clouds over cities, such as Philadelphia and Baltimore, are typically about 1000 m (3300 ft) above the ground (see Fig. 2). Farther west, in the Central Plains, where the air is drier and the spread between surface air temperature and dew point is greater, the cloud bases are higher. For example, west of Salina, Kansas, the cumulus cloud bases are generally greater than 1500 m (about 5000 ft) above the surface. On a summer afternoon in central Nevada, it is not uncommon to observe cumulus forming at 2400 m (about 8000 ft). In the Central Valley of California, where the summer afternoon spread between air temperature and dew point usually exceeds 22°C (40°F), the air must rise to almost 2700 m (about 9000 ft) before a cloud forms. Due to sinking air aloft, thermals in this area are unable to rise to that elevation, and afternoon cumulus clouds are seldom observed forming overhead.

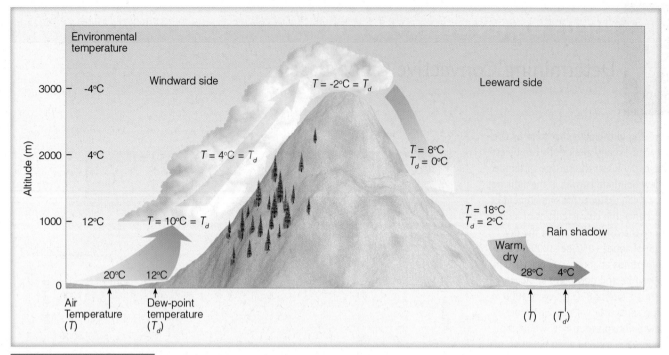

ACTIVE FIGURE 6.22 Orographic uplift, cloud development, and the formation of a rain shadow. Visit the Meteorology Resource Center to view this and other active figures at academic.cengage.com/login

uses this information and a simple formula to determine the bases of convective clouds.)

TOPOGRAPHY AND CLOUDS Horizontally moving air obviously cannot go through a large obstacle, such as a mountain, so the air must go over it. Forced lifting along a topographic barrier is called **orographic uplift.** Often, large masses of air rise when they approach a long chain of mountains like the Sierra Nevada or Rockies. This lifting produces cooling, and, if the air is humid, clouds form. Clouds produced in this manner are called *orographic clouds.* The type of cloud that forms will depend on the air's stability and moisture content. On the leeward (downwind) side of the mountain, as the air moves downhill, it warms. This sinking air is now drier, since much of its moisture was removed in the form of clouds and precipitation on the windward side. This region on the leeward side, where precipitation is noticeably less, is called a **rain shadow.**

An example of orographic uplift and cloud development is given in ● Fig. 6.22. Before rising up and over the barrier, the air at the base of the mountain (0 m) on the windward side has an air temperature of 20°C (68°F) and a dew-point temperature of 12°C (54°F). Notice that the atmosphere is conditionally unstable, as indicated by the environmental lapse rate of 8°C per 1000 m. (Remember from our earlier discussion that the atmosphere is conditionally unstable

● FIGURE 6.23 Satellite view of wave clouds forming many kilometers downwind of the mountains in Scotland and Ireland.

when the environmental lapse rate falls between the dry adiabatic rate and the moist adiabatic rate.)

As the unsaturated air rises, the air temperature decreases at the dry adiabatic rate (10°C per 1000 m) and the dew-point temperature decreases at 2°C per 1000 m. Notice that the rising, cooling air reaches its dew point and becomes saturated at 1000 m. This level (called the **lifting condensation level,** or **LCL**) marks the base of the cloud that has formed as air is lifted (in this case by the mountain). As the rising saturated air condenses into many billions of liquid cloud droplets, and as latent heat is liberated by the condensing vapor, both the air temperature and dew-point temperature decrease at the moist adiabatic rate.

At the top of the mountain, the air temperature and dew point are both −2°C. Note in Fig. 6.22 that this temperature (−2°C) is higher than that of the surrounding air (−4°C). Consequently, the rising air at this level is not only warmer, but unstable with respect to its surroundings. Therefore, the rising air should continue to rise and build into a much larger cumuliform cloud.

Suppose, however, that the air at the top of the mountain (temperature and dew point of −2°C) is forced to descend to the base of the mountain (0 m) on the leeward side. If we assume that the cloud remains on the windward side and does not extend beyond the mountain top, the temperature of the sinking air will increase at the dry adiabatic rate (10°C per 1000 m) all the way down to the base of the mountain. (The dew-point temperature increases at a much lower rate of 2°C per 1000 m.)

We can see in Fig. 6.22 that on the leeward side, after descending 3000 m, the air temperature is 28°C (82°F) and the dew-point temperature is 4°C (39°F). The air is now 8°C (14°F) warmer than it was before being lifted over the barrier. The higher air temperature on the leeward side is the result of latent heat being converted into sensible heat during condensation on the windward side. (In fact, the rising air at the *top* of the mountain is considerably warmer than it would have been had condensation not occurred.) The lower dew-point temperature and, hence, drier air on the leeward side are the result of water vapor condensing and then remaining as liquid cloud droplets and precipitation on the windward side. (A graphic representation of the preceding example is given in the Focus section on adiabatic charts, pp. 158-159.)

Although clouds are more prevalent on the windward side of mountains, they may, under certain atmospheric conditions, form on the leeward side as well. For example, stable air flowing over a mountain often moves in a series of waves that may extend for several hundred kilometers on the leeward side (see Fig. 6.23). These waves resemble the waves

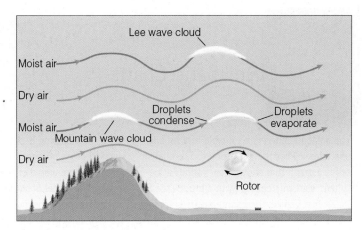

● FIGURE 6.24 Clouds that form in the wave directly over the mountains are called *mountain wave clouds,* whereas those that form downwind of the mountain are called *lee wave clouds.*

that form in a river downstream from a large boulder. Recall from Chapter 5 that *wave clouds* often have a characteristic lens shape and are commonly called *lenticular clouds.*

The formation of lenticular clouds is shown in Fig. 6.24. As moist air rises on the upwind side of the wave, it cools and condenses, producing a cloud. On the downwind side of the wave, the air sinks and warms; the cloud evaporates. Viewed from the ground, the clouds appear motionless as the air rushes through them; hence, they are often referred to as *standing wave clouds.* Since they most frequently form at altitudes where middle clouds form, they are called *altocumulus standing lenticulars.*

When the air between the cloud-forming layers is too dry to produce clouds, lenticular clouds will form one above the other. Actually, when a strong wind blows almost perpendicular to a high mountain range, mountain waves may extend into the stratosphere, producing a spectacular display, sometimes resembling a fleet of hovering spacecraft (see Fig. 6.25).

● FIGURE 6.25 Lenticular clouds forming one on top of the other over the Sierra Nevada.

FOCUS ON AN ADVANCED TOPIC

Adiabatic Charts

The adiabatic chart is a valuable tool for anyone who studies the atmosphere. The chart itself is a graph that shows how various atmospheric elements change with altitude (see Fig. 7). At first glance, the chart appears complicated because of its many lines. We will, therefore, construct these lines on the chart step by step.

Figure 3 shows horizontal lines of pressure decreasing with altitude, and vertical lines of temperature in °C increasing toward the right. The height values on the far right are approximate elevations that have been computed assuming that the air temperature decreases at a standard rate of 6.5°C per kilometer.

In Fig. 4, the slanted solid red lines are called *dry adiabats*. They show how the air temperature would change inside a rising or descending *unsaturated* air parcel. Suppose, for example, that an unsaturated air parcel at the surface (pressure 1013 mb) with a temperature of 10°C rises and cools at the dry adiabatic rate (10°C per km). What would be the parcel temperature at a pressure of 900 mb? To find out, simply follow the dry adiabat from the surface temperature of 10°C up to where it crosses the 900-mb line. Answer: about 0°C. If the same parcel returns to the surface, follow the dry adiabat back to the surface and read the temperature, 10°C.

On some charts, the dry adiabats are expressed as a potential temperature in Kelvins. The *potential temperature* is the temperature an air parcel would have if it were moved dry adiabatically to a pressure of 1000 mb. Moving parcels to the same level allows them to be observed under identical conditions. Thus, it can be determined which parcels are potentially warmer than others.

The sloping dashed blue lines in Fig. 5 are called *moist adiabats*. They show how the air temperature would change inside a rising or descending parcel of *saturated* air. In other words, they represent the moist adiabatic rate for a rising or sinking saturated air parcel, such as in a cloud.

The sloping gray lines in Fig. 6 are lines of constant *mixing ratio*. At any given temperature and pressure, they show how much water vapor the air could hold if it were saturated— the *saturation mixing ratio* (w_s) in grams of water vapor per kilogram of dry air (g/kg). At a given dew-point temperature, they show how much water vapor the air is actually holding— the *actual mixing ratio (w)* in g/kg. Hence, given the air temperature and dew-point temperature at some level, we can compute the relative humidity of the air.* For example, suppose at the surface (pressure 1013 mb) the air temperature and dew-point temperature are 29°C and 15°C, respectively. In Fig. 6, observe that at 29°C the saturation mixing ratio (w_s) is 26 g/kg, and with a dew-point temperature of 15°C, the actual mixing ratio *(w)* is 11 g/kg. This produces a relative humidity of $^{11}/_{26} \times 100$ percent, or 42 percent.

The mixing ratio lines also show how the dew-point temperature changes in a rising or sinking unsaturated air parcel. If an unsaturated air parcel with a dew point of 15°C rises from the surface (pressure 1013 mb) up to where the pressure is 700 mb (approximately 3 km), notice in Fig. 6 that the dew-point temperature inside the parcel would have dropped to a temperature near 10°C.

---

*The relative humidity (RH) of the air can be expressed as: RH = $w/w_s \times 100\%$.

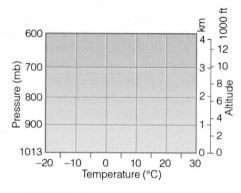

● FIGURE 3

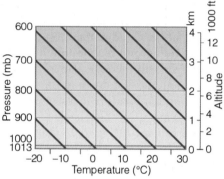

● FIGURE 4

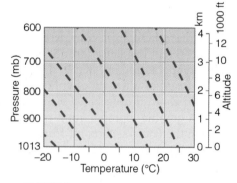

● FIGURE 5

Notice in Fig. 6.24 that beneath the lenticular cloud downwind of the mountain range, a large swirling eddy forms. The rising part of the eddy may cool enough to produce **rotor clouds.** The air in the rotor is extremely turbulent and presents a major hazard to aircraft in the vicinity. Dangerous flying conditions also exist near the leeside of the mountain, where strong downwind air motions are present. (These types of winds will be treated in more detail in Chapter 9.)

Now, having examined the concept of stability and the formation of clouds, we are ready to see what role stability might play in changing a cloud from one type into another.

CHANGING CLOUD FORMS Under certain conditions, a layer of altostratus may change into altocumulus. This happens if the top of the original cloud deck cools while the bottom warms. Because clouds are such good absorbers and

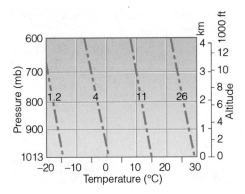

● FIGURE 6

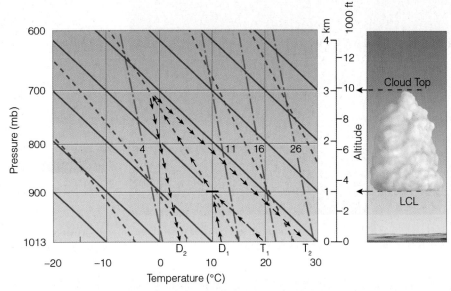

● FIGURE 7 The adiabatic chart. The arrows illustrate the example given in the text. The cloud on the right side represents the base and height of the cloud given in the example.

Figure 7 shows all of the lines described thus far on a single chart. We have already seen that the chart can be used to obtain graphically a number of atmospheric mathematical relationships. Therefore, let's use the chart to obtain information on air that rises up and over a mountain range.

Suppose we use the example given in Fig. 6.22 on p. 156. Air at an elevation of 0 m (pressure 1013 mb), with a temperature of 20°C (T_1) and a dew-point temperature of 12°C (D_1), first ascends, then descends a 3000-meter-high mountain range. Look at Fig. 7 closely and observe that the surface air with a temperature of 20°C indicates a saturation mixing ratio of about 15 g/kg, and at 12°C the dew-point temperature indicates an actual mixing ratio of about 9 g/kg. Hence, the relative humidity of the air before rising over the mountain is $^9/_{15}$, or 60 percent.

Now, as the unsaturated air rises (as indicated by arrows in Fig. 7), the air temperature follows a dry adiabat (solid red line), and the dew-point temperature follows a line of constant mixing ratio (gray line). Carefully follow the mixing ratio line in Fig. 7 from 12°C up to where it intersects the dry adiabat that slopes upward from 20°C. Notice that the intersection occurs at an elevation near 1 km. This, of course, marks the base of the cloud — the *lifting condensation level (LCL)* — where the relative humidity is 100 percent and condensation begins. Above this level, the rising air is saturated. Consequently, the air temperature and dew-point temperature together follow a moist adiabat (dashed blue line) to the top of the mountain.

Notice in Fig. 7 that, at the top of the mountain (at 3 km or about 700 mb), both the air temperature and dew point are −2°C. If we assume that the cloud stays on the windward side, then from 3 km (700 mb) the descending air follows a dry adiabat all the way to the surface (1013 mb). Notice that, after descending, the air has a temperature of 28°C (T_2). From the mountaintop, the dew-point temperature follows a line of mixing ratio and reaches the surface (1013 mb) with a temperature of 4°C (D_2). Observe in Fig. 7 that, with an air temperature of 28°C, the saturation mixing ratio is about 25 g/kg and, with a dew point of 4°C, the actual mixing ratio is about 5 g/kg. Thus, the relative humidity of the air after descending is about $^5/_{25}$, or 20 percent. A more complete adiabatic chart is provided in Appendix J.

emitters of infrared radiation, the top of the cloud will often cool as it radiates infrared energy to space more rapidly than it absorbs solar energy. Meanwhile, the bottom of the cloud will warm as it absorbs infrared energy from below more quickly than it radiates this energy away. This process makes the cloud layer conditionally unstable to the point that small convection cells begin within the cloud itself. The up and down motions in a layered cloud produce globular elements that give the cloud a lumpy appearance. The cloud forms in the rising part of a cell, and clear spaces appear where descending currents occur.*

Cirrocumulus and stratocumulus may form in a similar way. When the wind is fairly uniform throughout a cloud

---

*An example of an altocumulus cloud with a lumpy appearance is given in Fig. 5.16 on p. 125.

NASA

● **FIGURE 6.26** Satellite view of cloud streets, rows of stratocumulus clouds forming over the warm Georgia landscape.

© C. Donald Ahrens

● **FIGURE 6.27** Billow clouds forming in a region of rapidly changing wind speed, called *wind shear*.

layer, these new cloud elements appear evenly distributed across the sky. However, if the wind speed or direction changes with height, the horizontal axes of the convection cells align with the average direction of the wind. The new cloud elements then become arranged in rows and are given the name **cloud streets** (see ● Fig. 6.26). When the changes in wind speed and direction reach a critical value, and an inversion caps the cloud-forming layers, wavelike clouds called **billows** may form along the top of the cloud layer (see ● Fig. 6.27).

Occasionally, altocumulus show vertical development and produce towerlike extensions. The clouds often resemble

© C. Donald Ahrens

● **FIGURE 6.28** An example of altocumulus castellanus.

floating castles and, for this reason, they are called *altocumulus castellanus* (see ● Fig. 6.28). They form when rising currents within the cloud extend into conditionally unstable air above the cloud. Apparently, the buoyancy for the rising air comes from the latent heat released during condensation within the cloud. This process can occur in cirrocumulus clouds, producing *cirrocumulus castellanus*. When altocumulus castellanus appear, they indicate that the mid-level of the troposphere is becoming more unstable (destabilizing). This destabilization is often the precursor to shower activity. So a morning sky full of altocumulus castellanus will likely become afternoon showers and even thunderstorms.

Occasionally, the stirring of a moist layer of stable air will produce a deck of stratocumulus clouds. In ● Fig. 6.29, the air is stable and close to saturation. Suppose a strong wind mixes the layer from the surface up to an elevation of 600 m (2000 ft). As we saw earlier, the lapse rate will steepen as the upper part of the layer cools and the lower part warms. At the same time, mixing will make the moisture distribution in the layer more uniform. The warmer temperature and decreased moisture content cause the lower part of the layer to dry out. On the other hand, the decrease in temperature and increase in moisture content saturate the top of the mixed layer, producing a layer of stratocumulus clouds. Figure 6.29 indicates that the air above the region of mixing is still stable and inhibits further mixing. In some cases, an inversion may actually form above the clouds. However, if the surface warms substantially, rising thermals may penetrate the stable region and the stratocumulus clouds may change into more widely separated clouds, such as cumulus or cumulus congestus. A stratocumulus layer changing to a sky dotted with growing cumulus clouds often occurs as surface heating increases on a warm, humid summer day.

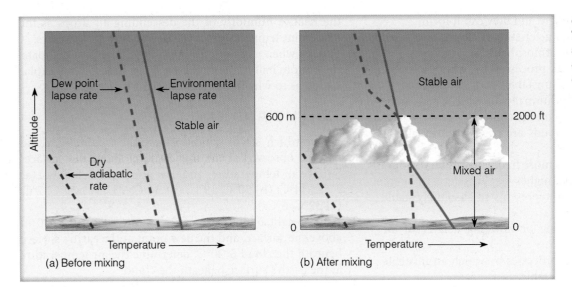

● FIGURE 6.29
The mixing of a moist layer of air near the surface can produce a deck of stratocumulus clouds.

SUMMARY

In this chapter, we tied together the concepts of stability and the formation of clouds. We learned that rising unsaturated air cools at the dry adiabatic rate and, due to the release of latent heat, rising saturated air cools at the moist adiabatic rate. In a stable atmosphere, a lifted parcel of air will be colder (heavier) than the air surrounding it at each new level, and it will sink back to its original position. Because stable air tends to resist upward vertical motions, clouds forming in a stable atmosphere often spread horizontally and have a stratified appearance, such as cirrostratus and altostratus. A stable atmosphere may be caused by either cooling the surface air, warming the air aloft, or by the sinking (subsidence) of an entire layer of air, in which case a very stable subsidence inversion usually forms.

In an unstable atmosphere, a lifted parcel of air will be warmer (lighter) than the air surrounding it at each new level, and it will continue to rise upward away from its original position. In a conditionally unstable atmosphere, an unsaturated parcel of air can be lifted to a level where condensation begins, latent heat is released, and instability results, as the temperature inside the rising parcel becomes warmer than the air surrounding it. In a conditionally unstable atmosphere, rising air tends to form clouds that develop vertically, such as cumulus congestus and cumulonimbus. Instability may be caused by warming the surface air, cooling the air aloft, or by the lifting or mixing of an entire layer of air.

On warm humid days the instability generated by surface heating can produce cumulus clouds at a height determined by the temperature and moisture content of the surface air. Instability may cause changes in existing clouds as convection changes an altostratus into an altocumulus. Also, mixing can change a clear day into a cloudy one.

KEY TERMS

The following terms are listed (with page numbers) in the order they appear in the text. Define each. Doing so will aid you in reviewing the material covered in this chapter.

parcel of air, 142
adiabatic process, 142
dry adiabatic rate, 142
moist adiabatic rate, 142
environmental lapse rate, 143
absolutely stable atmosphere, 143
subsidence inversion, 145
neutral stability, 145
absolutely unstable atmosphere, 146
conditionally unstable atmosphere, 146
condensation level, 153
entrainment, 154
orographic uplift, 156
rain shadow, 156
lifting condensation level (LCL), 157
rotor clouds, 158
cloud streets, 160
billow clouds, 160

QUESTIONS FOR REVIEW

1. What is an adiabatic process?
2. Why are moist and dry adiabatic rates of cooling different?
3. Under what conditions would the moist adiabatic rate of cooling be almost equal to the dry adiabatic rate?
4. Explain the difference between environmental lapse rate and dry adiabatic rate.
5. How would one normally obtain the environmental lapse rate?

6. What is a stable atmosphere and how can it form?
7. Describe the general characteristics of clouds associated with stable and unstable atmospheres.
8. List and explain several processes by which a stable atmosphere can be made unstable.
9. If the atmosphere is conditionally unstable, what condition is necessary to bring on instability?
10. Explain why cumulus clouds are conspicuously absent over a cool water surface.
11. Why are cumulus clouds more frequently observed during the afternoon than at night?
12. Explain why an inversion represents an absolutely stable atmosphere.
13. How and why does lifting or lowering a layer of air change its stability?
14. List and explain several processes by which an unstable atmosphere can be made stable.
15. Why do cumulonimbus clouds often have flat tops?
16. Why are there usually large spaces of blue sky between cumulus clouds?
17. List four primary ways clouds form, and describe the formation of one cloud type by each method.
18. (a) Why are lenticular clouds also called standing wave clouds? (b) On which side of a mountain (windward or leeward) would lenticular clouds most likely form?
19. Explain why rain shadows form on the leeward side of mountains.
20. How can a layer of altostratus change into one of altocumulus?
21. Describe the conditions necessary to produce stratocumulus clouds by mixing.
22. Briefly describe how each of the following clouds forms:
    (a) lenticular
    (b) rotor
    (c) billow
    (d) castellanus

QUESTIONS FOR THOUGHT

1. How is it possible for a layer of air to be convectively unstable and absolutely stable at the same time?
2. Are the bases of convective clouds generally higher during the day or the night? Explain.
3. Where would be the safest place to build an airport in a mountainous region? Why?
4. Use Fig. 4.14b, p. 99 (Chapter 4) to help you explain why the bases of cumulus clouds, which form from rising thermals during the summer, increase in height above the surface as you move due west of a line that runs north-south through central Kansas.
5. For least polluted conditions, what would be the best time of day for a farmer to burn agricultural debris?
6. Suppose that surface air on the windward side of a mountain rises and descends on the leeward side. Recall from Chapter 4 that the dew-point temperature is a measure of the amount of water vapor in the air. Explain, then, why

the relative humidity of the descending air drops as the dew-point temperature of the descending air increases.
7. Usually when a cumulonimbus cloud begins to dissipate, the bottom half of the cloud dissipates first. Give an explanation as to why this situation might happen.

PROBLEMS AND EXERCISES

1. Under which set of conditions would a cumulus cloud base be observed at the highest level above the surface? Surface air temperatures and dew points are as follows: (a) 35°C, 14°C; (b) 30°C, 19°C; (c) 34°C, 9°C; (d) 29°C, 7°C; (e) 32°C, 6°C.
2. If the height of the base of a cumulus cloud is 1000 m above the surface, and the dew point at the earth's surface beneath the cloud is 20°C, determine the air temperature at the earth's surface beneath the cloud.
3. The condensation level over New Orleans, Louisiana, on a warm muggy afternoon is 2000 ft. If the dew-point temperature of the rising air at this level is 73°F, what is the approximate dew-point temperature and air temperature at the surface? Determine the surface relative humidity. (Hint: See Chapter 4, p. 102.)
4. Suppose the air pressure outside a conventional jet airliner flying at an altitude of 10 km (about 33,000 ft) is 250 mb. Further, suppose the air inside the aircraft is pressurized to 1000 mb. If the outside air temperature is −50°C (−58°F), what would be the temperature of this air if brought inside the aircraft and compressed at the dry adiabatic rate to a pressure of 1000 mb? (Assume that a pressure of 1000 mb is equivalent to an altitude of 0 m.)
5. In ● Fig. 6.30, a radiosonde is released and sends back temperature data as shown in the diagram. (This is the environment temperature.)
   (a) Calculate the environmental lapse rate from the surface up to 3000 m.

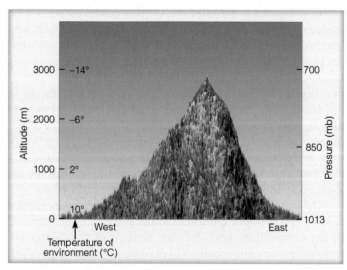

● FIGURE 6.30 The information in this illustration is to be used in answering questions 5 and 6 in the Problems and Exercises section.

(b) What type of atmospheric stability does the sounding indicate?

Suppose the wind is blowing from the west and a parcel of surface air with a temperature of 10°C and a dew point of 2°C begins to rise upward along the western (windward) side of the mountain.

(c) What is the relative humidity of the air parcel at 0 m (pressure 1013 mb) before rising? (Hint: See Chapter 4, p. 102.)

(d) As the air parcel rises, at approximately what elevation would condensation begin and a cloud start to form?

(e) What is the air temperature and dew point of the rising air at the base of the cloud?

(f) What is the air temperature and dew point of the rising air inside the cloud at an elevation of 3000 m? (Use moist adiabatic rate of 6°C per 1000 m.)

(g) At an altitude of 3000 m, how does the air temperature inside the cloud compare with the air temperature outside the cloud, as measured by the radiosonde? What type of atmospheric stability (stable or unstable) does this suggest? Explain.

(h) At an elevation of 3000 m, would you expect the cloud to continue to develop vertically? Explain.

6. Answer the same questions in problem 5 except, this time, use the adiabatic chart provided in Appendix J.

(i) What would be the name of the cloud that is forming?

Suppose that a parcel of air inside the cloud descends from the top of the mountain at 3000 m (pressure 700 mb) down the eastern (leeward) side of the mountain to an elevation of 0 m (pressure 1013 mb).

(j) If the descending air warms at the dry adiabatic rate from the top of the mountain all the way down to 0 m, what is the sinking air's temperature and dew point when it reaches 0 m?

(k) What would be the relative humidity of the sinking air at 0 m? (Hint: See Chapter 4, p. 102.)

(l) What accounts for the sinking air being warmer at the base of the mountain on the eastern side?

(m) Explain why the sinking air is drier (its dew point is lower) on the eastern side at 0 m.

Heavy snowfall blankets the field, while wet snowflakes cling to trees in Saarland, Germany.
© Ray Juno/CORBIS

CHAPTER 7

Precipitation

By an unfortunate coincidence, as I write, the New Jersey countryside around me is in the grip of an ice storm — "the worst ice storm in a generation" so the papers tell me, and a look at my garden suffices to convince me. A 150-year-old tulip tree has already lost enough limbs to keep us in firewood for the rest of the winter; a number of black locusts stand beheaded; the silver birches are bent double to the ground; and almost every twig of every bush and tree is encased in a translucent cylinder of ice one to two inches in diameter. There is beauty in the sight, to be sure, for the sun has momentarily transmuted the virginal whites and grays into liquid gold. And there is hope, too, for some of the trees are still unbowed and look as though they had every intention of living to tell the tale.

George H. T. Kimble, *Our American Weather*

CONTENTS

The young boy pushed his nose against the cold windowpane, hoping to see snowflakes glistening in the light of the street lamp across the way. Perhaps if it snowed, he thought, accumulations would be deep enough to cancel school—maybe for a day, possibly a week, or, perhaps, forever. But a full moon with a halo gave little hope for snow on this evening. Nor did the voice from the back room that insisted, "Don't even think about snow. You know it won't snow tonight—it's too cold to snow."

Is it ever "too cold to snow"? Although many believe in this expression, the fact remains that it is *never* too cold to snow. True, colder air cannot "hold" as much water vapor as warmer air; but, no matter how cold the air becomes, it always contains some water vapor that could produce snow. At Fort Yellowstone, Wyoming, for example, 3 inches of snow fell on February 2, 1899, when the maximum temperature reached only −28°C (−18°F). In fact, tiny ice crystals have been observed falling at temperatures as low as −47°C (−53°F). We usually associate extremely cold air with "no snow" because the coldest winter weather occurs on clear, calm nights—conditions that normally prevail with strong high-pressure areas that have few, if any, clouds.

This chapter raises a number of interesting questions to consider regarding **precipitation.*** Why, for example, does the largest form of precipitation—hail—often fall during the warmest time of the year? Why does it sometimes rain on one side of the street but not on the other? What is "sleet" and how does it differ from hail? First, we will examine the processes that produce rain and snow; then, we will look closely at the other forms of precipitation. Our discussion will conclude with a section on how precipitation is measured.

Precipitation Processes

As we all know, cloudy weather does not necessarily mean that it will rain or snow. In fact, clouds may form, linger for many days, and never produce precipitation. In Eureka, California, the August daytime sky is overcast more than 50 percent of the time, yet the average precipitation there for August is merely one-tenth of an inch. We know that clouds form by condensation, yet apparently condensation alone is not sufficient to produce rain. Why not? To answer this question we need to closely examine the tiny world of cloud droplets.

HOW DO CLOUD DROPLETS GROW LARGER?   An ordinary cloud droplet is extremely small, having an average diameter of 20 μm† or 0.002 cm. Notice in • Fig. 7.1 that a typical cloud droplet is 100 times smaller than a typical raindrop. If a cloud droplet is in equilibrium with its surround-

---

*Recall from Chapter 4 that precipitation is any form of water (liquid or solid) that falls from a cloud and reaches the ground.

†Remember from Chapter 2 that one micrometer (μm) equals one-millionth of a meter.

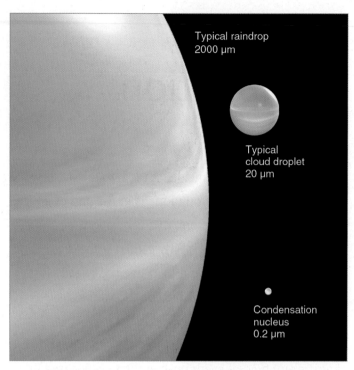

• FIGURE 7.1 Relative sizes of raindrops, cloud droplets, and condensation nuclei in micrometers (μm).

ings, the size of the droplet does not change because the water molecules condensing onto the droplet will be exactly balanced by those evaporating from it. If, however, it is not in equilibrium, the droplet size will either increase or decrease, depending on whether condensation or evaporation predominates.

Consider a cloud droplet in equilibrium with its environment. The total number of vapor molecules around the droplet remains fairly constant and defines the droplet's *saturation vapor pressure*. Since the droplet is in equilibrium, the saturation vapor pressure is also called the **equilibrium vapor pressure.** • Figure 7.2 shows a cloud droplet and a flat water surface, both of which are in equilibrium. Because more vapor molecules surround the droplet, it has a greater equilibrium vapor pressure. The reason for this fact is that water molecules are less strongly attached to a curved (convex) water surface; hence, they evaporate more readily.

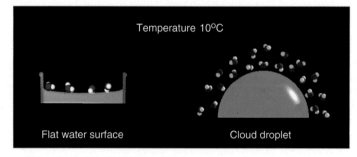

• FIGURE 7.2 At equilibrium, the vapor pressure over a curved droplet of water is greater than that over a flat surface.

To keep the droplet in equilibrium, more vapor molecules are needed around it to replace those molecules that are constantly evaporating from its surface. Smaller cloud droplets exhibit a greater curvature, which causes a more rapid rate of evaporation. As a result of this process (called the **curvature effect**), smaller droplets require an even greater vapor pressure to keep them from evaporating away. Therefore, *when air is saturated with respect to a flat surface, it is unsaturated with respect to a curved droplet of pure water,* and the droplet evaporates. So, to keep tiny cloud droplets in equilibrium with the surrounding air, the air must be *supersaturated;* that is, the relative humidity must be greater than 100 percent. The smaller the droplet, the greater its curvature, and the higher the supersaturation needed to keep the droplet in equilibrium.

• Figure 7.3 shows the curvature effect for pure water. The dark blue line represents the relative humidity needed to keep a droplet with a given diameter in equilibrium with its environment. Note that when the droplet's size is less than 2 μm, the relative humidity (measured with respect to a flat surface) must be above 100.1 percent for the droplet to survive. As droplets become larger, the effect of curvature lessens; for a droplet whose diameter is greater than 20 μm, the curvature effect is so small that the droplet behaves as if its surface were flat.

Just as relative humidities less than that required for equilibrium permit a water droplet to evaporate and shrink, those greater than the equilibrium value allow the droplet to grow by condensation. From Fig. 7.3, we can see that a droplet whose diameter is 1 μm will grow larger as the relative humidity approaches 101 percent. But relative humidities, even in clouds, rarely become greater than 101 percent. How, then, do tiny cloud droplets of less than 1 μm grow to the size of an average cloud droplet?

Recall from Chapter 5 (fog formation discussion) that condensation begins on tiny particles called *cloud condensation nuclei.* Because many of these nuclei are *hygroscopic* (that is, they have an affinity for water vapor), condensation may begin on such particles when the relative humidity is well below 100 percent. When condensation begins on hygroscopic salt particles, for example, they dissolve, forming a solution. Since the salt ions in solution bind closely with water molecules, it is more difficult for the water molecules to evaporate. This condition reduces the equilibrium vapor pressure, an effect known as the **solute effect.** Due to the solute effect, once an impurity (such as a salt particle) replaces a water molecule in the lattice structure of the droplet, the equilibrium vapor pressure surrounding the droplet is lowered. As a result of the solute effect, a droplet containing salt can be in equilibrium with its environment when the atmospheric relative humidity is much lower than 100 percent. Should the relative humidity of the air increase, water vapor molecules would attach themselves to the droplet at a faster rate than they would leave, and the droplet would grow larger in size.

Imagine that we place cloud condensation nuclei of varying sizes into moist but unsaturated air. As the air cools, the

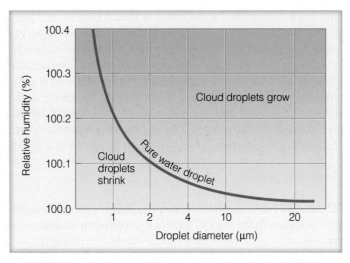

• **FIGURE 7.3** The curved line represents the relative humidity needed to keep a droplet in equilibrium with its environment. For a given droplet size, the droplet will evaporate and shrink when the relative humidity is less than that given by the curve. The droplet will grow by condensation when the relative humidity is greater than the value on the curve.

relative humidity increases. When the relative humidity reaches a value near 78 percent, condensation occurs on the majority of nuclei. As the air cools further, the relative humidity increases, with the droplets containing the most salt reaching the largest sizes. And since the smaller nuclei are more affected by the curvature effect, only the larger nuclei are able to become cloud droplets.

Over land masses where large concentrations of nuclei exist, there may be many hundreds of droplets per cubic centimeter, all competing for the available supply of water vapor. Over the oceans where the concentration of nuclei is less, there are normally fewer (typically less than 100 per cubic centimeter) but larger cloud droplets. So, in a given volume we tend to find more cloud droplets in clouds that form over land and fewer, but larger, cloud droplets in clouds that form over the ocean.

We now have a cloud composed of many small droplets—too small to fall as rain. These minute droplets require only slight upward air currents to keep them suspended. Those droplets that do fall descend slowly and evaporate in the drier air beneath the cloud. It is evident, then, that most clouds cannot produce precipitation. The condensation process by itself is entirely too slow to produce rain. Even under ideal conditions, it would take several days for this process alone to create a raindrop. However, observations show that clouds can develop and begin to produce rain in less than an hour. Since it takes about 1 million average size (20 μm) cloud droplets to make an average size (2000 μm) raindrop, there must be some other process by which cloud droplets grow large and heavy enough to fall as precipitation. Even though all of the intricacies of how rain is produced are not yet fully understood, two important processes stand out: (1) the collision-coalescence process and (2) the ice-crystal (Bergeron) process.

COLLISION AND COALESCENCE PROCESS In clouds with tops warmer than −15°C (5°F), the **collision-coalescence process** can play a significant role in producing precipitation. To produce the many collisions necessary to form a raindrop, some cloud droplets must be larger than others. Larger drops may form on large condensation nuclei, such as salt particles, or they may form through random collisions of droplets. Studies suggest that turbulent mixing between the cloud and its drier environment may play a role in producing larger droplets.

As cloud droplets fall, air retards the falling drops. The amount of air resistance depends on the size of the drop and on its rate of fall: The greater its speed, the more air molecules the drop encounters each second. The speed of the falling drop increases until the air resistance equals the pull of gravity. At this point, the drop continues to fall, but at a constant speed, which is called its **terminal velocity.** Because larger drops have a smaller surface-area-to-weight ratio, they must fall faster before reaching their terminal velocity. Thus *larger drops fall faster than smaller drops* (see ▼ Table 7.1). Note in Table 7.1 that, in calm air, a typical raindrop falls over 600 times faster than a typical cloud droplet!

Large droplets overtake and collide with smaller drops in their path. This merging of cloud droplets by collision is called **coalescence.** Laboratory studies show that collision

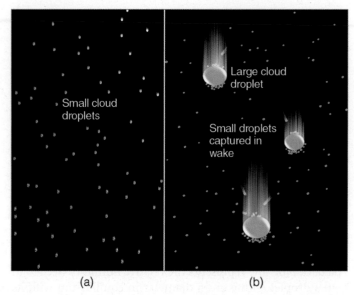

● FIGURE 7.4 Collision and coalescence. (a) In a warm cloud composed only of small cloud droplets of uniform size, the droplets are less likely to collide as they all fall very slowly at about the same speed. Those droplets that do collide, frequently do not coalesce because of the strong surface tension that holds together each tiny droplet. (b) In a cloud composed of different size droplets, larger droplets fall faster than smaller droplets. Although some tiny droplets are swept aside, some collect on the larger droplet's forward edge, while others (captured in the wake of the larger droplet) coalesce on the droplet's backside.

does not always guarantee coalescence; sometimes the droplets actually bounce apart during collision. For example, the forces that hold a tiny droplet together (surface tension) are so strong that if the droplet were to collide with another tiny droplet, chances are they would not stick together (coalesce) (see ● Fig. 7.4). Coalescence appears to be enhanced if colliding droplets have opposite (and, hence, attractive) electrical charges.* An important factor influencing cloud droplet growth by the collision process is the amount of time the droplet spends in the cloud. A very large cloud droplet of 200 μm falling in still air takes about 12 minutes to travel through a cloud 500 m (1640 ft) thick and over an hour if the cloud thickness is 2500 m (8200 ft). Rising air currents in a forming cloud slow the rate at which droplets fall toward the ground. Consequently, a thick cloud with strong updrafts maximizes the time cloud droplets spend in the cloud and, hence, the size to which they can grow.

A warm stratus cloud is typically less than 500 m thick and has slow upward air movement (generally less than 0.1 m/sec). Under these conditions, a large droplet would be in the cloud for a relatively short time and grow by coalescence

▼ TABLE 7.1   Terminal Velocity of Different-Size Particles Involved in Condensation and Precipitation Processes

Diameter (μm)	TERMINAL VELOCITY		
	m/sec	ft/sec	Type of Particle
0.2	0.0000001	0.0000003	Condensation nuclei
20	0.01	0.03	Typical cloud droplet
100	0.27	0.9	Large cloud droplet
200	0.70	2.3	Large cloud droplet or drizzle
1000	4.0	13.1	Small raindrop
2000	6.5	21.4	Typical raindrop
5000	9.0	29.5	Large raindrop

*It was once thought that atmospheric electricity played a significant role in the production of rain. Today, many scientists feel that the difference in electrical charge that exists between cloud droplets results from the bouncing collisions between them. It is felt that the weak separation of charge and the weak electrical fields in developing, relatively warm clouds are not significant in initiating precipitation. However, studies show that coalescence is often enhanced in thunderstorms where strongly charged droplets exist in a strong electrical field.

to only about 200 μm. If the air beneath the cloud is moist, the droplets may reach the ground as *drizzle,* the lightest form of rain. If, however, the stratus cloud base is fairly high above the ground, the drops will evaporate before reaching the surface, even when the relative humidity is 90 percent.

Clouds that have above-freezing temperatures at all levels are called *warm clouds.* In such clouds, precipitation forms by the collision and coalescence process. For example, in tropical regions, where warm cumulus clouds build to great heights, convective updrafts of at least 1 m/sec (and some exceeding many tens of meters per second) occur. Look at the warm cumulus cloud in ● Fig. 7.5. Suppose a cloud droplet of 100 μm is caught in an updraft whose velocity is 6.5 m/sec (about 15 mi/hr). As the droplet rises, it collides with and captures smaller drops in its path and grows until it reaches a size of about 1000 μm. At this point, the updraft in the cloud is just able to balance the pull of gravity on the drop. Here, the drop remains suspended until it grows just a little bigger. Once the fall velocity of the drop is greater than the updraft velocity in the cloud, the drop slowly descends. As the drop falls, larger cloud droplets are captured by the falling drop, which then grows larger. By the time this drop reaches the bottom of the cloud, it will be a large raindrop with a diameter of over 5000 μm (5 mm). Because raindrops of this size fall faster and reach the ground first, they typically occur at the beginning of a rain shower originating in these warm, convective cumulus clouds.

Raindrops that reach the earth's surface are seldom larger than about 5 mm. The collisions between raindrops (whether glancing or head-on) tend to break them up into many smaller drops. Additionally, a large drop colliding with another large drop may result in oscillations within the combined drop. As the resultant drop grows, these oscillations may tear the drop apart into many fragments, all smaller than the original drop.

So far, we have examined the way cloud droplets in warm clouds (that is, those clouds with temperatures above freezing) grow large enough by the collision-coalescence process to fall as raindrops. Rain that falls from warm clouds is sometimes called *warm rain.* The most important factor in the production of raindrops is the cloud's *liquid water content.* In a cloud with sufficient water, other significant factors are:

1. the range of droplet sizes
2. the cloud thickness
3. the updrafts of the cloud
4. the electric charge of the droplets and the electric field in the cloud

Relatively thin stratus clouds with slow, upward air currents are, at best, only able to produce drizzle, whereas the towering cumulus clouds associated with rapidly rising air can cause heavy showers. Now, let's turn our attention to see how clouds with temperatures below freezing are able to produce precipitation.

ICE-CRYSTAL PROCESS The **ice-crystal** (or **Bergeron**)* **process** of rain formation is extremely important in middle and high latitudes, where clouds extend upward into regions where the air temperature is well below freezing. Such clouds are called *cold clouds.* ● Figure 7.6 illustrates a typical cold cloud that has formed over the Great Plains, where the "cold" part is well above the 0°C isotherm.

Suppose we take an imaginary balloon flight up through the cumulonimbus cloud in Fig. 7.6. Entering the cloud, we observe cloud droplets growing larger by processes described in the previous section. As expected, only water droplets exist here, for the base of the cloud is warmer than 0°C. Surprisingly, in the cold air just above the 0°C isotherm, almost all of the cloud droplets are still composed of liquid water. Water droplets existing at temperatures below freezing are referred to as **supercooled.** Even at higher levels, where the air temperature is −10°C (14°F), there is only one ice crystal for every million liquid droplets. Near 5500 m (18,000 ft), where the temperature becomes −20°C (−4°F), ice crystals become more numerous, but are still outnumbered by water droplets.† The distribution of ice crystals, however, is not uniform, as the downdrafts contain more ice than the updrafts.

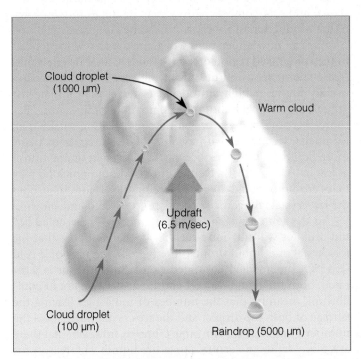

Cloud droplet
(1000 μm)

Warm cloud

Updraft
(6.5 m/sec)

Cloud droplet
(100 μm)

Raindrop (5000 μm)

● FIGURE 7.5 A cloud droplet rising then falling through a warm cumulus cloud can grow by collision and coalescence, and emerge from the cloud as a large raindrop.

---

*The ice-crystal process is also known as the *Bergeron process* after the Swedish meteorologist Tor Bergeron, who proposed that essentially all raindrops begin as ice crystals.

†In continental clouds, such as the one in Fig. 7.6, where there are many small cloud droplets less than 20 μm in diameter, the onset of ice-crystal formation begins at temperatures between −9°C and −15°C. In clouds where larger but fewer cloud droplets are present, ice crystals begin to form at temperatures between −4°C and −8°C. In some of these clouds, glaciation can occur at −8°C, which may be only 2500 m (8200 ft) above the surface.

● FIGURE 7.6 The distribution of ice and water in a cumulonimbus cloud.

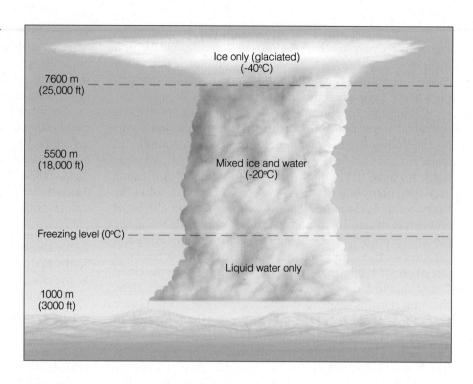

Not until we reach an elevation of 7600 m (25,000 ft), where temperatures drop below −40°C (also −40°F), do we find *only* ice crystals. The region of a cloud where only ice particles exist is called *glaciated*. Why are there so few ice crystals in the middle of the cloud, even though temperatures there are well below freezing? Laboratory studies reveal that the smaller the amount of pure water, the lower the temperature at which water freezes. Since cloud droplets are extremely small, it takes very low temperatures to turn them into ice. (More on this topic is given in the Focus section on p. 171.)

Just as liquid cloud droplets form on condensation nuclei, ice crystals may form in subfreezing air on particles called **ice nuclei.** The number of ice-forming nuclei available in the atmosphere is small, especially at temperatures above −10°C (14°F). However, as the temperature decreases, more particles become active and promote freezing. Although some uncertainty exists regarding the principal source of ice nuclei, it is known that clay minerals, such as kaolinite, become effective nuclei at temperatures near −15°C (5°F). Some types of bacteria in decaying plant leaf material and ice crystals themselves are also excellent ice nuclei. Moreover, particles serve as excellent ice-forming nuclei if their geometry resembles that of an ice crystal. However, it is difficult to find substances in nature that have a lattice structure similar to ice, since there are so many possible lattice structures. In the atmosphere, it is easy to find hygroscopic ("water seeking") particles. Consequently, ice-forming nuclei are rare compared to cloud condensation nuclei.

In a cold cloud, there may be several types of ice-forming nuclei present. For example, certain ice nuclei allow water vapor to deposit as ice directly onto their surfaces in cold, saturated air. These are called *deposition nuclei* because, in this situation, water vapor changes directly into ice without going through the liquid phase. Ice nuclei that promote the freezing of supercooled liquid droplets are called *freezing nuclei.* Some freezing nuclei cause freezing after they are immersed in a liquid drop; some promote condensation, then freezing; yet others cause supercooled droplets to freeze if they collide with them. This last process is called **contact freezing,** and the particles involved are called *contact nuclei.* Studies suggest that contact nuclei can be just about any substance and that contact freezing may be the dominant force in the production of ice crystals in some clouds.

We can now understand why there are so few ice crystals in the cold mixed region of some clouds. Cloud droplets may freeze spontaneously, but only at the very low temperatures usually found at high altitudes. Ice nuclei may initiate the growth of ice crystals, but they do not abound in nature. Because there are many more cloud condensation nuclei than ice nuclei, we are left with a cold cloud that contains many more liquid droplets than ice particles, even at temperatures as low as −10°C (14°F). Neither the tiny liquid nor solid particles are large enough to fall as precipitation. How, then, does the ice-crystal (Bergeron) process produce rain and snow?

In the subfreezing air of a cloud, many supercooled liquid droplets will surround each ice crystal. Suppose that the ice crystal and liquid droplet in ● Fig. 7.7 are part of a cold (−15°C) supercooled, saturated cloud. Since the air is saturated, both the liquid droplet and the ice crystal are in equilibrium, meaning that the number of molecules leaving the surface of both the droplet and the ice crystal must equal the number of molecules returning. Observe, however, that there are more vapor molecules above the liquid. The reason for this fact is that molecules escape the surface of water much easier than they escape the surface of ice. Consequently, more molecules escape the water surface at a given temperature,

FOCUS ON A SPECIAL TOPIC

The Freezing of Tiny Cloud Droplets

Over large bodies of fresh water, ice ordinarily forms when the air temperature drops slightly below 0°C. Yet, a cloud droplet of pure water about 25 μm in diameter will not freeze spontaneously until the air temperature drops to about −40°C(−40°F) or below.

The freezing of pure water (without the benefit of some nucleus) is called *spontaneous* or *homogeneous freezing*. For this type of freezing to occur, enough molecules within the water droplet must join together in a rigid pattern to form a tiny ice structure, or *ice embryo*. When the ice embryo grows to a critical size, it acts as a nucleus. Other molecules in the droplet then attach themselves to the nucleus of ice and the water droplet freezes.

Tiny ice embryos form in water at temperatures just below freezing, but at these temperatures thermal agitations are large enough to weaken their structure. The ice embryos simply form and then break apart. At lower temperatures, thermal motion is reduced, making it eas-

● FIGURE I
This cirrus cloud is probably composed entirely of ice crystals, because any liquid water droplet, no matter how small, must freeze spontaneously at the very low temperature (below −40°C) found at this altitude, 9 km (29,500 ft).

© C. Donald Ahrens

ier for bigger ice embryos to form. Hence, freezing is more likely.

The chances of an ice embryo growing large enough to freeze water before the embryo is broken up by thermal agitation increase with larger volumes of water. Consequently, only larger cloud droplets will freeze by homogeneous freezing at air temperatures higher than −40°C. In air colder than −40°C, however, it

is almost certain that an ice embryo will grow to critical size in even the smallest cloud droplet. Thus, any cloud that forms in extremely cold air (below −40°C), such as cirrus clouds (see Fig. I), will almost certainly be composed of ice, since any cloud droplets that form will freeze spontaneously.

---

requiring more in the vapor phase to maintain saturation. This situation reflects the important fact discussed briefly in Chapter 4: At the same subfreezing temperature, *the saturation vapor pressure just above a water surface is greater than the saturation vapor pressure above an ice surface.* This difference in saturation vapor pressure between water and ice is illustrated in ● Fig. 7.8.

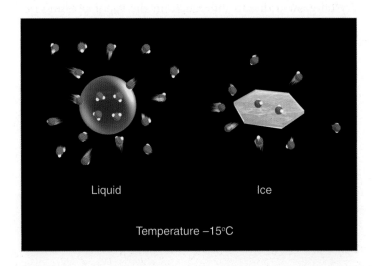

Liquid

Ice

Temperature −15°C

This difference in vapor pressure causes water vapor molecules to move (diffuse) from the water droplet toward the ice crystal. The removal of vapor molecules reduces the vapor pressure above the water droplet. Since the droplet is now out of equilibrium with its surroundings, it evaporates to replenish the diminished supply of water vapor above it. This process provides a continuous source of moisture for the ice crystal, which absorbs the water vapor and grows rapidly (see ● Fig. 7.9). Hence, during the *ice-crystal (Bergeron) process, ice crystals grow larger at the expense of the surrounding water droplets.*

The constant supply of moisture to the ice crystal allows it to enlarge rapidly. At some point, the ice crystal becomes heavy enough to overcome updrafts in the cloud and begins to fall. But a single falling ice crystal does not comprise a snowstorm; consequently, other ice crystals must quickly form.

● FIGURE 7.7 In a saturated environment, the water droplet and the ice crystal are in equilibrium, as the number of molecules leaving the surface of each droplet and ice crystal equals the number returning. The greater number of vapor molecules above the liquid indicates, however, that the saturation vapor pressure over water is greater than it is over ice.

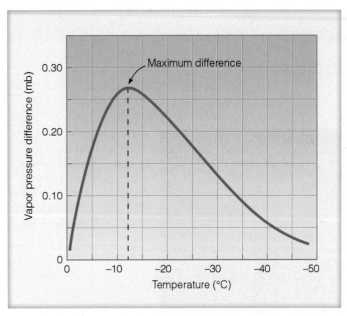

● FIGURE 7.8 The difference in saturation vapor pressure between supercooled water and ice at different temperatures.

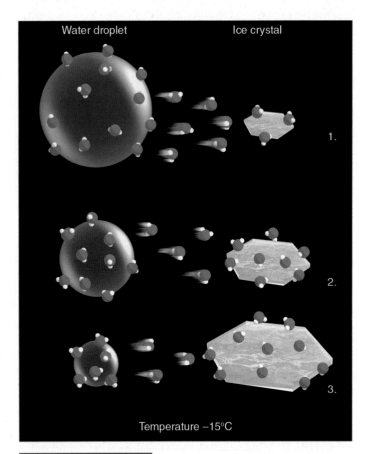

ACTIVE FIGURE 7.9   The ice-crystal (Bergeron) process.
(1) The greater number of water vapor molecules around the liquid droplet causes water molecules to diffuse from the liquid droplet toward the ice crystal. (2) The ice crystal absorbs the water vapor and grows larger, while (3) the water droplet grows smaller.

In some clouds, especially those with relatively warm tops, ice crystals might collide with supercooled droplets. Upon contact, the liquid droplets freeze into ice and stick together. This process of ice crystals growing larger as they collide with supercooled cloud droplets is called **accretion.** The icy matter that forms is called **graupel** (or *snow pellets*). As the graupel falls, it may fracture or splinter into tiny ice particles when it collides with cloud droplets. These splinters may grow to become new graupel, which, in turn, may produce more splinters.

In colder clouds, the delicate ice crystals may collide with other crystals and fracture into smaller ice particles, or tiny seeds, which freeze hundreds of supercooled droplets on contact. In both cases a chain reaction may develop, producing many ice crystals. As the ice crystals fall, they may collide and stick to one another. The process of ice crystals colliding then sticking together is called **aggregation.**[*] The end product of this clumping together of ice crystals is a **snowflake** (see ● Fig. 7.10). If the snowflake melts before reaching the ground, it continues its fall as a raindrop. Therefore, much of the rain falling in middle and high latitudes—even in summer— begins as snow.

For ice crystals to grow large enough to produce precipitation there must be many, many times more water droplets than ice crystals. Generally, the ratio of ice crystals to water droplets must be on the order of 1:100,000 to 1:1,000,000. When there are too few ice crystals in the cloud, each crystal grows large and falls out of the cloud, leaving the majority of cloud behind (unaffected). Since there are very few ice crystals, there is very little precipitation. If, on the other hand, there are too many ice crystals (such as an equal number of crystals and droplets), then each ice crystal receives the mass of one droplet. This would create a cloud of many tiny ice crystals, each too small to fall to the ground, and no precipitation. Now, if the ratio of crystals to droplets is on the order of 1:100,000, then each ice crystal would receive the mass of 100,000 droplets. Most of the cloud would convert to precipitation, as the majority of ice crystals would grow large enough to fall to the ground as precipitation.

The first person to formally propose the theory of ice-crystal growth due to differences in the vapor pressure between ice and supercooled water was Alfred Wegener (1880– 1930), a German climatologist who also proposed the geological theory of continental drift. In the early 1930s, important additions to this theory were made by the Swedish meteorologist Tor Bergeron. Several years later, the German meteorologist Walter Findeisen made additional contributions to Bergeron's theory; hence, the ice-crystal theory of rain formation has come to be known as the *Wegener-Bergeron-Findeisen process,* or, simply, the *Bergeron process.*

View this concept in action on the Meteorology Resource Center at academic.cengage.com/login

---

[*]Significant aggregation seems possible only when the air is relatively warm, usually warmer than −10°C (14°F).

(a) Falling ice crystals may freeze supercooled droplets on contact (accretion), producing larger ice particles.

(b) Falling ice particles may collide and fracture into many tiny (secondary) ice particles.

(c) Falling ice crystals may collide and stick to other ice crystals (aggregation), producing snowflakes.

● FIGURE 7.10 Ice particles in clouds.

CLOUD SEEDING AND PRECIPITATION The primary goal in many **cloud seeding** experiments is to inject (or seed) a cloud with small particles that will act as nuclei, so that the cloud particles will grow large enough to fall to the surface as precipitation. The first ingredient in any seeding project is, of course, the presence of clouds, as seeding does not generate clouds. However, not just any cloud will do. For optimum results, the cloud must be cold; that is, at least a portion of it (preferably the upper part) must be supercooled because cloud seeding uses the ice-crystal (Bergeron) process to cause the cloud particles to grow.

The idea in cloud seeding is to first find clouds that have *too low* a ratio of ice crystals to droplets and then to add enough artificial ice nuclei so that the ratio of crystals to droplets is about 1:100,000. However, it should be noted that the natural ratio of ice nuclei to cloud condensation nuclei in a typical cold cloud is about 1:100,000, just about optimal for producing precipitation.

Some of the first experiments in cloud seeding were conducted by Vincent Schaefer and Irving Langmuir during the late 1940s. To seed a cloud, they dropped crushed pellets of *dry ice* (solid carbon dioxide) from a plane. Because dry ice has a temperature of $-78°C$ ($-108°F$), it acts as a cooling agent. As the extremely cold, dry ice pellets fall through the cloud, they quickly cool the air around them. This cooling causes the air around the pellet to become supersaturated. In this supersaturated air, water vapor forms directly into many tiny cloud droplets. In the very cold air created by the falling pellets (below $-40°C$), the tiny droplets instantly freeze into tiny ice crystals. The newly formed ice crystals then grow larger by deposition as the water vapor molecules attach themselves to the ice crystals at the expense of the nearby liquid droplets. Upon reaching a sufficiently large size, they fall as precipitation.

In 1947, Bernard Vonnegut demonstrated that silver iodide (AgI) could be used as a cloud-seeding agent. Because silver iodide has a crystalline structure similar to an ice crystal, it acts as an effective ice nucleus at temperatures of $-4°C$ ($25°F$) and lower. Silver iodide causes ice crystals to form in two primary ways:

1. Ice crystals form when silver iodide crystals come in contact with supercooled liquid droplets.
2. Ice crystals grow in size as water vapor deposits onto the silver iodide crystal.

Silver iodide is much easier to handle than dry ice, since it can be supplied to the cloud from burners located either on the ground or on the wing of a small aircraft. Although other substances, such as lead iodide and cupric sulfide, are also effective ice nuclei, silver iodide still remains the most commonly used substance in cloud-seeding projects. (Additional information on the controversial topic, the effectiveness of cloud seeding, is given in the Focus section on p. 175.)

Under certain conditions, clouds may be seeded naturally. For example, when cirriform clouds lie directly above a lower cloud deck, ice crystals may descend from the higher cloud and seed the cloud below (see ● Fig. 7.11). As the ice crystals mix into the lower cloud, supercooled droplets are converted to ice crystals, and the precipitation process is enhanced. Sometimes the ice crystals in the lower cloud may

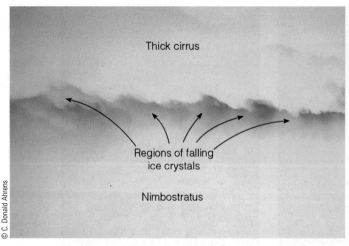

● FIGURE 7.11 Ice crystals falling from a dense cirriform cloud into a lower nimbostratus cloud. This photo was taken at an altitude near 6 km (19,700 ft) above western Pennsylvania. At the surface, moderate rain was falling over the region.

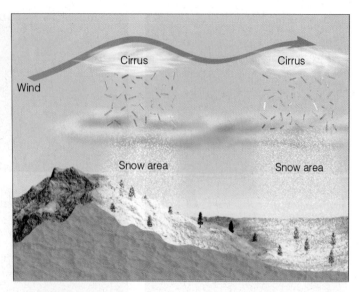

● FIGURE 7.12 Natural seeding by cirrus clouds may form bands of precipitation downwind of a mountain chain.

settle out, leaving a clear area or "hole" in the cloud. When the cirrus clouds form waves downwind from a mountain chain, bands of precipitation often form (see ● Fig. 7.12).

PRECIPITATION IN CLOUDS   In cold, strongly convective clouds, precipitation may begin only minutes after the cloud forms and may be initiated by either the collision-coalescence or the ice-crystal (Bergeron) process. Once either process begins, most precipitation growth is by accretion. Although precipitation is commonly absent in warm-layered clouds, such as stratus, it is often associated with such cold-layered clouds as nimbostratus and altostratus. This precipitation is thought to form principally by the ice-crystal (Bergeron) process because the liquid-water content of these clouds is generally lower than that in convective clouds, thus making

the collision-coalescence process much less effective. Nimbostratus clouds are normally thick enough to extend to levels where air temperatures are quite low, and they usually last long enough for the ice-crystal (Bergeron) process to initiate precipitation. ● Figure 7.13 illustrates how ice crystals produce precipitation in clouds of both low and high liquid-water content.

━━━━━━━━━━━━━━━━━━━━━━━━━

BRIEF REVIEW

In the last few sections we encountered a number of important concepts and ideas about how cloud droplets can grow large enough to fall as precipitation. Before examining the various types of precipitation, here is a summary of some of the important ideas presented so far:

● FIGURE 7.13
How ice crystals grow and produce precipitation in clouds with a low liquid-water content and a high liquid-water content.

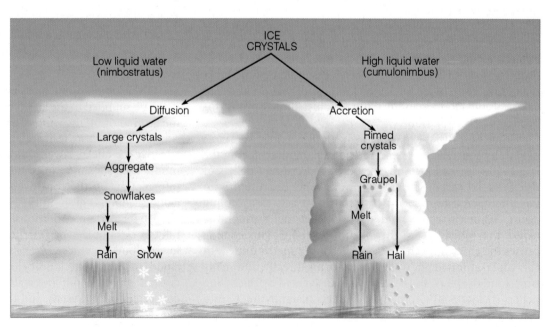

FOCUS ON AN ENVIRONMENTAL ISSUE

Does Cloud Seeding Enhance Precipitation?

Just how effective is artificial seeding with silver iodide in increasing precipitation? This is a much-debated question among meteorologists. First of all, it is difficult to evaluate the results of a cloud-seeding experiment. When a seeded cloud produces precipitation, the question always remains as to how much precipitation would have fallen had the cloud not been seeded.

Other factors must be considered when evaluating cloud-seeding experiments: the type of cloud, its temperature, moisture content, droplet size distribution, and updraft velocities in the cloud.

Although some experiments suggest that cloud seeding does not increase precipitation, others seem to indicate that seeding *under the right conditions* may enhance precipitation between 5 percent and 20 percent. And so the controversy continues.

Some cumulus clouds show an "explosive" growth after being seeded. The latent heat given off when the droplets freeze functions to warm the cloud, causing it to become more buoyant. It grows rapidly and becomes a longer-lasting cloud, which may produce more precipitation.

The business of cloud seeding can be a bit tricky, since overseeding can produce too many ice crystals. When this phenomenon occurs, the cloud becomes glaciated (all liquid droplets become ice) and the ice particles, being very small, do not fall as precipitation. Since few liquid droplets exist, the ice crystals cannot grow by the ice-crystal (Bergeron) process; rather, they evaporate, leaving a clear area in a thin, stratified cloud. Because dry ice can produce the most ice crystals in a supercooled cloud, it is the substance most suitable for deliberate overseeding. Hence, it is the substance most commonly used to dissipate cold fog at airports (see Chapter 5, p. 122).

Warm clouds with temperatures above freezing have also been seeded in an attempt to produce rain. Tiny water drops and particles of hygroscopic salt are injected into the base (or top) of the cloud. These particles (called *seed drops*), when carried into the cloud by updrafts, create large cloud droplets, which grow even larger by the collision-coalescence process. Apparently, the seed drop size plays a major role in determining the effectiveness of seeding with hygroscopic particles. To date, however, the results obtained using this method are inconclusive.

Cloud seeding may be inadvertent. Some industries emit large concentrations of condensation nuclei and ice nuclei into the air. Studies have shown that these particles are at least partly responsible for increasing precipitation in, and downwind of, cities. On the other hand, studies have also indicated that the burning of certain types of agricultural waste may produce smoke containing many condensation nuclei. These produce clouds that yield less precipitation because they contain numerous, but very small, droplets.

In summary, cloud seeding in certain instances may lead to more precipitation; in others, to less precipitation, and, in still others, to no change in precipitation amounts. Many of the questions about cloud seeding have yet to be resolved.

- Cloud droplets are very small, much too small to fall as rain.
- The smaller the cloud droplet, the greater its curvature, and the more likely it will evaporate.
- Cloud droplets form on cloud condensation nuclei. Hygroscopic nuclei, such as salt, allow condensation to begin when the relative humidity is less than 100 percent.
- Cloud droplets, in above-freezing air, can grow larger as faster-falling, bigger droplets collide and coalesce with smaller droplets in their path.
- In the ice-crystal (Bergeron) process of rain formation, both ice crystals and liquid cloud droplets must coexist at below-freezing temperatures. The difference in saturation vapor pressure between liquid and ice causes water vapor to diffuse from the liquid droplets (which shrink) toward the ice crystals (which grow).
- Most of the rain that falls over middle latitudes results from melted snow that formed from the ice-crystal (Bergeron) process.
- Cloud seeding with silver iodide can only be effective in coaxing precipitation from a cloud if the cloud is supercooled and the proper ratio of cloud droplets to ice crystals exists.

Precipitation Types

Up to now, we have seen how cloud droplets are able to grow large enough to fall to the ground as rain or snow. While falling, raindrops and snowflakes may be altered by atmospheric conditions encountered beneath the cloud and transformed into other forms of precipitation that can profoundly influence our environment.

RAIN Most people consider **rain** to be any falling drop of liquid water. To the meteorologist, however, that falling drop must have a diameter equal to, or greater than, 0.5 mm to be considered rain. Fine uniform drops of water whose diameters are smaller than 0.5 mm (which is a diameter about one-half the width of the letter "o" on this page) are called **drizzle.** Most drizzle falls from stratus clouds; however, small raindrops may fall through air that is unsaturated, partially evaporate, and reach the ground as drizzle. Occasionally, the rain falling from a cloud never reaches the surface because the low humidity causes rapid evaporation. As the drops become smaller, their rate of fall decreases, and they appear to

• FIGURE 7.14 The streaks of falling precipitation that evaporate before reaching the ground are called *virga*.

© Ross DePaola

hang in the air as a rain streamer. These evaporating streaks of precipitation are called **virga*** (see • Fig. 7.14).

Raindrops may also fall from a cloud and not reach the ground, if they encounter rapidly rising air. Large raindrops have a terminal velocity of about 9 m/sec (20 mi/hr), and, if they encounter rising air whose speed is greater than 9 m/sec, they will not reach the surface. If the updraft weakens or changes direction and becomes a downdraft, the suspended drops will fall to the ground as a sudden rain **shower.** The showers falling from cumuliform clouds are usually brief and sporadic, as the cloud moves overhead and then drifts on by. If the shower is excessively heavy, it is termed a *cloudburst.* Beneath a cumulonimbus cloud, which normally contains large convection currents of rising and descending air, it is entirely possible that one side of a street may be dry (updraft side), while a heavy shower is occurring across the street (downdraft side). Continuous rain, on the other hand, usually falls from a layered cloud that covers a large area and has smaller vertical air currents. These are the conditions normally associated with nimbostratus clouds.

Raindrops that reach the earth's surface are seldom larger than about 6 mm (0.2 in.), the reason being that the collisions (whether glancing or head-on) between raindrops tend to break them up into many smaller drops. Additionally, when raindrops grow too large they become unstable and break apart. What is the shape of the falling raindrop? Is it tear-shaped, or is it round? If you are unsure of the answer, read the Focus section on p. 177.

After a rainstorm, visibility usually improves primarily because precipitation removes (scavenges) many of the suspended particles. When rain combines with gaseous pollutants, such as oxides of sulfur and nitrogen, it becomes acidic. *Acid rain,* which has an adverse effect on plants and water resources, is becoming a major problem in many industrialized regions of the world.

It is important to know the interval of time over which rain falls. Did it fall over several days, gradually soaking into the soil? Or did it come all at once in a cloudburst, rapidly eroding the land, clogging city gutters, and causing floods along creeks and rivers unable to handle the sudden increased flow? The *intensity* of rain is the amount that falls in a given period; intensity of rain is always based on the accumulation during a certain interval of time (see ▼ Table 7.2).

▼ TABLE 7.2   Rainfall Intensity

Rainfall Description	Rainfall Rate (in./hr)*
Light	0.01 to 0.10
Moderate	0.11 to 0.30
Heavy	>0.30

*In the United States, the National Weather Service measures rainfall in inches.

---

*Studies suggest that the "rain streamer" is actually caused by ice (which is more reflective) changing to water (which is less reflective). Apparently, most evaporation occurs below the virga line.

FOCUS ON A SPECIAL TOPIC

Are Raindrops Tear-Shaped?

As rain falls, the drops take on a characteristic shape. Choose the shape in Fig. 2 that you feel most accurately describes that of a falling raindrop. Did you pick number 1? The tear-shaped drop has been depicted by artists for many years. Unfortunately, *raindrops are not tear-shaped.* Actually, the shape depends on the drop size. Raindrops less than 2 mm in diameter are nearly spherical and look like raindrop number 2. The attraction among the molecules of the liquid (surface tension) tends to squeeze the drop into a shape that has the smallest surface area for its total volume — a sphere.

Large raindrops, with diameters exceeding 2 mm, take on a different shape as they fall. Believe it or not, they look like number 3, slightly elongated, flattened on the bottom, and rounded on top. As the larger drop falls, the air pressure against the drop is greatest on the bottom and least on the sides. The pressure of the air on the bottom flattens the drop, while the lower pressure on its sides allows it to expand a little. This mushroom shape has been described as everything from a falling parachute to a loaf of bread, or even a hamburger bun. You may call it what you wish, but remember: It is not tear-shaped.

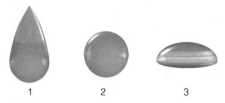

● FIGURE 2 Which of the three drops drawn here represents the real shape of a falling raindrop?

SNOW We know that much of the precipitation reaching the ground actually begins as **snow.** In summer, the freezing level is usually above 3600 m (12,000 ft), and the snowflakes falling from a cloud melt before reaching the ground. However, in winter, the freezing level is much lower, and falling snowflakes have a better chance of survival. Snowflakes can generally fall about 300 m (1000 ft) below the freezing level before completely melting. Occasionally, you can spot the melting level when you look in the direction of the sun, if it is near the horizon. Because snow scatters incoming sunlight better than rain, the darker region beneath the cloud contains falling snow, while the lighter region is falling rain. The melting zone, then, is the transition between the light and dark areas (see ● Fig. 7.15).

The sky will look different, however, if you are looking directly up at the precipitation. Because snowflakes are such effective scatterers of light, they redirect the light beneath the cloud in all directions — some of it eventually reaching your eyes, making the region beneath the cloud appear a lighter shade of gray. Falling raindrops, on the other hand, scatter very little light toward you, and the underside of the cloud appears dark. It is this change in shading that enables some observers to predict with uncanny accuracy whether falling precipitation will be in the form of rain or snow.

When ice crystals and snowflakes fall from high cirrus clouds they are called **fallstreaks.** Fallstreaks behave in much the same way as virga. As the ice particles fall into drier air, they usually sublimate (that is, change from ice into vapor). Because the winds at higher levels move the cloud and ice particles horizontally more quickly than do the slower winds at lower levels, fallstreaks appear as dangling white streamers (see ● Fig. 7.16). Moreover, fallstreaks descending into lower, supercooled clouds may actually seed them.

Snowflakes and Snowfall Snowflakes that fall through moist air that is slightly above freezing slowly melt as they descend. A thin film of water forms on the edge of the flakes, which acts like glue when other snowflakes come in contact with it. In this way, several flakes join to produce giant snowflakes often measuring several centimeters or more in diameter. These large, soggy snowflakes are associated with moist air and temperatures near freezing. However, when snowflakes fall through extremely cold air with a low moisture content, small, powdery flakes of "dry" snow accumulate on the ground. (To understand how snowflakes can survive in air that is above freezing, read the Focus section "Snowing When the Air Temperature Is Well Above Freezing" on p. 179.)

● FIGURE 7.15 Snow scatters sunlight more effectively than rain. Consequently, when you look toward the sun, the region of falling precipitation looks darker above the melting level than below it.

● **FIGURE 7.16** The dangling white streamers of ice crystals beneath these cirrus clouds are known as *fallstreaks*. The bending of the streaks is due to the changing wind speed with height.

If you catch falling snowflakes on a dark object and examine them closely, you will see that the most common snowflake form is a fernlike branching star shape called a *dendrite*. Since many types of ice crystals grow (see ● Fig. 7.17), why is the dendrite crystal the most common shape for snowflakes?

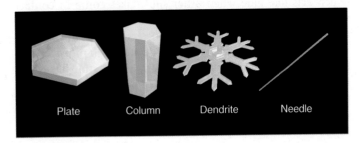

Plate Column Dendrite Needle

● **FIGURE 7.17** Common ice crystal forms (habits).

▼ **TABLE 7.3** Ice Crystal Habits That Form at Various Temperatures*

ENVIRONMENTAL TEMPERATURE		CRYSTAL HABIT
(°C)	(°F)	
0 to −4	32 to 25	Thin plates
−4 to −6	25 to 21	Needles
−6 to −10	21 to 14	Columns
−10 to −12	14 to 10	Plates
−12 to −16	10 to 3	Dendrites, plates
−16 to −22	3 to −8	Plates
−22 to −40	−8 to −40	Hollow columns

*Note that at each temperature, the type of crystal that forms (e.g., hollow columns versus solid columns) will depend on the difference in saturation vapor pressure between ice and supercooled water.

The type of crystal that forms, as well as its growth rate, depends on the air temperature and relative humidity (the degree of supersaturation between water and ice). ▼ Table 7.3 summarizes the crystal forms (habits) that develop when supercooled water and ice coexist in a saturated environment. Note that dendrites are common at temperatures between −12°C and −16°C. The maximum growth rate of ice crystals depends on the difference in saturation vapor pressure between water and ice, and this difference reaches a maximum in the temperature range where dendrite crystals are most likely to grow. (Look back at Fig. 7.8, p. 172.) Therefore, this type of crystal grows more rapidly than the other crystal forms. As ice crystals fall through a cloud, they are constantly exposed to changing temperature and moisture conditions. Since many ice crystals can join together (aggregate) to form a much larger snowflake, snow crystals may assume many complex patterns (see ● Fig. 7.18 on p. 180).

Snow falling from developing cumulus clouds is often in the form of **flurries**. These are usually light showers that fall intermittently for short durations and produce only light accumulations. A more intense snow shower is called a **snow squall**. These brief but heavy falls of snow are comparable to summer rain showers and, like snow flurries, usually fall from cumuliform clouds. A more continuous snowfall (sometimes steadily, for several hours) accompanies nimbostratus and altostratus clouds. The intensity of snow is based on its reduction of horizontal visibility at the time of observation (see ▼ Table 7.4).

When a strong wind is blowing at the surface, snow can be picked up and deposited into huge drifts. Drifting snow is usually accompanied by *blowing snow*; that is, snow lifted from the surface by the wind and blown about in such quantities that horizontal visibility is greatly restricted. The combination of drifting and blowing snow, after falling snow has ended, is called a *ground blizzard*. A true **blizzard** is a weather

FOCUS ON A SPECIAL TOPIC

Snowing When the Air Temperature Is Well Above Freezing

In the beginning of this chapter, we learned that it is never too cold to snow. So when is it too warm to snow? A person who has never been in a snowstorm might answer, "When the air temperature rises above freezing." However, anyone who lives in a climate that experiences cold winters knows that snow may fall when the air temperature is considerably above freezing (see Fig. 3). In fact, in some areas, snowstorms often begin with a surface air temperature near 2°C (36°F). Why doesn't the falling snow melt in this air? Actually, it does melt, at least to some degree. Let's examine this in more detail.

In order for falling snowflakes to survive in air with temperatures much above freezing, the air must be unsaturated (relative humidity is less than 100 percent), and the wet-bulb temperature must be at freezing or below. You may recall from our discussion on humidity in Chapter 4 that the wet-bulb temperature is the lowest temperature that can be attained by evaporating water into the air. Consequently, it is a measure of the amount of cooling that can occur in the atmosphere as water evaporates into the air. When rain falls into a layer of dry air with a low wet-bulb temperature, rapid evaporation and cooling occurs, which is why the air temperature often decreases when it begins to rain. During the winter, as raindrops evaporate in this dry air, rapid cooling may actually change a rainy day into a snowy one. This same type of cooling allows snowflakes to survive above freezing (melting) temperatures.

Suppose it is winter and the sky is overcast. At the surface, the air temperature is 2°C (36°F), the dew point is −6°C (21°F), and the

● FIGURE 3
It is snowing at 40°F during the middle of July near the summit of Beartooth Mountain, Montana.

© C. Donald Ahrens

wet-bulb temperature is 0°C (32°F).* The air temperature drops sharply with height, from the surface up to the cloud deck. Soon, flakes of snow begin to fall from the clouds into the unsaturated layer below. In the above-freezing temperatures, the snowflakes begin to partially melt. The air, however, is dry, so the water quickly evaporates, cooling the air. In addition, evaporation cools the falling snowflake to the wet-bulb temperature, which retards the flake's rate of melting. As snow continues to fall, evaporative cooling causes the air temperature to continue to drop. The addition of water vapor to the air increases the dew point, while the wet-bulb temperature remains essentially unchanged. Eventually, the entire layer of air cools to the wet-bulb temperature and becomes saturated at 0°C. As

*The wet-bulb temperature is always higher than the dew point, except when the air is saturated. At that point, the air temperature, wet-bulb temperature, and dew point are all the same.

long as the wind doesn't bring in warmer air, the precipitation remains as snow.

We can see that when snow falls into warmer air (say at 8°C or 46°F), the air must be extremely dry in order to have a wet-bulb temperature at freezing or below. In fact, with an air temperature of 8°C (46°F) and a wet-bulb temperature of 0°C (32°F), the dew point would be −23°C (−9°F) and the relative humidity 11 percent. Conditions such as these are extremely unlikely at the surface before the onset of precipitation. Actually, the highest air temperature possible with a below-freezing wet-bulb temperature is about 10°C (50°F). Hence, snowflakes will melt rapidly in air with a temperature above this value. However, it is still possible to see flakes of snow at temperatures greater than 10°C (50°F), especially if the snowflakes are swept rapidly earthward by the cold, relatively dry downdraft of a thunderstorm.

---

condition characterized by low temperatures and strong winds (greater than 30 knots) bearing large amounts of fine, dry, powdery particles of snow, which can reduce visibility to only a few meters.

A Blanket of Snow A mantle of snow covering the landscape is much more than a beautiful setting — it is a valuable resource provided by nature. A blanket of snow is a good insulator (poor heat conductor). In fact, the more air spaces

there are between the individual snowflake crystals, the better insulator they become. A light, fluffy covering of snow protects sensitive plants and their root systems from damaging low temperatures by retarding the loss of ground heat.

On winter nights, ground that is covered with dry snow maintains a higher temperature than ground that is exposed to the cold air. In this way, snow can prevent the ground from freezing downward to great depths. In cold climates that receive little snow, it is often difficult to grow certain crops be-

● FIGURE 7.18 Computer color-enhanced image of dendrite snowflakes.

▼ TABLE 7.4 Snowfall Intensity

SNOWFALL DESCRIPTION	VISIBILITY
Light	Greater than ½ mile*
Moderate	Greater than ¼ mile, less than or equal to ½ mile
Heavy	Less than or equal to ¼ mile

*In the United States, the National Weather Service determines visibility (the greatest distance you can see) in miles.

cause the frozen soil makes spring cultivation almost impossible. Frozen ground also prevents early spring rains from percolating downward into the soil, leading to rapid water runoff and flooding. If subsequent rains do not fall, the soil could even become moisture-deficient. If you become lost in a cold and windy snowstorm, build a snow cave and climb inside. It not only will protect you from the wind, but it also will protect you from the extreme cold by slowing the escape of heat your body generates.

The accumulation of snow in mountains provides for winter recreation, and the melting snow in spring and summer is of great economic value in that it supplies streams and reservoirs with much-needed water.

Winter snows may be beautiful, but they are not without hardships and potential hazards. As spring approaches, rapid melting of the snowpack may flood low-lying areas. Too much snow on the side of a steep hill or mountain may become an avalanche as the spring thaw approaches. The added weight of snow on the roof of a building may cause it to collapse, leading to costly repairs and even loss of life. Each winter, heavy snows clog streets and disrupt transportation. To keep traffic moving, streets must be plowed and sanded, or salted to lower the temperature at which the snow freezes (melts). This effort can be expensive, especially if the snow is heavy and wet. Cities unaccustomed to snow are usually harder hit by a moderate snowstorm than cities that frequently experience snow. A January snowfall of several centimeters in New Orleans, Louisiana, can bring traffic to a standstill, while a snowfall of several centimeters in Buffalo, New York, would go practically unnoticed.
● Figure 7.19 gives the annual average snowfall across the United States. (A blanket of snow also has an effect on the way sounds are transmitted. More on this subject is given in the Focus section on p.181.)

● FIGURE 7.19
Average annual snowfall over the United States. (NOAA)

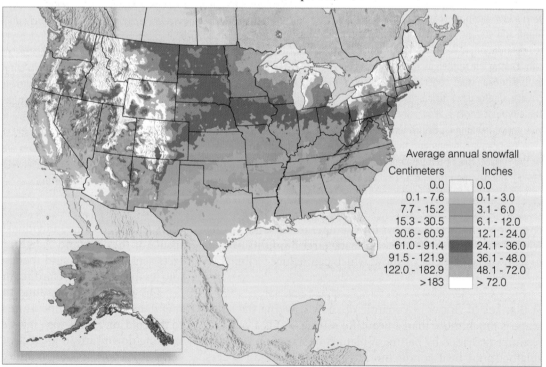

Average annual snowfall

Centimeters	Inches
0.0	0.0
0.1 - 7.6	0.1 - 3.0
7.7 - 15.2	3.1 - 6.0
15.3 - 30.5	6.1 - 12.0
30.6 - 60.9	12.1 - 24.0
61.0 - 91.4	24.1 - 36.0
91.5 - 121.9	36.1 - 48.0
122.0 - 182.9	48.1 - 72.0
>183	> 72.0

FOCUS ON A SPECIAL TOPIC

Sounds and Snowfalls

A blanket of snow is not only beautiful, but it can affect what we hear. You may have noticed that, after a snowfall, it seems quieter than usual: Freshly fallen snow can absorb sound—just like acoustic tiles. As the snow gets deeper, this absorption increases. Anyone who has walked through the woods on a snowy evening knows the quiet created by a thick blanket of snow. As snow becomes older and more densely packed, its ability to absorb sound is reduced. That's why sounds you couldn't hear right after a snowstorm become more audible several days later.

New snow covering a pavement will sometimes squeak as you walk in it. The sound produced is related to the snow's temperature. When the air and snow are only slightly below freezing, the pressure from the heel of a boot

© C Donald Ahrens

partially melts the snow. The snow can then flow under the weight of the boot and no sound is made. However, on cold days when the snow temperature drops below −10°C (14°F), the heel of the boot will not melt the snow, and the ice crystals are crushed. The crunching of the crystals produces the creaking sound.

● FIGURE 4 This freshly fallen blanket of snow absorbs sound waves so effectively that even the water flowing in the tiny stream is difficult to hear.

SLEET AND FREEZING RAIN Consider the falling snowflake in ● Fig. 7.20. As it falls into warmer air, it begins to melt. When it falls through the deep subfreezing surface layer of air, the partially melted snowflake or cold raindrop turns back into ice, not as a snowflake, but as a *tiny ice pellet* called

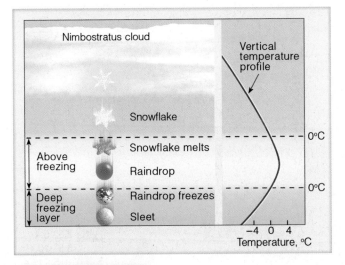

ACTIVE FIGURE 7.20 Sleet forms when a partially melted snowflake or a cold raindrop freezes into a pellet of ice before reaching the ground. Visit the Meteorology Resource Center to view this and other active figures at academic.cengage.com/login

sleet.* Generally, these ice pellets are transparent (or translucent), with diameters of 5 mm (0.2 in.) or less. They bounce when striking the ground and produce a tapping sound when they hit a window or piece of metal.

The cold surface layer beneath a cloud may be too shallow to freeze raindrops as they fall. In this case, they reach the surface as supercooled liquid drops. Upon striking a cold object, the drops spread out and almost immediately freeze, forming a thin veneer of ice. This form of precipitation is called **freezing rain,** or **glaze.** If the drops are small (less than 0.5 mm in diameter), the precipitation is called **freezing drizzle.** When small supercooled cloud or fog droplets strike an object whose temperature is below freezing, the tiny droplets freeze, forming an accumulation of white or milky granular ice called **rime** (see ● Fig. 7.21).†

Freezing rain can create a beautiful winter wonderland by coating everything with silvery, glistening ice. At the same time, highways turn into skating rinks for automobiles, and the destructive weight of the ice—which can be many tons on a single tree—breaks tree branches, power lines, and telephone cables (see ● Fig. 7.22). A case in point is the huge ice

---

*Occasionally, the news media incorrectly use the term *sleet* to represent a mixture of rain and snow. The term used in this manner is, however, the British meaning.

†When a sheet of ice covering a road surface or pavement appears relatively dark, it is often referred to as *black ice.* Black ice commonly forms when light rain, drizzle, or supercooled fog droplets come in contact with surfaces (especially those of bridges and overpasses) that have cooled to a temperature below freezing.

NCAR/UCAR/NSF

● FIGURE 7.21 An accumulation of rime forms on tree branches as supercooled fog droplets freeze on contact in the below-freezing air.

storm of January, 1998, which left millions of people without power in Northern New England and Canada and caused over $1 billion in damages. In the worst ice storm to hit Kansas and Missouri in 100 years, 5 cm (2 in.) of ice covered sections of these states in January, 2002, causing over 300,000 people to be without power. The area most frequently hit by these storms extends over a broad region from Texas into Minnesota and eastward into the middle Atlantic states and New England. Such storms are extremely rare in most of California and Florida (see ● Fig. 7.23). (For additional information on freezing rain and its effect on aircraft, read the Focus section on p. 183.)

In summary, ● Fig. 7.24 shows various winter temperature profiles and the type of precipitation associated with each. In profile (a), the air temperature is below freezing at all levels, and snowflakes reach the surface. In (b), a zone of above-

© Dick Blume/The Image Works

● FIGURE 7.22 A heavy coating of freezing rain (glaze) covers Syracuse, New York, during January, 1998, causing tree limbs to break and power lines to sag.

freezing air causes snowflakes to partially melt; then, in the deep, subfreezing air at the surface, the liquid freezes into sleet. In the shallow subfreezing surface air in (c), the melted snowflakes, now supercooled liquid drops, freeze on contact, producing freezing rain. In (d), the air temperature is above freezing in a sufficiently deep layer so that precipitation reaches the

● FIGURE 7.23 Average annual number of days with freezing rain and freezing drizzle over the United States. (NOAA)

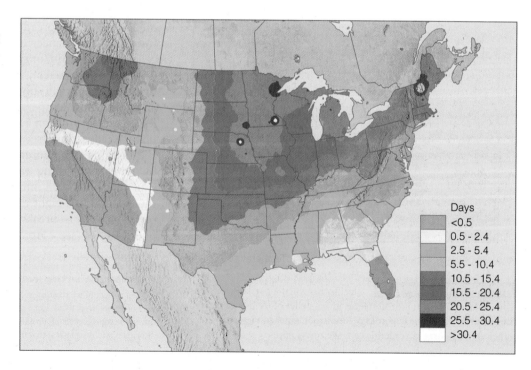

Days
<0.5
0.5 - 2.4
2.5 - 5.4
5.5 - 10.4
10.5 - 15.4
15.5 - 20.4
20.5 - 25.4
25.5 - 30.4
>30.4

Aircraft Icing

Consider an aircraft flying through an area of freezing rain or through a region of large super-cooled droplets in a cumuliform cloud. As the large, supercooled drops strike the leading edge of the wing, they break apart and form a film of water, which quickly freezes into a solid sheet of ice. This smooth, transparent ice — called *clear ice* — is similar to the freezing rain or glaze that coats trees during ice storms. Clear ice can build up quickly; it is heavy and difficult to remove, even with modern de-icers.

When an aircraft flies through a cloud composed of tiny, supercooled liquid droplets, *rime ice* may form. Rime ice forms when some of the cloud droplets strike the wing and freeze before they have time to spread, thus leaving a rough and brittle coating of ice on the wing. Because the small, frozen droplets trap air between them, rime ice usually appears white (see Fig. 7.21). Even though rime ice redistributes the flow of air over the wing more than clear ice does, it is lighter in weight and is more easily removed with de-icers.

• **FIGURE 5** An aircraft undergoing de-icing during inclement winter weather.

Because the raindrops and cloud droplets in most clouds vary in size, a mixture of clear and rime ice usually forms on aircraft. Also, be-

cause concentrations of liquid water tend to be greatest in warm air, icing is usually heaviest and most severe when the air temperature is between 0°C and −10°C (32°F and 14°F).

A major hazard to aviation, icing reduces aircraft efficiency by increasing weight. Icing has other adverse effects, depending on where it forms. On a wing or fuselage, ice can disrupt the air flow and decrease the plane's flying capability. When ice forms in the air intake of the engine, it robs the engine of air, causing a reduction in power. Icing may also affect the operation of brakes, landing gear, and instruments. Because of the hazards of ice on an aircraft, its wings are usually sprayed with a type of antifreeze before taking off during cold, inclement weather (see Fig. 5).

---

surface as rain. (Weather symbols for these and other forms of precipitation are presented in Appendix B.)

SNOW GRAINS AND SNOW PELLETS **Snow grains** are small, opaque grains of ice, the solid equivalent of drizzle. They are fairly flat or elongated, with diameters generally less than 1 mm (0.04 in.). They fall in small quantities from stra-

tus clouds, and never in the form of a shower. Upon striking a hard surface, they neither bounce nor shatter. **Snow pellets,** on the other hand, are white, opaque grains of ice, with diameters less than 5 mm (0.2 in.). They are sometimes confused with snow grains. The distinction is easily made, however, by remembering that, unlike snow grains, snow pellets are brittle, crunchy, and bounce (or break apart) upon hitting a hard

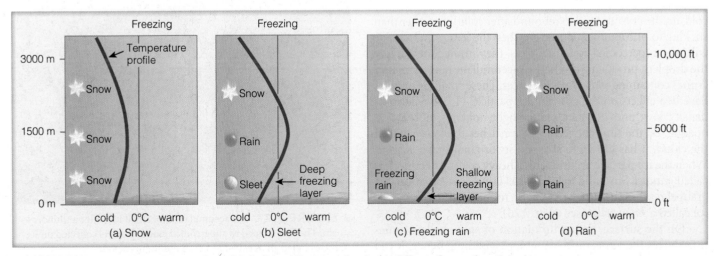

• FIGURE 7.24 Vertical temperature profiles (solid red line) associated with different forms of precipitation.

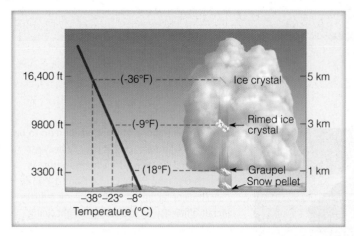

● FIGURE 7.25 The formation of snow pellets. In the cold air of a convective cloud, with a high liquid-water content, ice particles collide with supercooled cloud droplets, freezing them into clumps of icy matter called *graupel*. Upon reaching the relatively cold surface, the graupel is classified as snow pellets.

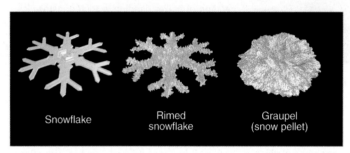

● FIGURE 7.26 A snowflake becoming a rimed snowflake, then finally graupel (a snow pellet).

surface. They usually fall as showers, especially from cumulus congestus clouds.

To understand how snow pellets form consider the cumulus congestus cloud with a high liquid-water content in ● Fig. 7.25. The freezing level is near the surface and, since the atmosphere is conditionally unstable, the air temperature drops quickly with height. An ice crystal falling into the cold (−23°C) middle region of the cloud would be surrounded by many supercooled cloud droplets and ice crystals. In the very cold air, the crystals tend to rebound after colliding rather than sticking to one another. However, when the ice crystals collide with the supercooled water droplets, they immediately freeze the droplets, producing a spherical accumulation of icy matter (rime) containing many tiny air spaces. These small air bubbles have two effects on the growing ice particle: (1) They keep its density low; and (2) they scatter light, making the particle opaque. By the time the ice particle reaches the lower half of the cloud, it has grown in size and its original shape is gone. When the ice particle accumulates a heavy coating of rime, it is called *graupel*. Since the freezing level is at a low elevation, the graupel reaches the surface as a light, round clump of snowlike ice called a *snow pellet* (see ● Fig. 7.26).

On the surface, the accumulation of snow pellets sometimes gives the appearance of tapioca pudding; hence, it can

be referred to as *tapioca snow.* In a thunderstorm, when the freezing level is well above the surface, graupel that reaches the ground is sometimes called *soft hail.* During summer, the graupel may melt and reach the surface as a large raindrop. In vigorously convective clouds, however, the graupel may develop into full-fledged hailstones.

HAIL **Hailstones** are pieces of ice, either transparent or partially opaque, ranging in size from that of small peas to that of golf balls or larger (see ● Fig. 7.27). Some are round; others take on irregular shapes. The largest authenticated hailstone in the United States fell on Aurora, Nebraska, during June, 2003. This giant hailstone had a measured diameter of 17.8 cm (7 in.) and a circumference of 47.6 cm (18.7 in.) (see ● Fig. 7.28). Although an accurate weight was difficult to obtain, the hailstone (being almost as large as a soccer ball) probably weighed over 1.75 lbs. Canada's record hailstone fell on Cedoux, Saskatchewan, during August, 1973. It weighed 290 grams (0.6 lb) and measured about 10 cm (4 in.) in diameter. Needless to say, large hailstones are quite destructive. They can break windows, dent cars, batter roofs of homes, and cause extensive damage to livestock and crops. In fact, a single hailstorm can destroy a farmer's crop in a matter of minutes, which is why farmers sometimes call it "the white plague."

Estimates are that, in the United States alone, hail damage amounts to hundreds of millions of dollars annually. The costliest hailstorm on record in the United States battered the Front Range of the Rocky Mountains in Colorado with golf ball- and baseball-size hail on July 11, 1990. The storm damaged thousands of roofs and tens of thousands of cars, trucks, and streetlights, causing an estimated $625 million in damage. Although hailstones are potentially lethal, only two fatalities due to falling hail have been documented in the United States during the twentieth century.

© C. Donald Ahrens

● FIGURE 7.27 The accumulation of small hail after a thunderstorm. The hail formed as supercooled cloud droplets collected on ice particles called *graupel* inside a cumulonimbus cloud.

● FIGURE 7.28 This giant hailstone—the largest ever reported in the United States with a diameter of 17.8 cm (7 in.)—fell on Aurora, Nebraska, during June, 2003.

Hail is produced in a cumulonimbus cloud—usually an intense thunderstorm—when graupel, or large frozen raindrops, or just about any particles (even insects) act as *embryos* that grow by accumulating supercooled liquid droplets—*accretion*. It takes a million cloud droplets to form a single raindrop, but it takes about 10 billion cloud droplets to form a golf ball–size hailstone. For a hailstone to grow to this size, it must remain in the cloud between 5 and 10 minutes. Violent, upsurging air currents within the storm carry small ice particles high above the freezing level where the ice particles grow by colliding with supercooled liquid cloud droplets. Violent rotating updrafts in severe thunderstorms are even capable of sweeping the growing ice particles latterly through the cloud. In fact, it appears that the best trajectory for hailstone growth is one that is nearly horizontal through the storm (see ● Fig. 7.29).

As growing ice particles pass through regions of varying liquid water content, a coating of ice forms around them, causing them to grow larger and larger. In a strong updraft, the larger hailstones ascend very slowly, and may appear to "float" in the updraft, where they continue to grow rapidly by colliding with numerous supercooled liquid droplets. When winds aloft carry the large hailstones away from the updraft or when the hailstones reach appreciable size, they become too heavy to be supported by the rising air, and they begin to fall.

In the warmer air below the cloud, the hailstones begin to melt. Small hail often completely melts before reaching the ground, but in the violent thunderstorms of late spring and summer, hailstones often grow large enough to reach the surface before completely melting. Strangely, then, we find the largest form of frozen precipitation occurring during the warmest time of the year.

● Figure 7.30 shows a cut section of a very large hailstone. Notice that it has distinct concentric layers of milky white and clear ice. We know that a hailstone grows by accumulating supercooled water droplets. If the growing hailstone enters a region inside the storm where the liquid water content is relatively low (called the *dry growth regime*), supercooled droplets will freeze immediately on the stone, producing a coating of white or opaque rime ice containing many air bubbles. As supercooled water droplets freeze onto the hailstone's surface, the liquid-to-ice transformation releases latent heat, which keeps the hailstone's surface temperature (which is below freezing) warmer than that of its environment. As long as the hailstone's surface temperature remains below freezing, liquid supercooled droplets freeze on contact, producing a coating of rime.

Should, however, the hailstone get swept into a region of the storm where the liquid-water contact is higher (called the

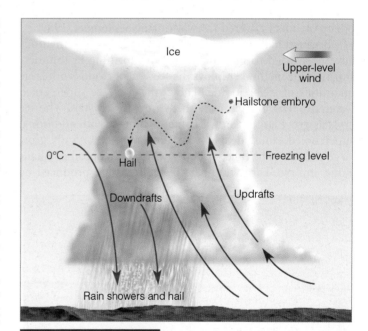

ACTIVE FIGURE 7.29 Hailstones begin as embryos (usually ice particles called *graupel*) that remain suspended in the cloud by violent updrafts. When the updrafts are tilted, the ice particles are swept horizontally through the cloud, producing the optimal trajectory for hailstone growth. Along their path, the ice particles collide with supercooled liquid droplets, which freeze on contact. The ice particles eventually grow large enough and heavy enough to fall toward the ground as hailstones. Visit the Meteorology Resource Center to view this and other active figures at academic.cengage.com/login

wet growth regime), supercooled water droplets will collect so rapidly on the stone that, due to the release of latent heat, the stone's surface temperature will remain at 0°C. Now the supercooled droplets no longer freeze on impact; instead, they spread a coating of water around the hailstone, filling in the porous regions. As the water coating the hailstone slowly freezes, air bubbles are able to escape, leaving a layer of clear ice around the stone. Therefore, as a hailstone passes through a thunderstorm of changing liquid water content (the dry and wet growth regimes) alternating layers of opaque and clear ice form, as illustrated in Fig. 7.30.

As a thunderstorm moves along, it may deposit its hail in a long narrow band (often several kilometers wide and about 10 kilometers long) known as a **hailstreak.** If the storm should remain almost stationary for a period of time, substantial accumulation of hail is possible. For example, in June, 1984, a devastating hailstorm lasting over an hour dumped knee-deep hail on the suburbs of Denver, Colorado. In addition to its destructive effect, accumulation of hail on a roadway is a hazard to traffic, as when, for example, four people lost their lives near Soda Springs, California, in a 15-vehicle pileup on a hail-covered freeway during September, 1989.

Because hailstones are so damaging, various methods have been tried to prevent them from forming in thunderstorms. One method employs the seeding of clouds with large quantities of silver iodide. These nuclei freeze supercooled water droplets and convert them into ice crystals. The ice crystals grow larger as they come in contact with additional supercooled cloud droplets. In time, the ice crystals grow large enough to be called graupel, which then becomes a hailstone embryo. Large numbers of embryos are produced by seeding in hopes that competition for the remaining supercooled droplets may be so great that none of the embryos would be able to grow into large and destructive hailstones. Russian scientists claim great success in suppressing hail using ice nuclei, such as silver iodide and lead iodide. In the United States, the results of most hail-suppression experiments have been inconclusive.

Up to this point, we have examined the various types of precipitation. The different types (from drizzle to hail) are summarized in ▼ Table 7.5.

Measuring Precipitation

INSTRUMENTS   Any instrument that can collect and measure rainfall is called a *rain gauge.* A **standard rain gauge** consists of a funnel-shaped collector attached to a long measuring tube (see ● Fig. 7.31). The cross-sectional area of the collector is 10 times that of the tube. Hence, rain falling into the collector is amplified tenfold in the tube, permitting measurements of great precision. A wooden scale, calibrated to allow for the vertical exaggeration, is inserted into the tube and withdrawn. The wet portion of the scale indicates the depth of water. So, 10 inches of water in the tube would be measured as 1 inch of rainfall. Because of this amplification, rainfall measurements can be made when the amount is as small as one-hundredth (0.01) of an inch. An amount of rainfall less than one-hundredth of an inch is called a **trace.**

The measuring tube can only collect 5 cm or 2 in. of rain. Rainfall of more than this amount causes an overflow into an outer cylinder. Here, the excess rainfall is stored and protected from appreciable evaporation. When the gauge is emptied, the overflow is carefully poured into the tube and measured.

Another instrument that measures rainfall is the **tipping bucket rain gauge.** In ● Fig. 7.32, notice that this gauge has a receiving funnel leading to two small metal collectors (buckets). The bucket beneath the funnel collects the rain water. When it

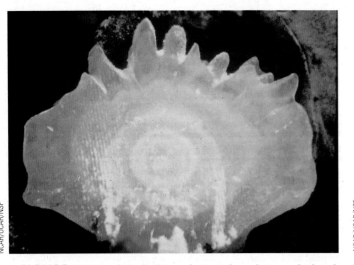

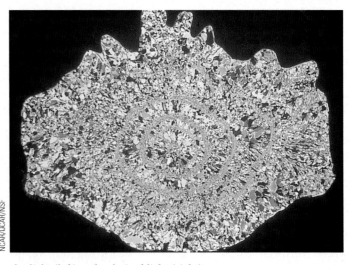

● FIGURE 7.30 A large hailstone first cut then photographed under regular light (left) and polarized light (right). This procedure reveals its layered structure.

▼ TABLE 7.5   Summary of Precipitation Types

PRECIPITATION TYPE	WEATHER SYMBOL	DESCRIPTION
Drizzle	'' (light)	Tiny water drops with diameters less than 0.5 mm that fall slowly, usually from a stratus cloud
Rain	•• (light)	Falling liquid drops that have diameters greater than 0.5 mm
Snow	✱✱ (light)	White (or translucent) ice crystals in complex hexagonal (six-sided) shapes that often join together to form snowflakes
Sleet (ice pellets)	△	Frozen raindrops that form as cold raindrops (or partially melted snowflakes) refreeze while falling through a relatively deep subfreezing layer
Freezing rain	∿ (light)	Supercooled raindrops that fall through a relatively shallow subfreezing layer and freeze upon contact with cold objects at the surface
Snow grains (granular snow)	△	White or opaque particles of ice less than 1 mm in diameter that usually fall from stratus clouds, and are the solid equivalent of drizzle
Snow pellets (graupel)	⚥ (light showers)	Brittle, soft white (or opaque), usually round particles of ice with diameters less than 5 mm that generally fall as showers from cumuliform clouds; they are softer and larger than snow grains
Hail	⚥ (moderate or heavy showers)	Transparent or partially opaque ice particles in the shape of balls or irregular lumps that range in size from that of a pea to that of a softball; the largest form of precipitation. *Large hail* has a diameter of ¾ in. or greater; hail almost always is produced in a thunderstorm

accumulates the equivalent of one-hundredth of an inch of rain, the weight of the water causes it to tip and empty itself. The second bucket immediately moves under the funnel to catch the water. When it fills, it also tips and empties itself, while the original bucket moves back beneath the funnel. Each time a bucket tips, an electric contact is made, causing a pen to register a mark on a remote recording chart. Adding up the total number of marks gives the rainfall for a certain time period.

A problem with the tipping bucket rain gauge is that during each "tip" it loses some rainfall and, therefore, undermeasures rainfall amounts, especially during heavy downpours. The tipping bucket is the rain gauge used in the automated (ASOS) weather stations.

Remote recording of precipitation can also be made with a **weighing-type rain gauge.** With this gauge, precipitation is caught in a cylinder and accumulates in a bucket. The bucket

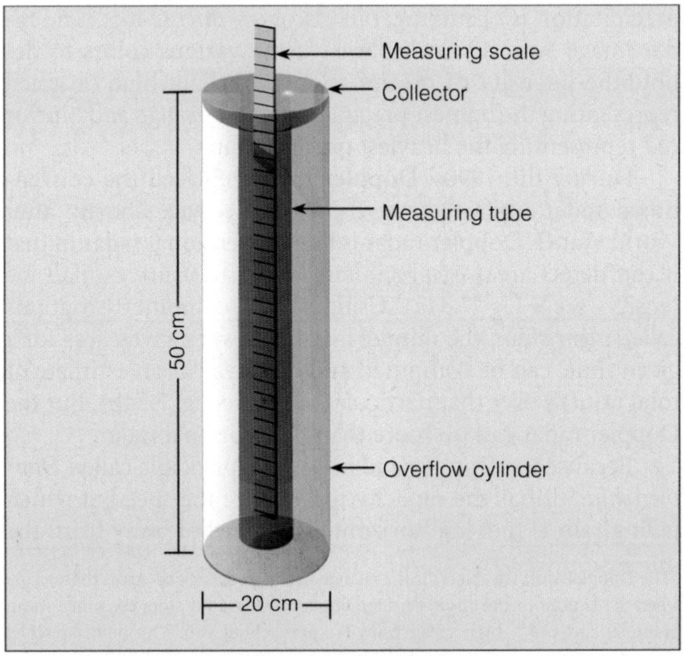

● FIGURE 7.31 Components of the standard rain gauge.

● FIGURE 7.32 The tipping bucket rain gauge. Each time the bucket fills with one-hundredth of an inch of rain, it tips, sending an electric signal to the remote recorder.

sits on a sensitive weighing platform. Special gears translate the accumulated weight of rain or snow into millimeters or inches of precipitation. The precipitation totals are recorded by a pen on chart paper, which covers a clock-driven drum. By using special electronic equipment, this information can be transmitted from rain gauges in remote areas to satellites or land-based stations, thus providing precipitation totals from previously inaccessible regions.

The depth of snow can be determined by measuring its depth at three or more representative areas. The amount of snowfall is defined as the average of these measurements. Since snow often blows around and accumulates into drifts, finding a representative area can be a problem. Determining the actual depth of snow can include considerable educated guesswork. Snow depth may also be measured by removing the collector and inner cylinder of a standard rain gauge and allowing snow to accumulate in the outer tube. Turbulent air around the edge of the tube often blows flakes away from the gauge. This makes the amount of snow collected less than the actual snowfall. To remedy this, slatted windshields are placed around the cylinder to block the wind and ensure a more correct catch.

The depth of water that would result from the melting of a snow sample is called the **water equivalent.** In a typical fresh snowpack, about 10 cm of snow will melt down to about 1 cm of water, giving a water equivalent ratio of 10:1. This ratio, however, will vary greatly, depending on the texture and packing of the snow. Very wet snow falling in air near freezing may have a water equivalent of 6:1. On the other hand, in dry powdery snow, the ratio may be as high as 30:1. Toward the end of the winter, large compacted drifts representing the accumulation of many storms may have a water equivalent of less than 2:1.

Determining the water equivalent of snow is a fairly straightforward process: The snow accumulated in a rain gauge is melted and its depth is measured. Another method uses a long, hollow tube pushed into the snow to a desired depth. This snow sample is then melted and poured into a rain gauge for measuring its depth. Knowing the water equivalent of snow can provide valuable information about spring runoff and the potential for flooding, especially in mountain areas.

Precipitation is a highly variable weather element. A huge thunderstorm may drench one section of a town while leaving another section completely dry. Given this variability, it should be apparent that a single rain gauge on top of a building cannot represent the total precipitation for any particular region.

DOPPLER RADAR AND PRECIPITATION   **Radar** (*radio detection and ranging*) has become an essential tool of the atmospheric scientist, for it gathers information about storms and precipitation in previously inaccessible regions. Atmospheric scientists use radar to examine the inside of a cloud much like physicians use X-rays to examine the inside of a human body. Essentially, the radar unit consists of a transmitter that sends out short, powerful microwave pulses. When this energy encounters a foreign object—called a *target*—a fraction of the energy is scattered back toward the transmitter and is detected by a receiver (see ● Fig. 7.33). The returning signal is amplified and displayed on a screen, producing an image, or *echo,* from the target. The elapsed time between transmission and reception indicates the target's distance.

Smaller targets require detection by shorter wavelengths. Cloud droplets are detected by radar using wavelengths of 1 cm, whereas longer wavelengths (between 3 and 10 cm) are only weakly scattered by tiny cloud droplets, but are strongly scattered by larger precipitation particles. The brightness of the echo is directly related to the amount (intensity) of rain falling in the cloud. So, the radar screen shows not only *where* precipitation is occurring, but also how *intense* it is. The radar image typically is displayed using various colors to denote the intensity of precipitation, with light blue or green representing the lightest precipitation and orange and blue or red representing the heaviest precipitation.

During the 1990s, **Doppler radar** replaced the conventional radar units that were put into service shortly after World War II. Doppler radar is like conventional radar in that it can detect areas of precipitation and measure rainfall intensity (see ● Fig. 7.34a). Using special computer programs called *algorithms,* the rainfall intensity, over a given area for a given time, can be computed and displayed as an estimate of total rainfall over that particular area (see Fig. 7.34b). But the Doppler radar can do more than conventional radar.

Because the Doppler radar uses the principle called *Doppler shift,** it has the capacity to measure the speed at which falling rain is moving horizontally toward or away from the

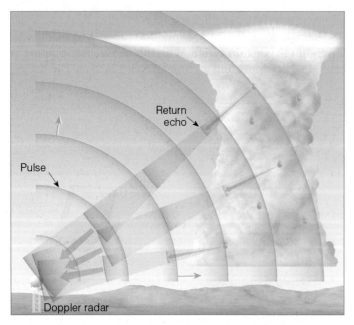

● FIGURE 7.33 A microwave pulse is sent out from the radar transmitter. The pulse strikes raindrops and a fraction of its energy is reflected back to the radar unit, where it is detected and displayed, as shown in Fig. 7.34.

---

*The Doppler shift (or effect) is the change in the frequency of waves that occurs when the emitter or the observer is moving toward or away from the other. As an example, suppose a high-speed train is approaching you. The higher-pitched (higher frequency) whistle you hear as the train approaches will shift to a lower pitch (lower frequency) after the train passes.

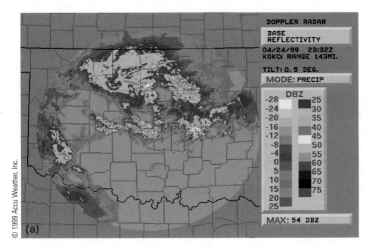

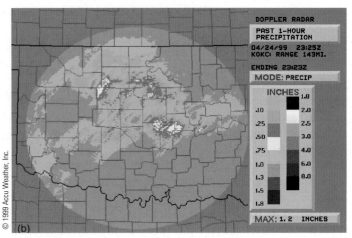

• FIGURE 7.34 (a) Doppler radar display showing precipitation intensity over Oklahoma for April 24, 1999. The lightest precipitation is shown as blue and green; heavier rainfall is indicated by the color yellow. The numbers under the letters DBZ represent the logarithmic scale for measuring the size and volume of precipitation particles. (b) Doppler radar display showing 1-hour rainfall amounts over Oklahoma for April 24, 1999.

radar antenna. Falling rain moves with the wind. Consequently, Doppler radar allows scientists to peer into a tornado-generating thunderstorm and observe its wind. We will investigate these ideas further in Chapter 14, when we consider the formation of severe thunderstorms and tornadoes.

In some instances, radar displays indicate precipitation where there is none reaching the surface. This situation happens because the radar beam travels in a straight line and the earth curves away from it. Hence, the return echo is not necessarily that of precipitation reaching the ground, but is that of raindrops in the cloud. So, if Doppler radar indicates that it's raining in your area, and outside you observe that it is not, remember it is raining, but the raindrops are probably evaporating before reaching the ground.

The next improvement for Doppler radar is *polarimetric radar*. This form of Doppler radar transmits both a vertical and horizontal pulse that will make it easier to determine whether falling precipitation is in the form of rain or snow.

MEASURING PRECIPITATION FROM SPACE As it circles the earth at an altitude of about 400 km, the *TRMM* (*Tropical Rainfall Measuring Mission*) satellite is able to measure rainfall intensity in previously inaccessible regions of the tropics and subtropics. The onboard Precipitation Radar is capable of detecting rainfall rates down to about 0.7 mm (0.03 in.) per hour, while at the same time providing vertical profiles of rain and snow intensity from the surface up to about 20 km (12 mi). The Microwave Imager complements the Precipitation Radar by measuring emitted microwave energy from the earth, the atmosphere, clouds, and precipitation, which is translated into rainfall rates.

The Visible and Infrared Scanner (VIS) onboard the satellite measures visible and infrared energy from the earth, the atmosphere, and clouds. This information is used to determine such things as the temperature of cloud tops, which can then be translated into rainfall rates. A *TRMM* satellite image of Hurricane Humberto and its pattern of precipitation is provided in • Fig. 7.35.

Launched in April 2006, the satellite *CloudSat* circles the earth in an orbit about 700 km above the surface. Onboard *CloudSat,* a very sensitive radar (called the Cloud Profiling Radar, or CPR) is able to peer into a cloud and provide a vertical view of its tiny cloud droplets and ice particles. Such vertical profiling of liquid water and ice will hopefully provide scientists with a better understanding of precipitation processes that go on inside the cloud and the role that clouds play in the earth's global climate system.

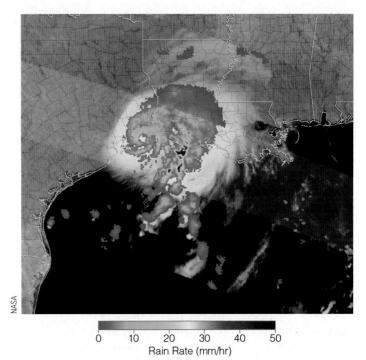

• FIGURE 7.35 A satellite and radar image of Hurricane Humberto obtained by the *TRMM* satellite on September 13, 2007. Precipitation rates (lowest in blue, highest in dark red) were obtained by the satellite's Precipitation Radar and Microwave Imager. The rainfall estimates are overlain on the infrared image of the storm.

SUMMARY

In this chapter, we have seen that cloud droplets are too small and light to reach the ground as rain. Cloud droplets do grow larger by condensation, but this process by itself is much too slow to produce substantial precipitation. Because larger cloud droplets fall faster and farther than smaller ones, they grow larger as they fall by coalescing with drops in their path. If the air temperature in a cloud drops below freezing, then ice crystals play an important role in producing precipitation. Some ice crystals may form directly on ice nuclei, or they may result when an ice nucleus makes contact with and freezes a supercooled water droplet. Because of differences in vapor pressures between water and ice, an ice crystal surrounded by water droplets grows larger at the expense of the droplets. As the ice crystal begins to fall, it grows even larger by colliding with supercooled liquid droplets, which freeze on contact. In an attempt to coax more precipitation from them, some clouds are seeded with silver iodide.

Precipitation can reach the surface in a variety of forms. In winter, raindrops may freeze on impact, producing freezing rain that can disrupt electrical service by downing power lines. Raindrops may freeze into tiny pellets of ice above the ground and reach the surface as sleet. Depending on conditions, snow may fall as pellets, grains, or flakes, all of which can influence how far we see and hear. Strong updrafts in a cumulonimbus cloud may keep ice particles suspended above the freezing level, where they acquire a further coating of ice and form destructive hailstones. Although the rain gauge is still the most commonly used method of measuring precipitation, Doppler radar has become an important instrument for determining precipitation intensity and estimating rainfall amount. In tropical regions, rainfall estimates can be obtained from radar and microwave scanners onboard satellites.

KEY TERMS

The following terms are listed in the order they appear in the text. Define each. Doing so will aid you in reviewing the material covered in this chapter.

precipitation, 166
equilibrium vapor pressure, 166
curvature effect, 167
solute effect, 167
collision-coalescence process, 168
terminal velocity, 168
coalescence, 168
ice-crystal (Bergeron) process, 169
supercooled (water droplets), 169
ice nuclei, 170
contact freezing, 170
accretion, 172
graupel, 172
aggregation, 172
snowflake, 172

cloud seeding, 173
rain, 175
drizzle, 175
virga, 176
shower (rain), 176
snow, 177
fallstreaks, 177
flurries (of snow), 178
snow squall, 178
blizzard, 178
sleet (ice pellets), 181
freezing rain (glaze), 181
freezing drizzle, 181

rime, 181
snow grains, 183
snow pellets, 183
hailstones, 184
hailstreak, 186
standard rain gauge, 186

trace (of precipitation), 186
tipping bucket rain gauge, 186
weighing-type rain gauge, 187
water equivalent, 188
radar, 188
Doppler radar, 188

QUESTIONS FOR REVIEW

1. What is the primary difference between a cloud droplet and a raindrop?
2. Why do typical cloud droplets seldom reach the ground as rain?
3. Describe how the process of collision and coalescence produces precipitation.
4. Would the collision-and-coalescence process work better at producing rain in (a) a warm, thick nimbostratus cloud or (b) a warm, towering cumulus congestus cloud? Explain.
5. List and describe three ways in which ice crystals can form in a cloud.
6. When the temperature in a cloud is −30°C, are larger cloud droplets more likely to freeze than smaller cloud droplets? Explain.
7. In a cloud where the air temperature is −10°C, why are there many more cloud droplets than ice crystals?
8. How does the ice-crystal (Bergeron) process produce precipitation? What is the *main* premise describing this process?
9. Why do heavy showers usually fall from cumuliform clouds? Why does steady precipitation normally fall from stratiform clouds?
10. Why are large snowflakes usually observed when the air temperature near the ground is just below freezing?
11. In a cloud composed of water droplets and ice crystals, is the saturation vapor pressure greater over the droplets or over the ice?
12. Why is it foolish to seed a clear sky with silver iodide?
13. When seeding a cloud to promote rainfall, is it possible to overseed the cloud so that it prevents rainfall? Explain.
14. Explain how clouds can be seeded naturally.
15. What atmospheric conditions are necessary for snow to fall when the air temperature is considerably above freezing?
16. List the advantages and disadvantages of heavy snowfall.
17. How do the atmospheric conditions that produce sleet differ from those that produce hail?
18. What is the difference between freezing rain and sleet?
19. Describe how hail might form in a cumulonimbus cloud.
20. Why is hail more common in summer than in winter?
21. List the common precipitation gauges that measure rain and snow.
22. (a) What is Doppler radar? (b) How does Doppler radar measure the intensity of precipitation?

QUESTIONS FOR THOUGHT

1. Ice crystals that form by accretion are fairly large. Explain why they fall slowly.
2. Why is a warm, tropical cumulus cloud more likely to produce precipitation than a cold, stratus cloud?
3. Explain why very small cloud droplets of pure water evaporate even when the relative humidity is 100 percent.
4. Suppose a thick nimbostratus cloud contains ice crystals and cloud droplets all about the same size. Which precipitation process will be most important in producing rain from this cloud? Why?
5. Clouds that form over water are usually more efficient in producing precipitation than clouds that form over land. Why?
6. Everyday in summer a blizzard occurs over the Great Plains. Explain where and why.
7. During a recent snowstorm, Denver, Colorado, received 7 cm (3 in.) of snow. Sixty kilometers east of Denver, a city received no measurable snowfall, while 150 km east of Denver another city received 10 cm (4 in.) of snow. Since Denver is located to the east of the Rockies, and the upper-level winds were westerly during the snowstorm, give an explanation as to what *could* account for this snowfall pattern.
8. Raindrops rarely grow larger than 5 mm. Two reasons were given on p. 169. Can you think of a third? (Hint: See the Focus section on p. 177, and look at the shape of a large drop.)
9. Lead iodide is an effective ice-forming nucleus. Why do you think it has *not* been used for that purpose?
10. When cirrus clouds are above a deck of altocumulus clouds, occasionally a clear area, or "hole," will appear in the altocumulus cloud layer. What do you suppose could cause this to happen?
11. It is −12°C (10°F) in Albany, New York, and freezing rain is falling. Can you explain why? Draw a vertical profile of the air temperature (a sounding) that illustrates why freezing rain is occurring at the surface.
12. When falling snowflakes become mixed with sleet, why is this condition often followed by the snowflakes changing into rain?
13. A major snowstorm occurred in a city in northern New Jersey. Three volunteer weather observers measured the snowfall. Observer #1 measured the depth of newly fallen snow every hour. At the end of the storm, Observer #1 added up the measurements and came up with a total of 12 inches of new snow. Observer #2 measured the depth of new snow twice: once in the middle of the storm and once at the end, and came up with a total snowfall of 10 inches. Observer #3 measured the new snowfall only once, after the storm had stopped, and reported 8.4 inches. Which of the three observers do you feel has the correct snowfall total? List *at least five* possible reasons why the snowfall totals were different.

PROBLEMS AND EXERCISES

1. In the daily newspaper, a city is reported as receiving 1.32 cm (0.52 in.) of precipitation over a 24-hour period. If all the precipitation fell as snow, and if we assume a normal water equivalent ratio of 10:1, how much snow did this city receive?
2. How many times faster does a large raindrop (diameter 5000 μm) fall than a cloud droplet (diameter 20 μm), if both are falling at their terminal velocity in still air?
3. (a) How many minutes would it take drizzle with a diameter of 200 μm to reach the surface if it falls at its terminal velocity from the base of a cloud 1000 m (about 3300 ft) above the ground? (Assume the air is saturated beneath the cloud, the drizzle does not evaporate, and the air is still.)
   (b) Suppose the drizzle in problem 3a evaporates on its way to the ground. If the drop size is 200 μm for the first 450 m of descent, 100 μm for the next 450 m, and 20 μm for the final 100 m, how long will it take the drizzle to reach the ground if it falls in still air?
4. Suppose a large raindrop (diameter 5000 μm) falls at its terminal velocity from the base of a cloud 1500 m (about 5000 ft) above the ground.
   (a) If we assume the raindrop does not evaporate, how long would it take the drop to reach the surface?
   (b) What would be the shape of the falling raindrop just before it reaches the ground?
   (c) What type of cloud would you expect this raindrop to fall from? Explain.
5. In ● Fig. 7.36, a drawing of a large hailstone, explain how the areas of clear ice and rime ice could have formed.

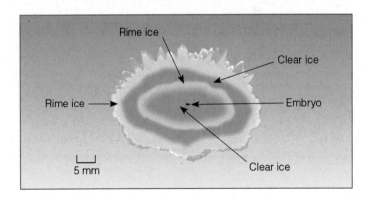

● FIGURE 7.36

Strong swirling winds help to develop these clouds over the Labrador Sea, off the coast of Newfoundland.

CHAPTER 8

Air Pressure and Winds

December 19, 1980, was a cool day in Lynn, Massachusetts, but not cool enough to dampen the spirits of more than 2000 people who gathered in Central Square — all hoping to catch at least one of the 1500 dollar bills that would be dropped from a small airplane at noon. Right on schedule, the aircraft circled the city and dumped the money onto the people below. However, to the dismay of the onlookers, a westerly wind caught the currency before it reached the ground and carried it out over the cold Atlantic Ocean. Had the pilot or the sponsoring leather manufacturer examined the weather charts beforehand, it might have been possible to predict that the wind would ruin the advertising scheme.

 CONTENTS

This opening scenario raises two questions: (1) Why does the wind blow? and (2) How can one tell its direction by looking at weather charts? Chapter 1 has already answered the first question: Air moves in response to horizontal differences in pressure. This phenomenon happens when we open a vacuum-packed can—air rushes from the higher pressure region outside the can toward the region of lower pressure inside. In the atmosphere, the wind blows in an attempt to equalize imbalances in air pressure. Does this mean that the wind always blows directly from high to low pressure? Not really, because the movement of air is controlled not only by pressure differences but by other forces as well.

In this chapter, we will first consider how and why atmospheric pressure varies, then we will look at the forces that influence atmospheric motions aloft and at the surface. Through studying these forces, we will be able to tell how the wind should blow in a particular region by examining surface and upper-air charts.

Atmospheric Pressure

In Chapter 1, we learned several important concepts about atmospheric pressure. One stated that **air pressure** is simply the mass of air above a given level. As we climb in elevation above the earth's surface, there are fewer air molecules above us; hence, atmospheric pressure always decreases with increasing height. Another concept we learned was that most of our atmosphere is crowded close to the earth's surface, which causes air pressure to decrease with height, rapidly at first, then more slowly at higher altitudes.

So one way to change air pressure is to simply move up or down in the atmosphere. But what causes the air pressure to change in the horizontal? And why does the air pressure change at the surface?

HORIZONTAL PRESSURE VARIATIONS—A TALE OF TWO CITIES   To answer these questions, we eliminate some of the complexities of the atmosphere by constructing *models*. • Figure 8.1 shows a simple atmospheric model—a column of air, extending well up into the atmosphere. In the column, the dots represent air molecules. Our model assumes: (1) that the air molecules are not crowded close to the surface and, unlike the real atmosphere, the air density remains constant from the surface up to the top of the column, (2) that the width of the column does not change with height and (3) that the air is unable to freely move into or out of the column.

Suppose we somehow force more air into the column in Fig. 8.1. What would happen? If the air temperature in the column does not change, the added air would make the column more dense, and the added weight of the air in the column would increase the surface air pressure. Likewise, if a great deal of air were removed from the column, the surface air pressure would decrease. Consequently, to change the surface air pressure, we need to change the mass of air

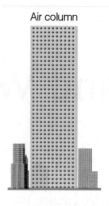

Air column

• FIGURE 8.1 A model of the atmosphere where air density remains constant with height. The air pressure at the surface is related to the number of molecules above. When air of the same temperature is stuffed into the column, the surface air pressure rises. When air is removed from the column, the surface pressure falls.

in the column above the surface. But how can this feat be accomplished?

Look at the air columns in •Fig. 8.2a.* Suppose both columns are located at the same elevation, both have the same air temperature, and both have the same surface air pressure. This condition, of course, means that there must be the same number of molecules (same mass of air) in each column above both cities. Further suppose that the surface air pressure for both cities remains the same, while the air above city 1 cools and the air above city 2 warms (see Fig. 8.2b).

As the air in column 1 cools, the molecules move more slowly and crowd closer together—the air becomes more dense. In the warm air above city 2, the molecules move faster and spread farther apart—the air becomes less dense. Since the width of the columns does not change (and if we assume an invisible barrier exists between the columns), the total number of molecules above each city remains the same, and the surface pressure does not change. Therefore, in the more-dense cold air above city 1, the column shrinks, while the column rises in the less-dense, warm air above city 2.

We now have a cold, shorter dense column of air above city 1 and a warm, taller less-dense air column above city 2. From this situation, we can conclude that *it takes a shorter column of cold, more-dense air to exert the same surface pressure as a taller column of warm, less-dense air.* This concept has a great deal of meteorological significance.

Atmospheric pressure decreases more rapidly with height in the cold column of air. In the cold air above city 1 (Fig. 8.2b), move up the column and observe how quickly you pass through the densely packed molecules. This activity indicates a rapid change in pressure. In the warmer, less-dense air, the pressure does not decrease as rapidly with height, simply because you climb above fewer molecules in the same vertical distance.

In Fig. 8.2c, move up the warm, red column until you come to the letter *H*. Now move up the cold, blue column the same distance until you reach the letter *L*. Notice that there are more molecules above the letter *H* in the warm column than above the letter *L* in the cold column. The fact that the

---

*We will keep our same assumption as in Fig. 8.1; that is, (1) the air molecules are not crowded close to the surface, (2) the width of the columns does not change, and (3) air is unable to move into or out of the columns.

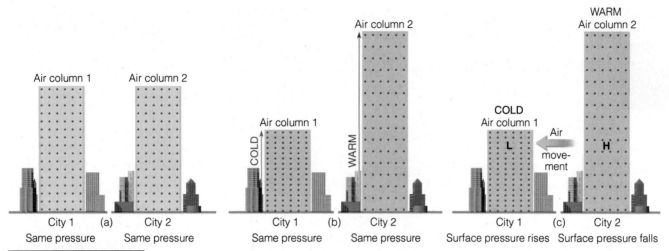

ACTIVE FIGURE 8.2 (a) Two air columns, each with identical mass, have the same surface air pressure. (b) Because it takes a shorter column of cold air to exert the same pressure as a taller column of warm air, as column 1 cools, it must shrink, and as column 2 warms, it must expand. (c) Because at the same level in the atmosphere there is more air above the H in the warm column than above the L in the cold column, warm air aloft is associated with high pressure and cold air aloft with low pressure. The pressure differences aloft create a force that causes the air to move from a region of higher pressure toward a region of lower pressure. The removal of air from column 2 causes its surface pressure to drop, whereas the addition of air into column 1 causes its surface pressure to rise. (The difference in height between the two columns is greatly exaggerated.) Visit the Meteorology Resource Center to view this and other active figures at academic.cengage.com/login

number of molecules above any level is a measure of the atmospheric pressure leads to an important concept: *Warm air aloft is normally associated with high atmospheric pressure, and cold air aloft is associated with low atmospheric pressure.*

In Fig. 8.2c, the horizontal difference in temperature creates a horizontal difference in pressure. The pressure difference establishes a force (called the *pressure gradient force*) that causes the air to move from higher pressure toward lower pressure. Consequently, if we remove the invisible barrier between the two columns and allow the air aloft to move horizontally, the air will move from column 2 toward column 1. As the air aloft leaves column 2, the mass of the air in the column decreases, and so does the surface air pressure. Meanwhile, the accumulation of air in column 1 causes the surface air pressure to increase.

Higher air pressure at the surface in column 1 and lower air pressure at the surface in column 2 causes the surface air to move from city 1 towards city 2 (see ● Fig. 8.3). As the surface air moves out away from city 1, the air aloft slowly sinks to replace this outwardly spreading surface air. As the surface air flows into city 2, it slowly rises to replace the depleted air aloft. In this manner, a complete circulation of air is established due to the heating and cooling of air columns.

In summary, we can see how heating and cooling columns of air can establish horizontal variations in air pressure both aloft and at the surface. It is these horizontal differences in air pressure that cause the wind to blow.

Air temperature, air pressure, and air density are all interrelated. If one of these variables changes, the other two usually change as well. The relationship among these three variables is expressed by the gas law, which is described in the Focus section on p. 196.

DAILY PRESSURE VARIATIONS From what we have learned so far, we might expect to see the surface pressure dropping as the air temperature rises, and vice versa. Over large continental areas, especially the southwestern United States in summer, hot surface air is accompanied by surface low pressure. Likewise, bitter cold arctic air in winter is often accom-

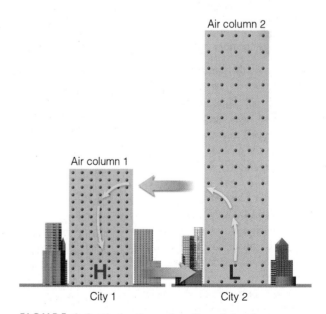

● **FIGURE 8.3** The heating and cooling of air columns causes horizontal pressure variations aloft and at the surface. These pressure variations force the air to move from areas of higher pressure toward areas of lower pressure. In conjunction with these horizontal air motions, the air slowly sinks above the surface high and rises above the surface low.

FOCUS ON A SPECIAL TOPIC

The Atmosphere Obeys the Gas Law

The relationship among the pressure, temperature, and density of air can be expressed by

Pressure = temperature
× density × constant.

This simple relationship, often referred to as the *gas law* (or *equation of state*), tells us that the pressure of a gas is equal to its temperature times its density times a constant. When we ignore the constant and look at the gas law in symbolic form, it becomes

$$p \sim T \times \rho,$$

where, of course, *p* is pressure, *T* is temperature, and ρ (the Greek letter rho, pronounced "row") represents air density.* The line ~ is a symbol meaning "is proportional to." A change in one variable causes a corresponding change in the other two variables. Thus, it will be easier to understand the behavior of a gas if we keep one variable from changing and observe the behavior of the other two.

Suppose, for example, we hold the temperature constant. The relationship then becomes

$$p \sim \rho \text{ (temperature constant)}.$$

This expression says that the pressure of the gas is proportional to its density, as long as its temperature does not change. Consequently, if the temperature of a gas (such as air) is held constant, as the pressure increases the density increases, and as the pressure decreases the density decreases. In other words, *at the same temperature, air at a higher pressure is more dense than air at a lower pressure.* If we apply this concept to the atmosphere, then with nearly the same temperature and elevation, air above a region of surface high pressure is more dense than air above a region of surface low pressure (see Fig. 1).

We can see, then, that for surface high-pressure areas (anticyclones) and surface low-pressure areas (mid-latitude cyclones) to form,

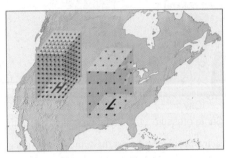

● FIGURE 1 Air above a region of surface high pressure is more dense than air above a region of surface low pressure (at the same temperature). (The dots in each column represent air molecules.)

the air density (mass of air) above these systems must change. As we will see later in this chapter, as well as in other chapters, surface air pressure increases when the wind causes more air to move into a column of air than is able to leave (called *net convergence*), and surface air pressure decreases when the wind causes more air to move out of a column of air than is able to enter (called *net divergence*).

Earlier, we considered how pressure and density are related when the temperature is not changing. What happens to the gas law when the pressure of a gas remains constant? In shorthand notation, the law becomes

(Constant pressure) × constant = $T \times \rho$.

This relationship tells us that when the pressure of a gas is held constant, the gas becomes less dense as the temperature goes up, and more dense as the temperature goes down. Therefore, *at a given atmospheric pressure, air that is cold is more dense than air that is warm.* Keep in mind that the idea that cold air is more dense than warm air applies only when we compare volumes of air at the same level, where pressure changes are small in any horizontal direction.

We can use the gas law to obtain information about the atmosphere. For example, at an altitude of about 5600 m (18,400 ft) above sea level, the atmospheric pressure is normally close to 500 millibars. If we obtain the average

density at this level, with the aid of the gas law we can calculate the average air temperature.

Recall that the gas law is written as

$$p = T \times \rho \times C.$$

With the pressure *(p)* in millibars (mb), the temperature *(T)* in Kelvins, and the density (ρ) in kilograms per cubic meter (kg/m^3), the numerical value of the constant (C) is about 2.87.*

At an altitude of 5600 m above sea level, where the average (or standard) air pressure is about 500 mb and the average air density is 0.690 kg/m^3, the average air temperature becomes

$$p = T \times \rho \times C$$
$$500 = T \times 0.690 \times 2.87$$
$$\frac{500}{0.690 \times 2.87} = T$$
$$252.5 \text{ K} = T.$$

To convert Kelvins into degrees Celsius, we subtract 273 from the Kelvin temperature and obtain a temperature of $-20.5°$C, which is the same as $-5°$F.

If we know the numerical values of temperature and density, with the aid of the gas law we can obtain the air pressure. For example, in Chapter 1 we saw that the average global temperature near sea level is 15°C (59°F), which is the same as 288 K. If the average air density at sea level is 1.226 kg/m^3, what would be the standard (average) sea-level pressure?

Using the gas law, we obtain

$$p = T \times \rho \times C$$
$$p = 288 \times 1.226 \times 2.87$$
$$p = 1013 \text{ mb}.$$

Since the air pressure is related to both temperature and density, a small change in either or both of these variables can bring about a change in pressure.

---

*This gas law may also be written as $p \times v = T \times$ constant. Consequently, pressure and temperature changes are also related to changes in volume.

---

*The constant is usually expressed as 2.87×10^6 erg/g K, or, in the SI system, as 287 J/kg K. (See Appendix A for information regarding the units used here.)

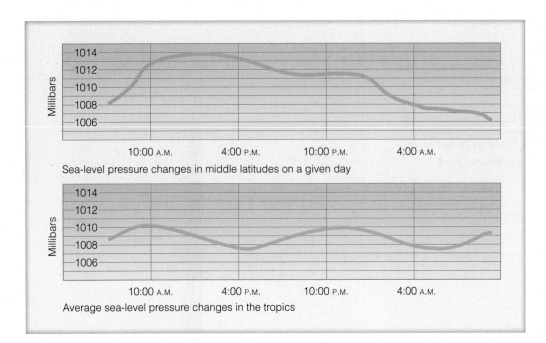

● FIGURE 8.4 Diurnal surface pressure changes in the middle latitudes and in the tropics.

Sea-level pressure changes in middle latitudes on a given day

Average sea-level pressure changes in the tropics

panied by surface high pressure. Yet, on a daily basis, any cyclic change in surface pressure brought on by daily temperature changes is concealed by the pressure changes created by the warming of the upper atmosphere.

In the tropics, for example, pressure rises and falls in a regular pattern twice a day (see ● Fig. 8.4). Maximum pressures occur around 10:00 A.M. and 10:00 P.M., minimum near 4:00 A.M. and 4:00 P.M. The largest pressure difference, about 2.5 mb, occurs near the equator. It also shows up in higher latitudes, but with a much smaller amplitude. This daily (*diurnal*) fluctuation of pressure appears to be due primarily to the absorption of solar energy by ozone in the upper atmosphere and by water vapor in the lower atmosphere. The warming and cooling of the air creates density oscillations known as *thermal* (or *atmospheric*) *tides* that show up as small pressure changes near the earth's surface.

In middle latitudes, surface pressure changes are primarily the result of large high- and low-pressure areas that move toward or away from a region. Generally, when an area of high pressure approaches a city, surface pressure usually rises. When it moves away, pressure usually falls. Likewise, an approaching low causes the air pressure to fall, and one moving away causes surface pressure to rise.

PRESSURE MEASUREMENTS   Instruments that detect and measure pressure changes are called **barometers,** which literally means an instrument that measures bars. You may recall from Chapter 1 that a *bar* is a unit of pressure that describes a force over a given area.* Because the bar is a relatively large unit, and because surface pressure changes are normally

---

*A bar is a force of 100,000 newtons acting on a surface area of 1 square meter. A *newton* (N) is the amount of force required to move an object with a mass of 1 kilogram so that it increases its speed at a rate of 1 meter per second each second. Additional pressure units and conversions are given in Appendix A.

Although 1013.25 mb (29.92 in.) is the *standard atmospheric pressure* at sea level, it is *not* the average sea-level pressure. The earth's average sea-level pressure is 1011.0 mb (29.85 in.). Because much of the earth's surface is above sea level, the earth's annual average *surface pressure* is estimated to be 984.43 mb (29.07 in.).

small, the unit of pressure commonly found on surface weather maps is, as we saw in Chapter 1, the **millibar** (mb), where 1 mb = 1/1000 bar or

$$1 \text{ bar} = 1000 \text{ mb}.$$

A common pressure unit used in aviation is *inches of mercury* (Hg). At sea level, *standard atmospheric pressure** is

$$1013.25 \text{ mb} = 29.92 \text{ in. Hg} = 76 \text{ cm}.$$

As a reference, ● Fig. 8.5 compares pressure readings in inches of mercury and millibars.

The unit of pressure designated by the International System (SI) of measurement is the *pascal,* named in honor of Blaise Pascal (1632–1662), whose experiments on atmospheric pressure greatly increased our knowledge of the atmosphere. A pascal (Pa) is the force of 1 newton acting on a surface area of 1 square meter. Thus, 100 pascals equals 1 millibar. The scientific community often uses the *kilopascal* (kPa) as the unit of pressure, where 1 kPa = 10 mb. However, a more convenient unit is the **hectopascal** (hPa), as

$$1 \text{ hPa} = 1 \text{ mb}.$$

---

*Standard atmospheric pressure at sea level is the pressure extended by a column of mercury 29.92 in. (760 mm) high, having a density of 1.36×10^4 kg/m^3, and subject to an acceleration of gravity of 9.80 m/sec^2.

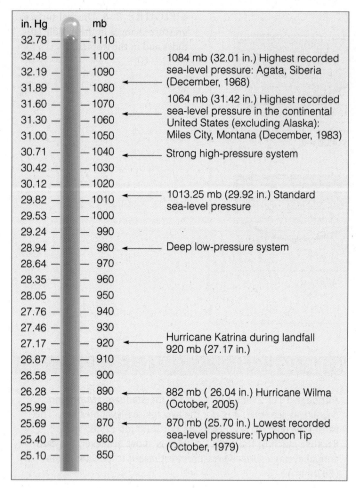

• FIGURE 8.5 Atmospheric pressure in inches of mercury and in millibars.

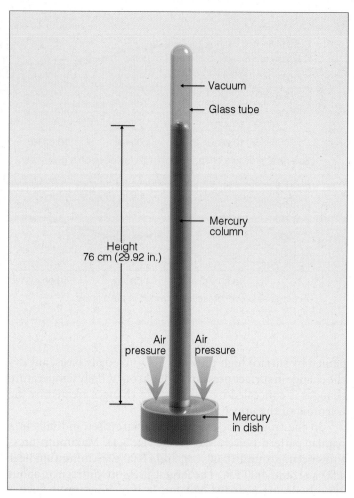

• FIGURE 8.6 The mercury barometer. The height of the mercury column is a measure of atmospheric pressure.

Presently, the hectopascal is gradually replacing the millibar as the preferred unit of pressure on surface weather maps.

Because we measure atmospheric pressure with an instrument called a *barometer*, atmospheric pressure is also referred to as *barometric pressure*. Evangelista Torricelli, a student of Galileo, invented the **mercury barometer** in 1643. His barometer, similar to those in use today, consisted of a long glass tube open at one end and closed at the other (see • Fig. 8.6). Removing air from the tube and covering the open end, Torricelli immersed the lower portion into a dish of mercury. He removed the cover, and the mercury rose up the tube to nearly 76 cm (or about 30 in.) above the level in the dish. Torricelli correctly concluded that the column of mercury in the tube was balancing the weight of the air above the dish, and hence its height was a measure of atmospheric pressure.

Why is mercury rather than water used in the barometer? The primary reason is convenience. (Also, water can evaporate in the tube.) Mercury seldom rises to a height above 80 cm (31.5 in.). A water barometer, however, presents a problem. Because water is 13.6 times less dense than mercury, an atmospheric pressure of 76 cm (30 in.) of mercury would be equivalent to 1034 cm (408 in.) of water. A water barom-

eter resting on the ground near sea level would have to be read from a ladder over 10 m (33 ft) tall.

The most common type of home barometer—the **aneroid barometer**—contains no fluid. Inside this instrument is a small, flexible metal box called an *aneroid cell*. Before the cell is tightly sealed, air is partially removed, so that small changes in external air pressure cause the cell to expand or contract. The size of the cell is calibrated to represent different pressures, and any change in its size is amplified by levers and transmitted to an indicating arm, which points to the current atmospheric pressure (see • Fig. 8.7).

Notice that the aneroid barometer often has descriptive weather-related words printed above specific pressure values. These descriptions indicate the most likely weather conditions when the needle is pointing to that particular pressure reading. Generally, the higher the reading, the more likely clear weather will occur, and the lower the reading, the better are the chances for inclement weather. This situation occurs because surface high-pressure areas are associated with sinking air and normally fair weather, whereas surface low-pressure areas are associated with rising air and usually cloudy, wet weather. A steady rise in atmospheric pressure (a

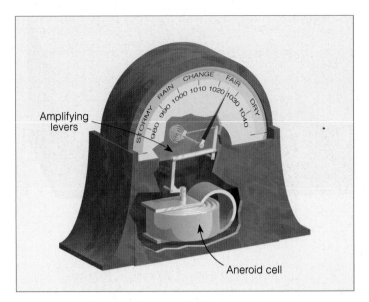

● FIGURE 8.7 The aneroid barometer.

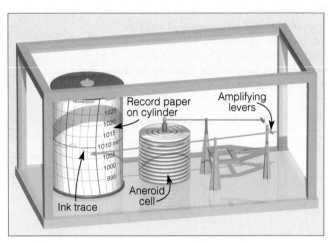

● FIGURE 8.8 A recording barograph.

rising barometer) usually indicates clearing weather or fair weather, whereas a steady drop in atmospheric pressure (a falling barometer) often signals the approach of a storm with inclement weather.

The *altimeter* and *barograph* are two types of aneroid barometers. Altimeters are aneroid barometers that measure pressure, but are calibrated to indicate altitude. Barographs are recording aneroid barometers. Basically, the barograph consists of a pen attached to an indicating arm that marks a continuous record of pressure on chart paper. The chart paper is attached to a drum rotated slowly by an internal mechanical clock (see ● Fig. 8.8).

PRESSURE READINGS The seemingly simple task of reading the height of the mercury column to obtain the air pressure is actually not all that simple. Being a fluid, mercury is sensitive to changes in temperature; it will expand when heated and contract when cooled. Consequently, to obtain accurate pressure readings without the influence of temperature, all mercury barometers are corrected as if they were read at the same temperature. Because the earth is not a perfect sphere, the force of gravity is not a constant. Since small gravity differences influence the height of the mercury column, they must be considered when reading the barometer. Finally, each barometer has its own "built-in" error, called *instrument error,* which is caused, in part, by the surface tension of the mercury against the glass tube. After being corrected for temperature, gravity, and instrument error, the barometer reading at a particular location and elevation is termed **station pressure.**

● Figure 8.9a gives the station pressure measured at four locations only a few hundred kilometers apart. The different station pressures of the four cities are due primarily to the cities being at different altitudes. This fact becomes even clearer when we realize that atmospheric pressure changes

much more quickly when we move upward than it does when we move sideways. As an example, the vertical change in air pressure from the base to the top of the Empire State Building—a distance of a little more than $\frac{1}{2}$ km—is typically much greater than the horizontal difference in air pressure from New York City to Miami, Florida—a distance of over 1600 km. Therefore, we can see that a small vertical difference between two observation sites can yield a large difference in station pressure. Thus, to properly monitor horizontal changes in pressure, barometer readings must be corrected for altitude.

Altitude corrections are made so that a barometer reading taken at one elevation can be compared with a barometer reading taken at another. Station pressure observations are normally adjusted to a level of *mean sea level*—the level representing the average surface of the ocean. The adjusted reading is called **sea-level pressure.** The size of the correction depends primarily on how high the station is above sea level.

Near the earth's surface, atmospheric pressure decreases on the average by about 10 mb for every 100 m increase in elevation (about 1 in. of mercury for each 1000-ft rise).* Notice in Fig. 8.9a that city A has a station pressure of 952 mb. Notice also that city A is 600 m above sea level. Adding 10 mb per 100 m to its station pressure yields a sea-level pressure of 1012 mb (Fig. 8.9b). After all the station pressures are adjusted to sea level (Fig. 8.9c), we are able to see the horizontal variations in sea-level pressure—something we were not able to see from the station pressures alone in Fig. 8.9a.

When more pressure data are added (see Fig. 8.9c), the chart can be analyzed and the pressure pattern visualized.

---

*This decrease in atmospheric pressure with height (10 mb/100 m) occurs when the air temperature decreases at the standard lapse rate of 6.5°C/1000 m. Because atmospheric pressure decreases more rapidly with height in cold (more-dense) air than it does in warm (less-dense) air, the vertical rate of pressure change is typically greater than 10 mb per 100 m in cold air and less than that in warm air.

● FIGURE 8.9 The top diagram (a) shows four cities (A, B, C, and D) at varying elevations above sea level, all with different station pressures. The middle diagram (b) represents sea-level pressures of the four cities plotted on a sea-level chart. The bottom diagram (c) shows sea-level pressure readings of the four cities plus other sea-level pressure readings, with isobars drawn on the chart (gray lines) at intervals of 4 millibars.

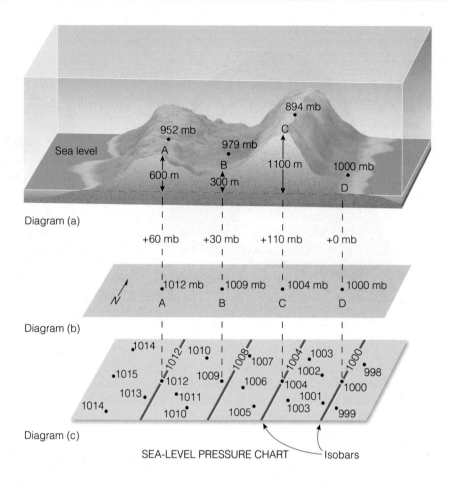

Diagram (a)

Diagram (b)

Diagram (c)

SEA-LEVEL PRESSURE CHART — Isobars

Isobars (lines connecting points of equal pressure) are drawn at intervals of 4 mb,* with 1000 mb being the base value. Note that the isobars do not pass through each point, but, rather, between many of them, with the exact values being interpolated from the data given on the chart. For example, follow the 1008-mb line from the top of the chart southward and observe that there is no plotted pressure of 1008 mb. The 1008-mb isobar, however, comes closer to the station with a sea-level pressure of 1007 mb than it does to the station with a pressure of 1010 mb. With its isobars, the bottom chart (Fig. 8.9c) is now called a *sea-level pressure chart* or simply a **surface map.** When weather data are plotted on the map it becomes a *surface weather map.*

Surface and Upper-Level Charts

The isobars on the surface map in ● Fig. 8.10a are drawn precisely, with each individual observation taken into account. Notice that many of the lines are irregular, especially in mountainous regions over the Rockies. The reason for the wiggle is due, in part, to small-scale local variations in pressure and to errors introduced by correcting observations that were taken at high-altitude stations. An extreme case of

*An interval of 2 mb would put the lines too close together, and an 8-mb interval would spread them too far apart.

this type of error occurs at Leadville, Colorado (elevation 3096 m), the highest city in the United States. Here, the station pressure is typically near 700 mb. This means that nearly 300 mb must be added to obtain a sea-level pressure reading! A mere 1 percent error in estimating the exact correction would result in a 3-mb error in sea-level pressure. For this reason, isobars are smoothed through readings from high-altitude stations and from stations that might have small observational errors. Figure 8.10b shows how the isobars appear on the surface map after they are smoothed.

The sea-level pressure chart described so far is called a *constant height chart* because it represents the atmospheric pressure at a constant level—in this case, sea level. The same type of chart could be drawn to show the horizontal variations in pressure at any level in the atmosphere; for example, at 3000 m (see ● Fig. 8.11).

Another type of chart commonly used in studying the weather is the *constant pressure chart,* or **isobaric chart.** Instead of showing pressure variations at a constant altitude, these charts are constructed to show height variations along an equal pressure *(isobaric)* surface. Constant pressure charts are convenient to use because the height variables they show are easier to deal with in meteorological equations than the variables of pressure. Since isobaric charts are in common use, let's examine them in detail.

Imagine that the dots inside the air column in ● Fig. 8.12 represent tightly packed air molecules from the surface up to

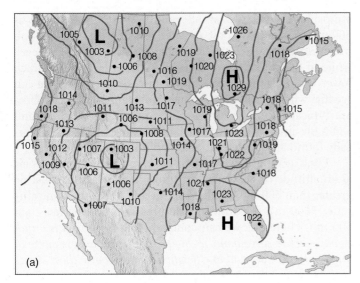

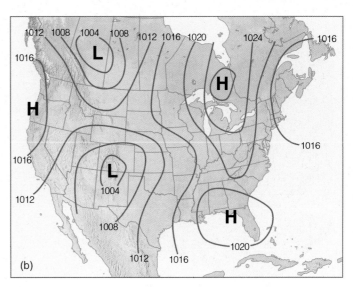

● FIGURE 8.10 (a) Sea-level isobars drawn so that each observation is taken into account. Not all observations are plotted. (b) Sea-level isobars after smoothing.

the tropopause. Assume that the air density is constant throughout the entire air layer and that all of the air molecules are squeezed into this layer. If we climb halfway up the air column and stop, then draw a sheetlike surface representing this level, we will have made a constant height surface. This altitude (5600 m) is where we would, under standard conditions, measure a pressure of 500 mb. Observe that ev-

erywhere along this surface (shaded tan in the diagram) there are an equal number of molecules above it. This condition means that the level of constant height also represents a level of constant pressure. At every point on this *isobaric surface*, the height is 5600 m above sea level and the pressure is 500 mb. Within the air column, we could cut any number of horizontal slices, each one at a different altitude, and each slice would represent both an isobaric and constant height surface. A map of any one of these surfaces would be blank, since there are no horizontal variations in either pressure or altitude.

● FIGURE 8.11 Each map shows isobars on a constant height chart. The isobars represent variations in horizontal pressure at that altitude. An average isobar at sea level would be about 1000 mb; at 3000 m, about 700 mb; and at 5600 m, about 500 mb.

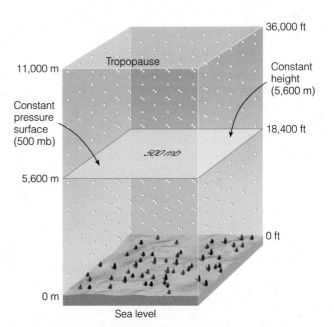

● FIGURE 8.12 When there are no horizontal variations in pressure, constant pressure surfaces are parallel to constant height surfaces. In the diagram, a measured pressure of 500 mb is 5600 m above sea level everywhere. (Dots in the diagram represent air molecules.)

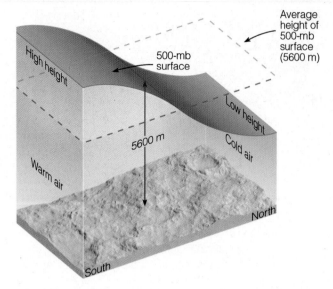

● FIGURE 8.13 The area shaded gray in the above diagram represents a surface of constant pressure, or isobaric surface. Because of the changes in air density, the isobaric surface rises in warm, less-dense air and lowers in cold, more-dense air. Where the horizontal temperature changes most quickly, the isobaric surface changes elevation most rapidly.

If the air temperature should change in any portion of the column, the air density and pressure would change along with it. Notice in ● Fig. 8.13 that we have colder air to the north and warmer air to the south. To simplify this situation, we will assume that the atmospheric pressure at the earth's surface remains constant. Hence, the total number of molecules in the column above each region must remain constant.

In Fig. 8.13, the area shaded gray at the top of the column represents a constant pressure (isobaric) surface, where the

atmospheric pressure at all points along this surface is 500 mb. Notice that in the warmer, less-dense air the 500-mb pressure surface is found at a higher (than average) level, while in the colder, more-dense air, it is observed at a much lower (than average) level. From these observations, we can see that when the air aloft is warm, constant pressure surfaces are typically found at higher elevations than normal, and when the air aloft is cold, constant pressure surfaces are typically found at lower elevations than normal.

Look again at Fig. 8.13 and observe that in the warm air at an altitude of 5600 m, the atmospheric pressure must be greater than 500 mb, whereas in the cold air, at the same altitude (5600 m), the atmospheric pressure must be less than 500 mb. Therefore, we can conclude that *high heights on an isobaric chart correspond to higher-than-normal pressures at any given altitude, and low heights on an isobaric chart correspond to lower-than-normal pressures.*

The variations in height of the isobaric surface in Fig. 8.13 are shown in ● Fig. 8.14. Note that where the constant altitude lines intersect the 500-mb pressure surface, **contour lines** (lines connecting points of equal elevation) are drawn on the 500-mb map. Each contour line, of course, tells us the altitude above sea level at which we can obtain a pressure reading of 500 mb. In the warmer air to the south, the elevations are high, while in the cold air to the north, the elevations are low. The contour lines are crowded together in the middle of the chart, where the pressure surface dips rapidly due to the changing air temperatures. Where there is little horizontal temperature change, there are also few contour lines. Although contour lines are height lines, keep in mind that they illustrate pressure as do isobars in that contour lines of low height represent a region of lower pressure and contour lines of high height represent a region of higher pressure.

● FIGURE 8.14 Changes in elevation of an isobaric surface (500 mb) show up as contour lines on an isobaric (500 mb) map. Where the surface dips most rapidly, the lines are closer together.

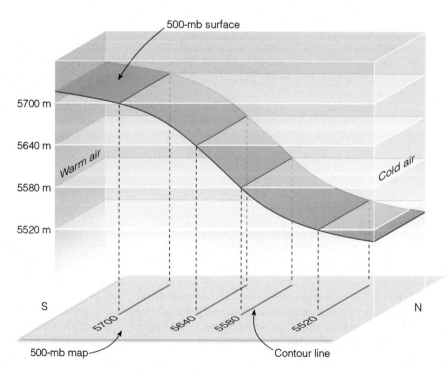

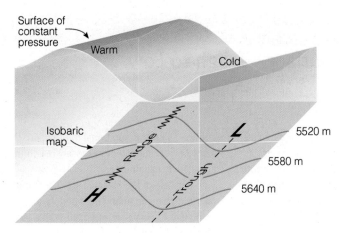

● FIGURE 8.15 The wavelike patterns of an isobaric surface reflect the changes in air temperature. An elongated region of warm air aloft shows up on an isobaric map as higher heights and a ridge; the colder air shows as lower heights and a trough.

Since cold air aloft is normally associated with low heights or low pressures, and warm air aloft with high heights or high pressures, on upper-air charts representing the Northern Hemisphere, contour lines and isobars usually decrease in value from south to north because the air is typically warmer to the south and colder to the north. The lines, however, are not straight; they bend and turn, indicating **ridges** *(elongated highs)* where the air is warm and indicating depressions, or **troughs** *(elongated lows)*, where the air is cold. In ● Fig. 8.15, we can see how the wavy contours on the map relate to the changes in altitude of the isobaric surface.

Although we have examined only the 500-mb chart, other isobaric charts are commonly used. ▼ Table 8.1 lists these charts and their approximate heights above sea level.

▼ TABLE 8.1 Common Isobaric Charts and Their Approximate Elevation above Sea Level

ISOBARIC SURFACE (MB) CHARTS	APPROXIMATE ELEVATION (M)	(FT)
1000	120	400
850	1,460	4,800
700	3,000	9,800
500	5,600	18,400
300	9,180	30,100
200	11,800	38,700
100	16,200	53,200

Upper-level charts are a valuable tool. As we will see, they show wind-flow patterns that are extremely important in forecasting the weather. They can also be used to determine the movement of weather systems and to predict the behavior of surface pressure areas. To the pilot of a small aircraft, a constant pressure chart can help determine whether the plane is flying at an altitude either higher or lower than its altimeter indicates. (For more information on this topic, read the Focus section "Flying on a Constant Pressure Surface—High to Low, Look Out Below," p. 204.)

● Figure 8.16a is a simplified surface map that shows areas of high and low pressure and arrows that indicate *wind direction*—the direction from which the wind is blowing. The large blue H's on the map indicate the centers of high pressure, which are also called **anticyclones**. The large L's represent centers of low pressure, also known as *depressions* or **mid-latitude cyclonic storms** because they form in the

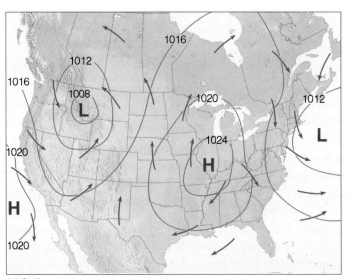

(a) Surface map

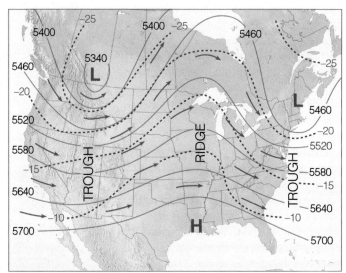

(b) Upper-air map (500 mb)

● FIGURE 8.16 (a) Surface map showing areas of high and low pressure. The solid lines are isobars drawn at 4-mb intervals. The arrows represent wind direction. Notice that the wind blows *across* the isobars. (b) The upper-level (500-mb) map for the same day as the surface map. Solid lines on the map are contour lines in meters above sea level. Dashed red lines are isotherms in °C. Arrows show wind direction. Notice that, on this upper-air map, the wind blows *parallel* to the contour lines.

Flying on a Constant Pressure Surface—High to Low, Look Out Below

Aircraft that use pressure altimeters typically fly along a constant pressure surface rather than a constant altitude surface. They do this because the *altimeter*, as we saw earlier, is simply an aneroid barometer calibrated to convert atmospheric pressure to an approximate elevation. The altimeter elevation indicated by an altimeter assumes a standard atmosphere where the air temperature decreases at the rate of 6.5°C/1000 m (3.6°F/1000 ft). Since the air temperature seldom, if ever, decreases at exactly this rate, altimeters generally indicate an altitude different from their true elevation.

Figure 2 shows a standard column of air bounded on each side by air with a different temperature and density. On the left side, the air is warm; on the right, it is cold. The orange line represents a constant pressure surface of 700 mb as seen from the side. In the standard air, the 700-mb surface is located at 10,000 ft above sea level.

In the warm air, the 700-mb surface rises; in the cold air, it descends. An aircraft flying along the 700-mb surface would be at an altitude less than 10,000 ft in the cold air, equal to 10,000 ft in the standard air, and greater than 10,000 ft in the warmer air. With no corrections for temperature, the altimeter would indicate the same altitude at all three positions because the air pressure does not change. We can see that, if no temperature corrections are made, an aircraft flying into warm air will increase in altitude and fly higher than its altimeter indicates. Put another way: The altimeter inside the plane will read an altitude lower than the plane's true elevation.

Flying from standard air into cold air represents a potentially dangerous situation. As an aircraft flies into cold air, it flies along a lowering pressure surface. If no correction for temperature is made, the altimeter shows no change in elevation even though the aircraft is losing altitude; hence, the plane will be flying lower than the altimeter indicates. This problem can be serious, especially for planes flying above mountainous terrain with poor visibility and where high winds and turbulence can reduce the air pressure drastically. To ensure adequate clearance under these conditions, pilots fly their aircraft higher than they normally would, consider air temperature, and compute a more realistic altitude by resetting their altimeters to reflect these conditions.

Even without sharp temperature changes, pressure surfaces may dip suddenly (see Fig. 3). An aircraft flying into an area of decreasing pressure will lose altitude unless corrections are made. For example, suppose a pilot has set the altimeter for sea-level pressure above station A. At this location, the plane is flying along an isobaric surface at a true altitude of 500 ft. As the plane flies toward station B, the pressure surface (and the plane) dips but the altimeter continues to read 500 ft, which is too high. To correct for such changes in pressure, a pilot can obtain a current altimeter setting from ground control. With this additional information, the altimeter reading will more closely match the aircraft's actual altitude.

Because of the inaccuracies inherent in the pressure altimeter, many high performance and commercial aircraft are equipped with a radio altimeter. This device is like a small radar unit that measures the altitude of the aircraft by sending out radio waves, which bounce off the terrain. The time it takes these waves to reach the surface and return is a measure of the aircraft's altitude. If used in conjunction with a pressure altimeter, a pilot can determine the variations in a constant pressure surface simply by flying along that surface and observing how the true elevation measured by the radio altimeter changes.

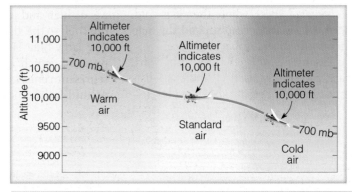

● FIGURE 2
An aircraft flying along a surface of constant pressure (orange line) may change altitude as the air temperature changes. Without being corrected for the temperature change, a pressure altimeter will continue to read the same elevation.

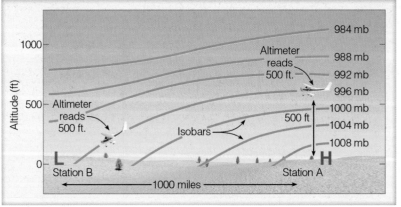

● FIGURE 3 In the absence of horizontal temperature changes, pressure surfaces can dip toward the earth's surface. An aircraft flying along the pressure surface will either lose or gain altitude, depending on the direction of flight.

middle latitudes, outside of the tropics. The solid dark lines are isobars with units in millibars. Notice that the surface winds tend to blow across the isobars toward regions of lower pressure. In fact, as we briefly observed in Chapter 1, in the Northern Hemisphere the winds blow counterclockwise and inward toward the center of the lows and clockwise and outward from the center of the highs.

Figure 8.16b shows an upper-air chart (a 500-mb isobaric map) for the same day as the surface map in Fig. 8.16a. The solid gray lines on the map are contour lines given in meters above sea level. The difference in elevation between each contour line (called the *contour interval*) is 60 meters. Superimposed on this map are dashed red lines, which represent lines of equal temperature (isotherms). Observe how the contour lines tend to parallel the isotherms. As we would expect, the contour lines tend to decrease in value from south to north.

The arrows on the 500-mb map show the wind direction. Notice that, unlike the surface winds that cross the isobars in Fig. 8.16a, the winds on the 500-mb chart tend to flow *parallel* to the contour lines in a wavy west-to-east direction. Why does the wind tend to cross the isobars on a surface map, yet blow parallel to the contour lines (or isobars) on an upper-air chart? To answer this question we will now examine the forces that affect winds.

Newton's Laws of Motion

Our understanding of why the wind blows stretches back through several centuries, with many scientists contributing to our knowledge. When we think of the movement of air, however, one great scholar stands out — Isaac Newton (1642–1727), who formulated several fundamental laws of motion.

Newton's first law of motion states that *an object at rest will remain at rest and an object in motion will remain in motion (and travel at a constant velocity along a straight line) as long as no force is exerted on the object.* For example, a baseball in a pitcher's hand will remain there until a force (a push) acts upon the ball. Once the ball is pushed (thrown), it would continue to move in that direction forever if it were not for the force of air friction (which slows it down), the force of gravity (which pulls it toward the ground), and the catcher's mitt (which exerts an equal but opposite force to bring it to a halt). Similarly, to start air moving, to speed it up, to slow it down, or even to change its direction requires the action of an external force. This brings us to Newton's second law.

Newton's second law states that *the force exerted on an object equals its mass times the acceleration produced.* In symbolic form, this law is written as

$$F = ma.$$

From this relationship we can see that, when the mass of an object is constant, the force acting on the object is directly related to the acceleration that is produced. A force in its simplest form is a push or a pull. *Acceleration is the speeding up, the slowing down, or the changing of direction of an object.* (More precisely, acceleration is the change in velocity* over a period of time.)

Because more than one force may act upon an object, Newton's second law always refers to the *net*, or total, force that results. An object will always accelerate in the direction of the total force acting on it. Therefore, to determine in which direction the wind will blow, we must identify and examine all of the forces that affect the horizontal movement of air. These forces include:

1. pressure gradient force
2. Coriolis force
3. centripetal force
4. friction

We will first study the forces that influence the flow of air aloft. Then we will see which forces modify winds near the ground.

Forces That Influence the Winds

We already know that horizontal differences in atmospheric pressure cause air to move and, hence, the wind to blow. Since air is an invisible gas, it may be easier to see how pressure differences cause motion if we examine a visible fluid, such as water.

In ● Fig. 8.17, the two large tanks are connected by a pipe. Tank A is two-thirds full and tank B is only one-half full. Since the water pressure at the bottom of each tank is proportional to the weight of water above, the pressure at the bottom of tank A is greater than the pressure at the bottom of tank B. Moreover, since fluid pressure is exerted equally in all directions, there is a greater pressure in the pipe directed from tank A toward tank B than from B toward A.

Since pressure is force per unit area, there must also be a net force directed from tank A toward tank B. This force causes the water to flow from left to right, from higher pressure toward lower pressure. The greater the pressure difference, the stronger the force, and the faster the water moves. In a similar way, horizontal differences in atmospheric pressure cause air to move.

PRESSURE GRADIENT FORCE ● Figure 8.18 shows a region of higher pressure on the map's left side, lower pressure on the right. The isobars show how the horizontal pressure is changing. If we compute the amount of pressure change that occurs over a given distance, we have the **pressure gradient;** thus

$$\text{Pressure gradient} = \frac{\text{difference in pressure}}{\text{distance}}.$$

---

*Velocity specifies both the speed of an object and its direction of motion.

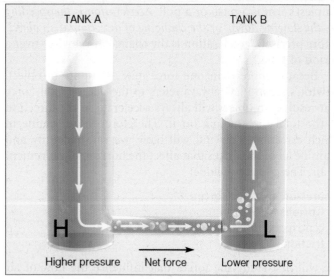

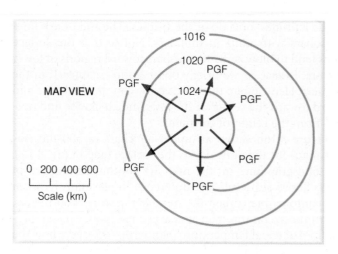

● F I G U R E 8.17 The higher water level creates higher fluid pressure at the bottom of tank A and a net force directed toward the lower fluid pressure at the bottom of tank B. This net force causes water to move from higher pressure toward lower pressure.

● F I G U R E 8.19 The closer the spacing of the isobars, the greater the pressure gradient. The greater the pressure gradient, the stronger the pressure gradient force (PGF). The stronger the PGF, the greater the wind speed. The red arrows represent the relative magnitude of the force, which is always directed from higher toward lower pressure.

If we let the symbol delta (Δ) mean "a change in," we can simplify the expression and write the pressure gradient as

$$PG = \frac{\Delta p}{d},$$

where Δp is the pressure difference between two places some horizontal distance (d) apart. In Fig. 8.18 the pressure gradient between points 1 and 2 is 4 mb per 100 km.

Suppose the pressure in Fig. 8.18 were to change and the isobars become closer together. This condition would produce a rapid change in pressure over a relatively short distance, or what is called a *steep* (or *strong*) *pressure gradient*. However, if the pressure were to change such that the isobars spread farther apart, then the difference in pressure would be small over a relatively large distance. This condition is called a *gentle* (or *weak*) *pressure gradient*.

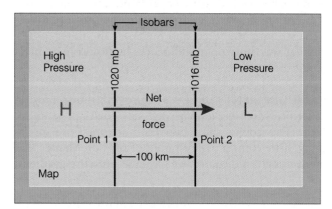

● F I G U R E 8.18 The pressure gradient between point 1 and point 2 is 4 mb per 100 km. The net force directed from higher toward lower pressure is the *pressure gradient force*.

Notice in Fig. 8.18 that when differences in horizontal air pressure exist there is a net force acting on the air. This force, called the **pressure gradient force** (PGF), *is directed from higher toward lower pressure at right angles to the isobars.* The magnitude of the force is directly related to the pressure gradient. Steep pressure gradients correspond to strong pressure gradient forces and vice versa. ● Figure 8.19 shows the relationship between pressure gradient and pressure gradient force.

The *pressure gradient force is the force that causes the wind to blow.* Because of this effect, closely spaced isobars on a weather map indicate steep pressure gradients, strong forces, and high winds. On the other hand, widely spaced isobars indicate gentle pressure gradients, weak forces, and light winds. An example of a steep pressure gradient and strong winds is given in ● Fig. 8.20. Notice that the tightly packed isobars along the green line are producing a steep pressure gradient of 32 mb per 500 km and strong surface winds of 40 knots.

If the pressure gradient force were the only force acting upon air, we would always find winds blowing directly from higher toward lower pressure. However, the moment air starts to move, it is deflected in its path by the *Coriolis force.*

CORIOLIS FORCE   The Coriolis force describes an apparent force that is due to the rotation of the earth. To understand how it works, consider two people playing catch as they sit opposite one another on the rim of a merry-go-round (see ● Fig. 8.21, platform A). If the merry-go-round is not moving, each time the ball is thrown, it moves in a straight line to the other person.

Suppose the merry-go-round starts turning counterclockwise—the same direction the earth spins as viewed from above the North Pole. If we watch the game of catch from

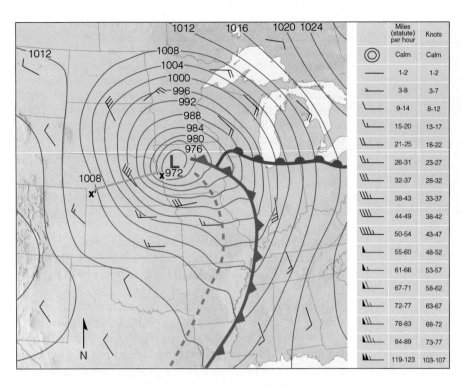

Miles (statute) per hour	Knots	
◎	Calm	Calm
—	1-2	1-2
⌐	3-8	3-7
⌐	9-14	8-12
⌐	15-20	13-17
⌐	21-25	18-22
⌐	26-31	23-27
⌐	32-37	28-32
⌐	38-43	33-37
⌐	44-49	38-42
⌐	50-54	43-47
⌐	55-60	48-52
⌐	61-66	53-57
⌐	67-71	58-62
⌐	72-77	63-67
⌐	78-83	68-72
⌐	84-89	73-77
⌐	119-123	103-107

● FIGURE 8.20 Surface weather map for 6 A.M. (CST), Tuesday, November 10, 1998. Dark gray lines are isobars with units in millibars. The interval between isobars is 4 mb. A deep low with a central pressure of 972 mb (28.70 in.) is moving over northwestern Iowa. The distance along the green line X-X′ is 500 km. The difference in pressure between X and X′ is 32 mb, producing a pressure gradient of 32 mb/500 km. The tightly packed isobars along the green line are associated with strong northwesterly winds of 40 knots, with gusts even higher. Wind directions are given by lines that parallel the wind. Wind speeds are indicated by barbs and flags. (A wind indicated by the symbol ⌐ would be a wind from the northwest at 10 knots. See blue insert.) The solid blue line is a cold front, the solid red line a warm front, and the solid purple line an occluded front. The dashed gray line is a trough.

above, we see that the ball moves in a straight-line path just as before. However, to the people playing catch on the merry-go-round, the ball seems to veer to its right each time it is thrown, always landing to the right of the point intended by the thrower (see Fig. 8.21, platform B). This perception is due to the fact that, while the ball moves in a straight-line path, the merry-go-round rotates beneath it; by the time the ball reaches the opposite side, the catcher has moved. To anyone on the merry-go-round, it seems as if there is some force causing the ball to deflect to the right. This apparent force is called the **Coriolis force** after Gaspard Coriolis, a nineteenth-century French scientist who worked it out mathematically. (Because it is an *apparent force* due to the rotation of the earth, it is also called the *Coriolis effect*.) This effect occurs on the rotating earth, too. All free-moving objects, such as ocean currents, aircraft, artillery projectiles, and air molecules seem to deflect from a straight-line path because the earth rotates under them.

The Coriolis force *causes the wind to deflect to the right of its intended path in the Northern Hemisphere and to the left of its intended path in the Southern Hemisphere.* To illustrate this, consider a satellite in polar circular orbit. If the earth were not rotating, the path of the satellite would be observed to move directly from north to south, parallel to the earth's meridian lines. However, the earth *does* rotate, carrying us and meridians eastward with it. Because of this rotation in the Northern Hemisphere, we see the satellite moving southwest instead of due south; it seems to veer off its path and move toward *its right*. In the Southern Hemisphere, the earth's direction of rotation is clockwise as viewed from above the South Pole. Consequently, a satellite moving northward from the South Pole would appear to move northwest and, hence, would veer to the *left* of its path.

The magnitude of the Coriolis force varies with the speed of the moving object and the latitude. ● Figure 8.22 shows this variation for various wind speeds at different latitudes. In each case, as the wind speed increases, the Coriolis force increases; hence, *the stronger the wind speed, the greater the de-*

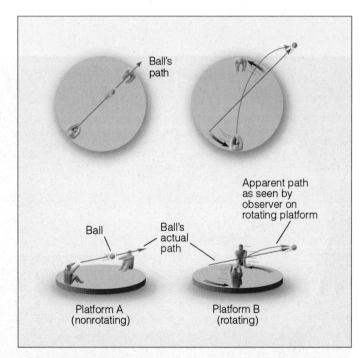

● FIGURE 8.21 On nonrotating platform A, the thrown ball moves in a straight line. On platform B, which rotates counterclockwise, the ball continues to move in a straight line. However, platform B is rotating while the ball is in flight; thus, to anyone on platform B, the ball appears to deflect to the right of its intended path.

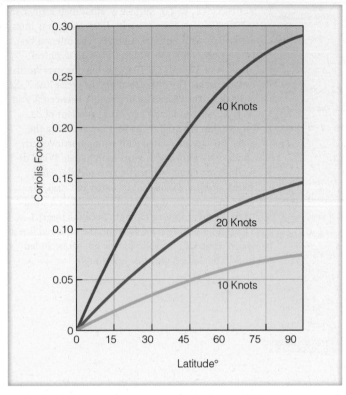

• FIGURE 8.22 The relative variation of the Coriolis force at different latitudes with different wind speeds.

The deep, low-pressure area illustrated in Fig. 8.20 was quite a storm. The intense low with its tightly packed isobars and strong pressure gradient produced extremely high winds that gusted over 90 knots in Wisconsin. The extreme winds caused blizzard conditions over the Dakotas, closed many interstate highways, shut down airports, and overturned trucks. The winds pushed a school bus off the road near Albert Lea, Minnesota, injuring two children, and blew the roofs off homes in Wisconsin. This notorious deep storm set an all-time record low pressure of 963 mb (28.43 in.) for Minnesota on November 10, 1998.

flection. Also, note that the Coriolis force increases for all wind speeds from a value of *zero at the equator to a maximum at the poles.* We can see this latitude effect better by examining • Fig. 8.23.

Imagine in Fig. 8.23 that there are three aircraft, each at a different latitude and each flying along a straight-line path, with no external forces acting on them. The destination of each aircraft is due east and is marked on the diagram (see Fig. 8.23a). Each plane travels in a straight path relative to an observer positioned at a fixed spot in space. The earth rotates beneath the moving planes, causing the destination points at latitudes 30° and 60° to change direction slightly (to the observer in space) (see Fig. 8.23b). To an observer standing on the earth, however, it is the plane that appears to deviate. The amount of deviation is greatest toward the pole and nonexistent at the equator. Therefore, the Coriolis force has a far

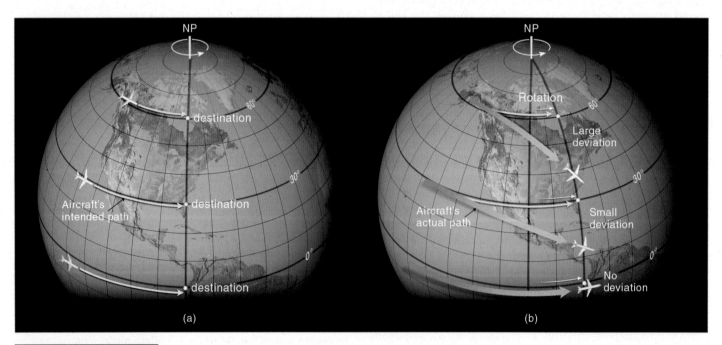

(a)   (b)

ACTIVE FIGURE 8.23 Except at the equator, a free-moving object heading either east or west (or any other direction) will appear from the earth to deviate from its path as the earth rotates beneath it. The deviation (Coriolis force) is greatest at the poles and decreases to zero at the equator. Visit the Meteorology Resource Center to view this and other active figures at academic.cengage.com/login

greater effect on the plane at high latitudes (large deviation) than on the plane at low latitudes (small deviation). On the equator, it has no effect at all. The same, of course, is true of its effect on winds.

In summary, to an observer on the earth, objects moving in *any direction* (north, south, east, or west) are deflected to the *right* of their intended path in the Northern Hemisphere and to the *left* of their intended path in the Southern Hemisphere. The amount of deflection depends upon:

1. the rotation of the earth
2. the latitude
3. the object's speed*

In addition, the *Coriolis force acts at right angles to the wind, only influencing wind direction and never wind speed.*

The Coriolis "force" behaves as a real force, constantly tending to "pull" the wind to its right in the Northern Hemisphere and to its left in the Southern Hemisphere. Moreover, this effect is present in all motions relative to the earth's surface. However, in most of our everyday experiences, the Coriolis force is so small (compared to other forces involved in those experiences) that it is negligible and, contrary to popular belief, does not cause water to turn clockwise or counterclockwise when draining from a sink. The Coriolis force is also minimal on small-scale winds, such as those that blow inland along coasts in summer. Here, the Coriolis force might be strong because of high winds, but the force cannot produce much deflection over the relatively short distances. Only where winds blow over vast regions is the effect significant.

> View this concept in action on the Meteorology Resource Center at academic.cengage.com/login

BRIEF REVIEW

In summary, we know that:

- Atmospheric (air) pressure is the pressure exerted by the mass of air above a region.
- A change in surface air pressure can be brought about by changing the mass (amount of air) above the surface.
- Heating and cooling columns of air can establish horizontal variations in atmospheric pressure aloft and at the surface.
- A difference in horizontal air pressure produces a horizontal pressure gradient force.
- The pressure gradient force is always directed from higher pressure toward lower pressure, and it is the pressure gradient force that causes the air to move and the wind to blow.

- Steep pressure gradients (tightly packed isobars on a weather map) indicate strong pressure gradient forces and high winds; gentle pressure gradients (widely spaced isobars) indicate weak pressure gradient forces and light winds.
- Once the wind starts to blow, the Coriolis force causes it to bend to the right of its intended path in the Northern Hemisphere and to the left of its intended path in the Southern Hemisphere.

WEATHER WATCH

As you drive your car along a highway (at the speed limit), the Coriolis force would "pull" your vehicle to the right about 1500 feet for every 100 miles you travel if it were not for the friction between your tires and the road surface.

With this information in mind, we will first examine how the pressure gradient force and the Coriolis force produce straight-line winds aloft. We will then see what influence the centripetal force has on winds that blow along a curved path.

STRAIGHT-LINE FLOW ALOFT — GEOSTROPHIC WINDS Earlier in this chapter, we saw that the winds aloft on an upper-level chart blow more or less parallel to the isobars or contour lines. We can see why this phenomenon happens by carefully looking at ●Fig. 8.24, which shows a map in the Northern Hemisphere, above the earth's frictional influence,* with horizontal pressure variations at an altitude of about 1 km above the earth's surface. The evenly spaced isobars indicate a constant pressure gradient force *(PGF)* directed from south toward north as indicated by the red arrow at the left. Why, then, does the map show a wind blowing from the west? We can answer this question by placing a parcel of air at position 1 in the diagram and watching its behavior.

At position 1, the *PGF* acts immediately upon the air parcel, accelerating it northward toward lower pressure. However, the instant the air begins to move, the Coriolis force deflects the air toward its right, curving its path. As the parcel of air increases in speed (positions 2, 3, and 4), the magnitude of the Coriolis force increases (as shown by the longer arrows), bending the wind more and more to its right. Eventually, the wind speed increases to a point where the Coriolis force just balances the *PGF*. At this point (position 5), the wind no longer accelerates because the net force is zero. Here the wind flows in a straight path, parallel to the isobars at a constant speed.† This flow of air is called a **geostrophic**

---

*These three factors are grouped together and shown in the expression

$$\text{Coriolis force} = 2m\,\Omega V \sin\phi,$$

where m is the object's mass, Ω is the earth's angular rate of spin (a constant), V is the speed of the object, and ϕ is the latitude.

---

*The friction layer (the layer where the wind is influenced by frictional interaction with objects on the earth's surface) usually extends from the surface up to about 1000 m (3300 ft) above the ground.

†At first, it may seem odd that the wind blows at a constant speed with no net force acting on it. But when we remember that the net force is necessary only to accelerate $(F = ma)$ the wind, it makes more sense. For example, it takes a considerable net force to push a car and get it rolling from rest. But once the car is moving, it only takes a force large enough to counterbalance friction to keep it going. There is no net force acting on the car, yet it rolls along at a constant speed.

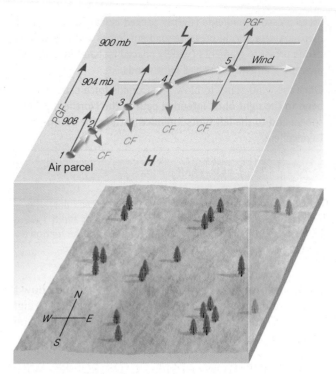

● FIGURE 8.24 Above the level of friction, air initially at rest will accelerate until it flows parallel to the isobars at a steady speed with the pressure gradient force (PGF) balanced by the Coriolis force (CF). Wind blowing under these conditions is called geostrophic.

(*geo:* earth; *strophic:* turning) **wind.** Notice that the geostrophic wind blows in the Northern Hemisphere with lower pressure to its left and higher pressure to its right.

When the flow of air is purely geostrophic, the isobars (or contours) are straight and evenly spaced, and the wind speed is constant. In the atmosphere, isobars are rarely straight or evenly spaced, and the wind normally changes speed as it flows along. So, the geostrophic wind is usually only an approximation of the real wind. However, the approximation is generally close enough to help us more clearly understand the behavior of the winds aloft.

As we would expect from our previous discussion of winds, the speed of the geostrophic wind is directly related to the pressure gradient. In ● Fig. 8.25, we can see that a geostrophic wind flowing parallel to the isobars is similar to

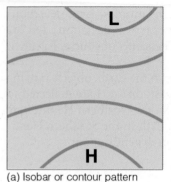 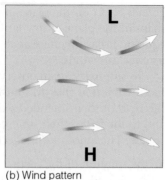

(a) Isobar or contour pattern   (b) Wind pattern

● FIGURE 8.26 By observing the orientation and spacing of the isobars (or contours) in diagram (a), the geostrophic wind direction and speed can be determined in diagram (b).

water in a stream flowing parallel to its banks. At position 1, the wind is blowing at a low speed; at position 2, the pressure gradient increases and the wind speed picks up. Notice also that at position 2, where the wind speed is greater, the Coriolis force is greater and balances the stronger pressure gradient force. (A more mathematical approach to the concept of geostrophic wind is given in the Focus section on p. 211.)

In ● Fig. 8.26, we can see that the geostrophic wind direction can be determined by studying the orientation of the isobars; its speed can be estimated from the spacing of the isobars. On an isobaric chart, the geostrophic wind direction and speed are related in a similar way to the contour lines. Therefore, if we know the isobar or contour patterns on an upper-level chart, we also know the direction and relative speed of the geostrophic wind, even for regions where no direct wind measurements have been made. Similarly, if we know the geostrophic wind direction and speed, we can estimate the orientation and spacing of the isobars, even if we don't have a current weather map. (It is also possible to estimate the wind flow and pressure patterns aloft by watching the movement of clouds. The Focus section on p. 212 illustrates this further.)

We know that the winds aloft do not always blow in a straight line; frequently, they curve and bend into meandering loops as they tend to follow the patterns of the isobars. In the Northern Hemisphere, winds blow counterclockwise around lows and clockwise around highs. The next section explains why.

● FIGURE 8.25 The isobars and contours on an upper-level chart are like the banks along a flowing stream. When they are widely spaced, the flow is weak; when they are narrowly spaced, the flow is stronger. The increase in winds on the chart results in a stronger Coriolis force (CF), which balances a larger pressure gradient force (PGF).

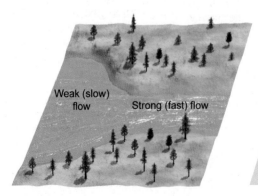

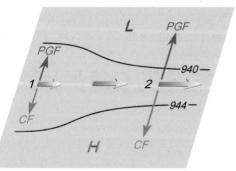

FOCUS ON AN ADVANCED TOPIC

A Mathematical Look at the Geostrophic Wind

We know from an earlier discussion that the geostrophic wind gives us a good approximation of the real wind above the level of friction, about 500 to 1000 m above the earth's surface. Above the friction layer, the winds tend to blow parallel to the isobars, or contours. We know that, for any given latitude, the speed of the geostrophic wind is proportional to the pressure gradient. This may be represented as

$$V_g \sim \frac{\Delta p}{d},$$

where V_g is the geostrophic wind and Δp is the pressure difference between two places some horizontal distance (d) apart. From this, we can see that the greater the pressure gradient, the stronger the geostrophic wind.

When we consider a unit mass of moving air, we must take into account the air density (mass per unit volume) expressed by the symbol ρ. The geostrophic wind is now directly proportional to the pressure gradient force; thus

$$V_g \sim \frac{1}{\rho}\frac{\Delta p}{d}.$$

We can see from this expression that, with the same pressure gradient (at the same latitude), the geostrophic wind will increase with increasing elevation because air density decreases with height.

In a previous section, we saw that the geostrophic wind represents a balance of forces between the Coriolis force and the pressure gradient force. Here, it should be noted that the Coriolis force (per unit mass) can be expressed as

Coriolis force = $2\Omega V \sin \phi$,

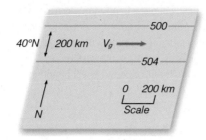

● FIGURE 4 A portion of an upper-air chart for part of the Northern Hemisphere at an altitude of 5600 meters above sea level. The lines on the chart are isobars, where 500 equals 500 millibars. The air temperature is $-25°C$ and the air density is 0.70 kg/m³.

where Ω is the earth's angular spin (a constant), V is the speed of the wind, and ϕ is the latitude. The sin ϕ is a trigonometric function that takes into account the variation of the Coriolis force with latitude. At the equator (0°), sin ϕ is 0; at 30° latitude, sin ϕ is 0.5, and, at the poles (90°), sin ϕ is 1.

This balance between the Coriolis force and the pressure gradient force can be written as

$$CF = PGF$$
$$2\Omega V_g \sin \phi = \frac{1}{\rho}\frac{\Delta p}{d}. \quad (1)$$

Solving for V_g, the geostrophic wind, the equation becomes

$$V_g = \frac{1}{2\Omega \sin \phi \, \rho}\frac{Dp}{d}.$$

Customarily, the rotational (2Ω) and latitudinal (sin ϕ) factors are combined into a single value f, called the *Coriolis parameter*. Thus, we have the geostrophic wind equation written as

$$V_g \sim \frac{1}{f\rho}\frac{\Delta p}{d}. \quad (2)$$

Suppose we compute the geostrophic wind for the example given in Fig. 4. Here the wind is blowing parallel to the isobars in the Northern Hemisphere at latitude 40°. The spacing between the isobars is 200 km and the pressure difference is 4 mb. The altitude is 5600 m above sea level, where the air temperature is $-25°C$ ($-13°F$) and the air density is 0.70 kg/m³. First, we list our data and put them in the proper units, as

$$\Delta p = 4 \text{ mb} = 400 \text{ Newtons/m}^2$$
$$d = 200 \text{ km} = 2 \times 10^5 \text{ m}$$
$$\sin \phi = \sin(40°) = 0.64$$
$$\rho = 0.70 \text{ kg/m}^3$$
$$2\Omega = 14.6 \times 10^{-5} \text{ radian/sec.*}$$

When we use equation (1) to compute the geostrophic wind, we obtain

$$V_g = \frac{1}{2\Omega \sin \phi \rho}\frac{\Delta \rho}{d},$$

$$V_g = \frac{400}{14.6 \times 10^{-5} \times 0.64 \times 0.70 \times 2 \times 10^5},$$

$$V_g = 30.6 \text{ m/sec, or } 59.4 \text{ knots.}$$

---

*The rate of the earth's rotation (Ω) is 360° in one day, actually a sidereal day consisting of 23 hr, 56 min, 4 sec, or 86,164 seconds. This gives a rate of rotation of 4.18 $\times 10^{-3}$ degree per second. Most often, Ω is given in radians, where 2π radians equals 360° ($\pi = 3.14$). Therefore, the rate of the earth's rotation can be expressed as 2π radians/86,164 sec, or 7.29 $\times 10^{-5}$ radian/sec, and the constant 2Ω becomes 14.6 $\times 10^{-5}$ radian/sec.

View this concept in action on the Meteorology Resource Center at academic.cengage.com/login

CURVED WINDS AROUND LOWS AND HIGHS ALOFT— GRADIENT WINDS

Because lows are also known as cyclones, the counterclockwise flow of air around them is often called *cyclonic flow*. Likewise, the clockwise flow of air around a high, or anticyclone, is called *anticyclonic flow*. Look at the wind flow around the upper-level low (Northern Hemisphere) in ● Fig. 8.27. At first, it appears as though the wind is defying the Coriolis force by bending to the left as it moves counterclockwise around the system. Let's see why the wind blows in this manner.

FOCUS ON AN OBSERVATION

Estimating Wind Direction and Pressure Patterns Aloft by Watching Clouds

Both the wind direction and the orientation of the isobars aloft can be estimated by observing middle- and high-level clouds from the earth's surface. Suppose, for example, we are in the Northern Hemisphere watching clouds directly above us move from southwest to northeast at an elevation of about 3000 m or 10,000 ft (see Fig. 5a). This indicates that the geostrophic wind at this level is southwesterly. Looking downwind, the geostrophic wind blows parallel to the isobars with lower pressure on the left and higher pressure on the right. Thus, if we *stand with our backs to the direction from which the clouds are moving, lower pressure aloft will always be to our left and higher pressure to our right.* From this observation, we can draw a rough upper-level chart (see Fig. 5b), which shows isobars and wind direction for an elevation of approximately 10,000 ft.

The isobars aloft will not continue in a southwest-northeast direction indefinitely;

rather, they will often bend into wavy patterns. We may carry our observation one step farther, then, by assuming a bending of the lines (Fig. 5c). Thus, with a southwesterly wind aloft, a trough of low pressure will be found

to our west and a ridge of high pressure to our east. What would be the pressure pattern if the winds aloft were blowing *from* the northwest? Answer: A trough would be to the east and a ridge to the west.

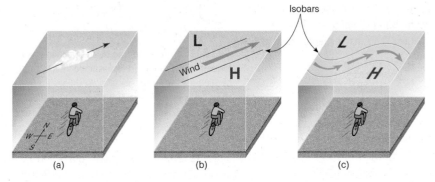

● FIGURE 5 This drawing of a simplified upper-level chart is based on cloud observations. Upper-level clouds moving from the southwest (a) indicate isobars and winds aloft (b). When extended horizontally, the upper-level chart appears as in (c), where a trough of low pressure is to the west and a ridge of high pressure is to the east.

Suppose we consider a parcel of air initially at rest at position 1 in Fig. 8.27a. The pressure gradient force accelerates the air inward toward the center of the low and the Coriolis force deflects the moving air to its right, until the air is moving parallel to the isobars at position 2. If the wind were geostrophic, at position 3 the air would move northward parallel to straight-line isobars at a constant speed. The wind is blowing at a con-

stant speed, but parallel to curved isobars. A wind that blows at a constant speed parallel to *curved isobars* above the level of frictional influence is termed a **gradient wind.**

Earlier in this chapter we learned that an object accelerates when there is a change in its speed or direction (or both). Therefore, the gradient wind blowing *around* the low-pressure center is constantly accelerating because it is con-

● FIGURE 8.27 Winds and related forces around areas of low and high pressure above the friction level in the Northern Hemisphere. Notice that the pressure gradient force (PGF) is in red, while the Coriolis force (CF) is in blue.

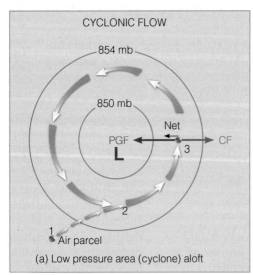

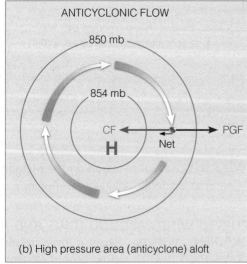

stantly changing direction. This acceleration, called the **centripetal acceleration,** is directed at right angles to the wind, inward toward the low center.

Remember from Newton's second law that, if an object is accelerating, there must be a *net force* acting on it. In this case, the net force acting on the wind must be directed toward the center of the low, so that the air will keep moving in a circular path. This inward-directed force is called the **centripetal force** (*centri:* center; *petal:* to push toward). The magnitude of the centripetal force is related to the wind velocity *(V)* and the radius of the wind's path *(r)* by the formula

$$\text{Centripetal force} = \frac{V^2}{r}.$$

Where wind speeds are light and there is little curvature (large radius), the centripetal force is weak and, compared to other forces, may be considered insignificant. However, where the wind is strong and blows in a tight curve (small radius), as in the case of tornadoes and tropical hurricanes, the centripetal force is large and becomes quite important.

The centripetal force results from an imbalance between the Coriolis force and the pressure gradient force.* Again, look closely at position 3 (Fig. 8.27a) and observe that the inward-direction pressure gradient force *(PGF)* is greater than the outward-directed Coriolis force *(CF)*. The difference between these two forces—the net force—is the inward-directed centripetal force. In Fig. 8.27b, the wind blows clockwise around the center of the high. The spacing of the isobars tells us that the magnitude of the *PGF* is the same as in Fig. 8.27a. However, to keep the wind blowing in a circle, the inward-directed Coriolis force must now be greater in magnitude than the outward-directed pressure gradient force, so that the centripetal force (again, the net force) is directed inward.

The greater Coriolis force around the high results in an interesting observation. Because the Coriolis force (at any given latitude) can increase only when the wind speed increases, we can see that for the same pressure gradient (the same spacing of the isobars), the winds around a high-pressure area (or a ridge) must be greater than the winds around a low-pressure area (or a trough). Normally, however, the winds blow much faster around an area of low pressure (a cyclonic storm)

---

*In some cases, it is more convenient to express the centripetal force (and the centripetal acceleration) as the *centrifugal force,* an apparent force that is equal in magnitude to the centripetal force, but directed outward from the center of rotation. The gradient wind is then described as a balance of forces between the centrifugal force

$$\frac{V^2}{r},$$

the pressure gradient force

$$\frac{1}{\rho}\frac{\Delta p}{d},$$

and the Coriolis force $2\Omega V \sin \phi$. Under these conditions, the *gradient wind equation* for a unit mass of air is expressed as

$$\frac{V^2}{r}+\frac{1}{\rho}\frac{\Delta p}{d}+2\Omega V \sin \phi = 0.$$

than they do around an area of high pressure because the isobars around the low are usually spaced much closer together, resulting in a much stronger pressure gradient.

In the Southern Hemisphere, the pressure gradient force starts the air moving, and the Coriolis force deflects the moving air to the *left*, thereby causing the wind to blow *clockwise around lows* and *counterclockwise around highs.* ● Figure 8.28 shows a satellite image of clouds and wind flow (dark arrows) around a low-pressure area in the Northern Hemisphere (8.28a) and in the Southern Hemisphere (8.28b).

Near the equator, where the Coriolis force is minimum, winds may blow around intense tropical storms with the centripetal force being almost as large as the pressure gradient force. In this type of flow, the Coriolis force is considered negligible, and the wind is called *cyclostrophic.*

So far we have seen how winds blow in theory, but how do they appear on an actual map?

WINDS ON UPPER-LEVEL CHARTS   On the upper-level 500-mb map (● Figure 8.29), notice that, as we would expect, the winds tend to parallel the contour lines in a wavy west-to-east direction. Notice also that the contour lines tend to decrease in elevation from south to north. This situation occurs because the air at this level is warmer to the south and colder to the north. On the map, where horizontal temperature contrasts are large there is also a large height gradient—the contour lines are close together and the winds are strong. Where the horizontal temperature contrasts are small, there is a small height gradient—the contour lines are spaced farther apart and the winds are weaker. In general, on maps such as this we find stronger north-to-south temperature contrasts in winter than in summer, which is why the winds aloft are usually stronger in winter.

In Fig. 8.29, the wind is geostrophic where it blows in a straight path parallel to evenly spaced lines; it is gradient where it blows parallel to curved contour lines. Where the wind flows in large, looping meanders, following a more or less north-south trajectory (such as along the west coast of North America), the wind-flow pattern is called **meridional.** Where the winds are blowing in a west-to-east direction (such as over the eastern third of the United States), the flow is termed **zonal.**

Because the winds aloft in middle and high latitudes generally blow from west to east, planes flying in this direction have a beneficial tail wind, which explains why a flight from San Francisco to New York City takes about thirty minutes less than the return flight. If the flow aloft is zonal, clouds, storms, and surface anticyclones tend to move more rapidly from west to east. However, where the flow aloft is meridional, as we will see in Chapter 12 surface storms tend to move more slowly, often intensifying into major storm systems.

We know that the winds aloft in the middle latitudes of the Northern Hemisphere tend to blow in a west-to-east pattern. Does this mean that the winds aloft in the Southern Hemisphere blow from east-to-west? If you are unsure of the answer, read the Focus section on p. 215.

(a) Northern Hemisphere

(b) Southern Hemisphere

● FIGURE 8.28 Clouds and related wind-flow patterns (black arrows) around low-pressure areas. (a) In the Northern Hemisphere, winds blow counterclockwise around an area of low pressure. (b) In the Southern Hemisphere, winds blow clockwise around an area of low pressure.

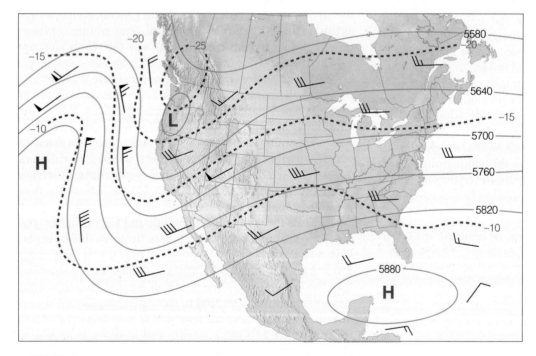

● FIGURE 8.29 An upper-level 500-mb map showing wind direction, as indicated by lines that parallel the wind. Wind speeds are indicated by barbs and flags. (See the blue insert.) Solid gray lines are contours in meters above sea level. Dashed red lines are isotherms in °C.

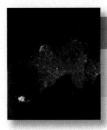

FOCUS ON AN OBSERVATION

Winds Aloft in the Southern Hemisphere

In the Southern Hemisphere, just as in the Northern Hemisphere, the winds aloft blow because of horizontal differences in pressure. The pressure differences, in turn, are due to variations in temperature. Recall from an earlier discussion of pressure that warm air aloft is associated with high pressure and cold air aloft with low pressure. Look at Fig. 6. It shows an upper-level chart that extends from the Northern Hemisphere into the Southern Hemisphere. Over the equator, where the air is warmer, the pressure aloft is higher. North and south of the equator, where the air is colder, the pressure aloft is lower.

Let's assume, to begin with, that there is no wind on the chart. In the Northern Hemisphere, the pressure gradient force directed northward starts the air moving toward lower pressure. Once the air is set in motion, the Coriolis force bends it to the right until it is a *west wind*, blowing parallel to the isobars. In the Southern Hemisphere, the pressure gradient

force directed southward starts the air moving south. But notice that the Coriolis force in the Southern Hemisphere bends the moving air to its *left*, until the wind is blowing parallel to the

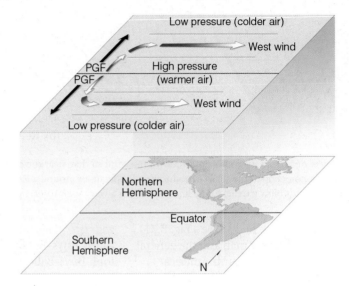

● FIGURE 6
Upper-level chart that extends over the Northern and Southern Hemispheres. Solid gray lines on the chart are isobars.

isobars *from the west*. Hence, in the middle and high latitudes of both hemispheres, we generally find westerly winds aloft.

Take a minute and look back at Fig. 8.20 on p. 207. Observe that the winds on this surface map tend to cross the isobars, blowing from higher pressure toward lower pressure. Observe also that along the green line, the tightly packed isobars are producing a steady surface wind of 40 knots. However, this same pressure gradient (with the same air temperature) would, on an upper-level chart, produce a much stronger wind. Why do surface winds normally cross the isobars and why do they blow more slowly than the winds aloft? The answer to both of these questions is *friction*.

SURFACE WINDS   The frictional drag of the ground slows the wind down. Because the effect of friction decreases as we move away from the earth's surface, wind speeds tend to increase with height above the ground. The atmospheric layer that is influenced by friction, called the **friction layer** (or *planetary boundary layer*), usually extends upward to an altitude near 1000 m (3300 ft) above the surface, but this altitude may vary due to strong winds or irregular terrain. (We will examine the planetary boundary layer winds more thoroughly in Chapter 9.)

In ● Fig. 8.30a, the wind aloft is blowing at a level above the frictional influence of the ground. At this level, the wind is approximately geostrophic and blows parallel to the isobars, with

the pressure gradient force *(PGF)* on its left balanced by the Coriolis force *(CF)* on its right. At the earth's surface, the same pressure gradient will not produce the same wind speed, and the wind will not blow in the same direction.

Near the surface, *friction reduces the wind speed, which in turn reduces the Coriolis force.* Consequently, the weaker Coriolis force no longer balances the pressure gradient force, and the wind blows across the isobars toward lower pressure. The angle (α) at which the wind crosses the isobars varies, but averages about 30°.* As we can see in Fig. 8.30a, at the surface the pressure gradient force is now balanced by the sum of the frictional force and the Coriolis force. Therefore, in the Northern Hemisphere, we find surface winds blowing counterclockwise and *into* a low; they flow clockwise and *out* of a high (see Fig. 8.30b). In the Southern Hemisphere, winds blow clockwise and inward around surface lows; counter-

---

*The angle at which the wind crosses the isobars to a large degree depends upon the roughness of the terrain. Everything else being equal, the rougher the surface, the larger the angle. Over hilly land, the angle might average between 35° and 40°, while over an open body of relatively smooth water it may average between 10° and 15°. Taking into account all types of surfaces, the average is near 30°. This angle also depends on the wind speed. Typically, the angle is smallest for high winds and largest for gentle breezes. As we move upward through the friction layer, the wind becomes more and more parallel to the isobars.

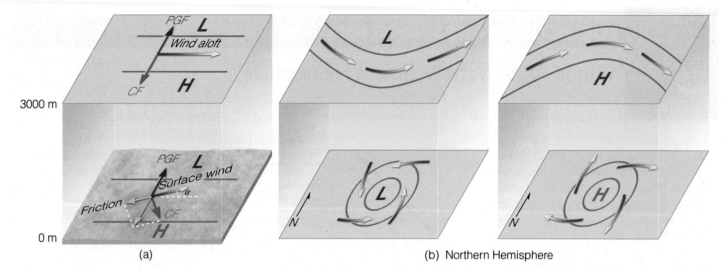

● FIGURE 8.30 (a) The effect of surface friction is to slow down the wind so that, near the ground, the wind crosses the isobars and blows toward lower pressure. (b) This phenomenon at the surface produces an inflow of air around a low and an outflow of air around a high. Aloft, away from the influence of friction, the winds blow parallel to the lines, usually in a wavy west-to-east pattern.

clockwise and outward around surface highs (see ● Fig. 8.31). ● Figure 8.32 illustrates a surface weather map and the general wind-flow pattern on a particular day in South America.

We know that, because of friction, surface winds move more slowly than do the winds aloft with the same pressure gradient. Surface winds also blow across the isobars toward lower pressure. The angle at which the winds cross the isobars depends upon surface friction, wind speed, and the height above the surface. Aloft, however, the winds blow parallel to contour lines, with lower pressure (in the Northern Hemisphere) to their left. Consequently, because of this fact, if you (in the Northern Hemisphere) stand with the wind aloft to your back, lower pressure will be to your left and higher pressure to your right (see ● Fig. 8.33a). The same rule applies to the surface wind, but with a slight modification due to the fact that here the wind crosses the isobars. Look at Fig. 8.33b and notice that, at the surface, *if you stand with your back to the wind, then turn clockwise about 30°, the center of lowest pressure will be to*

your left.* This relationship between wind and pressure is often called **Buys-Ballot's law,** after the Dutch meteorologist Christoph Buys-Ballot (1817–1890), who formulated it.

Winds and Vertical Air Motions

Up to this point, we have seen that surface winds blow inward the center of low pressure and outward away from the center of high pressure. Notice in ● Fig. 8.34 that as air moves inward toward the center of low pressure, it must go somewhere. Since this converging air cannot go into the ground, it slowly rises. Above the surface low (at about 6 km or so), the air begins to diverge (spread apart).

As long as the upper-level diverging air balances the converging surface air, the central pressure in the surface low

---

*In the Southern Hemisphere, stand with your back to the wind, then turn counterclockwise about 30°—the center of lowest pressure will then be to your right.

● FIGURE 8.31 Winds around an area of (a) low pressure and (b) high pressure in the Southern Hemisphere.

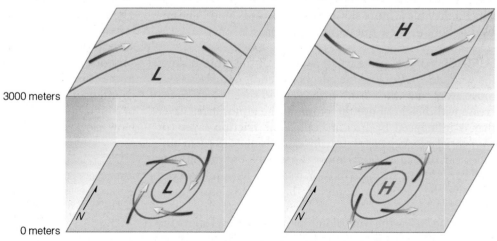

Southern Hemisphere

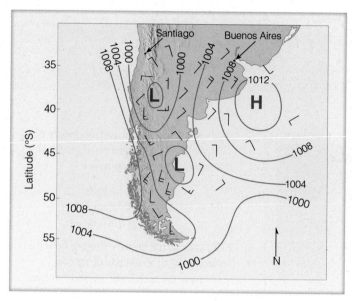

● FIGURE 8.32 Surface weather map showing isobars and winds on a day in December in South America.

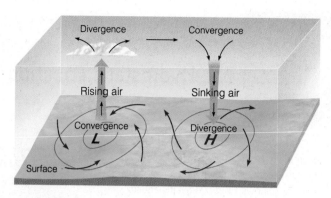

● FIGURE 8.34 Winds and air motions associated with surface highs and lows in the Northern Hemisphere.

does not change. However, the surface pressure *will change* if upper-level divergence and surface convergence are not in balance. For example, as we saw earlier in this chapter (when we examined the air pressure above two cities), the surface pressure will change if the mass of air above the surface changes. Consequently, if upper-level divergence exceeds surface convergence (that is, more air is removed at the top than

is taken in at the surface), the air pressure at the center of the surface low will decrease, and isobars around the low will become more tightly packed. This situation increases the pressure gradient (and, hence, the pressure gradient force), which, in turn, increases the surface winds.

Surface winds move outward (diverge), away from the center of a high-pressure area. To replace this laterally spreading air, the air aloft converges and slowly descends as shown in Fig. 8.34. Again, as long as upper-level converging air balances surface diverging air, the air pressure in the center of the high will not change. (Convergence and divergence of air are so important to the development or weakening of surface pressure systems that we will examine this topic again when

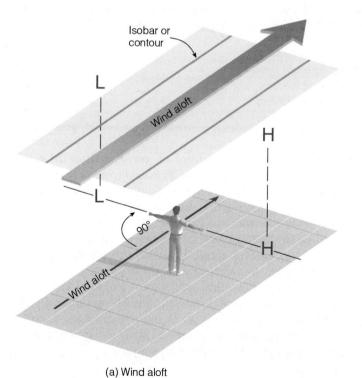

(a) Wind aloft

● FIGURE 8.33 (a) In the Northern Hemisphere, if you stand with the wind aloft at your back, lower pressure aloft will be to your left and higher pressure to your right. (b) At the surface, the center of lowest pressure will be to your left if, with your back to the surface wind, you turn clockwise about 30°.

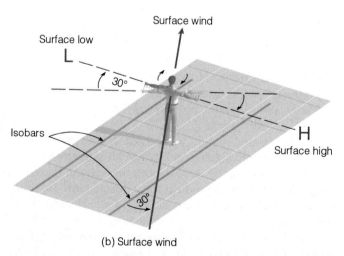

(b) Surface wind

Northern Hemisphere

FOCUS ON AN ADVANCED TOPIC

The Hydrostatic Equation

Air is in hydrostatic equilibrium when the upward-directed pressure gradient force is exactly balanced by the downward force of gravity. Figure 7 shows air in hydrostatic equilibrium. Since there is no net vertical force acting on the air, there is no net vertical acceleration, and the sum of the forces is equal to zero, all of which is represented by

$$PGF_{vertical} + g = 0$$
$$\frac{1}{\rho}\frac{\Delta p}{\Delta z} + g = 0,$$

where ρ is the air density, Δp is the decrease in pressure along a small change in height (Δz), and g is the force of gravity. This expression is usually given as

$$\frac{\Delta p}{\Delta z} = -\rho g.$$

This equation is called the *hydrostatic equation*. The hydrostatic equation tells us that the rate at which the pressure decreases with height is equal to the air density times the acceleration of gravity (where ρg is actually the force of gravity per unit volume). The minus sign indicates that, as the air pressure decreases, the height increases. When the hydrostatic equation is given as

$$\Delta p = -\rho g\,\Delta z,$$

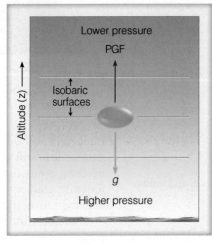

● **FIGURE 7** When the vertical pressure gradient force (PGF) is in balance with the force of gravity (g), the air is in hydrostatic equilibrium.

it tells us something important about the atmosphere that we learned earlier: The air pressure decreases more rapidly with height in cold (more-dense) air than it does in warm (less-dense) air. In addition, we can use the hydrostatic equation to determine how rapidly the air pressure decreases with increasing height above the surface. For example, suppose at the surface a 1000 meter-thick layer of air (under stan-

dard conditions) has an average density of 1.1 kg/m³ and an acceleration of gravity of 9.8 m/sec². Therefore, we have

$$\rho = 1.1 \text{ kg/m}^3$$
$$g = 9.8 \text{ m/sec}^2$$
$$\Delta z = 1000 \text{ m}$$

(This value is the height difference from the surface [0 m] to an altitude of 1000 m.)

Using the hydrostatic equation to compute Δp, the difference in pressure in a 1000-meter-thick layer of air, we obtain

$$\Delta p = \rho g\,\Delta z$$
$$\Delta p = (1.1)\,(9.8)\,(1000)$$
$$\Delta p = 10{,}780 \text{ Newtons/m}^2.$$

Since 1 mb = 100 Newtons/m²,

$$\Delta p = 108 \text{ mb}.$$

Hence, air pressure decreases by about 108 mb in a standard 1000-meter layer of air near the surface. This closely approximates the pressure change of 10 mb per 100 meters we used in converting station pressure to sea-level pressure earlier in this chapter.

---

we look more closely at the vertical structure of pressure systems in Chapter 12.)

The rate at which air rises above a low or descends above a high is small compared to the horizontal winds that spiral about these systems. Generally, the vertical motions are usually only about several centimeters per second, or about 1.5 km (or 1 mi) per day.

Earlier in this chapter we learned that air moves in response to pressure differences. Because air pressure decreases rapidly with increasing height above the surface, there is always a strong pressure gradient force directed upward, much stronger than in the horizontal. Why, then, doesn't the air rush off into space?

Air does not rush off into space because the upward-directed pressure gradient force is nearly always exactly bal-

anced by the downward force of gravity. When these two forces are in exact balance, the air is said to be in **hydrostatic equilibrium.** When air is in hydrostatic equilibrium, there is no net vertical force acting on it, and so there is no net vertical acceleration. Most of the time, the atmosphere approximates hydrostatic balance, even when air slowly rises or descends at a constant speed. However, this balance does not exist in violent thunderstorms and tornadoes, where the air shows appreciable vertical acceleration. But these occur over relatively small vertical distances, considering the total vertical extent of the atmosphere. (A more mathematical look at hydrostatic equilibrium, expressed by the *hydrostatic equation,* is given in the Focus section above.)

SUMMARY

This chapter gives us a broad view of how and why the wind blows. We examined constant pressure charts and found that low heights correspond to low pressure and high heights to high pressure. In regions where the air aloft is cold, the air pressure is normally lower than average; where the air aloft is warm, the air pressure is normally higher than average. Where horizontal variations in temperature exist, there is a corresponding horizontal change in pressure. The difference in pressure establishes a force, the pressure gradient force, which starts the air moving from higher toward lower pressure.

Once the air is set in motion, the Coriolis force bends the moving air to the right of its intended path in the Northern Hemisphere and to the left in the Southern Hemisphere. Above the level of surface friction, the wind is bent enough so that it blows nearly parallel to the isobars, or contours. Where the wind blows in a straight-line path, and a balance exists between the pressure gradient force and the Coriolis force, the wind is termed geostrophic. Where the wind blows parallel to curved isobars (or contours), the centripetal acceleration becomes important, and the wind is called a gradient wind. When the wind-flow pattern aloft is west-to-east, the flow is called *zonal;* where the wind flow aloft is more north-south, the flow is called *meridional.*

The interaction of the forces causes the wind in the Northern Hemisphere to blow clockwise around regions of high pressure and counterclockwise around areas of low pressure. In the Southern Hemisphere, the wind blows counterclockwise around highs and clockwise around lows. The effect of surface friction is to slow down the wind. This causes the surface air to blow across the isobars from higher pressure toward lower pressure. Consequently, in both hemispheres, surface winds blow outward, away from the center of a high, and inward, toward the center of a low.

When the upward-directed pressure gradient force is in balance with the downward force of gravity, the air is in hydrostatic equilibrium. Since there is no net vertical force acting on the air, it does not rush off into space.

KEY TERMS

The following terms are listed (with page numbers) in the order they appear in the text. Define each. Doing so will aid you in reviewing the material covered in this chapter.

air pressure, 194
barometer, 197
millibar, 197
hectopascal, 197
mercury barometer, 198
aneroid barometer, 198
station pressure, 199
sea-level pressure, 199

isobars, 200
surface map, 200
isobaric chart, 200
contour lines (on isobaric
  charts), 202
ridges, 203
troughs, 203
anticyclones, 203

mid-latitude cyclonic
  storms, 203
pressure gradient, 205
pressure gradient force, 206
Coriolis force, 207
geostrophic wind, 209
gradient wind, 212

centripetal acceleration, 213
centripetal force, 213
meridional flow, 213
zonal flow, 213
friction layer, 215
Buys-Ballot's law, 216
hydrostatic equilibrium, 218

QUESTIONS FOR REVIEW

1. Why does air pressure decrease with height more rapidly in cold air than in warm air?
2. What can cause the air pressure to change at the bottom of a column of air?
3. What is considered standard sea-level atmospheric pressure in millibars? In inches of mercury? In hectopascals?
4. How does an aneroid barometer differ from a mercury barometer?
5. How does sea-level pressure differ from station pressure? Can the two ever be the same? Explain.
6. On an upper-level chart, is cold air aloft generally associated with low or high pressure? What about warm air aloft?
7. What do Newton's first and second laws of motion tell us?
8. Explain why, in the Northern Hemisphere, the average height of contour lines on an upper-level isobaric chart tend to decrease northward.
9. What is the force that initially sets the air in motion?
10. What does the Coriolis force do to moving air (a) in the Northern Hemisphere? (b) in the Southern Hemisphere?
11. Explain how each of the following influences the Coriolis force: (a) rotation of the earth; (b) wind speed; (c) latitude.
12. How does a steep (or strong) pressure gradient appear on a weather map?
13. Explain why on a map, closely spaced isobars (or contours) indicate strong winds, and widely spaced isobars (or contours) indicate weak winds.
14. What is a geostrophic wind? Why would you *not* expect to observe a geostrophic wind at the equator?
15. Why do upper-level winds in the middle latitudes of both hemispheres generally blow from the west?
16. Describe how the wind blows around highs and lows aloft and near the surface (a) in the Northern Hemisphere and (b) in the Southern Hemisphere.
17. What are the forces that affect the horizontal movement of air?
18. What factors influence the angle at which surface winds cross the isobars?
19. Describe the type of vertical air motions associated with surface high- and low-pressure areas.
20. Since there is always an upward-directed pressure gradient force, why doesn't the air rush off into space?

21. How does Buys-Ballot's law help to locate regions of high and low pressure aloft and at the surface?
22. Explain the effect surface friction has on wind speed and direction.
23. Explain how on a 500-mb chart you would be able to distinguish a trough from a ridge?

QUESTIONS FOR THOUGHT

1. Explain why, on a sunny day, an aneroid barometer would indicate "stormy" weather when carried to the top of a hill or mountain.
2. The gas law states that pressure is proportional to temperature times density. Use the gas law to explain why a balloon will deflate when placed inside a refrigerator. Use the gas law to explain why the same balloon will inflate when removed from the refrigerator and placed in a warm room.
3. In ●Fig. 8.35 suppose the air column above city Q is completely saturated with water vapor, and the air column above city T is completely dry. If the temperature of the air in both columns is the same, which column will have the highest atmospheric pressure at the surface? Explain. (Hint: Refer back to the Focus section on p. 106 in Chapter 4, "Is Humid Air Heavier Than Dry Air?")

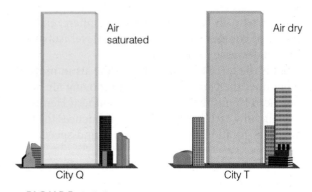

Air saturated   Air dry

City Q   City T

● FIGURE 8.35

4. Could station pressure ever *exceed* sea-level pressure? Explain.
5. Suppose you are in the Northern Hemisphere watching altocumulus clouds 4000 m (13,000 ft) above you drift from the northeast. Draw the orientation of the isobars above you. Locate and mark regions of lowest and highest pressure on this map. Finish the map by drawing isobars and the upper-level wind-flow pattern hundreds of kilometers in all directions from your position. Would this type of flow be zonal or meridional? Explain.
6. Pilots often use the expression "high to low, look out below." In terms of upper-level temperature and pressure, explain what this can mean.

7. Suppose an aircraft using a pressure altimeter flies along a constant pressure surface from standard temperature into warmer-than-standard air without any corrections. Explain why the altimeter would indicate an altitude lower than the aircraft's true altitude.
8. If the earth were not rotating, how would the wind blow with respect to centers of high and low pressure?
9. Why are surface winds that blow over the ocean closer to being geostrophic than those that blow over the land?
10. If the wind aloft is blowing parallel to curved isobars, with the horizontal pressure gradient force being of greater magnitude than the Coriolis force, would the wind flow be cyclonic or anticyclonic? In this example, what would be the relative magnitude of the centripetal acceleration, and how would it be directed?
11. With your present outside surface wind, use Buys-Ballot's law to determine where regions of surface high- and low-pressure areas are located. If clouds are moving overhead, use the relationship to locate regions of higher and lower pressure aloft.
12. If you live in the Northern Hemisphere and a region of surface low pressure is directly west of you, what would probably be the surface wind direction at your home? If an upper-level low is also directly west of your location, describe the probable wind direction aloft and the direction in which middle-type clouds would move. How would the wind direction and speed change from the surface to where the middle clouds are located?
13. In the Northern Hemisphere, you observe surface winds shift from N to NE to E, then to SE. From this observation, you determine that a west-to-east moving high-pressure area (anticyclone) has passed north of your location. Describe with the aid of a diagram how you were able to come to this conclusion.
14. The Coriolis force causes winds to deflect to the right of their intended path in the Northern Hemisphere, yet around a surface low-pressure area, winds blow counterclockwise, appearing to bend to their left. Explain why.
15. Why is it that, on the equator, winds may blow either counterclockwise or clockwise with respect to an area of low pressure?
16. Use the gas law in the Focus section on p. 196 to explain why a car with tightly closed windows will occasionally have a window "blow out" or crack when exposed to the sun on a hot day.
17. Consider wind blowing over a land surface that crosses a coastline and then blows over a lake. How will the wind speed and direction change as it moves from the land surface to the lake surface?
18. As a cruise ship crosses the equator, the entertainment director exclaims that water in a tub will drain in the opposite direction now that the ship is in the Southern Hemisphere. Give *two* reasons to the entertainment director why this assertion is not so.

PROBLEMS AND EXERCISES

1. A station 300 m above sea level reports a station pressure of 994 mb. What would be the sea-level pressure for this station, assuming standard atmospheric conditions? If the observation were taken on a hot summer afternoon, would the sea-level pressure be greater or less than that obtained during standard conditions? Explain.

2. ● Figure 8.36 is a sea-level pressure chart (Northern Hemisphere), with isobars drawn for every 4 mb. Answer the following questions, which refer to this map.

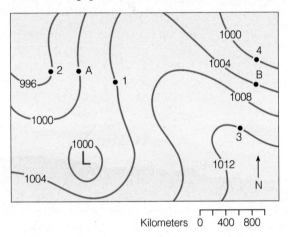

Kilometers 0   400 800

● FIGURE 8.36 Map for problem 2.

(a) What is the lowest possible pressure in whole millibars that there can be in the center of the closed low? What is the highest pressure possible?

(b) Place a dashed line through the ridge and a dotted line through the trough.

(c) What would be the wind direction at point A and at point B?

(d) Where would the stronger wind be blowing, at point A or B? Explain.

(e) Compute the pressure gradient between points 1 and 2, and between points 3 and 4. How do the computed pressure gradients relate to the pressure gradient force?

(f) If point A and point B are located at 30°N, and if the air density is 1.2 kg/m³, use the geostrophic wind equation in the Focus section on p. 211 to compute the geostrophic wind at point A and point B. (Hint: Be sure to convert km to m and mb to Newtons/m², where 1 mb = 100 Newtons/m².)

(g) Would the actual winds at point A and point B be greater than, less than, or equal to the wind speeds computed in problem (f)? Explain.

3. (a) Suppose the atmospheric pressure at the bottom of a deep air column 5.6 km thick is 1000 mb. If the average air density of the column is 0.91 kg/m³, and the acceleration of gravity is 9.8 m/sec², use the hydrostatic equation on p. 218 to determine the atmospheric pressure at the top of the column. (Hint: Be sure to convert km to m and mb to Newtons/m², where 1 mb = 100 N/m².)

(b) If the air in the column of problem (a) becomes much colder than average, would the atmospheric pressure at the top of the new column be greater than, less than, or equal to the pressure computed in problem (a)? Explain.

(c) Determine the atmospheric pressure at the top of the air column in problem (a) if the air in the column is quite cold and has an average density of 0.97 kg/m³.

4. Use the gas law in the Focus section on p. 196 to calculate the air pressure in millibars when the air temperature is −23°C and the air density is 0.700 kg/m³. (Hint: Be sure to use the Kelvin temperature.) At approximately what elevation would you expect to observe this pressure?

5. Suppose air in a closed container has a pressure of 1000 mb and a temperature of 20°C.

(a) Use the gas law to determine the air density in the container.

(b) If the density in the container remains constant, but the pressure doubles, what would be the new temperature?

6. A large balloon is filled with air so that the air pressure inside just equals the air pressure outside. The volume of the filled balloon is 3 m³, the mass of air inside is 3.6 kg, and the temperature inside is 20°C. What is the air pressure? (Hint: Density = mass/volume.)

7. If the clouds overhead are moving from north to south, would the upper-level center of low pressure be to the east or west of you? Draw a simplified map to explain.

Wind skims a rock across wet ground at Death Valley National Park, California.

© Michael Melford, National Geographic Image Collection

CHAPTER 9

Wind: Small Scale and Local Systems

On December 30, 1997, a United Airlines' Boeing 747 carrying 374 passengers was en route to Hawaii from Japan. Dinner had just been served, and the aircraft had reached a cruising altitude of 33,000 feet. Suddenly, east of Tokyo and over the Pacific Ocean, this routine, uneventful flight turned tragic. Without warning, the aircraft entered a region of severe air turbulence and a vibration ran through the aircraft. The plane nosed upward, then plunged toward the earth for about 1000 feet before stabilizing. Screaming, terrified passengers not fastened to their seats were flung against the walls of the aircraft, then dropped. Bags, serving trays, and luggage that slipped out from under the seats were flying about inside the plane. Within seconds, the entire ordeal was over. At least 110 people were injured, 12 seriously. Tragically, there was one fatality: a 32-year-old woman, who had been hurled against the ceiling of the plane, died of severe head injuries. What sort of atmospheric phenomenon could cause such turbulence?

CONTENTS

The aircraft in our opening vignette encountered a turbulent eddy—an "air pocket"—in perfectly clear air. Such violent eddies are not uncommon, especially in the vicinity of jet streams. In this chapter, we will examine a variety of eddy circulations. First, we will see how eddies form and how eddies and other small-scale winds interact with our environment. Then, we will examine slightly larger circulations—local winds—such as the sea breeze and the chinook, describing how they form and the type of weather they generally bring.

Small-Scale Winds Interacting with the Environment

The air in motion—what we commonly call *wind*—is invisible, yet we see evidence of it nearly everywhere we look. It sculptures rocks, moves leaves, blows smoke, and lifts water vapor upward to where it can condense into clouds. The wind is with us wherever we go. On a hot day, it can cool us off; on a cold day, it can make us shiver. A breeze can sharpen our appetite when it blows the aroma from the local bakery in our direction. The wind is a powerful element. The workhorse of weather, it moves storms and large fair-weather systems around the globe. It transports heat, moisture, dust, insects, bacteria, and pollens from one area to another.

Circulations of all sizes exist within the atmosphere. Little whirls form inside bigger whirls, which encompass even larger whirls—one huge mass of turbulent, twisting *eddies.** For clarity, meteorologists arrange circulations according to their size. This hierarchy of motion from tiny gusts to giant storms is called the **scales of motion.**

*Eddies are spinning globs of air that have a life history of their own.

SCALES OF MOTION   Consider smoke rising from a chimney into the otherwise clean air in the industrial section of a large city (see ● Fig. 9.1a). Within the smoke, small chaotic motions—tiny eddies—cause it to tumble and turn. These eddies constitute the smallest scale of motion—the **microscale.** At the microscale, eddies with diameters of a few meters or less not only disperse smoke, they also sway branches and swirl dust and papers into the air. They form by convection or by the wind blowing past obstructions and are usually short-lived, lasting only a few minutes at best.

In Fig. 9.1b observe that, as the smoke rises, it drifts toward the center of town. Here the smoke rises even higher and is carried many kilometers downwind. This circulation of city air constitutes the next larger scale—the **mesoscale** (meaning middle scale). Typical mesoscale winds range from a few kilometers to about a hundred kilometers in diameter. Generally, they last longer than microscale motions, often many minutes, hours, or in some cases as long as a day. Mesoscale circulations include local winds (which form along shorelines and mountains), as well as thunderstorms, tornadoes, and small tropical storms.

When we look at the smoke stack on a surface weather map (see Fig. 9.1c), neither the smoke stack nor the circulation of city air shows up. All that we see are the circulations around high- and low-pressure areas. We are now looking at the **synoptic scale,** or weather map scale. Circulations of this magnitude dominate regions of hundreds to even thousands of square kilometers and, although the life spans of these features vary, they typically last for days and sometimes weeks.

The largest wind patterns are seen at the **planetary (global) scale.** Here, we have wind patterns ranging over the entire earth. Sometimes, the synoptic and global scales are combined and referred to as the *macroscale.* (● Figure 9.2 summarizes the various scales of motion and their average life span.)

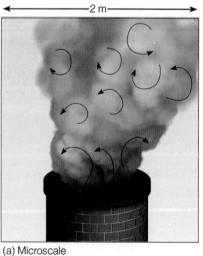 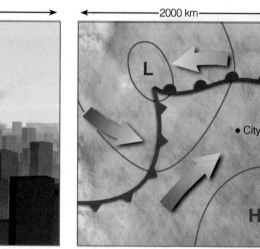

(a) Microscale   (b) Mesoscale   (c) Synoptic scale

● FIGURE 9.1   Scales of atmospheric motion. The tiny microscale motions constitute a part of the larger mesoscale motions, which, in turn, are part of the much larger synoptic scale. Notice that as the scale becomes larger, motions observed at the smaller scale are no longer visible.

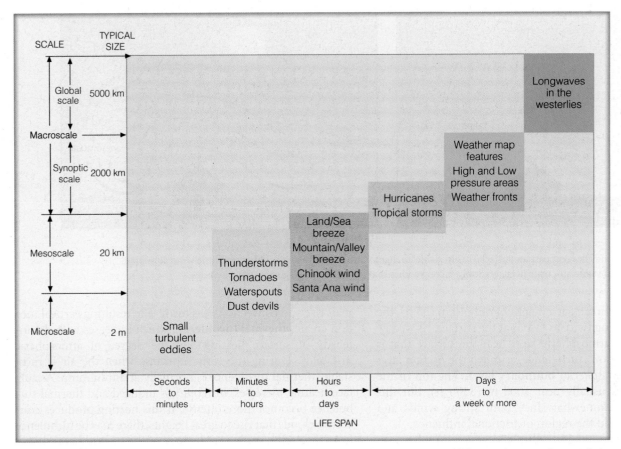

● FIGURE 9.2 The scales of atmospheric motion with the phenomenon's average size and life span. (Because the actual size of certain features may vary, some of the features fall into more than one category.)

In the next several sections, we will concentrate primarily on microscale winds and the effect they have on our environment.

FRICTION AND TURBULENCE IN THE BOUNDARY LAYER We are all familiar with friction. If we rub our hand over the top of a table, friction tends to slow its movement because of irregularities in the table's surface. On a microscopic level, friction arises as atoms and molecules of the two surfaces seem to adhere, then snap apart as the hand slides over the table. Friction is not restricted to solid objects; it occurs in moving fluids as well. Consider, for example, a steady flow of water in a stream. When a paddle is placed in the stream, turbulent whirls *(eddies)* form behind it. These eddies create fluid friction by draining energy from the main stream flow, slowing it down. Let's examine the idea of fluid friction in more detail.

The friction of fluid flow is called **viscosity.** When the slowing of a fluid—such as air—is due to the random motion of the gas molecules, the viscosity is referred to as *molecular viscosity.* Consider a mass of air gliding horizontally and smoothly *(laminar flow)* over a stationary mass of air. Even though the molecules in the stationary air are not moving horizontally, they are darting about and colliding with each other. At the boundary separating the air layers, there is a constant exchange of molecules between the stationary air and flowing air. The overall effect of this molecular exchange is to slow down the moving air. If molecular viscosity were the only type of friction acting on moving air, the effect of friction would disappear in a thin layer just above the surface. There is, however, another frictional effect that is far more important in reducing wind speeds.

When laminar flow gives way to irregular turbulent motion, there is an effect similar to molecular viscosity, but which occurs throughout a much larger portion of the moving air. The internal friction produced by turbulent whirling eddies is called *eddy viscosity.* Near the surface, it is related to the roughness of the ground. As wind blows over a landscape dotted with trees and buildings, it breaks into a series of irregular, twisting eddies that can influence the air flow for hundreds of meters above the surface. Within each eddy, the wind speed and direction fluctuate rapidly, producing the irregular air motion known as *wind gusts.* Eddy motions created by obstructions are commonly referred to as **mechanical turbulence.*** Mechanical turbulence creates a drag on the flow of air, one far greater than that caused by molecular viscosity.

The frictional drag of the ground normally decreases as we move away from the earth's surface. Because of the reduced friction, wind speeds tend to increase with height

---

*Keep in mind that *turbulence* represents any irregular or disturbed flow in the atmosphere that produces gusts and eddies.

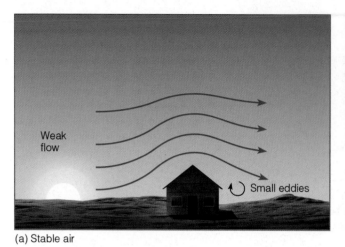

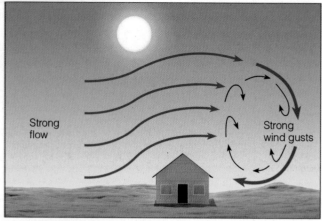

(a) Stable air

(b) Unstable air

● FIGURE 9.3 Winds flowing past an obstacle. (a) In stable air, light winds produce small eddies and little vertical mixing. (b) Greater winds in unstable air create deep, vertically mixing eddies that produce strong, gusty surface winds.

above the ground. In fact, at a height of only 10 m (33 ft), the wind is often moving twice as fast as at the surface. As we saw in Chapter 8, the atmospheric layer near the surface that is influenced by friction (turbulence) is called the *friction layer* or **planetary** *(atmospheric)* **boundary layer.** The top of the boundary layer is usually near 1000 m (3300 ft), but this height may vary somewhat since both strong winds and rough terrain extend the region of frictional influence.

Surface heating and instability also cause turbulence to extend to greater altitudes. As the earth's surface heats, ther-

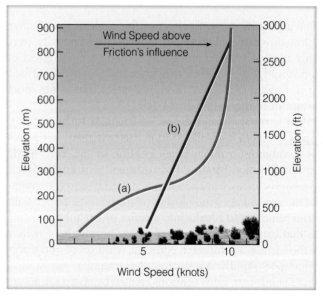

● FIGURE 9.4 When the air is stable and the terrain fairly smooth (a), vertical mixing is at a minimum, and the effect of surface friction only extends upward a relatively short distance above the surface. When the air is unstable and the terrain rough (b), vertical mixing is at a maximum, and the effect of surface friction extends upward through a much greater depth of atmosphere. Within the region of frictional influence, vertical mixing increases the wind speed near the ground and decreases it aloft. (Wind at the surface is measured at 10 m above the surface.)

mals rise and convection cells form. The resulting vertical motion creates **thermal turbulence,** which increases with the intensity of surface heating and the degree of atmospheric instability. During the early morning, when the air is most stable, thermal turbulence is normally at a minimum. As surface heating increases, instability is induced and thermal turbulence becomes more intense. If this heating produces convective clouds that rise to great heights, there may be turbulence from the earth's surface to the base of the stratosphere.

Although we have treated thermal and mechanical turbulence separately, they occur together in the atmosphere—each magnifying the influence of the other. Let's consider a simple example: the eddy forming behind the barn in ● Fig. 9.3. In stable air with weak winds, the eddy is nonexistent or small. As wind speed and surface heating increase, instability develops, and the eddy becomes larger and extends through a greater depth. The rising side of the eddy carries slow-moving surface air upward, causing a frictional drag on the faster flow of air aloft. Some of the faster moving air is brought down with the descending part of the eddy, producing a momentary gust of wind. Because of the increased depth of circulating eddies in unstable air, strong, gusty surface winds are more likely to occur when the atmosphere is unstable. Greater instability also leads to a greater exchange of faster moving air from upper levels with slower moving air at lower levels. In general, this exchange increases the average wind speed near the surface and decreases it aloft, producing the distribution of wind speed with height shown in ● Fig. 9.4.

We can now see why surface winds are usually stronger in the afternoon. Vertical mixing during the middle of the day links surface air with the faster moving air aloft. The result is that the surface air is pulled along more quickly. At night, when convection is reduced, the interchange between the air at the surface and the air aloft is at a minimum. Hence, the wind near the ground is less affected by the faster wind flow above, and so it blows more slowly.

In summary, the friction of air flow (viscosity) is a result of the exchange of air molecules moving at different speeds.

The exchange brought about by random molecular motions (molecular viscosity) is quite small in comparison with the exchange brought about by turbulent motions (eddy viscosity). Therefore, the frictional effect of the surface on moving air depends largely upon mechanical and thermal turbulent mixing. The depth of mixing and, hence, frictional influence (in the boundary layer) depend primarily upon three factors:

1. surface heating—producing a steep lapse rate and strong thermal turbulence
2. strong wind speeds—producing strong mechanical turbulent motions
3. rough or hilly landscape—producing strong mechanical turbulence

When these three factors occur simultaneously, the frictional effect of the ground is transferred upward to considerable heights, and the wind at the surface is typically strong and gusty.

EDDIES—BIG AND SMALL   When the wind encounters a solid object, a whirl of air—an eddy—forms on the object's leeward side (see •Fig. 9.5). The size and shape of the eddy often depend upon the size and shape of the obstacle and on the speed of the wind. Light winds produce small stationary eddies. Wind moving past trees, shrubs, and even your body produces small eddies. (You may have had the experience of dropping a piece of paper on a windy day only to have it carried away by a swirling eddy as you bend down to pick it up.) Air flowing over a building produces larger eddies that will, at best, be about the size of the building. Strong winds blowing past an open sports stadium can produce eddies that may rotate in such a way as to create surface winds on the playing field that move in a direction opposite to the wind flow above the stadium. Wind blowing over a fairly smooth surface produces few eddies, but when the surface is rough, many eddies form.

The eddies that form downwind from obstacles can produce a variety of interesting effects. For instance, wind moving over a mountain range in a stable atmosphere with a speed greater than 40 knots usually produces waves and eddies, such as those shown in •Fig. 9.6. We can see that eddies form both close to the mountain and beneath each wave crest. These are called *roll eddies,* or **rotors,** and have violent vertical motions that produce extreme turbulence and hazardous flying conditions. Strong winds blowing over a mountain in stable air sometimes provide a *mountain wave eddy* on the downwind side, with a reverse flow near the ground.

The largest atmospheric eddies form as the flow of air becomes organized into huge spiraling whirls—the cyclones

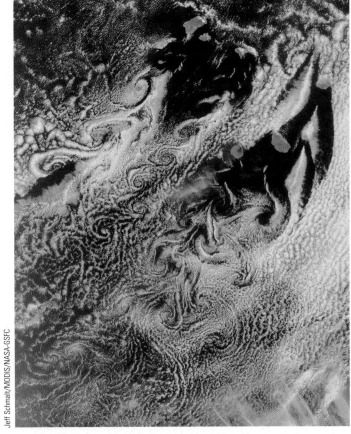

Jeff Schmalt/MODIS/NASA-GSFC

•FIGURE 9.5 Satellite image of eddies forming on the leeward (downwind) side of the Cape Verde Islands during April, 2004. As the air moves past the islands, it breaks into a variety of swirls as indicated by the cloud pattern. (The islands are situated in the Atlantic Ocean, off Africa's western coast.)

and anticyclones of middle latitudes—which can have diameters greater than 1000 km (600 mi). Since it is these migrating systems that make our middle latitude weather so changeable, we will examine the formation and movement of these systems in Chapters 11 and 12.

Turbulent eddies form aloft as well as near the surface. Turbulence aloft can occur suddenly and unexpectedly, especially where the wind changes its speed or direction (or both) abruptly. Such a change is called **wind shear.** The shearing

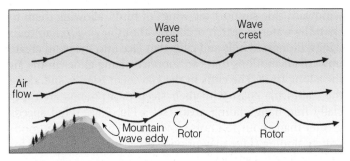

•FIGURE 9.6 Under stable conditions, air flowing past a mountain range can create eddies many kilometers downwind of the mountain itself.

creates forces that produce eddies along a mixing zone. If the eddies form in clear air, this form of turbulence is called **clear air turbulence** (or **CAT**). (Additional information on this topic is given in the Focus section on p. 229.)

THE FORCE OF THE WIND   Because a small increase in wind speed can greatly increase the wind force on an object, strong winds may blow down trees, overturn mobile homes, and even move railroad cars. For example, in February, 1965, the wind presented people in North Dakota with a "ghost train" as it pushed five railroad cars from Portal to Minot (about 125 km) without a locomotive. And, while the people in Mount Pleasant, Iowa, awaited the 1979 Fourth of July celebration, a phenomenally strong wind—estimated at 90 knots—blew the Goodyear blimp from its mooring and rolled it 300 m into a corn field, where it came to rest in ruins.

Wind blowing with sufficient force to demolish the Goodyear blimp is uncommon. However, wind blowing with enough force to move an automobile is very common, especially when the automobile is exposed to a strong crosswind. On a normal road, the force of a crosswind is usually insufficient to move a car sideways because of the reduced wind flow near the ground. However, when the car crosses a high bridge, where the frictional influence of the ground is reduced, the increased wind speed can be felt by the driver. Near the top of a high bridge, where the wind flow is typically strongest, complicated eddies pound against the car's side as the air moves past obstructions, such as guard railings and posts. In a strong wind, these eddies may even break into extremely turbulent whirls that buffet the car, causing difficult handling as it moves from side to side. If there is a wall on the bridge, the wind may swirl around and strike the car from the side opposite the wind direction.

A similar effect occurs where the wind moves over low hills paralleling a highway (see • Fig. 9.7). When the vehicle moves by the obstruction, a wind gust from the opposite direction can suddenly and without warning push it to the opposite side. This wind hazard is a special problem for trucks, campers, and trailers, and highway signs warning of gusty wind areas are often posted.

Up to now we've seen that, when the wind meets a barrier, it exerts a force upon it. If the barrier doesn't move, the wind moves around, up, and over it. When the barrier is long and low like a water wave, the slight updrafts created on the windward side support the wings of birds, allowing them to skim the water in search of food without having to flap their wings. Elongated hills and cliffs that face into the wind create upward air motions that can support a hang glider in the air for a long time. The cliffs in the Hawaiian Islands and along the California coast with their steep escarpments are especially fine hang-gliding areas (see • Fig. 9.8). Wind speeds greater than about 15 knots blowing over a smooth yet moderately sloping ridge may provide excellent ridge-soaring for the sailplane enthusiast.

As stable air flows over a ridge, it increases in speed. Thus, winds blowing over mountains tend to be stronger

• FIGURE 9.7 Strong winds flowing past an obstruction, such as these hills, can produce a reverse flow of air that strikes an object from the side opposite the general wind direction.

than winds blowing at the same level on either side. In fact, one of the greatest wind speeds ever recorded near the ground occurred at the summit of Mt. Washington, New Hampshire, elevation 1909 m (6262 ft), where the wind gusted to 200 knots (230 mi/hr) on April 12, 1934. A similar increase in wind speed occurs where air accelerates as it funnels through a narrow constriction, such as a low pass or saddle in a mountain crest.

MICROSCALE WINDS BLOWING OVER THE EARTH'S SURFACE   Where the wind blows over exposed soil, it takes an active role in shaping the landscape. This is especially noticeable in deserts. As tiny, loose particles of sand, silt, and dust are lifted from the surface and carried away by the wind, the ground level gradually lowers. The removal of this fine material leaves the surface covered with gravel and pebbles,

• FIGURE 9.8 With the prevailing wind blowing from off the ocean, the steep cliffs along the coast of Southern California promote rising air and good hang-gliding conditions.

FOCUS ON AN OBSERVATION

Eddies and "Air Pockets"

To better understand how eddies form along a zone of wind shear, imagine that, high in the atmosphere, there is a stable layer of air having vertical wind speed shear (changing wind speed with height) as depicted in Fig. 1a. The top half of the layer slowly slides over the bottom half, and the relative speed of both halves is low. As long as the wind shear between the top and bottom of the layer is small, few if any eddies form. However, if the shear and the corresponding relative speed of these layers increase (Figs. 1b and 1c), wavelike undulations may form. When the shearing exceeds a certain value (Fig. 1d), the waves break into large swirls, with significant vertical movement. Eddies such as these often form in the upper troposphere near jet streams, where large wind speed shears exist. They also occur in conjunc-tion with mountain waves, which may extend upward into the stratosphere (see Fig. 2). As we learned earlier, when these huge eddies de-velop in clear air, this form of turbulence is re-ferred to as *clear air turbulence,* or *CAT.*

The eddies that form in clear air may have diameters ranging from a couple of meters to several hundred meters. An unsuspecting air-craft entering such a region may be in for more than just a bumpy ride. If the aircraft flies into a zone of descending air, it may drop suddenly, producing the sensation that there is no air to support the wings. Consequently, these regions have come to be known as *air pockets.*

Commercial aircraft entering an air pocket have dropped hundreds of meters, injuring pas-sengers and flight attendants not strapped into their seats. For example, a DC-10 jetliner flying at 11,300 m (37,000 ft) over central Illinois dur-ing April, 1981, encountered a region of severe clear air turbulence and reportedly plunged about 600 m (2000 ft) toward the earth before stabilizing. Twenty-one of the 154 people aboard were injured; one person sustained a fractured hip and another person, after hitting the ceiling, jabbed himself in the nose with a fork, then landed in the seat in front of him.* Clear air turbulence has occasionally caused structural damage to aircraft by breaking off vertical stabilizers and tail structures. Fortu-nately, the effects are usually not this dramatic.

---

*Another example of an aircraft that experienced severe turbulence as it flew into an air pocket is given in the opening vignette on p. 223.

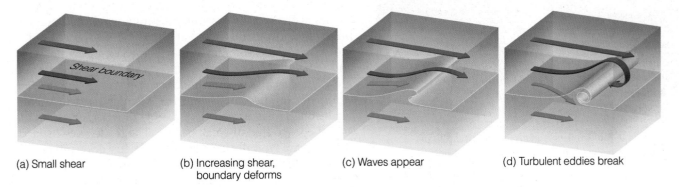

(a) Small shear

(b) Increasing shear, boundary deforms

(c) Waves appear

(d) Turbulent eddies break

● FIGURE 1 The formation of clear air turbulence (CAT) along a boundary of increasing wind speed shear. The wind in the top layer increases in speed from (a) through (d) as it flows over the bottom layer.

● FIGURE 2 Turbulent eddies forming downwind of a mountain chain in a wind shear zone produce these waves called *Kelvin Helmholtz waves.* The visible clouds that form are called *billow clouds.*

NCAR/UCAR/NSF

• FIGURE 9.9 The shape of this sand dune reveals that the wind was blowing from left to right when it formed. Note also the shape of the sand ripples on the dune.

• FIGURE 9.10 Snow rollers—natural cylindrical rolls of snow—grow larger as the wind blows them down a hillside.

• FIGURE 9.11 Snow drifts accumulating behind snow fences in Wyoming.

which are too large to be transported by the wind. Such a landscape is termed *desert pavement.*

Blowing sand eventually comes to rest *behind* obstacles, which can be anything from a rock to a clump of vegetation. As the sand grains accumulate, they pile into a larger heap that, when high enough, acts as an obstacle itself. If the wind speed is strong and continues to blow in the same direction for a sufficient time, the sand piles up higher and eventually becomes a *sand dune.* On the dune's surface, the sand rolls, slides, and gradually creeps along, producing wavelike patterns called *sand ripples.* Each ripple forms perpendicular to the wind direction, with a gentle slope on the upwind side and a steeper slope on the downwind side. (If the wind direction frequently changes, the ripple becomes more symmetric.) On a larger scale, the dune itself may take on a more symmetric shape. Sand is carried forward and up the dune until it reaches the top. Here, the air flow is strongest, and the sand continues its forward movement and cascades down the backside of the dune into quieter air. The effect of this migration is to create a dune whose windward slope is more gentle than its leeward slope. Therefore, the shape of a sand dune reveals the prevailing wind direction during its formation (see • Fig. 9.9).

Wind blowing over a snow-covered landscape may also create wavelike patterns several centimeters high and oriented at right angles to the wind. These *snow ripples* are similar to sand ripples. On a larger scale, winds may create *snow dunes,* which are quite similar to sand dunes. Irregularities at the surface can cause a strong wind (40 knots) to break into turbulent eddies. If the snow on the ground is moist and sticky, some of it may be picked up by the wind and sent rolling. As it rolls along, it collects more snow and grows bigger. If the wind is sufficiently strong, the moving clump of snow becomes cylindrical, often with a hole extending through it lengthwise. These **snow rollers** range from the size of eggs to that of small barrels. The tracks they make in the snow are typically less than 1 centimeter deep and several meters long. Snow rollers are rare, but, when they occur, they create a striking winter scene (see • Fig. 9.10). In populated areas, they may escape notice as they are often mistaken as having been made by children rather than by nature.

Strong winds blowing over a vast region of open plains can alter the landscape in a different way. Consider, for example, a light snowfall several centimeters deep covering a large portion of central South Dakota. After the snow stops falling, strong winds may whip it into the air, leaving fields barren of snow. The cold, dry wind also robs the soil of any remaining moisture and freezes it solid. Meanwhile, the snow settles out of the air when the wind encounters obstacles. Since the greatest density of such obstructions is normally in towns, municipal snowfall measurements may show an accumulation of many centimeters, while the surrounding countryside, which may desperately need the snow, has practically none.

To help remedy this situation, *snow fences* are constructed in open spaces (see • Fig. 9.11). Behind the snow fence, the wind speed is reduced because the air is broken into small eddies, which allow the snow to settle to the ground. Added snow cover is important for open areas because it acts like an insulating blanket that protects the ground from the bitter cold air, which often follows in the wake of a major snowstorm. In regions of low rainfall, moisture from the spring snowmelt can be a critical factor during long, dry summers. Snow fences are also built to protect major highways in these areas. Hopefully, the snow will accumulate behind the fence rather than in huge drifts on the road.

Strong winds can have an effect on vegetation, too. Armed with sand, winds can damage or destroy tender new vegetation, decreasing crop productivity. Most plants increase their rate of transpiration as wind speed increases.* This leads to rapid water loss, especially in warmer areas having low humidities, and may actually dry out plants. If sustained, this drying-out effect may stunt plant growth, and, in some windy, dry regions, mature trees that should be many meters tall grow only to the height of a small shrub.

Wind-dried vegetation can result in an area of high fire danger. If a fire should begin here, any additional wind helps it along, directing its movement, adding oxygen for combustion, and carrying burning embers elsewhere to start new fires. On the open plains, where the wind blows practically unimpeded, wind-whipped prairie fires can imperil homes and livestock as the fires burn out of control over large areas.

Wind erosion is greatly reduced by a continuous cover of vegetation. The vegetation screens the surface from the direct force of the wind and anchors the soil. Soil moisture also helps to resist wind erosion by cementing particles together, which increases their cohesiveness. From this fact, we can see that land where natural vegetation has been removed for farming purposes—followed by several years of drought—is ripe for wind erosion. This situation happened in parts of the Great Plains in the middle 1930s, when winds carried millions of tons of dust into the air, creating vast duststorms that buried whole farmhouses, reduced millions of acres to unproductive wasteland, and financially ruined thousands of families. Because of these disastrous effects of the wind, portions of the western plains became known as the "dust bowl."

To protect crops and soil, *windbreaks*—commonly called **shelterbelts**—are planted. Shelterbelts usually consist of a series of mixed conifer and deciduous trees or shrubs planted in rows perpendicular to the prevailing wind flow. They greatly reduce the wind speed behind them (see Fig. 9.12). As air filters through the belt, the flow is broken into small eddies, which have little mixing effect on the air near the surface. However, if trees are planted too close together, several unwanted effects may result. For one thing, the air moving past the belt may be broken into larger, more turbulent eddies, which swirl soil about. Furthermore, in high winds, strong downdrafts may damage the crops.

The use of properly designed shelterbelts has benefited agriculture. In some parts of the Central Plains, these belts have stabilized the soil and increased wheat yield. Despite

*This effect actually drops above a certain wind speed and varies greatly among plant species.

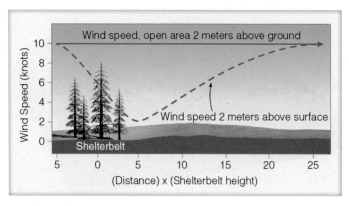

● FIGURE 9.12 A properly designed shelterbelt can reduce the air flow downwind for a distance of 25 times the height of the belt. The minimum wind flow behind the belt is typically measured downwind at a distance of about 4 times the belt's height.

their advantages, many of the shelterbelts planted during the mid-1930s drought years have been removed. Some are economically unfeasible because they occupy valuable crop land. Others interfere with the large center pivot sprinkler systems now in use. At any rate, one wonders how the absence of these shelterbelts would affect this region if it were struck by a drought similar to that experienced in the 1930s.

The impact of the wind on the earth's surface is not limited to land; wind also influences water—it makes waves. Waves forming by wind blowing over the surface of the water are known as **wind waves.** Just as air blowing over the top of a water-filled pan creates tiny ripples, so waves are created as the frictional drag of the wind transfers energy to the water. In general, the greater the wind speed, the greater the amount of energy added, and the higher will be the waves. Actually, the amount of energy transferred to the water (and thus the height to which a wave can build) depends upon three factors:

1. the wind speed
2. the length of time that the wind blows over the water
3. the *fetch,* or distance, of deep water over which the wind blows

A sustained 50-knot wind blowing steadily for nearly three days over a minimum distance of 2600 km (1600 mi) can generate waves with an average height of 15 m (49 ft). Thus, a stationary storm system centered somewhere over the open sea is capable of creating large waves with wave heights occasionally measuring over 31 m (100 ft).

Microscale winds actually help waves grow taller. Consider, for example, the wind blowing over the small wave depicted in Fig. 9.13. Observe that both the wind and the wave are moving in the same direction, and that the wave crest deflects the wind upward, producing an undulation in the air flow just above the water. This looping air motion establishes a small eddy of air between the two crests. The upward and downward motion of the eddy reinforces the upward and downward motion of the water. Consequently, the eddy helps the wave to build in height.

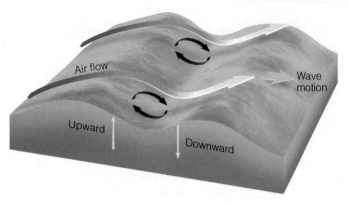

● FIGURE 9.13 Wind blowing over a wave creates a small eddy of air that helps to reinforce the up-and-down motion of the water.

Traveling in the open ocean, waves represent a form of energy. As they move into a region of weaker winds, they gradually change: Their crests become lower and more rounded, forming what are commonly called *swells*. When waves reach a shoreline they transfer their energy—sometimes catastrophically—to the coast and structures along it. High, storm-induced waves can hurl thousands of tons of water against the shore. If this happens during an unusually high tide, resort homes overlooking the ocean can be pounded into a twisted mass of board and nails by the surf. Bear in mind that the storms creating these waves may be thousands of kilometers away and, in fact, may never reach the shore. Some of the largest, most damaging waves ever to strike the beach communities of Southern California arrived on what was described as "one of the clearest days imaginable." On the more positive side, in the Hawaiian Islands these high waves are excellent for surfboarding.

Up to this point, we have seen how the wind blowing over the surface can produce a variety of features, from snow rollers to large ocean waves. How the force of the wind can influence someone riding a bicycle is found in the Focus section on p. 233.

Determining Wind Direction and Speed

Wind—the horizontal movement of air—is characterized by its direction, speed, and gustiness. If we imagine air molecules as being a swarm of bees, the wind may be seen as the movement of the entire swarm. This analogy can be carried a little further: On a calm day, the swarm will remain in one spot with each bee randomly darting about; while on a windy day the entire swarm will move quickly from one place to another. The swarm's speed would be the rate at which it moves past you. In like manner, wind speed is the rate at which air moves by a stationary observer. This movement can be expressed as the distance in nautical miles traveled in one hour (knots) or as the number of meters traveled in one second (m/sec).

Unlike a swarm of bees, air is invisible; we cannot really see it. Rather, we see things being moved by it. Therefore, we can determine wind direction by watching the movement of objects as air passes them. For example, the rustling of small leaves, smoke drifting near the ground, and flags waving on a pole all indicate wind direction. In a light breeze, a tried and true method of determining wind direction is to raise a wet finger into the air. The dampness quickly evaporates on the wind-facing side, cooling the skin. Traffic sounds carried from nearby railroads or airports can be used to help figure out the direction of the wind. Even your nose can alert you to the wind direction as the smell of fried chicken or broiled hamburgers drifts with the wind from a local restaurant.

We already know that *wind direction* is given as the direction *from which* it is blowing—a north wind blows from the north toward the south. However, near large bodies of water and in hilly regions, wind direction may be expressed differently. For example, wind blowing from the water onto the land is referred to as an **onshore wind,** whereas wind blowing from land to water is called an **offshore wind** (see ● Fig. 9.14). Consequently, a sea breeze is an onshore wind and a land

● FIGURE 9.14 An onshore wind blows from water to land; whereas an offshore wind blows from land to water.

FOCUS ON A SPECIAL TOPIC

Pedaling into the Wind

Anyone who rides a bicycle knows that it is much easier to pedal with the wind than against it. The reason is obvious: As the wind blows against an object, it exerts a force upon it. The amount of force exerted by the wind over an area increases as the square of the wind velocity. This relationship is shown by

$$F \sim V^2,$$

where F is the wind force and V is the wind velocity. From this we can see that, if the wind velocity doubles, the force goes up by a factor of 2^2, or 4, which means that pedaling into a 40-knot wind requires 4 times as much effort as pedaling into a 20-knot wind.

Wind striking an object exerts a pressure on it. The amount of pressure depends upon the object's shape and size, as well as on the amount of reduced pressure that exists on the object's downwind side. Without concern for all the complications, we can approximate the wind pressure on an object with a simple formula. For example, if the wind velocity (V) is in miles per hour, and the wind force (F) is in pounds, and the object's surface area (A) is measured in square feet, the wind pressure (P), in pounds per square feet, is

$$\frac{F}{A} = P = 0.004 \ V^2.$$

We can look at a practical example of this expression if we consider a bicycle rider going 10 mi/hr into a head wind of 40 mi/hr. With

© Richard R. Hansen/Photo Researchers, Inc.

● FIGURE 3 Pedaling into a 15-knot wind requires nine times as much effort as pedaling into a 5-knot wind.

the total velocity of the wind against the rider (wind speed plus bicycle speed) being 50 mi/hr, the pressure of the wind is

$$P = 0.004 \ V^2$$
$$P = 0.004 \ (50^2)$$
$$P = 10 \ \text{lb/ft}^2.$$

If the rider has a surface body area of 5 ft², the total force exerted by the wind becomes

$$F = P \times A$$
$$F = 10 \ \text{lb/ft}^2 \times 5 \ \text{ft}^2$$
$$F = 50 \ \text{lb}.$$

This force is enough to make pedaling into the wind extremely difficult. To remedy this adverse effect, cyclists—especially racers—bend forward as low as possible in order to expose a minimum surface area to the wind. It should also be obvious why track records are asterisked with "wind-aided" when the runners race with a tail wind of more than 3 mi/hr.

---

breeze, an offshore wind. Air moving uphill is an *upslope wind;* air moving downhill is a *downslope wind.* Hence, valley breezes are upslope winds, and mountain breezes are downslope winds. The wind direction may also be given as degrees about a 360° circle. These directions are expressed by the numbers shown in ● Fig. 9.15. For example: A wind direction of 360° is a north wind; an east wind is 90°; a south wind is 180°; and calm is expressed as zero. It is also common practice to express the wind direction in terms of compass points, such as N, NW, NE, and so on. (Helpful hints for estimating wind speeds from surface observations may be found in the *Beaufort Wind Scale,* located in Appendix C, toward the back of the book.)

THE INFLUENCE OF PREVAILING WINDS At many locations, the wind blows more frequently from one direction than from any other. The **prevailing wind** is the name given to the wind direction most often observed during a given time period. Prevailing winds can greatly affect the climate of a region. For example, where the prevailing winds are upslope, the rising, cooling air makes clouds, fog, and precipitation more likely than where the winds are downslope. Prevailing onshore winds in summer carry moisture, cool air, and fog into coastal regions, whereas prevailing offshore breezes carry warmer and drier air into the same locations.

In city planning, the prevailing wind can help decide where industrial centers, factories, and city dumps should be

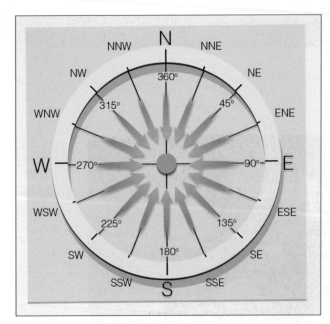

• FIGURE 9.15 Wind direction can be expressed in degrees about a circle or as compass points.

built. All of these, of course, must be located so that the wind will not carry pollutants into populated areas. Sewage disposal plants must be situated downwind from large housing developments, and major runways at airports must be aligned with the prevailing wind to assist aircraft in taking off or landing. In the high country, strong prevailing winds can bend and twist tree branches toward the downwind side, producing *wind-sculptured "flag trees"* (see • Fig. 9.16).

• FIGURE 9.16 In the high country, trees standing unprotected from the wind are often sculpted into "flag" trees, such as these trees in Wyoming.

• FIGURE 9.17 Was the wind blowing from the right or from the left when this cinder cone in Iceland erupted? (Answer given in the footnote below.)

The prevailing wind can even be a significant factor in building an individual home. In the northeastern half of the United States, the prevailing wind in winter is northwest and in summer it is southwest. Thus, houses built in the northeastern United States should have windows facing southwest to provide summertime ventilation and few, if any, windows facing the cold winter winds from the northwest. The northwest side of the house should be thoroughly insulated and even protected by a windbreak.

From the prevailing wind, biologists can predict the direction disease-carrying insects and plant spores will move and, hence, how a disease may spread. Geologists use the prevailing wind to predict where ejected debris from potentially active volcanoes will land.

Many local ground and landscape features show the effect of a prevailing wind. For example, smoke particles from an industrial stack settle to the ground on its downwind side. From the air, the prevailing wind direction can be seen as a discolored landscape on the downwind side of the stack. Wind blowing over surfaces of snow and sand produces ripples with a more gentle slope facing into the wind. As previously mentioned, sand dunes have similar shapes and, thus, show the prevailing wind direction. Look at • Fig. 9.17 and see if you can determine the prevailing wind when this cinder cone in Iceland erupted.*

The prevailing wind can be represented by a **wind rose,** which indicates the percentage of time the wind blows from different directions. Extensions from the center of a circle point to the wind direction, and the length of each extension indicates the percentage of time the wind blew from that direction.

---

*During eruption, the prevailing wind was from left to right. We can tell this by the volcano's shape. Particles ejected from the volcano were blown by the wind to the right, where they accumulated, producing a more gentle slope.

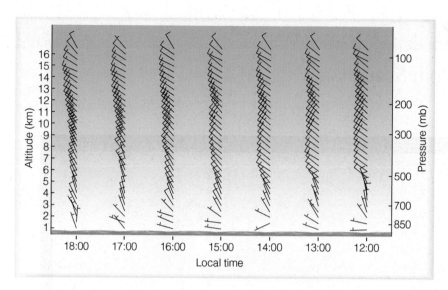

● FIGURE 9.21 A profile of wind direction and speed above Hillsboro, Kansas, on June 28, 2006.

Above about 30 km, rockets and radar provide information about the wind flow. One type of rocket ejects an instrument attached to a parachute that drifts with the wind as it slowly falls to earth. While descending, the instrument is tracked by a ground-based radar unit that determines wind information for that region of the atmosphere. Other rockets eject metal strips at some desired level. Again, radar tracks these drifting pieces of chaff, which provide valuable wind speed and direction data for elevations outside the normal radiosonde range.

A device similar to radar called **lidar** (*light detection and ranging*) uses infrared or visible light in the form of a laser beam to determine wind information. Basically, it sends out a narrow beam of light that is reflected from particles, such as smoke or dust—it measures wind velocity by measuring the movement of these particles.

Doppler radar has been employed to obtain a vertical profile of wind speed and direction up to an altitude of 16 km or so above the ground. Such a profile is called a *wind sounding*, and the radar, a **wind profiler** (or simply a *profiler*). Doppler radar, like conventional radar, emits pulses of microwave radiation that are returned (backscattered) from a target, in this case the irregularities in moisture and temperature created by turbulent, twisting eddies that move with the wind. Doppler radar works on the principle that, as these eddies move toward or away from the receiving antenna, the returning radar pulse will change in frequency. The Doppler radar wind profilers are so sensitive that they can translate the backscattered energy from these eddies into a vertical picture of wind speed and direction in a column of air 16 km (10 mi) thick (see ● Fig. 9.21). Presently, there is a network of wind profilers scattered across the central United States.

In remote regions of the world where upper-air observations are lacking, wind speed and direction can be obtained from satellites. Geostationary satellites positioned above a particular location show the movement of clouds. The direction of cloud movement indicates wind direction, and the horizontal distance the cloud moves during a given time period indicates the wind speed. Satellites now measure surface winds above the ocean by observing the roughness of the sea. (More information on this topic is given in the Focus section on p. 238.)

BRIEF REVIEW

Up to this point we've been examining microscale winds and how they affect our environment. Before we turn our attention to winds on a larger scale, here is a brief review of some of the main points presented so far:

- Viscosity is the friction of fluid flow. The small-scale fluid friction that is due to the random motion of the molecules is called *molecular viscosity*. The larger scale internal friction produced by turbulent flow is called *eddy viscosity*.

- Mechanical turbulence is created by twisting eddies that form as the wind blows past obstructions. Thermal turbulence results as rising and sinking air forms when the earth's surface is heated unevenly by the sun.

- The planetary boundary layer (or friction layer) is usually given as the first 1000 m (3300 ft) above the surface.

- Wind shear is a sudden change in wind speed or wind direction (or both).

- Onshore winds blow from water to land; offshore winds blow from land to water.

- The wind can shape a landscape, influence crop production, transport material from one area to another, and generate waves.

- The prevailing wind is the wind direction most frequently observed during a given time.

- The wind rose gives the percent of time the wind blows from different directions.

- Wind speed and direction above the earth's surface can be obtained with pilot balloons, radiosondes, rockets, satellites, and Doppler radar.

Observing Winds from Space

The oceans cover more than 70 percent of the earth's surface. For many years, our only observations of surface winds over the open seas came from a few ships and buoys. Today, however, NASA's *QuickScat* satellite, equipped with a sophisticated onboard instrument, is able to provide a clear picture of wind speed and wind direction over the open ocean. This instrument called *SeaWinds*, a scatterometer, is actually capable of obtaining wind information during all types of weather.

The scatterometer (a type of radar) gathers wind data in this manner: From the satellite, the scatterometer sends out a microwave pulse of energy that travels through the clouds, down to the sea surface. A portion of this energy is scattered (bounced) back to the satellite. The amount of energy returning to the scatterometer (called the *echo*) depends on the roughness of the sea — rougher seas have a stronger echo because they scatter back more incoming energy. Since the sea's roughness depends upon the strength of the wind blowing over it, the echo's intensity can be translated into surface wind speed and direction (see Fig. 5).

Surface wind information of this nature can be extremely valuable to the shipping industry, as well as to coastal communities.

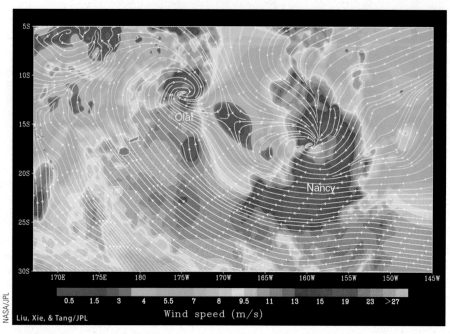

• FIGURE 5 A *QuickScat* satellite image of wind direction and wind speed associated with tropical cyclones Olaf and Nancy over the South Pacific Ocean on February 15, 2005. Wind direction is shown with arrows. Wind speed is indicated by colors, where purple represents the lightest winds and light pink the strongest winds.

Storms over the open ocean can be carefully monitored to see how their winds are changing. And incorporating sea surface wind information into computer forecast models may have the benefit of improving weather forecasts.

Local Wind Systems

Every summer, millions of people flock to the New Jersey shore, hoping to escape the oppressive heat and humidity of the inland region. On hot, humid afternoons, these travelers often encounter thunderstorms about 30 km or so from the ocean, thunderstorms that invariably last for only a few minutes. In fact, by the time the vacationers arrive at the beach, skies are generally clear and air temperatures are much lower, as cool ocean breezes greet them. If the travelers return home in the afternoon, these "mysterious" showers often occur at just about the same location as before.

The showers are not really mysterious. Actually, they are caused by a local wind system—the sea breeze. As cooler ocean air pours inland, it forces the warmer, conditionally unstable humid air to rise and condense, producing majestic clouds and rainshowers along a line where the air with contrasting temperatures meets.

The sea breeze forms as part of a thermally driven circulation. Consequently, we will begin our study of local winds by examining the formation of thermal circulations.

THERMAL CIRCULATIONS   Consider the vertical distribution of pressure shown in •Fig. 9.22a. The isobaric surfaces all lie parallel to the earth's surface; thus, there is no horizontal variation in pressure (or temperature), and there is no pressure gradient and no wind. Suppose in Fig. 9.22b the atmosphere is cooled to the north and warmed to the south. In the cold, dense air above the surface, the isobars bunch closer together, while in the warm, less-dense air, they spread farther apart. This dipping of the isobars produces a horizontal pressure gradient force aloft that causes the air to move from higher pressure (warm air) toward lower pressure (cold air).

At the surface, the air pressure changes as the air aloft begins to move. As the air aloft moves from south to north, air leaves the southern area and "piles up" above the northern area. This redistribution of air reduces the surface air pressure to the south and raises it to the north. Consequently, a pressure gradient force is established at the earth's surface from north to south and, hence, surface winds begin to blow from north to south.

We now have a distribution of pressure and temperature and a circulation of air, as shown in Fig. 9.22c. As the cool surface air flows southward, it warms and becomes less dense. In the region of surface low pressure, the warm air slowly rises, expands, cools, and flows out the top at an elevation of about 1 km above the surface. At this level, the air flows horizontally northward toward lower pressure, where it completes the circulation by slowly sinking and flowing out the bottom of the surface high. Circulations brought on by changes in air temperature, in which warmer air rises and colder air sinks, are termed **thermal circulations.**

The regions of surface high and low atmospheric pressure created as the atmosphere either cools or warms are called **thermal** (cold-core) **highs** and **thermal** (warm-core) **lows.** In general, they are shallow systems, usually extending no more than a few kilometers above the ground. These systems weaken with height. For example, at the surface, atmospheric pressure is lowest in the center of the warm thermal low in ●Fig. 9.23. In the warm air above the low, the isobars spread apart, and, at some intermediate level, the thermal low disappears and actually changes into a high. A similar phenomenon happens above the cold thermal high. The surface pressure is greatest in its center, but because the isobars aloft are crowded together due to the cold dense air, the surface thermal high becomes a low a kilometer or so above the ground. We can summarize the typical characteristics of thermal pressure systems as being shallow, weakening with height, and being maintained, for the most part, by local surface heating and cooling.

SEA AND LAND BREEZES The sea breeze is a type of thermal circulation. The uneven heating rates of land and water (described in Chapter 3) cause these mesoscale coastal winds. During the day, the land heats more quickly than the adjacent water, and the intensive heating of the air above produces a shallow thermal low. The air over the water remains cooler than the air over the land; hence, a shallow thermal high exists above the water. The overall effect of this pressure distribution is a **sea breeze** that blows at the surface from the sea toward the land (see ●Fig. 9.24a). Since the strongest gradients of temperature and pressure occur near the land-water boundary, the strongest winds typically occur right near the beach and diminish inland. Further, since the greatest contrast in temperature between land and water usually occurs in the afternoon, sea breezes are strongest at this time. (The

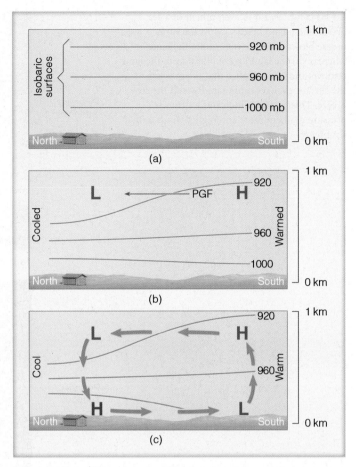

●FIGURE 9.22 A thermal circulation produced by the heating and cooling of the atmosphere near the ground. The H's and L's refer to atmospheric pressure. The lines represent surfaces of constant pressure (isobaric surfaces), where 1000 is 1000 millibars. For more information on isobaric surfaces, see Chapter 8, p. 200.

same type of breeze that develops along the shore of a large lake is called a **lake breeze.**)

At night, the land cools more quickly than the water. The air above the land becomes cooler than the air over the water,

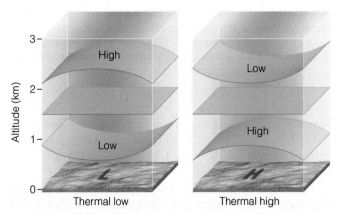

●FIGURE 9.23 The vertical distribution of pressure with thermal highs and thermal lows.

● FIGURE 9.24 Development of a sea breeze and a land breeze. (a) At the surface, a sea breeze blows from the water onto the land, whereas (b) the land breeze blows from the land out over the water. Notice that the pressure at the surface changes more rapidly with the sea breeze. This situation indicates a stronger pressure gradient force and higher winds with a sea breeze.

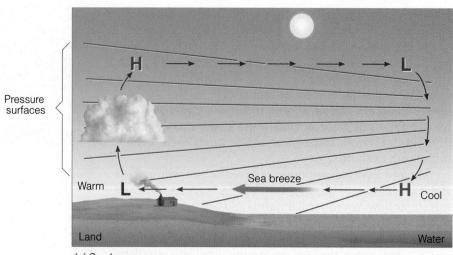

(a) Sea breeze

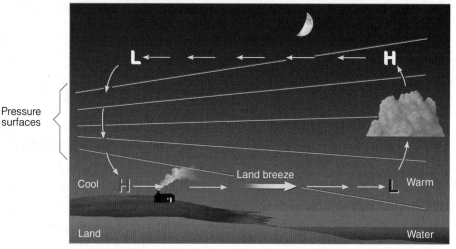

(b) Land breeze

producing a distribution of pressure such as the one shown in Fig. 9.24b. With higher surface pressure now over the land, the surface wind reverses itself and becomes a **land breeze** — a breeze that flows from the land toward the water. Temperature contrasts between land and water are generally much smaller at night; hence, land breezes are usually weaker than their daytime counterpart, the sea breeze. In regions where greater nighttime temperature contrasts exist, stronger land breezes occur over the water, off the coast. They are not usually noticed much on shore, but are frequently observed by ships in coastal waters.

Look at Fig. 9.24 again and observe that the rising air is over the land during the day and over the water during the night. Therefore, along the humid East Coast, daytime clouds tend to form over land and nighttime clouds over water. This explains why, at night, distant lightning flashes are sometimes seen over the ocean.

Sea breezes are best developed where large temperature differences exist between land and water. Such conditions prevail year-round in many tropical regions. In middle latitudes, however, sea breezes are invariably spring and summer phenomena.

During the summer, a sea breeze usually sets in about mid-morning after the land has been warmed. By early afternoon, the breeze has increased in strength and depth. By late afternoon, the cool ocean air may reach a depth of more than 300 m (1000 ft) and extend inland for more than 20 km (12 mi).

The leading edge of the sea breeze is called the **sea breeze front.** As the front moves inland, a rapid drop in temperature usually occurs just behind it. In some locations, this temperature change may be 5°C (9°F) or more during the first hours — a refreshing experience on a hot, sultry day. In regions where the water temperature is warm, the cooling effect of the sea breeze is hardly evident. Since cities near the ocean usually experience the sea breeze by noon, their highest temperature usually occurs much earlier than in inland cities. Along the East Coast, the passage of the sea breeze front is marked by a wind shift, usually from west to east. In the cool ocean air, the relative humidity rises as the temperature drops. If the relative humidity increases to above 70 percent, water vapor begins to condense upon particles of sea salt or industrial smoke, producing haze. When the ocean air is highly concentrated with pollutants, the sea breeze front may meet relatively clear air and thus appear as a *smoke front,* or a *smog front.* If the ocean air becomes saturated, a mass of low clouds and fog will mark the leading edge of the marine air.

When there is a sharp contrast in air temperature across the frontal boundary, the warmer, lighter air will converge and rise. In many regions, this makes for good sea breeze glider soaring. If this rising air is sufficiently moist, a line of cumulus clouds will form along the sea breeze front, and, if the air is also conditionally unstable, thunderstorms may form. As previously mentioned, on a hot, humid day one can drive toward the shore, encounter heavy showers several kilometers from the ocean, and arrive at the beach to find it sunny with a steady onshore breeze.

A sea breeze moving over a forest fire can be dangerous. First of all, gusty surface winds often make the fire difficult to control. Another problem is the return flow aloft. Along the sea breeze frontal boundary, air can rise to elevations where it becomes part of the return flow. Should burning embers drift seaward with this flow and drop to the ground behind the fire, they could start new fires. Flames from these fires pushed on by surface winds can trap firefighters between the two blazes.

When cool, dense, stable marine air encounters an obstacle, such as a row of hills, the heavy air tends to flow around them rather than over them. When the opposing breezes meet on the opposite side of the obstruction, they form what is called a *sea breeze convergence zone.* Such conditions are common along the Pacific Coast of North America.

Sea breezes in Florida help produce that state's abundant summertime rainfall. On the Atlantic side of the state, the sea

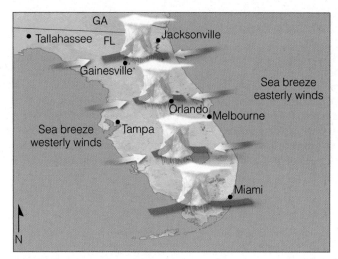

● FIGURE 9.25 Typically, during the summer over Florida, converging sea breezes in the afternoon produce uplift that enhances thunderstorm development and rainfall. However, when westerly surface winds dominate and a ridge of high pressure forms over the area, thunderstorm activity diminishes, and dry conditions prevail.

breeze blows in from the east; on the Gulf shore, it moves in from the west (see ● Fig. 9.25). The convergence of these two moist wind systems, coupled with daytime convection, produces cloudy conditions and showery weather over the land (see ● Fig. 9.26). Over the water (where cooler, more stable air lies close to the surface), skies often remain cloud-free. On many days during June and July of 1998, however, Florida's converging wind system did not materialize. The lack of con-

● FIGURE 9.26 Surface heating and lifting of air along a converging sea breeze combine to form thunderstorms almost daily during the summer in southern Florida.

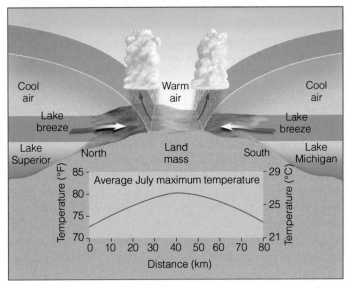

• FIGURE 9.27 The convergence of two lake breezes and their influence on the maximum temperature during July in upper Michigan.

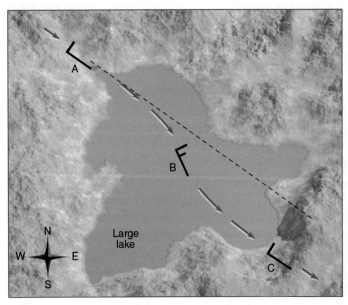

• FIGURE 9.28 Wind can change in both speed and direction when crossing a large lake.

verging surface air and its accompanying showers left much of the state parched. Huge fires broke out over northern and central Florida, which left hundreds of people homeless and burned many thousands of acres of grass and woodlands. A weakened sea breeze and dry condition produced wild fires on numerous other occasions, including the spring of 2006.

Convergence of coastal breezes is not restricted to ocean areas, as large lakes are capable of producing well-defined lake breezes. For example, both Lake Superior and Lake Michigan can produce strong lake breezes. In upper Michigan, these large bodies of water are separated by a narrow strip of land about 80 km (50 mi) wide. As can be seen from ● Fig. 9.27, the two breezes push inland and converge near the center of the peninsula, creating afternoon clouds and showers, while the lakeshore area remains sunny, pleasantly cool, and dry.

LOCAL WINDS AND WATER   Frequently, local winds will change speed and direction as they cross a large body of water. ● Figure 9.28 shows the wind speed and direction as air flows over a large lake. At position A, on the upwind side, the wind is blowing at 10 knots from the WNW; at position B, the wind speed is 15 knots from the NW; at position C, the wind is again blowing at 10 knots from the WNW. Why does the wind blow faster and from a slightly different direction in the center of the lake? As the air moves from the rough land over the relatively smooth lake, friction with the surface lessens, and the wind speed increases. The increase in wind speed, however, increases the Coriolis force, which turns the wind flow to the right, as shown by the wind report at position B. When the air reaches the opposite side of the lake, it again encounters rough land, and its speed slows. This process reduces the Coriolis force, and the wind responds by shifting to a more westerly direction, as shown by the report at position C.

Changes in wind speed along the shore of a large lake can inhibit cloud formation on one side and enhance it on the other. Suppose warm, moist air flows over a lake, as illustrated in ● Fig. 9.29. Observe that clouds are forming on the downwind side, but not on the upwind side. The lake is slightly cooler than the air. Consequently, by the time the air reaches the downwind side of the lake, it will be cooler, denser, and less likely to rise. Why, then, are clouds forming on this side of the lake? As air moves from the land over the water, it travels from a region of greater friction into a region of less friction, so it increases in speed, which causes the surface air to diverge—to spread apart. Such spreading of air forces air from above to slowly sink, which, of course, inhibits the formation of clouds. Hence, there are no clouds on the upwind side of the lake. Out over the lake, the separation between air temperature and dew point lessens. As this nearly saturated air moves onshore, friction with the rougher ground slows it down, causing it to "bunch up" or converge (which forces the air upward). This slight upward motion coupled with surface heating is often sufficient to initiate the formation of clouds along the downwind side of the lake.

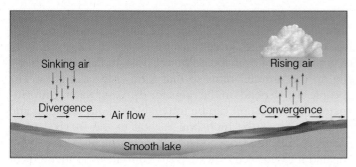

• FIGURE 9.29 Sinking air develops where surface winds move offshore, speed up, and diverge. Rising air develops as surface winds move onshore, slow down, and converge.

Strong winds blowing over an open body of water, such as a lake, can cause the water to slosh back and forth rhythmically. This sloshing causes the water level to periodically rise and fall, much like water does at both ends of a bathtub when the water is disturbed. Such water waves that oscillate back and forth are called **seiches** (pronounced "sayshes"). In addition to strong winds, seiches may also be generated by sudden changes in atmospheric pressure or by earthquakes.* Around the Great Lakes, seiche applies to any sudden rise in water level whether or not it oscillates. During December, 1986, seiches generated by strong easterly winds caused extensive coastal flooding along the southwestern shores of Lake Michigan. More recently, in November, 2003, strong westerly winds, gusting to more than 50 knots, created a seiche on Lake Erie that caused a 4-m (12-ft) difference in lake level between Toledo, Ohio (on its western shore) and Buffalo, New York (on its eastern shore).

SEASONALLY CHANGING WINDS—THE MONSOON

The word *monsoon* derives from the Arabic *mausim*, which means seasons. A **monsoon wind system** is one that *changes*

---

*Earthquakes and other disturbances on a lake floor can cause the water to slosh back and forth, producing a seiche. Earthquakes on the ocean basin floor can cause a tsunami, a Japanese word meaning "harbor waves" because these waves build in height as they enter a bay or harbor.

direction seasonally, blowing from one direction in summer and from the opposite direction in winter. This seasonal reversal of winds is especially well developed in eastern and southern Asia.

In some ways, the monsoon is similar to a large-scale sea breeze. During the winter, the air over the continent becomes much colder than the air over the ocean. A large, shallow high-pressure area develops over continental Siberia, producing a *clockwise* circulation of air that flows out over the Indian Ocean and South China Sea (see • Fig. 9.30a). Subsiding air of the anticyclone and the downslope movement of northeasterly winds from the inland plateau provide eastern and southern Asia with generally fair weather. Hence, the *winter monsoon,* which lasts from about December through February, means clear skies (*dry season*), with surface winds that blow from land to sea.

In summer, the wind flow pattern reverses itself as air over the continents becomes much warmer than air above the water. A shallow thermal low develops over the continental interior. The heated air within the low rises, and the surrounding air responds by flowing *counterclockwise* into the low center. This condition results in moisture-bearing winds sweeping into the continent from the ocean. The humid air converges with a drier westerly flow, causing it to rise; further lifting is provided by hills and mountains. Lifting cools the air to its saturation point, resulting in heavy showers and thunderstorms. Thus, the *summer monsoon* of southeastern Asia, which lasts from about June through September, means wet, rainy weather (*wet season*) with winds that blow from sea to land (see Fig. 9.30b). Although the majority of rain falls during the wet season, it does not rain all the time. In fact, rainy periods of between 15 to 40 days are often followed by several weeks of hot, sunny weather.

Many factors help create the monsoon wind system. The latent heat given off during condensation aids in the warming of the air over the continent and strengthens the summer monsoon circulation. Rainfall is enhanced by weak, westward moving low-pressure areas called *monsoon depressions.* The formation of these depressions is aided by an upper-level jet

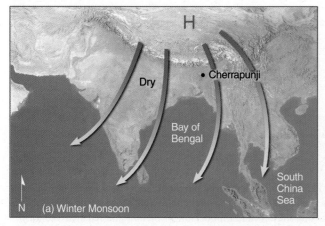

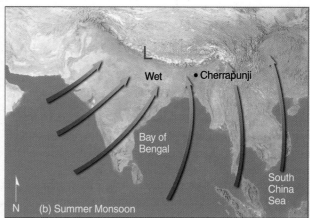

• FIGURE 9.30 Changing annual wind-flow patterns associated with the winter and summer Asian monsoon.

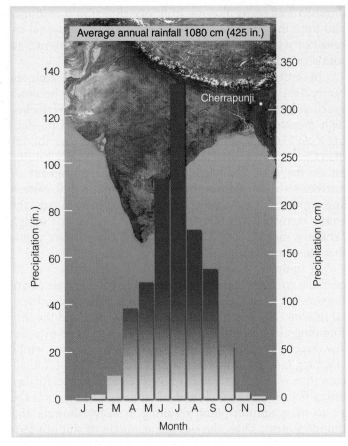

● FIGURE 9.31 Average annual precipitation for Cherrapunji, India. Note the abundant rainfall during the summer monsoon (April through October) with the lack of rainfall during the winter monsoon (November through March).

stream. Where winds in the jet diverge, surface pressures drop, the monsoon depressions intensify, and surface winds increase. The greater inflow of moist air supplies larger quantities of latent heat, which, in turn, intensifies the summer monsoon circulation.

The strength of the Indian monsoon appears to be related to the reversal of surface air pressure that occurs at irregular intervals about every two to seven years at opposite ends of the tropical South Pacific Ocean. As we will see in Chapter 10, this reversal of pressure (which is known as the *Southern Oscillation*) is linked to an ocean warming phenomenon known as *El Niño*. During a major El Niño event, surface water near the equator becomes much warmer over the central and eastern Pacific. Over the region of warm water we find rising air, huge convective clouds, and heavy rain. Meanwhile, to the west of the warm water (over the region influenced by the summer monsoon), sinking air inhibits cloud formation and convection. Hence, during El Niño years, monsoon rainfall is likely to be deficient.

Summer monsoon rains over southern Asia can reach record amounts. Located about 300 km inland on the southern slopes of the Khasi Hills in northeastern India, Cherrapunji receives an average of 1080 cm (425 in.) of rainfall each

year, most of it during the summer monsoon between April and October (see ● Fig. 9.31). The summer monsoon rains are essential to the agriculture of that part of the world. With a population of over 900 million people, India depends heavily on the summer rains so that food crops will grow. The people also depend on the rains for drinking water. Unfortunately, the monsoon can be unreliable in both duration and intensity. Since the monsoon is vital to the survival of so many people, it is no wonder that meteorologists have investigated it extensively. They have tried to develop methods of accurately forecasting the intensity and duration of the monsoon. With the aid of current research projects and the latest climate models, there is hope that monsoon forecasts will begin to improve in accuracy.

Monsoon wind systems exist in other regions of the world, such as Australia, Africa, and North and South America, where large contrasts in temperature develop between oceans and continents. (Usually, however, these systems are not as pronounced as in southeast Asia.) For example, a monsoonlike circulation exists in the southwestern United States, especially in Arizona, New Mexico, Nevada, and the southern part of California where spring and early summer are normally dry, as warm westerly winds sweep over the region. By mid-July, however, humid southerly or southeasterly winds are more common, and so are afternoon showers and thunderstorms (see ● Fig. 9.32 and ● Fig. 9.33).

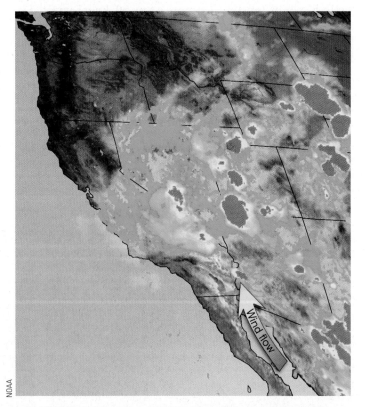

● FIGURE 9.32 Enhanced infrared satellite image with heavy arrow showing strong monsoonal circulation. Moist, southerly winds are causing showers and thunderstorms (yellow and red areas) to form over the southwestern section of the United States during July, 2001.

Clouds and thunderstorms forming over Arizona, as humid monsoonal air flows northward over the region during July, 2007.

© C. Donald Ahrens

MOUNTAIN AND VALLEY BREEZES   Mountain and valley breezes develop along mountain slopes. Observe in ● Fig. 9.34 that, during the day, sunlight warms the valley walls, which in turn warm the air in contact with them. The heated air, being less dense than the air of the same altitude above the valley, rises as a gentle upslope wind known as a **valley breeze.** At night, the flow reverses. The mountain slopes cool quickly, chilling the air in contact with them. The cooler, more-dense air glides downslope into the valley, providing a **mountain breeze.** (Because gravity is the force that directs these winds downhill, they are also referred to as *gravity winds,* or *nocturnal drainage winds.*) This daily cycle of wind flow is best developed in clear summer weather when prevailing winds are light.

In many areas, the upslope winds begin early in the morning, reach a peak speed of about 6 knots by midday, and re-

WEATHER WATCH

Cherrapunji, India, received 26.5 m (87 ft) of rain in 1861, most of which fell between April and October—the summer monsoon. In fact, during July, 1861, Cherrapunji recorded a whopping 9.3 m (30.5 ft) of rain.

verse direction by late evening. The downslope mountain breeze increases in intensity, reaching its peak in the early morning hours, usually just before sunrise. In the Northern Hemisphere, valley breezes are particularly well developed on south-facing slopes, where sunlight is most intense. On partially shaded north-facing slopes, the upslope breeze may be weak or absent. Since upslope winds begin soon after the sun's

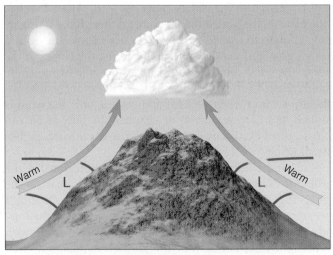

Valley breeze

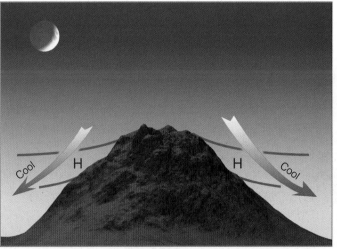

Mountain breeze

● FIGURE 9.34 Valley breezes blow uphill during the day; mountain breezes blow downhill at night. (The L's and H's represent pressure, whereas the purple lines represent surfaces of constant pressure.)

● FIGURE 9.35 As mountain slopes warm during the day, air rises and often condenses into cumuliform clouds, such as these.

rays strike a hill, valley breezes typically begin first on the hill's east-facing side. In the late afternoon, this side of the mountain goes into shade first, producing the onset of downslope winds at an earlier time than experienced on west-facing slopes. Hence, it is possible for campfire smoke to drift downslope on one side of a mountain and upslope on the other side.

When the upslope winds are well developed and have sufficient moisture, they can reveal themselves as building cumulus clouds above mountain summits (see ● Fig. 9.35). Since valley breezes usually reach their maximum strength in the early afternoon, cloudiness, showers, and even thunderstorms are common over mountains during the warmest part

of the day—a fact well known to climbers, hikers, and seasoned mountain picnickers.

KATABATIC WINDS Although any downslope wind is technically a **katabatic wind,** the name is usually reserved for downslope winds that are much stronger than mountain breezes. Katabatic (or *fall*) winds can rush down elevated slopes at hurricane speeds, but most are not that intense and many are on the order of 10 knots or less.

The ideal setting for a katabatic wind is an elevated plateau surrounded by mountains, with an opening that slopes rapidly downhill. When winter snows accumulate on the plateau, the overlying air grows extremely cold and a shallow dome of high pressure forms near the surface (see ● Fig. 9.36). Along the edge of the plateau, the horizontal pressure gradient force is usually strong enough to cause the cold air to flow across the isobars through gaps and saddles in the hills. Along the slopes of the plateau, the wind continues downhill as a gentle or moderate cold breeze. If the horizontal pressure gradient increases substantially, such as when a storm approaches, or if the wind is confined to a narrow canyon or channel, the flow of air can increase, often destructively, as cold air rushes downslope like water flowing over a fall.

Katabatic winds are observed in various regions of the world. For example, along the northern Adriatic coast in the former Yugoslavia, a polar invasion of cold air from Russia descends the slopes from a high plateau and reaches the lowlands as the *bora*—a cold, gusty, northeasterly wind with speeds sometimes in excess of 100 knots. A similar, but often less violent, cold wind known as the *mistral* descends the western mountains into the Rhone Valley of France, and then out over the Mediterranean Sea. It frequently causes frost damage

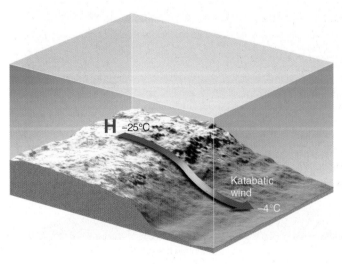

● FIGURE 9.36 Strong katabatic winds can form where cold winds rush downhill from an elevated plateau covered with snow.

to exposed vineyards and makes people bundle up in the otherwise mild climate along the Riviera. Strong, cold katabatic winds also blow downslope off the icecaps in Greenland and Antarctica, occasionally with speeds greater than 100 knots.

In North America, when cold air accumulates over the Columbia plateau,* it may flow westward through the Columbia River Gorge as a strong, gusty, and sometimes violent wind. Even though the sinking air warms by compression, it is so cold to begin with that it reaches the ocean side of the Cascade Mountains much colder than the marine air it replaces. The *Columbia Gorge wind* (called the *coho*) is often the harbinger of a prolonged cold spell.

Strong downslope katabatic-type winds funneled through a mountain canyon can do extensive damage. During January, 1984, a ferocious downslope wind blew through Yosemite National Park in California at speeds estimated at 100 knots. The wind toppled many trees and, unfortunately, caused a fatality when a tree fell on a park employee sleeping in a tent.

CHINOOK (FOEHN) WINDS   The **chinook wind** is a warm, dry wind that descends the eastern slope of the Rocky Mountains. The region of the chinook is rather narrow (only several hundred kilometers wide) and extends from northeastern New Mexico northward into Canada. Similar winds occur along the leeward slopes of mountains in other regions of the world. In the European Alps, for example, such a wind is called a **foehn** and, in Argentina, a *zonda*. When these winds move through an area, the temperature rises sharply, sometimes 20°C (36°F) or more in one hour, and a corresponding sharp drop in the relative humidity occurs, occasionally to less than 5 percent. (More information on temperature changes associated with chinooks is given in the Focus section on p. 248.)

Chinooks occur when strong westerly winds aloft flow over a north-south-trending mountain range, such as the

___
*Information on geographic features and their location in North America is provided in the back of the book.

Rockies and Cascades. Such conditions (described in Chapter 12) can produce a trough of low pressure on the mountain's eastern side, a trough that tends to force the air downslope. As the air descends, it is compressed and warms at the dry adiabatic rate (10°C/km). So the main source of warmth for a chinook is *compressional heating,* as potentially warmer (and drier) air is brought down from aloft.

When clouds and precipitation occur on the mountain's windward side, they can enhance the chinook. For example, as the cloud forms on the upwind side of the mountain in ● Fig. 9.37a. The release of latent heat inside the cloud supplements the compressional heating on the downwind side. This phenomenon makes the descending air at the base of the mountain on the downwind side warmer than it was before it started its upward journey on the windward side. The air is also drier, since much of its moisture was removed as precipitation on the windward side (see Fig. 9.37b).

Along the front range of the Rockies, a bank of clouds forming over the mountains is a telltale sign of an impending chinook. This *chinook wall cloud* (which looks like a wall of clouds) usually remains stationary as air rises, condenses, and then rapidly descends the leeward slopes, often causing strong winds in foothill communities. ● Figure 9.38 shows how a chinook wall cloud appears as one looks west toward the Rockies from the Colorado plains. The photograph was taken on a winter afternoon with the air temperature about −7°C (20°F). That evening, the chinook moved downslope at high speeds through foothill valleys, picking up sand and pebbles (which dented cars and cracked windshields). The chinook spread out over the plains like a warm blanket, raising the air temperature the following day to a mild 15°C (59°F). The chinook and its wall of clouds remained for several days, bringing with it a welcomed break from the cold grasp of winter.

View this concept in action on the Meteorology Resource Center at academic.cengage.com/login

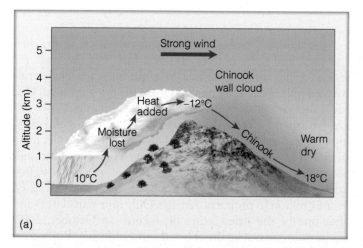

(a)

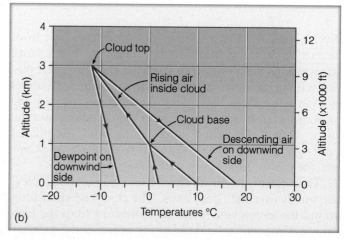

(b)

● FIGURE 9.37 (a) A chinook wind can be enhanced when clouds form on the mountain's windward side. Heat added and moisture lost on the upwind side produce warmer and drier air on the downwind sides. (b) A graphic representation of the rising and sinking air as it moves over the mountain.

FOCUS ON A SPECIAL TOPIC

Snow Eaters and Rapid Temperature Changes

Chinooks are thirsty winds. As they move over a heavy snow cover, they can melt and evaporate a foot of snow in less than a day. This situation has led to some tall tales about these so-called "snow eaters." Canadian folklore has it that a sled-driving traveler once tried to outrun a chinook. During the entire ordeal his front runners were in snow while his back runners were on bare soil.

Actually, the chinook is important economically. It not only brings relief from the winter cold, but it uncovers prairie grass, so that livestock can graze on the open range. Also, these warm winds have kept railroad tracks clear of snow, so that trains can keep running. On the other hand, the drying effect of a chinook can create an extreme fire hazard. And when a chinook follows spring planting, the seeds may die in the parched soil. Along with the dry air comes a buildup of static electricity, making a simple handshake a shocking experience. These warm dry winds have sometimes adversely affected human behavior. During periods of chinook winds some people feel irritable and depressed and others become ill. The exact reason for this phenomenon is not clearly understood.

Chinook winds have been associated with rapid temperature changes. In fact, on January

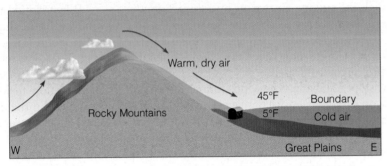

● FIGURE 6 Cities near the warm air–cold air boundary can experience sharp temperature changes if cold air should rock up and down like water in a bowl.

11, 1980, due to a chinook wind, the air temperature in Great Falls, Montana, rose from −32°F to 17°F (a 49°F rise in temperature) in just seven minutes. How such rapid changes in temperature can occur is illustrated in Fig. 6. Notice that a shallow layer of extremely cold air has moved out of Canada and is now resting against the Rocky Mountains. The cold air behaves just as any fluid, and, in some cases, atmospheric conditions may cause the air to move up and down much like water does when a bowl is rocked back and forth. This rocking motion can cause extreme temperature variations for cities located at the base of the hills along the periphery of the cold air–warm air boundary, as they are alternately in and then

out of the cold air. Such a situation is held to be responsible for the extremely rapid two-minute temperature change of 49°F recorded at Spearfish, South Dakota, during the morning of January 22, 1943. On the same morning, in nearby Rapid City, the temperature fluctuated from −4°F at 5:30 A.M. to 54°F at 9:40 A.M., then down to 11°F at 10:30 A.M. and up to 55°F just 15 minutes later. At nearby cities, the undulating cold air produced similar temperature variations that lasted for several hours.

SANTA ANA WINDS A warm, dry wind that blows from the east or northeast into southern California is the **Santa Ana wind.** As the air descends from the elevated desert plateau, it funnels through mountain canyons in the San Gabriel and San Bernardino Mountains, finally spreading over the Los Angeles basin and San Fernando Valley and out over the Pacific Ocean. The wind often blows with exceptional speed—occasionally over 90 knots—in the Santa Ana Canyon (the canyon from which it derives its name).

These warm, dry winds develop as a region of high pressure builds over the Great Basin. The clockwise circulation around the anticyclone forces air downslope from the high plateau. Thus, *compressional heating* provides the primary source of warming. The air is dry, since it originated in the desert, and it dries out even more as it is heated. ● Figure 9.39 shows a typical wintertime Santa Ana situation.

As the wind rushes through canyon passes, it lifts dust and sand and dries out vegetation, which sets the stage for serious brush fires, especially in autumn, when chaparral-covered hills are already parched from the dry summer.* One such fire in November of 1961—the infamous *Bel Air fire*—burned for three days, destroying 484 homes and causing over $25 million in damage. During October, 2003, massive wildfires driven by strong Santa Ana winds swept through Southern California. The fires charred more than 740,000 acres, destroyed over 2800 homes, took 20 lives, and caused over $1 billion in property damage. Only four years later (and after one of the driest years on record) in October, 2007, wildfires broke out again in Southern California. Pushed on by hellacious Santa Ana winds that gusted to over 80 knots,

*Chaparral denotes a shrubby environment in which many of the plant species contain highly flammable oils.

© C. Donald Ahrens

● FIGURE 9.38 A chinook wall cloud forming over the Colorado Rockies (viewed from the plains).

the fires raced through dry vegetation, scorching everything in their paths. The fires, which extended from north of Los Angeles to the Mexican border (see ● Fig. 9.40), burned over 500,000 acres, destroyed more than 1800 homes, and took 8 lives. The total costs of the fires exceeded $1.5 billion.

Four hundred miles to the north of Los Angeles in Oakland, California, a ferocious Santa Ana-type wind was responsible for the disastrous Oakland hills fire during October, 1991. The fire started in the parched Oakland hills, just east of San Francisco, where a firestorm driven by strong northeast winds blackened almost 2000 acres, damaged or destroyed over 3000 dwellings, caused almost $5 billion in damage, and took 25 lives. With the protective vegetation cover removed, the land is ripe for erosion, as winter rains may wash away topsoil and, in some areas, create serious mudslides such as those that occurred in Southern California during May, 2005. The adverse effects of a wind-driven Santa Ana fire may be felt long after the fire itself has been put out.

A similar downslope-type wind called a *California norther* can produce unbearably high temperatures in the northern half of California's Central Valley. On August 8, 1978, for example, a ridge of high pressure formed to the north of this region, while a thermal low was well entrenched to the south. This pressure pattern produced a north wind in the area. A summertime north wind in most parts of the country means cooler weather and a welcome relief from a hot spell, but not in Red Bluff, California, where the winds moved downslope off the mountains. Heated by compression, these winds increased the air temperature in Red Bluff

to an unbelievable 48°C (119°F) for two consecutive days—amazing when you realize that Red Bluff is located at about the same latitude as Philadelphia, Pennsylvania.

DESERT WINDS   Winds of all sizes develop over the deserts. Huge *dust storms* form in dry regions, where strong winds are able to lift and fill the air with particles of fine dust. An exceptionally large dust storm formed over the African Sahara, during February, 2001. The storm—about the size of

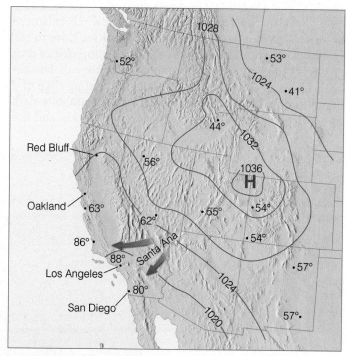

● FIGURE 9.39 Surface weather map showing Santa Ana conditions in January. Maximum temperatures for this particular day are given in °F. Observe that the downslope winds blowing into southern California raised temperatures into the upper 80s, while elsewhere temperature readings were much lower.

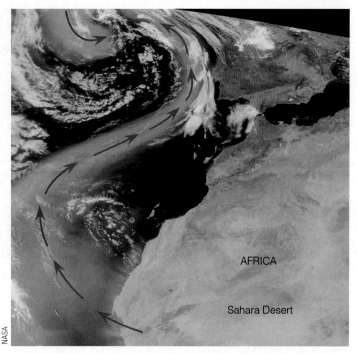

● FIGURE 9.40 Strong northeasterly Santa Ana winds on October 23, 2007, blew the smoke from massive wild fires (red dots) across southern California out over the Pacific Ocean.

● FIGURE 9.41 A large dust storm over the African Sahara Desert during February, 2001, sweeps westward off the coast, then northward into a mid-latitude cyclonic storm west of Spain, as indicated by red arrow.

Spain—swept westward off the African coast, then northeastward (see ● Fig. 9.41). In desert areas where loose sand is more prevalent, *sandstorms* develop, as high winds enhanced by surface heating rapidly carry sand particles close to the ground. A spectacular example of a storm composed of dust or sand is the **haboob** (from Arabic *hebbe:* blown). The haboob forms as cold downdrafts along the leading edge of a thunderstorm lift dust or sand into a huge, tumbling dark cloud that may extend horizontally for over 150 km and rise

vertically to the base of the thunderstorm (see ● Fig. 9.42). Spinning whirlwinds of dust frequently form along the turbulent cold air boundary. Haboobs are most common in the African Sudan (where about 24 occur each year) and in the desert southwest of the United States, especially in southern Arizona.

On a smaller scale, the spinning vortices so commonly seen on hot days in dry areas are called **dust devils** or *whirlwinds.* (In Australia, the Aboriginal word *willy-willy* refers to a dust devil.) Generally, dust devils form on clear, hot days

● FIGURE 9.42 An haboob approaching Phoenix, Arizona. The dust cloud is rising to a height of about 450 m (1475 ft) above the valley floor.

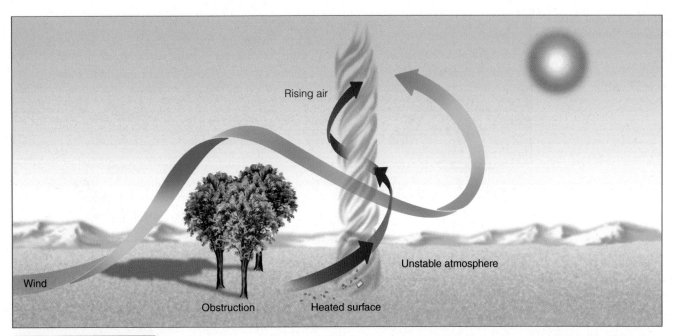

ACTIVE FIGURE 9.43 The formation of a dust devil. On a hot, dry day, the atmosphere next to the ground becomes unstable. As the heated air rises, wind blowing past an obstruction twists the rising air, forming a rotating air column or *dust devil*. Air from the sides rushes into the rising column, lifting sand, dust, leaves, or any other loose material from the surface. Visit the Meteorology Resource Center to view this and other active figures at academic.cengage.com/login

over a dry surface where most of the sunlight goes into heating the surface, rather than evaporating water from vegetation. The atmosphere directly above the hot surface becomes unstable, convection sets in, and the heated air rises. Wind, often deflected by small topographic barriers, flows into this region, rotating the rising air (see ● Fig. 9.43). Depending on the nature of the topographic feature, the spin of a dust devil around its central eye may be cyclonic or anticyclonic, and both directions occur with about equal frequency.

Having diameters of only a few meters and heights of less than a hundred meters, most dust devils are small and last only a short time (see ● Fig. 9.44). There are, however, some dust devils of sizable dimension, extending upward from the surface for many hundreds of meters. Such whirlwinds are capable of considerable damage; winds exceeding 75 knots may overturn mobile homes and tear the roofs off buildings. Fortunately, the majority of dust devils are small. Also keep in mind that dust devils *are not* tornadoes. The circulation of many tornadoes (as we will see in Chapter 14) descends downward from the base of a thunderstorm, whereas the circulation of a dust devil begins at the surface, normally in sunny weather, although some form beneath convective-type clouds.

There are other desert winds that should be mentioned. Winds originating over the Sahara Desert are given local names as they move into different regions. For example, the normal flow of surface air over North Africa is from the north; however, when a storm system is located west of Africa or southern Spain (position 1, ● Fig. 9.45), a hot, dry, and dusty easterly or southeasterly wind—the *leste*—blows over Morocco and out into the Atlantic. If the wind crosses the Mediterranean, it becomes the *leveche* when it enters south-

© Sherwood Idso

● **FIGURE 9.44**

A dust devil forming on a clear, hot summer day just south of Phoenix, Arizona.

● FIGURE 9.45 Local winds that originate over North Africa.

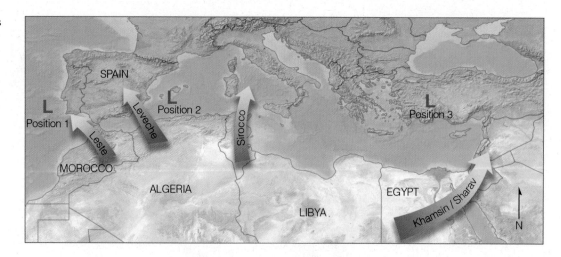

ern Spain. Because of the short time it is over water, the leveche remains hot and dry.

When a low-pressure center is located at position 2, a warm, dry, dust-laden south or southeast wind originates over the Sahara Desert and blows across North Africa. This wind is known as the *sirocco*. As it moves over the Mediterranean, it picks up moisture and arrives in Sicily and southern Italy as a warm but more humid wind.

A storm located still farther to the east (position 3) can cause a dry, hot southerly wind — the *khamsin* — to blow over Egypt, the Red Sea, and Saudi Arabia. In Israel, this wind is called the *sharav*. These winds are exceedingly hot and can raise the air temperature to 50°C (122°F), while lowering the relative humidity to less than 10 percent. Because storm systems are not common over the Mediterranean in summer, scorching breezes such as these occur in spring or fall.

Desert winds are not confirmed to planet Earth; they form on the planet Mars as well. Most of the Martian dust storms are small, and only cover a relatively small portion of

that planet. However, during 2001 an enormous dust storm developed that actually wrapped itself around the entire planet (see ● Fig. 9.46). Dust storms of this size may form as the reddish-colored airborne dust particles absorb sunlight and warm the air around them. On the extremely cold Martian planet, the air in the immediate vicinity of the dust becomes much warmer than the surrounding air. In the thin air of Mars, this variation in temperature sets up a strong pressure gradient that causes high winds that may exceed 100 knots. Dust devils often form as the twisting air sweeps over the uneven landscape.

OTHER LOCAL WINDS OF INTEREST   Up to now, we have examined a number of wind systems recognized more than just locally. Other winds have various names in different locales. Let's look at some examples.

In winter, when an intense storm tracks east across the Great Plains of North America, cold northerly winds often plunge southward behind it. As the cold air moves through Texas, it may drop temperatures tens of degrees in a few hours. Such a cold wind is called a **Texas norther,** or *blue norther,* especially if accompanied by snow. If the cold air penetrates into Central America, it is known as a *norte.* Meanwhile, if the strong, cold winds over the plains states are accompanied by drifting, blowing, or falling snow, the term *blizzard* is applied to this weather situation.

Along the eastern slope of the Rocky Mountains, strong down-mountain winds occasionally blow during chinook conditions. Such winds are especially notorious in winter in Boulder, Colorado, where the average yearly windstorm damage is about $1 million. These *Boulder winds* have been recorded at over 100 knots, damaging roofs, uprooting trees, overturning mobile homes and trucks, and sandblasting car windows. Although the causes of these high winds are not completely understood, some meteorologists believe that they may be associated with large vertically oriented spinning whirls of air that some scientists call *mountainadoes.* How these rapidly rotating vortices form is presently being investigated.

▼ Table 9.1 lists winds of local significance observed in other regions of the world.

● FIGURE 9.46 A huge dust storm (dark red region) covers Mars during July, 2001.

▼ TABLE 9.1   Local Winds of the World

NAME	DESCRIPTION
Cold Winds	
Buran	A strong, cold wind that blows over Russia and central Asia
Purga	A buran accompanied by strong winds and blowing snow
Pampero	A cold wind blowing from the south over Argentina, Uruguay, and into the Amazon Basin
Burga	A cold northeasterly wind in Alaska usually accompanied by snow; similar to the buran and purga of Russia
Bise	Generally a cold north or northeast wind that blows over southern France; often brings damaging spring frosts
Papagayo	A cold northeasterly wind along the Pacific coast of Nicaragua and Guatemala; occurs when a cold air mass overrides the mountains of Central America
Tehuantepecer	A strong wind from the north or northwest funneled through the gap between the Mexican and Guatemalan mountains and out into the Gulf of Tehuantepec
Mild Winds	
Levanter	A mild, humid, and often rainy east or northeast wind that blows across southern Spain
Harmattan	A dry, dusty but mild wind from the northeast or east that originates over the cool Sahara in winter and blows over the west coast of Africa; brings relief from the hot, humid weather along the coastal region
Hot Winds	
Simoom	A strong, dry, and dusty desert wind that blows over the African and Arabian deserts; name means "poison wind" because it is often accompanied by temperatures in excess of 52°C (125°F), which may cause heat stroke

SUMMARY

In this chapter, we concentrated on microscale and mesoscale winds. In the beginning of the chapter, we considered both how our environment influences the wind and how the wind influences our environment. We saw that the friction of airflow—viscosity—can be brought about by the random motion of air molecules (molecular viscosity) or by turbulent whirling eddies of air (eddy viscosity). The depth of the atmospheric layer near the surface that is influenced by surface friction (the boundary layer) depends upon atmospheric stability, the wind speed, and the roughness of the terrain. Although it may vary, the top of the boundary layer is usually near 1000 meters.

Winds blowing past obstructions can produce a number of effects, from gusty winds at a sports stadium to howling winds on a blustery night. Aloft, winds blowing over a mountain range may generate hazardous rotors downwind of the range. And the eddies that form in a region of strong wind shear, especially in the vicinity of a jet stream, can produce extreme turbulence, even in clear air.

Wind blowing over the earth's surface can create a variety of features. In deserts, we see sand dunes, and desert pavement. Over a snow surface, the wind produces snow ripples and snow rollers. Where high winds blow over a ridge,

trees may be sculpted into "flag" trees. In unprotected areas, shelterbelts are planted to protect crops and soil from damaging winds.

We also examined winds on a slightly larger scale. Land and sea breezes are true mesoscale winds that blow in response to local pressure differences created by the uneven heating and cooling rates of land and water. When winds move across a large body of water, they often change in speed and direction. Where the winds change direction seasonally, they are termed monsoon winds. Monsoon winds exist in many parts of the world, including North America, Asia, Australia, and Africa.

Local winds that blow uphill during the day are called valley breezes and those that blow downhill at night, mountain breezes. A strong, cold downslope wind is the katabatic (or fall) wind.

A warm, dry wind that descends the eastern side of the Rocky Mountains is the chinook. The same type of wind in the Alps is called a foehn. A warm, dry, usually strong downslope wind that blows into Southern California from the east or northeast is the Santa Ana wind.

Local intense heating of the surface can produce small rotating winds, such as the dust devil, while downdrafts in a thunderstorm are responsible for the desert haboob. Some winds, such as the blizzard, are snow-bearing, whereas others, such as the sirocco, are dust-bearing. Finally, the wind's name may express the direction from which it blows (the Texas norther), or it may represent the region that it blows from (the Santa Ana).

KEY TERMS

The following terms are listed (with page numbers) in the order they appear in the text. Define each. Doing so will aid you in reviewing the material covered in this chapter.

scales of motion, 224
microscale, 224
mesoscale, 224
synoptic scale, 224
planetary (global) scale, 224
viscosity, 225
mechanical turbulence, 225
planetary boundary layer, 226
thermal turbulence, 226
rotors, 227
wind shear, 227
clear air turbulence
  (CAT), 228
snow rollers, 230
shelterbelts (windbreaks), 231
wind waves, 231
onshore wind, 232
offshore wind, 232
prevailing wind, 233
wind rose, 234
wind vane, 235
anemometer, 235

aerovane (skyvane), 236
pilot balloon, 236
lidar, 237
wind profiler, 237
thermal circulations, 239
thermal highs, 239
thermal lows, 239
sea breeze, 239
lake breeze, 239
land breeze, 240
sea breeze front, 240
seiches, 243
monsoon wind system, 243
valley breeze, 245
mountain breeze, 245
katabatic (or fall) wind, 246
chinook wind, 247
foehn wind, 247
Santa Ana wind, 248
haboob, 250
dust devil (whirlwind), 250
Texas norther, 252

QUESTIONS FOR REVIEW

1. Describe the various scales of motion, and give an example of each.

2. How does the earth's surface influence the flow of air above it?

3. What causes wind gusts?

4. How does mechanical turbulence differ from thermal turbulence?

5. Why are winds near the surface typically stronger and more gusty in the afternoon?

6. Describe several ways in which an eddy might form.

7. A friend has just returned from a trans-Atlantic jet aircraft flight and reported that the plane dropped about 1000 m when it entered an "air pocket." Explain to your friend what apparently happened to cause this drop.

8. Explain why the car in the diagram in ● Fig. 9.47 may experience a west wind as it travels past the wall.

● FIGURE 9.47

9. What is wind shear and how does it relate to clear air turbulence?

10. Explain how shelterbelts protect sensitive crops from wind damage.

11. With the same wind speed, explain why a camper is more easily moved by the wind than a car.

12. What are the necessary conditions for the development of large wind waves?

13. How can a coastal area have heavy waves on a clear, nonstormy day?

14. If you are standing directly south of a smoke stack and the wind from the stack is blowing over your head, what would be the wind direction?

15. An upper wind direction is reported as 315°. From what compass direction is the wind blowing?

16. List as many ways as you can of determining the wind direction and the wind speed.

17. Name and describe three instruments used to measure wind speed and direction.

18. Using a diagram, explain how a thermal circulation develops.

19. Why do winds usually change direction and speed when moving over a large body of water?

20. Discuss the factors that contribute to the formation of the summer monsoon and the winter monsoon in India.

21. You are fly fishing in a mountain stream during the early morning; would you expect the wind to be blowing upstream or downstream? Explain.

22. Which wind will most likely produce clouds: a valley breeze or a mountain breeze? Why?

23. Explain why chinook winds are warm and dry.

24. Name some of the benefits of a chinook wind.

25. What atmospheric conditions contribute to the development of a strong Santa Ana condition? Why is a Santa Ana wind warm?

26. How do strong katabatic winds form?

27. Why are haboobs more prevalent in Arizona than in Oklahoma?

28. Describe how dust devils usually form.

29. In what part of the world would you expect to encounter each of the following winds, and what type of weather would each wind bring?

    (a) foehn   (e) chinook
    (b) California norther (f) Columbia Gorge wind
    (c) Santa Ana   (g) sirocco
    (d) zonda   (h) mistral

QUESTIONS FOR THOUGHT

1. A pilot enters the weather service office and wants to know what time of the day she can expect to encounter the least turbulent winds at 760 m (2500 ft) above central Kansas. If you were the weather forecaster, what would you tell her?

2. Why is it dangerous during hang gliding to enter the downwind side of the hill when the wind speed is strong?

3. After a winter snowstorm, Cheyenne, Wyoming, reports a total snow accumulation of 48 cm (19 in.), while the maximum depth in the surrounding countryside is only 28 cm (11 in.). If the storm's intensity and duration were practically the same for a radius of 50 km around Cheyenne, explain why Cheyenne received so much more snow.

4. Why is the difference in surface wind speed between morning and afternoon typically greater on a clear, sunny day than on a cloudy, overcast day?

5. Might it be possible to have a city/suburb breeze? If so, would you expect it to be more prominent during the day or night? Describe how it would form. Use a diagram to help you.

6. Average annual wind speed information in knots is given here for two cities located on the Great Plains. Which city would probably be the best site for a wind turbine? Why?

	Time									
	MID-NIGHT	3	6	9	NOON	3	6	9	AVERAGE ANNUAL WIND SPEED (KNOTS)
City A:	12	7	8	13	15	18	14	13	12.5
City B:	8	6	6	13	20	22	15	10	12.5

7. Which of the sites in ● Fig. 9.48 would probably be the best place to construct a wind turbine? A, B, or C? Which would be the worst? Explain.

● FIGURE 9.48

8. Explain why cities near large bodies of cold water in summer experience well-developed sea breezes, but only poorly developed land breezes.

9. Why do clouds tend to form over land with a sea breeze and over water with a land breeze?

10. The convergence of two sea breezes in Florida frequently produces rain showers; the convergence of two sea breezes in California does not. Explain.

11. If campfire smoke is blowing uphill along the east-facing side of the hill and downhill along the west-facing side of the same hill, are the fires cooking breakfast or dinner? From the drift of the smoke, how were you able to tell?

12. Why don't chinook winds form on the east side of the Appalachians?

13. Show, with the aid of a diagram, what atmospheric and topographic conditions are necessary for an area in the Northern Hemisphere to experience hot summer breezes from the north.

14. The prevailing winds in southern Florida are northeasterly. Knowing this, would you expect the strongest sea breezes to be along the east or west coast of southern Florida? What about the strongest land breezes?

PROBLEMS AND EXERCISES

1. A model city is to be constructed in the middle of an uninhabited region. The wind rose seen here (●Fig. 9.49) shows the annual frequency of wind directions for this region. With the aid of the wind rose, on a square piece of paper determine where the following should be located:

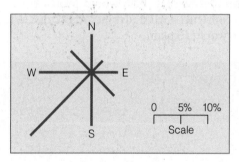

● FIGURE 9.49

(a) industry
(b) parks
(c) schools
(d) shopping centers
(e) sewage disposal plants
(f) housing development
(g) an airport with two runways

2. What would be the total force exerted on a camper 15 ft long and 8 ft high, if a wind of 40 mi/hr blows perpendicular to one of its sides?

3. On the map of the United States (●Fig. 9.50), label where each of the following winds might be observed, then show with arrows the general direction of air flow that occurs with each of the winds.

(a) Santa Ana wind
(b) chinook wind
(c) California norther
(d) northeaster
(e) Columbia Gorge wind (downslope)
(f) Texas norther (blue norther)
(g) sea breeze along the New Jersey shore
(h) sea breeze in Los Angeles, California

4. On the same map of the United States (Fig. 9.50), show where the centers of atmospheric pressure should be located in order to produce the following winds. Place a large *L* on the map for the center of low pressure and a large *H* for the center of high pressure. (Be sure to place the letter representing the wind next to the *L* or the *H*.)

(a) high-pressure area for a Santa Ana wind
(b) low-pressure area for a chinook wind
(c) high-pressure area for a California norther
(d) high-pressure area for a Columbia Gorge wind (downslope wind)
(e) high- and low-pressure areas for a Texas norther (blue norther)
(f) high-pressure area for a sea breeze along the New Jersey shore
(g) low-pressure area for a sea breeze in Los Angeles, California

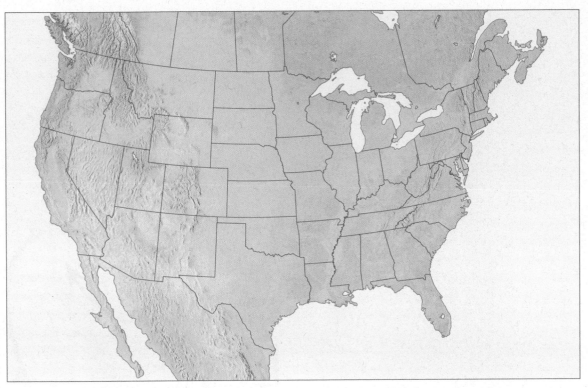

●FIGURE 9.50

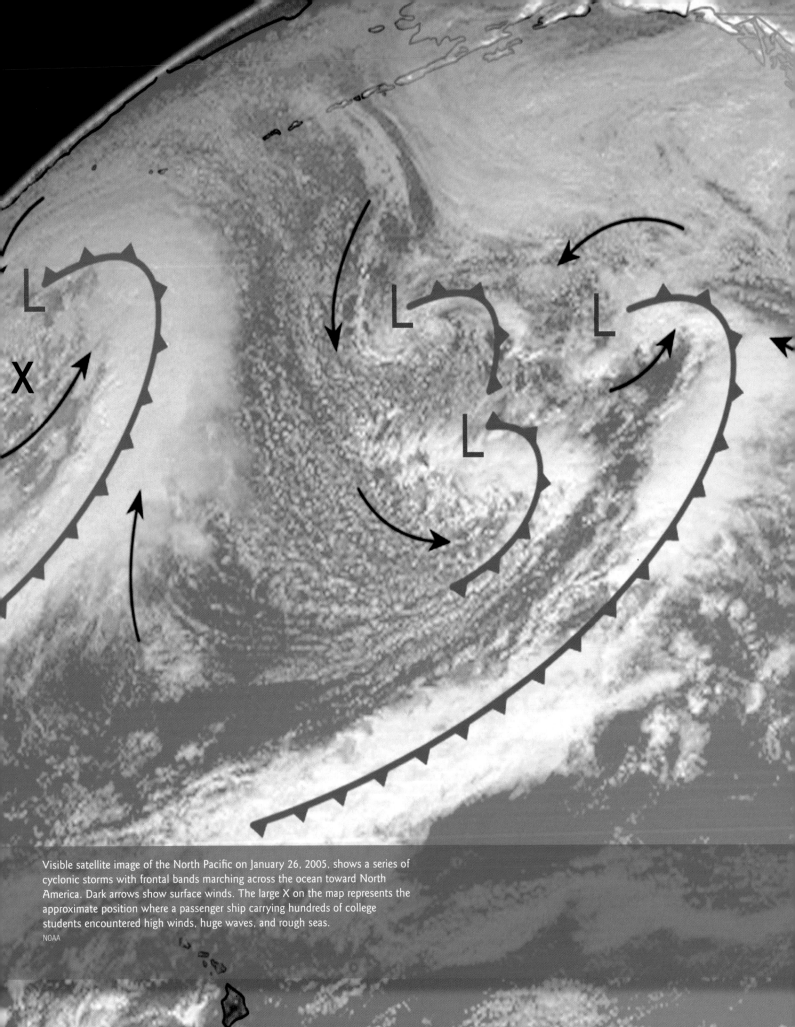

Visible satellite image of the North Pacific on January 26, 2005, shows a series of cyclonic storms with frontal bands marching across the ocean toward North America. Dark arrows show surface winds. The large X on the map represents the approximate position where a passenger ship carrying hundreds of college students encountered high winds, huge waves, and rough seas.

NOAA

CHAPTER 10

Wind: Global Systems

M ost of us know someone who at one time was "sick of school." But what about someone who was "sick in school"? That's what happened to many of the nearly 700 college students who took off from Vancouver, British Columbia, on January 18, 2005, for a 100-day "semester at sea."

After days of riding the huge waves of the north Pacific, the weather turned real ugly, as a storm greeted the ship (see the X on the chapter opening satellite image). In the wee hours of the morning on January 26, hurricane-force winds and huge waves—one estimated at 50 feet high—smashed the glass on the bridge and shorted out the ship's electrical and navigational systems. With three of the four engines disabled, the 590-foot vessel swayed from side to side, knocking students from their bunks onto the floor, where they dodged flying books, television sets, coffee pots, and furniture.

The captain ordered everyone to put on their life jackets and get into the ship's narrow hallways. As the students huddled together, the ship continued to roll from side to side. Students found themselves tumbling over one another as they slid across the floor. The storm gradually subsided and the ship was soon under control. The vessel with its cargo of anxious students then limped into Honolulu, Hawaii, for repairs. Fortunately, except for a few bumps and bruises and many nauseated passengers, no one was seriously injured, and many of the students continued their "semester at sea" in the air, flying from one destination to another.

What these students learned firsthand on this adventure was how the interaction between the atmosphere and ocean can have a rather exciting, if not violent, outcome.

CONTENTS

In Chapter 9, we learned that local winds vary considerably from day to day and from season to season. As you may suspect, these winds are part of a much larger circulation—the little whirls within larger whirls that we spoke of before. Indeed, if the rotating high- and low-pressure areas we see on a weather map are like spinning eddies in a huge river, then the flow of air around the globe is like the meandering river itself. When winds throughout the world are averaged over a long period of time, the local wind patterns vanish, and what we see is a picture of the winds on a global scale—what is commonly called the **general circulation of the atmosphere.** Just as the eddies in a river are carried along by the overall flow of water, so the highs and lows in the atmosphere are swept along by the general circulation. We will examine this large-scale circulation of air, its effects and its features, in this chapter.

General Circulation of the Atmosphere

Before we study the general circulation, we must remember that it only represents the *average* air flow around the world. Actual winds at any one place and at any given time may vary considerably from this average. Nevertheless, the average can answer why and how the winds blow around the world the way they do—why, for example, prevailing surface winds are northeasterly in Honolulu, Hawaii, and westerly in New York City. The average can also give a picture of the driving mechanism behind these winds, as well as a model of how heat and momentum are transported from equatorial regions poleward, keeping the climate in middle latitudes tolerable.

The underlying cause of the general circulation is the unequal heating of the earth's surface. We learned in Chapter 2 that, averaged over the entire earth, incoming solar radiation is roughly equal to outgoing earth radiation. However, we also know that this energy balance is not maintained for each latitude, since the tropics experience a net gain in energy, while polar regions suffer a net loss. To balance these inequities, the atmosphere transports warm air poleward and cool air equatorward. Although seemingly simple, the actual flow of air is complex; certainly not everything is known about it. In order to better understand it, we will first look at some models (that is, artificially constructed simulations) that eliminate some of the complexities of the general circulation.

SINGLE-CELL MODEL The first model is the single-cell model, in which we assume that:

1. The earth's surface is uniformly covered with water (so that differential heating between land and water does not come into play).
2. The sun is always directly over the equator (so that the winds will not shift seasonally).
3. The earth does not rotate (so that the only force we need to deal with is the pressure gradient force).

With these assumptions, the general circulation of the atmosphere on the side of the earth facing the sun would look much like the representation in ● Fig. 10.1a, a huge thermally driven convection cell in each hemisphere. (For reference, the names of the different regions of the world and their approximate latitudes are given in Figure 10.1b.)

The circulation of air described in Fig. 10.1a is the **Hadley cell** (named after the eighteenth-century English meteorologist George Hadley, who first proposed the idea). It is referred to as a *thermally direct cell* because it is driven by energy from the sun as warm air rises and cold air sinks. Excessive heating of the equatorial area produces a broad region of surface low pressure, while at the poles excessive cooling creates a region of surface high pressure. In response to the horizontal pressure gradient, cold surface polar air flows equatorward, while at higher levels air flows toward the poles. The entire circulation consists of a closed loop with rising air near the equator, sinking air over the poles, an equatorward flow of air near the surface, and a return flow aloft. In this manner, some of the excess energy of the tropics is transported as sensible and latent heat to the regions of energy deficit at the poles.*

Such a simple cellular circulation as this does not actually exist on the earth. For one thing, the earth rotates, so the Coriolis force would deflect the southward-moving surface air in the Northern Hemisphere to the right, producing easterly surface winds at practically all latitudes. These winds would be moving in a direction opposite to that of the earth's rotation and, due to friction with the surface, would slow down the earth's spin. We know that this does not happen and that prevailing winds in middle latitudes actually blow from the west. Therefore, observations alone tell us that a closed circulation of air between the equator and the poles is not the proper model for a rotating earth. But this model does show us how a nonrotating planet would balance an excess of energy at the equator and a deficit at the poles. How, then, does the wind blow on a rotating planet? To answer, we will keep our model simple by retaining our first two assumptions—that is, that the earth is covered with water and that the sun is always directly above the equator.

THREE-CELL MODEL If we allow the earth to spin, the simple convection system breaks into a series of cells as shown in ● Fig. 10.2. Although this model is considerably more complex than the single-cell model, there are some similarities. The tropical regions still receive an excess of heat and the poles a deficit. In each hemisphere, three cells instead of one have the task of energy redistribution. A surface high-pressure area is located at the poles, and a broad trough of surface low pressure still exists at the equator. From the equator to latitude 30°, the circulation is the *Hadley cell*. Let's look at this model more closely by examining what happens to the air above the equator. (Refer to Fig. 10.2, as you read the following section.)

Over equatorial waters, the air is warm, horizontal pressure gradients are weak, and winds are light. This region is

---

*Additional information on thermal circulations is found on p. 238 in Chapter 9.

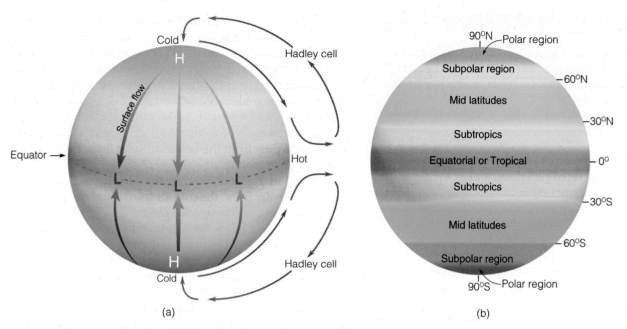

(a)   (b)

● FIGURE 10.1 Diagram (a) shows the general circulation of air on a nonrotating earth uniformly covered with water and with the sun directly above the equator. (Vertical air motions are highly exaggerated in the vertical.) Diagram (b) shows the names that apply to the different regions of the world and their approximate latitudes.

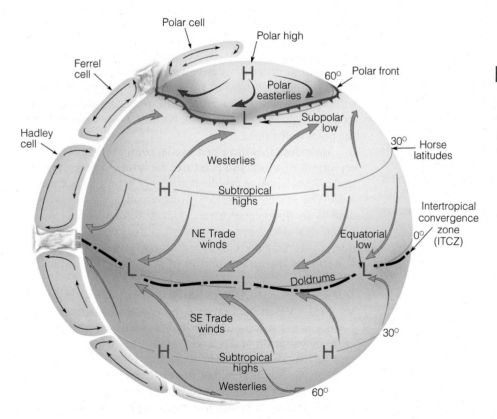

ACTIVE FIGURE 10.2 The idealized wind and surface-pressure distribution over a uniformly water-covered rotating earth. Visit the Meteorology Resource Center to view this and other active figures at academic.cengage.com/login

referred to as the **doldrums.** (The monotony of the weather in this area has given rise to the expression "down in the doldrums.") Here, warm air rises, often condensing into huge cumulus clouds and thunderstorms called *convective "hot" towers* because of the enormous amount of latent heat they liberate. This heat makes the air more buoyant and provides energy to drive the Hadley cell. The rising air reaches the

tropopause, which acts like a barrier, causing the air to move laterally toward the poles. The Coriolis force deflects this poleward flow toward the right in the Northern Hemisphere and to the left in the Southern Hemisphere, providing westerly winds aloft in both hemispheres. (We will see later that these westerly winds reach maximum velocity and produce jet streams near 30° and 60° latitudes.)

As air moves poleward from the tropics, it constantly cools by giving up infrared radiation, and at the same time it also begins to converge, especially as it approaches the middle latitudes.* This convergence (piling up) of air aloft increases the mass of air above the surface, which in turn causes the air pressure at the surface to increase. Hence, at latitudes near 30°, the convergence of air aloft produces belts of high pressure called **subtropical highs** (or anticyclones). As the converging, relatively dry air above the highs slowly descends, it warms by compression. This subsiding air produces generally clear skies and warm surface temperatures; hence, it is here that we find the major deserts of the world, such as the Sahara. Over the ocean, the weak pressure gradients in the center of the high produce only weak winds. According to legend, sailing ships traveling to the New World were frequently becalmed in this region, and, as food and supplies dwindled, horses were either thrown overboard or eaten. As a consequence, this region is sometimes called the **horse latitudes.**

From the horse latitudes, some of the surface air moves back toward the equator. It does not flow straight back, however, because the Coriolis force deflects the air, causing it to blow from the northeast in the Northern Hemisphere and from the southeast in the Southern Hemisphere. These steady winds provided sailing ships with an ocean route to the New World; hence, these winds are called the **trade winds.** Near the equator, the *northeast trades* converge with the *southeast trades* along a boundary called the **intertropical convergence zone (ITCZ).** In this region of surface convergence, air rises and continues its cellular journey.

Meanwhile, at latitude 30°, not all of the surface air moves equatorward. Some air moves toward the poles and deflects toward the east, resulting in a more or less westerly air flow—called the *prevailing westerlies,* or, simply, **westerlies**—in both hemispheres. Consequently, from Texas northward into Canada, it is much more common to experience winds blowing out of the west than from the east. The westerly flow in the real world is not constant as migrating areas of high and low pressure break up the surface flow pattern from time to time. In the middle latitudes of the Southern Hemisphere, where the surface is mostly water, winds blow more steadily from the west.

As this mild air travels poleward, it encounters cold air moving down from the poles. These two air masses of contrasting temperature do not readily mix. They are separated

*You can see why the air converges if you have a globe of the world. Put your fingers on meridian lines at the equator and then follow the meridians poleward. Notice how the lines and your fingers bunch together in the middle latitudes.

by a boundary called the **polar front,** a zone of low pressure—the **subpolar low**—where surface air converges and rises, and storms and clouds develop. Some of the rising air returns at high levels to the horse latitudes, where it sinks back to the surface in the vicinity of the subtropical high. In this model, the middle cell (a *thermally indirect cell,* in which cool air rises and warm air sinks, called the *Ferrel cell,* after the American meteorologist William Ferrel) is completed when surface air from the horse latitudes flows poleward toward the polar front.

Notice in Fig 10.2 that, in the Northern Hemisphere, behind the polar front, the cold air from the poles is deflected by the Coriolis force, so that the general flow of air is from the northeast. Hence, this is the region of the **polar easterlies.** In winter, the polar front with its cold air can move into middle and subtropical latitudes, producing a cold polar outbreak. Along the front, a portion of the rising air moves poleward, and the Coriolis force deflects the air into a westerly wind at high levels. Air aloft eventually reaches the poles, slowly sinks to the surface, and flows back toward the polar front, completing the weak *polar cell.*

We can summarize all of this by referring back to Fig. 10.2 and noting that, at the surface, there are two major areas of high pressure and two major areas of low pressure. Areas of high pressure exist near latitude 30° and the poles; areas of low pressure exist over the equator and near 60° latitude in the vicinity of the polar front. By knowing the way the winds blow around these systems, we have a generalized picture of surface winds throughout the world. The trade winds extend from the subtropical high to the equator, the westerlies from the subtropical high to the polar front, and the polar easterlies from the poles to the polar front.

How does this three-cell model compare with actual observations of winds and pressure? We know, for example, that upper-level winds at middle latitudes generally blow from the west. The middle Ferrel cell, however, suggests an east wind aloft as air flows equatorward. Hence, discrepancies exist between this model and atmospheric observations. This model does, however, agree closely with the winds and pressure distribution at the *surface,* and so we will examine this next.

AVERAGE SURFACE WINDS AND PRESSURE: THE REAL WORLD When we examine the real world with its continents and oceans, mountains and ice fields, we obtain an average distribution of sea-level pressure and winds for January and July, as shown in ●Fig.10.3a and 10.3b. Look closely at both maps and observe that there are regions where pressure systems appear to persist throughout the year. These systems are referred to as **semipermanent highs and lows** because they move only slightly during the course of a year.

In Fig. 10.3a, we can see that there are four semipermanent pressure systems in the Northern Hemisphere during January. In the eastern Atlantic, between latitudes 25° and 35°N is the *Bermuda–Azores high,* often called the **Bermuda high,** and, in the Pacific Ocean, its counterpart, the **Pacific high.** These are the subtropical anticyclones that develop in response to the convergence of air aloft near an upper-level

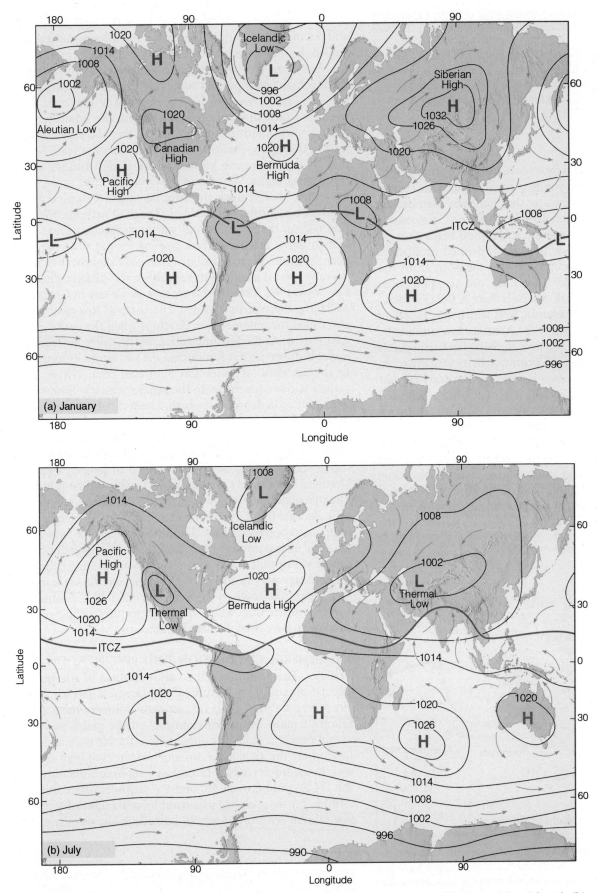

● FIGURE 10.3 Average sea-level pressure distribution and surface wind-flow patterns for January (a) and for July (b). The solid red line represents the position of the ITCZ.

jet stream. Since surface winds blow clockwise around these systems, we find the trade winds to the south and the prevailing westerlies to the north. In the Southern Hemisphere, where there is relatively less land area, there is less contrast between land and water, and the subtropical highs show up as well-developed systems with a clearly defined circulation.

Where we would expect to observe the polar front (between latitudes 40° and 65°), there are two semipermanent subpolar lows. In the North Atlantic, there is the *Greenland-Icelandic low,* or simply **Icelandic low,** which covers Iceland and southern Greenland, while the **Aleutian low** sits over the Gulf of Alaska and Bering Sea near the Aleutian Islands in the North Pacific. These zones of cyclonic activity actually represent regions where numerous storms, having traveled eastward, tend to converge, especially in winter.* In the Southern Hemisphere, the subpolar low forms a continuous trough that completely encircles the globe.

On the January map (Fig. 10.3a), there are other pressure systems, which are not semipermanent in nature. Over Asia, for example, there is a huge (but shallow) thermal anticyclone called the **Siberian high,** which forms because of the intense cooling of the land. South of this system, the winter monsoon

---
*For a better picture of these cyclonic storms in the Gulf of Alaska during the winter, see the opening satellite image on p. 258.

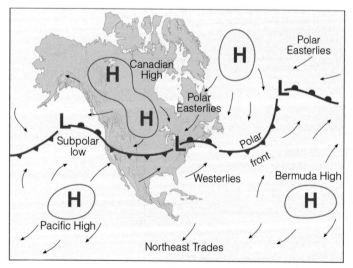

● FIGURE 10.4 A winter weather map depicting the main features of the general circulation over North America. Notice that the Canadian high, polar front, and subpolar lows have all moved southward into the United States, and that the prevailing westerlies exist south of the polar front. The arrows on the map illustrate wind direction.

shows up clearly, as air flows away from the high across Asia and out over the ocean. A similar (but less intense) anticyclone (called the *Canadian high*) is evident over North America.

As summer approaches, the land warms and the cold shallow highs disappear. In some regions, areas of surface low pressure replace areas of high pressure. The lows that form over the warm land are *thermal lows.* On the July map (Fig. 10.3b), warm thermal lows are found over the desert southwest of the United States and over the plateau of Iran. Notice that these systems are located at the same latitudes as the subtropical highs. We can understand why they form when we realize that, during the summer, the subtropical high-pressure belt girdles the world *aloft* near 30° latitude.* Within this system, the air sinks and warms, producing clear skies (which allow intense surface heating by the sun). This air near the ground warms rapidly, rises only slightly, then flows laterally several hundred meters above the surface. The outflow lowers the surface pressure and, as we saw in Chapter 9, a shallow thermal low forms. The thermal low over India, also called the *monsoon low,* develops when the continent of Asia warms. As the low intensifies, warm, moist air from the ocean is drawn into it, producing the wet summer monsoon so characteristic of India and Southeast Asia. Where these surface winds converge with the general westerly flow, rather weak monsoon depressions form. These enhance the position of the monsoon low on the July map.

When we compare the January and July maps, we can see several changes in the semipermanent pressure systems. The strong subpolar lows so well developed in January over the Northern Hemisphere are hardly discernible on the July map. The subtropical highs, however, remain dominant in both seasons. Because the sun is overhead in the Northern Hemisphere in July and overhead in the Southern Hemisphere in January, the zone of maximum surface heating shifts seasonally. In response to this shift, the major pressure systems, wind belts, and ITCZ (heavy red line in Fig.10.3) *shift toward the north in July and toward the south in January.†* ● Figure 10.4 illustrates a winter weather map where the main features of the general circulation have been displaced southward.

THE GENERAL CIRCULATION AND PRECIPITATION PATTERNS The position of the major features of the general circulation and their latitudinal displacement (which annually averages about 10° to 15°) strongly influence the precipitation of many areas. For example, on the global scale, we would expect abundant rainfall where the air rises and very little where the air sinks. Consequently, areas of high rainfall exist in the tropics, where humid air rises in conjunction with the ITCZ, and between 40° and 55° latitude, where middle-latitude storms and the polar front force air upward. Areas of low rainfall are found near 30° latitude in the vicinity of the

---
*An easy way to remember the seasonal shift of pressure systems is to think of birds—in the Northern Hemisphere, they migrate south in the winter and north in the summer.

†This belt of high pressure aloft shows up well in Fig. 10.8b, p. 266, the average 500-mb map for July.

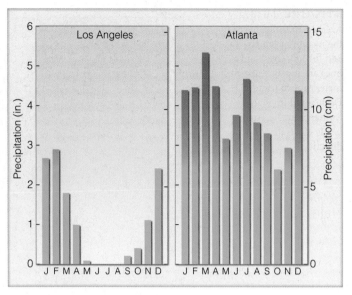

● FIGURE 10.7 Average annual precipitation for Los Angeles, California, and Atlanta, Georgia.

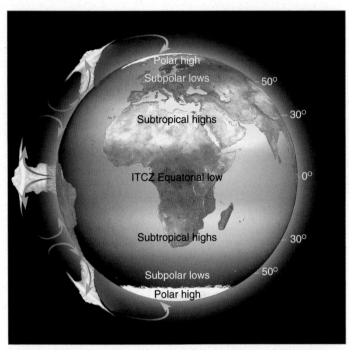

● FIGURE 10.5 Rising and sinking air associated with the major pressure systems of the earth's general circulation. Where the air rises, precipitation tends to be abundant (blue shade); where the air sinks, drier regions prevail (tan shade). Note that the sinking air of the subtropical highs produces the major desert regions of the world.

subtropical highs and in polar regions where the air is cold and dry (see ● Fig. 10.5).

Poleward of the equator, between the doldrums and the horse latitudes, the area is influenced by both the ITCZ and the subtropical high. In summer (high sun period), the subtropical high moves poleward and the ITCZ invades this area, bringing with it ample rainfall. In winter (low sun period), the subtropical high moves equatorward, bringing with it clear, dry weather.

During the summer, the Pacific high drifts northward to a position off the California coast (see ● Fig.10.6). Sinking air on its eastern side produces a strong upper-level subsidence inversion, which tends to keep summer weather along the West Coast relatively dry. The rainy season typically occurs in winter when the high moves south and the polar front and

storms are able to penetrate the region. Observe in Fig. 10.6 that along the East Coast, the clockwise circulation of winds around the Bermuda high brings warm tropical air northward into the United States and southern Canada from the Gulf of Mexico and the Atlantic Ocean. Because subsiding air is not as well developed on this side of the high, the humid air can rise and condense into towering cumulus clouds and thunderstorms. So, in part, it is the air motions associated with the subtropical highs that keep summer weather dry in California and moist in Georgia. (Compare the rainfall patterns for Los Angeles, California, and Atlanta, Georgia in ● Fig. 10.7.)

AVERAGE WIND FLOW AND PRESSURE PATTERNS ALOFT

● Figures 10.8a and 10.8b are average global 500-mb charts for the months of January and July, respectively. Look at both charts carefully and observe that some of the surface features of the general circulation are reflected on these upper-air charts. On the January map, for example, both the Icelandic low and Aleutian low are located to the west of their surface counterparts. On the July map, the subtropical high-pressure areas of the Northern Hemisphere appear as belts of high

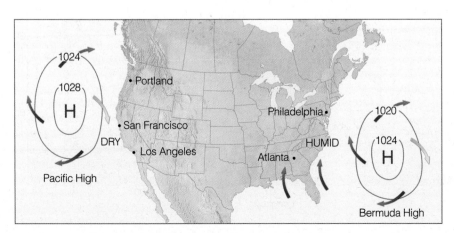

● FIGURE 10.6 During the summer, the Pacific high moves northward. Sinking air along its eastern margin (over California) produces a strong subsidence inversion, which causes relatively dry weather to prevail. Along the western margin of the Bermuda high, southerly winds bring in humid air, which rises, condenses, and produces abundant rainfall.

● FIGURE 10.8
Average 500-mb chart for the month of January (a) and for July (b). Solid lines are contour lines in meters above sea level. Dashed red lines are isotherms in °C. Arrowheads illustrate wind direction.

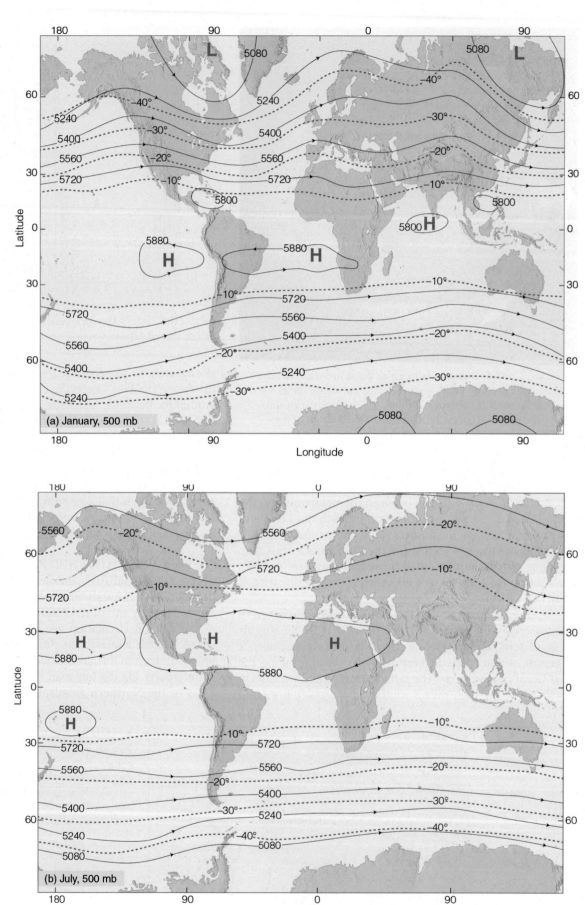

FOCUS ON AN OBSERVATION

The "Dishpan" Experiment

We know that the primary cause of the atmosphere's general circulation is the unequal heating that occurs between tropical and polar regions. A laboratory demonstration that tries to replicate this situation is the *"dishpan"* experiment.

The dishpan experiment consists of a flat, circular pan, filled with water several centimeters deep. The pan is positioned on a rotating table (see Fig. 1a). Around the edge of the pan, a heating coil supplies heat to the pan's "equator." In the center of the pan, a cooling cylinder represents the "pole," and ice water is continu-

ally supplied here. When the pan is rotated counterclockwise, the temperature difference between "equator" and "pole" produces a thermally driven circulation that transports heat poleward.

Aluminum powder (or dye) is added to the water so that the motions of the fluid can be seen. If the pan rotates at a speed that corresponds to the rotation of the earth, the flow develops into a series of waves and rotating eddies similar to those shown in Fig. 1b. The atmospheric counterpart of these eddies are the cyclones and anticyclones of the middle

latitudes. At the earth's surface, they occur as winds circulating around centers of low- and high-pressure. Aloft, the waves appear as a series of troughs and ridges that encircle the globe and slowly migrate from west to east (see Fig. 2). The waves represent a fundamental feature of the atmosphere's circulation. Later in this chapter we will see how they transfer momentum, allowing the atmosphere to maintain its circulation. In Chapter 12, we will see that these waves are instrumental in the development of surface mid-latitude cyclonic storms.

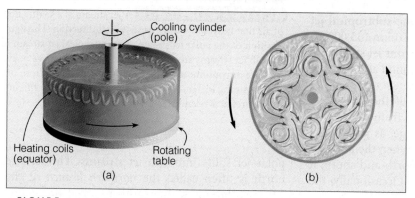

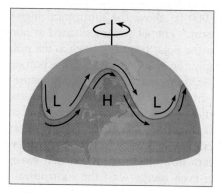

● FIGURE 1 (a) A "dishpan" with a hot "equator" and a cold "pole" rotating at a speed corresponding to that of the earth (b) produces troughs, ridges, and eddies, which appear (when viewed from above) very similar to the patterns we see on an upper-level chart.

● FIGURE 2 The circulation of the air aloft is in the form of waves — troughs and ridges — that encircle the globe.

---

height (high pressure) that tend to circle the globe south of 30°N. In both hemispheres, the air is warmer over low latitudes and colder over high latitudes. This horizontal temperature gradient establishes a horizontal pressure (contour) gradient that causes the winds to blow from the west, especially in middle and high latitudes.* Notice that the temperature gradients and the contour gradients are steeper in January than July. Consequently, the winds aloft are stronger in winter than in summer. The westerly winds, however, do not extend all the way to the equator, as easterly winds appear on the equatorward side of the upper-level subtropical highs.

In middle and high latitudes, the westerly winds continue to increase in speed above the 500-mb level. We already know that the wind speed increases up through the friction layer, but

why should it continue to increase at higher levels? You may remember from Chapter 8 that the geostrophic wind at any latitude is directly related to the pressure gradient and inversely related to the air density. Therefore, a greater pressure gradient will result in stronger winds, and so will a decrease in air density. Owing to the fact that air density decreases with height, the same pressure gradient will produce stronger winds at higher levels. In addition, the north-to-south temperature gradient causes the horizontal pressure (contour) gradient to increase with height up to the tropopause. As a result, the winds increase in speed up to the tropopause. Above the tropopause, the temperature gradients reverse. This changes the pressure gradients and reduces the strength of the westerly winds. Where strong winds tend to concentrate into narrow bands at the tropopause, we find rivers of fast-flowing air — *jet streams.* (In the following section, you will read about a wavy jet stream. The Focus section above describes an experiment that illustrates how these waves form.)

*Remember that, at this level (about 5600 m or 18,000 ft above sea level), the winds are approximately geostrophic, and tend to blow more or less parallel to the contour lines.

Jet Streams

Atmospheric **jet streams** are swiftly flowing air currents thousands of kilometers long, a few hundred kilometers wide, and only a few kilometers thick. Wind speeds in the central core of a jet stream often exceed 100 knots and occasionally exceed 200 knots. Jet streams are usually found at the tropopause at elevations between 10 and 15 km (6 and 9 mi), although they may occur at both higher and lower altitudes.

Jet streams were first encountered by high-flying military aircraft during World War II, but their existence was suspected before the war. Ground-based observations of fast-moving cirrus clouds had revealed that westerly winds aloft must be moving rapidly.

● Figure 10.9 illustrates the average position of the jet streams, tropopause, and general circulation of air for the Northern Hemisphere in winter. From this diagram, we can see that there are two jet streams, both located in tropopause gaps, where mixing between tropospheric and stratospheric air takes place. The jet stream situated near 30° latitude at about 13 km (43,000 ft) above the subtropical high is the **subtropical jet stream.*** The jet stream situated at about 10 km (33,000 ft) near the polar front is known as the **polar front jet stream** or, simply, the *polar jet stream.* Since both are found at the tropopause, they are referred to as *tropopause jets.*

In Fig. 10.9, the wind in the center of the jet stream would be flowing as a westerly wind away from the viewer. This direction, of course, is only an average, as jet streams often flow in a wavy west-to-east pattern. When the polar jet stream flows in broad loops that sweep north and south, it may even merge with the subtropical jet. Occasionally, the

---

*The subtropical jet stream is normally found between 20° and 30° latitude.

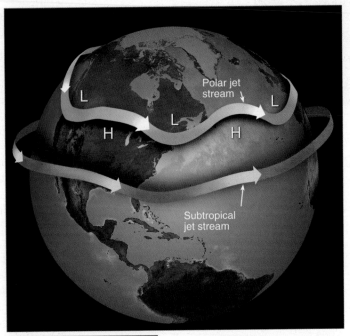

ACTIVE FIGURE 10.10 A jet stream is a swiftly flowing current of air that moves in a wavy west-to-east direction. The figure shows the position of the polar jet stream and subtropical jet stream in winter. Although jet streams are shown as one continuous river of air, in reality they are discontinuous, with their position varying from one day to the next. Visit the Meteorology Resource Center to view this and other active figures at academic.cengage.com/login

polar jet splits into two jet streams. The jet stream to the north is often called the *northern branch* of the polar jet, whereas the one to the south is called the *southern branch.* ● Figure 10.10 illustrates how the polar jet stream and the subtropical jet stream might appear as they sweep around the earth in winter.

● FIGURE 10.9 Average position of the polar jet stream and the subtropical jet stream, with respect to a model of the general circulation in winter. Both jet streams are flowing from west to east.

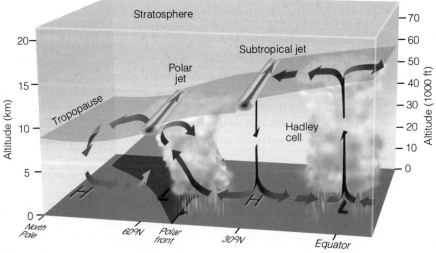

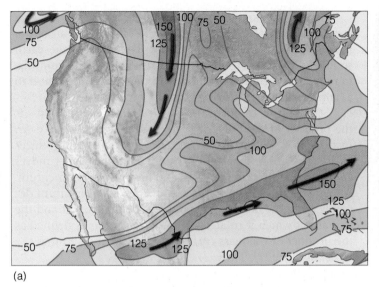

(a)

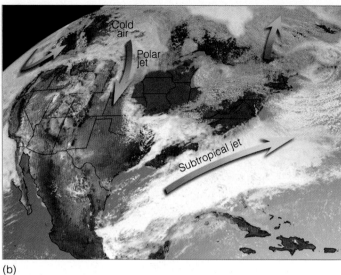

(b)

● FIGURE 10.11 (a) Position of the polar jet stream (blue arrows) and the subtropical jet stream (orange arrows) at the 300-mb level (about 9 km or 30,000 ft above sea level) on March 9, 2005. Solid lines are lines of equal wind speed (isotachs) in knots. (b) Satellite image showing clouds and positions of the jet streams for the same day.

We can better see the looping pattern of the jet by studying ● Fig. 10.11a, which shows the position of the polar jet stream and the subtropical jet stream at the 300-mb level (near 9 km or 30,000 ft) on March 9, 2005. The fastest flowing air, or *jet core,* is represented by the heavy dark arrows. The map shows a strong polar jet stream sweeping south over the Great Plains with an equally strong subtropical jet over the Gulf states. Notice that the polar jet has a number of loops, with one off the west coast of North America and another over eastern Canada. Observe in the satellite image (Fig. 10.11b) that the polar jet stream (blue arrows) is directing cold, polar air into the Plains States, while the subtropical jet stream (orange arrow) is sweeping subtropical moisture, in the form of a dense cloud cover, over the southeastern states.

The looping (meridional) pattern of the polar jet stream has an important function. In the Northern Hemisphere, where the air flows southward, swiftly moving air directs cold air equatorward; where the air flows northward, warm air is carried toward the poles. Jet streams, therefore, play a major role in the global transfer of heat. Moreover, since jet streams tend to meander around the world, we can easily understand how pollutants or volcanic ash injected into the atmosphere in one part of the globe could eventually settle to the ground many thousands of kilometers downwind. And, as we will see in Chapter 12, the looping nature of the polar jet stream has an important role in the development of mid-latitude cyclonic storms.

The ultimate cause of jet streams is the energy imbalance that exists between high and low latitudes. How, then, do jet streams actually form?

THE FORMATION OF THE POLAR FRONT JET AND THE SUBTROPICAL JET Horizontal variations in temperature and pressure offer clues to the existence of the polar jet

stream. ● Figure 10.12 is a 3-D model that shows a side view of the atmosphere in the region of the polar front. Since the polar front is a boundary separating cold polar air to the north from warm subtropical air to the south, the greatest contrast in air temperature occurs along the frontal zone. We can see this contrast as the −20°C isotherm dips sharply crossing the front. This rapid change in temperature produces a rapid change in pressure (as shown by the sharp bending of the constant pressure (isobaric) 500 mb surface as it passes through the front). In Figure 10.12b we can see that *the sudden change in pressure along the front sets up a steep pressure (contour) gradient that intensifies the wind speed and causes the jet stream.* Observe in Figure 10.12b that the wind is blowing along the front (from the west), parallel to the contour lines, with cold air on its left side.* The north-south temperature contrast along the polar front is strongest in winter and weakest in summer. This situation explains why the polar-front jet shows seasonal variations. In winter, the winds blow stronger and the jet moves farther south as the leading edge of the cold air extends into subtropical regions. In summer, the jet is weaker, and is usually found over more northern latitudes.

The subtropical jet stream, which is usually strongest slightly above the 200-mb level (above 12 km), tends to form along the poleward side of the Hadley cell as shown in Fig. 10.9, p. 268. Here, warm air carried poleward by the Hadley cell produces sharp temperature contrasts along a boundary sometimes called the *subtropical front.* In the vicinity of

_____

*Recall from Chapter 8 that any horizontal change in temperature causes the isobaric surfaces to dip or slant. The greater the temperature difference, the greater the slanting, and the stronger the winds. The changing wind speed with height due to horizontal temperature variations is referred to as the *thermal wind.* The thermal wind always blows with cold air on its left side in the Northern Hemisphere and on its right side in the Southern Hemisphere.

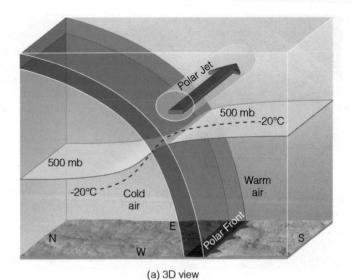

(a) 3D view

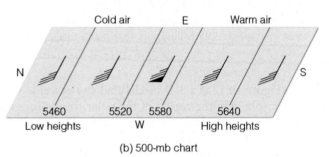

(b) 500-mb chart

● FIGURE 10.12 Diagram (a) is a model that shows a vertical 3-D view of the polar front in association with a sharply dipping 500-mb pressure surface, an isotherm (dashed line), and the position of the polar front jet stream in winter. The diagram is highly exaggerated in the vertical. Diagram (b) represents a 500-mb chart that cuts through the polar front as illustrated by the dipping 500-mb surface in (a). Sharp temperature contrasts along the front produce tightly packed contour lines and strong winds (contour lines are in meters above sea level).

the subtropical front (which does not have a frontal structure extending to the surface), sharp contrasts in temperature produce sharp contrasts in pressure and strong winds.

When we examine jet streams carefully, we see that another mechanism (other than a steep temperature gradient) causes a strong westerly flow aloft. The cause appears to be the same as that which makes an ice skater spin faster when the arms are pulled in close to the body—the *conservation of angular momentum.*

WEATHER WATCH

An unusually strong jet stream (and strong upper-level westerlies) during February, 2006, disrupted westbound transcontinental air travel over North America. These strong winds caused jet aircraft to make unscheduled fuel stops that added 45 minutes to some flights. To add insult to injury, many disgruntled passengers missed their connecting flights. On the bright side, these same winds cut many minutes from east-bound trips.

At the equator, the earth rotates toward the east at a speed close to 1000 knots. On a windless day, the air above moves eastward at the same speed. If somehow the earth should suddenly stop rotating, the air above would continue to move eastward until friction with the surface brought it to a halt; the air keeps moving because it has momentum.

Straight-line momentum—called *linear momentum*—is the product of the mass of the object times its velocity. An increase in either the mass or the velocity (or both) produces an increase in momentum. Air on a spinning planet moves about an axis in a circular path and has angular momentum. Along with the mass and the speed, angular momentum depends upon the distance (*r*) between the mass of air and the axis about which it rotates. *Angular momentum* is defined as the product of the mass (*m*) times the velocity (*v*) times the radial distance (*r*). Thus

$$\text{Angular momentum} = mvr.$$

As long as there are no external twisting forces (torques) acting on the rotating system, the angular momentum of the system does not change. We say that angular momentum is *conserved;* that is, the product of the quantity *mvr* at one time will equal the numerical quantity *mvr* at some later time. Hence, a decrease in radius must produce an increase in speed and vice versa. An ice skater, for instance, with arms fully extended rotates quite slowly. As the arms are drawn in close to the body, the radius of the circular path (*r*) decreases, which causes an increase in rotational velocity (*v*), and the skater spins faster. As arms become fully extended again, the skater's speed decreases. The conservation of angular momentum, when applied to moving air, will help us to understand the formation of fast-flowing air aloft.

Consider heated air parcels rising from the equatorial surface on a calm day. As the parcels approach the tropopause, they spread laterally and begin to move poleward. If we follow the air moving northward (● Fig. 10.13), we see that, because of the curvature of the earth, air constantly moves closer to its axis of rotation (*r* decreases). Because angular momentum is conserved (and since the mass of air is unchanged), the decrease in radius must be compensated for by an increase in speed. The air must, therefore, move faster to the east than a point on the earth's surface does. To an observer, this is a west wind. Hence, the conservation of angular momentum of northward-flowing air leads to the generation of strong westerly winds and the formation of a jet stream. (A more detailed look at the general circulation and the exchange of momentum between the earth and the atmosphere is provided in the Focus section on p. 271.)

OTHER JET STREAMS There is another jet stream that forms in summer near the tropopause above Southeast Asia, India, and Africa. Here, the altitude of the summer tropopause and the jet stream is near 15 km. Because the jet forms on the equatorward side of the upper-level subtropical high, its winds are easterly and, hence, it is known as the **tropical easterly jet stream.** Although the exact causes of this jet have yet to be

FOCUS ON AN ADVANCED TOPIC

Momentum—A Case of Give and Take

We know that energy from the sun drives the atmospheric circulation. Although the circulation pattern is complex, the general flow aloft is westerly. However, the westerly flow is not constant, for it breaks into eddies, cyclones, and anticyclones that transfer heat and momentum poleward. This transfer process feeds energy to the jet stream and maintains the westerly winds aloft. Therefore, this transfer mechanism is responsible for maintaining the general circulation of the atmosphere. How, then, does the momentum exchange take place?

There is a constant exchange of momentum between the earth and the atmosphere. Since momentum is simply mass times velocity, the momentum of an object whose mass is 1 is represented by velocity only. For such an object (a mass of air, for example), a change in momentum represents a change in velocity.

Consider the earth to be a rotating globe and the atmosphere to be your hand. When your hand rests on the globe and rotates with it at the same speed, there is, for all practical purposes, no transfer of momentum. But if your hand is on the equator and moves more slowly to the east than the rotating globe, then the friction between your hand and the globe will reduce the globe's rate of spin. Hence, there is a transfer of momentum from your hand to the globe. Because the actual winds in the tropics blow from east to west (and the earth rotates from west to east), there is a transfer of momentum from the moving air to the earth below, which should slow down the earth's rotation. But the earth's spin does not slow because of the westerly winds in middle latitudes.

To see what effect westerly winds have on the earth, place your hand on a rotating globe, then move it faster to the east than the globe rotates. Momentum transferred from your hand to the globe will cause the globe to spin faster. Similarly, the westerly winds of middle latitudes should increase the earth's rate of spin, but they do not because they are compensated for by the easterly winds in the tropics.

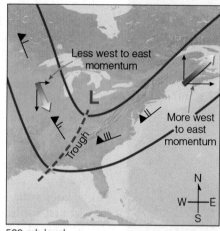

● FIGURE 3 A well-developed surface storm usually shows up as a wave with a tilted trough (dashed line) on a 500-mb chart. The wave transports westerly momentum poleward because the winds east of the trough have a greater westerly component than do the winds west of the trough.

The rotating earth affects the momentum of the atmosphere as well. We know that the northeast trades blow from about 30°N latitude toward the equator. As the air moves closer to the equator, it also moves farther away from the earth's axis of rotation (r increases in Fig. 10.13). Therefore, to conserve *angular momentum*, the increase in r must be offset by a decrease in velocity (v). However, the slower the air moves eastward on the rotating earth, the faster the wind appears to be blowing from the east to an observer on the earth's surface. (For a calm wind, the air is moving eastward at the same rate that the earth spins.) As a consequence, the trades should be strong easterly winds near the equator. In fact, however, the trades are fairly steady, but weak. The reason for these weak winds is that friction with the earth drags the air along more rapidly toward the east; thus, the apparent westward motion of the air decreases. The net result of this frictional interaction is that the earth imparts some of its momentum to the tropical air above. So, in low latitudes, the at-

mosphere *gains* momentum from the earth.

Meanwhile, in middle latitudes, the prevailing westerlies curve slightly northward, away from the subtropical high. As they move northward, they also move closer to the earth's axis of rotation (r decreases). The conservation of angular momentum requires that, as r decreases, surface wind velocity should increase and eventually reach the speed of a fast-flowing westerly jet. Surface winds do not blow that fast; again, the reason is surface friction, which slows the westerlies and reduces the air's angular momentum. Therefore, in middle latitudes, the air near the surface *loses* momentum to the earth.

If the atmosphere were to continually lose momentum in middle latitudes and gain momentum in low latitudes, both the prevailing westerlies and northeast trades would slow until the air is calm. Since we know this does not happen, there must be a net transfer of momentum from low latitudes toward high latitudes in order to maintain our wind systems. It is the large low- and high-pressure areas, the cyclones and anticyclones of the middle latitudes, that are primarily responsible for this transfer of momentum.

For instance, a storm in the upper troposphere might look like a trough similar to the one seen in Fig. 3. Notice that it has an asymmetric shape and tilts in a southwest-northeast direction. The winds on the east side of the trough are generally stronger than the winds on the west side. Also, the northward-moving winds on the east side have a stronger west-to-east component than do the southward-moving winds on the west side. This situation means that there is a net west-to-east transfer of momentum from lower latitudes to the middle latitude westerlies. Therefore, the next time a mid-latitude cyclonic storm causes you some discomfort with its high winds, remember that, without such storms, the atmosphere could not maintain its circulation.

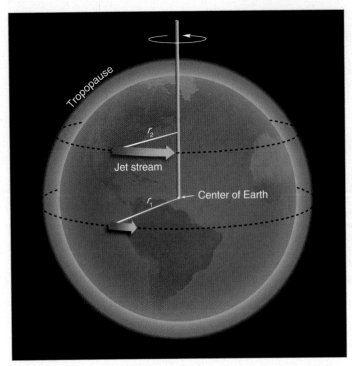

● FIGURE 10.13 Air flowing poleward at the tropopause moves closer to the rotational axis of the earth (r_2 is less than r_1). This decrease in radius is compensated for by an increase in velocity and the formation of a jet stream.

completely resolved, its formation appears to be, at least in part, related to the warming of the air over large elevated land masses, such as Tibet. During the summer, the air above this region (even at high elevations) is warmer than the air above the ocean to the south. This contrast in temperature produces a north-to-south pressure gradient and strong easterly winds that usually reach a maximum speed near 15°N latitude.

Not all jet streams form at the tropopause. For example, there is a jet stream that forms near the top of the stratosphere over polar latitudes. Because little, if any, sunlight reaches the polar region during the winter, air in the upper stratosphere is able to cool to low temperatures. By comparison, in equatorial regions, sunlight prevails all year long, allowing stratospheric ozone to absorb solar energy and warm the air. The horizontal temperature gradients between the cold poles and the warm tropics create steep horizontal pressure gradients, and a strong westerly jet forms in polar regions at an elevation near 50 km (30 mi). Because this wind maximum occurs in the stratosphere during the dark polar winter, it is known as the *stratospheric polar night jet stream.*

In summer, the polar regions experience more hours of sunlight than do tropical areas. Stratospheric temperatures over the poles increase more than at the same altitude above the equator, which causes the horizontal temperature gradient to reverse itself. The jet stream disappears, and in its place there are weaker easterly winds.

Jet streams also form in the upper mesosphere and in the thermosphere. Not much is known about the winds at these high levels, but they are probably related to the onslaught of charged particles that constantly bombard this region of the atmosphere.

Jet streams form near the earth's surface as well. One such jet develops over the central plains of the United States, where it occasionally attains speeds of 60 knots several hundred meters above the surface. This wind speed maximum, which usually flows from the south or southwest, is known as a **low-level jet.** It typically forms at night above a temperature inversion, and so it is sometimes called a *nocturnal jet stream.* Apparently, the stable air reduces the interaction between the air within the inversion and the air directly above. Consequently, the air in the vicinity of the jet is able to flow faster because it is not being slowed by the lighter winds below. Also, the north-south trending Rocky Mountains tend to funnel the air northward. Another important element contributing to the formation of the low-level jet is the downward sloping of the land from the Rockies to the Mississippi Valley, which causes nighttime air above regions to the west to be cooler than air at the same elevation to the east. This horizontal contrast in temperature causes pressure surfaces to dip toward the west. The dipping of pressure surfaces produces strong pressure gradient forces directed from east to west which, in turn, cause strong southerly winds.

During the summer, these strong southerly winds carry moist air from the Gulf of Mexico into the Central Plains. This moisture, coupled with converging, rising air of the low-level jet, enhances thunderstorm formation. Therefore, on warm, moist, summer nights, when the low-level jet is present, it is common to have nighttime thunderstorms over the plains.

BRIEF REVIEW

Before going on to the next section, which describes the many interactions between the atmosphere and the ocean, here is a review of some of the important concepts presented so far:

● The two major semipermanent subtropical highs that influence the weather of North America are the Pacific high situated off the west coast and the Bermuda high situated off the southeast coast.

● The polar front is a zone of low pressure where cyclonic storms often form. It separates the mild westerlies of the middle latitudes from the cold, polar easterlies of the high latitudes.

● In equatorial regions, the intertropical convergence zone (ITCZ) is a boundary where air rises in response to the convergence of the northeast trades and the southeast trades.

● In the Northern Hemisphere, the major global pressure systems and wind belts shift northward in summer and southward in winter.

● The northward movement of the Pacific high in summer tends to keep summer weather along the west coast of North America relatively dry.

● Jet streams exist where strong winds become concentrated in narrow bands. The polar front jet stream is associated with the polar front. The polar jet meanders in a wavy, west-to-east pattern, becoming strongest in winter when the contrast in temperature along the front is greatest.

- The subtropical jet stream is found on the poleward side of the Hadley cell, between 20° and 30° latitude.

- The conservation of angular momentum plays a role in producing strong westerly winds aloft. As air aloft moves from lower latitudes toward higher latitudes, its axis of rotation decreases, which results in an increase in its speed.

Atmosphere-Ocean Interactions

Although scientific understanding of all the interactions between the oceans and the atmosphere is far from complete, there are some relationships that deserve mentioning here.

GLOBAL WIND PATTERNS AND SURFACE OCEAN CURRENTS   As the wind blows over the oceans, it causes the surface water to drift along with it. The moving water gradually piles up, creating pressure differences within the water itself. This leads to further motion several hundreds of me-

ters down into the water. In this manner, the general wind flow around the globe starts the major surface ocean currents moving. The relationship between the general circulation and ocean currents can be seen by comparing Fig. 10.3, p. 263 and • Fig. 10.14.

Because of the larger frictional drag in water, ocean currents move more slowly than the prevailing wind. Typically, these currents range in speed from several kilometers per day to several kilometers per hour. As we can see in Fig. 10.14, major ocean currents do not follow the wind pattern exactly; rather, they spiral in semiclosed circular whirls called *gyres*. In the North Atlantic and North Pacific, the prevailing winds blow clockwise and outward from the subtropical highs. As the water moves beneath the wind, the Coriolis force deflects the water to the right in the Northern Hemisphere (to the left in the Southern Hemisphere). This deflection causes the surface water to move at an angle between 20° and 45° to the direction of the wind. Hence, surface water tends to move in a circular pattern as winds blow outward, away from the center of the subtropical highs.

Important interactions between the atmosphere and the ocean can be seen by examining the huge gyre in the North Atlantic. Flowing northward along the east coast of the

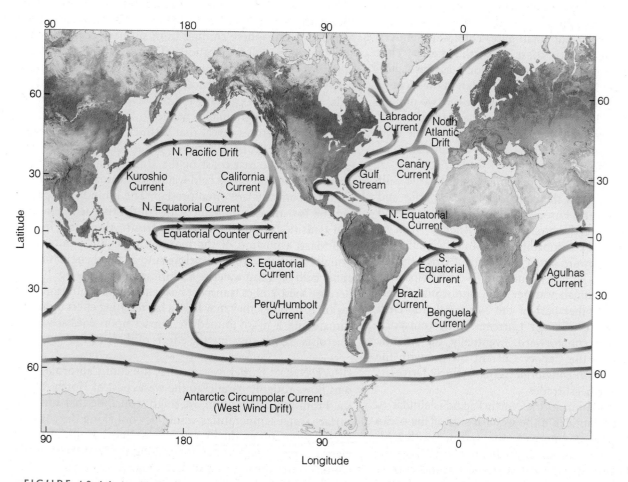

• FIGURE 10.14 Average position and extent of the major surface ocean currents. Cold currents are shown in blue; warm currents are shown in red.

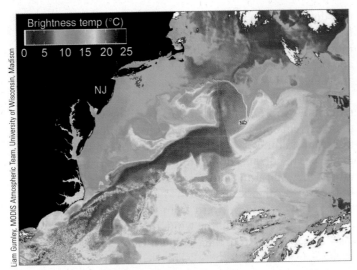

Liam Gumley, MODIS Atmospheric Team, University of Wisconsin, Madison

Brightness temp (°C)
0 5 10 15 20 25

NJ

• FIGURE 10.15 The Gulf Stream (dark red band) and its eddies are revealed in this satellite mosaic of sea surface temperatures of the western North Atlantic during May, 2001. Bright red shows the warmest water, followed by orange and yellow. Green, blue, and purple represent the coldest water.

United States is a tremendous warm water current called the *Gulf Stream*. The Gulf Stream carries vast quantities of warm tropical water into higher latitudes. To the north, on the western side of the smaller subpolar gyre, cold water moves southward along the Atlantic coast of North America. This *Labrador Current* brings cold water as far south as Massachusetts in summer and North Carolina in winter. In the vicinity of the Grand Banks of Newfoundland, where the two opposing currents flow side by side, there is a sharp temperature gradient. When warm Gulf Stream air blows over the cold Labrador Current water, the stage is set for the formation of the fog so common to this region.

Meanwhile, steered by the prevailing westerlies, the Gulf Stream swings away from the coast of North America and moves eastward toward Europe. Gradually, it widens and slows as it merges into the broader *North Atlantic Drift*. As this current approaches Europe, it divides into two currents. A portion flows northward along the coasts of Great Britain and Norway, bringing with it warm water (which helps keep winter temperatures much warmer than one would expect this far north). The other part of the North Atlantic Drift flows southward as the *Canary Current*, which transports cool northern water equatorward. Eventually, the Atlantic gyre is completed as the Canary Current merges with the westward-moving *North Equatorial Current*, which derives its energy from the northeast trades.

The ocean circulation in the North Pacific is similar to that in the North Atlantic. On the western side of the ocean is the Gulf Stream's counterpart, the warm, northward-flowing *Kuroshio Current*, which gradually merges into the slower-moving *North Pacific Drift*. A portion of this current flows southward along the coastline of the western United

States as the cool *California Current*. In the Southern Hemisphere, surface ocean circulations are much the same except that the gyres move counterclockwise in response to the winds around the subtropical highs. Notice that the ocean currents at higher latitudes tend to move in a more west-to-east pattern than do the currents of the Northern Hemisphere. This zonal pattern limits, to some extent, the poleward transfer of warm tropical water. Hence, there is a much smaller temperature difference between the ocean's surface and the atmosphere than exists over Northern Hemisphere oceans. This situation tends to limit the development of vigorous convective activity over the oceans of the Southern Hemisphere. In the Indian Ocean, monsoon circulations tend to complicate the general pattern of ocean currents.

To sum up: On the eastern edge of continents there usually is a warm current that carries huge quantities of warm water from the equator toward the pole; whereas on the western side of continents a cool current typically flows from the pole toward the equator.

Up to now, we have seen that atmospheric circulations and ocean circulations are closely linked; wind blowing over the oceans produces surface ocean currents. The currents, along with the wind, transfer heat from tropical areas, where there is a surplus of energy, to polar regions, where there is a deficit. This helps to equalize the latitudinal energy imbalance with about 40 percent of the total heat transport in the Northern Hemisphere coming from surface ocean currents. The environmental implications of this heat transfer are tremendous. If the energy imbalance were to go unchecked, yearly temperature differences between low and high latitudes would increase greatly, and, as we will see in Chapter 16, the climate would gradually change.

Satellite pictures reveal that distinct temperature gradients exist along the boundaries of surface ocean currents. For example, off the east coast of the United States, where the warm Gulf Stream meets cold waters to the north, sharp temperature contrasts are often present. The boundary separating the two masses of water with contrasting temperatures and densities is called an **oceanic front**. Along this frontal boundary, a portion of the meandering Gulf Stream occasionally breaks away and develops into a closed circulation of either cold or warm water—a whirling eddy (see • Fig. 10.15). Because these eddies transport heat and momentum from one region to another, they may have a far-reaching effect upon the climate and a more immediate impact upon coastal waters. Scientists are investigating the effects of these eddies.

UPWELLING Earlier, we saw that the cool California Current flows roughly parallel to the west coast of North America. From this observation, we might conclude that summer surface water temperatures would be cool along the coast of Washington and gradually warm as we move south. A quick glance at the water temperatures along the west coast of the United States during August (see • Fig. 10.16) quickly alters that notion. The coldest water is observed along the northern

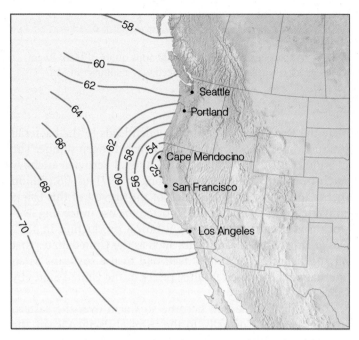

• FIGURE 10.16 Average sea surface temperatures (°F) along the west coast of the United States during August.

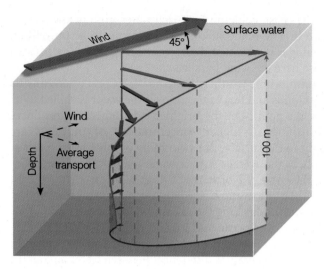

• FIGURE 10.17 The Ekman Spiral. Winds move the water, and the Coriolis force deflects the water to the right (Northern Hemisphere). Below the surface each successive layer of water moves more slowly and is deflected to the right of the layer above. The average transport of surface water in the Ekman layer is at right angles to the prevailing winds.

California coast near Cape Mendocino. Why there? To answer this, we need to examine how the wind influences the movement of surface water.

As the wind blows over an open stretch of ocean, the surface water beneath it is set in motion. The Coriolis force bends the moving water to the right in the Northern Hemisphere. Thus, if we look at a shallow surface layer of water, we see in ● Fig. 10.17 that it moves at an average angle of about 45° to the direction of the wind. If we imagine the top layer of ocean water to be broken into a series of layers, then each layer will exert a frictional drag on the layer below. Each successive layer will not only move a little slower than the one above, but (because of the Coriolis effect) each layer will also rotate slightly to the right of the layer above. (The rotation of each layer is to the right in the Northern Hemisphere and to the left in the Southern Hemisphere.) Consequently, descending from the surface, we would find water slowing and turn-

ing to the right until, at some depth (usually about 100 m), the water actually moves in a direction opposite to the flow of water at the surface. This turning of water with depth is known as the **Ekman Spiral.*** The Ekman Spiral in Fig. 10.17 shows us that the average movement of surface water down to a depth of about 100 m is at right angles (90°) to the surface wind direction. The Ekman Spiral helps to explain why, in summer, surface water is cold along the west coast of North America.

The summertime position of the Pacific high and the low coastal mountains cause winds to blow parallel to the California coastline (see ● Fig. 10.18). The net transport of surface water (called the **Ekman transport**) is at right angles to the wind, in this case, out to sea. As surface water drifts away

*The Ekman Spiral is also present in the atmospheric boundary layer, from the surface up to the top of the friction layer, which is usually about 1000 m above the surface.

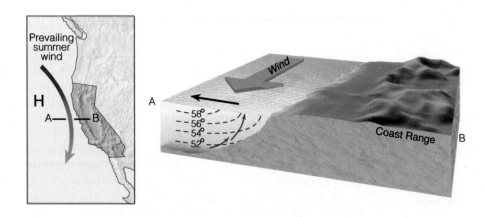

• FIGURE 10.18 As winds blow parallel to the west coast of North America, surface water is transported to the right (out to sea). Cold water moves up from below (upwells) to replace the surface water.

from the coast, cold, nutrient-rich water from below rises to replace it. The rising of cold water is known as **upwelling.** Upwelling is strongest and surface water is coolest in this area because here the wind parallels the coast.

Summertime weather along the West Coast often consists of low clouds and fog, as the air over the water is chilled to its saturation point. On the brighter side, upwelling produces good fishing, as higher concentrations of nutrients are brought to the surface. But swimming is only for the hardiest of souls, as the average surface water temperature in summer along the coast of Northern California is nearly 10°C (18°F) colder than the average coastal water temperature found at the same latitude along the Atlantic Coast.

Between the ocean surface and the atmosphere, there is an exchange of heat, moisture, and momentum that depends, in part, on temperature differences between water and air. In winter, when air-water temperature contrasts are greatest, there is a substantial transfer of sensible and latent heat from the ocean surface into the atmosphere. This energy helps to maintain the global air flow. Because of the difference in heat capacity between water and air, even a relatively small change in surface ocean temperatures could modify atmospheric circulations. Such change could have far-reaching effects on global weather patterns. One ocean-atmospheric phenomenon that is linked to worldwide weather events is a warming of the tropical Pacific Ocean known as *El Niño.*

EL NIÑO AND THE SOUTHERN OSCILLATION
Along the west coast of South America, where the cool Peru Current sweeps northward, southerly winds promote upwelling of cold, nutrient-rich water that gives rise to large fish populations, especially anchovies. The abundance of fish supports a large population of sea birds whose droppings (called *guano*) produce huge phosphate-rich deposits, which support the fertilizer industry. Near the end of the calendar year, a warm current of nutrient-poor tropical water often moves southward, replacing the cold, nutrient-rich surface water. Because this condition frequently occurs around Christmas, local residents call it *El Niño* (Spanish for *boy child*), referring to the Christ child.

In most years, the warming lasts for only a few weeks to a month or more, after which weather patterns usually return to normal and fishing improves. However, when El Niño conditions last for many months, and a more extensive ocean warming occurs, the economic results can be catastrophic. This extremely warm episode, which occurs at irregular intervals of two to seven years and covers a large area of the tropical Pacific Ocean, is now referred to as a *major El Niño event,* or simply **El Niño.***

During a major El Niño event, large numbers of fish and marine plants may die. Dead fish and birds may litter the water and beaches of Peru; their decomposing carcasses de-

---

*It was once thought that El Niño was a local event that occurs along the west coast of Peru and Ecuador. It is now known that the ocean-warming associated with a major El Niño can cover an area of the tropical Pacific much larger than the continental United States.

plete the water's oxygen supply, which leads to the bacterial production of huge amounts of smelly hydrogen sulfide. The El Niño of 1972–1973 reduced the annual Peruvian anchovy catch from 10.3 million metric tons in 1971 to 4.6 million metric tons in 1972. Since much of the harvest of this fish is converted into fishmeal and exported for use in feeding livestock and poultry, the world's fishmeal production in 1972 was greatly reduced. Countries such as the United States that rely on fishmeal for animal feed had to use soybeans as an alternative. This raised poultry prices in the United States by more than 40 percent.

Why does the ocean become so warm over the eastern tropical Pacific? Normally, in the tropical Pacific Ocean, the trades are persistent winds that blow westward from a region of higher pressure over the eastern Pacific toward a region of lower pressure centered near Indonesia (see ● Fig. 10.19a). The trades create upwelling that brings cold water to the surface. As this water moves westward, it is heated by sunlight and the atmosphere. Consequently, in the Pacific Ocean, surface water along the equator usually is cool in the east and warm in the west. In addition, the dragging of surface water by the trades raises sea level in the western Pacific and lowers it in the eastern Pacific, which produces a thick layer of warm water over the tropical western Pacific Ocean and a weak ocean current (called the *countercurrent*) that flows slowly eastward toward South America.

Every few years, the surface atmospheric pressure patterns break down, as air pressure rises over the region of the western Pacific and falls over the eastern Pacific (see Fig. 10.19b). This change in pressure weakens the trades, and, during strong pressure reversals, east winds are replaced by west winds that strengthen the countercurrent. Surface water warms over a broad area of the tropical Pacific and heads eastward toward South America in a surge known as a *Kelvin wave,* which is an enormous wave perhaps 15 centimeters high but extending for hundreds of kilometers north and south of the equator (see ● Fig. 10.20). Toward the end of the warming period, which may last between one and two years, atmospheric pressure over the eastern Pacific reverses and begins to rise, whereas, over the western Pacific, it falls. This seesaw pattern of reversing surface air pressure at opposite ends of the Pacific Ocean is called the **Southern Oscillation.** Because the pressure reversals and ocean warming are more or less simultaneous, scientists call this phenomenon the *El Niño/Southern Oscillation* or **ENSO** for short. Although most ENSO episodes follow a similar evolution, each event has its own personality, differing in both strength and behavior.

During especially strong ENSO events (such as in 1982–83 and 1997–98) the easterly trades may actually be-

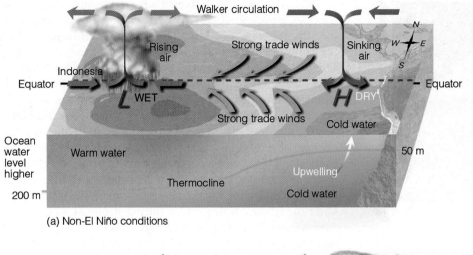

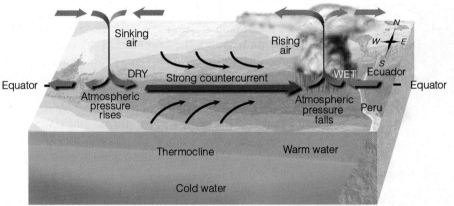

● FIGURE 10.19 In diagram (a), under ordinary conditions higher pressure over the southeastern Pacific and lower pressure near Indonesia produce easterly trade winds along the equator. These winds promote upwelling and cooler ocean water in the eastern Pacific, while warmer water prevails in the western Pacific. The trades are part of a circulation (called the *Walker circulation*) that typically finds rising air and heavy rain over the western Pacific and sinking air and generally dry weather over the eastern Pacific. When the trades are exceptionally strong, water along the equator in the eastern Pacific becomes quite cool. This cool event is called La Niña. During El Niño conditions—diagram (b)—atmospheric pressure decreases over the eastern Pacific and rises over the western Pacific. This change in pressure causes the trades to weaken or reverse direction. This situation enhances the countercurrent that carries warm water from the west over a vast region of the eastern tropical Pacific. The thermocline, which separates the warm water of the upper ocean from the cold water below, changes as the ocean conditions change from non-El Niño to El Niño.

come westerly winds, as shown in Fig. 10.19b. As these winds push eastward, they drag surface water with them. This dragging raises sea level in the eastern Pacific and lowers sea level in the western Pacific. The eastward-moving water gradually warms under the tropical sun, becoming as much as 6°C (11°F) warmer than normal in the eastern equatorial Pacific.

Gradually, a thick layer of warm water pushes into coastal areas of Ecuador and Peru, choking off the upwelling that supplies cold, nutrient-rich water to South America's coastal region. The unusually warm water may extend from South America's coastal region for many thousands of kilometers westward along the equator (see ● Fig. 10.21a).

● FIGURE 10.20 These three images depict the evolution of a warm water Kelvin wave moving eastward in the equatorial Pacific Ocean during March and April, 1997. The white areas near the equator represent ocean levels about 20 cm (8 in.) higher than average, while the red areas represent ocean levels about 10 cm (4 in.) higher than average. Notice how the wave (high region) moves eastward across the tropical Pacific Ocean. These data were collected by the altimeter on board the joint United States/French *TOPEX/Poseidon* satellite.

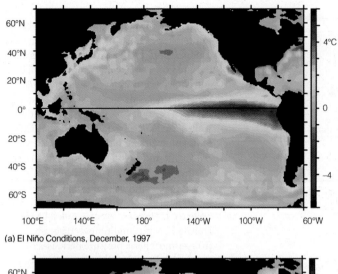

(a) El Niño Conditions, December, 1997

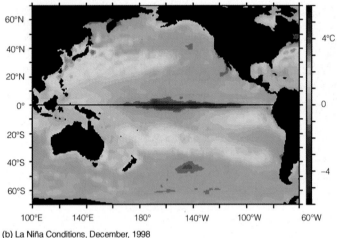

(b) La Niña Conditions, December, 1998

● FIGURE 10.21 (a) Average sea surface temperature departures from normal as measured by satellite. During El Niño conditions upwelling is greatly diminished and warmer than normal water (deep red color) extends from the coast of South America westward, across the Pacific. (b) During La Niña conditions, strong trade winds promote upwelling, and cooler than normal water (dark blue color) extends over the eastern and central Pacific. (NOAA/PHEL/TAO)

Such a large area of abnormally warm water can have an effect on global wind patterns. The warm tropical water fuels the atmosphere with additional warmth and moisture, which the atmosphere turns into additional storminess and rainfall. The added warmth from the oceans and the release of latent heat during condensation apparently influence the westerly winds aloft in such a way that certain regions of the world experience too much rainfall, whereas others have too little. Meanwhile, over the warm tropical central Pacific, the frequency of typhoons usually increases. However, over the tropical Atlantic, between Africa and Central America, the winds aloft tend to disrupt the organization of thunderstorms that is necessary for hurricane development; hence, there are fewer hurricanes in this region during strong El Niño events. And, as we saw in Chapter 9, during El Niño events there is a tendency for monsoon conditions over India to weaken.

Although the actual mechanism by which changes in surface ocean temperatures influence global wind patterns is not fully understood, the by-products are plain to see. For example, during exceptionally warm El Niños, drought is normally felt in Indonesia, southern Africa, and Australia, while heavy rains and flooding often occur in Ecuador and Peru. In the Northern Hemisphere, a strong subtropical westerly jet stream normally directs storms into California and heavy rain into the Gulf Coast states. The total damage worldwide due to flooding, winds, and drought may exceed many billions of dollars.

Following an ENSO event, the trade winds usually return to normal. However, if the trades are exceptionally strong, unusually cold surface water moves over the central and eastern Pacific, and the warm water and rainy weather is confined mainly to the western tropical Pacific (see Fig. 10.21b). This cold-water episode, which is the opposite of El Niño conditions, has been termed **La Niña** (the girl child).

● Figure 10.22 shows warm events, El Niño years, in red and cold events, or La Niña years, in blue. Notice that the two strongest El Niños were 1982–83 and 1997–98. The weaker El Niño events also have an effect on the Northern Hemisphere's weather patterns. For example, during the El Niño of 1986–87,

● FIGURE 10.22 The Ocean Niño Index (ONI). The numbers on the left side of the diagram represent a running 3-month mean for sea surface temperature variations (from normal) over the tropical Pacific Ocean from latitude 5°N to 5°S and from longitude 120°W to 170°W. Warm El Niño episodes are in red; cold La Niña episodes are in blue. Warm and cold events occur when the deviation from the normal is 0.5 or greater. An index value between 0.5 and 0.9 is considered weak; an index value between 1.0 and 1.4 is considered moderate, and an index value of 1.5 or greater is considered strong. (Courtesy of NOAA and Jan Null.)

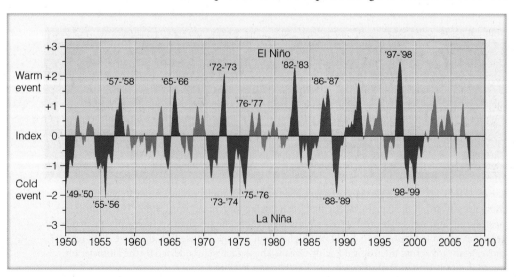

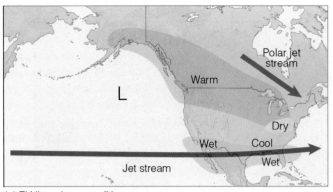 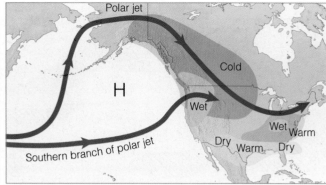

(a) El Niño winter conditions   (b) La Niña winter conditions

● FIGURE 10.23 Typical winter weather patterns across North America during an El Niño warm event (a) and during a La Niña cold event (b). During El Niño conditions, a persistent trough of low pressure forms over the north Pacific and, to the south of the low, the jet stream (from off the Pacific) steers wet weather and storms into California and the southern part of the United States. During La Niña conditions, a persistent high-pressure area forms south of Alaska forcing the polar jet stream and accompanying cold air over much of western North America. The southern branch of the polar jet stream directs moist air from the ocean into the Pacific Northwest, producing a wet winter for that region.

the subtropical jet stream (being fueled by the warm tropical waters and huge thunderstorms) curved its way over the southeastern United States, where it brought abundant rainfall to a region that, during the previous summer, had suffered through a devastating drought. During the El Niño of 1991–92, the subtropical jet stream once again swung over North America. Water evaporating from the warm tropical oceans fueled huge thunderstorms. The subtropical jet stream initially swept this moisture into Texas, where it caused extensive flooding.

● Figure 10.23a illustrates typical winter weather patterns over North America during El Niño conditions. Notice that a persistent trough of low pressure forms over the north Pacific and, to the south of the low, the jet stream (from off the Pacific) steers wet weather and storms into California and the southern part of the United States. A weak polar jet stream forms over eastern Canada allowing warmer than normal weather to prevail over a large part of North America.

Figure 10.23b shows typical winter weather patterns with a La Niña. Notice that a persistent high-pressure area (called a *blocking high*) forms south of Alaska forcing the polar jet stream into Alaska, then southward into Canada and the western United States. The southern branch of the polar jet, which forms south of the high, directs moist air from the ocean into the Pacific northwest, producing a wet winter for that region. Meanwhile, winter months in the southern part of the United States tend to be warmer and drier than normal.

As we have seen, El Niño and the Southern Oscillation are part of a large-scale ocean-atmosphere interaction that can take several years to run its course. During this time, there are certain regions in the world where significant climatic responses to an ENSO event are likely. Using data from previous ENSO episodes, scientists at the National Oceanic and Atmospheric Administration's Climate Prediction Center have obtained a global picture of where climatic

abnormalities are most likely (see ● Fig. 10.24). Such ocean-atmosphere interactions where warm or cold surface ocean temperatures can influence precipitation patterns in a distant part of the world are called **teleconnections.**

Some scientists feel that the trigger necessary to start an ENSO event lies within the changing of the seasons, especially the transition periods of spring and fall. Others feel that the winter monsoon plays a major role in triggering a major El Niño event. As noted earlier, it appears that an ENSO episode and the monsoon system are intricately linked, so that a change in one brings about a change in the other.

Is there a similar pattern in the Atlantic that compares to the Southern Oscillation in the Pacific? Typically in the eastern Atlantic (off the coast of Africa), the water is cool and the weather is drier than in the tropical western Atlantic. Periodically, the cool water along the African coast is replaced by warm water and heavy rainfall. This Atlantic warming, however, occurs more sporadically and is not as strong as the warming in the tropical Pacific.

Presently, scientists (with the aid of general circulation models) are trying to simulate atmospheric and oceanic conditions, so that El Niño and the Southern Oscillation can be anticipated. At this point, several models have been formulated that show promise in predicting the onset and life history of an ENSO event. In addition, an in-depth study known as TOGA (*Tropical Ocean and Global Atmosphere*), which ended in 1994, is providing scientists with valuable information about the interactions that occur between the ocean and the atmosphere. The primary aim of TOGA, a major component of the *World Climate Research Program* (WCRP), is to provide enough scientific information so that researchers can better predict climatic fluctuations (such as ENSO) that occur over periods of months and years. The hope is that a better understanding of El Niño and the Southern Oscillation will provide improved long-range forecasts of weather and climate.

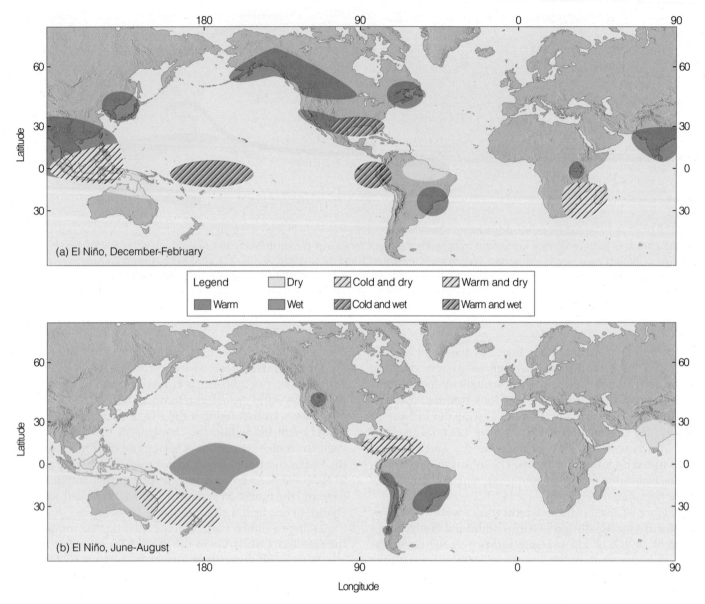

● FIGURE 10.24 Regions of climatic abnormalities associated with El Niño–Southern Oscillation conditions (a) during December through February and (b) during June through August. A strong ENSO event may trigger a response in nearly all indicated areas, whereas a weak event will likely play a role in only some areas. (After NOAA Climate Prediction Center.)

Up to this point, we have looked at El Niño and the Southern Oscillation and how the reversal of surface ocean temperatures and atmospheric pressure combine to influence regional and global weather and climate patterns. There are other atmosphere-ocean interactions that can have an effect on large-scale weather patterns. Some of these are described in the following sections.

PACIFIC DECADAL OSCILLATION   Over the Pacific Ocean, changes in surface water temperatures appear to influence winter weather along the west coast of North America. In the mid 1990s, scientists at the University of Washington, while researching connections between Alaskan salmon production and Pacific climate, identified a long-term Pacific Ocean

temperature fluctuation, which they called the **Pacific Decadal Oscillation (PDO)** because the ocean surface temperature reverses every 20 to 30 years. The Pacific Decadal Oscillation is like ENSO in that it has a warm phase and a cool phase, but its temperature behavior is much different from that of El Niño.

During the warm (or positive) phase, unusually warm surface water exists along the west coast of North America, whereas over the central North Pacific, cooler than normal surface water prevails (see ● Fig. 10.25a). At the same time, the Aleutian low in the Gulf of Alaska strengthens, which causes more Pacific storms to move into Alaska and California. This situation causes winters, as a whole, to be warmer and drier over northwestern North America. Elsewhere, win-

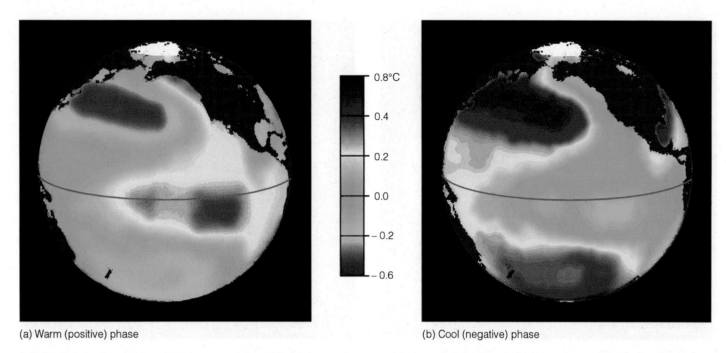

(a) Warm (positive) phase (b) Cool (negative) phase

0.8°C
0.4
0.2
0.0
-0.2
-0.6

● FIGURE 10.25 Typical winter sea-surface temperature departure from normal in °C during the Pacific Decadal Oscillation's warm phase (a) and cool phase (b). (*Source:* JISAO, University of Washington, obtained via http://www.tao.atmos.washington.edu/pdo. Used with permission of N. Mantua.)

ters tend to be drier over the Great Lakes, and cooler and wetter in the southern United States. Meanwhile, during this warm phase, salmon populations increase in Alaska and diminish along the Pacific Northwest coast. The latest warm phase began in 1977 and ended in the late 1990s.

The present cool (or negative) phase finds cooler-than-average surface water along the west coast of North America and an area of warmer-than-normal surface water extending from Japan into the central North Pacific (see Fig. 10.25b). Winters in the cool phase tend to be cooler and wetter than average over northwestern North America, wetter over the Great Lakes, and warmer and drier in the southern United States. Salmon fishing diminishes in Alaska and picks up along the Pacific Northwest Coast.

The climate patterns described so far only represent average conditions, as individual years within either phase may vary considerably. In fact, the Pacific Ocean temperature pattern in a particular phase may even reverse for a few years, as it did from 1958 to 1960. These small reversals can make it difficult to decipher exactly when the ocean temperature changes from one phase to another. Hopefully, as our understanding of the interactions between the ocean and atmosphere improves, climate forecasts across North America and elsewhere will improve as well.

NORTH ATLANTIC OSCILLATION   Over the Atlantic there is a reversal of pressure called the **North Atlantic Oscillation (NAO)** that has an effect on the weather in Europe and along the east coast of North America. For example, in winter if the atmospheric pressure in the vicinity of the Icelandic low

drops, and the pressure in the region of the Bermuda-Azores high rises, there is a corresponding large difference in atmospheric pressure between these two regions that strengthens the westerlies. The strong westerlies in turn direct strong storms on a more northerly track into northern Europe, where winters tend to be wet and mild. During this *positive phase* of the NAO, winters in the eastern United States tend to be wet and relatively mild, while northern Canada and Greenland are usually cold and dry (see ● Fig. 10.26a).

The *negative phase* of the NAO occurs when the atmospheric pressure in the vicinity of the Icelandic low rises, while the pressure drops in the region of the Bermuda high (see Fig. 10.26b). This pressure change results in a reduced pressure gradient and weaker westerlies that steer fewer and weaker winter storms across the Atlantic in a more westerly path. These storms bring wet weather to southern Europe and to the region around the Mediterranean Sea. Meanwhile, winters in Northern Europe are usually cold and dry, as are the winters along the east coast of North America. Greenland and northern Canada usually experience mild winters.

Although the NAO varies from year to year (and sometimes from month to month), it may exhibit a tendency to remain in one phase for several years. It is interesting to note that the NAO during the past 30 years or so has been trending toward a more positive phase.

ARCTIC OSCILLATION   Closely related to the North Atlantic Oscillation is the **Arctic Oscillation (AO),** where changes in atmospheric pressure between the Arctic and regions to the south cause changes in upper-level westerly winds. Dur-

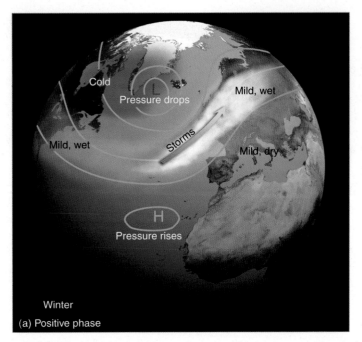

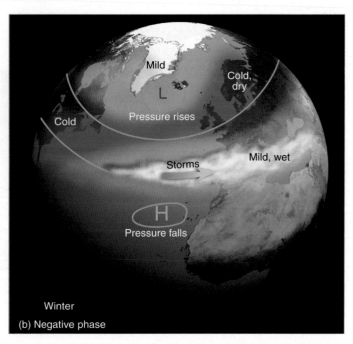

● FIGURE 10.26 Change in surface atmospheric pressure and typical winter weather patterns associated with the (a) positive phase and (b) negative phase of the North Atlantic Oscillation.

ing the *positive warm phase* of the *AO* (see ● Fig. 10.27a), strong pressure differences produce strong westerly winds aloft that prevent cold arctic air from invading the United States, and so winters in this region tend to be warmer than normal. With cold arctic air in place to the north, winters over Newfoundland and Greenland tend to be very cold.

Meanwhile, strong winds over the Atlantic direct storms into northern Europe, bringing with them wet, mild weather.

During the *negative cold phase* of the *AO* (Fig. 10.27b), small pressure differences between the arctic and regions to the south produce weaker westerly winds aloft. Cold arctic air is now able to penetrate farther south, producing colder than

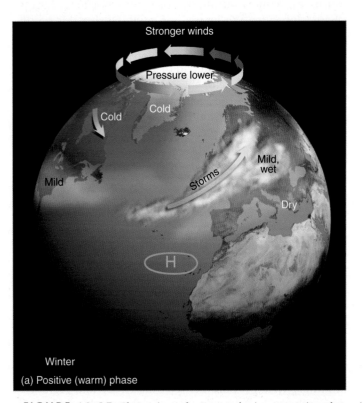

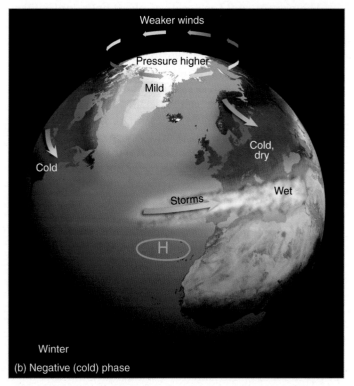

● FIGURE 10.27 Change in surface atmospheric pressure in polar regions and typical winter weather patterns associated with the (a) positive (warm) phase and the (b) negative (cold) phase of the Arctic Oscillation.

normal winters over much of the United States. Cold air also invades northern Europe and Asia, while Newfoundland and Greenland experience warmer than normal winters.

So when Greenland has mild winters, northern Europe has cold winters and vice versa. This seesaw in winter temperatures between Greenland and northern Europe has been known for many years. What was not known until recently is that during the warm Arctic Oscillation phase relatively warm, salty water from the Atlantic is able to move into the Arctic Ocean, where it melts sea ice, causing it to thin by more than 40 centimeters. During the cold phase, surface winds tend to keep warmer Atlantic water to the south, which promotes thicker sea ice. Although the Arctic Oscillation switches from one phase to another on an irregular basis, one phase may persist for several years in a row, bringing with it a succession of either cold or mild winters.

SUMMARY

In this chapter, we described the large-scale patterns of wind and pressure that persist around the world. We found that, at the surface in both hemispheres, the trade winds blow equatorward from the semipermanent high-pressure areas centered near 30° latitude. Near the equator, the trade winds converge along a boundary known as the intertropical convergence zone (ITCZ). On the poleward side of the subtropical highs are the prevailing westerly winds. The westerlies meet cold polar easterly winds along a boundary called the polar front, a zone of low pressure where middle-latitude cyclonic storms often form. The annual shifting of the major pressure areas and wind belts—northward in July and southward in January—strongly influences the annual precipitation of many regions.

Warm air aloft (high pressure) over low latitudes and cold air aloft (low pressure) over high latitudes produce westerly winds aloft in both hemispheres, especially at middle and high latitudes. Near the equator, easterly winds exist. The jet streams are located where strong winds concentrate into narrow bands. The polar jet stream forms in response to temperature contrasts along the polar front, while the subtropical jet stream forms at higher elevations above the subtropics, along an upper-level boundary called the subtropical front.

Near the surface, we examined the interaction between the atmosphere and oceans. We found the interaction to be an ongoing process where everything, in one way or another, seems to influence everything else. On a large scale, winds blowing over the surface of the water drive the major ocean currents; the oceans, in turn, release energy to the atmosphere, which helps to maintain the general circulation of winds. Where winds and the Ekman Spiral move surface water away from a coastline, cold, nutrient-rich water upwells to replace it, creating good fishing and cooler surface water.

When atmospheric circulation patterns change over the tropical Pacific, and the trade winds weaken or reverse direction, warm tropical water is able to flow eastward toward South America, where it chokes off upwelling and produces disastrous economic conditions. When the warm water extends over a vast area of the tropical Pacific, the warming is called a major El Niño event and the associated reversal of pressure over the Pacific Ocean is called the Southern Oscillation. The large-scale interaction between the atmosphere and the ocean during El Niño and the Southern Oscillation (ENSO) affects global atmospheric circulation patterns. The sweeping winds aloft provide too much rain in some areas and not enough in others. La Niña is the name given to the situation where the surface water of the central and eastern tropical Pacific turns cooler than normal.

Over the north-central Pacific and along the west coast of North America there is a reversal of surface water temperature, called the Pacific Decadal Oscillation that occurs every 20 to 30 years. Over the Atlantic Ocean there is a reversal of pressure called the North Atlantic Oscillation that influences weather in various regions of the world. Surface atmospheric pressure changes over the Arctic also seem to influence weather patterns over regions of the Northern Hemisphere. Studies now in progress are designed to determine how the interaction between the atmosphere and the ocean can influence climate patterns in various regions of the world.

KEY TERMS

The following terms are listed (with page numbers) in the order they appear in the text. Define each. Doing so will aid you in reviewing the material covered in this chapter.

general circulation of the atmosphere, 260
Hadley cell, 260
doldrums, 261
subtropical highs, 262
horse latitudes, 262
trade winds, 262
intertropical convergence zone (ITCZ), 262
westerlies, 262
polar front, 262
subpolar low, 262
polar easterlies, 262
semipermanent highs and lows, 262
Bermuda high, 262
Pacific high, 262
Icelandic low, 264
Aleutian low, 264
Siberian high, 264
jet streams, 268

subtropical jet stream, 268
polar front jet stream, 268
tropical easterly jet stream, 270
low-level jet, 272
oceanic front, 274
Ekman Spiral, 275
Ekman transport, 275
upwelling, 276
El Niño, 276
Southern Oscillation, 276
ENSO, 276
La Niña, 278
teleconnections, 279
Pacific Decadal Oscillation (PDO), 280
North Atlantic Oscillation (NAO), 281
Arctic Oscillation (AO), 281

QUESTIONS FOR REVIEW

1. Draw a large circle. Now, place the major surface pressure and wind belts of the world at their appropriate latitudes.

2. Explain how and why the average surface pressure features shift from summer to winter.

3. Why is it impossible on the earth for a Hadley cell to extend from the pole to the equator?

4. Along a meridian line running from the equator to the poles, how does the general circulation help to explain zones of abundant and sparse precipitation?

5. Most of the United States is in what wind belt?

6. Explain why the winds in the middle and upper troposphere tend to blow from west to east in both the Northern and the Southern Hemispheres.

7. How does the polar front influence the development of the polar front jet stream?

8. Describe how the conservation of angular momentum plays a role in the formation of a jet stream.

9. Why is the polar front jet stream stronger in winter than in summer?

10. Explain the relationship between the general circulation of air and the circulation of ocean currents.

11. List at least four important interactions that exist between the ocean and the atmosphere.

12. Describe how the Ekman Spiral forms.

13. What conditions are necessary for upwelling to occur along the west coast of North America? The east coast of North America?

14. (a) What is a major El Niño event?
    (b) What happens to the surface pressure at opposite ends of the Pacific Ocean during the Southern Oscillation?
    (c) Describe how the Southern Oscillation influences a major El Niño event.

15. What are the conditions over the tropical eastern and central Pacific Ocean during the phenomenon known as La Niña?

16. What type of weather (cold/warm, wet/dry) would you expect over North America during a strong El Niño? During a strong La Niña?

17. Describe the ocean surface temperatures associated with the Pacific Decadal Oscillation. What climate patterns (cool/warm, wet/dry) tend to exist during the warm phase and the cool phase?

18. How does the positive phase of the North Atlantic Oscillation differ from the negative phase?

19. During the negative cold phase of the Arctic Oscillation, when Greenland is experiencing mild winters, what type of winters (cold or mild) is northern Europe usually experiencing?

QUESTIONS FOR THOUGHT

1. What effect would continents have on the circulation of air in the single-cell model?

2. How would the general circulation of air appear in summer and winter if the earth were tilted on its axis at an angle of 45° instead of 23½°?

3. Summer weather in the southwestern section of the United States is influenced by a subtropical high-pressure cell, yet Fig. 10.3b (p. 263) shows an area of low pressure at the surface. Explain.

4. Explain why icebergs tend to move at right angles to the direction of the wind.

5. Over the open ocean in the vicinity of the Pacific high, observations have indicated that ozone concentrations hundreds of meters above the surface are greater than expected. Give a possible explanation for this.

6. Give *two* reasons why pilots would prefer to fly in the core of a jet stream rather than just above or below it.

7. Why do the major ocean currents in the North Indian Ocean reverse direction between summer and winter?

8. Explain why the surface water temperature along the northern California coast is warmer in winter than it is in summer.

9. You are given an upper-level map that shows the position of two jet streams. If one is the polar front jet and the other the subtropical jet, how would you be able to tell which is which?

10. The Coriolis force deflects moving water to the right of its intended path in the Northern Hemisphere and to the left of its intended path in the Southern Hemisphere. Why, then, does upwelling tend to occur along the western margin of continents in both hemispheres?

PROBLEMS AND EXERCISES

1. Locate the following cities on a world map. Then, based on the general circulation of surface winds, predict the prevailing wind for each one during July and January.
   (a) Nashville, Tennessee
   (b) Oklahoma City, Oklahoma
   (c) Melbourne, Australia
   (d) London, England
   (e) Paris, France
   (f) Reykjavik, Iceland
   (g) Fairbanks, Alaska
   (h) Seattle, Washington

2. In the next column is a list of average weather conditions that prevail during the month of July at San Francisco, California, and Atlantic City, New Jersey. Both cities lie adjacent to an ocean at nearly the same latitude; however, the average weather conditions vary greatly. In terms of the average surface winds and pressure systems (see

Fig. 10.3b, p. 263) and the interaction between the atmosphere and ocean, explain what accounts for the variation between the two cities of each weather element.

Atlantic City, New Jersey
(latitude 39°N)
Average weather, July

Temperature maximum	84°F
minimum	66°F
Dew point	64°F
Precipitation	3.72 in.
Prevailing wind	S
Weather: Clear to partly cloudy with the possibility of afternoon showers	
Water temperature	70°F

San Francisco, California
(latitude 37°N)
Average weather, July

Temperature maximum	64°F
minimum	53°F
Dew point	53°F
Precipitation	0.01 in.
Prevailing wind	NW
Weather: Fog and low clouds during the night and morning with afternoon clearing	
Water temperature	53°F

Clouds developing along an approaching cold front.

© C. Donald Ahrens

CHAPTER 11

Air Masses and Fronts

About two o'clock in the afternoon it began to grow dark from a heavy, black cloud which was seen in the northwest. Almost instantly the strong wind, traveling at the rate of 70 miles an hour, accompanied by a deep bellowing sound, with its icy blast, swept over the land, and everything was frozen hard. The water in the little ponds in the roads froze in waves, sharp edged and pointed, as the gale had blown it. The chickens, pigs and other small animals were frozen in their tracks. Wagon wheels ceased to roll, froze to the ground. Men, going from their barns or fields a short distance from their homes, in slush and water, returned a few minutes later walking on the ice. Those caught out on horseback were frozen to their saddles, and had to be lifted off and carried to the fire to be thawed apart. Two young men were frozen to death near Rushville. One of them was found with his back against a tree, with his horse's bridle over his arm and his horse frozen in front of him. The other was partly in a kneeling position, with a tinder box in one hand and a flint in the other, with both eyes wide open as if intent on trying to strike a light. Many other casualties were reported. As to the exact temperature, however, no instrument has left any record; but the ice was frozen in the stream, as variously reported, from six inches to a foot in thickness in a few hours.

John Moses, *Illinois: Historical and Statistical*

  CONTENTS

The opening details the passage of a spectacular cold front as it moved through Illinois on December 21, 1836. Although no reliable temperature records are available, estimates are that, as the front swept through, air temperatures dropped almost instantly from the balmy 40s (°F) to 0 degrees. Fortunately, temperature changes of this magnitude with cold fronts are quite rare.

In this chapter, we will examine the more typical weather associated with cold fronts and warm fronts. We will address questions such as: Why are cold fronts usually associated with showery weather? How can warm fronts during the winter cause freezing rain and sleet to form over a vast area? And how can one read the story of an approaching warm front by observing its clouds? But, first, so that we may better understand fronts, we will examine air masses. We will look at where and how they form and the type of weather usually associated with them.

Air Masses

An **air mass** is an extremely large body of air whose properties of temperature and humidity are fairly similar in any horizontal direction at any given altitude. Air masses may cover many thousands of square kilometers. In ●Fig. 11.1, a large winter air mass, associated with a high-pressure area, covers over half of the United States. Note that, although the surface air temperature and dew point vary somewhat, everywhere the air is cold and dry, with the exception of the zone of snow showers on the eastern shores of the Great Lakes. This cold, shallow anticyclone will drift eastward, carrying

with it the temperature and moisture characteristic of the region where the air mass formed; hence, in a day or two, cold air will be located over the central Atlantic Ocean. Part of weather forecasting is, then, a matter of determining air mass characteristics, predicting how and why they change, and in what direction the systems will move.

SOURCE REGIONS   Regions where air masses originate are known as **source regions**. In order for a huge mass of air to develop uniform characteristics, its source region should be generally flat and of uniform composition with light surface winds. The longer the air remains stagnant over its source region, or the longer the path over which the air moves, the more likely it will acquire properties of the surface below. Consequently, ideal source regions are usually those areas dominated by surface high pressure. They include the ice- and snow-covered arctic plains in winter and subtropical oceans in summer. The middle latitudes, where surface temperatures and moisture characteristics vary considerably, are not good source regions. Instead, this region is a transition zone where air masses with different physical properties move in, clash, and produce an exciting array of weather activity.

CLASSIFICATION   Air masses are usually classified according to their temperature and humidity, both of which usually remain fairly uniform in any horizontal direction.* There are cold and warm air masses, humid and dry air masses. Air

---

*In classifying air masses, it is common to use the *potential temperature* of the air. The potential temperature is the temperature that unsaturated (dry) air would have if moved from its original level to a pressure of 1000 millibars at the dry adiabatic rate (10°C/1000 m).

● FIGURE II.I Here, a large, extremely cold winter air mass is dominating the weather over much of the United States. At almost all cities, the air is cold and dry. Upper number is air temperature (°F); bottom number is dew point (°F).

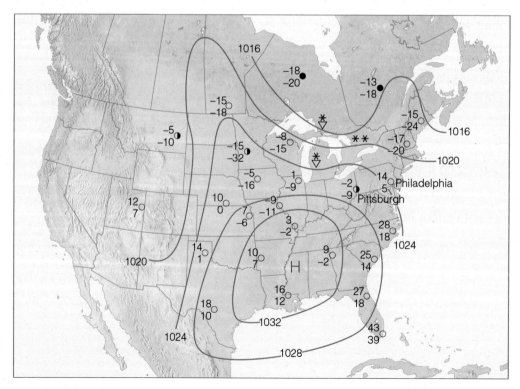

SOURCE REGION	ARCTIC REGION (A)	POLAR (P)	TROPICAL (T)
Land	cA	cP	cT
Continental(c)	extremely cold, dry stable; ice- and snow-covered surface	cold, dry, stable	hot, dry, stable air aloft; unstable surface air
Water		mP	mT
Maritime (m)		cool, moist, unstable	warm, moist; usually unstable

masses are grouped into five general categories according to their source region. Air masses that originate in polar latitudes are designated by the capital letter "P" (for *polar*); those that form in warm tropical regions are designated by the capital letter "T" (for *tropical*). If the source region is land, the air mass will be dry and the lowercase letter "c" (for *continental*) precedes the P or T. If the air mass originates over water, it will be moist—at least in the lower layers—and the lowercase letter "m" (for *maritime*) precedes the P or T. We can now see that polar air originating over land will be classified cP on a surface weather map, whereas tropical air originating over water will be marked as mT. In winter, an extremely cold air mass that forms over the arctic is designated as cA, *continental arctic*. Sometimes, however, it is difficult to distinguish between arctic and polar air masses, especially when the arctic air mass has traveled over warmer terrain. ▼ Table 11.1 lists the five basic air masses.

After the air mass spends some time over its source region, it usually begins to move in response to the winds aloft.

As it moves away from its source region, it encounters surfaces that may be warmer or colder than itself. When the air mass is colder than the underlying surface, it is warmed from below, which produces instability at low levels. In this case, increased convection and turbulent mixing near the surface usually produce good visibility, cumuliform clouds, and showers of rain or snow. On the other hand, when the air mass is warmer than the surface below, the lower layers are chilled by contact with the cold earth. Warm air above cooler air produces stable air with little vertical mixing. This situation causes the accumulation of dust, smoke, and pollutants, which restricts surface visibilities. In moist air, stratiform clouds accompanied by drizzle or fog may form.

AIR MASSES OF NORTH AMERICA   The principal air masses (with their source regions) that enter the United States are shown in ● Fig. 11.2. We are now in a position to study the formation and modification of each of these air masses and the variety of weather that accompanies them.

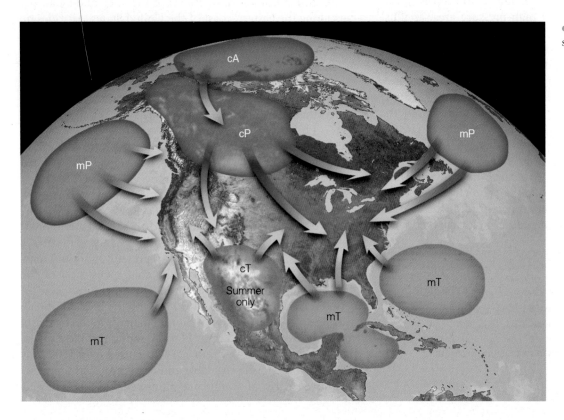

● FIGURE 11.2 Air mass source regions and their paths.

● FIGURE 11.3 A shallow but large dome of extremely cold air—a continental arctic air mass—moves slowly southeastward across the upper plains. The leading edge of the air mass is marked by a cold front. (Numbers represent air temperature, °F.)

cP (Continental Polar) and cA (Continental Arctic) Air Masses

The bitterly cold weather that invades southern Canada and the United States in winter is associated with **continental polar** and **continental arctic air masses.** These air masses originate over the ice- and snow-covered regions of the arctic, northern Canada and Alaska where long, clear nights allow for strong radiational cooling of the surface. Air in contact with the surface becomes quite cold and stable. Since little moisture is added to the air, it is also quite dry, and dew-point temperatures are often less than −30°C (−22°F). Eventually a portion of this cold air breaks away and, under the influence of the air flow aloft, moves southward as an enormous shallow high-pressure area, as illustrated in ● Fig. 11.3.

As the cold air moves into the interior plains, there are no topographic barriers to restrain it, so it continues southward, bringing with it cold wave warnings and frigid temperatures. The infamous Texas norther is associated with cP (and cA) air. As the air mass moves over warmer land to the south, the air temperature moderates slightly as it is heated from below. However, even during the afternoon, when the surface air is most unstable, cumulus clouds are rare because of the extreme dryness of the air mass. At night, when the winds die down, rapid radiational surface cooling and clear skies combine to produce low minimum temperatures. If the cold air moves as far south as central or southern Florida, the winter

vegetable crop may be severely damaged. When the cold, dry air mass moves over a relatively warm body of water, such as the Great Lakes, heavy snow showers—called **lake-effect snows**—often form on the eastern shores. (More information on lake-effect snows is provided in the Focus section on p. 291.)

In winter, the generally fair weather accompanying cP and cA air is due to the stable nature of the atmosphere aloft. Sinking air develops above the large dome of high pressure. The subsiding air warms by compression and creates warmer air, which lies above colder surface air. Therefore, a strong upper-level subsidence inversion often forms. Should the anticyclone stagnate over a region for several days, the visibility gradually drops as pollutants become trapped in the cold air near the ground. Usually, however, winds aloft move the cold air mass either eastward or southeastward.

The Rockies, Sierra Nevada, and Cascades normally protect the Pacific Northwest from the onslaught of cP and cA air, but, occasionally, very cold air masses do invade these regions. When the upper-level winds over Washington and Oregon blow from the north or northeast on a trajectory beginning over northern Canada or Alaska, cold cP (and cA) air can slip over the mountains and extend its icy fingers all the way to the Pacific Ocean. As the air moves off the high plateau, over the mountains, and on into the lower valleys, compressional heating of the sinking air causes its temperature to rise, so that by the time it reaches the lowlands, it is considerably warmer than it was originally. However, in no way would this air be considered warm. In some cases, the subfreezing temperatures slip over the Cascades and extend southward into the coastal areas of southern California.

A similar but less dramatic warming of cP and cA air occurs along the eastern coast of the United States. Air rides up and over the lower Appalachian Mountains. Turbulent mixing and compressional heating increase the air temperatures on the downwind side. Consequently, cities located to the east of the Appalachian Mountains usually do not experience temperatures as low as those on the west side. In Fig. 11.1, notice that for the same time of day—in this case 7 A.M. EST—Philadelphia, with an air temperature of 14°F, is 16°F warmer than Pittsburgh, with an air temperature of −2°F.

● Figure 11.4 shows two upper-air wind patterns that led to extremely cold outbreaks of arctic air during December 1989 and 1990. Upper-level winds typically blow from west to east, but, in both of these cases, the flow, as given by the heavy, dark arrows, had a strong north-south (meridional) trajectory. The H represents the positions of the cold surface anticyclones. Numbers on the map represent minimum temperatures (°F) recorded during the cold spells. East of the Rocky Mountains, over 350 record low temperatures were set between December 21 and 24, 1989, with the arctic outbreak causing an estimated $480 million in damage to the fruit and vegetable crops in Texas and Florida. Along the West Coast, the frigid air during December, 1990, caused over $300 million in damage to the vegetable and citrus crops, as temperatures over parts of California plummeted to their lowest

FOCUS ON A SPECIAL TOPIC

Lake-Effect (Enhanced) Snows

During the winter, when the weather in the Midwest is dominated by clear and cold polar or arctic air, people living on the eastern shores of the Great Lakes brace themselves for heavy snow showers. Snowstorms that form on the downwind side of one of these lakes are known as *lake-effect snows*. Since the lakes are responsible for enhancing the amount of snow that falls on its downwind side, these snowstorms are also called *lake-enhanced snows*, especially when the snow accompanies a cold front or mid-latitude cyclone. These storms are highly localized, extending from just a few kilometers to more than 100 km inland. The snow usually falls as a heavy shower or squall in a concentrated zone. So centralized is the region of snowfall, that one part of a city may accumulate many inches of snow, while, in another part, the ground is bare.

Lake-effect snows are most numerous from November to January. During these months, cold air moves over the lakes when they are relatively warm and not quite frozen. The contrast in temperature between water and air can be as much as 25°C (45°F). Studies show that the greater the contrast in temperature, the greater the potential for snow showers. In Fig. 1 we can see that, as the cold air moves over the warmer water, the air mass is quickly warmed from below, making it more buoyant and less stable. Rapidly, the air sweeps up moisture, soon becoming saturated. Out over the water, the vapor condenses into steam fog. As the air continues to warm, it rises and forms billowing cumuliform clouds, which continue to grow as the air becomes more unstable. Eventually, these clouds produce heavy showers of snow, which make the lake seem like a snow factory. Once the air and clouds reach the downwind side of the lake, additional lifting is provided by low hills and the convergence of air as it slows down over the rougher terrain. In late winter, the frequency and intensity of lake-effect snows often taper off as the temperature contrast between water and air diminishes and larger portions of the lakes freeze.

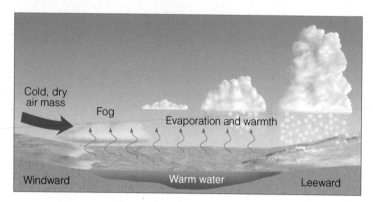

● FIGURE 1 The formation of lake-effect snows. Cold, dry air crossing the lake gains moisture and warmth from the water. The more buoyant air now rises, forming clouds that deposit large quantities of snow on the lake's leeward (downwind) shores.

Generally, the longer the stretch of water over which the air mass travels (the longer the fetch), the greater the amount of warmth and moisture derived from the lake, and the greater the potential for heavy snow showers. In fact, studies show that, for significant snowfall to occur, the air must move across 80 km (50 mi) of open water. Consequently, forecasting lake-effect snowfalls depends to a large degree on determining the trajectory of the air as it flows over the lake. Regions that experience heavy lake-effect snowfalls are shown in Fig. 2.*

As the cold air moves farther east, the heavy snow showers usually taper off; however, the western slope of the Appalachian Mountains produces further lifting, enhancing the possibility of more and heavier showers. The heat given off during condensation warms the air and, as the air descends the eastern slope, compressional heating warms it even more. Snowfall ceases, and by the time the air arrives in Philadelphia, New York, or Boston, the only remaining trace of the snow showers occurring

● FIGURE 2 Areas shaded white show regions that experience heavy lake-effect snows.

on the other side of the mountains are the puffy cumulus clouds drifting overhead.

Lake-effect snows are not confined to the Great Lakes. In fact, any large unfrozen lake (such as the Great Salt Lake) can enhance snowfall when cold, relatively dry air sweeps over it. Moreover, a type of lake-effect snow occurs when cold air moves over a relatively warm ocean, then lifts slightly as it moves over a landmass. Such *ocean-effect snows* are common over Cape Cod, Massachusetts, in winter.

---

*Buffalo, New York, is a city that experiences heavy lake-effect snows. Visit the National Weather Service website in Buffalo at http://www.erh.noaa.gov/buf/lakeffect/indexlk.html and read about lake-effect snowstorms measured in feet, as well as interesting weather stories.

● FIGURE 11.4 Average upper-level wind flow (heavy arrows) and surface position of anticyclones (H) associated with two extremely cold outbreaks of arctic air during December. Numbers on the map represent minimum temperatures (°F) measured during each cold snap.

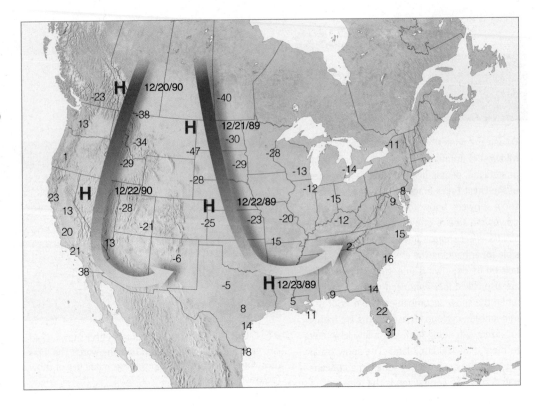

readings in more than fifty years. Notice in both cases how the upper-level wind directs the paths of the air masses.

The cP air that moves into the United States in summer has properties much different from its winter counterpart. The source region remains the same but is now characterized by long summer days that melt snow and warm the land. The

air is only moderately cool, and surface evaporation adds water vapor to the air. A summertime cP air mass usually brings relief from the oppressive heat in the central and eastern states, as cooler air lowers the air temperature to more comfortable levels. Daytime heating warms the lower layers, producing surface instability. The water vapor in the rising air may condense and create a sky dotted with fair weather cumulus clouds (cumulus humilis). A typical profile of temperatures for a summer and a winter cP air mass is given in ●Fig. 11.5. Notice that the strong inversion so prevalent in winter is absent in summer.

When an air mass moves over a large body of water, its original properties may change considerably. For instance, cold, dry cP air moving over the Gulf of Mexico warms rapidly and gains moisture. The air quickly assumes the qualities of a maritime air mass. Notice in ●Fig. 11.6 that rows of cumulus clouds (cloud streets) are forming over the Gulf of Mexico parallel to northerly surface winds as cP air is being warmed by the water beneath it, causing the air mass to destabilize. As the air continues its journey southward into Mexico and Central America, strong, moist northerly winds build into heavy clouds and showers along the northern coast. Hence, a once cold, dry, and stable air mass can be modified to such an extent that its original characteristics are no longer discernible. When this happens, the air mass is given a new designation.

Notice also in Fig. 11.6 that a similar modification of cP air is occurring along the Atlantic Coast, as northwesterly winds are blowing over the mild Atlantic. When this air encounters the much warmer Gulf Stream water, it warms rapidly and becomes conditionally unstable. Vertical mixing brings down faster-flowing cold air from aloft. This mixing

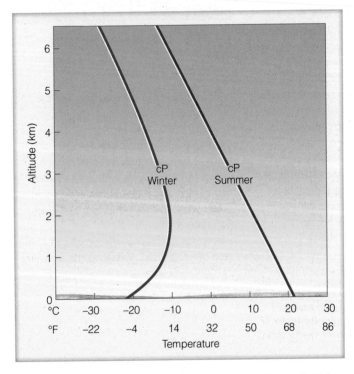

● FIGURE 11.5 Typical vertical temperature profile over land for a summer and a winter cP air mass.

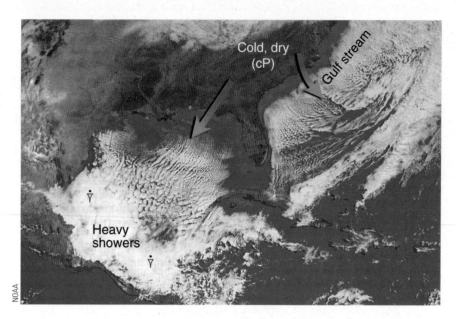

● FIGURE 11.6 Visible satellite image showing the modification of cold continental polar air as it moves over the warmer Gulf of Mexico and the Atlantic Ocean.

creates strong, gusty surface winds and choppy seas, which can be hazardous to shipping.

In summary, polar and arctic air masses are responsible for the bitter cold winter weather that can cover wide sections of North America. When the air mass originates over the Canadian Northwest Territories, frigid air can bring record-breaking low temperatures. Such was the case on Christmas Eve, 1983, when arctic air covered most of North America. (A detailed look at this air mass and its accompanying record-setting low temperatures is given in the Focus section on p. 294.)

mP (Maritime Polar) Air Masses   During the winter, cP and cA air originating over Asia and frozen polar regions is carried eastward and southward over the Pacific Ocean by the circulation around the Aleutian low. The ocean water modifies these cold air masses by adding warmth and moisture to them. Since this air travels across many hundreds or even thousands of kilometers of water, it gradually changes into *maritime polar air.*

By the time this **maritime polar air mass** reaches the Pacific Coast, it is cool, moist, and conditionally unstable. The ocean's effect is to keep air near the surface warmer than the air aloft. Temperature readings in the 40s and 50s (°F) are common near the surface, while air at an altitude of about a kilometer or so above the surface may be at the freezing point. Within this colder air, characteristics of the original cold, dry air mass may still prevail. As the air moves inland, coastal mountains force it to rise, and much of its water vapor condenses into rain-producing clouds. In the colder air aloft, the rain changes to snow, with heavy amounts accumulating in mountain regions. Over the relatively warm open ocean, the cool moist air mass produces cumulus clouds that show up as tiny white splotches on a visible satellite image (see ● Fig. 11.7).

When the maritime polar air moves inland, it loses much of its moisture as it crosses a series of mountain ranges. Be-

yond these mountains, it travels over a cold, elevated plateau that chills the surface air and slowly transforms the lower level into dry, stable continental polar air. East of the Rockies this air mass is referred to as *Pacific air* (see ● Fig. 11.8). Here, it often brings fair weather and temperatures that are cool but not nearly as cold as the continental polar and arctic air that invades this region from northern Canada. In fact, when

● FIGURE 11.7
Clouds and air flow aloft (large blue arrow) associated with maritime polar air moving into California. The large L shows the position of an upper-level low. Regions experiencing precipitation are also shown. The small, white clouds over the open ocean are cumulus clouds forming in the conditionally unstable air mass. (Precipitation symbols are given in Appendix B.)

The Return of the Siberian Express

The winter of 1983–1984 was one of the coldest on record across North America. Unseasonably cold weather arrived in December, which, for much of the United States, was one of the coldest Decembers since records have been kept. During the first part of the month, continental polar air covered most of the northern and central plains. As the cold air moderated slightly, far to the north a huge mass of bitter cold arctic air was forming over the frozen reaches of the Canadian Northwest Territories.

By midmonth, the frigid air, associated with a massive high-pressure area, covered all of northwest Canada. Meanwhile, an upper-level ridge was forming over Alaska. On the eastern side of the ridge, strong northerly winds associated with the polar jet stream directed the frigid air southward over the prairie provinces of Canada. A portion of the extraordinarily cold air broke away, and, like a large swirling bubble, moved as a cold, shallow anticyclone southward into the United States. Because the frigid air was accompanied in some regions by winds gusting to 45 knots, at least one news reporter dubbed the onslaught of this arctic blast, "the Siberian Express."

The express dropped temperatures to some of the lowest readings ever recorded during the month of December. On December 22, Elk Park, Montana, recorded an unofficial low of −64°F, only 6°F higher than the all-time low of −70°F for the United States (excluding Alaska) recorded at Rogers Pass, Montana, on January 20, 1954.

The center of the massive anticyclone gradually pushed southward out of Canada. By December 24, its center was over eastern Montana (Fig. 3), where the sea-level pressure at Miles City reached an incredible 1064 mb (31.40 in.)—a new United States' record excluding Alaska. An enormous ridge of high pressure stretched from the Canadian arctic coast to the Gulf of Mexico. On the east side of the ridge, cold westerly winds brought lake-effect snows to the eastern shores of the Great Lakes. To the south of the high-pressure center, cold easterly winds, rising along the elevated plains, brought light amounts of *upslope*

---

*Upslope snow forms as cold air moving from east to west over the Great Plains gradually rises (and cools even more) as it approaches the Rocky Mountains.

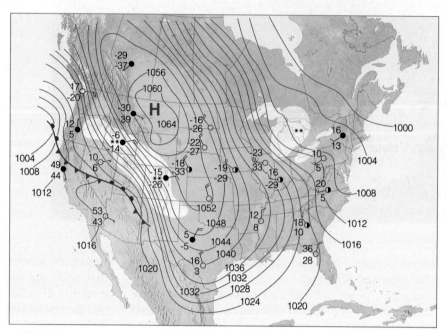

● FIGURE 3 Surface weather map for 7 A.M., EST, December 24, 1983. Solid lines are isobars. Areas shaded white represent snow. An extremely cold arctic air mass covers nearly 90 percent of the United States. (Weather symbols for the surface map are given in Appendix B.)

*snow** to sections of the Rocky Mountain states. Notice in Fig. 3 that, on Christmas Eve, arctic air covered almost 90 percent of the United States. As the cold air swept eastward and southward, a hard freeze caused hundreds of millions of dollars in damage to the fruit and vegetable crops in Texas, Louisiana, and Florida. On Christmas Day, 125 record low temperature readings were set in twenty-four states. That afternoon, at 1:00 P.M., it was actually colder in Atlanta, Georgia, at 9°F than it was in Fairbanks, Alaska (10°F). One of the worst cold waves to occur in December during the twentieth century continued through the week, as many new record lows were established in the Deep South from Texas to Louisiana.

By January 1, the extreme cold had moderated, as the upper-level winds became more westerly. These winds brought milder Pacific air eastward into the Great Plains. The warmer pattern continued until about January 10, when the Siberian Express decided to make a return visit. Driven by strong upper-level northerly winds, impulse after impulse of arctic air from Canada swept across the United States. On January 18, an all-time record low of −65°F

was recorded for the state of Utah at Middle Sinks. On January 19, temperatures plummeted to a new low of −7°F for the airports in Philadelphia and Baltimore. Toward the end of the month, the upper-level winds once again became more westerly. Over much of the nation, the cold air moderated. But the express was to return at least one more time.

The beginning of February saw relatively warm air covering much of the nation from California to the Atlantic coast. On February 4, an arctic outbreak spread southward and eastward across the United States. Although freezing air extended southward into central Florida, the express ran out of steam, and a February heat wave soon engulfed most of the states east of the Rocky Mountains as warm, humid air from the Gulf of Mexico spread northward.

Even though February was one of the warmest months on record over parts of the United States, the winter of 1983–1984 (December, January, and February) will go down in the record books as one of the coldest winters for the United States as a whole since reliable record keeping began in 1931.

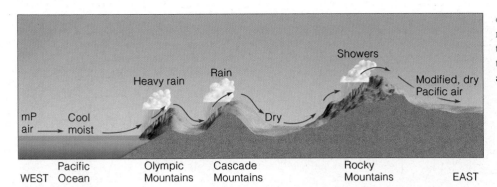

● FIGURE 11.8 After crossing several mountain ranges, cool moist mP air from off the Pacific Ocean descends the eastern side of the Rockies as modified, relatively dry Pacific air.

Pacific air from the west replaces retreating cold air from the north, chinook winds often develop. Furthermore, when the modified maritime polar air replaces moist tropical air, storms can form along the boundary separating the two air masses.

Along the East Coast, mP air originates in the North Atlantic as continental polar air moves southward some distance off the Atlantic Coast. (Look back at Fig. 11.2, p. 289.) Steered by northeasterly winds, mP air then swings southwestward toward the northeastern states. Because the water of the North Atlantic is very cold and the air mass travels only a short distance, wintertime Atlantic mP air masses are usually much colder than their Pacific counterparts. Because the prevailing winds aloft are westerly, Atlantic mP air masses are also much less common.

● Figure 11.9 illustrates a typical late winter or early spring surface weather pattern that carries mP air from the Atlantic into the New England and middle Atlantic states. A slow-moving, cold anticyclone drifting to the east (north of New England) causes a northeasterly flow of mP air to the south. The boundary separating this invading colder air from warmer air even farther south is marked by a stationary front. North of this front, northeasterly winds provide generally undesirable weather, consisting of damp air and low, thick clouds from which light precipitation falls in the form of rain, drizzle, or snow. When upper atmospheric conditions are right, storms may develop along the stationary front, move eastward, and intensify near the shores of Cape Hatteras. Such storms, called *Hatteras lows,* sometimes swing northeastward along the coast, where they become *northeasters* (commonly called *nor'easters*) bringing with them strong northeasterly winds, heavy rain or snow, and coastal flooding. (Such developing storms will be treated in detail in Chapter 12.)

mT (Maritime Tropical) Air Masses The wintertime source region for Pacific **maritime tropical** air masses is the subtropical east Pacific Ocean. (Look back at Fig. 11.2, p. 289.) Air from this region must travel over 1600 km of water before it reaches the southern California coast. Consequently, these air masses are very warm and moist by the time they arrive along the West Coast. In winter, the warm air produces heavy precipitation usually in the form of rain, even at high elevations. Melting snow and rain quickly fill rivers,

which overflow into the low-lying valleys. The rapid snow-melt leaves local ski slopes barren, and the heavy rain can cause disastrous mud slides in the steep canyons.

● Figure 11.10 shows maritime tropical air (usually referred to as *subtropical air*) streaming into northern California on January 1, 1997. The humid, subtropical air, which originated near the Hawaiian Islands, was termed by at least one forecaster as "*the pineapple express.*" After battering the Pacific Northwest with heavy rain, the pineapple express roared into northern and central California, causing catastrophic floods that sent over 100,000 people fleeing from their homes, mud slides that closed roads, property damage

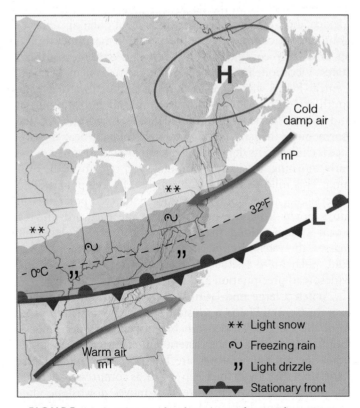

● FIGURE 11.9 Winter and early spring surface weather pattern that usually prevails during the invasion of cold, moist mP air into the mid-Atlantic and New England states. (Green-shaded area represents light rain and drizzle; pink-shaded region represents freezing rain and sleet; white-shaded area is experiencing snow.)

● FIGURE 11.10
An infrared satellite image
that shows maritime
tropical air (heavy yellow
arrow) moving into
northern California on
January 1, 1997. The
warm, humid airflow
(sometimes called "the
pineapple express")
produced heavy rain and
extensive flooding in
northern and central
California.

(including crop losses) that amounted to more than $1.5 billion, and eight fatalities. Yosemite National Park, which sustained over $170 million in damages due mainly to flooding, was forced to close for more than two months.

The warm, humid subtropical air that influences much of the weather east of the Rockies originates over the Gulf of Mexico and Caribbean Sea. In winter, cold polar air tends to dominate the continental weather scene, so maritime tropical air is usually confined to the Gulf and extreme southern states. Occasionally, a slow-moving storm system over the Central Plains draws warm, humid air northward. Gentle south or southwesterly winds carry this air into the central and eastern parts of the United States in advance of the storm. Since the land is still extremely cold, air near the surface is chilled to its dew point. Fog and low clouds form in the early morning, dissipate by midday, and reform in the evening. This mild winter weather in the Mississippi and Ohio valleys lasts, at best, only a few days. Soon cold polar air will move down from the north behind the eastward-moving storm system. Along the boundary between the two air masses, the warm, humid air is lifted above the more dense cold, polar air—a situation that often leads to heavy and widespread precipitation and storminess.

When a large mid-latitude cyclonic storm system stalls over the Central Plains, a constant supply of warm, humid air from the Gulf of Mexico can bring record-breaking maximum temperatures to the eastern half of the country. Sometimes the air temperatures are higher in the mid-Atlantic states than they are in the Deep South, as compressional heating warms the air even more as it moves downslope after crossing the Appalachian Mountains.

● Fig. 11.11 shows a surface weather map and the associated upper airflow (heavy arrow) that brought unseasonably warm maritime tropical air into the central and eastern states during April, 1976. A large surface high-pressure area centered off the southeast coast coupled with a strong southwest-

erly flow aloft carried warm moist air into the Midwest and East, causing a record-breaking April heat wave. The flow aloft prevented the surface low and the cold polar (cP) air behind it from making much eastward progress, so that the warm spell lasted for five days. Providence, Rhode Island, experienced a record-breaking high temperature for April of 96°F. Note that, on the west side of the surface low, the winds aloft funneled cold air from the north into the western states, creating unseasonably cold weather from California to the Rockies. Hence, while people in the Southwest were huddled around heaters, others several thousand kilometers away in the Northeast were turning on air conditioners. We can see that it is the upper-level meridional flow, directing cold polar air southward and warm subtropical air northward that makes these contrasts in temperature possible.

In summer, the circulation of air around the Bermuda High (which sits off the southeast coast of North America—see Fig. 10.3b, p. 263) pumps warm, humid (mP) air northward from off the Gulf of Mexico and from off the Atlantic Ocean into the eastern half of the United States. As this humid air moves inland, it warms even more, rises, and frequently condenses into cumuliform clouds, which produce afternoon showers and thunderstorms. You can almost count on thunderstorms developing along the Gulf Coast every summer afternoon. As evening approaches, thunderstorm activity typically dies off. Nighttime cooling lowers the temperature of this hot, muggy air only slightly. Should the air become saturated, fog or low clouds usually form, and these normally dissipate by late morning as surface heating warms the air again.

A weak, but often persistent, flow around an upper-level anticyclone in summer will spread warm, humid air from the Gulf of Mexico and from the Gulf of California into the southern and central Rockies, where it causes afternoon thunderstorms. Occasionally, this easterly flow may work its way even farther west, producing shower activity in the otherwise dry southwestern desert.

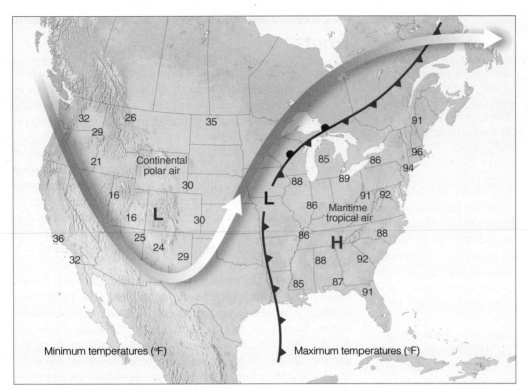

Minimum temperatures (°F)   Maximum temperatures (°F)

● FIGURE 11.11 Weather conditions during an unseasonably hot spell in the eastern portion of the United States that occurred between the 15th and 20th of April, 1976. The surface low-pressure area and fronts are shown for April 17. Numbers to the east of the surface low (in red) are maximum temperatures recorded during the hot spell, while those to the west of the low (in blue) are minimum temperatures reached during the same time period. The heavy arrow is the average upper-level flow during the period. The purple L and H show average positions of the upper-level trough and ridge.

During the summer, humid subtropical air originating over the southeastern Pacific and Gulf of California normally remains south of California. Occasionally, an upper-level southerly flow will spread this humid air northward into the southwestern United States, most often Arizona, Nevada, and the southern part of California. In many places, the moist, conditionally unstable air aloft only shows up as middle and high cloudiness, especially altocumulus and cirrocumulus castellanus. However, where the moist flow meets a mountain barrier, it usually rises and condenses into towering, shower-producing clouds. (For an exceptionally strong flow of subtropical air into this region, look at Fig. 9.32 on p. 244.)

cT (Continental Tropical) Air Masses The only real source region for hot, dry **continental tropical** air masses in North America is found during the summer in northern Mexico and the adjacent arid southwestern United States. Here, the air mass is hot, dry, and conditionally unstable at low levels, with frequent dust devils forming during the day. Because of the low relative humidity (typically less than 10 percent during the afternoon), air must rise over 3000 m (10,000 ft) before condensation begins. Furthermore, an upper-level ridge usually produces weak subsidence over the region, tending to make the air aloft rather stable and the surface air even warmer. Consequently, skies are generally clear, the weather is hot, and rainfall is practically nonexistent where continental tropical air masses prevail. If this air mass moves outside its source region and into the Great Plains and stagnates over that region for any length of time, a severe drought may result. ●Figure 11.12 shows a weather map situation where continental tropical air produces hot, dry weather over a large portion of the southwestern United States during July, 2005.

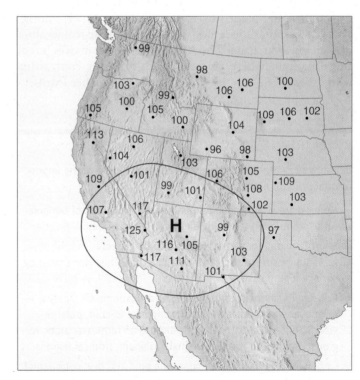

● FIGURE 11.12 From July 14 through July 22, 2005, continental tropical air covered a large area of the southwestern United States. Numbers on the map represent maximum temperatures (°F) during this period. The large H with the isobar shows the upper-level position of the subtropical high. Sinking air associated with the high contributed to the hot weather. Winds aloft were weak, with the main flow over central Canada.

So far, we have examined the various air masses that enter North America annually. The characteristics of each depend upon the air mass source region and the type of surface over which the air mass moves. The winds aloft determine the trajectories of these air masses. Occasionally, an air mass will control the weather in a region for some time. These persistent weather conditions are sometimes referred to as *airmass weather.*

Airmass weather is especially common in the southeastern United States during summer as, day after day, humid subtropical air from the Gulf brings sultry conditions and afternoon thunderstorms. It is also common in the Pacific Northwest in winter when conditionally unstable, cool moist air accompanied by widely scattered showers dominates the weather for several days or more. The real weather action, however, usually occurs not within air masses but at their margins, where air masses with sharply contrasting properties meet—in the zone marked by weather fronts.*

BRIEF REVIEW

Before we examine fronts, here is a review of some of the important facts about air masses:

- An air mass is a large body of air whose properties of temperature and humidity are fairly similar in any horizontal direction.

- Source regions for air masses tend to be generally flat, of uniform composition, and in an area of light winds, dominated by surface high pressure.

- Continental air masses form over land. Maritime air masses form over water. Polar air masses originate in cold, polar regions, and extremely cold arctic air masses form over arctic regions. Tropical air masses originate in warm, tropical regions.

- Continental polar (cP) air masses are cold and dry; continental arctic (cA) air masses are extremely cold and dry; continental tropical (cT) air masses are hot and dry; maritime tropical (mT) air masses are warm and moist; maritime polar (mP) air masses are cold and moist.

---

*The word *front* is used to denote the clashing or meeting of two air masses, probably because it resembles the fighting in Western Europe during World War I, when the term originated.

Fronts

Although we briefly looked at fronts in Chapter 1, we are now in a position to study them in depth, which will aid us in forecasting the weather. We will now learn about the general nature of fronts—how they move and what weather patterns are associated with them.

A **front** is the transition zone between two air masses of different densities. Since density differences are most often caused by temperature differences, fronts usually separate air masses with contrasting temperatures. Often, they separate air masses with different humidities as well. Remember that air masses have both horizontal and vertical extent; consequently, the upward extension of a front is referred to as a *frontal surface,* or a *frontal zone.*

● Figure 11.13 illustrates the vertical extent of two frontal zones—the *polar front* and the *arctic front.* The polar front boundary, which extends upward to over 5 km, separates warm, humid air to the south from cold polar air to the north. The arctic front, which separates cold air from extremely cold arctic air, is much more shallow than the polar front and only extends upward to an altitude of about one or two kilometers. In the next several sections, as we examine fronts on a flat surface weather map, keep in mind that all fronts have horizontal and vertical extent.

● Figure 11.14 shows a surface weather map illustrating four different fronts. Notice that the fronts are associated with lower pressure and that the fronts separate differing air masses. As we move from west to east across the map, the fronts appear in the following order: a stationary front between points A and B; a cold front between points B and C; a warm front between points C and D; and an occluded front between points C and L. Let's examine the properties of each of these fronts.

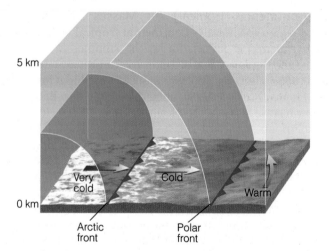

● FIGURE II.13 The polar front represents a cold frontal boundary that separates colder air from warmer air at the surface and aloft. The more shallow arctic front separates cold air from extremely cold air.

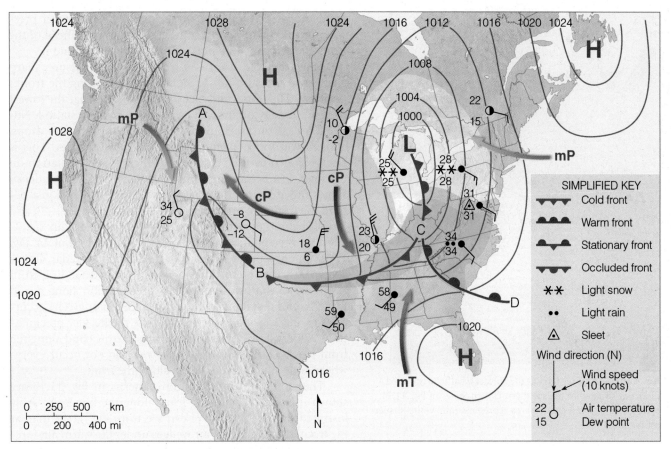

● FIGURE 11.14 A surface weather map showing surface-pressure systems, air masses, fronts, and isobars (in millibars) as solid gray lines. Large arrows in color show air flow. (Green-shaded area represents rain; pink-shaded area represents freezing rain and sleet; white-shaded area represents snow.)

STATIONARY FRONTS   A **stationary front** has essentially no movement.* On a colored weather map, it is drawn as an alternating red and blue line. Red semicircles face toward colder air on the red line and blue triangles point toward warmer air on the blue line. The stationary front between points A and B in Fig. 11.14 marks the boundary where cold, dense continental polar (cP) air from Canada butts up against the north-south trending Rocky Mountains. Unable to cross the barrier, the cold air shows little or no westward movement. The stationary front is drawn along a line separating the cP air from the milder, more humid maritime polar (mP) air to the west. Notice that the surface winds tend to blow parallel to the front, but in opposite directions on either side of it. Upper-level winds often blow parallel to a stationary front.

The weather along the front is clear to partly cloudy, with much colder air lying on its eastern side. Because both air masses are relatively dry, there is no precipitation. This is not, however, always the case. When warm moist air rides up and over the cold air, widespread cloudiness with light precipita-

tion can cover a vast area. These are the conditions that prevail north of the east-west running stationary front depicted in Fig. 11.9 (p. 295).

If the warmer air to the west begins to move and replace the colder air to the east, the front in Fig. 11.14 will no longer remain stationary; it will become a warm front. If, on the other hand, the colder air slides up over the mountain and replaces the warmer air on the other side, the front will become a cold front. If either a cold front or a warm front stop moving, they become a stationary front.

COLD FRONTS   The **cold front** between points B and C on the surface weather map (in Fig. 11.14) represents a zone where cold, dry stable polar air is replacing warm, moist, conditionally unstable subtropical air. The front is drawn as a solid blue line with the triangles along the front showing its direction of movement. How did the meteorologist know to draw the front at that location? A closer look at the front will give us the answer.

The weather in the immediate vicinity of this cold front in the southern United States is shown in ● Fig. 11.15. The data plotted on the map represent the current weather at selected cities. The station model used to represent the data at

---
*They are usually called *quasi-stationary fronts* because they can show some movement.

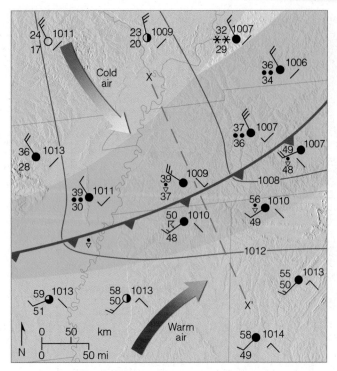

● FIGURE 11.15 A closer look at the surface weather associated with the cold front situated in the southern United States in Fig. 11.14. (Gray lines are isobars. Green-shaded area represents rain; white-shaded area represents snow.)

each reporting station is a simplified one that shows temperature, dew point, present weather, cloud cover, sea-level pressure, wind direction and speed. The little line in the lower right-hand corner of each station shows the *pressure tendency*—the pressure change, whether rising (/) or falling (\)—during the last three hours. With all of this information, the front can be properly located.* (Appendix B explains the weather symbols and the station model more completely.)

The following criteria are used to locate a front on a surface weather map:

1. sharp temperature changes over a relatively short distance
2. changes in the air's moisture content (as shown by marked changes in the dew point)
3. shifts in wind direction
4. pressure and pressure changes
5. clouds and precipitation patterns

In Fig. 11.15, we can see a large contrast in air temperature and dew point on either side of the front. There is also a wind shift from southwesterly ahead of the front, to northwesterly behind it. Notice that each isobar kinks as it crosses the front, forming an elongated area of low pressure—a *trough*—which accounts for the wind shift. Since surface

*Locating any front on a weather map is not always a clear-cut process. Even meteorologists can disagree on an exact position.

winds normally blow across the isobars toward lower pressure, we find winds with a southerly component ahead of the front and winds with a northerly component behind it.

Since the cold front is a trough of low pressure, sharp changes in pressure can be significant in locating the front's position. One important fact to remember is that the lowest pressure usually occurs just as the front passes a station. Notice that, as you move toward the front, the pressure drops, and, as you move away from it, the pressure rises. This is clearly shown by the pressure tendencies for each station on the map. Just before the front passes, the pressure tendency shows the atmospheric pressure is falling (\), while just behind the front, the pressure is now beginning to rise (✓), and farther behind the front, the pressure is rising steadily (/).

The precipitation pattern along the cold front in Fig. 11.15 might appear similar to the Doppler radar image shown in ● Fig. 11.16. The region in color extending from northeast to southwest represents precipitation along a cold front. Notice that light-to-moderate rain (color green) occurs over a wide area along the front, while the heavier precipitation (color yellow) tends to occur in a narrow band along the front itself. Thunderstorms (color red) do not occur everywhere, but only in certain areas along the front.

The cloud and precipitation patterns in Fig. 11.15 are shown in a side view of the front along the line X–X' as illustrated in ● Fig. 11.17. We can see from Fig. 11.17 that, at the front, the cold, dense air wedges under the warm air, forcing the warm air upward, much like a snow shovel forces snow upward as the shovel glides through the snow. As the moist, conditionally unstable air rises, it condenses into a series of cumuliform clouds. Strong, upper-level westerly winds blow the delicate ice crystals (which form near the top

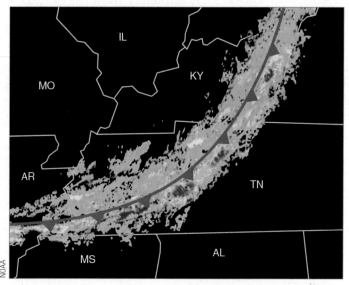

● FIGURE 11.16 A Doppler radar image showing precipitation patterns along a cold front similar to the cold front in Fig. 11.15. Green represents light-to-moderate precipitation; yellow represents heavier precipitation; and red the most likely areas for thunderstorms. (The cold front is superimposed on the radar image.)

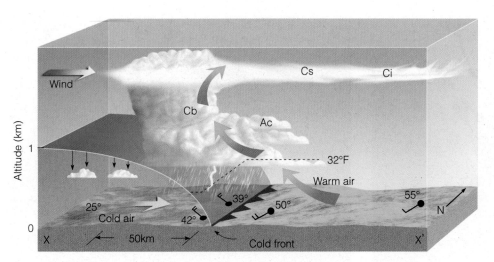

● FIGURE 11.17 A vertical view of the weather across the cold front in Fig. 11.15 along the line X–X'.

of the cumulonimbus) into cirrostratus (Cs) and cirrus (Ci). These clouds usually appear far in advance of the approaching front. At the front itself, a relatively narrow band of thunderstorms (Cb) produces heavy showers with gusty winds. Behind the front, the air cools quickly. (Notice how the freezing level dips as it crosses the front.) The winds shift from southwesterly to northwesterly, pressure rises, and precipitation ends. As the air dries out, the skies clear, except for a few lingering fair weather cumulus clouds.

Observe that the leading edge of the front is steep. The steepness is due to friction, which slows the airflow near the ground. The air aloft pushes forward, blunting the frontal surface. If we could walk from where the front touches the surface back into the cold air, a distance of 50 km, the front would be about 1 km above us. Thus, the slope of the front—the ratio of vertical rise to horizontal distance—is 1:50. This is typical for a fast-moving cold front—those that move about 25 knots. In a slower-moving cold front—one that moves about 15 knots—the slope is much more gentle.

With slow-moving cold fronts, clouds and precipitation usually cover a broad area behind the front. When the ascending warm air is stable, stratiform clouds, such as nimbostratus, become the predominate cloud type and even fog may develop in the rainy area. Occasionally, out ahead of a fast-moving front, a line of active showers and thunderstorms, called a *squall line,* develops parallel to and often ahead of the advancing front, producing heavy precipitation and strong gusty winds.

As the temperature contrast across a front lessens, the front will often weaken and dissipate. Such a condition is known as **frontolysis.** On the other hand, an increase in the temperature contrast across a front can cause it to strengthen and regenerate into a more vigorous frontal system, a condition called **frontogenesis.**

An example of a regenerated front is shown in the infrared satellite images in ● Fig. 11.18. The cold front in Fig. 11.18a is weak, as indicated by the low clouds (gray tones) along the front. As the front moves offshore, over the warm Gulf Stream (Fig. 11.18b), it intensifies into a more vigorous fron-

tal system as surface air becomes conditionally unstable and convective activity develops. Notice that the area of cloudiness is more extensive and thunderstorms are now forming along the frontal zone.

So far, we have considered the general weather patterns of "typical" cold fronts. There are, of course, many exceptions. In fact, no two fronts are exactly alike. In some, the cold air is very shallow; in others, it is much deeper. If the rising warm air is dry and stable, scattered clouds are all that form, and there is no precipitation. In extremely dry weather, a marked change in the dew point, accompanied by a slight wind shift, may be the only clue to a passing cold front.

During the winter, a series of cold polar outbreaks may travel across the United States so quickly that warm air is unable to develop ahead of the front. In this case, frigid arctic air associated with an arctic front usually replaces cold polar air, and a drop in temperature is the only indication that a front has moved through your area. Along the West Coast, the Pacific Ocean modifies the air so much that cold fronts, such as those described in the previous section, are never seen. In fact, as a cold front moves inland from the Pacific Ocean, the surface temperature contrast across the front may be quite small. Topographic features usually distort the wind pattern so much that locating the position of the front and the time of its passage is exceedingly difficult. In this case, the pressure tendency is the most reliable indication of a frontal passage.

In some instances along the West Coast, an approaching cold front (or upper-level trough) will cause cool marine air at the surface to *surge* into coastal and inland valleys. The cool air (which is often accompanied by a wind shift) may

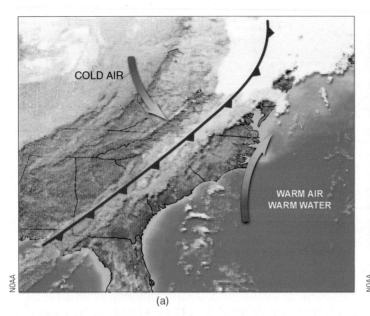

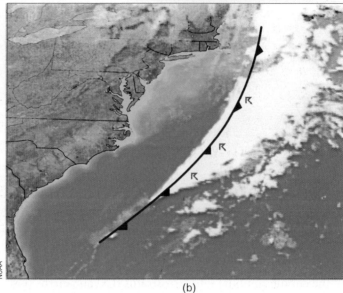

NOAA

NOAA

(a)   (b)

● FIGURE 11.18 The infrared satellite image (a) shows a weakening cold front over land on Tuesday morning, November 21, intensifying into (b) a vigorous front over warm Gulf Stream water on Wednesday morning, November 22.

produce a sharp drop in air temperature. This may give the impression that a rather strong cold front has moved through, when in reality, the front may be many kilometers offshore.

Cold fronts usually move toward the south, southeast, or east. But sometimes they will move southwestward. In New England, this movement occurs when northeasterly surface winds, blowing clockwise around an anticyclone centered to the north over Canada, push a cold front southwestward often as far south as Boston. Because the cold front moves in from the east, or northeast, it is known as a **"back door" cold**

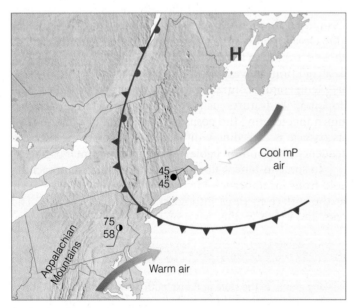

● FIGURE 11.19 A "back door" cold front moving into New England during the spring. Notice that, behind the front, the weather is cold and damp with drizzle, while to the south, ahead of the front, the weather is partly cloudy and warm.

front. As the front passes, westerly surface winds usually shift to easterly or northeasterly and temperatures drop as mP air flows in off the Atlantic Ocean.

An example of a "back door" cold front is shown in ● Fig. 11.19. This is a springtime situation where, behind the front, the weather is cold and damp with drizzle, as northeasterly winds sweep into the region from off the chilly Atlantic. To the south of the front, the weather is much warmer. Should the front move through this area, the more summer-like weather would change, in a matter of hours, to more winterlike. The cold, dense air behind the front is rather shallow. Consequently, the Appalachian Mountains act as a dam to the front's forward progress, halting its westward movement. This situation, where the cold, damp air is confined to the eastern side of the mountains, is called **cold air damming.** The stalled cold front now becomes a stationary front. The cool air behind the front may linger for some time as warmer, less-dense air to the south rides up and over it. Forecasting how far south the "back door" cold front will move and when the entrenched cold air will leave can be a bit tricky.

Even though cold-front weather patterns have many exceptions, learning these patterns can be to your advantage if you live in an area that experiences well-defined cold fronts. Knowing them improves your own ability to make short-range weather forecasts. For your reference, ▼ Table 11.2 summarizes idealized cold-front weather in the Northern Hemisphere.

View this concept in action on the Meteorology Resource Center at academic.cengage.com/login

WARM FRONTS In Fig. 11.14, p. 299, a **warm front** is drawn along the solid red line running from points C to D.

▼ TABLE 11.2   Typical Weather Conditions Associated with a Cold Front in the Northern Hemisphere

WEATHER ELEMENT	BEFORE PASSING	WHILE PASSING	AFTER PASSING
Winds	South or southwest	Gusty, shifting	West or northwest
Temperature	Warm	Sudden drop	Steadily dropping
Pressure	Falling steadily	Minimum, then sharp rise	Rising steadily
Clouds	Increasing Ci, Cs, then either Tcu* or Cb*	Tcu or Cb	Often Cu, Sc* when ground is warm
Precipitation	Short period of showers	Heavy showers of rain or snow, sometimes with hail, thunder, and lightning	Decreasing intensity of showers, then clearing
Visibility	Fair to poor in haze	Poor, followed by improving	Good, except in showers
Dew point	High; remains steady	Sharp drop	Lowering

*Tcu stands for towering cumulus, such as cumulus congestus; whereas Cb stands for cumulonimbus. Sc stands for stratocumulus.

Here, the leading edge of advancing warm, moist, subtropical (mT) air from the Gulf of Mexico replaces the retreating cold maritime polar air from the North Atlantic. The direction of frontal movement is given by the half circles, which point into the cold air; this front is heading toward the northeast. As the cold air recedes, the warm front slowly advances. The average speed of a warm front is about 10 knots, or about half that of an average cold front. During the day, as mixing occurs on both sides of the front, its movement may be much faster. Warm fronts often move in a series of rapid jumps, which show up on successive weather maps. At night, however, radiational cooling creates cool dense surface air behind the front. This inhibits both lifting and the front's forward progress. When the forward surface edge of the warm front passes a station, the wind shifts, the temperature rises, and the overall weather conditions improve. To see why, we will examine the weather commonly associated with the warm front both at the surface and aloft.

• Figure 11.20 is a surface weather map showing the position of a warm front and its associated weather. • Figure 11.21 is a vertical view of the warm front in Fig. 11.20. Look at these two figures and observe that the warmer, less-dense air rides up and over the colder, more-dense surface air. This rising of warm air over cold, called **overrunning,** produces clouds and precipitation well in advance of the front's surface boundary. The warm front that separates the two air masses has an average slope of about 1:300*—a much more gentle or inclined shape than that of a typical cold front. Warm air overriding the cold air creates a stable atmosphere (see the vertical temperature profile in Fig. 11.21b). Notice that a temperature inversion—called a **frontal inversion**—exists in the region of the upper-level front at the boundary where the warm air overrides the cold air. Another fact to notice in Fig. 11.21b is that the wind *veers* (shifts clockwise) with altitude, so that the

southeasterly (SE) surface winds become southwesterly (SW) and westerly (W) aloft.

Suppose we are standing at the position marked P′ in Figs. 11.20 and 11.21. Note that we are over 1200 km (750 mi) ahead of where the warm front is touching the surface. Here, the surface winds are light and variable. The air is cold and about the only indication of an approaching warm front is the high cirrus clouds overhead. We know the front is moving slowly toward us and that within a day or so it will pass our area. Suppose that, instead of waiting for the front to pass us, we drive toward it, observing the weather as we go.

Heading toward the warm front, we notice that the cirrus (Ci) clouds gradually thicken into a thin, white veil of cirrostratus (Cs) whose ice crystals cast a halo around the sun.* Almost imperceptibly, the clouds thicken and lower, becoming altocumulus (Ac) and altostratus (As) through which the sun shows only as a faint spot against an overcast gray sky. Snowflakes begin to fall, and we are still over 600 km (370 mi) from the surface front. The snow increases, and the clouds thicken into a sheetlike covering of nimbostratus (Ns). The winds become brisk and out of the southeast, while the atmospheric pressure slowly falls. Within 400 km (250 mi) of the front, the cold surface air mass is now quite shallow. The surface air temperature moderates and, as we approach the front, the light snow changes first into sleet. It then becomes freezing rain and finally rain and drizzle as the air temperature climbs above freezing. Overall, the precipitation remains light or moderate but covers a broad area. Moving still closer to the front, warm, moist air mixes with cold, moist air producing ragged wind-blown stratus (St) and fog. (Thus, flying in the vicinity of a warm front is quite hazardous.)

Finally, after a trip of over 1200 km, we reach the warm front's surface boundary. As we cross the front, the weather changes are noticeable, but much less pronounced than those

---

*The slope of 1:300 is a much more gentle slope than that of most warm fronts. Typically, the slope of a warm front is on the order of 1:150 to 1:200.

*If the warm air is relatively unstable, ripples or waves of cirrocumulus clouds will appear as a "mackerel sky."

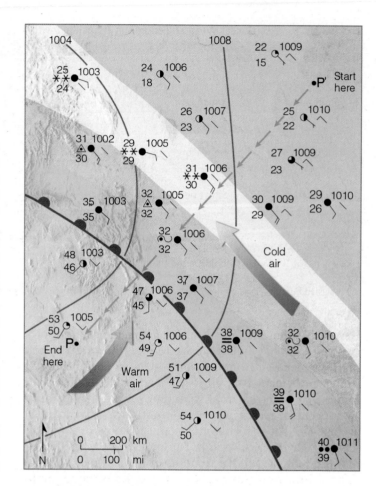

ACTIVE FIGURE 11.20 Surface weather associated with a typical warm front. A vertical view along the dashed line P-P′ is shown in Fig. 11.21. (Green-shaded area represents rain; pink-shaded area represents freezing rain and sleet; white-shaded area represents snow.) Visit the Meteorology Resource Center to view this and other active figures at academic.cengage.com/login

experienced with the cold front; they show up more as a gradual transition rather than a sharp change. On the warm side of the front, the air temperature and dew point rise, the wind shifts from southeast to south or southwest, and the air pressure stops falling. The light rain ends and, except for a few stratocumulus (Sc), the fog and low clouds vanish.

This scenario of an approaching warm front represents average (if not idealized) warm-front weather in winter. In some instances, the weather can differ from this dramatically. For example, if the overrunning warm air is relatively dry and stable, only high and middle clouds will form, and no precipitation will occur. On the other hand, if the warm air is relatively moist and conditionally unstable (as is often the case during the summer), heavy showers can develop as thunderstorms become embedded in the cloud mass. In the southern Great Plains warm, humid air may be separated from warm, dry air along a boundary called a **dryline.** More on drylines is given in the Focus section on p. 305.

Along the West Coast, the Pacific Ocean significantly modifies the surface air so that warm fronts are difficult to locate on a surface weather map. Also, not all warm fronts

ACTIVE FIGURE 11.21 Vertical view of clouds, precipitation, and winds across the warm front in Fig. 11.20 along the line P–P′. Visit the Meteorology Resource Center to view this and other active figures at academic.cengage.com/login

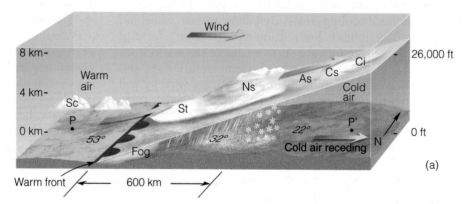

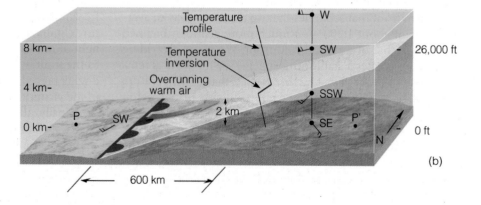

FOCUS ON A SPECIAL TOPIC

Drylines

Drylines are not warm fronts or cold fronts, but represent a narrow boundary where there is a steep horizontal change in moisture, so drylines separate moist air from dry air. Because dew-point temperatures may drop along this boundary by as much as 9°C (16°F) per km, drylines have been referred to as *dew-point fronts.** Although drylines can occur in the United States as far north as the Dakotas, and as far east as the Texas-Louisiana border, they are most frequently observed in the western half of Texas, Oklahoma, and Kansas, especially during spring and early summer. In these locations, drylines tend to move eastward during the day, then westward toward evening. Drylines are observed in other regions of the world, too. They occur, for example, in Central West Africa and in India before the onset of the summer monsoon.

Figure 4 shows a dryline moving across Texas during May, 2001. Notice that the dryline is represented as a line with brown half circles. Notice also that, to the west of the dryline, warm, dry continental tropical air is moving in from the southwest. Consequently, on this side, the weather is usually hot and dry with gusty southwesterly winds. To the east of the dryline, warm, very humid maritime tropical air is sweeping northward from the Gulf of Mexico. Here we typically find air temperatures to

*Recall from Chapter 4 that the dew-point temperature is a measure of the amount of water vapor in the air.

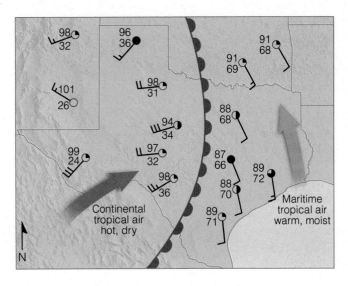

● FIGURE 4 A dryline represents a narrow boundary where there is a steep horizontal change in moisture as indicated by a rapid change in dew-point temperature. Here, a dryline moving across Texas and Oklahoma separates warm, moist air from warm, dry air during an afternoon in May.

be slightly lower and the humidity (as indicated by the higher dew points) considerably higher than on the western side. The semicircles of the dryline point toward this humid air.

Even though the dryline represents a moisture boundary, its actual position on a weather map is plotted according to a shift in surface winds. When insects and insect-eating birds congregate along the dryline, Doppler radar may be able to locate it. On the radar screen, the echo from insects and birds shows up as a thin line, called a *fine line*.

Sometimes drylines are associated with mid-latitude cyclones, sometimes they are not. Cumulus clouds and thunderstorms often form along or to the east of the dryline. This cloud

development is caused in part by daytime convection and a sloping terrain. The Central Plains area of North America is higher to the west and lower to the east. Convection over the elevated western Plains carries dry air high above the surface. Westerly winds sweep this dry air eastward over the lower plains where it overrides the slightly cooler but more humid air at the surface. This situation sets up a potentially unstable atmosphere that finds warm, dry air above warm, moist air. In regions where the air rises, cumulus clouds and organized bands of thunderstorms can form. We will examine in more detail the development of these storms in Chapter 14.

---

move northward or northeastward. On rare occasions, a front will move into the eastern seaboard from the Atlantic Ocean as the front spins all the way around a deep storm positioned off the coast. Cold northeasterly winds ahead of the front usually become warm northeasterly winds behind it. Even with these exceptions, knowing the normal sequence of warm-front weather will be useful, especially if you live where warm fronts become well developed. You can look for certain cloud and weather patterns and make reasonably accurate short-range forecasts of your own. ▼ Table 11.3 summarizes typical warm-front weather. (Before going on to the next section, you may wish to read the Focus section on p. 306, which gives additional information about warm fronts.)

View this concept in action on the Meteorology Resource Center at academic.cengage.com/login

OCCLUDED FRONTS   If a cold front catches up to and overtakes a warm front, the frontal boundary created between the two air masses is called an **occluded front,** or, simply, an **occlusion** (meaning "closed off"). On the surface weather map, it is represented as a purple line with alternating cold-front triangles and warm-front half circles; both symbols point in the direction toward which the front is moving. Look back at Fig. 11.14, p. 299, and notice that the air behind the occluded front is colder than the air ahead of it. This is known as a

FOCUS ON A SPECIAL TOPIC

The Wavy Warm Front

Up to this point, we've examined idealized warm fronts on a surface weather map — like the one shown in Fig. 11.14, p. 299. Some warm fronts do look like this example; others, however, have an entirely different appearance. For instance, look at the warm front in Fig. 5. Notice that it has a wavelike shape as it approaches North Carolina from three different directions. So what causes the warm front to bend in this manner?

Look carefully at Fig. 5 and notice that at the surface cold air is flowing southwestward around a high-pressure area centered over southern Canada. As cold, dense, surface air pushes south into the southern states, it flows up against the Appalachian Mountains, which impede its westward progress. Since the shallow layer of cold air is unable to ride up and over the mountains, it becomes wedged along the mountains' eastern foothills. Recall from an earlier discussion that this trapping of cold air is called *cold air damming*.

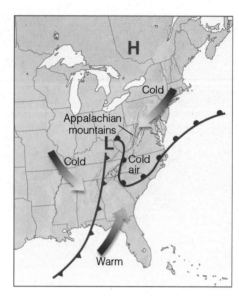

● FIGURE 5 Surface weather map for 11:00 P.M. (EST), February 13, 2007.

Warm air pushing northward from the Gulf of Mexico rides up and over the cold, dry, surface air. Clouds and precipitation often form in this rising warm air. When rain falls into the shallow, cold air, it may evaporate, chilling the air even more. Sometimes the rain freezes before reaching the ground, producing sleet; other times, the rain freezes on impact, producing freezing rain. If the frozen precipitation falls for many hours, severe ice storms may result, with heavy accumulations of ice causing treacherous driving conditions and downed power lines.

The shallow layer of cold air usually becomes entrenched in low-lying areas and therefore retreats northward very slowly. As the cold air slowly recedes northward, warmer air pushes in from different directions, and the leading edge of the warm air (the warm front) no longer has a nice curved shape, but begins to take on a move wavy shape, such as the warm front in Fig. 5.

---

cold-type occluded front, or **cold occlusion.** Let's see how this front develops.

The development of a cold occlusion is shown in ● Fig. 11.22. Along line A–A′, the cold front is rapidly approaching the slower-moving warm front. Along line B–B′, the cold front overtakes the warm front, and, as we can see in the vertical view across C–C′, underrides and lifts off the

ground both the warm front and the warm air mass. As a cold-occluded front approaches, the weather sequence is similar to that of a warm front, with high clouds lowering and thickening into middle and low clouds, with precipitation forming well in advance of the surface front. Since the front represents a trough of low pressure, southeasterly winds and falling atmospheric pressure occur ahead of it. The fron-

▼ TABLE 11.3   Typical Weather Conditions Associated with a Warm Front in the Northern Hemisphere

WEATHER ELEMENT	BEFORE PASSING	WHILE PASSING	AFTER PASSING
Winds	South or southeast	Variable	South or southwest
Temperature	Cool to cold, slow warming	Steady rise	Warmer, then steady
Pressure	Usually falling	Leveling off	Slight rise, followed by fall
Clouds	In this order: Ci, Cs, As, Ns, St, and fog; occasionally Cb in summer	Stratus type	Clearing with scattered Sc, especially in summer; occasionally Cb in summer
Precipitation	Light-to-moderate rain, snow, sleet, or drizzle; showers in summer	Drizzle or none	Usually none; sometimes light rain or showers
Visibility	Poor	Poor, but improving	Fair in haze
Dew point	Steady rise	Steady	Rise, then steady

tal passage, however, brings weather similar to that of a cold front: heavy, often showery precipitation with winds shifting to west or northwest. After a period of wet weather, the sky begins to clear, atmospheric pressure rises, and the air turns colder. The most violent weather usually occurs where the cold front is just overtaking the warm front, at the point of occlusion, where the greatest contrast in temperature occurs. Cold occlusions are the most prevalent type of front that moves into the Pacific coastal states and into interior North America. Occluded fronts frequently form over the North Pacific and North Atlantic, as well as in the vicinity of the Great Lakes.

Continental polar air over eastern Washington and Oregon may be much colder than milder maritime polar air moving inland from the Pacific Ocean. ● Figure 11.23 illustrates this situation. Observe that the air ahead of the warm front is colder than the air behind the cold front. Consequently, when the cold front catches up to and overtakes the warm front, the milder, lighter air behind the cold front is unable to lift the colder, heavier air off the ground. As a result, the cold front rides "piggyback" along the sloping warm front. This produces a *warm-type occluded front,* or a **warm occlusion.** The surface weather associated with a warm occlusion is similar to that of a warm front.*

Contrast Fig. 11.22 and Fig. 11.23. Note that the primary difference between the warm- and cold-type occluded front is the location of the upper-level front. In a warm occlusion, the upper-level cold front *precedes* the surface occluded front, whereas in a cold occlusion the upper warm front *follows* the surface occluded front.

In the world of weather fronts, occluded fronts are the mavericks. In our discussion, we treated occluded fronts as forming when a cold front overtakes a warm front. Some may form in this manner, but others apparently form as new fronts, which develop when a surface mid-latitude cyclonic storm intensifies in a region of cold air after its trailing cold and warm fronts have broken away and moved eastward. The new occluded front shows up on a surface chart as a trough of low pressure separating two cold air masses. Because of this, locating and defining occluded fronts at the surface is often difficult for the meteorologist. Similarly, you too may find it hard to recognize an occlusion. In spite of this, we will assume that the weather associated with occluded fronts behaves in a similar way to that shown in ▼ Table 11.4.

The frontal systems described in this chapter are actually part of a much larger storm system—the middle-latitude cyclone. ● Figure 11.24 shows the cold front, warm front, and occluded front in association with a mid-latitude cyclonic storm. Notice that, as we would expect, clouds and precipitation form in a rather narrow band along the cold front, and in a much wider band with the warm and occluded fronts. In Chapter 12, we will look more closely at middle-latitude cy-

---

*Due to the relatively mild winter air that moves into Europe from the North Atlantic, many of the occlusions that move into this region in winter are of the warm occlusion variety.

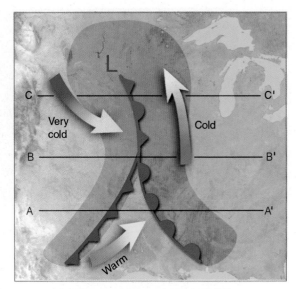

(d)

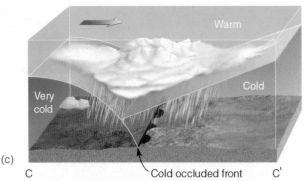

(c)

C   Cold occluded front   C'

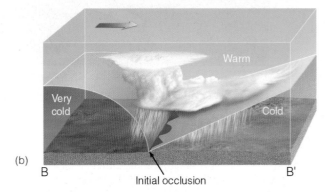

(b)

B   Initial occlusion   B'

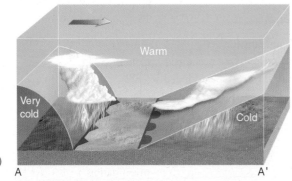

(a)

A   A'

● FIGURE 11.22 The formation of a cold-occluded front. The faster-moving cold front (a) catches up to the slower-moving warm front (b) and forces it to rise off the ground (c). (Green-shaded area in (d) represents precipitation.)

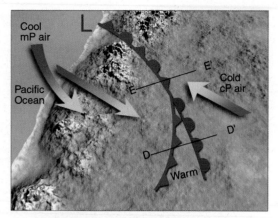

(c)

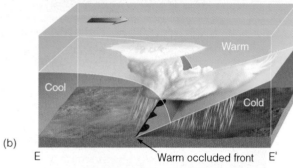

(b)

E — Warm occluded front — E'

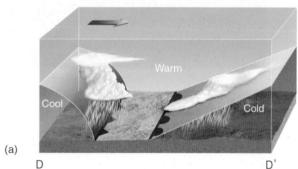

(a)

D — D'

● **FIGURE 11.23** (*at left*) The formation of a warm-type occluded front. The faster-moving cold front in (a) overtakes the slower-moving warm front in (b). The lighter air behind the cold front rises up and over the denser air ahead of the warm front. Diagram (c) shows a surface map of the situation.

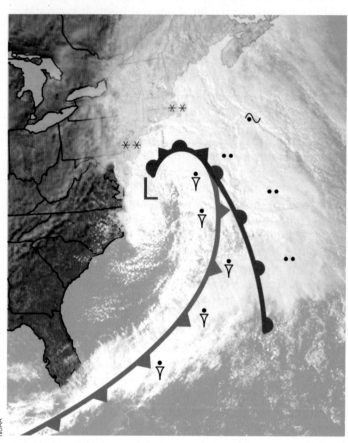

● **FIGURE 11.24** A visible satellite image showing a mid-latitude cyclonic storm with its weather fronts over the Atlantic Ocean during March, 2005. Superimposed on the photo is the position of the surface cold front, warm front, and occluded front. Precipitation symbols indicate where precipitation is reaching the surface.

▼ **TABLE 11.4** Typical Weather Most Often Associated with Occluded Fronts in North America

WEATHER ELEMENT	BEFORE PASSING	WHILE PASSING	AFTER PASSING
Winds	East, southeast, or south	Variable	West or northwest
Temperature (a) Cold-type occluded (b) Warm-type occluded	Cold or cool / Cold	Dropping / Rising	Colder / Milder
Pressure	Usually falling	Low point	Usually rising
Clouds	In this order: Ci, Cs, As, Ns	Ns, sometimes Tcu and Cb	Ns, As, or scattered Cu
Precipitation	Light, moderate, or heavy precipitation	Light, moderate, or heavy continuous precipitation or showers	Light-to-moderate precipitation followed by general clearing
Visibility	Poor in precipitation	Poor in precipitation	Improving
Dew point	Steady	Usually slight drop, especially if cold-occluded	Slight drop, although may rise a bit if warm-occluded

clonic storms, examining where, why, and how they form. Before we move on, however, we need to look at fronts that form in the upper troposphere—that may, or may not, show up at the surface.

UPPER-AIR FRONTS   An **upper-air front** (which is also known as *upper front,* or *upper-tropospheric front*) is a front that is present aloft. It may or may not extend down to the surface. ● Figure 11.25 shows a north-to-south side view of an idealized upper-air front. Notice that the front forms when the tropopause—the boundary separating the troposphere from the stratosphere—dips downward and folds under the polar jet stream. In the fold, the isotherms are tightly packed, marking the position of the upper front. Although the upper front may not connect with a surface front, the position of the surface front is shown in the diagram.

The small arrows in Fig. 11.25 show air motion associated with the upper front. On the north side of the front (and the north side of the jet stream), the air is slowly sinking. Here, in the folded troposphere, ozone-rich air from the stratosphere descends into the troposphere. To the south of the front (and south of the jet stream), the air slowly rises. These rising and descending air motions can aid in the development of middle-latitude cyclonic storms described in the next chapter.

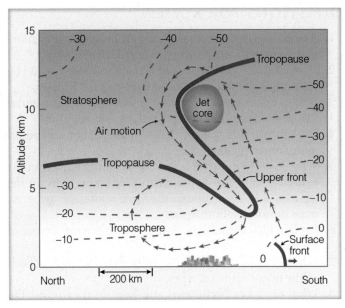

● FIGURE 11.25 An idealized vertical view of an upper-air front showing tropopause (heavy red line), isotherms in °C (dashed gray lines), and vertical air motions. The polar jet stream core (maximum winds) is flowing into the page (from west to east).

SUMMARY

In this chapter, we considered the different types of air masses and the various weather each brings to a particular region. Continental arctic air masses are responsible for the extremely cold (arctic) outbreaks of winter, whereas continental polar air masses are responsible for cold, dry weather in winter and cool, pleasant weather in summer. Maritime polar air, having traveled over an ocean for a considerable distance, brings cool, moist weather to an area. The hot, dry weather of summer is associated with continental tropical air masses, while warm, humid conditions are due to maritime tropical air masses. Where air masses with sharply contrasting properties meet, we find weather fronts.

A front is a boundary between two air masses of different densities. Stationary fronts have essentially no movement, with cold air on one side and warm air on the other. Winds tend to blow parallel to the front, but in opposite directions on either side of it. Along the leading edge of a cold front, where colder air replaces warmer air, showers are prevalent, especially if the warmer air is moist and conditionally unstable. Along a warm front, warmer air rides up and over colder surface air, producing widespread cloudiness and light-to-moderate precipitation that can cover thousands of square kilometers. When the rising air is conditionally unstable (such as it often is in summer), showers and thunderstorms may form ahead of the advancing warm front. Cold fronts typically move faster and are more steeply sloped than

warm fronts. Occluded fronts, which are often difficult to locate and define on a surface weather map, may have characteristics of both cold and warm fronts. Fronts that form in the upper troposphere, in the vicinity of the polar-front jet stream, are called upper-air fronts.

KEY TERMS

The following terms are listed (with page numbers) in the order they appear in the text. Define each. Doing so will aid you in reviewing the material covered in this chapter.

air mass, 288
source regions (for air masses), 288
continental polar (air mass), 290
continental arctic (air mass), 290
lake-effect snows, 290
maritime polar (air mass), 293
maritime tropical (air mass), 295
continental tropical (air mass), 297
front, 298
stationary front, 299

cold front, 299
frontolysis, 301
frontogenesis, 301
"back door" cold front, 302
cold air damming, 302
warm front, 302
overrunning, 303
frontal inversion, 303
dryline, 304
occluded front (occlusion), 305
cold occlusion, 306
warm occlusion, 307
upper-air front, 309

QUESTIONS FOR REVIEW

1. If an area is described as a "good air mass source region," what information can you give about it?
2. It is summer. What type of afternoon weather would you expect from an air mass designated as mT? Explain.
3. Why is continental polar air not welcome to the Central Plains in winter and yet very welcome in summer?
4. Explain why the central United States is not a good air mass source region.
5. Why do air temperatures tend to be a little higher on the eastern side of the Appalachian Mountains than on the western side, even though the same winter cP or cA air mass dominates both areas?
6. Explain how the airflow aloft regulates the movement of air masses.
7. List the temperature and moisture characteristics of each of the major air mass types.
8. What are lake-effect snows and how do they form? On which side of a lake do they typically occur?
9. Why are maritime polar air masses along the East Coast of the United States usually colder than those along the nation's West Coast? Why are they also *less* prevalent?
10. The boundaries between neighboring air masses tend to be more distinct during the winter than during the summer. Explain why.
11. What type of air mass would be responsible for the weather conditions listed as follows?
    (a) heavy snow showers and low temperatures at Buffalo, New York
    (b) hot, muggy summer weather in the Midwest and the East
    (c) daily afternoon thunderstorms along the Gulf Coast
    (d) heavy snow showers along the western slope of the Rockies
    (e) refreshing, cool, dry breezes after a long summer hot spell on the Central Plains
    (f) heavy summer rainshowers in southern Arizona
    (g) drought with high temperatures over the Great Plains
    (h) persistent cold, damp weather with drizzle along the East Coast
    (i) summer afternoon thunderstorms forming along the eastern slopes of the Sierra Nevada
    (j) record low winter temperatures in South Dakota
12. On a surface weather map, what do you know about a region where the word *frontogenesis* is marked?
13. Explain why barometric pressure usually falls with the approach of a cold front or occluded front.
14. How does the weather usually change along a dryline?
15. Based on the following weather forecasts, what type of front will most likely pass the area?
    (a) Light rain and cold today, with temperatures just above freezing. Southeasterly winds shifting to westerly tonight. Turning colder with rain becoming heavy and possibly changing to snow.
    (b) Cool today with rain becoming heavy at times by this afternoon. Warmer tomorrow. Winds southeasterly becoming westerly by tomorrow morning.
    (c) Increasing cloudiness and warm today, with the possibility of showers by evening. Turning much colder tonight. Winds southwesterly, becoming gusty and shifting to northwesterly by tonight.
    (d) Increasing high cloudiness and cold this morning. Clouds increasing and lowering this afternoon, with a chance of snow or rain tonight. Precipitation ending tomorrow morning. Turning much warmer. Winds light easterly today, becoming southeasterly tonight and southwesterly tomorrow.
16. Sketch side views of a typical cold front, warm front, and cold-occluded front. Include in each diagram cloud types and patterns, areas of precipitation, surface winds, and relative temperature on each side of the front.
17. During the spring, on a warm, sunny day in Boston, Massachusetts, the wind shifts from southwesterly to northeasterly and the weather turns cold, damp, and overcast. What type of front moved through the Boston area? From what direction did the front apparently approach Boston?
18. How does the tropopause show where an upper-level front is located?

QUESTIONS FOR THOUGHT

1. Suppose an mP air mass moving eastward from the Pacific Ocean travels across the United States. Describe all of the modifications that could take place as this air mass moves eastward in winter. In summer.
2. Explain how an anticyclone during autumn can bring record-breaking low temperatures and continental polar air to the southeastern states, and only a day or so later very high temperatures and maritime tropical air to the same region.
3. In Fig. 11.5 (p. 292), there is a temperature inversion. How does this inversion differ from the frontal inversion illustrated in Fig. 11.21b (p. 304)?
4. For Chicago, Illinois, to experience heavy lake-effect snows, from what direction would the wind have to be blowing?
5. When a very cold air mass covers half of the United States, a very warm air mass often covers the other half. Explain how this happens.
6. Explain why freezing rain more commonly occurs with warm fronts than with cold fronts.
7. In winter, cold-front weather is typically more violent than warm-front weather. Why? Explain why this is not necessarily true in summer.
8. When a cold front passes a station in the Northern Hemisphere, the wind shifts in a clockwise manner. How would the winds shift during the passage of a cold front in the Southern Hemisphere?

9. Why does the same cold front typically produce more rain over Kentucky than over western Kansas?

10. You are in upstate New York and observe the wind shift from easterly to southerly. This shift in wind is accompanied by a sudden rise in both the air temperature and dew-point temperature. What type of front passed?

11. If Lake Erie freezes over in January, is it still possible to have lake-effect snows off Lake Erie in February? Why or why not?

12. Why are ocean-effect snow storms (described on p. 291) fairly common when a persistent cold northeasterly wind blows over Cape Cod, Massachusetts, but are not common when a cold northeasterly wind blows over Long Island, New York?

PROBLEMS AND EXERCISES

1. Make a sketch of North America and show the upper-air wind-flow pattern that would produce:
   (a) very cold cA air moving into the far western states in winter
   (b) cold cP air over the Central Plains in winter
   (c) warm mT air over the Midwest in winter
   (d) warm, moist mT air over southern California and Arizona during the summer

2. You are presently taking a weather observation. The sky is full of wispy cirrus clouds estimated to be about 6 km (20,000 ft) overhead. If a warm front is approaching from the south, about how far away is it (assuming a slope of 1:200)? If it is moving toward you at an average warm-front speed of about 10 knots, how long will it take before it passes your area?

Visit the
Meteorology Resource Center
at
academic.cengage.com/login
for more assets, including questions for exploration, animations, videos, and more.

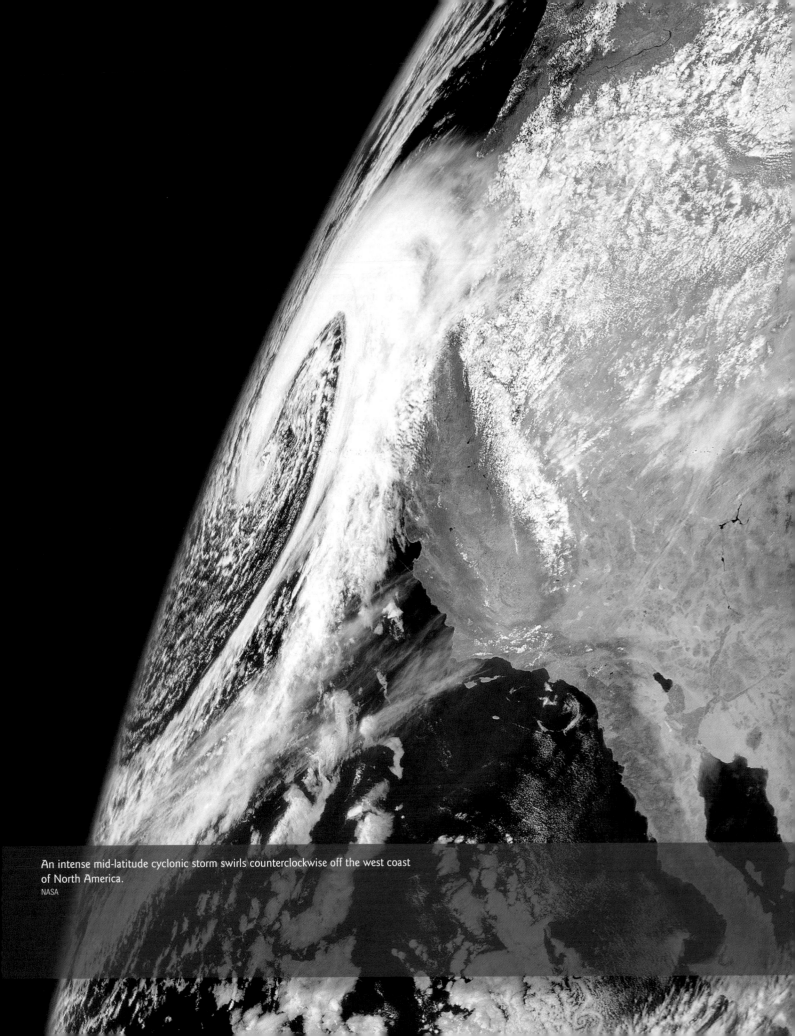

An intense mid-latitude cyclonic storm swirls counterclockwise off the west coast of North America.
NASA

CHAPTER 12

Middle-Latitude Cyclones

It never for a moment occurred to me to regard this storm as a thing to prevent me from getting to New York. The Jackson Avenue station of the New Jersey Central was only five or six blocks from the house, and I anticipated not the slightest trouble in getting there. It was still snowing and blowing . . . and not a print had been made in the snow about the house. I got out to the gate and into the "street," and was then able to discover the real snow.

It was everywhere. Great piles of it rose up like gigantic arctic graves . . . in all directions. Every way that I turned I was confronted with these awful mounds. I took my bearings and steered for the Jackson Avenue station. Every step I took I went into my knees in snow and every other step I fell over on my face and tried to see how much of the stuff I could swallow. The wind was at my back and its accompanying snowflakes cut the back of my head and ears like a million icy lashes. I . . . plowed my way, jumping, falling, and crawling over the drifts, some of which were nine or ten feet high . . . and after an hour and ten minutes I got to the end of my six blocks. There were trains there, two of them, but they were stuck.

I gave up the idea of going to New York. My trip back to the house was simply awful. The wind was straight in my face and beat so in my eyes that I couldn't see a rod before me. My mustache was frozen stiff, and over my eyebrows were cakes of frozen snow. I stumbled along, falling down at almost every step, burying myself in the snow. Then, I began to feel like a crazy man. Every time I fell down, I shouted and cursed and beat the snow with my fists. Then it got dark, the wind howled and tore along, hurling the ice flakes in my face, and the very snow on the ground seemed to rise up and fling itself upon me.

In one of my crazy efforts to force ahead, I caught just a glimpse of the welcome gate posts, and then I laid down on my back and hollered. Somebody heard my cries, and just as I was going off comfortably to sleep my friend came plowing out through the snow, and he and this man dragged me into the house.

U.S. Department of Commerce, *American Weather Stories*

CONTENTS

The storm described in our opening is now referred to as the "Blizzard of '88." This legendary storm of March, 1888, was accompanied by high winds of 60 miles per hour at Atlantic City, New Jersey, a severe cold wave, and unprecedented snowfall—up to 50 inches over portions of southeastern New York and southern New England, with drifts 30 to 40 feet high. In New York City, people died in the street, trapped in snowdrifts up to their hips. What atmospheric conditions are needed for such a monstrous storm to develop?

Early weather forecasters were aware that precipitation generally accompanied falling barometers and areas of low pressure. However, it was not until the early part of the twentieth century that scientists began to piece together the information that yielded the ideas of modern meteorology and storm development.

Working largely from surface observations, a group of scientists in Bergen, Norway, developed a model explaining the life cycle of an *extratropical,* or *middle-latitude cyclonic storm;* that is, a storm that forms at middle and high latitudes outside of the tropics. This extraordinary group of meteorologists included Vilhelm Bjerknes, his son Jakob, Halvor Solberg, and Tor Bergeron. They published their *Norwegian cyclone model* shortly after World War I. It was widely acclaimed and became known as the "polar front theory of a developing wave cyclone" or, simply, the **polar front theory.** What these meteorologists gave to the world was a working model of how a mid-latitude cyclone progresses through the stages of birth, growth, and decay. An important part of the model involved the development of weather along the polar front. As new information became available, the original work was modified, so that, today, it serves as a convenient way to describe the structure and weather associated with a migratory middle-latitude cyclonic storm system.

In the following sections we will first examine, from a surface perspective, how a mid-latitude cyclone develops along the polar front. Then we will examine how the winds aloft influence the developing surface storm. Later on, we will obtain a three-dimensional view of a mid-latitude cyclone by observing how ribbons of air glide through the storm system.

Polar Front Theory

The development of a mid-latitude cyclone, according to the Norwegian model, begins along the polar front. Remember from the discussion of the general circulation in Chapter 10 that the polar front is a semicontinuous global boundary separating cold polar air from warm subtropical air. Because the mid-latitude cyclonic storm forms and moves along the polar front in a wavelike manner, the developing storm is called a **wave cyclone.** The stages of a developing wave cyclone are illustrated in the sequence of surface weather maps shown in ● Fig. 12.1.

Figure 12.1a shows a segment of the polar front as a stationary front. It represents a trough of lower pressure with higher pressure on both sides. Cold air to the north and warm air to the south flow parallel to the front, but in opposite directions. This type of flow sets up a cyclonic wind shear. You can conceptualize the shear more clearly if you place a pen between the palms of your hands and move your left hand toward your body; the pen turns counterclockwise, cyclonically.

Under the right conditions (described later in this chapter), a wavelike kink forms on the front, as shown in Fig. 12.1b. The wave that forms is known as a **frontal wave** or an *incipient cyclone.* Watching the formation of a frontal wave on a weather map is like watching a water wave from its side as it approaches a beach: It first builds, then breaks, and finally dissipates, which is why a mid-latitude cyclonic storm system is known as a *wave cyclone.* Figure 12.1b shows the newly formed wave with a cold front pushing southward and a warm front moving northward. The region of lowest pressure (called the *central pressure*) is at the junction of the two fronts. As the cold air displaces the warm air upward along the cold front, and as *overrunning* occurs ahead of the warm front, a narrow band of precipitation forms (shaded green area). Steered by the winds aloft, the system typically moves east or northeastward and gradually becomes a fully developed *open wave* in 12 to 24 hours (see Fig. 12.1c). The central

● FIGURE 12.1 The idealized life cycle of a mid-latitude cyclone (a through f) in the Northern Hemisphere based on the polar front theory. As the life cycle progresses, the system moves eastward in a dynamic fashion. The small arrow next to each L shows the direction of storm movement.

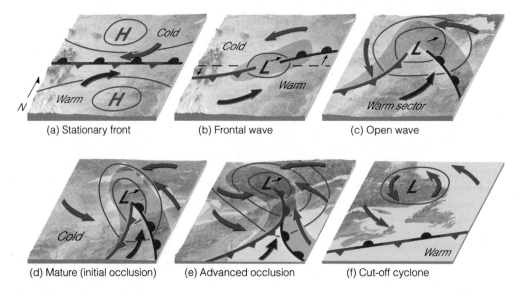

(a) Stationary front
(b) Frontal wave
(c) Open wave
(d) Mature (initial occlusion)
(e) Advanced occlusion
(f) Cut-off cyclone

pressure is now much lower, and several isobars encircle the wave's apex. These more tightly packed isobars create a stronger cyclonic flow, as the winds swirl counterclockwise and inward toward the low's center. Precipitation forms in a wide band *ahead* of the warm front and *along a narrow band* of the cold front. The region of warm air between the cold and warm fronts is known as the **warm sector.** Here, the weather tends to be partly cloudy, although scattered showers may develop if the air is conditionally unstable.

Energy for the storm is derived from several sources. As the air masses try to attain equilibrium, warm air rises and cold air sinks, transforming potential energy into kinetic energy—energy of motion. Condensation supplies energy to the system in the form of latent heat. And, as the surface air converges toward the low's center, wind speeds may increase, producing an increase in kinetic energy.

As the open wave moves eastward, its central pressure continues to decrease, and the winds blow more vigorously as the wave quickly develops into a *mature cyclone.* The faster-moving cold front constantly inches closer to the warm front, squeezing the warm sector into a smaller area, as shown in Fig. 12.1d. In this model, the cold front eventually overtakes

the warm front and the system becomes occluded. At this point, the storm is usually most intense, with clouds and precipitation covering a large area.

The point of occlusion where the cold front, warm front, and occluded front all come together in Fig. 12.1e is referred to as the *triple point.* Notice that in this region the cold and warm fronts appear similar to the open-wave cyclone in Fig. 12.1c. It is here where a new wave (called a **secondary low**) will occasionally form, move eastward, and intensify into a cyclonic storm. The center of the intense storm system shown in Fig. 12.1e gradually dissipates, because cold air now lies on both sides of the occluded front. The warm sector is still present, but is far removed from the center of the storm. Without the supply of energy provided by the rising warm, moist air, the old storm system dies out and gradually disappears (see Fig. 12.1f). We can think of the sequence of a developing wave cyclone as a whirling eddy in a stream of water that forms behind an obstacle, moves with the flow, and gradually vanishes downstream. The entire life cycle of a wave cyclone can last from a few days to over a week.

• Figure 12.2 shows a series of wave cyclones in various stages of development along the polar front in winter. Such a succession of storms is known as a *"family" of cyclones.* Observe that to the north of the front are cold anticyclones; to the south over the Atlantic Ocean is the warm, semipermanent Bermuda high. The polar front itself has developed into a series of loops, and at the apex of each loop is a cyclonic storm system. The cyclone over the northern plains (Low 1) is just forming; the one along the East Coast (Low 2) is an open wave; and the occluded system near Iceland (Low 3) is

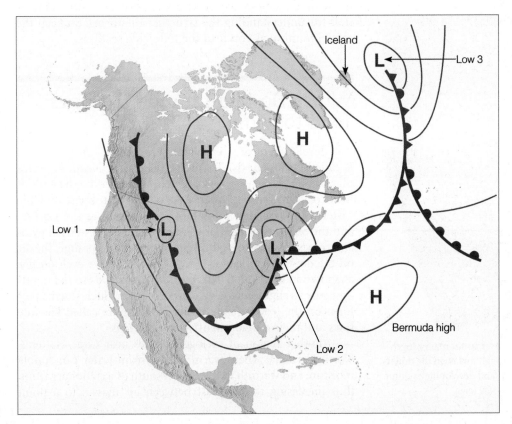

• FIGURE 12.2 A series of wave cyclones (a "family" of cyclones) forming along the polar front.

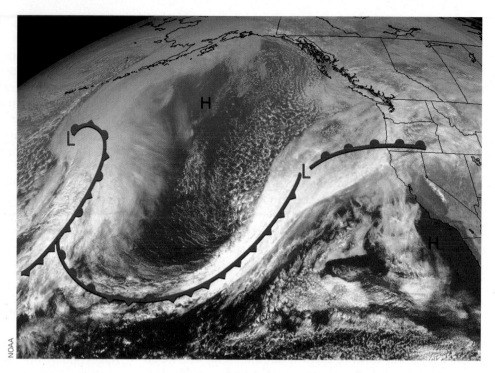

● FIGURE 12.3 Visible satellite image of the north Pacific with two mid-latitude cyclones in different stages of development during February, 2000.

NOAA

dying out. If the average rate of movement of a wave cyclone from birth to decay is 25 knots, then it is entirely possible for a storm to develop over the central part of the United States, intensify into a large storm over New England, become occluded over the ocean, and reach the coast of England in its dissipating stage less than a week after it formed. ● Figure 12.3 is a visible satellite image of clouds and two mid-latitude cyclones in different stages of development along the polar front. Superimposed on the image are the weather fronts. Look again at Fig. 12.1 and determine what stages of development the two cyclones are in.

Up to now, we have considered the polar front model of a developing wave cyclone, which represents a rather simplified version of the stages that an extratropical cyclonic storm system must go through. In fact, few (if any) storms adhere to the model exactly. Nevertheless, it serves as a good foundation for understanding the structure of storms. So keep the model in mind as you read the following sections.

Where Do Mid-Latitude Cyclones Tend to Form?

Any development or strengthening of a mid-latitude cyclone is called **cyclogenesis.** There are regions of North America that show a propensity for cyclogenesis, including the Gulf of Mexico, the Atlantic Ocean east of the Carolinas, and the eastern slope of high mountain ranges, such as the Rockies and the Sierra Nevada. For example, when a westerly flow of air crosses a north-to-south trending mountain range, the air on the downwind (leeward) side tends to curve cyclonically, as shown in ● Fig. 12.4. This curving of air adds to the developing or strengthening of a cyclonic storm. Such storms that form on the leeward side of a mountain are called **lee-side lows** and their development, *lee cyclogenesis.*

Another region of cyclogenesis lies near Cape Hatteras, North Carolina, where warm Gulf Stream water can supply moisture and warmth to the region south of a stationary front, thus increasing the contrast between air masses to a point

● FIGURE 12.4 As westerly winds blow over a mountain range, the airflow is deflected in such a way that a trough forms on the downwind (leeward) side of the mountain. Troughs and developing cyclonic storms that form in this manner are called *lee-side lows.*

where storms may suddenly spring up along the front. These cyclones, called **northeasters** or *nor'easters,* normally move northeastward along the Atlantic Coast, bringing high winds and heavy snow or rain to coastal areas. Before the age of modern satellite imagery and weather prediction, such coastal storms would often go undetected during their formative stages, and sometimes an evening weather forecast of "fair and colder" along the eastern seaboard would have to be changed to "heavy snowfall" by morning. Fortunately, with today's weather information-gathering and forecasting techniques, these storms rarely strike by surprise. (Additional information on northeasters is given in the Focus section on p. 318.)

● Figure 12.5 shows the typical paths taken in winter by mid-latitude cyclones and anticyclones. Notice in Fig. 12.5a that some of the lows are named after the region where they form, such as the *Hatteras low* which develops off the coast near Cape Hatteras, North Carolina. The *Alberta Clipper* forms (or redevelops) on the eastern side of the Rockies in Alberta, Canada, then rapidly skirts across the northern tier states. The *Colorado low,* in contrast, forms (or redevelops) on the eastern side of the Rockies. Notice that the lows generally move eastward or northeastward, whereas the highs typically move southeastward, then eastward.

When mid-latitude cyclones deepen rapidly (in excess of 24 mb in 24 hours), the term *explosive cyclogenesis,* or "*bomb,*" is sometimes used to describe them. As an example, explosive cyclogenesis occurred in a storm that developed over the warm Atlantic just east of New Jersey on September 10, 1978. As the central pressure of the storm dropped nearly 60 mb (1.8 in.) in 24 hours, hurricane force winds battered the ocean liner *Queen Elizabeth II* and sank the fishing vessel *Captain Cosmo.*

Some frontal waves form suddenly, grow in size, and develop into huge cyclonic storms. They slowly dissipate with the entire process taking several days to a week to complete. Other frontal waves remain small and never grow into a giant weather-producer. Why is it that some frontal waves develop into huge cyclonic storms, whereas others simply dissipate in a day or so?

This question poses one of the real challenges in weather forecasting. The answer is complex. Indeed, there are many surface conditions that do influence the formation of a mid-latitude cyclone, including mountain ranges and land-ocean temperature contrasts. However, the real key to the development of a wave cyclone is found in the *upper-wind flow,* in the region of the high-level westerlies. Therefore, before we can arrive at a reasonable answer to our question, we need to see how the winds aloft influence surface pressure systems.

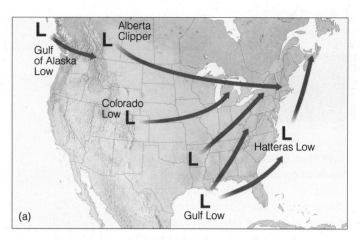

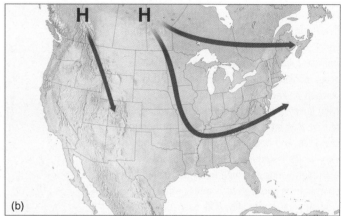

● **FIGURE 12.5** (a) Typical paths of winter mid-latitude cyclones. The lows are named after the region where they form. (b) Typical paths of winter anticyclones.

Vertical Structure of Deep Dynamic Lows

In Chapter 9, we learned that thermal pressure systems are shallow systems that weaken with increasing height above the surface. (Look back at Fig. 9.23, p. 239.) On the other hand, developing surface middle-latitude cyclones are deep *dynamic lows* that usually intensify with height. Hence, they appear on an upper-level chart as either a closed low or a trough.

Suppose the upper-level low is directly above the surface low as illustrated in ● Fig. 12.6. Notice that only at the surface (because of friction) do the winds blow inward toward the low's center. As these winds converge (flow together), the air "piles up." This piling up of air, called **convergence,** causes air density to increase directly above the surface low. This increase in mass causes surface pressures to rise; gradually, the low fills and the surface low dissipates. The same reasoning can be applied to surface anticyclones. Winds blow outward, away from the center of a surface high. If a closed high or ridge lies directly over the surface anticyclone, **divergence** (the spreading out of air) at the surface will remove air from the column directly above the high. The decrease in mass

Northeasters

Northeasters (commonly called *nor'easters*) are mid-latitude cyclonic storms that develop or intensify off the eastern seaboard of North America then move northeastward along the coast. They often bring gale force northeasterly winds to coastal areas, along with heavy rain, snow, or sleet. They usually deepen and become most intense off the coast of New England. The ferocious northeaster of December, 1992 (shown in Fig. 1), produced strong northeasterly winds from Maryland to Massachusetts. Huge waves accompanied by hurricane-force winds that reached 78 knots (90 mi/hr) in Wildwood, New Jersey, pounded the shoreline, causing extensive damage to beaches, beachfront homes, sea walls, and boardwalks. Heavy snow and rain, which lasted for several days, coupled with high winds and high tides, put many coastal areas and highways under water, including parts of the New York City subway. Another strong northeaster dumped between one and three feet of snow over portions of the northeast during late March, 1997.

Studies suggest that some of the northeasters, which batter the coastline in winter, may actually possess some of the characteristics of a tropical hurricane. For example, the northeaster shown in Fig. 1 actually developed something like a hurricane's "eye" as the winds at its center went calm when it moved over Atlantic City, New Jersey. (We will examine hurricanes and their characteristics in more detail in Chapter 15.)

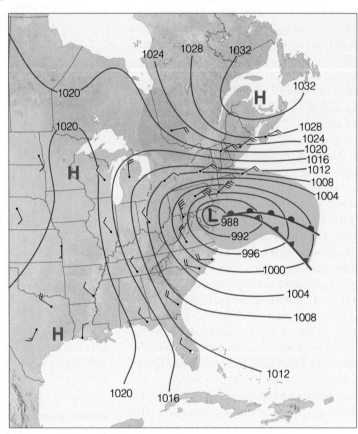

● FIGURE 1 The surface weather map for 7:00 A.M. (EST) December 11, 1992, shows an intense low-pressure area (central pressure 988 mb, or 29.18 in.), which is generating strong northeasterly winds and heavy precipitation (area shaded green) from the mid-Atlantic states into New England. This northeaster devastated a wide area of the eastern seaboard, causing damage in the hundreds of millions of dollars.

causes the surface pressure to fall and the surface high-pressure area to weaken. Consequently, it appears that, if upper-level pressure systems were always located directly above those at the surface (such as shown in Fig. 12.6), cyclones and anticyclones would die out soon after they form (if they could form at all). What, then, is it that allows these systems to develop and intensify?

● Figure 12.7 is an idealized model of the vertical structure of a middle-latitude cyclone and anticyclone in the Northern Hemisphere. Note that behind the cold front there is cold air both at the surface and aloft. This cold, dense air is helping to maintain the surface anticyclone. But remember from Chapter 8, that aloft, in a region of cold air, constant pressure surfaces are squeezed closer together. This squeezing is due to the fact that in the cold, dense air the atmospheric pressure decreased rapidly with height, causing cold air aloft to be associated with low pressure. Consequently, in the cold air aloft we find the upper low; and it is located *behind*, or to the *west* of the surface low. Observe also how the surface low tilts toward the northwest as we move up from the surface, showing up as a closed system on the 500-mb chart and as a trough on the 300-mb chart. Directly above the surface low, at 300 mb, the air spreads out and diverges (as indicated by the wind flow). This allows the converging surface air to rise and flow out of the top of the air column just below the tropopause, which acts as a constraint to vertical motions. We now have a mechanism for developing mid-latitude cyclonic storms. *When upper-level divergence is stronger than surface convergence (more air is taken out at the top than is brought in at the bottom), surface pressures drop, and the low intensifies (deepens). By the same token, when upper-level divergence is less than surface convergence (more air flows in at the bottom than is removed at the top), surface pressures rise, and the system weakens.*

We can also use Fig. 12.7 to explain the structure of the anticyclone. Notice that at the surface and aloft, warm air lies to the southwest of the surface high. Again, in Chapter 8 we saw that warm air aloft causes the isobaric surfaces to spread farther apart, which results in warm air aloft being associated with higher pressure. This situation causes the surface anticyclone to tilt toward the southwest—toward the warmer air—at higher altitudes. As we move upward from the surface, we observe that the closed area of high pressure at 500 mb becomes a ridge at the 300-mb level. Also notice that directly above the surface high at 300 mb there is convergence of air (as indicated by the wind flow lines). Convergence causes an accumulation of air above the surface high, which allows the air to sink slowly and replace the diverging surface air. Hence, *when upper-level convergence of air exceeds low-level divergence (inflow at top is greater than outflow near the surface), surface pressures rise, and the anticyclone builds.* On the other hand, *when upper-level convergence of air is less than low-level divergence, the anticyclone weakens as surface pressures fall.* (Additional information on the subject of convergence and divergence is provided in the Focus section on p. 320.)

Look at the wind direction at the 500-mb level in Fig. 12.7. Winds at this altitude tend to steer surface systems in the same direction that the winds are moving. Thus, the surface mid-latitude cyclone will move toward the northeast, while the surface anticyclone will move toward the southeast. As we can see in Fig. 12.5, these paths indicate the average movement of surface pressure systems in the eastern two-thirds of the United States. In general, surface storms travel across the United States at about 16 knots in summer and about 27 knots in winter. The faster winter velocity reflects the stronger upper-level flow during this time of year.*

So far, we have seen that deep pressure systems exist at the surface and aloft throughout much of the troposphere. When the upper-level trough lies to the west of the surface mid-latitude cyclone, the atmosphere is able to redistribute its mass. Regions of low-level converging air are compensated for by regions of upper-level diverging air and vice versa. Cyclones and anticyclones can intensify and, steered by the winds aloft, move away from their region of formation.

Since regions of strong upper-level divergence and convergence typically occur when deep troughs and ridges—waves—exist in the flow aloft, the next section examines these waves and their influence on a developing mid-latitude cyclonic storm.

---

*As a forecasting rule of thumb, surface pressure systems tend to move in the same direction as the wind at the 500-mb level. The speed at which the surface systems move is about half the speed of the 500-mb winds.

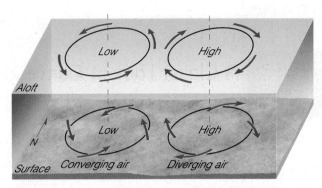

● **FIGURE 12.6** If lows and highs aloft were always directly above lows and highs at the surface, the surface systems would quickly dissipate.

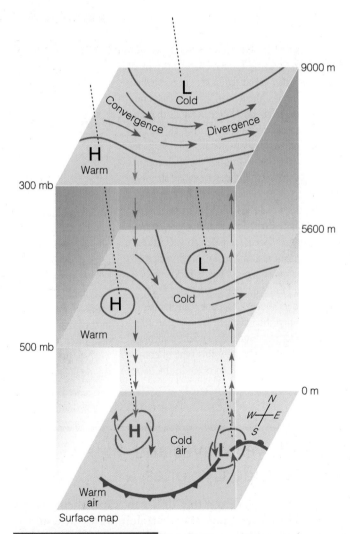

ACTIVE FIGURE 12.7 An idealized vertical structure of a middle-latitude cyclone and anticyclone. Visit the Meteorology Resource Center to view this and other active figures at academic. cengage.com/login

FOCUS ON A SPECIAL TOPIC

A Closer Look at Convergence and Divergence

We know that *convergence* is the piling up of air above a region, while *divergence* is the spreading out of air above some region. Convergence and divergence of air may result from changes in wind direction and wind speed. For example, convergence occurs when moving air is funneled into an area, much in the way cars converge when they enter a crowded freeway. Divergence occurs when moving air spreads apart, much as cars spread out when a congested two-lane freeway becomes three lanes. On an upper-level chart, this type of convergence (also called *confluence*) occurs when contour lines move closer together, as a steady wind flows parallel to them (see the upper-level chart in Fig. 2). On the same chart, this type of divergence (also called *diffluence*) occurs when the contour lines move apart as a steady wind flows parallel to them. Notice in Fig. 2 that below the area of convergence lies the surface anticyclone, whereas below the area of divergence lies the surface middle-latitude cyclonic storm.

Convergence and divergence may also result from changes in wind speed. *Speed convergence* occurs when the wind slows down as it moves along, whereas *speed divergence* occurs when the wind speeds up. We can grasp these relationships more clearly if we imagine air molecules to be marching in a band. When the marchers in front slow down, the rest of the band members squeeze together, causing convergence; when the marchers in front start to run, the band members spread apart, or diverge.

Figure 3 illustrates how this type of convergence and divergence can occur in the upper

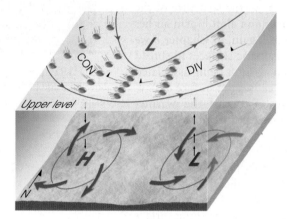

• FIGURE 2 The formation of convergence (CON) and divergence (DIV) of air with a constant wind speed (indicated by flags) in the upper troposphere. Circles represent air parcels that are moving parallel to the contour lines on a constant pressure chart. Below the area of convergence the air is sinking, and we find the surface high (H). Below the area of divergence the air is rising, and we find the surface low (L).

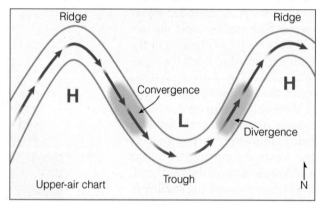

• FIGURE 3 As the faster-flowing air in the ridge moves toward the slower-flowing air in the trough, the air piles up and converges. As the slower-moving air in the trough moves toward the faster-flowing air in the ridge, the air spreads apart and diverges.

troposphere. The upper-air chart shows a trough and two ridges with evenly spaced contour lines. Notice that even though the contours are evenly spaced, the winds blow faster in the ridge than they do in the trough (a concept discussed in Chapter 8 on p. 213).

As the faster-flowing air moves away from the ridge and approaches the slower-moving air

in the trough, the air piles up, producing convergence. Where the slower-moving air in the trough approaches the faster-moving air in the ridge, the air spreads out, producing divergence in the air flow.

Upper-Level Waves and Mid-Latitude Cyclones

You may remember from the "dish-pan" experiment (found in the Focus section on p. 267) that, aloft, waves are a fundamental feature of an unevenly heated, rotating sphere, such as the earth. If we examine an upper-level chart that shows almost the entire Northern (or Southern) Hemisphere, such as
• Fig. 12.8, the waves appear as a series of troughs and ridges

with significant amplitude that encircle the globe. The distance from trough to trough (or ridge to ridge), is known as the *wavelength*. When the wavelength is on the order of many thousands of kilometers, the wave is called a longwave. At any one time there are usually between three and six longwaves looping around the earth. The fewer the number of waves, the longer their wavelengths. Since mountain ranges tend to disturb the upper-level wind flow, these waves are often found to the east of such topographic barriers as the Rockies and Tibetan Plateau. Sometimes longwaves exhibit a relatively small amplitude (or north-to-south extent) and the flow is mainly zonal, or west to east. On other occasions, the waves exhibit considerable amplitude and the flow has a strong north-to-south (meridional) component to it.

Longwaves are also known as *planetary waves* and as **Rossby waves,** after C. G. Rossby, a famous meteorologist who carefully studied their motion. Imbedded in longwaves are **shortwaves,** which are small disturbances, or ripples that move with the wind flow (see ● Fig. 12.9a). Rossby found that the shorter the wavelength of a particular wave, the faster it moved downstream. Shortwaves tend to move eastward at a speed proportional to the average wind flow near the 700-mb level, about 3 km above sea level. Longwaves, on the other hand, often remain stationary, move eastward very slowly at less than 4° of longitude per day (about 8 knots), or even move westward *(retrograde).** We can obtain a better idea of this wave movement if we think of longwaves as being huge meanders (loops) in a swiftly flowing stream of water. Water moves through the loops quickly, while the loops themselves move eastward very slowly, as the fast-flowing water cuts away at one bank and deposits material on the other. Suppose debris tumbles into the stream, disturbing the flow. The disturbed flow appears as a small wrinkle that travels downstream through the loops at a speed near the average stream flow. This wrinkle in the flow is analogous to a shortwave in the atmosphere.

Notice in Fig. 12.9b that, while the longwaves move eastward very slowly, the shortwaves move fairly quickly around the longwaves. Notice also that the shortwaves tend to deepen (that is, increase in size) when they approach a longwave trough and weaken (become smaller) when they approach a

---

*Retrograde wave motion means that the wave is actually moving in the opposite direction of the wind flow.

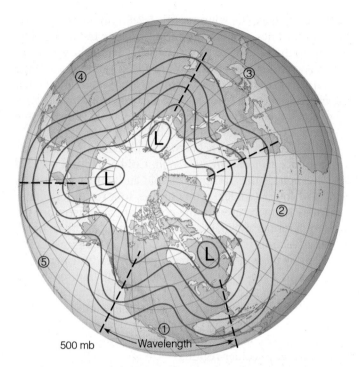

● FIGURE 12.8 A 500-mb map of the Northern Hemisphere from a polar perspective shows five longwaves encircling the globe. Note that the wavelength of wave number 1 is as great as the width of the United States. Solid lines are contours. Dashed lines show the position of longwave troughs.

ridge. Moreover, when a shortwave moves into a longwave trough, the trough tends to deepen. (Look at shortwave 3 in Fig. 12.9b.)

Where the contour lines in Fig. 12.9b are roughly parallel to the isotherms (dashed lines), the atmosphere is said to be

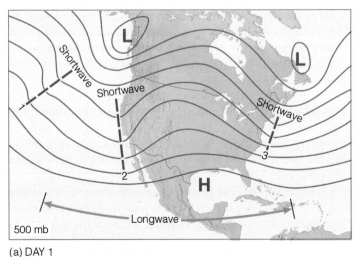

(a) DAY 1

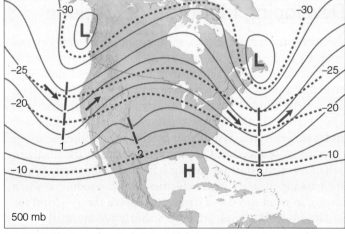

(b) DAY 2 (24 hours later)

ACTIVE FIGURE 12.9   (a) Upper-air chart showing a longwave with three shortwaves (heavy dashed lines) embedded in the flow. (b) Twenty-four hours later the shortwaves have moved rapidly around the longwave. Notice that the shortwaves labeled 1 and 3 tend to deepen the longwave trough, while shortwave 2 has weakened as it moves into a ridge. Notice also that as the longwave deepens in diagram (b), its length actually shortens. Dashed lines are isotherms in °C. Solid lines are contours. Blue arrows indicate cold advection and red arrows, warm advection. Visit the Meteorology Resource Center to view this and other active figures at academic.cengage.com/login

barotropic. Since the winds at this level more or less parallel the contour lines, in a barotropic atmosphere the winds blow parallel to the isotherms. By comparison, where the isotherms *cross* the contour lines, temperature advection occurs and the atmosphere is said to be **baroclinic.*** Notice in Fig. 12.9b that the baroclinic region tends to be in a narrow zone in the vicinity of shortwaves 1 and 3. The shortwaves actually disturb the flow and accentuate the region of baroclinicity.

In the region of baroclinicity, winds cross the isotherms and produce *temperature advection.* **Cold advection** (or *cold air advection*) is the transport of cold air by the wind from a region of lower (colder) temperatures to a region of higher (warmer) temperatures. In the region of cold advection, the air temperature normally decreases. On the other hand, **warm advection** (or *warm air advection*) is the transport of warm air by the wind from a region of higher (warmer) temperatures to a region of lower (colder) temperatures. In the region of warm advection, the air temperature normally increases. For cold advection to occur, the wind must blow across the isotherms from colder to warmer regions, whereas for warm advection, the wind must blow across the isotherms from warmer to colder regions.

In the baroclinic region in Fig. 12.9b, observe that strong winds cross the isotherms, producing cold advection (blue arrows) on the trough's west side and warm advection (red arrows) on its east side. Below the baroclinic zone lies the polar front; above it flows the polar-front jet stream. The disturbed flow created by the shortwaves is now capable of aiding in the development or intensification of a surface mid-latitude cyclonic storm. The theory explaining how this phenomenon occurs is known as the *baroclinic wave theory of developing cyclones.*

The Necessary Ingredients for a Developing Mid-Latitude Cyclone

To better understand how a wave cyclone may develop and intensify into a huge mid-latitude cyclonic storm, we need to examine atmospheric conditions at the surface and aloft. Suppose that a portion of a longwave trough at the 500-mb level lies directly above a surface stationary front, as illustrated in ● Fig. 12.10a. On the 500-mb chart, contour lines (solid lines) and isotherms (dashed lines) parallel each other and are crowded close together. Colder air is located in the northern half of the map, while warmer air is located to the south. Winds are blowing at fairly high velocities, which produce a sharp change in wind speed—a strong *wind speed shear*—from the surface up to this level. Suppose a shortwave moves through this region, disturbing the flow as shown in Fig. 12.10b. This sets up a kind of instability in the flow (as warmer air rises and colder air sinks) known as **baroclinic instability.**

UPPER-AIR SUPPORT   With the onset of baroclinic instability, horizontal and vertical air motions begin to enhance the formation of a cyclonic storm. For example, as the flow aloft becomes disturbed, it begins to lend support for the intensification of surface pressure systems, as a region of converging air forms above position 1 in Fig. 12.10b and a region of diverging air forms above position 2.* The converging air aloft causes the surface air pressure to rise in the region marked *H* in Fig. 12.10b. Surface winds begin to blow out away from the region of higher pressure, and the air aloft gradually sinks to replace it. Meanwhile, diverging air aloft causes the surface air pressure to decrease beneath position 2, in the region marked *L* on the surface map. This initiates rising air, as the surface winds blow in toward the region of lower pressure. As the converging surface air develops cyclonic spin, cold air flows southward and warm air northward. We can see in Fig. 12.10b that the western half of the stationary front is now a cold front and the eastern half a warm front. Cold air moves in behind the cold front, while warm air slides up along the warm front. These regions of cold and warm advection occur all the way up to the 500-mb level.

On the 500-mb chart in Fig. 12.10b, cold advection is occurring at position 1 (blue arrow) as the wind crosses the isotherms, bringing cold air into the trough. The cold advection makes the air more dense and lowers the height of the air column from the surface up to the 500-mb level. (Recall that, on a 500-mb chart, lower heights mean the same as lower pressures.) Consequently, the pressure in the trough lowers and the trough deepens. The deepening of the upper trough causes the contour lines to crowd closer together and the winds aloft to increase. Meanwhile, at position 2 (red arrow) warm advection is taking place, which has the effect of raising the height of a column of air; here, the 500-mb heights increase and a ridge builds (strengthens). Therefore, *the overall effect of differential temperature advection is to amplify the upper-level wave.* As the trough aloft deepens, its curvature increases, which in turn increases the region of divergence above the developing surface storm. At this point, the surface mid-latitude cyclone rapidly develops as surface pressures fall.

Regions of cold and warm advection are associated with vertical motions. Where there is cold advection, some of the cold, heavy air sinks; where there is warm advection, some of the warm, light air rises. Hence, due to advection, air must be sinking in the vicinity of position 1 and rising in the vicinity of position 2.

The sinking of cold air and the rising of warm air provide energy for a developing cyclone, as potential energy is transformed into kinetic energy. Further, if clouds form, condensation in the ascending air releases latent heat, which warms the air. The warmer air lowers the surface pressure, which strengthens the surface low even more. So, we now have a

---

*Actually, on a constant pressure surface, baroclinic conditions exist where the air density varies, and barotropic conditions exist where the air density does not vary.

*Look back at Fig. 12.7, p. 319, and the upper-air chart in Fig. 2 and Fig. 3 on p. 320, and note the regions of converging air and diverging air on these maps.

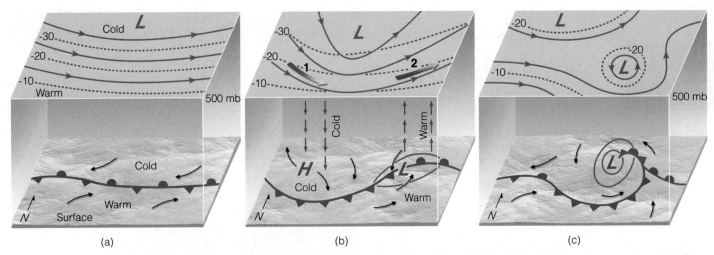

● FIGURE 12.10 An idealized view of the formation of a mid-latitude cyclone during baroclinic instability. (a) A longwave trough at 500 mb lies parallel to and directly above the surface stationary front. (b) A shortwave (not shown) disturbs the flow aloft, initiating temperature advection (blue arrow, cold advection; red arrow, warm advection). The upper trough intensifies and provides the necessary vertical motions (as shown by vertical arrows) for the development of the surface cyclone. (c) The surface storm occludes, and without upper level divergence to compensate for surface converging air, the storm system dissipates.

full-fledged middle-latitude cyclone with all of the necessary ingredients for its development.

Eventually, the warm air curls around the north side of the low, and the storm system occludes (see Fig. 12.10c).* Some storms may continue to deepen, but most do not as they move out from under the region of upper-level divergence. Additionally, at the surface the storm weakens as the supply of warm air is cut off and cold, dry air behind the cold front (called a *dry slot*) is drawn in toward the surface low.

Sometimes, an upper-level pool of cold air (which has broken away from the main flow) lies almost directly above the surface low. Occasionally the upper low will break away entirely from the main flow, producing a **cut-off low,** which often appears as a single contour line on an upper-level chart. When the upper low lies directly above the surface low (as in Fig. 12.10c), the storm system is said to be *vertically stacked.* Usually the isotherms around the upper low parallel the contour lines, which indicates that no significant temperature advection is occurring. Without the necessary energy transformations, the surface system gradually dissipates. As its winds slacken and its central pressure gradually rises, the low is said to be *filling.* The upper-level low, however, may remain stationary for many days. If air is forced to ascend into this cold pocket, widespread clouds and precipitation may persist for some time, even though the surface storm system itself has moved east out of the picture.

THE ROLE OF THE JET STREAM   As we have seen, in order for middle-latitude cyclones to develop and intensify there must be upper-level diverging air above the surface storm. The polar jet stream can provide such areas of divergence. In

Chapter 10, we learned that the axis of the polar-front jet stream pretty much coincides with the position of the polar front. The region of strongest winds in the jet stream is known as a *jet stream core,* or **jet streak.** When the polar jet stream flows in a wavy west-to-east pattern, a jet streak tends to form in the trough of the jet, where pressure gradients are tight. The curving of the jet stream coupled with the changing wind speeds around the jet streak produces regions of strong convergence and divergence of air along the flanks of the jet. (More on this topic is given in the Focus section on p. 324.)

Notice that the region of diverging air (marked *D* in ● Fig. 12.11a) draws warm surface air upward to the jet, which quickly sweeps the air downstream. Since the air above the mid-latitude cyclone is being removed more quickly than converging surface winds can supply air to the storm's center, the central pressure of the storm drops rapidly. As surface pressure gradients increase, the wind speed increases. Above the high-pressure area, a region of converging air (marked *C* in Fig. 12.11a) feeds air downward into the anticyclone. Hence, we find *the polar jet stream removing air above the surface cyclone and supplying air to the surface anticyclone.*

As the jet stream steers the mid-latitude cyclonic storm along—toward the northeast in this case—the surface cyclone occludes, and cold air surrounds the surface low (see Fig. 12.11b). Since the surface low has moved out from under the pocket of diverging air aloft, the occluded storm gradually fills as surface air flows into the system.

Since the polar jet stream is strongest and moves farther south in winter, we can see why mid-latitude cyclonic storms are better developed and move more quickly during the colder months. During the summer when the polar jet shifts northward, developing mid-latitude storm activity shifts northward and occurs principally over the Canadian provinces of Alberta and the Northwest Territories.

*If the occluded front should extend west of the low's center, it is sometimes referred to as a *bentback occlusion.*

FOCUS ON A SPECIAL TOPIC

Jet Streaks and Storms

Figure 4 shows an area of maximum winds, a *jet streak*, on a 300-mb chart. Jet streaks have winds of at least 50 knots, and represent small segments (ranging in length from a few hundred kilometers to over 3000 kilometers) within the meandering jet stream flow.

Jet streaks are important in the development of surface mid-latitude cyclones because areas of convergence and divergence form at specific regions around them. To understand why, consider air moving through a straight jet streak (shaded area) in Fig. 5. As the air enters the front of the streak (known as the *entrance region*), it increases in speed; as it leaves the rear of the streak (known as the *exit region*), it decreases in speed. At this elevation in the atmosphere (about 10,000 m or 33,000 ft above the surface), the wind flow is nearly in geostrophic balance with the pressure gradient force (directed north) and the Coriolis force (directed south). As the air enters the jet streak, it increases in speed because the contour lines are closer together, causing an increase in the pressure gradient force. The greater force temporarily exceeds the Coriolis force, and the air swings slightly to the north across the contour lines, which causes a piling up of air (called a *bottleneck effect*) and *strong convergence at point 1*. Weak divergence occurs at point 2.

Toward the middle of the jet streak, the increase in wind speed causes the Coriolis force to increase and the wind to become nearly geostrophic again. However, as the air exits the jet streak, the pressure gradient force is reduced as the contour lines spread farther apart. Hence, the Coriolis force temporarily exceeds the pressure gradient force, causing the air to cross the contour lines and swing slightly to the south. This process produces *strong divergence at point 3* and weak convergence at point 4.

The conditions described so far exist in a straight jet streak that shows no curvature. When the jet stream becomes wavy, and the jet streak exhibits cyclonic curvature (as it does in Fig. 4), the areas of weak divergence (at point 2) and weak convergence (at point 4) all

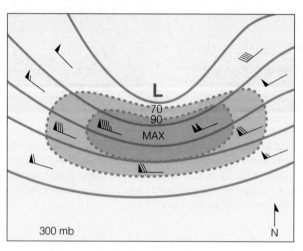

● FIGURE 4 A portion of a 300-mb chart (about 33,000 ft above sea level) that shows the core of the jet — the region of maximum winds (MAX)—called a *jet streak*.

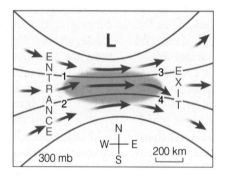

● FIGURE 5 Changing air motions within a straight jet streak (shaded area) cause strong convergence of air at point 1 (left entrance region) and strong divergence at point 3 (left exit region).

but disappear. What we are left with is a curving jet streak that exhibits strong divergence at point 3 (in the left exit region) and strong convergence at point 1 (in the left entrance region). Notice in Fig. 6 that below the area of strong divergence the air rises, cools and, if sufficiently moist, condenses into clouds. Moreover, the removal of air in the region of strong divergence causes surface pressures to fall, which results in the development of an area of surface low pressure.

● FIGURE 6 An area of strong divergence (DIV) can form with a curving jet streak. Below the area of divergence are rising air, clouds, and the developing mid-latitude cyclonic storm.

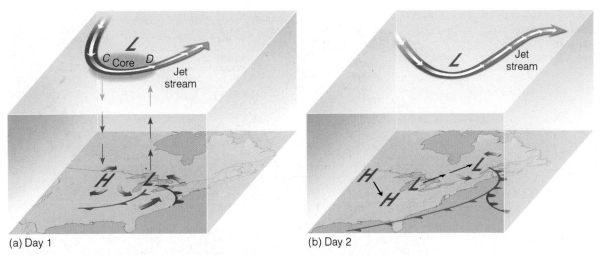

(a) Day 1   (b) Day 2

● FIGURE 12.11 (a) As the polar jet stream and its area of maximum winds (the jet streak, or core) swings over a developing mid-latitude cyclone, an area of divergence *(D)* draws warm surface air upward, and an area of convergence *(C)* allows cold air to sink. The jet stream removes air above the surface storm, which causes surface pressures to drop and the storm to intensify. (b) When the surface storm moves northeastward and occludes, it no longer has the upper-level support of diverging air, and the surface storm gradually dies out.

In general, we now have a fairly good picture as to why some surface lows intensify into huge mid-latitude cyclones while others do not. For a surface cyclonic storm to intensify, there must be an upper-level counterpart—a trough of low pressure—that lies to the *west* of the surface low. As short-waves disturb the flow aloft, they cause regions of differential temperature advection to appear, leading to an intensification of the upper-level trough. At the same time, the polar jet forms into waves and swings slightly south of the developing storm. When these conditions exist, zones of converging and diverging air, along with rising and sinking air, provide energy conversions for the storm's growth. With this atmospheric situation, storms may form even where there are no pre-existing fronts.* In regions where the upper-level flow is not disturbed by shortwaves or where no upper trough or jet stream exists, the necessary vertical and horizontal motions are insufficient to enhance cyclonic storm development and we say that the surface storm does not have the proper *upper-air support.* The horizontal and vertical motions, cloud patterns, and weather that typically occur with a developing open-wave cyclone are summarized in ● Fig. 12.12.

CONVEYOR BELT MODEL OF MID-LATITUDE CYCLONES A three-dimensional model of a developing mid-latitude cyclone is illustrated in ● Fig. 12.13. The model describes rising and sinking air as traveling along three main "conveyor belts." Just as people ride escalators to higher levels in a department

store, so air glides along through a constantly evolving mid-latitude cyclone. According to the **conveyor belt model,** a warm air stream (known as the *warm conveyor belt*—orange arrow in Fig. 12.13) originates at the surface in the warm sector, ahead of the cold front. As the warm air stream moves northward, it slowly rises along the sloping warm front, up and over the cold air below. As the rising air cools, water vapor condenses, and clouds form well out ahead of the surface low and its surface warm front. From these clouds, steady precipitation usually falls in the form of rain or snow. Aloft, the warm air flow gradually turns toward the northeast, parallel to the upper-level winds.

Directly below the warm conveyor belt, a cold, relatively dry airstream—the *cold conveyor belt*—moves slowly westward (see Fig. 12.13). As the air moves west ahead of the warm front, precipitation and surface moisture evaporates into the cold air, making it moist. As the cold, moist airstream moves into the vicinity of the surface low, rising air gradually forces the cold conveyor belt upward. As the cold, moist air sweeps northwest of the surface low, it often brings heavy winter snowfalls to this region of the cyclone. The rising airstream usually turns counterclockwise, around the surface low, first heading south, then northeastward, when it gets caught in the upper air flow. It is the counterclockwise turning of the cold conveyor belt that produces the comma-shaped cloud similar to the one shown in ● Fig. 12.14.

The last conveyor belt is a dry one that forms in the cold, very dry region of the upper troposphere. Called the *dry conveyor belt,* and shaded yellow in Fig. 12.13, this airstream slowly descends from the northwest behind the surface cold front, where it brings generally clear, dry weather and, occasionally, blustery winds. If a branch of the dry air sweeps into the storm, it produces a clear area called a **dry slot,** which ap-

---

*It is interesting to note that the beginning stage of a wave cyclone almost always takes place when an area of upper-level divergence passes over a surface front. However, even if initially there are no fronts on the surface map, they may begin to form where air masses having contrasting properties are brought together in the region where the surface air rises and the surrounding air flows inward.

● FIGURE 12.12
Summary of clouds, weather, vertical motions, and upper-air support associated with a developing mid-latitude cyclone.

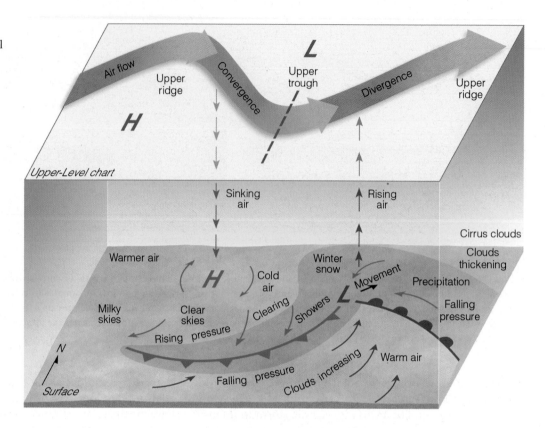

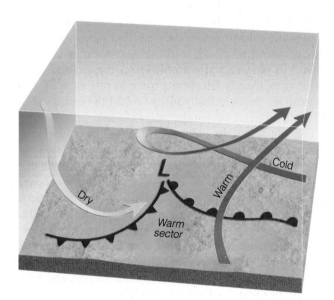

● FIGURE 12.13 The conveyor belt model of a developing mid-latitude cyclone. The warm conveyor belt (in orange) rises along the warm front, causing clouds and precipitation to cover a vast area. The cold conveyor belt (in blue) slowly rises as it carries cold, moist air westward ahead of the warm front but under the rising warm air. The cold conveyor belt lifts rapidly and wraps counterclockwise around the center of the surface low. The dry conveyor belt (in yellow) brings very dry, cold air downward from the upper troposphere.

pears to pinch off the comma cloud's head from its tail. This phenomenon tends to show up on satellite images as the mid-latitude storm becomes more fully developed (see Fig. 12.14).

We are now in a position to tie together many of the concepts we have learned about developing mid-latitude cyclones by examining a monstrous storm that formed during March, 1993.

A DEVELOPING MID-LATITUDE CYCLONE — THE MARCH STORM OF 1993   A color-enhanced infrared satellite image of a developing mid-latitude cyclone on the morning of March 13, 1993, is shown in ● Fig. 12.15. Notice that its cloud band is in the shape of a comma that covers the entire eastern seaboard. Such **comma clouds** indicate that the storm is still developing and intensifying. But this storm is not an ordinary wave cyclone — this storm intensified into a superstorm, which some forecasters dubbed "the storm of the century."

The surface weather map for the morning of March 13 (see ● Fig. 12.16) shows that the center of the open wave is over northern Florida. Observe in Fig. 12.15 that this position is in the head of the comma cloud. The central pressure of the storm is 975 mb (28.79 in.), which indicates an incredibly deep system, considering a typical open wave would have a central pressure closer to 996 mb (29.41 in.). A strong cold front stretches from the storm's center through western Florida. Behind the front, cold arctic air pours into the Deep South. Ahead of the advancing front, a band of heavy thunderstorms (along a squall line) is pounding Florida with

● FIGURE 12.14 Visible satellite image of a mature mid-latitude cyclone with the three conveyor belts superimposed on the storm. As in Fig. 12.13, the warm conveyor belt is in orange, the cold conveyor belt is in blue, and the dry conveyor belt (forming the *dry slot*) is in yellow.

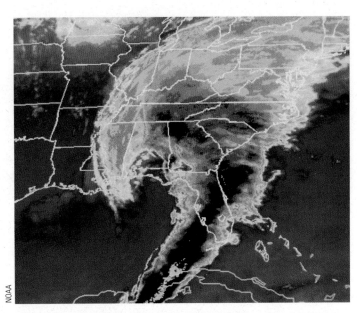

● FIGURE 12.15 A color-enhanced infrared satellite image that shows a developing mid-latitude cyclone at 2 A.M. (EST) on March 13, 1993. The darkest shades represent clouds with the coldest and highest tops. The dark cloud band moving through Florida represents a line of severe thunderstorms. Notice that the cloud pattern is in the shape of a comma.

heavy rain, high winds, and tornadoes. Warm humid air in the warm sector is streaming northward, overrunning cold surface air ahead of the warm front, which is causing precipitation in the form of rain, snow, and sleet to fall over a vast area extending from Florida to New York.

The 500-mb chart for the morning of March 13 (see ● Fig. 12.17) shows that a deep trough extending southward out of Canada lies to the west of the surface low. A baroclinic atmosphere exists around the trough as isotherms intersect contour lines and strong winds produce differential tempera-

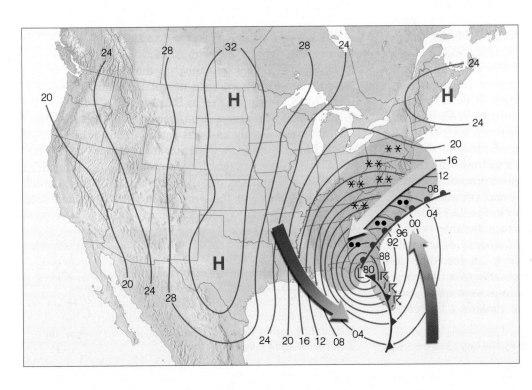

● FIGURE 12.16 Surface weather map for 4 A.M (EST) on March 13, 1993. Lines on the map are isobars. A reading of 96 is 996 mb and a reading of 00 is 1000 mb. (To obtain the proper pressure in millibars, place a 9 before those readings between 80 and 96, and place a 10 before those readings of 00 or higher.) Green shaded areas are receiving precipitation. Heavy arrows represent surface winds. The orange arrow represents warm, humid air; the light blue arrow, cold, moist air; and the dark blue arrow, cold, arctic air.

328 CHAPTER 12

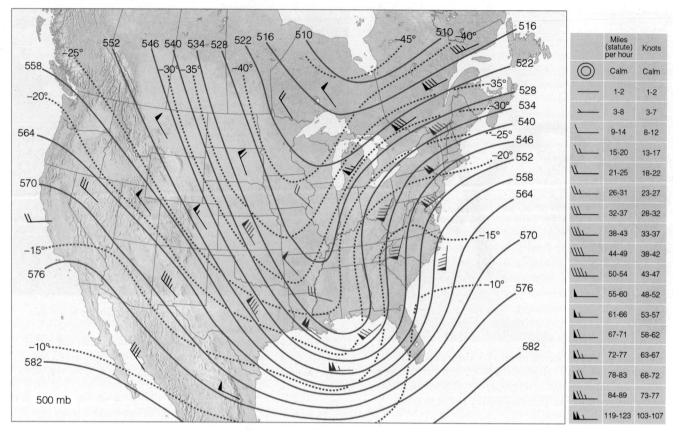

● FIGURE 12.17 The 500-mb chart for 7 A.M. (EST) March 13, 1993. Solid lines are contours where 564 equals 5640 meters. Dashed lines are isotherms in °C. Wind entries in red show warm advection. Those in blue show cold advection. Those in black indicate no appreciable temperature advection is occurring.

ture advection. Notice that warm advection (indicated by red barbs) is occurring on the trough's eastern side, ahead of the surface warm front shown in Fig. 12.16. Cold advection (indicated by blue barbs) is occurring on the trough's western side, behind the position of the surface cold front. As temperature advection deepens the trough, rising and sinking air provide energy for the developing surface storm.

In ● Fig. 12.18, we can see that the storm began on March 12 as a frontal wave off the Texas coast. In the upper air, a shortwave, moving rapidly around a longwave, disturbed the flow, setting up the necessary ingredients for the surface storm's development. By the morning of March 13, the storm had intensified into a deep open-wave cyclone centered over Florida. In the upper air, a region of diverging air positioned above the storm caused the storm's surface pressures to drop rapidly. Upper-level southwesterly winds (Fig. 12.17) directed the surface low northeastward, where it became occluded over Virginia during the afternoon of March 13. At this point, the storm's central pressure dropped to an incredibly low 960 mb (28.35 in.), a pressure comparable to a Category 3 hurricane.* Although the surface winds were quite strong and gusty, they were not as strong as those in a Category 3

hurricane because the isobars around the storm were spread farther apart than those in a hurricane and because surface friction slowed the winds. Higher up, away from the influence of the surface, the winds were much stronger as a wind of 125 knots was reported at the top of 1900-meter-high Mount Washington, New Hampshire.

The upper trough remained to the west of the surface low, and the storm continued its northeastward movement. In Fig. 12.18, we can see that by the morning of March 14 the storm (which was now a deep, bentback occluded system) had weakened slightly and was centered along the coast of Maine. Moving out from under its area of upper-level divergence, the storm weakened even more as it continued its northeastward journey, out over the North Atlantic. In all, the storm was one of the strongest ever. It blanketed deep snow from Alabama to Canada. Fierce winds piled the snow

WEATHER WATCH

The great March storm of 1993 set record low barometric pressure readings in a dozen states, produced wind gusts exceeding 90 knots from New England to Florida, and deposited 50 billion tons of snow over the east coast of the United States. When melted, this quantity of snow yielded enough water to equal the flow over Niagara Falls for 100 days.

---

*As we will see in Chapter 15, a pressure of 960 mb is equivalent to the pressure in a Category 3 hurricane on the Saffir-Simpson scale, which ranges from 1 to 5, with 5 being the strongest.

into huge drifts that closed roads, leaving motorists stranded. The storm shut down every major airport along the East Coast and more than 3 million people lost electric power at some point during the storm. "The storm of the century" damaged or destroyed hundreds of homes, produced 27 tornadoes, stranded hikers in North Carolina and Tennessee, caused an estimated $800 million in damage, and claimed the lives of at least 270 people.

BRIEF REVIEW

Up to this point, we have looked at the structure and development of mid-latitude cyclones. Before going on, here is a summary of a few of the important ideas presented so far:

- The polar front (or Norwegian) model of a developing mid-latitude cyclonic storm represents a simplified but useful model of how an ideal storm progresses through the stages of birth, maturity, and dissipation.

- For a surface mid-latitude cyclone to develop or intensify (deepen), the upper-level low must be located to the west of (behind) the surface low.

- For a surface mid-latitude cyclonic storm to form, there must be an area of upper-level diverging air above the surface low. For the surface storm to intensify, the region of upper-level diverging air must be greater than surface converging air (that is, more air must be removed above the storm than is brought in at the surface).

- When the upper-air flow develops into waves, winds often cross the isotherms, producing regions of cold advection and warm advection, which tend to amplify the wave. At the same time, vertical air motions begin to enhance the formation of the surface storm as the rising of warm air and the sinking of cold air provide the proper energy conversion for the storm's growth.

- When the polar-front jet stream develops into a looping wave, it provides an area of upper-level diverging air for the development of surface mid-latitude cyclonic storms.

- The curving nature of the polar-front jet stream tends to direct surface mid-latitude cyclonic storms northeastward and surface anticyclones southeastward.

Vorticity, Divergence, and Developing Mid-Latitude Cyclones

We know that for a surface mid-latitude cyclone to develop into a deep low-pressure area there must be an area of strong upper-level divergence situated above the developing storm. Meteorologists, then, are interested in locating regions of divergence on upper-level charts so that developing storms can be accurately predicted. We know from an earlier discussion that divergence and convergence of air are due to changes in either wind speed or wind direction. The problem is that it is a difficult task to measure divergence (or convergence) with

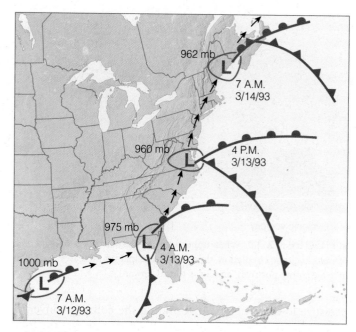

• FIGURE 12.18 The development of the ferocious mid-latitude cyclonic storm of March, 1993. A small wave in the western Gulf of Mexico intensifies into a deep open-wave cyclone over Florida. It moves northeastward and becomes occluded over Virginia where its central pressure drops to 960 mb (28.35 in.). As the occluded storm continues its northeastward movement, it gradually fills and dissipates. The number next to the storm is its central pressure in millibars. Arrows show direction of movement. Time is Eastern Standard Time.

any degree of accuracy using upper-level wind information. Therefore, meteorologists must look for something else that can be measured and, at the same time, can be related to regions of diverging (and converging) air. That something is called *vorticity*.

When something spins, it has vorticity. The faster it spins, the greater its vorticity. In meteorology, **vorticity** is a measure of the spin of small air parcels. Although the spin can be in any direction, our concern will be with the spin of horizontally flowing air about a vertical axis, much like an ice skater spins about an imaginary vertical axis. Because our goal is to see how vorticity can be used to identify regions of divergence and convergence, we must give vorticity some quantitative value. When viewed from above, air that spins cyclonically (counterclockwise) has *positive vorticity* and air that spins anticyclonically (clockwise) has *negative vorticity*.

We can see in • Fig. 12.19 how divergence aloft and the vorticity of surface air are related. Suppose the air column in Fig. 12.19 represents an area of low pressure with weak cyclonic (positive) spin. Further, suppose that an invisible barrier separates the column (and the ice skater inside) from the surrounding air. Now suppose an area of divergence aloft (at the 300-mb level) moves directly over the air column. Divergence aloft means that within this region more air is leaving than is entering. This removal of air above the column lowers the atmospheric pressure at the surface. As the surface

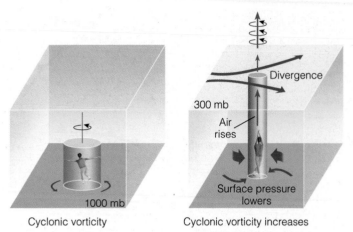

● FIGURE 12.19 When upper-level divergence moves over an area of weak cyclonic circulation, the cyclonic circulation increases (that is, it becomes more positive), and air is forced upward.

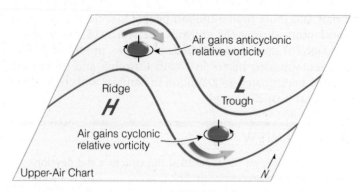

● FIGURE 12.21 In a region where the contour lines curve, air moving through a ridge spins clockwise and gains anticyclonic relative vorticity. In the trough, the air spins counterclockwise and gains cyclonic relative vorticity.

pressure lowers, the air surrounding the column converges on it, pushing inward on its sides. If we assume that the total mass of air in the column does not change, then squeezing the column causes it to shrink horizontally and stretch vertically, forcing the air in the column upward. Meanwhile, as the column stretches, the ice skater's arms are pulled in close, and the skater spins much faster. Hence, the rate at which air flows around the center of the column must also increase. Consequently, the vorticity of the column increases, becoming more positive. So we can see that divergence aloft causes an increase in the cyclonic (positive) vorticity of surface cyclones, which usually results in cyclogenesis and upward air motions.

Before we consider how vorticity on an upper-level chart ties in with developing mid-latitude cyclones, we need to ex-

amine two important types of this phenomenon: the earth's vorticity and relative vorticity.

VORTICITY ON A SPINNING PLANET   Because the earth spins, it has vorticity. In the Northern Hemisphere, the **earth's vorticity** (also called *planetary vorticity*) is always positive because the earth spins counterclockwise about its vertical North Pole axis. The amount of earth vorticity imparted to any object—even to those that are not moving relative to the earth's surface—depends upon the latitude. In ● Fig. 12.20, an observer standing on the equator would not spin about his or her own *vertical axis;* farther north, the observer would spin very slowly, while at the North Pole the observer would spin at a maximum rate of one revolution per day. It is now apparent that any object on the earth has vorticity simply because the earth is spinning, and the amount of this earth vorticity increases from zero at the equator to a maximum at the poles.*

Moving air will generally have additional vorticity relative to the earth's surface. This type of vorticity, called **relative vorticity,** is the sum of two effects: the curving of the air flow *(curvature)* and the changing of the wind speed over a horizontal distance *(shear)*. ● Figure 12.21 illustrates vorticity due to curvature. Air moving through a trough tends to spin cyclonically (counterclockwise), increasing its relative vorticity. In the ridge, the spin tends to be anticyclonic (clockwise), and the relative vorticity of the air increases, but in a negative direction. Whenever the wind blows faster on one side of an air parcel than on the other, a shear force is imparted on the parcel and it will spin and gain (or lose) relative vorticity, as illustrated in ● Fig. 12.22.

The sum of the *earth's vorticity* and the *relative vorticity* is called the **absolute vorticity.** To further illustrate this concept,

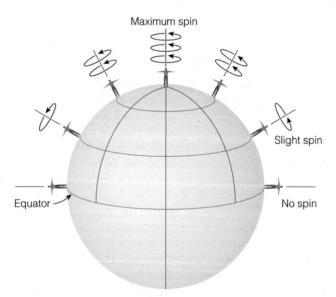

● FIGURE 12.20 Due to the rotation of the earth, the rate of spin of observers about their vertical axes increases from zero at the equator to a maximum at the poles.

---

*The earth's vorticity at any latitude is equal to the product of twice the earth's angular rate of spin (2Ω) and the sine of the latitude (ϕ); that is, $2\Omega \sin(\phi)$. This expression is referred to as the *Coriolis parameter,* and it is usually expressed by the letter *f*.

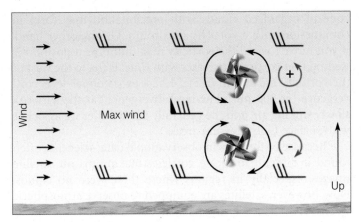

● FIGURE 12.22 Areas of cyclonic (positive) relative vorticity and anticyclonic (negative) relative vorticity can form in a region of strong wind-speed shear. Notice that the pinwheel changes its direction of spin when placed above and below the region of maximum winds.

suppose you are watching an ice skater spin counterclockwise on a frozen lake. The ice skater possesses positive relative vorticity. If you could suddenly leave the earth and watch the same spinning skater from space, you would see the ice skater spinning on a rotating platform — the earth. The combination of the skater spinning about her vertical axis (her relative vorticity) plus the small spin imparted to her from the earth (earth's vorticity) yields the absolute vorticity of the skater. We are now in a position to see how vorticity aloft ties in with divergence and the development of middle-latitude cyclones. (The concept of absolute vorticity can explain why the westerly flow aloft tends to form into waves. This topic is presented in the Focus section on p. 333.)

● Figure 12.23 shows an air parcel — a blob of air in this case — moving through a ridge and a trough in the upper troposphere. (For simplicity, we will assume that the parcel has a constant speed and that there is no shear acting on it.) Notice that at every position the parcel has cyclonic spin. This situation occurs because the parcel's total, or absolute vorticity, is a combination of the earth's vorticity plus its relative vorticity. In middle and high latitudes, the earth's vorticity is great enough to make the parcel's spin cyclonic everywhere on the earth.

At position 1 (in the ridge), the air parcel has only a slight cyclonic spin because the relative vorticity, due to curvature, is anticyclonic and subtracts from the earth's vorticity. At position 2, the relative vorticity due to curvature is zero, which allows the earth's vorticity to spin the parcel faster. At position 3, the parcel spins even faster as the curvature is cyclonic, which adds to the earth's vorticity. At position 4, the parcel spins more slowly as the curvature is once again zero.

We can see in Fig. 12.23 that as the parcel moves from position 1 in the ridge to position 3 in the trough, its absolute vorticity *increases* as it moves along. Within this region is typically found an area of upper-level convergence. As the parcel moves from position 3 in the trough to position 5 in

the ridge, its absolute vorticity *decreases* as it moves along. Within this region is typically found an area of upper-level divergence. We can now summarize the information in Fig. 12.23 by stating that as a parcel of air moves with the upper-level flow, an *increase in its absolute vorticity with respect to time is related to upper-level converging air, and a decrease in its absolute vorticity with respect to time is related to upper-level diverging air.*[*]

On upper-air charts that show absolute vorticity (such as on a 500-mb chart), we find that even though the clockwise flow around a high-pressure area produces negative relative vorticity, because of the earth's (positive) vorticity, the absolute vorticity is normally positive everywhere on the map. However, there are regions on the map (about the size of the state of Iowa) where the absolute vorticity is considerably greater. An area of high absolute vorticity is referred to as a *vorticity maximum* or *vort max*. A region of low absolute vorticity is known as a *vorticity minimum.*

● Figure 12.24 illustrates how vorticity, divergence, vertical air motions, and surface mid-latitude cyclonic storm development are linked together. The 500-mb chart (middle chart) in Fig. 12.24 shows a vorticity maximum. If you move from the max eastward toward point 1, notice that, as you move along, vorticity decreases with time. Associated with this region we find upper-level divergence (at 300 mb) and low-level convergence (at the surface), which infers rising air and the possibility of clouds, precipitation, and cyclonic storm development. It follows that when a vorticity maximum aloft moves toward a stationary front, a wave will form along the front, and a mid-latitude cyclonic storm will have a good chance of developing. Even in the absence of fronts, a

---

[*]Viewed another way, in the upper troposphere, an area of converging air increases the total spin — the absolute vorticity — of air parcels, whereas an area of diverging air decreases the absolute vorticity of air parcels.

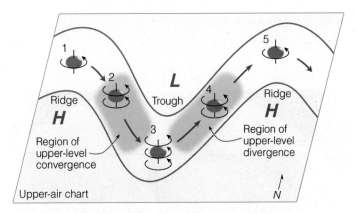

● FIGURE 12.23 The vorticity of an air parcel changes as we follow it through a wave. From position 1 to position 3, the parcel's absolute vorticity increases with time. In this region (shaded blue), we normally experience an area of upper-level converging air. As the air parcel moves from position 3 to position 5, its absolute vorticity decreases with time. In this region (shaded green), we normally experience an area of upper-level diverging air.

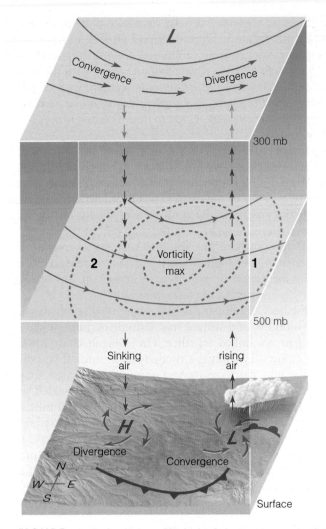

● FIGURE 12.24 A region of high absolute vorticity—a vorticity maximum—on its downwind (eastern) side has diverging air aloft, converging surface air, and ascending air motions. On its upwind (western) side, there is converging air aloft, diverging surface air, and descending air motions.

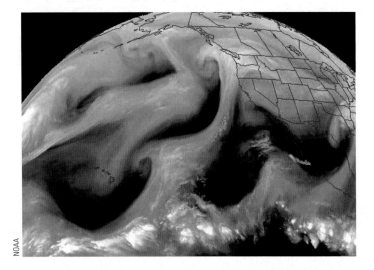

zone of organized clouds with precipitation may form in conjunction with a vorticity maximum. On the other hand, if you move toward the vorticity maximum from position 2, we find that vorticity increases with time. Here, to the west of the vorticity maximum in Fig. 12.24, exists upper-level convergence (at 300 mb), low-level divergence (at the surface), slowly sinking air, and the generally fair weather we associate with surface high-pressure areas.

Because of the gaps in observational data, scientists often found it difficult to locate vorticity maximums on satellite images, especially in regions where there were no clouds. Now, however, satellites are equipped to observe atmospheric water vapor. The swirling patterns of moisture clearly identify the position of vorticity centers. In ● Fig. 12.25, observe the cyclonic swirl of water vapor associated with a region of maximum vorticity just off the coast of Oregon and Washington.

PUTTING IT ALL TOGETHER—A MONSTROUS SNOW-STORM   We are now in a position to place much of what we have learned into a real-life weather situation. ● Figure 12.26 illustrates the atmospheric conditions that turned an open-wave cyclone into a ferocious storm. The lower map shows a developing mid-latitude cyclone off the North Carolina coast on February 11, 1983. The counterclockwise circulation around the low is causing warm, moist air from subtropical waters to ride up and over very cold surface air (heavy red arrow) that is entrenched over the eastern seaboard. Snow is falling in the cold air in advance of a warm front that extends from the low northeastward over the Atlantic Ocean.

Above the surface, the 500-mb chart (middle chart, Fig. 12.26) shows a broad longwave trough with a shortwave (heavy dashed line) moving through it. Notice that the shortwave is just to the west of the surface low. Apparently, the shortwave has disturbed the flow, as a strong area of baroclinicity exists to the west of the shortwave. Here, isotherms (dashed lines) are intersecting contour lines (solid lines), and cold advection is occurring.

At the 200-mb level (upper chart, Fig. 12.26), the polar-front jet stream (blue arrow) swings just to the south of the surface low. The region in orange represents the zone of strongest winds, the jet streak. The area of strong divergence (marked by the letters *DIV* on the 200-mb chart) is almost over the surface low.

● Figure 12.27 is a 500-mb chart that shows absolute vorticity for the same date and time as Fig. 12.26. Notice that the vorticity maximum of 14 (which is actually 14×10^{-5}/sec) is located in the same position as the shortwave in Fig. 12.26. Notice also that aloft, to the east of the vorticity maximum,

● FIGURE 12.25 This infrared water vapor image shows regions of maximum vorticity as cyclonic swirls of moisture off the coast of Oregon and Washington and out over the Pacific. The stretched-out band of clouds toward the bottom of the picture is the intertropical convergence zone.

FOCUS ON A SPECIAL TOPIC

Vorticity and Longwaves

We know that longwaves develop aloft in the atmosphere and that longwaves are important, because they transfer heat and momentum from one latitude to another. But how do these waves form in the first place? The concept of vorticity can help explain longwave formation. For example, let's assume that in the atmosphere there is no divergence or convergence of air so that air columns cannot stretch or contract. Where these conditions prevail, the absolute vorticity of air will be conserved, meaning that the numerical value of the sum of the earth's vorticity and the relative vorticity will not change with time. Thus

Absolute vorticity =
earth's vorticity + relative vorticity =
constant.

If ζ_a* is the absolute vorticity, ζ_r the relative vorticity, and f the earth's vorticity, then the expression becomes

$$\zeta_a = \zeta_r + f = \text{constant.}$$

Hence, any decrease in the earth's vorticity must be compensated for by an increase in the relative vorticity and vice versa.

Consider, for example, that air in Fig. 7 is moving horizontally at a constant speed at an altitude near 5.5 km (about 18,000 ft) above

---

*The symbol ζ is the Greek letter zeta. Meteorologists often use ζ to represent vorticity.

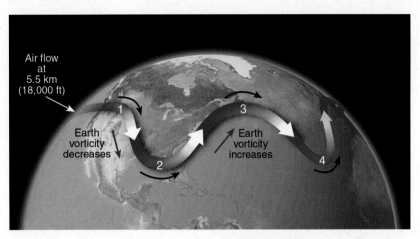

● FIGURE 7 The wavy path of air aloft due to the conservation of absolute vorticity.

sea level.* At this level, divergence is usually near zero, so our initial assumption approaches a real situation. Since there is no wind speed shear, any change in the relative vorticity will be due to curvature.

Suppose that the air flow is disturbed by a mountain range such that the air at position 1 is flowing southeastward. Heading equatorward, the air moves into a region of decreasing earth vorticity. To keep the absolute vorticity of the air constant, there must be a corresponding increase in the relative vorticity. Since increas-

---

*You may recall that the atmospheric pressure at this level is about 500 millibars.

ing relative vorticity implies cyclonic curvature, the air turns counterclockwise at position 2 and heads northeastward. But now the air is moving into a region where the earth's vorticity steadily increases. To offset this increase, the relative vorticity must decrease. The relative vorticity will decrease if the curvature becomes anticyclonic, so at position 3 air turns clockwise and heads toward the equator once again. This again brings the air into a region where the earth's vorticity decreases. To compensate, the air must now turn cyclonically at position 4, and so on. In this manner, a series of upper-level longwaves may develop, encircling the entire earth.

---

lies the region of strong upper-level divergence associated with the jet stream. Directly below this region is the open-wave cyclone. Hence, the stage is set, and the necessary ingredients are in place for the surface low to develop into a major storm system.

The winds aloft steered the surface low northeastward, but the strong blocking high to the north over southern Canada (surface map, Fig. 12.26) slowed the storm's forward pace, allowing the region of upper-level divergence to intensify the system. With the deep low just off the coast and the strong high to the north, pressure gradients increased, and strong, howling winds in excess of 30 knots battered New Jersey, eastern Pennsylvania, and southeastern New York. Meanwhile, huge quantities of moist ocean air produced co-

pious snowfalls along the Atlantic coastal states. High above the surface, as cold air from the west moved over the region, the air became conditionally unstable; strong convective cells developed, and heavy "downpours" of snow were accompanied by lightning and thunder. At one point in the storm, Allentown, Pennsylvania, reported snow falling at the rate of five inches per hour.

The storm left a buried landscape from northern Virginia to Connecticut, where snow drifts in some areas rose to housetops. The storm shut down entire cities. It crippled travel on the ground and in the air; closed businesses, government agencies, and schools; knocked out power to many thousands of homes; and caused several deaths. In addition, many cities reported their greatest 24-hour snowfall ever.

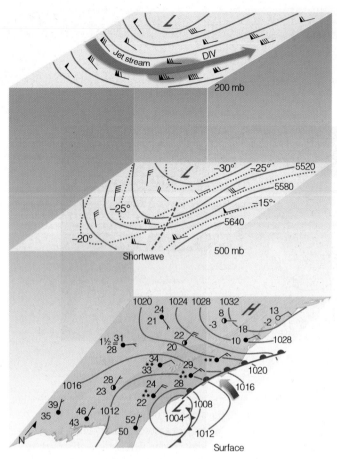

● FIGURE 12.26 The atmospheric conditions for February 11, 1983, at 7 A.M., EST. The bottom chart is the surface weather map. The middle chart is the 500-mb chart that shows contour lines (solid lines) in meters above sea level, isotherms (dashed lines) in °C, and the position of a shortwave (heavy dashed line). The upper chart is the 200-mb chart that illustrates contours, winds, and the position of the polar jet stream (dark blue arrow). The letters *DIV* represent an area of strong divergence. The region shaded orange represents the jet stream core—the jet streak.

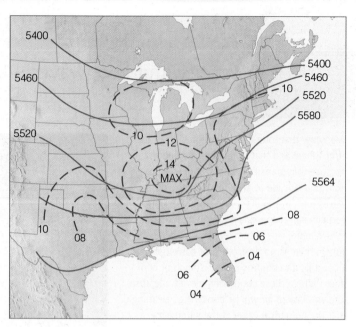

● FIGURE 12.27 The 500-mb chart for February 11, 1983, at 7 A.M., EST. Solid lines are height contours in meters above sea level. Dashed lines are lines of constant absolute vorticity × 10^{-5}/sec.

Because the heavy snow was accompanied by high winds and low temperatures, the storm has come to be known as "the Blizzard of '83."

POLAR LOWS Up to now, we have concentrated on middle-latitude cyclones, especially those storms that form along the polar front. There are storms, however, that develop over polar water behind (or poleward of) the main polar front. Such storms are called **polar lows**. Although polar lows develop in both hemispheres, our discussion will center on those storms that form in the Northern Hemisphere in the cold polar air of the North Pacific, North Sea, and North Atlantic, especially in the region south of Iceland.

With diameters of 1000 to 500 km (600 to 300 mi) or less, polar lows are generally smaller in size than their mid-latitude cousin, the wave cyclone that tends to form along the

polar front. Some polar lows have a comma-shaped cloud band. Others have a tight spiral of convective clouds that swirls counterclockwise about a clear area, or "eye," which resembles the eye of a tropical hurricane (see ● Fig. 12.28). In fact, like hurricanes, these smaller intense storms normally have a warmer central core, strong winds (often gale force or higher), and heavy showery precipitation that, unlike a hurricane, is in the form of snow.*

Polar lows typically form during the winter, from November through March. During this time, the sun is low on the horizon and absent for extended periods. This situation allows the air next to snow-and-ice-covered surfaces to cool rapidly and become incredibly cold. As this frigid air sweeps off the winter ice that covers much of the Arctic Ocean, it may come in contact with warmer air that is resting above a relatively warm ocean current. Where these two masses of contrasting air meet, the boundary separating them is called an **arctic front.**

Along the arctic front, the warmer (less-dense) air rises, while the much colder (more-dense) air slowly sinks beneath it. Recall from our discussion of developing middle-latitude cyclones (p. 322) that the rising of warm air and the sinking of cold air establishes a condition known as *baroclinic instability*. As the warm air rises, some of its water vapor condenses, resulting in the formation of clouds and the release of

---

*Tropical storms such as hurricanes are covered in Chapter 15. As we will see, the input of heat from the ocean surface into the hurricane increases as the wind speed increases, causing a positive feedback. Moreover, in hurricane environments, the ocean and air temperatures are about the same, whereas in the Arctic, the transfer of sensible heat from the surface is large because the air-sea temperature difference is large.

© Steven Businger

● FIGURE 12.28 An enhanced infrared satellite image of an intense polar low situated over the Norwegian Sea, north of the Arctic Circle. Notice that convective clouds swirl counterclockwise about a clear area, or eye. Surprising similarities exist between polar lows and tropical hurricanes described in Chapter 15.

latent heat, which warms the atmosphere. The warmer air has the effect of lowering the surface air pressure. Meanwhile, at the ocean surface there is a transfer of sensible heat from the relatively warm water to the cold air above. This transfer drives convective updrafts directly from the surface. In addition, it tends to destabilize the atmosphere, as heat is gained at the surface and lost to space at the top of the clouds as they radiate infrared energy upward.

The storm's development is enhanced if an upper trough lies to the west of the surface system and a shortwave disturbs the flow aloft. Similarly, the storm may intensify if a band of maximum winds—a jet streak—moves over the surface storm and a region of upper-level divergence draws the surface air upward. The developing cyclonic storm may attain a central pressure of 980 mb (28.94 in.) or lower. Generally, polar lows dissipate rapidly when they move over land.

SUMMARY

In this chapter, we discussed where, why, and how a mid-latitude cyclone forms. We began by examining the early polar front theory proposed by Norwegian scientists after World War I. We saw that the wave cyclone goes through a series of stages from birth to maturity to finally decay as an occluded storm.

We looked at the important effect that the upper air flow has on the intensification and movement of surface mid-latitude cyclones. We saw that when an upper-level trough lies to the west of a surface low-pressure area and when a shortwave disturbs the flow aloft, horizontal and vertical air motions begin to enhance the formation of the surface storm. Aloft, regions of divergence remove air from above surface mid-latitude cyclones, and regions of convergence supply air to surface anticyclones. The rising of warm air and the sinking of cold air provide the proper energy conversions for the storm's growth, as potential energy is transformed into kinetic energy. A region of maximum winds—a jet streak—associated with the polar jet stream provides additional support as an area of divergence removes air above the surface mid-latitude cyclone, allowing it to develop into a deep low-pressure area. The curving nature of the jet stream tends to direct mid-latitude cyclones northeastward and anticyclones southeastward.

We looked at the concept of vorticity and how it relates to developing mid-latitude cyclones. We found that on the downwind (eastern) side of a vorticity maximum, there is normally diverging air aloft, converging air at the surface, and cyclonic storm development. Finally, we examined polar lows—those storms that form over water in the cold air of the polar regions.

KEY TERMS

The following terms are listed (with page numbers) in the order they appear in the text. Define each. Doing so will aid you in reviewing the material covered in this chapter.

QUESTIONS FOR REVIEW

1. On a piece of paper, draw the different stages of a mid-latitude cyclonic storm as it goes through birth to decay according to the polar front (Norwegian) model.
2. Why do mid-latitude cyclones "die out" after they become occluded?
3. List four regions in North America where mid-latitude cyclones tend to develop or redevelop.
4. Explain this fact: Without upper-level divergence, a surface open wave would probably persist for less than a day.
5. Why do middle latitude surface low-pressure areas tilt westward with increasing height?
6. If upper-level diverging air above a surface area of low pressure exceeds converging air around the surface low, will the surface low-pressure area weaken or intensify? Explain.
7. What is an Alberta Clipper? Where does it form and how does it move?
8. What are northeasters? Why are they given that name?
9. How are longwaves in the upper-level westerlies different from shortwaves?
10. What are the necessary ingredients for a mid-latitude cyclonic storm to develop into a huge storm system?
11. Why do surface storms tend to dissipate "fill" when the upper-level low and the surface low become vertically stacked?
12. How does the polar-front jet stream influence the formation of a mid-latitude cyclone?
13. Explain why, even though the polar-front jet stream coincides with the polar front, some surface regions are more favorable for the development of mid-latitude cyclones than others.
14. Using a diagram, explain why a surface high-pressure area over North Dakota will typically move southeastward while, at the same time, a deep mid-latitude cyclone over the Great Lakes will generally move northeastward.
15. What are the sources of energy for a developing cyclone?
16. How does warm and cold advection aid in the development of a surface mid-latitude cyclone?
17. What are the roles of warm, cold, and dry conveyor belts in the development of a mid-latitude cyclonic storm system?
18. Explain why surface lows tend to deepen when a vorticity maximum aloft moves in from the west.
19. What are polar lows? How and where do they form? What do some of them have in common with tropical hurricanes?

QUESTIONS FOR THOUGHT

1. An English friend of yours says that last night's rain over northern Great Britain was caused by a storm that originally formed east of the Colorado Rockies. Explain how this could happen.
2. Would a mid-latitude cyclone intensify or dissipate if the upper trough were located to the *east* of the surface disturbance? Explain your answer with the aid of a diagram.
3. Explain why, at 500 millibars, when cold advection is occurring, the air temperature does not drop as fast as it should. (Hint: What type of vertical air motions are also occurring?)
4. Over the earth as a whole, would you expect the atmosphere to be mainly barotropic or baroclinic? Explain.
5. Baroclinic waves seldom form in the tropics. Why not?
6. Suppose that the earth stops rotating. How would this affect the earth's vorticity? What would happen to the absolute vorticity of a moving air parcel? If the parcel were initially moving southwestward, how would its direction change, if at all?
7. Why do Pacific storms often redevelop on the eastern side of the Sierra Nevada?
8. If you only had isotherms on an upper-level chart, how would a cut-off low appear?
9. If polar lows form in cold polar air over water, how is the atmosphere made conditionally unstable so that towering convective cumulus clouds can form?
10. The 500-mb level (at about 5600 m above the surface) is referred to as the "level of non-divergence." Give an explanation as to what this statement means with reference to what is taking place above and below this level.

PROBLEMS AND EXERCISES

1. On the 300-mb chart (Fig. 12.29, p. 337), suppose the winds are blowing parallel to the contour lines.
    (a) On the chart, mark where regions of convergence and divergence are occurring.
    (b) Put an *L* on the chart where you might expect to observe a developing mid-latitude cyclone at the surface.
    (c) Put an *H* on the chart where you might expect to observe a surface anticyclone.
    (d) In which directions would the surface cyclonic storm and anticyclone most likely move?
    (e) In terms of convergence and divergence, what are the necessary conditions for the intensification of the surface mid-latitude storm? For the building of the anticyclone?

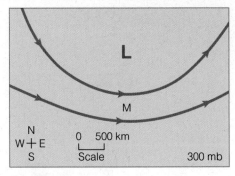

• FIGURE 12.29

2. Suppose there is a region of maximum absolute vorticity at position M on the chart.
   (a) How would the absolute vorticity be changing with time as an air parcel moves downstream (from position M) with the flow?
   (b) Based on absolute vorticity, circle on the map where you would expect to find upper-level divergence, surface convergence, and a developing mid-latitude cyclone.

Corpus Christi ●

Brownsville ●

Satellite view of South Texas along the Gulf Coast on Christmas day, 2004. The white area covering Corpus Christi and Brownsville is snow. The probability of measurable snow on the ground in either of these two cities on Christmas day is less than one percent. Yet, Corpus Christi received over 4 inches of snow and Brownsville about 1.5 inches, making it the first snowfall in Brownsville in 109 years. Just days later, the temperature climbed into the 80s (°F). NASA/GSFC

CHAPTER 13

Weather Forecasting

Sometimes there is no job security in weather forecasting. In fact, a weather forecaster actually lost his job for not altering his forecast. On April 15, 2001, a function honoring a well-known conservative radio talk show host was scheduled outdoors at the Madera, California, fairgrounds. The story goes that a local forecaster at the radio station that sponsored the event had called for a "chance of rain" on April 15th. Upset that such a forecast might discourage people from attending the function, the station manager told the forecaster to alter his forecast and predict a greater possibility of sunshine. The forecaster refused and was promptly fired. Apparently, retribution reigned supreme—it poured on the event.

Weather forecasts are issued to save lives, to save property and crops, and to tell us what to expect in our atmospheric environment. In addition, knowing what the weather will be like in the future is vital to many human activities. For example, a summer forecast of extended heavy rain and cool weather would have construction supervisors planning work under protective cover, department stores advertising umbrellas instead of bathing suits, and ice cream vendors vacationing as their business declines. The forecast would alert farmers to harvest their crops before their fields became too soggy to support the heavy machinery needed for the job. And the commuter? Well, the commuter knows that prolonged rain could mean clogged gutters, flooded highways, stalled traffic, blocked railway lines, and late dinners.

On the other side of the coin, a forecast calling for extended high temperatures with low humidity has an entirely different effect. As ice cream vendors prepare for record sales, the dairy farmer anticipates a decrease in milk and egg production. The forest ranger prepares warnings of fire danger in parched timber and grasslands. The construction worker is on the job outside once again, but the workday begins in the early morning and ends by early afternoon to avoid the oppressive heat. And the commuter prepares for increased traffic stalls due to overheated car engines.

Put yourself in the shoes of a weather forecaster: It is your responsibility to predict the weather accurately so that thousands (possibly millions) of people will know whether to carry an umbrella, wear an overcoat, or prepare for a winter storm. Since weather forecasting is not an exact science, your predictions will occasionally be incorrect. If your erroneous forecast misleads many people, you may become the target of jokes, insults, and even anger. There are even people who expect you to be able to predict the unpredictable. For example, on Monday you may be asked whether two Mondays from now will be a nice day for a picnic. And, of course, what about next winter? Will it be bitterly cold?

Unfortunately, accurate answers to such questions are beyond meteorology's present technical capabilities, but "useful" answers may be possible by applying different techniques to current forecast methods. Will forecasters ever be able to answer such questions confidently? If so, what steps are being taken to improve the forecasting art? How are forecasts made, and why do they sometimes go wrong? These are just a few of the questions we will address in this chapter.

Acquisition of Weather Information

Weather forecasting basically entails predicting how the present state of the atmosphere will change. Consequently, if we wish to make a weather forecast, present weather conditions over a large area must be known. To obtain this information, a network of observing stations is located throughout the world. Over 10,000 land-based stations and hundreds of ships and buoys provide surface weather information four times a day.* Most airports observe conditions hourly. Additional information, especially upper-air data, is supplied by radiosondes, aircraft, and satellites. Radiosonde data are usually available only at 0000 and 1200 UTC, while aircraft and satellite observations may occur throughout the day.

A United Nations agency—the *World Meteorological Organization* (WMO)—consists of over 175 nations. The WMO is responsible for the international exchange of weather data and certifies that the observation procedures do not vary among nations, an extremely important task, since the observations must be comparable.

After an observation is taken, it is immediately sent to a communication substation by electronic means, usually over dedicated phone lines or by satellite relay. From there, the data collected at many observation stations are sent to World Meteorological Centers (located in Melbourne, Australia; Moscow, Russia; and Washington, D.C.). Then, worldwide weather information is transmitted electronically to the National Center for Environmental Prediction (NCEP), a branch of the National Weather Service (NWS), located in Camp Springs, Maryland, just outside Washington, D.C. Here, the massive job of analyzing the data, preparing weather maps and charts, and predicting the weather on a global and national basis begins.† From NCEP, this information is transmitted to public and private agencies worldwide.

The compiled charts, maps, and forecasts are sent electronically to *Weather Forecast Offices* (WFO). These stations use the data for preparing regional weather forecasts, as well as for advisories and warnings of impending severe weather. The region serviced by one of these offices is a state or a large portion of a state.

The public hears weather forecasts over radio or television. Many stations hire private meteorological companies or professional meteorologists to make their own forecasts aided by NCEP material or to modify a WFO forecast. Other stations hire meteorologically untrained announcers who paraphrase or read the forecasts of the National Weather Service word for word.

Today, the forecaster has access to many hundreds of maps and charts, as well as vertical profiles (called *soundings*) of temperature, dew point, and winds. Also available are visible and infrared satellite images, as well as Doppler radar information that can detect and monitor the severity of precipitation and thunderstorms.

When hazardous weather is likely, the National Weather Service issues advisories in the form of weather watches and warnings. A **watch** indicates that atmospheric conditions fa-

---

*Observations are usually taken at 0000, 0600, 1200, and 1800 Coordinated Universal Time (UTC), which is also called *Greenwich Mean Time (GMT)*—local time at the Greenwich Observatory in England. To convert from UTC to your local time, see Appendix F.

†By international agreement, data are plotted using symbols illustrated in Appendix B.

FOCUS ON A SPECIAL TOPIC

Watches, Warnings, and Advisories

As we have seen, where severe or hazardous weather is either occurring or possible, the National Weather Service issues a forecast in the form of a watch or warning. The public, however, is not always certain as to what this forecast actually means. For example, a *high wind warning* indicates that there will be high winds — but how high and for how long? The following describes a few of the various watches, warnings, and advisories issued by the National Weather Service and the necessary precautions that should be taken during the event.

Wind advisory Issued when sustained winds reach 25 to 39 mi/hr or wind gusts are up to 57 mi/hr.

High wind warning Issued when sustained winds are at least 40 mi/hr, or wind gusts exceed 57 mi/hr. Caution should be taken when driving high-profile vehicles, such as trucks, trailers, and motor homes.

Wind-chill advisory Issued for wind-chill temperatures of $-30°$ to $-35°F$ or below.*

Heat advisory/warning Advisory issued when the daytime Heat Index is expected to reach 105°F for 3 hours or more and nighttime lows do not drop below 80°F. Warning issued when Heat Index reaches 115°F or above.

*It should be noted that watches, warnings, or advisories for wind chill or for snowfall-related events (e.g., winter storms, etc.) may use different criteria in different regions. For example, mountainous areas that experience frequent heavy snow may have higher snowfall criteria, whereas areas with infrequent snow, may have lower snowfall criteria. Similarly, in areas that experience frequent extreme cold, wind chills may have lower (colder) criteria for advisories.

• **FIGURE 1** Flags indicating advisories and warnings in maritime areas.

Flash-flood watch Heavy rains may result in flash flooding in the specified area. Be alert and prepared for the possibility of a flood emergency that will require immediate action.

Flash-flood warning Flash flooding is occurring or is imminent in the specified area. Move to safe ground immediately.

Urban and small stream advisory Issued when flooding is occurring in small streams, streets, or in low-lying areas, such as railroad underpasses and urban storm drains.

Severe thunderstorm watch Thunderstorms (with winds exceeding 57 mi/hr and/or hail three-fourths of an inch or more in diameter) are possible.

Severe thunderstorm warning Severe thunderstorms have been visually sighted or indicated by Doppler radar. Be prepared for lightning, heavy rains, strong winds, and large hail. (Tornadoes can form with severe thunderstorms.)

Tornado watch Issued to alert people that tornadoes may develop within a specified area during a certain time period.

Tornado warning Issued to alert people that a tornado has been spotted either visually or by Doppler radar. Take shelter immediately.

Snow advisory In nonmountainous areas, expect a snowfall of 2 in. or more in 12 hours, or 3 in. or more in 24 hours.*

Winter storm warning (formerly heavy snow warning) In nonmountainous areas, expect a snowfall of 4 in. or more in 12 hours or 6 in. or more in 24 hours. (Where heavy snow is infrequent, a snowfall of several inches may justify a warning.)*

Blizzard warning Issued when falling or blowing snow and winds of at least 35 mi/hr frequently restrict visibility to less than $\frac{1}{4}$ mile for several hours.

Dense fog advisory Issued when fog limits visibility to less than $\frac{1}{4}$ mile, or in some parts of the country to less than $\frac{1}{8}$ mile.

WARNINGS OVER THE WATER

Small craft advisories Issued to alert mariners that weather or sea conditions might be hazardous to small boats. Expect winds of 18 to 34 knots (21 to 39 mi/hr). Figure 1 displays the posted advisory and warning flags.

Gale warning Winds will range between 34 and 47 knots (39 to 54 mi/hr) in the forecast area.

Storm warning Winds in excess of 47 knots (54 mi/hr) are to be expected in the forecast area.

Hurricane watch Issued when a tropical storm or hurricane becomes a threat to a coastal area. Be prepared to take precautionary action in case hurricane warnings are issued.

Hurricane warning Issued when it appears that the storm will strike an area within 24 hours. Expect wind speeds in excess of 64 knots (74 mi/hr).

---

vor hazardous weather occurring over a particular region during a specified time period, but the actual location and timing of the occurrence is uncertain. A **warning,** on the other hand, indicates that hazardous weather is either imminent or actually occurring within the specified forecast area. *Advisories* are issued to inform the public of less hazardous conditions caused by wind, dust, fog, snow, sleet, or freezing rain. (Additional information on watches, warnings, and advisories is given in the Focus section above.)

Weather Forecasting Tools

To help forecasters handle all the available charts and maps, high-speed data modeling systems using computers are employed by the National Weather Service. The communication system in use today is known as **AWIPS** (*Advanced Weather Interactive Processing System*). The AWIPS system is shown in •Fig. 13.1.

● FIGURE 13.1 The AWIPS computer workstation provides various weather maps and overlays on different screens.

The AWIPS system has data communications, storage, processing, and display capabilities (including graphical overlays) to better help the individual forecaster extract and assimilate information from the mass of available data. In addition, AWIPS is able to integrate information received from the Doppler radar system (the WSR-88D), satellite imagery, and the Automated Surface Observing Systems (ASOS) that are operational at selected airports and other sites throughout the United States. The ASOS system is designed to provide nearly continuous information about wind, temperature, pressure, cloud-base height, and runway visibility at various airports. Meteorologists are hopeful that information from all of these sources will improve the accuracy of weather forecasts by providing previously unobtainable data for integration into numerical models. Moreover, much of the information from ASOS and Doppler radar is processed by software according to predetermined formulas, or *algorithms*, before it goes to the forecaster. Certain criteria or combinations of measurements can alert the forecaster to an impending weather situation, such as the severe weather illustrated in ● Fig. 13.2.

A software component of AWIPS (called the Interactive Forecast Preparation System) allows forecasters to look at the daily prediction of weather elements, such as temperature and dew point, in a grid format with spacing as small as 2 km. Presenting the data in this format allows the forecaster to predict the weather more precisely over a relatively small area.

With so much information at the forecaster's disposal, it is essential that the data be easily accessible and in a format that allows several weather variables to be viewed at one time. The **meteogram** is a chart that shows how one or more weather variables has changed at a station over a given period of time. As an example, the chart may represent how air temperature, dew point, and sea-level pressure have changed over the past five days, or it may illustrate how these same variables are projected to change over the next five days (see ● Fig. 13.3).

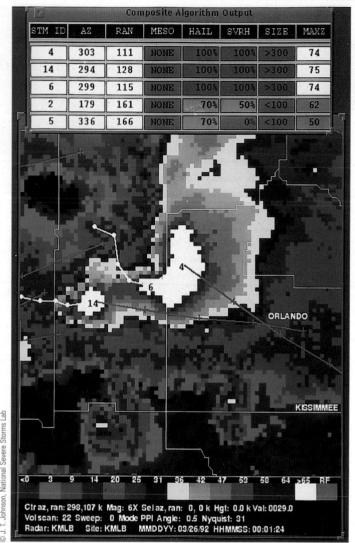

● FIGURE 13.2 Doppler radar data from Melbourne, Florida, during the time of a severe hailstorm in the Orlando area. In the table near the top of the display, the hail algorithm determined that there was 100 percent probability that the storm was producing hail and severe hail. The algorithm also estimated the maximum size of the hailstones to be greater than 3 inches. A forecaster can project the movement of the storm and adequately warn those areas in the immediate path of severe weather.

Another aid in weather forecasting is the use of *soundings*—a two-dimensional vertical profile of temperature, dew point, and winds (see ● Fig. 13.4).* The analysis of a sounding can be especially helpful when making a short-range forecast that covers a relatively small area, such as the mesoscale. The forecaster examines the sounding of the immediate area (or closest proximity), as well as the soundings of those sites upwind, to see how the atmosphere might be changing. Computer programs then automatically calculate

*A sounding is obtained from a radiosonde. For additional information on the radiosonde see the Focus section in Chapter 1 on p. 16.

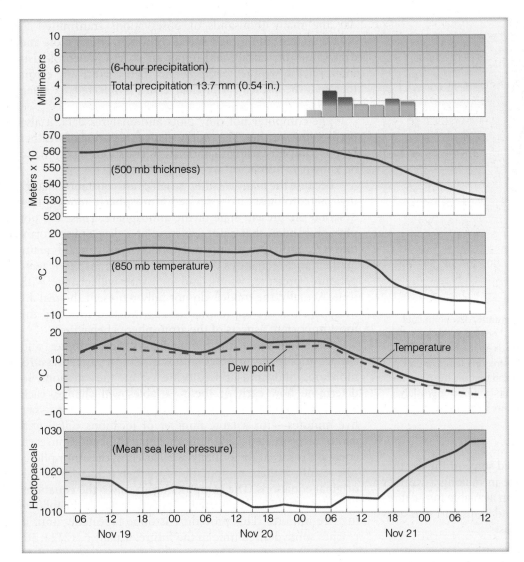

Meteogram illustrating predicted weather at the surface and aloft at St. Louis, Missouri, from 6 A.M., November 19, 2007, to noon on November 21, 2007. The forecast is derived from the Global Forecast System (GFS) model (NOAA)

from the sounding a number of meteorological *indexes* that can aid the forecaster in determining the likelihood of smaller-scale weather phenomena, such as thunderstorms, tornadoes, and hail. Soundings also provide information that can aid in the prediction of fog, air pollution alerts, and the downwind mixing of strong winds.

In the central United States, a network of *wind profilers* (see Chapter 9, p. 237) is providing forecasters with hourly wind speed and wind direction information at 72 different levels in a column of air 16 km thick. The almost continuous monitoring of winds is especially beneficial when briefing pilots on areas of strong headwinds and on regions of strong wind shear. Wind information from the profilers is also sent to the NCEP, where it is integrated into computer models used in preparing mid-range weather forecasts.

Satellite information is also a valuable tool for the forecaster. Visible, enhanced infrared, and water vapor images provide a wealth of information, some of which comes from inaccessible regions that can be plugged into forecast models. This added information provides a clearer representation of

Nightly news weather presentations have come a long way since the early days of television. The "weather girl," a fad that became popular during the 1960s, employed crazy gimmicks to attract viewers. Females gave weather forecasts in various attire (from bathing suits to bunny outfits), sometimes with the aid of hand puppets that resembled odd-looking turtles. One West Coast television station actually hired a woman to do the nightly news weather segment with the requirement that she be able to write backwards on a clear, Plexiglas screen.

the atmosphere.* When predicting temperatures, forecasters often look at a chart called the *thickness chart*. If you are interested in learning how this chart can be used to predict whether falling precipitation will be in the form of rain or snow, read the Focus section on p. 345.

*Information provided by satellites is located in various sections of this book. For example, see Chapter 5, p. 133, and Chapter 9, p. 238.

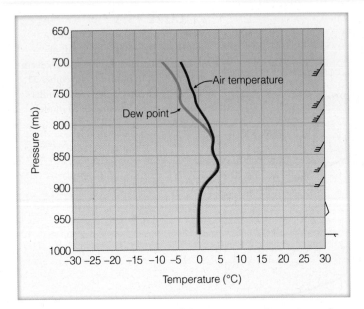

● FIGURE 13.4 A sounding of air temperature, dew point, and winds at Pittsburgh, PA, on January 14, 1999. Looking at this sounding, a forecaster would see that saturated air extends up to about 820 mb. The forecaster would also observe that below-freezing temperatures only exist in a shallow layer near the surface and that the freezing rain presently falling over the Pittsburgh area would continue or possibly change to rain, as cold easterly surface winds are swinging around to warmer southwesterly winds aloft.

Up to this point, we have examined some of the weather data and tools a forecaster might use in making a weather prediction. With all of this information available to the forecaster, including hundreds of charts and maps, just *how* does a meteorologist make a weather forecast?

Weather Forecasting Methods

As late as the mid-1950s, all weather maps and charts were plotted by hand and analyzed by individuals. Meteorologists predicted the weather using certain rules that related to the particular weather system in question. For short-range forecasts of six hours or less, surface weather systems were moved along at a steady rate. Upper-air charts were used to predict where surface storms would develop and where pressure systems aloft would intensify or weaken. The predicted positions of these systems were extrapolated into the future using linear graphical techniques and current maps. Experience played a major role in making the forecast. In many cases, these forecasts turned out to be amazingly accurate. They were good but, with the advent of modern computers, along with our present observing techniques, today's forecasts are even better.

THE COMPUTER AND WEATHER FORECASTING: NUMERICAL WEATHER PREDICTION Modern electronic computers can analyze large quantities of data extremely fast. Each day the many thousands of observations transmitted to NCEP are fed into a high-speed computer, which plots and draws lines on surface and upper-air charts. Meteorologists interpret the weather patterns and then correct any errors that may be present. The final chart is referred to as an **analysis.**

The computer not only plots and analyzes data, it also predicts the weather. The routine daily forecasting of weather by the computer using mathematical equations has come to be known as **numerical weather prediction.**

Because the many weather variables are constantly changing, meteorologists have devised **atmospheric models** that describe the present state of the atmosphere. These are not physical models that paint a picture of a developing storm; they are, rather, mathematical models consisting of many mathematical equations that describe how atmospheric temperature, pressure, winds, and moisture will change with time. Actually, the models do not fully represent the real atmosphere but are approximations formulated to retain the most important aspects of the atmosphere's behavior.

The models are programmed into the computer, and surface and upper-air observations of temperature, pressure, moisture, winds, and air density are fed into the equations. To determine how each of these variables will change, each equation is solved for a small increment of future time—say, five minutes—for a large number of locations called *grid points*, each situated a given distance apart.* In addition, each equation is solved for as many as 50 levels in the atmosphere. The results of these computations are then fed back into the original equations. The computer again solves the equations with the new "data," thus predicting weather over the following five minutes. This procedure is done repeatedly until it reaches some desired time in the future, usually 6, 12, 24, 36, and out to 84 hours. The computer then analyzes the data and draws the projected positions of pressure systems with their isobars or contour lines. The final forecast chart representing the atmosphere at a specified future time is called a **prognostic chart,** or, simply, a **prog.** Computer-drawn progs have come to be known as "machine-made" forecasts.

The computer solves the equations more quickly and efficiently than could be done by hand. For example, just to produce a 24-hour forecast chart for the Northern Hemisphere requires many hundreds of millions of mathematical calculations. It would, therefore, take a group of meteorologists working full time with hand calculators years to produce a single chart; by the time the forecast was available, the weather for that day would already be ancient history.

The forecaster uses the progs as a guide to predicting the weather. At present, there are a variety of models (and, hence, progs) from which to choose, each producing a slightly different interpretation of the weather for the same projected

---

*Some models have a grid spacing as small as 0.5 km, whereas the spacing in others exceeds 100 km. There are models that actually describe the atmosphere using a set of mathematical equations with wavelike characteristics rather than a set of discrete numbers associated with grid points.

FOCUS ON A SPECIAL TOPIC

The Thickness Chart — A Forecasting Tool

The thickness chart can be a valuable forecasting tool for the meteorologist. It can help identify air masses and locate fronts, and a prognostic thickness chart can help predict the daily max and min temperature. It can also help predict whether falling precipitation will be in the form of rain or snow. What, then, is a thickness chart?

A thickness chart shows the difference in height between two constant pressure surfaces. The vertical depth or thickness between any two pressure surfaces is related to the average air temperature between the two surfaces. Recall that air pressure decreases more rapidly with height in cold air than in warm air. This fact is illustrated in Fig. 2, which shows a 1000-mb pressure surface and a 500-mb pressure surface. The difference in height between these two pressure surfaces is called the *1000-mb to 500-mb thickness.** Notice that the vertical distance (thickness) between these two pressure surfaces is greater in warm air than in cold air. Consequently, warm air produces high thickness, and cold air low thickness.

When 1000-mb to 500-mb thickness lines are drawn on a chart, they may appear similar to those shown in Fig. 3. On the chart, regions of low thickness correspond to cold air, and regions of high thickness to warm air. There are a number of forecasting rules for predicting air temperature and precipitation using this chart. One forecasting rule is that the 5400-meter thickness line often represents the dividing line between rain and snow, especially for cities receiving precipitation east of the Rocky Mountains. If precipitation is falling, cities with a thickness greater than 5400 m should be receiving rain, whereas cities with a thickness less than 5400 m should be receiving snow. Look at Fig. 3. St. Louis is receiving precipitation. Would you expect it to

be in the form of rain or snow? What about Detroit? If you live east of the Rockies, find your city on the chart and see whether precipitation would be in the form of rain or snow on this day.

*Because the 1000-mb pressure surface is often close to the surface of the earth, the 1000-mb to 500-mb thickness is sometimes referred to as *the surface to 500-mb thickness.*

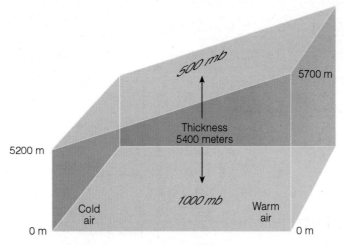

● FIGURE 2
The vertical separation (thickness) between the 1000-mb pressure surface and the 500-mb pressure surface is greater in warm air than in cold air.

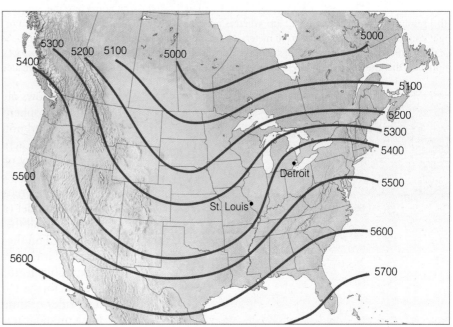

● FIGURE 3 A 1000-mb to 500-mb thickness chart for a January morning. The lines on the chart represent the vertical depth in meters between the 1000-mb and 500-mb pressure surfaces. Low thickness lines correspond to cold air, and high thickness lines to warm air. For reference, the 5400-meter thickness line represents a vertical layer of air 5400 meters thick with an average temperature of −7°C (19°F). The 5200 thickness line is roughly the boundary for arctic air.

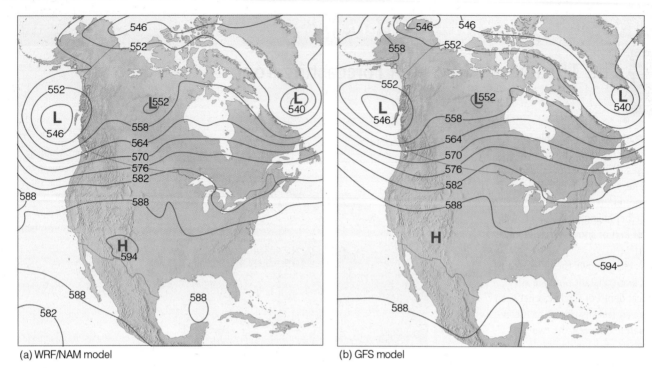

(a) WRF/NAM model   (b) GFS model

● FIGURE 13.5 Two 500-mb progs for 7 P.M. EST, July 12, 2006—48 hours into the future. Prog (a) is the WRF/NAM model, with a resolution (grid spacing) of 12 km, whereas prog (b) is the GFS model with a resolution of 60 km. Solid lines on each map are height contours, where 570 equals 5700 meters. Notice how the two progs (models) agree on the atmosphere's large scale circulation. The main difference between the progs is in the way the models handle the low off the west coast of North America. Model (a) predicts that the low will dig deeper along the coast, while model (b) predicts a more elongated west-to-east (zonal) low.

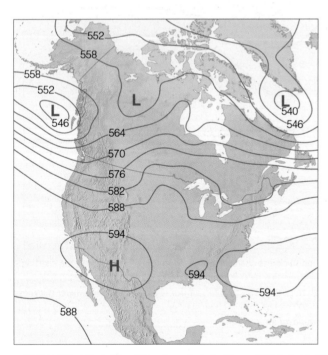

● FIGURE 13.6 The 500-mb analysis for 7 P.M. EST, July 12, 2006.

time and atmospheric level (see ● Fig. 13.5). The differences between progs may result from the way the models use the equations, or the distance between grid points, called *resolution*. Some models predict some features better than others: One model may work best in predicting the position of troughs on upper-level charts, whereas another forecasts the position of surface lows quite well. Some models even forecast the state of the atmosphere 384 hours (16 days) into the future. Look at ● Fig 13.6 and notice that model (b) in Fig. 13.5 with a resolution of 60 km actually did a better job of forecasting the structure of the low off the west coast of North America than did model (a) with a resolution of only 12 km.

A good forecaster knows the idiosyncrasies of each model [such as model (a) and model (b) in Fig. 13.5] and carefully scrutinizes all the progs. The forecaster then makes a prediction based on the *guidance* from the computer, a personalized practical interpretation of the weather situation and any local geographic features that influence the weather within the specific forecast area. ▼ Table 13.1 gives a brief summary of a few "rules of thumb" that a forecaster might use when making a prediction using a prog.

Currently, forecast models predict the weather reasonably well 4 to 6 days into the future. The models tend to do a better job at predicting temperature and jet-stream patterns than precipitation. However, even with all of the modern advances in weather forecasting provided by ever more powerful computers, National Weather Service (NWS) forecasts are sometimes wrong.

WHY NWS FORECASTS GO AWRY AND STEPS TO IMPROVE THEM Why do forecasts sometimes go wrong? There are a number of reasons for this unfortunate situation. For one, computer models have inherent flaws that limit the accuracy of weather forecasts. For example, computer-forecast models idealize the real atmosphere, meaning that

▼ TABLE 13.1   A Few Forecasting "Rules of Thumb"*

FORECAST QUESTION	USE OF FORECAST CHART
Cloudy or clear?	On the 700-mb forecast chart, the 70 percent relative humidity line usually encloses areas that are likely to have clouds.
Will it rain?	(a) On the 700-mb forecast chart, the 90 percent relative humidity line often encloses areas where precipitation is likely. If upward velocities are present, the chance of measurable precipitation is enhanced. (b) Along the west coast of North America, precipitation is much more likely north of the 5640-meter height contour on the 500-mb forecast chart.
Will it rain or snow?	On the 850-mb forecast chart, snow is likely north of the −5°C (23°F) isotherm, whereas rain is likely south of this line. On the 1000-mb to 500-mb thickness chart, the 5400-meter thickness line is widely used (east of the Rockies) as the dividing line between rain and snow.
Will the surface low intensify?	For the storm to intensify (deepen), an area of upper-level divergence must be over the surface cyclonic storm. On a 500-mb forecast chart that shows vorticity, look for a vorticity maximum (vort max) and remember from Chapter 12, p. 332, that to the east of an area of positive vorticity we usually find upper-level divergence, upward air motions, and cyclonic storm development.

*The forecast charts (progs) found in this table can be obtained from the World Wide Web.

each model makes certain assumptions about the atmosphere. These assumptions may be on target for some weather situations and be way off for others. Consequently, the computer may produce a prog that on one day comes quite close to describing the actual state of the atmosphere, and not so close on another. A forecaster who bases a prediction on an "off day" computer prog may find a forecast of "rain and windy" turning out to be a day of "clear and colder."

Another forecasting problem arises because the majority of models are not global in their coverage, and errors are able to creep in along the model's boundaries. For example, a model that predicts the weather for North America may not accurately treat weather systems that move in along its boundary from the western Pacific. This kind of inaccuracy is probably why model (b) in Fig. 13.5—a global model with a lower resolution—actually did a better job in predicting the low off the west coast than did model (a), which is a non-global model with a higher resolution. Obviously, a global model would usually be preferred. But a global model of similar sophistication with a high resolution requires an incredible number of computations.

Even though many thousands of weather observations are taken worldwide each day, there are still regions where observations are sparse, particularly over the oceans and at higher latitudes. To help alleviate this problem, the newest *GOES* satellite, with advanced atmospheric sounders, is providing a more accurate profile of temperature and humidity for the computer models. Wind information now comes from a variety of sources, such as Doppler radar, commercial aircraft, and satellites that translate ocean surface roughness into surface wind speed (see Chapter 9, p. 238).

Earlier, we saw that the computer solves the equations that represent the atmosphere at many locations called grid points, each spaced from 100 km to as low as 0.5 km apart. As a consequence, on computer models with large spacing between grid points (say 60 km), weather systems, such as extensive mid-latitude cyclones and anticyclones, show up on computer progs, whereas much smaller systems, such as thunderstorms, do not. The computer models that forecast for a large area such as North America are, therefore, better at predicting the widespread precipitation associated with a large cyclonic storm than local showers and thunderstorms. In summer, when much of the precipitation falls as local showers, a computer prog may have indicated fair weather, while outside it is pouring rain.

To capture the smaller-scale weather features as well as the terrain of the region, the distance between grid points on some models is being reduced. For example, the forecast model known as MM5 has a grid spacing as low as 0.5 km. This model predicts mesoscale atmospheric conditions over a limited region, such as a coastal area where terrain might greatly impact the local weather. The problem with models that have a small grid spacing (high resolution) is that, as the horizontal spacing between grid points decreases, the number of computations increases. When the distance is halved, there are 8 times as many computations to perform, and the time required to run the model goes up by a factor of 16.

Another forecasting problem is that many computer models cannot adequately interpret many of the factors that influence surface weather, such as the interactions of water, ice, surface friction, and local terrain on weather systems. Many large-scale models now take mountain regions and oceans into account. Some models (such as the MM5) take even smaller factors into account—features that large-scale computers miss due to their longer grid spacing. Given the effect of local terrain, as well as the impact of some of the other problems previously mentioned, computer models that forecast the weather over a vast area do an inadequate job of

● FIGURE 13.7 Ensemble 500-mb forecast chart for July 21, 2005 (48 hours into the future). The chart is constructed by running the model 15 different times, each time beginning with a slightly different initial condition. The blue lines represent the 5790-meter contour line; the red lines, the 5940-meter contour line; and the green line, the 500-mb 25-year average, called *climatology*.

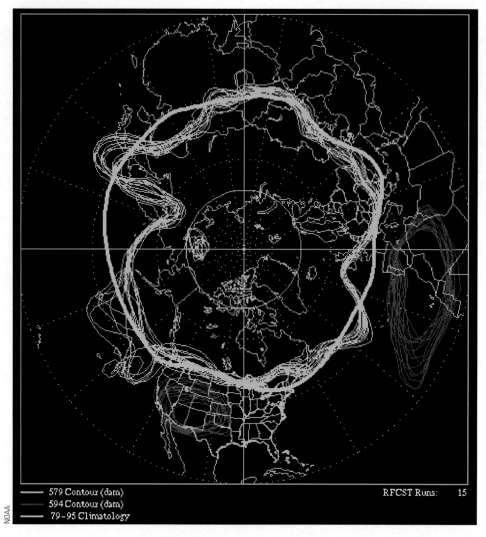

predicting local weather conditions, such as surface temperatures, winds, and precipitation.

Even with better observing techniques and near perfect computer models, there are countless small, unpredictable atmospheric fluctuations that fall under the heading of **chaos**. For example, tiny eddies are much smaller than the grid spacing on the computer model and, therefore, go unmeasured. These small disturbances, as well as small errors (uncertainties) in the data, generally amplify with time as the computer tries to project the weather farther and farther into the future. After a number of days, these initial imperfections tend to dominate, and the forecast shows little or no accuracy in predicting the behavior of the real atmosphere. In essence, what happens is that the small uncertainty in the initial atmospheric conditions eventually leads to a huge uncertainty in the model's forecast.

Because of the atmosphere's chaotic nature, meteorologists are turning to a technique called **ensemble forecasting** to improve short- and medium-range forecasts. The ensemble approach is based on running several forecast models — or different versions (simulations) of a single model — each beginning with slightly different weather information to reflect the errors inherent in the measurements.

Suppose, for example, a forecast model predicts the state of the atmosphere 24 hours into the future. For the ensemble forecast, the entire model simulation is repeated, but only after the initial conditions are "tweaked" just a little. The "tweaking," of course, represents the degree of uncertainty in the observations. Repeating this process several times creates an ensemble of forecasts for a range of small initial changes.

● Figure 13.7 shows an ensemble 500-mb forecast chart for July 21, 2005 (48 hours into the future) using the global atmospheric circulation model. The chart is constructed by running the model 15 different times, each time starting with slightly different initial conditions. Notice that the red contour line (which represents a height of 5940 meters) circles the southwestern United States, indicating a high degree of confidence in the model for that region. Here, a large upper-level high pressure area covers the region, and so a forecast for the southwestern United States would be "very hot and dry." The blue scrambled contour lines (representing a height of 5790 meters) off the west coast of North America indicate a great deal of uncertainty in the forecast model. As the forecast goes further and further into the future, the lines look more and more like scrambled spaghetti, which is why an

FOCUS ON AN OBSERVATION

TV Weathercasters — How Do They Do It?

As you watch the TV weathercaster, you typically see a person describing and pointing to specific weather information, such as satellite and radar images, and weather maps, as illustrated in Fig. 4. What you may not know is that the weathercaster is actually pointing to a blank board (usually green or blue) on which there is nothing (Fig. 5). This process of electronically superimposing weather information in the TV camera against a blank wall is called color-separation overlay, or *chroma key.*

The chroma key process works because the studio camera is constructed to pick up all colors except (in this case) blue. The various maps, charts, satellite photos, and other graphics are electronically inserted from a computer into this blue area of the color spectrum. The person in the TV studio should not wear blue clothes because such clothing would not be picked up by the camera — what you would see on your home screen would be a head and hands moving about the weather graphics!

How, then, does a TV weathercaster know where to point on the blank wall? Positioned on each side of the blue wall are TV monitors (look carefully at Fig. 5) that weathercasters watch so that they know where to point.

● FIGURE 4 On your home television, the weather forecaster Tom Loffman appears to be pointing to weather information directly behind him.

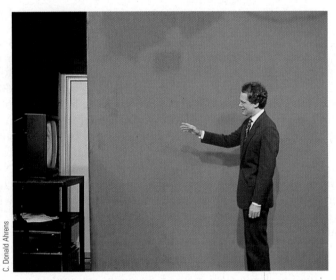

● FIGURE 5 In the studio, however, he is actually standing in front of a blank board.

ensemble forecast chart such as this one is often referred to as a *spaghetti plot.*

If, at the end of a specific time, the progs, or model runs, match each other fairly well, as they do over the southwestern United States in Fig. 13.7, the forecast is considered *robust.* This situation allows the forecaster to issue a prediction with a high degree of confidence. If the progs disagree, as they do off the west coast of North America in Fig. 13.7, the forecaster with little faith in the computer model prediction, issues a forecast with limited confidence. In essence, *the less agreement among the progs, or model runs, the less predictable the weather.* Consequently, it would not be wise to make outdoor plans for Saturday when on Monday the weekend forecast calls for "sunny and warm" with a low degree of confidence.

In summary, imperfect numerical weather predictions may result from flaws in the computer models, from errors that creep in along the models' boundaries, from the sparse-

ness of data, and/or from inadequate representation of many pertinent processes, interactions, and inherently chaotic behavior that occurs within the atmosphere.

Up to this point, we have looked primarily at weather forecasts made by high-speed computers using atmospheric models. There are, however, other forecasting methods, many of which have stood the test of time and are based mainly on the experience of the forecaster. Many of these techniques are of value, but often they give more of a general overview of what the weather should be like, rather than a specific forecast. (Before going on, you may wish to read the Focus section above, that describes how TV weather forecasters present weather visuals.)

OTHER FORECASTING METHODS Probably the easiest weather forecast to make is a **persistence forecast,** which is simply a prediction that future weather will be the same as

▼ TABLE 13.2   Forecast wording used by the National Weather Service to describe the percentage probability of measurable precipitation (0.01 inch or greater) for steady precipitation and for convective, showery precipitation

PERCENT PROBABILITY OF PRECIPITATION	FORECAST WORDING FOR STEADY PRECIPITATION	FORECAST WORDING FOR SHOWERY PRECIPITATION
>0 to 20 percent	*Slight chance* of precipitation	*Widely scattered* showers
30 to 50 percent	*Chance* of precipitation	*Scattered* showers
60 to 70 percent	Precipitation *likely*	*Numerous* showers
≥80 percent	Precipitation,* *rain, snow*	*Showers**

*A forecast that calls for an 80 percent chance of rain in the afternoon might read like this: ". . . cloudy today with rain this afternoon. . . ." For an 80 percent chance of rain showers, the forecast might read ". . . cloudy today with rain showers this afternoon. . . ."

present weather. If it is snowing today, a persistence forecast would call for snow through tomorrow. Such forecasts are most accurate for time periods of several hours and become less and less accurate after that.

Another method of forecasting is the **steady-state,** or **trend forecast.** The principle involved here is that surface weather systems tend to move in the same direction and at approximately the same speed as they have been moving, providing no evidence exists to indicate otherwise. Suppose, for example, that a cold front is moving eastward at an average speed of 30 km per hour and it is 90 km west of your home. Using the steady-state method, we might extrapolate and predict that the front should pass through your area in three hours.

The **analogue method** is yet another form of weather forecasting. Basically, this method relies on the fact that existing features on a weather chart (or a series of charts) may strongly resemble features that produced certain weather conditions sometime in the past. To the forecaster, the weather map "looks familiar," and for this reason the analogue method is often referred to as **pattern recognition.** A forecaster might look at a prog and say "I've seen this weather situation before, and this happened." Prior weather events can then be utilized as a guide to the future. The problem here is that, even though weather situations may appear similar, they are never *exactly* the same. There are always sufficient differences in the variables to make applying this method a challenge.

The analogue method can be used to predict a number of weather elements, such as maximum temperature. Suppose that in New York City the average maximum temperature on a particular date for the past 30 years is 10°C (50°F). By statistically relating the maximum temperatures on this date to other weather elements—such as the wind, cloud cover, and

humidity—a relationship between these variables and maximum temperature can be drawn. By comparing these relationships with current weather information, the forecaster can predict the maximum temperature for the day.

Presently, **statistical forecasts** are made routinely of weather elements based on the past performance of computer models. Known as *Model Output Statistics,* or MOS, these predictions, in effect, are statistically weighted analogue forecast corrections incorporated into the computer model output. For example, a forecast of tomorrow's maximum temperature for a city might be derived from a statistical equation that uses a numerical model's forecast of relative humidity, cloud cover, wind direction, and air temperature.

When the Weather Service issues a forecast calling for rain, it is usually followed by a probability. For example: "The chance of rain is 60 percent." Does this mean (a) that it will rain on 60 percent of the forecast area or (b) that there is a 60 percent chance that it will rain within the forecast area? Neither one! The expression means that there is a 60 percent chance that any random place in the forecast area, such as your home, will receive measurable rainfall.* Looking at the forecast in another way, if the forecast for 10 days calls for a 60 percent chance of rain, it should rain where you live on 6 of those days. The verification of the forecast (as to whether it actually rained or not) is usually made at the Weather Service office, but remember that the computer models forecast for a given region, not for an individual location. When the National Weather Service issues a forecast calling for a "slight chance of rain," what is the probability (percentage) that it will rain? ▼ Table 13.2 provides this information.

An example of a **probability forecast** using climatological data is given in ● Fig. 13.8. The map shows the probability of a "White Christmas"—1 inch or more of snow on the ground—across the United States. The map is based on the average of 30 years of data and gives the likelihood of snow in terms of a probability. For instance, the chances are greater than 90 percent (9 Christmases out of 10) that portions of northern Minnesota, Michigan, and Maine will experience a White Christmas. In Chicago, it is close to 50 percent; and in Washington, D.C., about 20 percent. Many places in the far west and south have probabilities less than 5 percent, but nowhere is the probability exactly 0, for there is always some chance (no matter how small) that a mantle of white will cover the ground on Christmas day. (Look at the opening photo in this chapter—p. 338—and notice that Brownsville, Texas, in the very southern part of the state, had a "White Christmas" during 2004.)

Predicting the weather by **weather types** employs the analogue method. In general, weather patterns are categorized into similar groups or "types," using such criteria as the position of the subtropical highs, the upper-level flow, and the prevailing storm track. As an example, when the Pacific

---

*The 60 percent chance of rain does not apply to a situation that involves rain showers. In the case of showers, the percentage refers to the expected area over which the showers will fall.

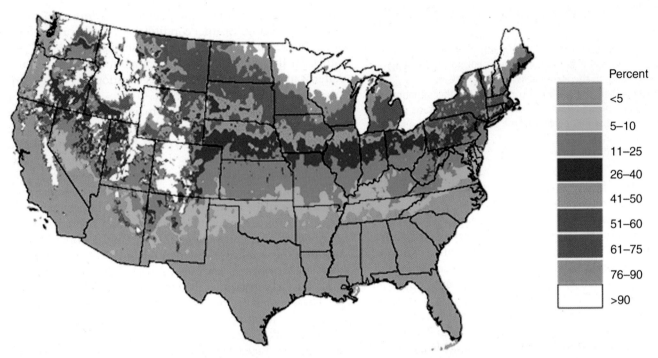

● FIGURE 13.8 Probability of a "White Christmas"—one inch or more of snow on the ground—based on a 30-year average. The probabilities do not include the mountainous areas in the western United States. (NOAA)

high is weak or depressed southward and the flow aloft is zonal (west-to-east), surface storms tend to travel rapidly eastward across the Pacific Ocean and into the United States without developing into deep systems. But when the Pacific high is to the north of its normal position and the upper airflow is meridional (north-south), looping waves form in the flow with surface lows usually developing into huge storms. As we saw in Chapter 12, these upper-level longwaves move slowly, usually remaining almost stationary for perhaps a few days to a week or more. Consequently, the particular surface weather at different positions around the wave is likely to persist for some time. ● Figure 13.9 presents an example of weather conditions most likely to prevail with a winter meridional weather type.

A forecast based on the climate* of a particular region is known as a **climatological forecast.** Anyone who has lived in Los Angeles for a while knows that July and August are practically rain-free. In fact, rainfall data for the summer months taken over many years reveal that rainfall amounts of more than a trace occur in Los Angeles about 1 day in every 90, or only about 1 percent of the time. Therefore, if we predict that it will not rain on some day next year during July or August in Los Angeles, our chances are nearly 99 percent that the forecast will be correct based on past records. Since it is unlikely that this pattern will significantly change in the near future, we can confidently make the same forecast for the year 2020.

---

*The climate of a region represents the total accumulation of daily and seasonal weather events for a specific interval of time, most often 30 years.

TYPES OF FORECASTS Weather forecasts are normally grouped according to how far into the future the forecast extends. For example, a weather forecast for up to a few hours (usually not more than 6 hours) is called a **very short-range forecast,** or *nowcast.* The techniques used in making such a forecast normally involve subjective interpretations of surface observations, satellite imagery, and Doppler radar information. Often weather systems are moved along by the steady

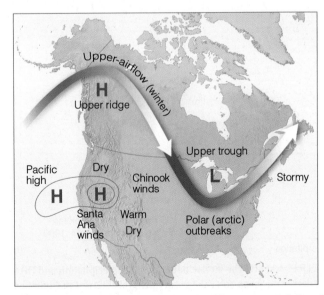

● FIGURE 13.9 Winter weather type showing upper-airflow (heavy arrow), surface position of Pacific high, and general weather conditions that should prevail.

state or trend method of forecasting, with human experience and pattern recognition coming into play.

Weather forecasts that range from about 12 hours to a few days (generally 2.5 days or 60 hours) are called **short-range forecasts.** The forecaster may incorporate a variety of techniques in making a short-range forecast, such as satellite imagery, Doppler radar, surface weather maps, upper-air winds, and pattern recognition. As the forecast period extends beyond about 12 hours, the forecaster tends to weight the forecast heavily on computer-drawn progs and statistical information, such as Model Output Statistics (MOS).

A **medium-range forecast** is one that extends from about 3 to 8.5 days (200 hours) into the future. Medium-range forecasts are almost entirely based on computer-derived products, such as forecast progs and statistical forecasts (MOS). A forecast that extends beyond 3 days is often called an *extended forecast.*

A forecast that extends beyond about 8.5 days (200 hours) is called a **long-range forecast.** Although computer progs are available for up to 16 days into the future, they are not accurate in predicting temperature and precipitation, and at best only show the broad-scale weather features. Pres-

ently, the Climate Prediction Center issues forecasts, called *outlooks,* of average weather conditions for a particular month or a season. These are not forecasts in the strict sense, but rather an overview of how precipitation and temperature patterns may compare with normal conditions. ● Figure 13.10 gives a typical 90-day outlook.

Initially, outlooks were based mainly on the relationship between the projected average upper-air flow and the surface weather conditions that the type of flow will create. Today, many of the outlooks are based on persistence statistics that carry over the general weather pattern from immediately preceding months, seasons, and years. In addition, long-range forecasts are made from models that link the atmosphere with the ocean surface temperature. As we saw in Chapter 10, a vast warming (*El Niño*) or cooling (*La Niña*) of the tropical Pacific can affect the weather in different regions of the world. These interactions, where a warmer tropical Pacific can influence rainfall in California, are called **teleconnections.***

These types of interactions between widely separated regions are identified through statistical correlations. For example, look back at Fig. 10.23, p. 279, and observe where seasonally averaged temperature and precipitation patterns over North America tend to depart from normal during El Niño and La Niña events. Using this type of information, the Climate Prediction Center can issue a seasonal outlook of an impending wetter or drier winter, months in advance. Forecasts using teleconnections have shown promise. For example, as the tropical equatorial Pacific became much warmer than normal during the spring and early summer of 1997, forecasters predicted a wet rainfall season over central and .

*Teleconnections include not only El Niño and La Niña but other indices, such as the Pacific Decadal Oscillation, the North Atlantic Oscillation, and the Arctic Oscillation. (See Chapter 10, p. 280.)

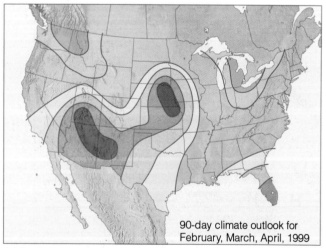

(a) Precipitation

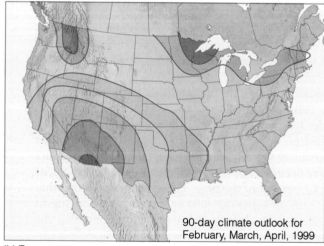

(b) Temperature

● FIGURE 13.10 The 90-day outlook for (a) precipitation and (b) temperature for February, March, and April, 1999. For precipitation (a), the darker the green color the greater the probability of precipitation being above normal, whereas the deeper the red color the greater the probability of precipitation being below normal. For temperature (b), the darker the orange/red colors the greater the probability of temperatures being above normal, whereas the darker the blue color, the greater the probability of temperatures being below normal. (National Weather Service/NOAA)

southern California. Although the heavy rains didn't begin until December, the weather during the winter of 1997–1998 was wet and wild: Storm after storm pounded the region producing heavy rains, mud slides, road closures, and millions of dollars in damages.

In most locations throughout North America, the weather is fair more often than rainy. Consequently, there is a forecasting bias toward fair weather, which means that, if you made a forecast of "no rain" where you live for each day of the year, your forecast would be correct more than 50 percent of the time. But did you show any *skill* in making your correct forecast? What constitutes skill, anyway? And how accurate are the forecasts issued by the National Weather Service?

ACCURACY AND SKILL IN FORECASTING   In spite of the complexity and ever-changing nature of the atmosphere, forecasts made for between 12 and 24 hours are usually quite accurate. Those made for between 2 and 5 days are fairly good. Beyond about 7 days, due to the chaotic nature of the atmosphere, computer prog forecast accuracy falls off rapidly. Although weather predictions made for up to 3 days are by no means perfect, they are far better than simply flipping a coin. But how accurate are they?

One problem with determining forecast accuracy is deciding what constitutes a right or wrong forecast. Suppose tomorrow's forecast calls for a minimum temperature of 35°F. If the official minimum turns out to be 37°F, is the forecast incorrect? Is it as incorrect as one 10 degrees off? By the same token, what about a forecast for snow over a large city, and the snow line cuts the city in half with the southern portion receiving heavy amounts and the northern portion none? Is the forecast right or wrong? At present, there is no clear-cut answer to the question of determining forecast accuracy.

How does forecast accuracy compare with forecast skill? Suppose you are forecasting the daily summertime weather in Los Angeles. It is not raining today and your forecast for tomorrow calls for "no rain." Suppose that tomorrow it doesn't rain. You made an accurate forecast, but did you show any skill in so doing? Earlier, we saw that the chance of measurable rain in Los Angeles on any summer day is very small indeed; chances are good that day after day it will not rain. For a forecast to show skill, it should be better than one based solely on the current weather *(persistence)* or on the "normal" weather *(climatology)* for a given region. Therefore, during the summer in Los Angeles, a forecaster will have many accurate forecasts calling for "no measurable rain," but will need skill to predict correctly on which summer days it will rain. So, if on a sunny July day in Los Angeles you forecast rain for tomorrow and it rains, you not only made an accurate forecast, you also showed skill in making your forecast because your forecast was better than both persistence and climatology.

Meteorological forecasts, then, show skill when they are more accurate than a forecast utilizing only persistence or climatology. Persistence forecasts are usually difficult to improve upon for a period of time of several hours or less. Weather forecasts ranging from 12 hours to a few days gener-

ally show much more skill than those of persistence. However, as the range of the forecast period increases, because of chaos the skill drops quickly. The 6- to 14-day mean outlooks both show some skill (which has been increasing over the last several decades) in predicting temperature and precipitation. However, the accuracy of precipitation forecasts is less than that for temperature. Presently, 7-day forecasts now show about as much skill as 3-day forecasts did a decade ago. Beyond 15 days, specific forecasts are only slightly better than climatology. However, skill in making forecasts of average monthly temperature and precipitation has approximately doubled from 1995 to 2006.

Forecasting large-scale weather events several days in advance (such as the blizzard of 1996 along the eastern seaboard of the United States) is far more accurate than forecasting the precise evolution and movement of small-scale, short-lived weather systems, such as tornadoes and severe thunderstorms. In fact, 3-day forecasts of the development and movement of a major low-pressure system show more skill today than 36-hour forecasts did 15 years ago.

Even though the *precise* location where a tornado will form is presently beyond modern forecasting techniques, the general area where the storm is *likely* to form can often be predicted up to 3 days in advance. With improved observing systems, such as Doppler radar and advanced satellite imagery, the lead time of watches and warnings for severe storms has increased. In fact, the lead time* for tornado warnings has more than doubled over the last decade, with the average lead time today being close to 15 minutes.

Although scientists may never be able to skillfully predict the weather beyond about 15 days using available observations, the prediction of *climatic trends* appears to be more promising. Whereas individual weather systems vary greatly and are difficult to forecast very far in advance, global-scale patterns of winds and pressure frequently show a high degree of persistence and predictable change over periods of a few weeks to a month or more. With the latest generation of high-speed supercomputers, general circulation models (GCMs) are doing a far better job at predicting large-scale atmospheric behavior than did the earlier models. (The GCMs are numerical computer models that simulate global patterns of wind, pressure, and temperature, and how these phenomena change over time.) In fact, the new GCMs are able to simulate a number of global patterns quite well, such as *blocking highs*† that can cause precipitation and temperature patterns to deviate considerably from average conditions. As new knowledge and methods of modeling are fed into the GCMs, it is hoped that they will become a reliable tool in the forecasting of weather and climate. (In Chapter 16, we will examine in more detail the climatic predictions based on numerical models.)

Lead time is the interval of time between the issue of the warning and actual observance of the event, in this case, the tornado.

†Blocking highs are high-pressure areas that tend to remain nearly stationary for some time, thus "blocking" the west-to-east movement of mid-latitude cyclonic storms.

BRIEF REVIEW

Up to this point, we have looked at the various methods of weather forecasting. Before going on, here is a review of some of the important ideas presented so far:

- Available to the forecaster are a number of tools that can be used when making a forecast, including surface and upper-air maps, computer progs, meteograms, soundings, Doppler radar, and satellite information.

- The forecasting of weather by high-speed computers is known as *numerical weather prediction*. Mathematical models that describe how atmospheric temperature, pressure, winds, and moisture will change with time are programmed into the computer. The computer then draws surface and upper-air charts, and produces a variety of forecast charts called *progs*.

- After a number of days, flaws in the computer models — atmospheric chaos and small errors in the data — greatly limit the accuracy of weather forecasts.

- Ensemble forecasting is a technique based on running several forecast models (or different versions of a single model), each beginning with slightly different weather information to reflect errors in the measurements.

- A *persistence forecast* is a prediction that future weather will be the same as the present weather, whereas a *climatological forecast* is based on the climatology of a particular region.

- For a forecast to show skill, it must be better than a persistence forecast or a climatological forecast.

- Weather forecasts for up to a few hours are called *very short-range forecasts;* those that range from about 6 hours to a few days are called *short-range forecasts; medium-range forecasts* extend from about 3 to 5 days into the future, whereas *long-range forecasts* extend beyond, to about 8.5 days.

- Seasonal outlooks provide an overview of how temperature and precipitation patterns may compare with normal conditions.

PREDICTING THE WEATHER FROM LOCAL SIGNS   Because the weather affects every aspect of our daily lives, attempts to predict it accurately have been made for centuries. One of the earliest attempts was undertaken by Theophrastus, a pupil of Aristotle, who in 300 B.C. compiled all sorts of weather indicators in his *Book of Signs*. A dominant influence in the field of weather forecasting for 2000 years, this work consists of ways to foretell the weather by examining natural signs, such as the color and shape of clouds, and the intensity at which a fly bites. Some of these signs have validity and are a part of our own weather folklore—"a halo around the moon portends rain" is one of these. Today, we realize that the halo is caused by the bending of light as it passes through ice crystals and that ice crystal-type clouds (cirrostratus) are often the forerunners of an approaching storm. (See ● Fig. 13.11.)

Weather predictions can be made by observing the sky and using a little weather wisdom. If you keep your eyes open and your senses keenly tuned to your environment, you

● FIGURE 13.11   A halo around the sun (or moon) means that rain is on the way. A weather forecast made by simply observing the sky.

should, with a little practice, be able to make fairly good short-range local weather forecasts by interpreting the messages written in the weather elements. ▼ Table 13.3 is designed to help you with this endeavor.

The movement of clouds at different levels can assist you in predicting changes in the temperature of the air above you. Knowing how the temperature is changing aloft can help you predict the stability of the air (Chapter 6), as well as whether falling snow will change to rain, or vice versa. (This topic is explored more extensively in the Focus section on p. 356.)

To further help you forecast the weather, the instant weather forecast chart (Appendix E) has been prepared by considering the relationship that the pressure and wind have to various weather systems. While the chart is applicable to much of the United States and southern Canada, local influences, such as mountains and large bodies of water, can affect the local weather to such an extent that the large-scale weather patterns on which the chart is based do not always show up clearly. (The chart works best during the fall, winter, and spring when the weather systems are active.)

Weather Forecasting Using Surface Charts

We are now in a position to forecast the weather, utilizing more sophisticated techniques. Suppose, for example, that we wish to make a *short-range weather prediction* and the only information available is a surface weather map. Can we make a forecast from such a chart? Most definitely. And our chances of that forecast being correct improve markedly if we have maps available from several days back. We can use these past maps to locate the previous position of surface features and predict their movement.

▼ TABLE 13.3   Forecast at a Glance–Forecasting the Weather from Local Weather Signs. Listed below are a few forecasting rules that may be applied when making a short-range local weather forecast.

OBSERVATION	INDICATION	LOCAL WEATHER FORECAST
Surface winds from the S or from the SW; clouds building to the west; warm (hot) and humid (pressure falling)	Possible cool front and thunderstorms approaching from the west	Possible showers; possibly turning cooler; windy
Surface winds from the E or from the SE, cool or cold; high clouds thickening and lowering; halo (ring of light) around the sun or moon (pressure falling)	Possible approach of a warm front	Possibility of precipitation within 12–24 hours; windy (rain with possible thunderstorms during the summer; snow changing to sleet or rain in winter)
Strong surface winds from the NW or W; cumulus clouds moving overhead (pressure rising)	A low-pressure area may be moving to the east, away from you; and an area of high pressure is moving toward you from the west	Continued clear to partly cloudy, cold nights in winter; cool nights with low humidity in summer
Winter night		
(a) If clear, relatively calm with low humidity (low dew-point temperature)	(a) Rapid radiational cooling will occur	(a) A very cold night
(b) If clear, relatively calm with low humidity and snow covering the ground	(b) Rapid radiational cooling will occur	(b) A very cold night with minimum temperatures lower than in (a)
(c) If cloudy, relatively calm with low humidity	(c) Clouds will absorb and radiate infrared (IR) energy to surface	(c) Minimum temperature will not be as low as in (a) or (b)
Summer night		
(a) Clear, hot, humid (high dew points)	(a) Strong absorption and emission of IR energy to surface by water vapor	(a) High minimum temperatures
(b) Clear and relatively dry	(b) More rapid radiational cooling	(b) Lower minimum temperatures
Summer afternoon		
(a) Scattered cumulus clouds that show extensive vertical growth by mid-morning	(a) Atmosphere is relatively unstable	(a) Possible showers or thunderstorms by afternoon with gusty winds
(b) Afternoon cumulus clouds with limited vertical growth and with tops at just about the same level	(b) Stable layer above clouds (region dominated by high pressure)	(b) Continued partly cloudy with no precipitation; probably clearing by nightfall

A simplified surface weather map is shown in ●Fig. 13.12. The map portrays early winter weather conditions on Tuesday morning at 6:00 A.M. A single isobar is drawn around the pressure centers to show their positions without cluttering the map. Note that an open-wave cyclone is developing over the Central Plains. The weather conforms to the cyclone model (see Fig. 12.12, p. 326), with showers forming along the cold front and light rain, snow, and sleet ahead of the warm front. The dashed lines on the map represent the position of the weather systems 6 hours ago. Our first question is: How will these systems move?

DETERMINING THE MOVEMENT OF WEATHER SYSTEMS   There are several methods we can use in forecasting the movement of surface pressure systems and fronts. The following are a few of these forecasting rules of thumb:

1. For short-time intervals, mid-latitude cyclonic storms and fronts tend to move in the same direction and at approximately the same speed as they did during the previous six hours (providing, of course, there is no evidence to indicate otherwise).

2. Low-pressure areas tend to move in a direction that parallels the isobars in the warm air (the warm air sector) ahead of the cold front.

3. Lows tend to move toward the region of greatest pressure drop, while highs tend to move toward the region of greatest rise.

4. Surface pressure systems tend to move in the same direction as the wind at 5500 m (18,000 ft)—the 500-mb level. The speed at which surface systems move is about half the speed of the winds at this level.

Forecasting Temperature Advection by Watching the Clouds

We know from Chapter 12 that when cold air is being brought into a region by the wind, we call this *cold advection*. When warm air is brought into a region, we call this *warm advection.*

A knowledge of temperature advection aloft is a valuable tool in forecasting the weather. In summer, when the surface is warm, cold advection aloft sets up instability and increases the likelihood of towering cumulus clouds and showers. On the other hand, warm advection aloft usually increases the temperature of the air, thus making it more stable. During the winter, this often leads to smoke and haze accumulating in the colder air near the surface.

By watching the movement of clouds, we get a good indication as to the wind direction at cloud level and also the type of advection. For instance, a cloud moving from the west indicates a west wind, a cloud from the south a south wind, and so on. Clouds at different levels frequently move in different directions, meaning that the wind direction is changing with height. Wind that changes direction in a clockwise sense (north to northeast to east, etc.) is a *veering wind*. Wind that changes direction in a counterclockwise sense (north to northwest to west, etc.) is a *backing wind*. There are two general rules that will help us determine whether cold or warm advection is occurring in a layer of air above us:

1. Winds that *back* with height (change counterclockwise) indicate *cold advection.*
2. Winds that *veer* with height (change clockwise) indicate *warm advection.*

As an example, suppose we observe lower clouds moving from a southerly direction (south wind) and higher clouds moving from a westerly direction (west wind) (see Fig. 6a). The wind direction is veering with height; warm advection is occurring between the cloud layers, and the air should be getting warmer. If, on another day, we see lower clouds moving from a southerly direction and higher clouds moving from an easterly direction (Fig. 6b), the wind is backing with height and the atmosphere between the cloud layers is probably becoming colder.*

An example of the relationship between winds and advection is seen in the vertically shifting winds that accompany weather fronts. Figure 7b is a 3-D model of a typical open-wave cyclone with its accompanying warm and cold front. Behind the cold front, swiftly moving cumulus clouds indicate a northwesterly wind exists about a kilometer or so above the surface. Ahead of the advancing warm front, stratiform clouds indicate that here southeasterly winds prevail about a kilometer above the ground. We know from Chapter 12 that warm advection takes place *ahead* of the warm front and cold advection *behind* the cold front. In both cases, the advection usually occurs in a layer from the surface up to at least the 500-mb level.

At the 500-mb level (Fig. 7b), the position of the upper trough and the region of coldest air is to the west of the surface low. The direction of the wind and also the cloud movement

---

*In both of these examples, we are assuming horizontal air motion only.

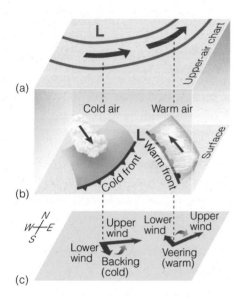

● FIGURE 7 Clouds, winds, and advection associated with a cold and a warm front.

is shown by arrows. Because of the upper trough's position, the winds aloft are westerly behind the cold front and southwesterly ahead of the warm front. Figure 7c shows how the wind direction changes from the surface to the 500-mb level. Behind the cold front, the winds back from northwesterly to westerly as we move upward. Cold advection is taking place as chilling air moves in from the west. Just ahead of the approaching warm front, the wind veers with height from southeasterly at the surface to southwesterly aloft as warm air glides up and over the cool surface air.

We can use this information to improve upon a weather forecast. For instance, if you happen to be located ahead of an advancing warm front and the winds above you are veering with height, the chances are that even if precipitation begins as snow it may change to rain as warm air moves in overhead. Behind a cold front where winds are backing with height, the influx of cold air may lower the temperature sufficiently so that rain first becomes mixed with snow, and then changes to snow before the storm moves eastward.

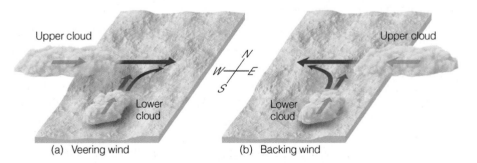

● FIGURE 6 (a) The wind veers with height, suggesting warm advection is occurring between the cloud layers. (b) The wind is backing with height, and cold advection is occurring between the cloud layers.

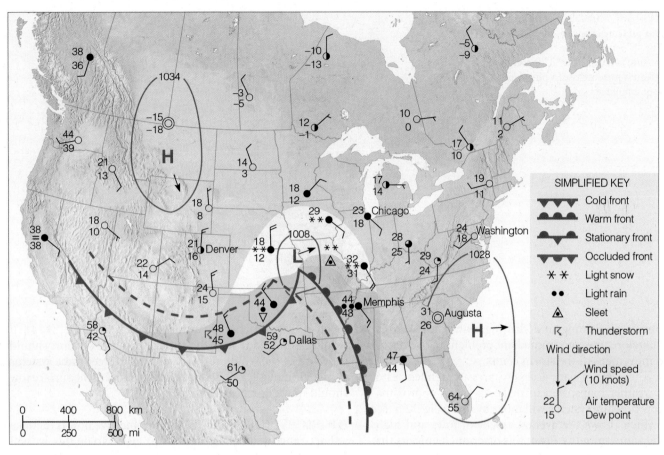

● **FIGURE 13.12** Surface weather map for 6:00 A.M. Tuesday. Dashed lines indicate positions of weather features six hours ago. Areas shaded green are receiving rain, while areas shaded white are receiving snow, and those shaded pink, freezing rain or sleet.

When the surface map (Fig. 13.12) is examined carefully and when rules of thumb 1 and 2 are applied, it appears that—based on present trends—the low pressure centered over the Central Plains should move northeast. If *pressure tendencies** are plotted on our map, we can draw lines connecting points of equal pressure change. These lines, called **isallobars,** help us to visualize the regions of falling and rising pressure. The distribution of pressure change for our map might look like the one in ● Fig. 13.13. Drawn at 2-mb intervals, the isallobars show a broad region of falling pressure ahead of the warm front, with the largest drop occurring to the northeast of the low-pressure area. This pattern fits with the previous observations and strengthens the prediction that the low center will move toward the northeast. The area of rising pressure immediately behind the cold front suggests that the high-pressure area over Montana will continue to move southeastward.

Pressure tendencies not only help predict the movement of highs and lows, they also indicate how the pressure systems are changing with time. The rapid fall in pressure in advance of the low indicates that the storm center is deepening as it moves. A deepening low means more closely spaced isobars,

a greater pressure gradient, and stronger winds—something to take into account when we make our weather forecast. A drop in pressure, on the other hand, in the vicinity of an anticyclone suggests that it is weakening, while a rise in pressure means that its central pressure is increasing. Hence, the high-pressure area over Montana is strong (1034 mb) and will remain so, whereas the high centered off the South Carolina coast is either moving eastward or weakening rapidly as indicated by the falling pressure in that area.

Before we complete our prediction about the movement of the pressure centers in Fig. 13.12, we need to look closely at the high-pressure area off the South Carolina coast. Strong highs, especially slow-moving ones, often retard the eastward progress of lows, deflecting them either north or south. From all indications—falling pressures and past movement—this

*The *pressure tendency* is the rate at which the pressure is changing during a given time, usually the past three hours.

● FIGURE 13.13 Isallobars—lines of equal 3-hour pressure change—for 6:00 A.M. Tuesday. The "F" represents the region of greatest pressure fall, while the "R" shows the region of greatest pressure rise. A plus 2 indicates a rise of 2 millibars.

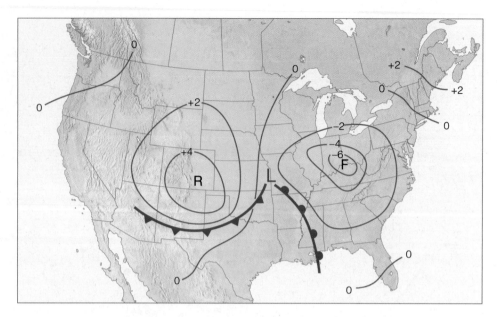

anticyclone is weakening and drifting slowly eastward. It should, therefore, pose no immediate problem to the northeastward movement of the storm center.

Even if we do not have access to pressure tendencies or previous weather maps, we can make an initial approximation of how pressure systems will move by using Fig. 12.5 (see p. 317), which shows the average tracks of lows and highs during the winter months. From this diagram, it appears that the cyclones and anticyclones in Fig. 13.12 are following rather typical trajectories.

If a 500-mb chart is available (such as ● Fig. 13.14), it would also indicate how the surface pressure systems should move, since the winds at this level tend to steer these systems along. From Fig. 13.14, it appears that, indeed, the surface low should move northeastward.

A FORECAST FOR SIX CITIES Our objective now is to make a short-range weather forecast for six cities. To do this, we will project the surface pressure systems, fronts, and current weather into the future by assuming steady-state conditions.

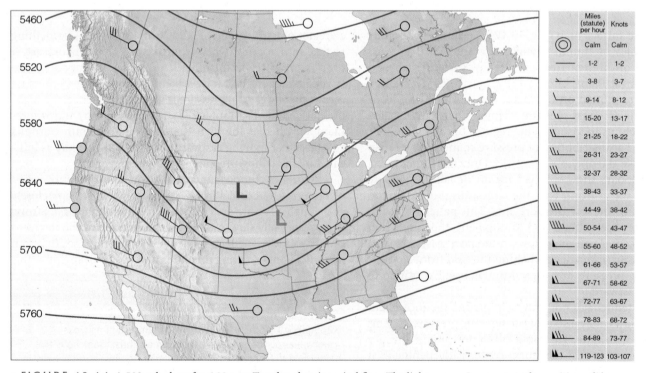

● FIGURE 13.14 A 500-mb chart for 6:00 A.M. Tuesday, showing wind flow. The light orange L represents the position of the surface low. The winds aloft tend to steer surface pressure systems along and, therefore, indicate that the surface low should move northeastward at about half the speed of the winds at this level, or 25 knots. Solid lines are contours in meters above sea level.

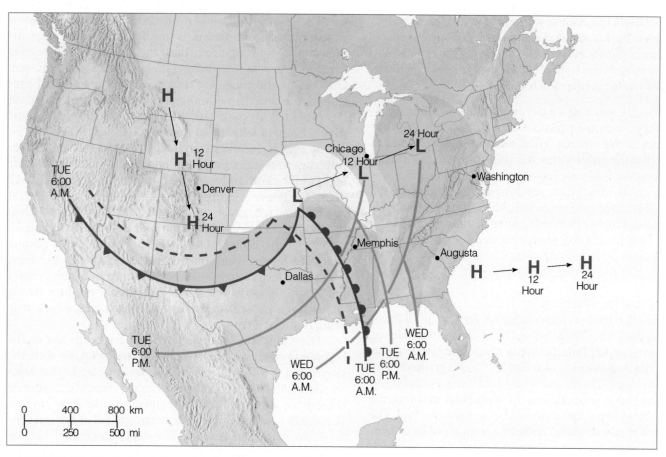

● FIGURE 13.15 Projected 12- and 24-hour movement of fronts, pressure systems, and precipitation from 6:00 A.M. Tuesday until 6:00 A.M. Wednesday. (The dashed lines represent frontal positions 6 hours ago.)

● Figure 13.15 gives the 12- and 24-hour projected positions of these features.

A word of caution before we make our forecasts. We are assuming that the pressure systems and fronts are moving at a constant rate. This may or may not occur. Low-pressure areas, for example, tend to accelerate until they occlude, after which their rate of movement slows. Furthermore, the direction of moving systems may change due to blocking-highs and -lows that exist in their path or because of shifting upper-level wind patterns. We will assume a constant rate of movement and forecast accordingly, always keeping in mind that the longer our forecasts extend into the future, the more susceptible they are to error.

If we move the low- and high-pressure areas eastward, as illustrated in Fig. 13.15, we can make a basic weather forecast for various cities. For example, the cold front moving into north Texas on Tuesday morning is projected to pass Dallas by that evening, so a forecast for the Dallas area would be "warm with showers, then turning colder." But we can do much better than this. Knowing the weather conditions that accompany advancing pressure areas and fronts, we can make more detailed weather forecasts that will take into account changes in temperature, pressure, humidity, cloud cover, precipitation, and winds. Our forecast will include the 24-hour

period from Tuesday morning to Wednesday morning for the cities of Augusta, Georgia; Washington, D.C.; Chicago, Illinois; Memphis, Tennessee; Dallas, Texas; and Denver, Colorado. We will begin with Augusta.

Weather Forecast for Augusta, Georgia

On Tuesday morning, continental polar air associated with a high-pressure area brought freezing temperatures and fair weather to the Augusta area (see Fig. 13.12). Clear skies, light winds, and low humidities allowed rapid nighttime cooling so that, by morning, temperatures were in the low thirties. Now look closely at Fig. 13.15 and observe that the anticyclone is moving slowly eastward, away from Augusta. Southerly winds on the western side of this system will bring warmer and more humid air to the region. Therefore, afternoon temperatures will be warmer than those of the day before. As the warm front approaches from the west, clouds will increase, appearing first as cirrus, then thickening and lowering into the normal sequence of warm-front clouds. Barometric pressure should fall. Clouds and high humidity should keep minimum temperatures well above freezing on Tuesday night. Note that the projected area of precipitation (green-shaded region) does not quite reach Augusta. With all of this in mind, our forecast might sound something like this:

Clear and cold this morning with moderating temperatures by afternoon. Increasing high clouds with skies becoming overcast by evening. Cloudy and not nearly as cold tonight and tomorrow morning. Winds will be light and out of the south or southeast. Barometric pressure will fall slowly.

Wednesday morning we discover that the weather in Augusta is foggy with temperatures in the upper 40s (°F). But fog was not in the forecast. What went wrong? We forgot to consider that the ground was still cold from the recent cold snap. The warm moist air moving over the cold surface was chilled below its dew point, resulting in fog. Above the fog were the low clouds we predicted. The minimum temperatures remained higher than anticipated because of the release of latent heat during fog formation and the absorption of infrared energy by the fog droplets. Not bad for a start. Now we will forecast the weather for Washington, D.C.

Rain or Snow for Washington, D.C.? Look at Fig. 13.15 and observe that the low-pressure area over the Central Plains is slowly approaching Washington, D.C., from the west. Hence, the clear weather, light southwesterly winds, and low temperatures on Tuesday morning (see Fig. 13.12) will gradually give way to increasing cloudiness, winds becoming southeasterly, and slightly higher temperatures. By Wednesday morning, the projected band of precipitation will be over the city. Will it be in the form of rain or snow? Without a vertical profile of temperature (a sounding) or a thickness chart, this question is difficult to answer. We can see in Fig. 13.12, however, that on Tuesday morning cities south of Washington, D.C.'s latitude are receiving snow. So a reasonable forecast would call for snow, possibly changing to rain as warm air moves in aloft in advance of the approaching fronts. A 24-hour forecast for Washington, D.C., might sound like this:

> Increasing clouds today and continued cold. Snow beginning by early Wednesday morning, possibly changing to rain. Winds will be out of the southeast. Pressures will fall.

Wednesday morning a friend in Washington, D.C., calls to tell us that the sleet began to fall but has since changed to rain. Sleet? Another fractured forecast! Well, almost. What we forgot to account for this time was the intensification of the storm. As the low-pressure area moved eastward, it deepened; central pressure lowered, pressure gradients tightened, and southeasterly winds blew stronger than anticipated. As air moved inland off the warmer Atlantic, it rode up and over the colder surface air. Snow falling into this warm layer at least partially melted; it then refroze as it entered the colder air near ground level. The advection of warmer air from the ocean slowly raised the surface temperatures, and the sleet soon became rain. Although we did not see this possibility when we made our forecast, a forecaster more familiar with local surroundings would have. Let's move on to Chicago.

Big Snowstorm for Chicago From Figs. 13.12 and 13.15, it appears that Chicago is in for a major snowstorm. Over-

running of warm air has produced a wide area of snow which, from all indications, is heading directly for the Chicago area. Since cold air north of the low center will be over Chicago, precipitation reaching the ground should be frozen. On Tuesday morning (Fig. 13.12), the leading edge of precipitation is less than 6 hours away from Chicago. Based on the projected path of the storm (Fig. 13.15), light snow should begin to fall around noon on Tuesday.

By evening, as the storm intensifies, snowfall should become heavy. It should taper off and finally end around midnight as the storm moves on east. If it snows for a total of 12 hours—6 hours as light snow (around 1 inch every 3 hours) and 6 hours as heavy snow (around 1 inch per hour)—then the total expected accumulation will be between 6 and 10 inches. As the center of the low moves eastward, passing south of Chicago, winds on Tuesday will gradually shift from southeasterly to easterly, then northeasterly by evening. Since the storm system is intensifying, it should produce strong winds that will swirl the snow into huge drifts, which may bring traffic to a crawl.

The winds will continue to shift to the north and finally become northwesterly by Wednesday morning. By then the storm center will probably be far enough east so that skies should begin to clear. Cold air advected from the northwest behind the storm will cause temperatures to drop further. Barometer readings during the storm will fall as the low center approaches and reach a low value sometime Tuesday night, after which they will begin to rise. A weather forecast for Chicago might be:

> Cloudy and cold with light snow beginning by noon, becoming heavy by evening and ending by Wednesday morning. Total accumulations will range between 6 and 10 inches. Winds will be strong and gusty out of the east or northeast today becoming northerly tonight and northwesterly by Wednesday morning. Barometric pressure will fall sharply today and rise tomorrow.

A call Wednesday morning to a friend in Chicago reveals that our forecast was correct except that the total snow accumulation so far is 13 inches. We were off in our forecast because the storm system slowed as it became occluded. We did not consider this because we moved the system by the steady-state forecast method. At this time of year (early winter), Lake Michigan is not quite frozen over, and the added moisture picked up from the lake by the strong easterly and northeasterly winds enhanced the snowfall. Again, a knowledge of the local surroundings would have helped make a more accurate forecast. The weather 500 miles south of Chicago should be much different from this.

Mixed Bag of Weather for Memphis Observe in Fig. 13.15 that, within 24 hours, both a warm and a cold front should move past Memphis, Tennessee. The light rain that began Tuesday morning (Fig. 13.12) should saturate the cool air, creating a blanket of low clouds and fog by midday. The warm front, as it moves through sometime Tuesday afternoon, should cause temperatures to rise slightly as winds shift to the south or southwest. At night, clear to partly cloudy skies should allow the ground and air above to cool, offsetting any tendency for a rapid rise in temperature. Falling pressures should level off in the warm air, then fall once again as the cold front approaches. According to the projection in Fig. 13.15, the cold front should arrive sometime before midnight on Tuesday, bringing with it gusty northwesterly winds, showers, the possibility of thunderstorms, rising pressures, and colder air. Taking all of this into account, our weather forecast for Memphis will be:

> Cloudy and cool with light rain, low clouds, and fog early today, becoming partly cloudy and warmer by late this afternoon. Clouds increasing with possible showers and thunderstorms later tonight and turning colder. Winds southeasterly this morning, becoming southerly or southwesterly this evening and shifting to northwesterly tonight. Pressures falling this morning, leveling off this afternoon, then falling again, but rising after midnight.

A friend who lives near Memphis calls Wednesday to inform us that our forecast was correct except that the thunderstorms did not materialize and that Tuesday night dense fog formed in low-lying valleys, but by Wednesday morning it had dissipated. Apparently, in the warm sector of the storm (the region ahead of the advancing cold front), winds were not strong enough to mix the cold, moist air that had settled in the valleys with the warm air above. It's on to Dallas.

Cold Wave for Dallas

From Fig. 13.15, it appears that our weather forecast for Dallas should be straightforward, since a cold front is expected to pass the area around noon on Tuesday. Weather along the front (Fig. 13.12) is showery with a few thunderstorms developing; behind the front the air is clear but cold. By Wednesday morning (Fig. 13.15) it looks as if the cold front will be far to the east and south of Dallas and an area of high pressure will be centered over southern Colorado. North or northwesterly winds on the east side of the high will bring cold arctic air into Texas, dropping temperatures as much as 40°F within a 24-hour period. With minimum temperatures well below freezing, Dallas will be in the grip of a cold wave. Our weather forecast should therefore sound something like this:

> Increasing cloudiness and mild this morning with the possibility of showers and thunderstorms this afternoon. Clearing and turning much colder tonight and tomorrow. Winds will be southwesterly today, becoming gusty north or northwesterly this afternoon and tonight. Pressures falling this morning, then rising later today.

How did our forecast turn out? A quick call to Dallas on Wednesday morning reveals that the weather there is cold but not as cold as expected, and the sky is overcast. Cloudy weather? How can this be?

The cold front moved through on schedule Tuesday afternoon, bringing showers, gusty winds, and cold weather with it. Moving southward, the front gradually slowed and became stationary along a line stretching from the Gulf of Mexico westward through southern Texas and northern Mexico. (From the surface map alone we had no way of knowing this would happen.) Along the stationary front a wave of low pressure formed. This disturbance caused warm, moist Gulf air to slide northward up and over the cold surface air. Clouds formed, minimum temperatures did not go as low as expected, and we are left with a fractured forecast. Let's give Denver a try.

Clear but Cold for Denver

In Fig. 13.15, we can see that, based on our projections, the cold high-pressure area will be centered slightly to the south of Denver by Wednesday morning. Sinking air aloft associated with this high-pressure area should keep the sky relatively free of clouds. Weak pressure gradients will produce only weak winds and this, coupled with dry air, will allow for intense radiational cooling. Minimum temperatures will probably drop to well below 0°F. Our forecast should therefore read:

> Clear and cold through tomorrow. Northerly winds today becoming light and variable by tonight. Low temperatures tomorrow morning will be below zero. Barometric pressure will continue to rise.

Wednesday morning, we (almost reluctantly) inquire about the weather conditions at Denver. "Clear and very cold" is the reply. A successful forecast at last! We are told, however, that the minimum temperature did not go below zero; in fact, 13°F was as cold as it got. A downslope wind coming off the mountains to the west of Denver kept the air mixed and the minimum temperature higher than expected. Again, a forecaster familiar with the local topography of the Denver area would have foreseen the conditions that lead to such downslope winds and would have taken this into account when making the forecast.

A complete picture of the surface weather systems for 6:00 A.M. Wednesday morning is given in Fig. 13.16. By comparing this chart with Fig. 13.15, we can summarize why our forecasts did not turn out exactly as we had predicted. For one thing, the storm center near the Great Lakes moved slower than expected. This slow movement allowed a southeasterly flow of mild Atlantic air to overrun cooler surface air ahead of the storm while, behind the low, cities remained in the snow area for a longer time. The weak wave that developed along the trailing cold front over south Texas brought cloudiness and precipitation to Texas and prevented the really cold air from penetrating deep into the south. Farther west, the high-pressure area originally over Montana moved more southerly than southeasterly, which set up a pressure gradient that brought westerly downslope winds to eastern Colorado.

The subjective forecasting techniques discussed so far are those you can use when making a short-range weather prediction. The following section describes how a meteorologist predicts the weather in a region where, to the west, surface weather features are extensively modified by a vast body of

● FIGURE 13.16
Surface weather map for
6:00 A.M. Wednesday.

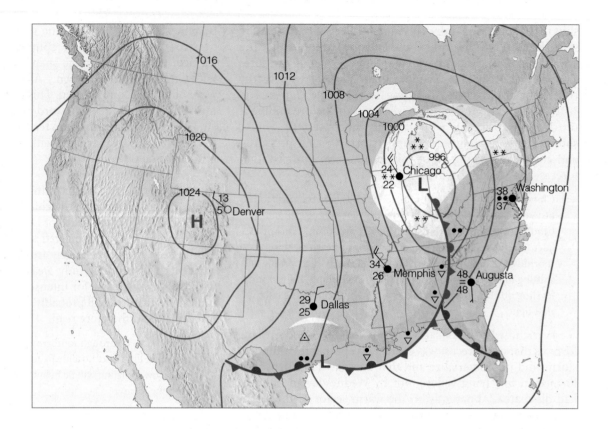

water and only scanty surface and upper-air data are available. Here, the forecaster must rely heavily on experience as well as more sophisticated tools, which include satellite data, upper-air charts, and computer progs.

A Meteorologist Makes a Prediction

It is late afternoon, and outside the weather forecast office near San Francisco the meteorologist mulls over what is going on in the sky. Overhead is a thin covering of cirrostratus; to the west, draped over the foothills, is the ever present stratus and fog. The air is cool and the winds are westerly. It is Sunday, March 25, and the forecaster's task is to predict the weather for the coastal area of central California.

What will tomorrow's weather be like? Will it be similar to today's or will it change markedly? A slowly falling barometer of 1016 mb (30 in.) and the high clouds moving in from the west point to an approaching storm system. A forecast of persistence might be good for the next several hours, but what about tomorrow morning or tomorrow afternoon?

The late afternoon surface analysis provides little assistance with these questions. The surface map for 4:00 P.M. (PST) Sunday, ● Fig. 13.17, shows there are no weather fronts approaching the West Coast. In fact, the nearest front is a stationary one that has stalled over the Rockies. There is, however, a region of low pressure centered about 1100 km (700 mi) west of San Francisco, which (according to previous maps) has been there for several days. With a central pressure

of only about 1012 mb (29.88 in.), the system is fairly weak. Could this weak storm system be causing the increase in high cloudiness and the falling barometer? And will this pattern lead to rain tomorrow? A look at the 500-mb chart may help with these questions.

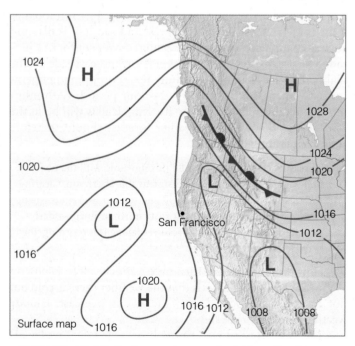

● FIGURE 13.17 Surface weather map for 4:00 P.M. Sunday, March 25.

HELP FROM THE 500-MB CHART • Figure 13.18 shows the 500-mb analysis for 4:00 P.M. Sunday afternoon. While examining the chart, the meteorologist recognizes certain clues that will aid in making the forecast. For one thing, the 5640-m height line is over northern California. The forecaster knows that when this contour line is situated here or farther south, the statistical probability of receiving measurable rainfall over central California increases greatly. There are, in fact, some forecasters who base their precipitation forecasts solely on the position of that line.

West of San Francisco the flow is meridional with a cutoff warm, upper high situated just south of Alaska. To the south both east and west of the high are troughs. Because the shape of this flow around the high resembles the Greek letter omega (Ω), the high and its accompanying ridge is known as an **omega high.** The forecaster recognizes the omega high as a *blocking high,* one that tends to persist in the same geographic location for many days. This blocking pattern also tends to keep the troughs in their respective positions, which has been the case for several days now. But the chart indicates that the cold upper trough located west of San Francisco may be changing somewhat.

Observe the spacing of the contour lines around this trough. Even with a limited number of actual wind observations, the close spacing of the contours to the west and northwest of the trough, and the more widely spaced contours to the east of the trough, hint that stronger winds exist to the west of the trough. The forecaster knows from past experience that this usually means the trough will deepen. Also note that west of the trough the winds are blowing across the isotherms (from colder regions), indicating that cold advection is occurring. The heavy dashed purple line on the west side of the trough represents the position of a shortwave trough, which is moving rapidly southward. The injection of cold air and the shortwave into the main trough should cause it to intensify. To the east of the main trough, the wind is crossing the isotherms and advecting warmer subtropical air northeastward. It is the lifting and condensing of this moist air that is producing the high clouds over San Francisco. All of these conditions—high wind speeds, regions of temperature advection, and a shortwave moving into a longwave trough—manifest themselves as a deepening of the longwave trough. As the upper trough deepens, it should be capable of providing the necessary conditions favorable for the development of the surface low into a major mid-latitude cyclonic storm. (You may remember from Chapter 12 that the generation of this type of dynamic instability is called *baroclinic instability.*)

One of the main ingredients necessary for the development and intensification of the surface low is divergence of the airflow aloft. The forecaster knows that divergence aloft is associated with a decrease in surface pressure. This decrease, in turn, causes surface air to converge, rise, and condense into widespread cloudiness. But where will regions of divergence, convergence, and rising air be found on tomorrow's map? And how will tomorrow's map be different from today's? This

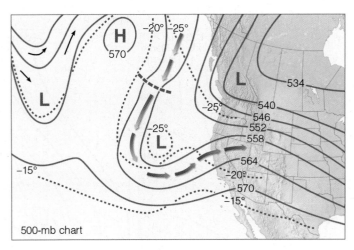

• FIGURE 13.18 The 500-mb chart for 4:00 P.M. Sunday, March 25. Arrows indicate wind flow. Red arrows indicate warm advection and blue arrows, cold advection. Solid lines are height contours where 564 equals 5640 meters. Dashed lines are isotherms in °C. Heavy dashed purple line shows position of shortwave trough.

is where the computer and the forecaster work together to come up with a prediction.

THE COMPUTER PROVIDES ASSISTANCE The computer progs predict the future positions of weather systems. Some of the progs also predict where shortwave troughs will be located. It is important to know where the shortwaves will be found, because to the east of them there is usually upper-level divergence, lower-level convergence, rising air, clouds, and precipitation. Hence, predicting the position of a shortwave means predicting regions of inclement weather. (The position of the shortwaves also pretty much corresponds to the position of the vorticity maximums discussed in Chapter 12.)

Three models that predict the positions of the shortwaves, upper-level pressure systems, and flow aloft at the 500-mb level for 12, 24, and 36 hours into the future are shown in • Fig. 13.19.* (Each prediction is made on Sunday afternoon.) Observe that there is good agreement among the models in that each model moves the upper trough slowly eastward and keeps it off the coast for the entire period. However, the actual positioning of the trough and the shortwaves (heavy dashed lines) differ for each model. For example, model A and model C move the upper trough eastward more quickly than does model B. Also, the 12-hour progs for model A and model C both show several shortwaves moving around the upper trough with one shortwave (labeled 1), positioned west of San Francisco by 4 A.M. Monday morning. Model B predicts that the same shortwave will be much farther to the west.

---

*Explaining the differences among the three models is beyond the scope of this book. Each model treats the atmosphere in a slightly different way. Some models have closer grid points. Some models have better resolution in the lower part of the atmosphere, whereas others have better resolution in the higher regions of the troposphere. The idea in this forecasting example is *not* to illustrate the different models in use, but rather to show how a forecaster might use *any* numerical computer model as a forecasting tool.

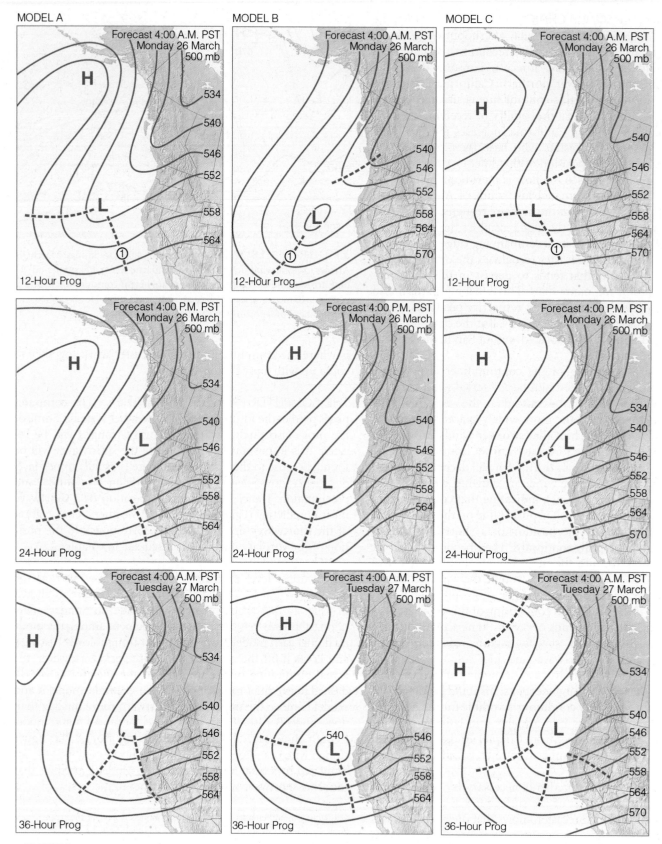

• FIGURE 13.19 Three computer-drawn progs (model A, model B, and model C) that show the 12-hour, 24-hour, and 36-hour projected 500-mb chart. Solid lines are contours. Dashed lines represent projected positions of shortwaves. (These predictions were made on Sunday, March 25, at 4:00 P.M., PST.)

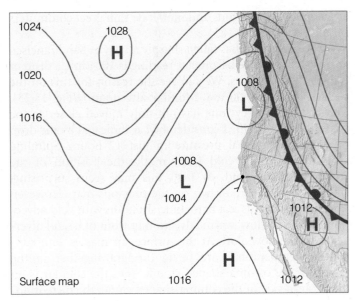

• FIGURE 13.20 Surface weather map for 4:00 A.M. (PST) Monday, March 26.

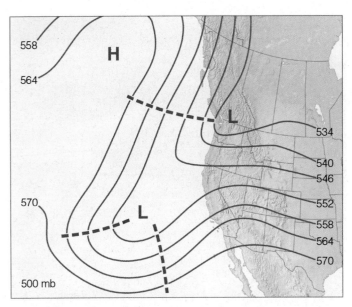

• FIGURE 13.21 The 500-mb analysis for 4:00 A.M. (PST) Monday, March 26. Heavy dashed lines show position of shortwaves. Solid lines are height contours where 564 equals 5640 m. (Compare with Fig. 13.19, the 12-hour progs for model A and model C.)

After examining each prog carefully, the forecaster must decide which model most accurately describes the future state of the atmosphere. Over the years, the forecaster knows that model A has performed well in predicting the positions of upper troughs that develop off the coast. Likewise, model C, because it uses more closely spaced grid points and a greater number of data points, has done an admirable job of forecasting the positions of upper troughs and shortwaves. Model B, on the other hand, tends to move the shortwaves along too slowly. Consequently, the forecaster puts more confidence into model A and model C.

Using experience and the progs, the meteorologist sets out to predict the weather. The 12-hour progs for model A and model C in Fig. 13.19 show a shortwave (labeled 1) approaching the California coast at 4 A.M. on Monday morning. As the shortwave approaches the coastline, clouds will increase and thicken and the likelihood of rain will increase. Because the main upper-level low is predicted to remain off the coast, southwesterly winds aloft will continue to pump moisture into the region and, as the 36-hour progs indicate, a series of shortwaves will move through the region, bringing a good chance of rain at least through Tuesday morning. Therefore, the precipitation forecast will sound like this:

> Increasing cloudiness Sunday night with rain beginning Monday morning. Periods of rain likely through Tuesday morning.

A VALID FORECAST   By early Monday morning, the maps begin to show the changes that the computer progs predicted. The surface map for 4:00 A.M. (PST) Monday morning (• Fig. 13.20) shows that the surface low in the Pacific has moved eastward and developed into a broad trough west of California. (Compare its position with Fig. 13.17.) The trough has deepened considerably as indicated by its central

pressure of 1004 mb (29.65 in.). The approach of the storm is evidenced in San Francisco by thick middle clouds, southerly winds, and a falling barometer, nearly 4 mb lower than 12 hours ago. All these signs suggest that rain is on the way.

On the 500-mb chart for 4:00 A.M. Monday morning (• Fig. 13.21), we can see that the injection of cold air and the shortwave into the main upper trough have caused the trough to deepen. Note that the height contours are now displaced farther south and that the contour in the middle of the trough is lower than on the previous 500-mb map (Fig. 13.18). Compare Fig. 13.21 with the 12-hour progs in model A and model C in Fig. 13.19 and notice how closely they match. The forecaster made a wise choice in showing confidence in these two computer models as they did a good job predicting the position of the upper-level low and shortwaves. Since the shortwaves are moving with the flow toward San Francisco, it should rain today. But at what time will the rain begin? Here is where satellite information assists the forecaster.

ASSISTANCE FROM THE SATELLITE   The infrared satellite image taken at 6:45 A.M. Monday (see • Fig. 13.22) shows that the middle clouds presently over California will soon give way to an organized band of cumuliform clouds in the shape of a comma. Such comma clouds tell the forecaster that the low-pressure area off the coast is developing into a mature mid-latitude cyclone. Observe that this comma-shaped cloud band lies almost beneath the shortwave approaching the coast, shown in Fig. 13.21. Also note that to the west a relatively unorganized mass of cumulus congestus clouds is beginning to form near the second shortwave in Fig. 13.21. Strong southwesterly winds aloft should carry the large comma-shaped cloud and its weather into California.

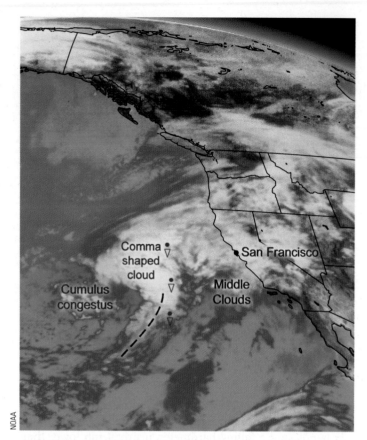

● FIGURE 13.22 Infrared satellite image taken at 6:45 A.M. (PST) Monday, March 26. The cloud in the shape of a comma indicates that the mid-latitude cyclonic storm is deepening. (The heavy dashed line shows the tail of the comma cloud.)

By examining the movement of the cloud mass on successive satellite images, the forecaster can predict its arrival time and, hence, when rainfall will begin. According to satellite photographs, the leading edge of the comma cloud should be just offshore by Monday afternoon. Also, Doppler radar indicates that, just off the coast, light rain is now falling from the middle cloud layer. Consequently, rainfall should begin sometime in the morning, becoming heavier by the afternoon as the cumuliform clouds move in. Because the surface mid-latitude cyclone is beneath the upper-level short-wave, the surface area of low pressure will probably continue to intensify, and pressure gradients around it will increase, creating strong and gusty winds from the south as the storm approaches. An amended forecast for San Francisco might read:

Rain beginning this morning becoming heavy by this afternoon. Strong and gusty southerly winds.

A DAY OF RAIN AND WIND The first raindrops falling from altostratus clouds dampen city streets near the end of the morning rush hour. Quickly, the rain spreads inland, and by late Monday afternoon, Doppler radar shows that precipitation is falling throughout northern and central California

as gusty southerly winds and moderate rain greet commuters on their way home.

The barometer has fallen sharply all day at San Francisco and by 4:00 P.M. the barometer reading is 1004 mb, a drop of 7 mb in just 6 hours. We can see the reason for this on the surface map for 4:00 P.M. Monday afternoon (● Fig. 13.23). The mid-latitude cyclone has not only moved closer to the coast, it has intensified considerably, as indicated by the drop of 11 mb in central pressure in just 12 hours. Spiraling around the low, a cold front marks the position of the comma-shaped cloud. At first, this may seem surprising, since no fronts were drawn on the previous map. However, remember that this is a baroclinic situation with cold advection behind the low, warm advection in front of it, and divergence in the flow aloft. At the surface, air masses with contrasting temperatures are being brought together in the region of the comma-shaped cloud. Since the cold air is on the western side of the comma-shaped cloud, the meteorologist saw fit to draw in a cold front. Notice that, to the north of the low, a stationary front marks the boundary between cold maritime polar air to the west and modified cool maritime polar air to the east.

As the surface low intensifies, it and the spiraling band of clouds move eastward more slowly. The front will, therefore, move through later than anticipated, sometime late Monday night or early Tuesday morning. The forecaster expects that the winds will remain strong and precipitation will be heavy as the front passes.

Early Tuesday morning, the front moves onshore, bringing with it heavy rain and winds with gusts exceeding 45 knots. Billowing cumulus clouds, and in some areas thunderstorms, drench the entire Pacific Coast with rain, and with snow at higher elevations. The storm center, constantly being drained of air by strong upper-level divergence, has

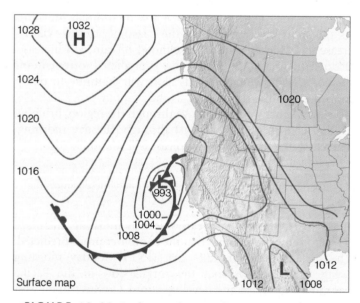

● FIGURE 13.23 Surface weather map for 4:00 P.M. (PST) Monday, March 26.

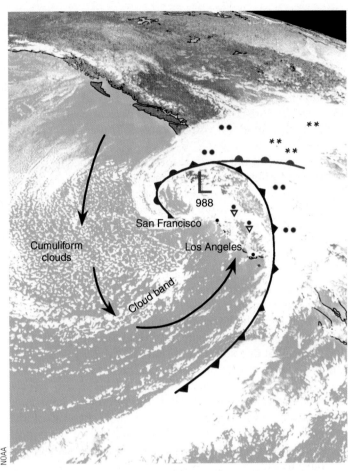

● **FIGURE 13.24** Visible satellite image for 9:00 A.M. (PST) Tuesday, March 27. Included in the picture are the positions of surface fronts, the upper-level flow (heavy arrows), and precipitation patterns.

Cumuliform clouds

Cloud band

San Francisco

Los Angeles

L 988

deepened into a furious system with a central pressure of 988 mb (29.17 in.).

The satellite image for 9:00 A.M. Tuesday morning (see ● Fig. 13.24) provides us with a visual interpretation of the storm. We see superimposed on the photograph the positions of the surface low, fronts, and the winds aloft (heavy arrow). Note that a front with its heavy band of clouds stretches from Idaho southward into Nevada and southern California, while the surface low is still positioned off the northern California coast. Moving through central California is a band of clouds and showers associated with a shortwave, spinning counterclockwise around the low.

The upper flow indicates that the trough aloft is still off the coast just as the 36-hour computer progs on Sunday (Fig. 13.19) had predicted. If we follow the flow southward out of Canada, we see a patch of clear weather just offshore. Here, the air pushing southward off the land is cold and dry, so clouds do not form. However, as the air moves farther south, it warms and picks up moisture from the water below. The rippled cloud pattern in the image is cumulus clouds, which form in the conditionally unstable air. Look closely at the image and notice that an organized band of clouds is developing to the southwest of the deep surface low. If you refer back to Fig. 13.19, you will see that the 36-hour prog for model A and model C had predicted that a shortwave would be located in this region on Tuesday morning. Is this cloud band organizing into another onslaught of high winds, heavy precipitation, and thunderstorms? If so, when will it arrive? And what about the storm off the coast? Will it deepen or fill, move inland or remain stationary? It's back to the drawing board—to the computer progs, the charts, Doppler radar, and the satellite images. The challenge and anticipation of making another forecast are at hand.

SUMMARY

Forecasting tomorrow's weather entails a variety of techniques and methods. Persistence, surface maps, satellite imagery, and Doppler radar are all useful when making a very short-range (0–6 hour) prediction. For short- and medium-range forecasts, the current analysis, satellite data, pattern recognition, meteorologist intuition, and experience, along with statistical information and guidance from the many computer progs supplied by the National Weather Service, all go into making a prediction. For monthly and seasonal long-range forecasts, meteorologists incorporate changes in sea surface temperature in the Pacific and Atlantic Oceans into seasonal outlooks of temperature and precipitation in North America.

Different computer progs are based upon different atmospheric models that describe the state of the atmosphere and how it will change with time. The atmosphere's chaotic behavior, along with flaws in the models and tiny errors (uncertainties) in the data, generally amplify as the computer tries to project weather further and further into the future. At present, computer progs that predict the weather over a vast region are better at forecasting the position of mid-latitude highs and lows and their development than at forecasting local showers and thunderstorms. To skillfully forecast smaller features, the grid spacing on some models is reduced to as low as 0.5 kilometers.

As new information from atmospheric research programs is fed into the latest generation of computers, it is hoped that the progs will be able to show more skill in predicting the weather up to 10 days in the future. More promising at this time is the simulation of large-scale climatic trends by the most recent general circulation models.

The latter part of this chapter does not cover all the methods of weather prediction, but, rather, conveys an understanding of the problems confronting anyone who attempts to predict the behavior of this churning mass of air we call our atmosphere.

Most of the forecasting methods in this chapter apply mainly to skill in predicting events associated with large-scale weather systems, such as fronts and mid-latitude cyclones. The next chapter on severe weather deals with the formation and forecasting of smaller scale (mesoscale) systems, such as thunderstorms, squall lines, and tornadoes.

KEY TERMS

The following terms are listed (with page numbers) in the order they appear in the text. Define each. Doing so will aid you in reviewing the material covered in this chapter.

weather watch, 340	analogue forecasting
weather warning, 341	method, 350
AWIPS, 341	pattern recognition, 350
meteogram, 342	statistical forecast, 350
analysis, 344	probability forecast, 350
numerical weather predic-	weather type forecasting, 350
tion, 344	climatological forecast, 351
atmospheric models, 344	very short-range forecast, 351
prognostic chart (prog), 344	short-range forecast, 352
chaos, 348	medium-range forecast, 352
ensemble forecasting, 348	long-range forecast, 352
persistence forecast, 349	teleconnections, 352
steady-state (trend) fore-	isallobars, 357
cast, 350	omega high, 363

QUESTIONS FOR REVIEW

1. What is the function of the National Center for Environmental Prediction?
2. How does a *weather watch* differ from a *weather warning*?
3. List some of the tools a weather forecaster might use when making a short-range forecast.
4. In what ways has the computer assisted the meteorologist in making weather forecasts?
5. How does a prog differ from an analysis?
6. How are computer-generated weather forecasts prepared?
7. What are some of the problems associated with computer-model forecasts?
8. Make a persistence forecast for your area for this same time tomorrow. Did you use any skill in making this prediction? Explain.
9. Describe four methods of forecasting the weather and give an example for each one.
10. How does pattern recognition aid a forecaster in making a prediction?
11. How can ensemble forecasts improve medium-range weather forecasts?

12. Explain how teleconnections are used in making a long-range seasonal forecast?
13. If today's weather forecast calls for a "chance of snow," what is the probability that it will snow today? (Hint: See Table 13.2, p. 350.)
14. Do all accurate forecasts show skill? Explain.
15. Would a forecast calling for a 20 percent chance of rain be high enough for you to cancel your plans for a picnic? Explain.
16. Do monthly and seasonal forecasts make specific predictions of rain or snow? Explain.
17. Use the instant weather forecast chart (Appendix E) as a guide to making a short-range weather forecast when the surface weather elements indicate those shown in ▼ Table 13.4, p. 369.
18. If low clouds at an elevation of 3000 ft above you are moving from the southeast, and clouds about 8000 ft higher are moving from the southwest, is cold or warm advection taking place between the cloud layers? Explain. (Hint: Read the Focus section found on p. 356.)
19. List four methods that you could use to predict the movement of a surface mid-latitude cyclone.
20. What is an omega high? What influence does it have on the movement of surface highs and lows?
21. Suppose that where you live, the middle of January is typically several degrees warmer than the rest of the month. If you forecast this "January thaw" for the middle of next January, what type of a weather forecast will you have made?
22. Given a map with isallobars, where will high- and low-pressure systems move?

QUESTIONS FOR THOUGHT

1. From Fig. 13.8, p. 351, determine the probability of a "White Christmas" for your area.
2. Suppose the chance for a "White Christmas" at your home is 10 percent. Last Christmas was a white one. If for next year you forecast a "nonwhite" Christmas, will you have shown any skill if your forecast turns out to be correct? Explain.
3. Suppose that it is presently warm and raining. A cold front will pass your area in 3 hours. Behind the front it is cold and snowing. Make a persistence forecast for your area 6 hours from now. Would you expect this forecast to be correct? Explain. Now, make a forecast for your area using the steady-state, or trend, method.
4. Since computer models have difficulty in adequately considering the effects of small-scale geographic features on a weather map, why don't numerical weather forecasts simply reduce the grid spacing to about one kilometer?
5. Explain how the phrase "sensitive dependence on initial conditions" relates to the final outcome of a computer-based weather forecast.

▼ TABLE 13.4   Surface Weather Elements (to be used in answering Review Question 17)

TIME	AIR PRESSURE (MB)	ANY CHANGES	AIR TEMPERATURE (°F)	WIND DIRECTION	CLOUD TYPE	PRECIPITATION
Morning	1017	Falling slowly	29	SE	As	None
Afternoon	1014	Rising rapidly	36	NW	Tcu	Rain showers
Afternoon	1025	Steady	86	NW	Ci	None
Afternoon	990	Falling rapidly	26	NE	Ns	Snow
Early morning	1020	Rising	13	NW	Clear	None
Nighttime	1012	Falling	31	SW	Sc	None
Afternoon	1006	Falling	52	SE	Ns	Rain
Nighttime	1005	Rising rapidly	45	W	Cb	Rain showers

6. You are in Calgary, AB, Canada, 100 km (62 mi) east of the Rockies in January. The current wind is from the north. Looking at a prog for tomorrow, you see that the wind will be from the west. Will tomorrow's temperature be warmer or cooler than today? Explain.

PROBLEMS AND EXERCISES

1. From the worldwide web, the newspaper, or the National Weather Service, obtain a copy of next month's 30-day outlook. Based on the temperature and precipitation patterns of this forecast, draw several contour lines representing the upper-air pattern that would be necessary for such a forecast to come true.

2. When a persistent winter pattern at 500 mb appears similar to that shown in ● Fig. 13.25, show on the map where you would forecast the following: (a) good chance of precipitation; (b) above seasonal temperatures; (c) below seasonal temperatures; and (d) generally dry weather. (e) What type of forecasting method did you use to figure out (a) through (d)?

3. In Fig. 13.12, p. 357, mark the position of the following cities: Cleveland, Ohio; Albuquerque, New Mexico; and New Orleans, Louisiana. Based on the projected movement of the surface weather systems in Fig. 13.15, p. 359, make a short-range 24-hour weather forecast for each of these cities. In your forecast, include temperature, pressure, cloud cover, humidity, winds, and precipitation (if any). Compare your forecasts with the actual weather at the end of the period in Fig. 13.16, p. 362.

4. Go outside and observe the weather. Make a weather forecast using the weather signs you observe. Explain the rationale for your forecast.

Visit the
Meteorology Resource Center
at
academic.cengage.com/login
for more assets, including questions for exploration, animations, videos, and more.

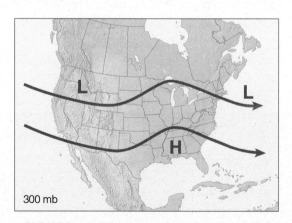
300 mb

● FIGURE 13.25 Diagram for Exercise 2.

A wall cloud associated with a supercell thunderstorm spins counterclockwise over the plains of Texas. Beneath the wall cloud, dust rising from the surface indicates that a tornado is about to form.

CHAPTER 14

Thunderstorms and Tornadoes

Wednesday, March 18, 1925, was a day that began uneventfully, but within hours turned into a day that changed the lives of thousands of people and made meteorological history. Shortly after 1:00 P.M., the sky turned a dark greenish-black and the wind began whipping around the small town of Murphysboro, Illinois. Arthur and Ella Flatt lived on the outskirts of town with their only son, Art, who would be four years old in two weeks. Arthur was working in the garage when he heard the roar of the wind and saw the threatening dark clouds whirling overhead.

Instantly concerned for the safety of his family, he ran toward the house as the tornado began its deadly pass over the area. With debris from the house flying in his path and the deafening thunder of destruction all around him, Arthur reached the front door. As he struggled in vain to get to his family, whose screams he could hear inside, the porch and its massive support pillars caved in on him. Inside the house, Ella had scooped up young Art in her arms and was making a panicked dash down the front hallway towards the door when the walls collapsed, knocking her to the floor, with Art cradled beneath her. Within seconds, the rest of the house fell down upon them. Both Arthur and Ella were killed instantly, but Art was spared, nestled safely under his mother's body.

As the dead and survivors were pulled from the devastation that remained, the death toll mounted. Few families escaped the grief of lost loved ones. The infamous tri-state tornado killed 234 people in Murphysboro and leveled 40 percent of the town.

 CONTENTS

The devastating tornado described in our opening cut a mile-wide path for a distance of more than 200 miles through the states of Missouri, Illinois, and Indiana. The tornado (which was most likely a series of tornadoes) totally obliterated 4 towns, killed an estimated 695 people, and left over 2000 injured. Tornadoes such as these, as well as much smaller ones, are associated with severe thunderstorms. Consequently, we will first examine the different types of thunderstorms. Later, we will focus on tornadoes, examining how and where they form, and why they are so destructive.

Thunderstorms

It probably comes as no surprise that a *thunderstorm* is merely a storm containing lightning and thunder. Sometimes a thunderstorm produces gusty surface winds with heavy rain and hail. The storm itself may be a single cumulonimbus cloud, or several thunderstorms may form into a cluster. In some cases, a line of thunderstorms will form that may extend for hundreds of kilometers.

Thunderstorms are *convective storms* that form with rising air. So the birth of a thunderstorm often begins when warm, moist air rises in a conditionally unstable environment.* The rising air may be a parcel of air ranging in size from a large balloon to a city block, or an entire layer, or slab of air, may be lifted. As long as a rising air parcel is warmer (less dense) than the air surrounding it, there is an upward-directed *buoyant force* acting on it. The warmer the parcel is compared to the air surrounding it, the greater the buoyant force and the stronger the convection. The trigger (or "forcing mechanism") needed to start air moving upward may be unequal heating at the surface, the effect of terrain, or the lifting of air along shallow boundaries of converging surface winds. Diverging upper-level winds, coupled with converging surface winds and rising air, also provide a favorable condition for thunderstorm development. Moreover, thunder-

storms often form when warm air rises along a frontal zone. Usually, several of these mechanisms work together with vertical wind shear to generate severe thunderstorms.

Although we often see thunderstorms forming where the surface air is quite buoyant (that is, warm and humid), they may also form when the surface air temperature is no more than 10°C (50°F). This latter situation often occurs in winter along the west coast of North America, when cold air aloft moves over the region. The cold air aloft destabilizes the atmosphere to the point where air parcels, given an initial push upwards, are able to continue their upward journey because they remain warmer (less dense) than the colder air surrounding them. The cold air aloft may even produce sufficient instability to generate thunderstorms in wintertime snowstorms.

Most thunderstorms that form over North America are short-lived, produce rain showers, gusty surface winds, thunder and lightning, and sometimes small hail. Many have an appearance similar to the mature thunderstorm shown in ● Fig. 14.1. The majority of these storms do not reach severe status. *Severe thunderstorms* are defined by the National Weather Service as having at least one of the following: large hail with a diameter of at least three-quarters of an inch and/or surface wind gusts of 50 knots (58 mi/hr) or greater, or produces a tornado.

Scattered thunderstorms (sometimes called "pop-up" storms) that typically form on warm, humid days are often referred to as *ordinary cell thunderstorms** or *air-mass thunderstorms* because they tend to form in warm, humid air masses away from significant weather fronts. Ordinary cell (air mass) thunderstorms can be considered "simple storms" because they rarely become severe, typically are less than a kilometer wide, and they go through a rather predictable life cycle from birth to maturity to decay that usually takes less than an hour to complete. However, under the right atmospheric conditions (described later in this chapter), more intense "complex thunderstorms" may form, such as the *multicell thunderstorm* and the *supercell thunderstorm*—a huge rotating storm that can last for hours and produce severe

---

*A conditionally unstable atmosphere exists when cold, dry air aloft overlies warm, moist surface air. Additional information on atmospheric instability is given in Chapter 6, beginning on p. 145.

*In convection, the cell may be a single updraft or a single downdraft, or a combination of the two.

● FIGURE 14.1 An ordinary thunderstorm in its mature stage. Note the distinctive anvil top.

© C. Donald Ahrens

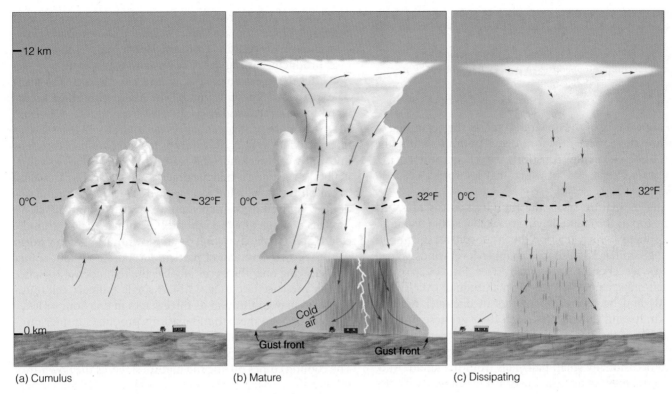

(a) Cumulus   (b) Mature   (c) Dissipating

● FIGURE 14.2 Simplified model depicting the life cycle of an ordinary cell thunderstorm that is nearly stationary as it forms in a region of low wind shear. (Arrows show vertical air currents. Dashed line represents freezing level, 0°C isotherm.)

weather such as strong surface winds, large damaging hail, flash floods, and violent tornadoes.

We will examine the development of ordinary cell (air mass) thunderstorms first, before we turn our attention to the more complex multicell and supercell storms.

ORDINARY CELL THUNDERSTORMS   **Ordinary cell (air mass) thunderstorms** or, simply, *ordinary thunderstorms,* tend to form in a region where there is limited wind shear—that is, where the wind speed and wind direction *do not* abruptly change with increasing height above the surface. Many ordinary thunderstorms appear to form as parcels of air are lifted from the surface by turbulent overturning in the presence of wind. Moreover, ordinary storms often form along shallow zones where surface winds converge. Such zones may be due to any number of things, such as topographic irregularities, sea-breeze fronts, or the cold outflow of air from inside a thunderstorm that reaches the ground and spreads horizontally. These converging wind boundaries are normally zones of contrasting air temperature and humidity and, hence, air density.

Extensive studies indicate that ordinary thunderstorms go through a cycle of development from birth to maturity to decay. The first stage is known as the **cumulus stage,** or *growth stage.* As a parcel of warm, humid air rises, it cools and condenses into a single cumulus cloud or a cluster of clouds (see ● Fig. 14.2a). If you have ever watched a thunderstorm develop, you may have noticed that at first the cumulus cloud grows upward only a short distance, then it dissipates. The

top of the cloud dissipates because the cloud droplets evaporate as the drier air surrounding the cloud mixes with it. However, after the water drops evaporate, the air is more moist than before. So, the rising air is now able to condense at successively higher levels, and the cumulus cloud grows taller, often appearing as a rising dome or tower.

As the cloud builds, the transformation of water vapor into liquid or solid cloud particles releases large quantities of latent heat, a process that keeps the rising air inside the cloud warmer (less dense) than the air surrounding it. The cloud continues to grow in the unstable atmosphere as long as it is constantly fed by rising air from below. In this manner, a cumulus cloud may show extensive vertical development and grow into a towering cumulus cloud (cumulus congestus) in just a few minutes. During the cumulus stage, there normally is insufficient time for precipitation to form, and the updrafts keep water droplets and ice crystals suspended within the cloud. Also, there is no lightning or thunder during this stage.

As the cloud builds well above the freezing level, the cloud particles grow larger. They also become heavier. Eventually, the rising air is no longer able to keep them suspended, and they begin to fall. While this phenomenon is taking place, drier air from around the cloud is being drawn into it in a process called *entrainment.* The entrainment of drier air causes some of the raindrops to evaporate, which chills the air. The air, now colder and heavier than the air around it, begins to descend as a *downdraft.* The downdraft may be enhanced as falling precipitation drags some of the air along with it.

The appearance of the downdraft marks the beginning of the **mature stage.** The downdraft and updraft within the mature thunderstorm now constitute the cell. In some storms, there are several cells, each of which may last for less than 30 minutes.

During its mature stage, the thunderstorm is most intense. The top of the cloud, having reached a stable region of the atmosphere (which may be the stratosphere), begins to take on the familiar anvil shape, as upper-level winds spread the cloud's ice crystals horizontally (see Fig. 14.2b). The cloud itself may extend upward to an altitude of over 12 km (40,000 ft) and be several kilometers in diameter near its base. Updrafts and downdrafts reach their greatest strength in the middle of the cloud, creating severe turbulence. Lightning and thunder are also present in the mature stage. Heavy rain (and occasionally small hail) falls from the cloud. And, at the surface, there is often a downrush of cold air with the onset of precipitation.

Where the cold downdraft reaches the surface, the air spreads out horizontally in all directions. The surface boundary that separates the advancing cooler air from the surrounding warmer air is called a *gust front*. Along the gust front, winds rapidly change both direction and speed. Look at Fig. 14.2b and notice that the gust front forces warm, humid air up into the storm, which enhances the cloud's updraft. In the region of the downdraft, rainfall may or may not reach the surface, depending on the relative humidity beneath the storm. In the dry air of the desert Southwest, for example, a mature thunderstorm may look ominous and contain all of the ingredients of any other storm, except that the raindrops evaporate before reaching the ground. However, intense downdrafts from the storm may reach the surface, producing strong, gusty winds and a gust front.

After the storm enters the mature stage, it begins to dissipate in about 15 to 30 minutes. The **dissipating stage** occurs when the updrafts weaken as the gust front moves away from the storm and no longer enhances the updrafts. At this stage, as illustrated in Fig. 14.2c, downdrafts tend to dominate throughout much of the cloud. The reason the storm does not normally last very long is that the downdrafts inside the cloud tend to cut off the storm's fuel supply by destroying the humid updrafts. Deprived of the rich supply of warm, humid air, cloud droplets no longer form. Light precipitation now falls from the cloud, accompanied by only weak downdrafts. As the storm dies, the lower-level cloud particles evaporate rapidly, sometimes leaving only the cirrus anvil as the reminder of the once mighty presence (see • Fig. 14.3). A single ordinary cell thunderstorm may go through its three stages in one hour or less.

Not only do these thunderstorms produce summer rainfall for a large portion of the United States but they also bring with them momentary cooling after an oppressively hot day. The cooling comes during the mature stage, as the downdraft reaches the surface in the form of a blast of welcome relief. Sometimes, the air temperature may lower as much as 10°C (18°F) in just a few minutes. Unfortunately, the cooling effect often is short-lived, as the downdraft diminishes or the thunderstorm moves on. In fact, after the storm has ended, the air temperature usually rises; and as the moisture from the rainfall evaporates into the air, the humidity increases, sometimes to a level where it actually feels more oppressive after the storm than it did before.

Up to this point, we've looked at ordinary cell thunderstorms that are short-lived, rarely become severe, and form in a region with weak vertical wind shear. As these storms develop, the updraft eventually gives way to the downdraft, and

• FIGURE 14.3 A dissipating thunderstorm near Naples, Florida. Most of the cloud particles in the lower half of the storm have evaporated.

© Howard B. Bluestein

● FIGURE 14.4 This multicell storm complex is composed of a series of cells in successive stages of growth. The thunderstorm in the middle is in its mature stage, with a well-defined anvil. Heavy rain is falling from its base. To the right of this cell, a thunderstorm is in its cumulus stage. To the left, a well-developed cumulus congestus cloud is about ready to become a mature thunderstorm. With new cells constantly forming, the multicell storm complex can exist for hours.

the storm ultimately collapses on itself. However, in a region where strong vertical wind shear exists, thunderstorms often take on a more complex structure. Strong, vertical wind shear can cause the storm to tilt in such a way that it becomes a multicell thunderstorm—a thunderstorm with more than one cell.

MULTICELL THUNDERSTORMS Thunderstorms that contain a number of cells, each in a different stage of development, are called **multicell thunderstorms** (see ● Fig. 14.4). Such storms tend to form in a region of moderate-to-strong vertical wind speed shear. Look at ● Fig. 14.5 and notice that on the left side of the illustration the wind speed increases rapidly with height, producing strong wind speed shear. This

type of shearing causes the cell inside the storm to tilt in such a way that the updraft actually rides up and over the downdraft. Note that the rising updraft is capable of generating new cells that go on to become mature thunderstorms. Notice also that precipitation inside the storm does not fall into the updraft (as it does in the ordinary cell thunderstorm), so the storm's fuel supply is not cut off and the storm complex can survive for a long time. Because the likelihood that a thunderstorm will become severe increases with the length of time the storm exists, long-lasting multicell storms can become intense and produce severe weather.

When convection is strong and the updraft intense (as it is in Fig. 14.5), the rising air may actually intrude well into the stable stratosphere, producing an **overshooting top.** As the air

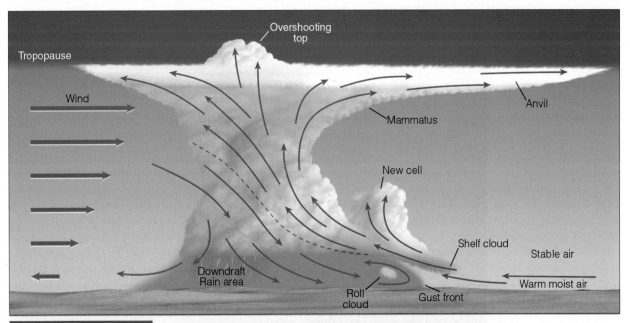

ACTIVE FIGURE 14.5 A simplified model describing air motions and other features associated with an intense multicell thunderstorm that has a tilted updraft. The severity depends on the intensity of the storm's circulation pattern. Visit the Meteorology Resource Center to view this and other active figures at academic.cengage.com/login

● FIGURE 14.6 A swirling mass of dust forms along the leading edge of a gust front as it moves across western Nebraska.

© Perry Samson

spreads laterally into the anvil, sinking air in this region of the storm can produce beautiful mammatus clouds. At the surface, below the thunderstorm's cold downdraft, the cold, dense air may cause the surface air pressure to rise—sometimes several millibars. The relatively small, shallow area of high pressure is called a *mesohigh* (meaning "mesoscale high").

The Gust Front When the cold downdraft reaches the earth's surface, it pushes outward in all directions, producing a strong **gust front** that represents the leading edge of the cold outflowing air. To an observer on the ground, the passage of the gust front resembles that of a cold front. During its passage, the temperature drops sharply and the wind shifts

and becomes strong and gusty, with speeds occasionally exceeding 55 knots. Along the leading edge of the gust front, the air is quite turbulent. Here, strong winds can pick up loose dust and soil and lift them into a huge tumbling cloud (see ● Fig. 14.6).* The cold surface air behind the gust front may even linger close to the ground for hours, well after thunderstorm activity has ceased.

As warm, moist air rises along the forward edge of the gust front, a **shelf cloud** (also called an *arcus cloud*) may form, such as the one shown in ● Fig. 14.7. These clouds are

---

*In dry, dusty areas or desert regions, the leading edge of the gust front is the haboob described in Chapter 9, p. 250.

● FIGURE 14.7 A dramatic example of a shelf cloud (or arcus cloud) associated with an intense thunderstorm. The photograph was taken in the Philippines as the thunderstorm approached from the northwest.

© Richard F. Picanso

● FIGURE 14.8 A roll cloud forming behind a gust front.

especially prevalent when the atmosphere is very stable near the base of the thunderstorm. Look again at Fig. 14.5 and notice that the shelf cloud is attached to the base of the thunderstorm. Occasionally, an elongated ominous-looking cloud forms just behind the gust front. These clouds, which appear to slowly spin about a horizontal axis, are called **roll clouds** (see ● Fig. 14.8).

When the atmosphere is conditionally unstable, the leading edge of the gust front may force the warm, moist air upward, producing a complex of multicell storms, each with new gust fronts. These gust fronts may then merge into a huge gust front called an **outflow boundary.** Along the outflow boundary, air is forced upward, often generating new thunderstorms (see ● Fig. 14.9).

Microbursts   Beneath an intense thunderstorm, the downdraft may become localized so that it hits the ground and spreads horizontally in a radial burst of wind, much like water pouring from a tap and striking the sink below. Such downdrafts are called **downbursts.** A downburst with winds extending only 4 km or less is termed a **microburst.** In spite of its small size, an intense microburst can induce damaging winds as high as 146 knots. (A larger downburst with winds extending more than 4 kilometers is termed a *macroburst.*) ● Figure 14.10 shows the dust clouds generated from a microburst north of Denver, Colorado. Since a microburst is an intense downdraft, its leading edge can evolve into a gust front.

Microbursts are capable of blowing down trees and inflicting heavy damage upon poorly built structures as well as upon sailing vessels that encounter microbursts over open water. In fact, microbursts may be responsible for some damage once attributed to tornadoes. Moreover, microbursts and their accompanying *wind shear* (that is, rapid changes in wind speed and wind direction) appear to be responsible for

several airline crashes. When an aircraft flies through a microburst at a relatively low altitude, say 300 m (1000 ft) above the ground, it first encounters a headwind that generates extra lift. This is position (a) in ● Fig. 14.11. At this point, the aircraft tends to climb (it gains lift), and if the pilot noses the aircraft downward there could be grave consequences, for in a matter of seconds the aircraft encounters the powerful downdraft (position b), and the headwind is replaced by a tail wind (position c). This situation causes a sudden loss of lift and a subsequent decrease in the performance of the aircraft, which is now accelerating toward the ground.

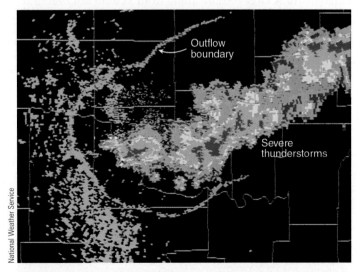

● FIGURE 14.9 Radar image of an outflow boundary. As cool (more-dense) air from inside the severe thunderstorms (red and orange colors) spreads outward, away from the storms, it comes in contact with the surrounding warm, humid (less-dense) air, forming a density boundary (blue line) called an *outflow boundary* between cool air and warm air. Along the outflow boundary, new thunderstorms often form.

● FIGURE 14.10 Dust clouds rising in response to the outburst winds of a microburst north of Denver, Colorado.

One accident attributed to a microburst occurred north of Dallas–Fort Worth Regional Airport during August, 1985. Just as an aircraft was making its final approach, it encountered severe wind shear beneath a small but intense thunderstorm. The aircraft then dropped to the ground and crashed, killing over 100 passengers. To detect the hazardous wind shear associated with microbursts, many major airports use a high resolution Doppler radar. The radar uses algorithms that are computer programmed to detect microbursts and low-level wind shear.

The leading edge of a microburst can contain an intense horizontally rotating vortex that is often filled with dust in a relatively dry region. In eastern Colorado, many microbursts emanate from virga—rain falling from a cloud but evaporating before reaching the ground. Apparently, in these "dry" microbursts (Fig. 14.10), evaporating rain cools the air. The

cooler heavy air then plunges downward through the warmer lighter air below. In humid regions, many microbursts are "wet" in that they are accompanied by blinding rain.

Microbursts can be associated with severe thunderstorms, producing strong, damaging winds. But studies show that they can also occur with ordinary cell thunderstorms and with clouds that produce only isolated showers—clouds that may or may not contain thunder and lightning.

Up to this point, you might think that thunderstorm downdrafts are always cool. Most are cool, but occasionally they can be extremely hot. For example, during the evening of May 22, 1996, in the town of Chickasha, Oklahoma, a blast of hot, dry air from a dissipating thunderstorm raised the surface air temperature from 88°F to 102°F in just 25 minutes. Such sudden warm downbursts are called **heat bursts.** Apparently, the heat burst originates high up in the thunderstorm and warms by compressional heating as it plunges toward the surface. The heat burst that hit Chickasha was exceptionally strong. Along with the hot air, it was accompanied by high winds that toppled trees, ripped down power lines, and lifted roofs off homes.

Squall-Line Thunderstorms Multicell thunderstorms may form as a line of thunderstorms, called a **squall line.** The line of storms may form directly along a cold front and extend for hundreds of kilometers, or the storms may form in the warm air 100 to 300 km out ahead of the cold front. These *pre-frontal squall-line thunderstorms* of the middle latitudes represent the largest and most severe type of squall line, with huge thunderstorms causing severe weather over much of its length (see ● Fig. 14.12).*

There is still debate as to exactly how pre-frontal squall lines form. Models that simulate their formation suggest that,

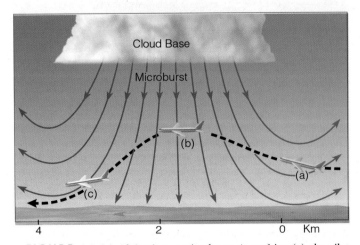

● FIGURE 14.11 Flying into a microburst. At position (a), the pilot encounters a headwind; at position (b), a strong downdraft; and at position (c), a tailwind that reduces lift and causes the aircraft to lose altitude.

*Within a squall line there may be multicell thunderstorms, as well as supercell storms—violent thunderstorms that contain a single rapidly rotating updraft. We will look more closely at supercells in the next section.

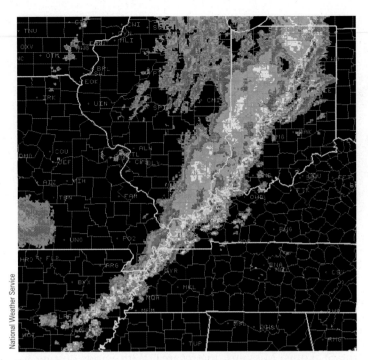

● FIGURE 14.12 A Doppler radar composite showing a pre-frontal squall line extending from Indiana southwestward into Arkansas. Severe thunderstorms (red and orange colors) associated with the squall line produced large hail and high winds during October, 2001.

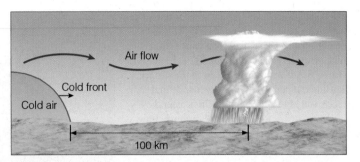

● FIGURE 14.13 Pre-frontal squall-line thunderstorms may form ahead of an advancing cold front as the upper-air flow develops waves downwind from the cold front.

initially, convection begins along the cold front, then reforms farther away. Moreover, the surging nature of the main cold front itself, or developing cumulus clouds along the front, may cause the air aloft to develop into waves (called *gravity waves*), much like the waves that form downwind of a mountain chain (see ● Fig. 14.13). Out ahead of the cold front, the rising motion of the wave may be the trigger that initiates the development of cumulus clouds and a pre-frontal squall line. In some instances, low-level converging air is better established out ahead of the advancing cold front.

Rising air along the frontal boundary (and along the gust front), coupled with the tilted nature of the updraft, promotes the development of new cells as the storm moves

along. Hence, as old cells decay and die out, new ones constantly form, and the squall line can maintain itself for hours on end. Occasionally, a new squall line will actually form out ahead of the front as the gust front pushes forward, beyond the main line of storms.

Squall lines that exhibit weaker updrafts and downdrafts tend to be more shallow than pre-frontal squall lines, and usually they have shorter life spans. These storms are referred to as *ordinary squall lines*. Severe weather may occur with them, but more typically they form as a line of thunderstorms that exhibit characteristics of ordinary cell thunderstorms. Ordinary squall lines may form along a gust front, with a stationary front, with a weak wave cyclone, or where no large-scale cyclonic storms are present. Many of the ordinary squall lines that form in the middle latitudes exhibit a structure similar to squall lines that form in the tropics.

In some squall lines, the leading area of thunderstorms and heavy precipitation is followed by a region of extensive stratified clouds and light precipitation (see ● Fig. 14.14). The stratiform clouds represent a region where the anvil cloud trails behind the thunderstorm. Slowly rising air within the region keeps the air saturated. Beneath the rising air, the air slowly descends in association with the falling rain.

The downdraft to the rear of the storm in Fig. 14.14 forms as some of the falling precipitation evaporates and chills the air. The heavy cooler air then descends, dragging

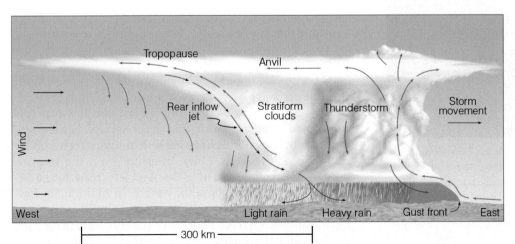

● FIGURE 14.14 A model describing air motions and precipitation associated with a squall line that has a trailing stratiform cloud layer.

● FIGURE 14.15 A side view of the lower half of a squall-line thunderstorm with the rear-inflow jet carrying strong winds from high altitudes down to the surface. These strong winds push forward along the surface, causing damaging straight-line winds that may reach 100 knots. If the high winds extend horizontally for a considerable distance, the wind storm is called a *derecho*.

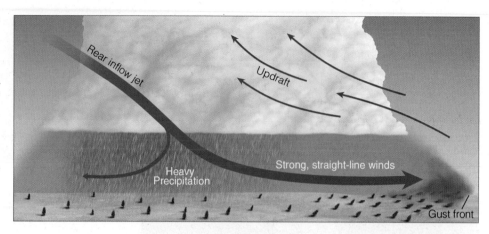

some of the surrounding air with it. If the cool air rapidly descends, it may concentrate into a rather narrow band of fast-flowing air called the *rear-inflow jet,* because it enters the storm from the west (see Fig. 14.14). Sometimes the rear-inflow jet will bring with it the stong upper-level winds from aloft. Should these winds reach the surface, they rush outward producing damaging *straight-line winds** that may exceed 90 knots (104 mi/hr). (See ● Fig. 14.15.)

As the strong winds rush forward along the ground, they sometimes push the squall line outward so that it appears as a *bow* (or a series of bows) on a radar screen. Such a bow-shaped squall line is called a **bow echo** (see ● Fig. 14.16). When the damage associated with the straight-line winds extends for several hundred kilometers along the squall line's path, the windstorm is called a **derecho** (day-ray-sho), after the Spanish word for "straight ahead." Typically, derechoes

---

*Straight-line winds are thunderstorm-generated winds that typically are not associated with rotation.

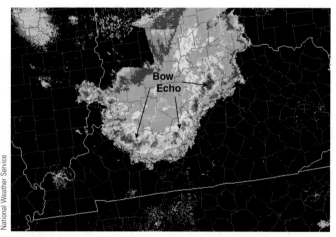

● FIGURE 14.16 The red and orange on this Doppler radar image show an intense squall line moving south southeastward into Kentucky. The thunderstorms are producing strong straight-line winds called a *derecho*. Notice that the line of storms is in the shape of a bow. Such *bow echos* are an indicator of strong, damaging surface winds near the center of the bow. Sometimes the left (usually northern) side of the bow will develop cyclonic rotation and produce a tornado.

form in the early evening and last throughout the night. An especially powerful derecho roared through New York State during the early morning of July 15, 1995, where it blew down millions of trees in Adirondack State Park. In an average year about 20 derechoes occur in the United States. During July, 2005, two derechoes within three days moved through the St. Louis, Missouri, metro area. With winds gusting to over 75 knots, they downed trees and power lines all across the region, leaving half a million residents without power.

Mesoscale Convective Complexes   Where conditions are favorable for convection, a number of individual multicell thunderstorms may occasionally grow in size and organize into a large circular convective weather system. These convectively driven systems, called **Mesoscale Convective Complexes (MCCs),** are quite large—they can be as much as 1000 times larger than an individual ordinary cell thunderstorm. In fact, they are often large enough to cover an entire state, an area in excess of 100,000 square kilometers (see ● Fig. 14.17).

Within the MCCs, the individual thunderstorms apparently work together to generate a long-lasting weather system that moves slowly (normally less than 20 knots) and often exists for periods exceeding 12 hours. The circulation of the MCCs supports the growth of new thunderstorms as well as a region of widespread precipitation. These systems are beneficial, as they provide a significant portion of the growing season rainfall over much of the corn and wheat belts of the United States. However, MCCs can also produce a wide variety of severe weather, including hail, high winds, destructive flash floods, and tornadoes.

Mesoscale Convective Complexes tend to form during the summer in regions where the upper-level winds are weak, which is often beneath a ridge of high pressure. If a weak cold front should stall beneath the ridge, surface heating and moisture may be sufficient to generate thunderstorms on the cool side of the front. Often moisture from the south is brought into the system by a low-level jet stream found between 1 km and 3 km above the surface. In addition, the low-level jet can provide shearing so that multicell storms can form. Most MCCs reach their maximum intensity in the early morning hours, which is partly due to the fact that the low-level jet reaches its maximum strength late at night or in the early morning. Moreover, at night, the cloud tops cool rapidly by emitting infrared energy to space. Gradually, the atmosphere destabilizes as a vast amount of latent heat is released in the lower and middle part of the clouds. Within the multicell storm complex new thunderstorms form as older ones dissipate. With only weak upper-level winds, most MCCs move southeastward very slowly.

SUPERCELL THUNDERSTORMS   In a region where there is strong vertical wind shear (both speed and direction shear), the thunderstorm may form in such a way that the outflow of cold air from the downdraft never undercuts the updraft. In such a storm, the wind shear may be so strong to create horizontal spin, which, when tilted into the updraft, causes it to rotate. A large, long-lasting thunderstorm with a single violently rotating updraft is called a **supercell**.* As we will see later in this chapter, it is the rotating aspect of the supercell that can lead to the formation of tornadoes.

 ●Figure 14.18 shows a supercell with a tornado. The internal structure of a supercell is organized in such a way that the storm may maintain itself as a single entity for hours. Storms of this type are capable of producing an updraft that may exceed 90 knots, damaging surface winds, and large tornadoes. Violent updrafts keep hailstones suspended in the cloud long enough for them to grow to considerable size—sometimes to the size of grapefruits. Once they are large enough, they may fall out the bottom of the cloud with the downdraft, or the violent spinning updraft may whirl them out the side of the cloud or even from the base of the anvil. Aircraft have actually encountered hail in clear air several kilometers from a storm.

*Smaller thunderstorms that occur with rotating updrafts are referred to as *mini supercells.*

● FIGURE 14.17 An enhanced infrared satellite image showing the cold cloud tops (dark red and orange colors) of a Mesoscale Convective Complex extending from central Kansas across western Missouri. This organized mass of multicell thunderstorms brought hail, heavy rain, and flooding to this area.

In some cases, the top of the storm may extend to as high as 18 km (60,000 ft) above the surface, and the width of the storm may exceed 40 kilometers.

Although no two supercells are exactly alike, for convenience they are often divided into three types. *Classic (CL) supercells,* for example, are well balanced storms that produce

● FIGURE 14.18 A supercell thunderstorm with a tornado sweeps over Texas.

heavy rain, large hail, high surface winds, and the majority of tornadoes. The classic supercell serves as an excellent model for all supercells, and is the one normally shown in diagrams. Supercells that produce heavy precipitation and large hail, which appears to fall in the center of the storm, are called *HP supercells*, (for *High Precipitation*). Such storms often produce extreme downdrafts (downburst) and flash flooding. If tornadoes are present, it is often difficult to see them, as they tend to form in the area of heavy precipitation. A supercell characterized by little precipitation, as the one shown in Fig. 14.18, is referred to as an *LP supercell* (for *Low Precipitation*). These storms, which are capable of producing tornadoes and large hail, often have a vertical tower that, due to the storm's rotation, resembles a corkscrew.

A model of a classic supercell with many of its features is given in • Fig. 14.19. In the diagram, we are viewing the storm from the southeast, and the storm is moving from southwest to northeast. The rotating air column on the south side of the storm, usually, 5 to 10 kilometers across, is called a **mesocyclone** (meaning "mesoscale cyclone"). The rotating updraft associated with the mesocyclone is so strong that precipitation cannot fall through it. This situation produces a rain-free area (called a *rain-free base*) beneath the updraft. Strong

southwesterly winds aloft usually blow the precipitation northeastward. Notice that large hail, having remained in the cloud for some time, usually falls just north of the updraft, and the heaviest rain occurs just north of the falling hail, with the lighter rain falling in the northeast quadrant of the storm. If low-level humid air is drawn into the updraft, a rotating cloud, called a **wall cloud**, may descend from the base of the storm (see • Fig. 14.20).

We can obtain a better picture of how wind shear plays a role in the development of supercell thunderstorms by observing • Fig 14.21. The illustration represents atmospheric conditions during the spring over the Central Plains. At the surface, we find an open-wave middle-latitude cyclone with cold, dry air moving in behind a cold front, and warm humid air pushing northward from the Gulf of Mexico behind a warm front. Above the warm surface air, a wedge or "tongue" of warm, moist air is streaming northward. It is in this region we find a relatively narrow band of strong winds called the *low-level jet*. Winds in the low-level jet may exceed 50 knots. Directly above the moist layer is a wedge of cooler, drier air moving in from the southwest. Higher up, at the 500-mb level, a trough of low pressure exists to the west of the surface low. At the 300-mb level, the polar front jet stream swings

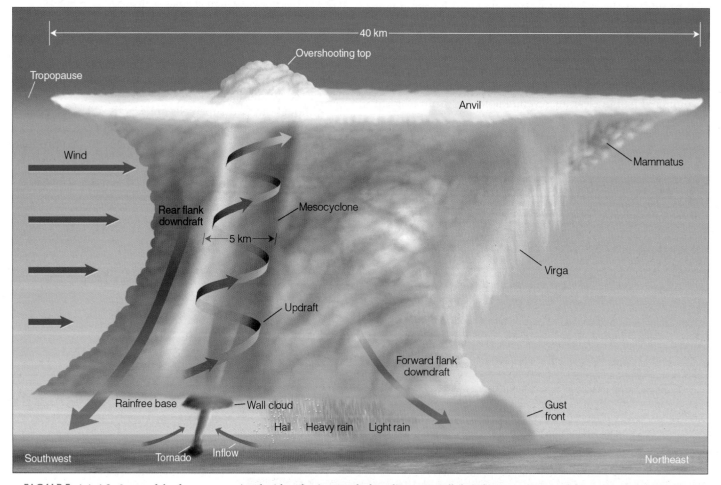

• FIGURE 14.19 Some of the features associated with a classic tornado-breeding supercell thunderstorm as viewed from the southeast. The storm is moving to the northeast.

• FIGURE 14.20 A wall cloud photographed southwest of Norman, Oklahoma.

over the region, often with an area of maximum wind (a jet streak) above the surface low. At this level, the jet stream provides an area of divergence that enhances surface convergence and rising air. The stage is now set for the development of supercell thunderstorms.

The light yellow area on the surface map (Fig. 14.21) shows where supercells are likely to form. They tend to form in this region because (1) the position of cold air above warm air produces a conditionally unstable atmosphere and because (2) strong vertical wind shear induces rotation.

Rapidly increasing wind speed from the surface up to the low-level jet provides strong wind speed shear. Within this region, wind shear causes the air to spin about a horizontal axis. You can obtain a better idea of this spinning by placing a pen (or pencil) in your left hand, parallel to the table. Now take your right hand and push it over the pen away from you. The pen rotates much like the air rotates. If you tilt the spinning pen into the vertical, the pen rotates counterclockwise from the perspective of looking down on it. A similar situation occurs with the rotating air. As the spinning air rotates counterclockwise about a horizontal axis, an updraft from a developing thunderstorm can draw the spinning air into the cloud, causing the updraft to rotate. It is this rotating updraft that is characteristic of all supercells. The increasing wind speed with height up to the 300-mb level, coupled with the changing wind direction with height from more southerly at low levels to more westerly at high levels, further induces storm rotation.*

Ahead of the advancing cold front, we might expect to observe many supercells forming as warm, conditionally unstable air rises from the surface. Often, however, numerous

---

*As we will see later in this chapter, it is this rotation that sets the stage for tornado development.

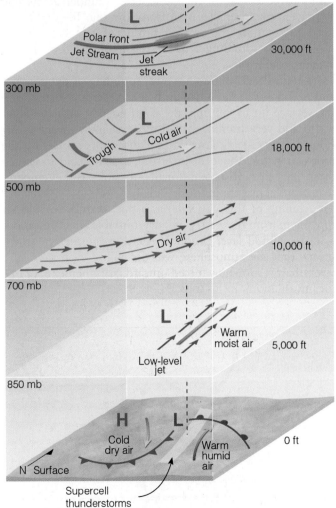

• FIGURE 14.21 Conditions leading to the formation of severe thunderstorms, and especially supercells. The area in yellow shows where supercell thunderstorms are likely to form.

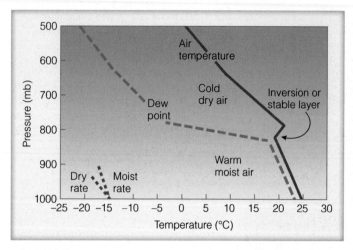

● FIGURE 14.22 A typical sounding of air temperature and dew point that frequently precedes the development of supercell thunderstorms.

supercells do not form as the atmospheric conditions that promote the formation of large supercell thunderstorms tend to prevent many smaller ones from forming. To see why, we need to examine the vertical profile of temperature and moisture—a sounding—in the warm air ahead of the advancing cold front.

● Figure 14.22 shows a typical sounding of temperature and dew point in the warm air before supercells form. From the surface up to 800 mb, the conditionally unstable air is warm and very humid. At 800 mb, a shallow inversion (or simply, a very stable layer) acts like a *cap* (or a *lid*) on the moist air below. Above the inversion, the air is cold and much drier. This air is also conditionally unstable, as the temperature drops at just about the dry adiabatic rate (10°C/1000 m). The cooling of this upper layer is due, mainly, to cold air moving in from the west. Cold, dry, unstable air sitting above a warm, humid layer produces *convective instability,* which means that the atmosphere will destabilize even more if a layer of air is somehow forced upward (see Chapter 6, p. 149, for information on this topic).

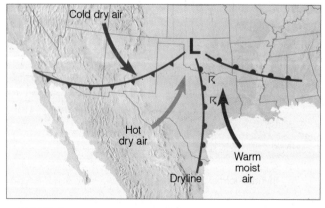

● FIGURE 14.23 Surface conditions that can produce a dryline with intense thunderstorms.

The lifting of warm surface air can occur at the frontal zones, but the air may also begin to rise anywhere in the region of warm air when the surface air heats up during the day. However, in the morning, the inversion acts as a lid on rising thermals and only small cumulus clouds form. As the day progresses (and the surface air heats even more), rising air breaks through the inversion at isolated places and clouds build rapidly, sometimes explosively, as the moist air is vented upward through the opening. Thus, we can see that the stable inversion prevents many small thunderstorms from forming. When the surface air is finally able to puncture the inversion, a jet streak associated with the upper-level jet stream (at the 300-mb level) rapidly draws the moist air up into the cold unstable air, and a large supercell quickly develops to great height.

Most thunderstorms move roughly in the direction of the winds in the middle troposphere. However, most supercell storms are *right-movers,* that is, they move to the right of the steering winds aloft. These right-movers tend to move about 30 degrees to the right of the mean wind in the middle troposphere. Apparently, the rapidly rising air of the storm's updraft interacts with increasing horizontal winds that change direction with height (from more southerly to more westerly) in such a way that vertical pressure gradients are able to generate new updrafts on the right side of the storm. Hence, as the storm moves along, new cells form to the right of the winds aloft.*

THUNDERSTORMS AND THE DRYLINE Thunderstorms may form along or just east of a boundary called a *dryline.* Recall from Chapter 11 that the dryline represents a narrow zone where there is a sharp horizontal change in moisture. In the United States, drylines are most frequently observed in the western half of Texas, Oklahoma, and Kansas. In this region, drylines occur most frequently during spring and early summer, where they are observed about 40 percent of the time.

● Figure 14.23 shows springtime weather conditions that can lead to the development of a dryline and intense thunderstorms. The map shows a developing mid-latitude cyclone with a cold front, a warm front, and three distinct air masses. Behind the cold front, cold dry continental polar (cP) air or modified cool dry Pacific air pushes in from the northwest. In the warm air, ahead of the cold front, warm dry continental tropical (cT) air moves in from the southwest. Farther east, warm but very humid maritime tropical (mT) air sweeps northward from the Gulf of Mexico. The dryline is the north–south oriented boundary that separates the warm, dry air and the warm, humid air.

Along the cold front—where cold, dry air replaces warm, dry air—there is insufficient moisture for thunderstorm development. The moisture boundary lies along the dryline. Because the Central Plains of North America are elevated to the west, some of the hot, dry air from the southwest is able

---

*Some thunderstorms move to the left of the steering winds aloft. This movement may happen as a thunderstorm splits into two storms, with the northern half of the storm often being a left-mover, and the southern half a right-mover.

to ride over the slightly cooler, more humid air from the Gulf. This condition sets up a potentially unstable atmosphere just east of the dryline. Converging surface winds in the vicinity of the dryline, coupled with upper-level outflow, may result in rising air and the development of thunderstorms. As thunderstorms form, the cold downdraft from inside the storm may produce a blast of cool air that moves along the ground as a gust front and initiates the uplift necessary for generating new (possibly more severe) thunderstorms.

BRIEF REVIEW

In the last several sections, we examined different types of thunderstorms. Listed below for your review are important concepts we considered:

- All thunderstorms need three basic ingredients: (1) moist surface air; (2) a conditionally unstable atmosphere; and (3) a mechanism "trigger" that forces the air to rise.
- Ordinary cell (air mass) thunderstorms tend to form where warm, humid air rises in a conditionally unstable atmosphere and where vertical wind shear is weak. They are usually short-lived and go through their life cycle of growth (cumulus stage), maturity (mature stage), and decay (dissipating stage) in less than an hour. They rarely produce severe weather.
- An ordinary cell thunderstorm dies because its downdraft falls into the updraft, which cuts off the storm's fuel supply.
- As wind shear increases (and the winds aloft become stronger), multicell thunderstorms are more likely to form as the storm's updraft rides up and over the downdraft. The tilted nature of the storm allows new cells to form as old ones die out.
- Multicell storms often form as a complex of storms, such as the squall line (a long line of thunderstorms that form along or out ahead of a frontal boundary) and the Mesoscale Convective Complex (a large circular cluster of thunderstorms).
- The stronger the convection and the longer a multistorm system exists, the greater the chances of the thunderstorm becoming severe.
- Supercell thunderstorms are large, long-lasting violent thunderstorms, with a single rotating updraft that forms in a region of strong vertical wind shear. A rotating supercell is more likely to develop when (a) the winds aloft are strong and change direction from southerly at the surface to more westerly aloft and (b) a low-level jet exists just above the earth's surface.
- Although supercells are likely to produce severe weather, such as strong surface winds, large hail, heavy rain, and tornadoes, not all do.
- A gust front, or outflow boundary, represents the leading edge of cool air that originates inside a thunderstorm, reaches the surface as a downdraft, and moves outward away from the thunderstorm.
- Strong downdrafts of a thunderstorm, called downbursts (or microbursts if the downdrafts are smaller than 4 km), have been responsible for several airline crashes, because upon strik-

ing the surface, these winds produce extreme wind shear—rapid changes in wind speed and wind direction.
- A derecho is a strong straight-line wind produced by strong downbursts from intense thunderstorms that often appear as a bow (bow echo) on a radar screen.
- Intense thunderstorms often form along a dryline, a narrow zone that separates warm, dry air from warm, humid air.

FLOODS AND FLASH FLOODS Intense thunderstorms are often associated with **flash floods**—floods that rise rapidly with little or no advance warning. Such flooding often results when thunderstorms stall or move very slowly, causing heavy rainfall over a relatively small area. Such flooding occurred over parts of New England and the mid-Atlantic states during June, 2006, when a stationary front stalled over the region, and tropical moist air, lifted by the front, produced heavy rainfall that caused extensive flooding and damage to thousands of homes. Flooding may also occur when thunderstorms move quickly, but keep passing over the same area, a phenomenon called *training*. (Like railroad cars, one after another, passing over the same tracks.) In recent years, flash floods in the United States have claimed an average of more than 100 lives a year, and have accounted for untold property and crop damage. (An example of a terrible flash flood that took the lives of more than 135 people is given in the Focus section on p. 386).

In some areas, flooding occurs primarily in the spring when heavy rain and melting snow cause rivers to overflow their banks. During March, 1997, heavy downpours over the Ohio River Valley caused extensive flooding that forced thousands from their homes along rivers and smaller streams in Ohio, Kentucky, Tennessee, and West Virginia. One month later, heavy rain coupled with melting snow caused the Red River to overflow its banks, inundating 75 percent of the city of Grand Forks, North Dakota. Flooding also occurs with tropical storms that deposit torrential rains over an extensive area. (Hundreds of people died in Algiers, Algeria, when a flash flood and mud flow roared through that city during November, 2001.)

During the summer of 1993, thunderstorm after thunderstorm rumbled across the upper Midwest, causing the worst flood ever in that part of the United States. What began as a wetter than normal winter and spring for most of the upper Midwest turned into "The Great Flood of 1993" by the end of July. In mid-June, thunderstorms began to form almost daily along a persistent frontal boundary that stretched across the upper Midwest. The front (which remained nearly stationary for days on end) was positioned beneath the polar jet stream that was situated much farther south than usual for this time of year (see Fig. 14.24).* The jet stream provided

---
*As a note, the position of the jet stream caused the weather to be cooler than normal in the Pacific Northwest and warmer than normal in the East. While the Midwest was deluged with rain, the southeastern section of the United States was experiencing an extensive dry period.

FOCUS ON A SPECIAL TOPIC

The Terrifying Flash Flood in the Big Thompson Canyon

July 31, 1976, was like any other summer day in the Colorado Rockies, as small cumulus clouds with flat bases and dome-shaped tops began to develop over the eastern slopes near the Big Thompson and Cache La Poudre rivers. At first glance, there was nothing unusual about these clouds, as almost every summer afternoon they form along the warm mountain slopes. Normally, strong upper-level winds push them over the plains, causing rainshowers of short duration. But the cumulus clouds on this day were different. For one thing, they were much lower than usual, indicating that the southeasterly surface winds were bringing in a great deal of moisture. Also, their tops were somewhat flattened, suggesting that an inversion aloft was stunting their growth. But these harmless-looking clouds gave no clue that later that evening in the Big Thompson Canyon more than 135 people would lose their lives in a terrible flash flood.

By late afternoon, a few of the cumulus clouds were able to puncture the inversion. Fed by moist southeasterly winds, these clouds soon developed into gigantic multicell thunderstorms with tops exceeding 18 km (60,000 ft). By early evening, these same clouds were producing incredible downpours in the mountains.

In the narrow canyon of the Big Thompson River, some places received as much as 30.5 cm (12 in.) of rain in the four hours between 6:30 P.M. and 10:30 P.M. local time. This is an incredible amount of precipitation, considering that the area normally receives about 40.5 cm (16 in.) for an entire year. The heavy downpours turned small creeks into raging torrents, and the Big Thompson River was quickly

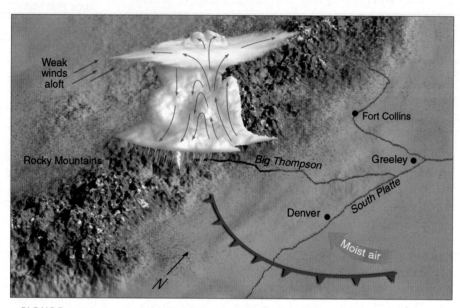

● FIGURE 1 Weather conditions that led to the development of intense thunderstorms that remained nearly stationary over the Big Thompson Canyon in the Colorado Rockies. The arrows within the thunderstorm represent air motions.

filled to capacity. Where the canyon narrowed, the river overflowed its banks and water covered the road. The relentless pounding of water caused the road to give way.

Soon cars, tents, mobile homes, resort homes, and campgrounds were being claimed by the river. Where the debris entered a narrow constriction, it became a dam. Water backed up behind it, then broke through, causing a wall of water to rush downstream.

Figure 1 shows the weather conditions during the evening of July 31, 1976. A cool front moved through earlier in the day and remained south of Denver. The weak inversion layer associated with the front kept the cumu-

lus clouds from building to great heights earlier in the afternoon. However, the strong southeasterly flow behind the cool front pushed unusually moist air upslope along the mountain range. Heated from below, the conditionally unstable air eventually punctured the inversion and developed into a huge multicell thunderstorm complex that remained nearly stationary for several hours due to the weak southerly winds aloft. The deluge may have deposited 19 cm (7.5 in.) of rain on the main fork of the Big Thompson River in about one hour. Of the approximately 2000 people in the canyon that evening, over 135 lost their lives, and property damage exceeded $35.5 million.

pockets of upper-level divergence for the development of weak surface waves that rippled along the front, while converging surface air provided uplift for thunderstorm growth.

Fed by warm, humid air from the Gulf of Mexico, thunderstorms almost daily rolled through an area that stretched eastward from Nebraska and South Dakota into Minnesota and Wisconsin, and southward into Iowa, Illinois, and Missouri. Torrential rains from these storms quickly saturated the soil and soon runoff began to raise the water level in creeks and

rivers. By the end of June, communities in the northern regions of the Mississippi River Valley were experiencing flooding.

As the thunderstorms continued into July, city after city was claimed by the rising waters (see ● Fig. 14.25). Between April and July, many areas had received twice their normal rainfall, and rivers continued to crest well above flood stage through July. By the time the water began to recede in August, more than 60 percent of the levees along the Mississippi River had been destroyed, and an area larger than the state of

Texas had been covered with water (the blue-shaded area in Fig. 14.24). Estimates are that $6.5 billion in crops was lost as millions of acres of valuable farmland were inundated by flood waters. The worst flooding this area had ever seen took 45 human lives, damaged or destroyed 45,000 homes, and forced the evacuation of 74,000 people.

DISTRIBUTION OF THUNDERSTORMS   It is estimated that more than 50,000 thunderstorms occur each day throughout the world. Hence, over 18 million occur annually. The combination of warmth and moisture make equatorial landmasses especially conducive to thunderstorm formation. Here, thunderstorms occur on about one out of every three days. Thunderstorms are also prevalent over water along the intertropical convergence zone, where the low-level convergence of air helps to initiate uplift. The heat energy liberated in these storms helps the earth maintain its heat balance by distributing heat poleward (see Chapter 10). Thunderstorms are much less prevalent in dry climates, such as the polar regions and the desert areas dominated by subtropical highs.

• Figure 14.26 shows the average annual number of days having thunderstorms in various parts of the United States. Notice that they occur most frequently in the southeastern states along the Gulf Coast with a maximum in Florida. A

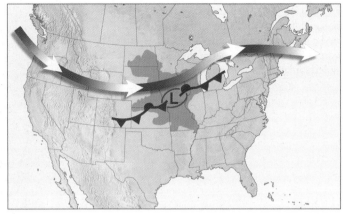

• FIGURE 14.24 The heavy arrow represents the average position of the upper-level jet stream from mid-June through July, 1993. The jet stream helped fuel thunderstorms that developed in association with a stationary front that seemed to oscillate back and forth over the region as an alternating cold front and warm front. The "L" marks the center of a frontal wave that is moving along the front. Many of the thunderstorms that formed in conjunction with this pattern were severe, and over a period of weeks produced "The Great Flood of 1993." Most of the counties within the blue-shaded area were declared "disaster areas" due to flooding.

secondary maximum exists over the central Rockies. The region with the fewest thunderstorms is the Pacific coastal and interior valleys.

In many areas, thunderstorms form primarily in summer during the warmest part of the day when the surface air is most unstable. There are some exceptions, however. During the summer in the valleys of central and southern California, dry, sinking air produces an inversion that inhibits the devel-

• FIGURE 14.25 Flooding during the summer of 1993 covered a vast area of the upper Midwest. Here, floodwaters near downtown Des Moines, Iowa, during July, 1993, inundate buildings of the Des Moines waterworks facility. Flood-contaminated water left 250,000 people without drinking water.

© Wide World Photos

• FIGURE 14.26 The average number of days each year on which thunderstorms are observed throughout the United States. (Due to the scarcity of data, the number of thunderstorms is underestimated in the mountainous far west.)

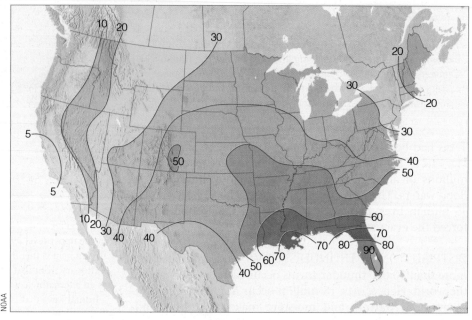

• FIGURE 14.27 The average number of days each year on which hail is observed throughout the United States.

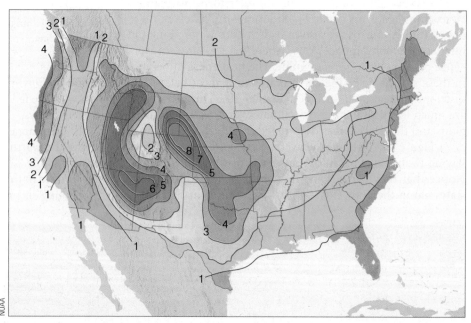

opment of towering cumulus clouds. In these regions, thunderstorms are most frequent in winter and spring, particularly when cold, moist, conditionally unstable air aloft moves over moist, mild surface air. The surface air remains relatively warm because of its proximity to the ocean. Over the Central Plains, thunderstorms tend to form more frequently at night. These storms may be caused by a low-level southerly jet stream that forms at night, and not only carries humid air northward but also initiates areas of converging surface air, which helps to trigger uplift. As the thunderstorms build, their tops cool by radiating infrared energy to space. This cooling process tends to destabilize the atmosphere, making it more suitable for nighttime thunderstorm development.

At this point, it is interesting to compare Fig. 14.26 and • Fig. 14.27. Notice that, even though the greatest frequency

of thunderstorms is near the Gulf Coast, the greatest frequency of hailstorms is over the western Great Plains. One reason for this situation is that conditions over the Great Plains are more favorable for the development of severe thunderstorms. We also find that, in summer along the Gulf Coast, a thick layer of warm, moist air extends upward from the surface. Most hailstones falling into this layer will melt before reaching the ground. Over the plains, the warm surface layer is much shallower and drier. Falling hailstones do begin to melt, but the water around their periphery quickly evaporates in the dry air. This cools the hailstones and slows the melting rate so that many survive as ice all the way to the surface.*

---

*The formation of hail is detailed in Chapter 7 on p. 184.

Now that we have looked at the development and distribution of thunderstorms, we are ready to examine an interesting, though yet not fully understood, aspect of all thunderstorms—lightning.

LIGHTNING AND THUNDER   **Lightning** is simply a discharge of electricity, a giant spark, which usually occurs in mature thunderstorms. Lightning may take place within a cloud, from one cloud to another, from a cloud to the surrounding air, or from a cloud to the ground (see ● Fig. 14.28). (The majority of lightning strikes occur within the cloud, while only about 20 percent or so occur between cloud and ground.) The lightning stroke can heat the air through which it travels to an incredible 30,000°C (54,000°F), which is 5 times hotter than the surface of the sun. This extreme heating causes the air to expand explosively, thus initiating a shock wave that becomes a booming sound wave—called **thunder**—that travels outward in all directions from the flash.

Light travels so fast that we see light instantly after a lightning flash. But the sound of thunder, traveling at only about 330 m/sec (1100 ft/sec), takes much longer to reach the ear. If we start counting seconds from the moment we see the lightning until we hear the thunder, we can determine how far away the stroke is. Because it takes sound about 3 seconds to travel 1 kilometer (5 seconds for each mile), if we see lightning and hear the thunder 15 seconds later, the lightning stroke occurred 5 km (3 mi) away.

When the lightning stroke is very close—100 m (330 ft) or less—thunder sounds like a clap or a crack followed immediately by a loud bang. When it is farther away, it often rumbles. The rumbling can be due to the sound emanating from different areas of the stroke. Moreover, the rumbling is accentuated when the sound wave reaches an observer after having bounced off obstructions, such as hills and buildings.

In some instances, lightning is seen but no thunder is heard. Does this mean that thunder was not produced by the lightning? Actually, there is thunder, but the atmosphere refracts (bends) and attenuates the sound waves, making the thunder inaudible. Sound travels faster in warm air than in cold air.* Because thunderstorms form in a conditionally unstable atmosphere, where the temperature normally drops rapidly with height, a sound wave moving outward away from a lightning stroke will often bend upward, away from an observer at the surface. Consequently, an observer closer than about 5 km to a lightning stroke will usually hear thunder, while an observer 15 km away will not.

However, even when a viewer is as close as several kilometers to a lightning flash, thunder may not be heard. For one thing, the complex interaction of sound waves and air molecules tends to attenuate the thunder. In addition, turbulent eddies of air less than 50 m in diameter scatter the sound waves. Hence, when thunder from a low-energy lightning flash travels several kilometers through turbulent air, it may become inaudible.

*The speed of sound in calm air is equal to 20 $\sqrt{T}$, where T is the air temperature in Kelvins.

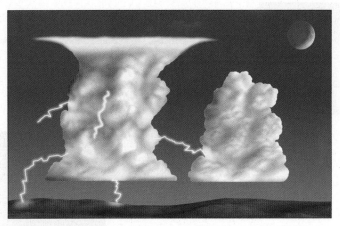

● FIGURE 14.28 The lightning stroke can travel in a number of directions. It can occur within a cloud, from one cloud to another cloud, from a cloud to the air, or from a cloud to the ground. Notice that the cloud-to-ground lightning can travel out away from the cloud, then turn downward, striking the ground many miles from the thunderstorm. When lightning behaves in this manner, it is often described as a *"bolt from the blue."*

A sound occasionally mistaken for thunder is the **sonic boom.** Sonic booms are produced when an aircraft exceeds the speed of sound at the altitude at which it is flying. The aircraft compresses the air, forming a shock wave that trails out as a cone behind the aircraft. Along the shock wave, the air pressure changes rapidly over a short distance. The rapid pressure change causes the distinct boom. (Exploding fireworks generate a similar shock wave and a loud bang.)

Earlier, we learned that lightning occurs with mature thunderstorms. But lightning may also occur in snowstorms, in dust storms, in the gas cloud of an erupting volcano, and on very rare occasions in nimbostratus clouds. Lightning may also shoot from the top of thunderstorms into the upper atmosphere. More on this topic is given in the Focus section on p. 390.

What causes lightning? The normal fair weather electric field of the atmosphere is characterized by a negatively charged surface and a positively charged upper atmosphere. For lightning to occur, separate regions containing opposite electrical charges must exist within a cumulonimbus cloud. Exactly how this charge separation comes about is not totally comprehended; however, there are many theories to account for it.

Electrification of Clouds   One theory proposes that clouds become electrified when graupel (small ice particles called *soft hail*) and hailstones fall through a region of supercooled liquid droplets and ice crystals. As liquid droplets collide with a hailstone, they freeze on contact and release latent heat. This process keeps the surface of the hailstone warmer than that of the surrounding ice crystals. When the warmer hailstone comes in contact with a colder ice crystal, an important phenomenon occurs: *There is a net transfer of positive ions (charged molecules) from the warmer object to the colder object.* Hence, the hailstone (larger, warmer particle) becomes negatively charged and the ice crystal (smaller, cooler particle)

ELVES in the Atmosphere

For many years airline pilots reported seeing strange bolts of light shooting upward, high above the tops of intense thunderstorms. These faint, mysterious flashes did not receive much attention, however, until they were first photographed in 1989. Photographs from sensitive, low–light-level cameras on board jet aircraft revealed that the mysterious flashes were actually a colorful display called *red sprites* and *blue jets*, which seemed to dance above the clouds.

Sprites are massive, but dim, light flashes that appear directly above an intense thunderstorm system (see Fig. 2). Usually red, and lasting but a few thousandths of a second, sprites tend to form almost simultaneously with lightning in the cloud below and with severe thunderstorms that have positive cloud-to-ground lightning strokes. (Most cloud-to-ground lightning is negative.) Although it is not entirely clear how they form, the thinking now is that sprites form when positive lightning disrupts the atmosphere's electrical field in such a way that charged particles in the upper atmosphere are accelerated downward toward the thunderstorm and upward to higher levels in the atmosphere.

Blue jets usually dart upward in a conical shape from the tops of thunderstorms that are experiencing vigorous lightning activity (Fig. 2). Although faint, blue jets can be seen with the

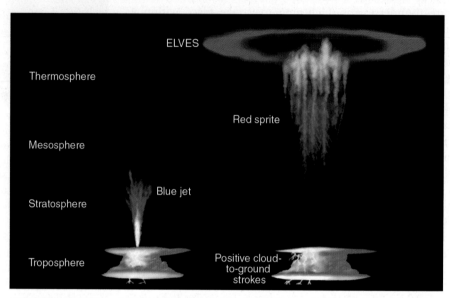

● FIGURE 2 Various electrical phenomena observed in the upper atmosphere.

naked eye. They are not well understood, but appear to transfer large amounts of electrical energy into the upper atmosphere.

ELVES as illustrated in Fig. 2 appear as a faint red halo — too faint to be seen with the naked eye, only with sensitive cameras. They occur in the ionized region of the upper atmo-

sphere. ELVES occur at night and are extremely short-lived. They appear to form when a lightning bolt from an intense thunderstorm gives off a strong electromagnetic pulse (EMP) that causes electrons in the ionosphere to collide with molecules that become excited and give off light.

The roles that red sprites, blue jets, and ELVES play in the earth's global electrical system have yet to be determined.

*The acronym ELVES is from *E*missions of *L*ight and *V*ery low frequency from lightning-induced electromagnetic pulsation sources.

positively charged, as the positive ions are incorporated into the ice crystal (see ● Fig. 14.29). The same effect occurs when colder, supercooled liquid droplets freeze on contact with a warmer hailstone and tiny splinters of positively charged ice break off. These lighter, positively charged particles are then carried to the upper part of the cloud by updrafts. The larger hailstones (or graupel), left with a negative charge, either remain suspended in an updraft or fall toward the bottom of the cloud. By this mechanism, the cold upper part of the cloud becomes positively charged, while the middle of the cloud becomes negatively charged. The lower part of the cloud is generally of negative and mixed charge except for an occasional positive region located in the falling precipitation near the melting level (see ● Fig. 14.30).

Another school of thought proposes that during the formation of precipitation, regions of separate charge exist within tiny cloud droplets and larger precipitation particles. In the upper part of these particles we find negative charge, while in the lower part we find positive charge. When falling precipitation collides with smaller particles, the larger precipitation particles become negatively charged and the smaller particles, positively charged. Updrafts within the cloud then sweep the smaller positively charged particles into the upper reaches of the cloud, while the larger negatively charged particles either settle toward the lower part of the cloud or updrafts keep them suspended near the middle of the cloud.

The Lightning Stroke Because unlike charges attract one another, the negative charge at the bottom of the cloud causes a region of the ground beneath it to become positively charged. As the thunderstorm moves along, this region of positive charge follows the cloud like a shadow. The positive charge is most dense on protruding objects, such as trees, poles, and buildings. The difference in charges causes an electric potential between the cloud and ground. In dry air, however, a flow of current does not occur because the air is a good electrical insulator. Gradually, the electrical potential gradient builds, and when it becomes sufficiently large (on the order of one million volts per meter), the insulating properties of the air break down, a current flows, and lightning occurs.

Cloud-to-ground lightning begins within the cloud when the localized electric potential gradient exceeds 3 million volts per meter along a path perhaps 50 m long. This situation causes a discharge of electrons to rush toward the cloud base and then toward the ground in a series of steps (see ● Fig. 14.31a). Each discharge covers about 50 to 100 m, then stops for about 50-millionths of a second, then occurs again over another 50 m or so. This **stepped leader** is very faint and is usually invisible to the human eye. As the tip of the stepped leader approaches the ground, the potential gradient (the voltage per meter) increases, and a current of positive charge starts upward from the ground (usually along elevated objects) to meet it (see Fig. 14.31b). After they meet, large numbers of electrons flow to the ground and a much larger, more luminous **return stroke** several centimeters in diameter surges upward to the cloud along the path followed by the stepped leader (Fig. 14.31c). Hence, the downward flow of electrons establishes the bright channel of upward propagating current. Even though the bright return stroke travels from the ground up to the cloud, it happens so quickly — in one ten-thousandth of a second — that our eyes cannot resolve the motion, and we see what appears to be a continuous bright flash of light (see ● Fig. 14.32).

Sometimes there is only one lightning stroke, but more often the leader-and-stroke process is repeated in the same ionized channel at intervals of about four-hundredths of a second. The subsequent leader, called a **dart leader,** proceeds from the cloud along the same channel as the original stepped leader; however, it proceeds downward more quickly because the electrical resistance of the path is now lower. As the leader approaches the ground, normally a less energetic return stroke than the first one travels from the ground to the cloud. Typically, a lightning flash will have three or four leaders, each followed by a return stroke. A lightning flash consisting of many strokes (one photographed flash had 26 strokes) usually lasts less than a second. During this short period of time, our eyes may barely be able to perceive the individual strokes, and the flash appears to flicker.

The lightning described so far (where the base of the cloud is negatively charged and the ground positively charged) is called *negative cloud-to-ground lightning,* because the stroke carries negative charges from the cloud to the ground. About 90 percent of all cloud-to-ground lightning is negative. However, when the base of the cloud is positively charged and the ground negatively charged, a *positive cloud-to-ground lightning* flash may result. Positive lightning, most common with supercell thunderstorms, has the potential to cause more damage because it generates a much higher current

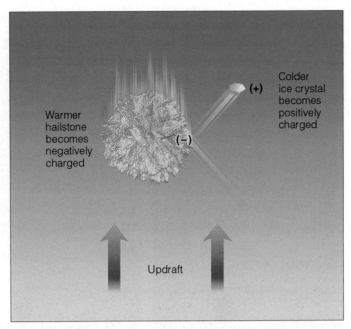

● FIGURE 14.29 When the tiny colder ice crystals come in contact with the much larger and warmer hailstone (or graupel), the ice crystal becomes positively charged and the hailstone negatively charged. Updrafts carry the tiny positively charged ice crystal into the upper reaches of the cloud, while the heavier hailstone falls through the updraft toward the lower region of the cloud.

● FIGURE 14.30 The generalized charge distribution in a mature thunderstorm.

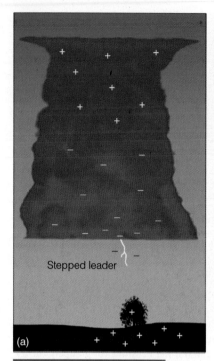
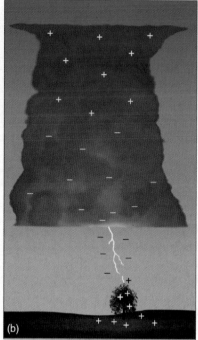
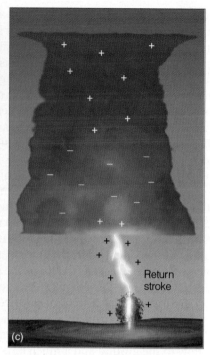

ACTIVE FIGURE 14.31 The development of a lightning stroke. (a) When the negative charge near the bottom of the cloud becomes large enough to overcome the air's resistance, a flow of electrons—the stepped leader—rushes toward the earth. (b) As the electrons approach the ground, a region of positive charge moves up into the air through any conducting object, such as trees, buildings, and even humans. (c) When the downward flow of electrons meets the upward surge of positive charge, a strong electric current—a bright return stroke—carries positive charge upward into the cloud. Visit the Meteorology Resource Center to view this and other active figures at academic.cengage.com/login

level and its flash lasts for a longer duration than negative lightning.

Notice in Fig. 14.32 that lightning may take on a variety of shapes and forms. When a dart leader moving toward the ground deviates from the original path taken by the stepped leader, the lightning appears crooked or forked, and it is called *forked lightning.* An interesting type of lightning is *ribbon lightning* that forms when the wind moves the ionized channel between each return stroke, causing the lightning to appear as a ribbon hanging from the cloud. If the lightning channel

● FIGURE 14.32 Time exposure of an evening thunderstorm with an intense lightning display near Denver, Colorado. The bright flashes are return strokes. The lighter forked flashes are probably stepped leaders that did not make it to the ground.

breaks up, or appears to break up, the lightning (called *bead lightning*) looks like a series of beads tied to a string. **Ball lightning** looks like a luminous sphere that appears to float in the air or slowly dart about for several seconds. Although many theories have been proposed, the actual cause of ball lightning remains an enigma. **Sheet lightning** forms when either the lightning flash occurs inside a cloud or intervening clouds obscure the flash, such that a portion of the cloud (or clouds) appears as a luminous white sheet. When cloud-to-ground lightning occurs with thunderstorms that do not produce rain, the lightning is often called **dry lightning.** Such lightning often starts forest fires in regions of dry timber.

Distant lightning from thunderstorms that is seen but not heard is commonly called **heat lightning** because it frequently occurs on hot summer nights when the overhead sky is clear. As the light from distant electrical storms is refracted through the atmosphere, air molecules and fine dust scatter the shorter wavelengths of visible light, often causing heat lightning to appear orange to a distant observer.

As the electric potential near the ground increases, a current of positive charge moves up pointed objects, such as antennas and masts of ships. However, instead of a lightning stroke, a luminous greenish or bluish halo may appear above them, as a continuous supply of sparks—a *corona discharge*—is sent into the air. This electric discharge, which can cause the top of a ship's mast to glow, is known as **St. Elmo's Fire,** named after the patron saint of sailors. St. Elmo's Fire is also seen around power lines and the wings of aircraft. When St. Elmo's Fire is visible and a thunderstorm is nearby, a lightning flash may occur in the near future, especially if the electric field of the atmosphere is increasing.

Lightning rods are placed on buildings to protect them from lightning damage. The rod is made of metal and has a pointed tip, which extends well above the structure (see Fig. 14.33). The positive charge concentration will be maximum on the tip of the rod, thus increasing the probability that the lightning will strike the tip and follow the metal rod harmlessly down into the ground, where the other end is deeply buried.

When lightning enters sandy soil, the extremely high temperature of the stroke may fuse sand particles together, producing a rootlike system of tubes called a **fulgurite,** after the Latin word for "lightning" (see Fig. 14.34). When lightning strikes an object such as a car, it normally leaves the passengers unharmed because it usually takes the quickest path to the ground along the outside metal casing of the vehicle. The lightning then jumps to the road through the air, or it enters the roadway through the tires (see Fig. 14.35). If you should be caught in the open in a thunderstorm, what should you do? Of course, seek shelter immediately, but under a tree? If you are not sure, please read the Focus section on p. 395.

Lightning Detection and Suppression

For many years, lightning strokes were detected primarily by visual observation. Today, cloud-to-ground lightning is located by means of an instrument called a *lightning direction-finder,* which works

The folks of Elgin, Manitoba, literally had their "goose cooked" during April, 1932, when a lightning bolt killed 52 geese that were flying overhead in formation. As the birds fell to the ground, they were reportedly gathered up and distributed to the townspeople for dinner.

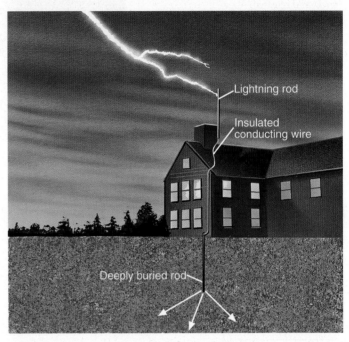

Lightning rod

Insulated conducting wire

Deeply buried rod

● FIGURE 14.33 The lightning rod extends above the building, increasing the likelihood that lightning will strike the rod rather than some other part of the structure. After lightning strikes the metal rod, it follows an insulated conducting wire harmlessly into the ground.

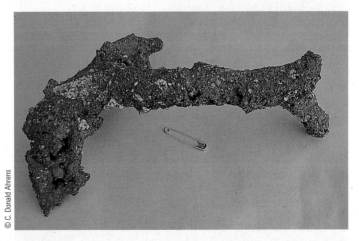

© C. Donald Ahrens

● FIGURE 14.34 A fulgurite that formed by lightning fusing sand particles.

by detecting the radio waves produced by lightning. Such waves are called *sferics,* a contraction from their earlier designation, *atmospherics.* A web of these magnetic devices is a valuable tool in pinpointing lightning strokes throughout the

● FIGURE 14.35 The four marks on the road surface represent areas where lightning, after striking a car traveling along south Florida's Sunshine State Parkway, entered the roadway through the tires. Lightning flattened three of the car's tires and slightly damaged the radio antenna. The driver and a six-year-old passenger were taken to a nearby hospital, treated for shock, and released.

© C. Donald Ahrens

contiguous United States, Canada, and Alaska. Lightning detection devices allow scientists to examine in detail the lightning activity inside a storm as it intensifies and moves (see ● Fig. 14.36). This gives forecasters a better idea where intense lightning strokes might be expected.* In addition, when this information is correlated with satellite images, a more complete and precise structure of a thunderstorm is

_____

*In fact, with the aid of these instruments and computer models of the atmosphere, the National Weather Service currently issues lightning probability forecasts for the western United States.

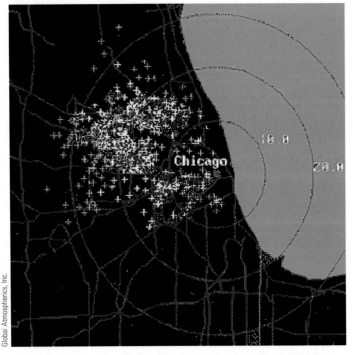

Global Atmospherics, Inc.

● FIGURE 14.36 Cloud-to-ground lightning strikes in the vicinity of Chicago, Illinois, as detected by the National Lightning Detection Network.

obtained. In the future, geostationary satellites will carry lightning detection sensors, which will expand our ability to detect types of lightning worldwide.

Each year, approximately 10,000 fires are started by lightning in the United States alone and around $50 million worth of timber is destroyed. For this reason, tests have been conducted to see whether the number of cloud-to-ground lightning discharges can be reduced. One technique that has shown some success in suppressing lightning involves seeding a cumulonimbus cloud with hair-thin pieces of aluminum about 10 centimeters long. The idea is that these pieces of metal will produce many tiny sparks, or *corona discharges,* and prevent the electrical potential in the cloud from building to a point where lightning occurs. While the results of this experiment are inconclusive, many forestry specialists point out that nature itself may use a similar mechanism to prevent excessive lightning damage. The long, pointed needles of pine trees may act as tiny lightning rods, diffusing the concentration of electric charges and preventing massive lightning strokes.

Now that we have looked at thunderstorms, we are ready to explore a product of a thunderstorm that is one of nature's most awesome phenomena: the tornado, a rapidly spiraling column of air that usually extends down from the base of a cumulonimbus cloud and can strike sporadically and violently.

Tornadoes

A **tornado** is a rapidly rotating column of air that blows around a small area of intense low pressure with a circulation that reaches the ground. A tornado's circulation is present on the ground either as a funnel-shaped cloud or as a swirling cloud of dust and debris. Sometimes called *twisters* or *cyclones,* tornadoes can assume a variety of shapes and forms that range from twisting rope-like funnels, to cylindrical-shaped funnels, to massive black funnels, to funnels that re-

FOCUS ON AN OBSERVATION

Don't Sit Under the Apple Tree

Because a single lightning stroke may involve a current as great as 100,000 amperes, animals and humans can be electrocuted when struck by lightning. The average yearly death toll in the United States attributed to lightning is nearly 100, with Florida accounting for the most fatalities. Many victims are struck in open places, riding on farm equipment, playing golf, attending sports events, or sailing in a small boat. Some live to tell about it, as did the retired champion golfer Lee Trevino. Others are less fortunate. About 10 percent of people struck by lightning are killed. Most die from cardiac arrest. Consequently, when you see someone struck by lightning, immediately give CPR (cardiopulmonary resuscitation), as lightning normally leaves its victims unconscious without heartbeat and without respiration. Those who do survive often suffer from long-term psychological disorders, such as personality changes, depression, and chronic fatigue.

Many lightning fatalities occur in the vicinity of relatively isolated trees (see Fig. 3). As a tragic example, during June, 2004, three people were killed near Atlanta, Georgia, seeking shelter under a tree. Because a positive charge tends to concentrate in upward projecting objects, the upward return stroke that meets the

© Johnny Autery

● FIGURE 3 A cloud-to-ground lightning flash hitting a 65-foot sycamore tree. It should be apparent why one should *not* seek shelter under a tree during a thunderstorm.

stepped leader is most likely to originate from such objects. Clearly, sitting under a tree during an electrical storm is not wise. What *should* you do?

When caught in a thunderstorm, the best protection, of course, is to get inside a building. But stay away from electrical appliances and corded phones, and avoid taking a shower. Automobiles with metal frames and trucks (but not golf carts) may also provide protection. If no such shelter exists, be sure to avoid elevated places and isolated trees. If you are on level ground, try to keep your head as low as possible, but do not lie down. Because lightning channels usually emanate outward through the ground at the point of a lightning strike, a surface current may travel through your body and injure or kill you. Therefore, crouch down as low as possible and minimize the contact area you have with the ground by touching it with only your toes or your heels.

There are some warning signs to alert you to a strike. If your hair begins to stand on end or your skin begins to tingle and you hear clicking sounds, beware — lightning may be about to strike. And if you are standing upright, you may be acting as a lightning rod.

---

semble an elephant's trunk hanging from a large cumulonimbus cloud. A **funnel cloud** is a tornado whose circulation has not reached the ground. When viewed from above, the majority of North American tornadoes rotate counterclockwise about their central core of low pressure. A few have been seen rotating clockwise, but those are rare.

The diameter of most tornadoes is between 100 and 600 m (about 300 to 2000 ft), although some are just a few meters wide and others have diameters exceeding 1600 m (1 mi). In fact, on May 22, 2004, one of the largest tornadoes on record touched down near Hallam, Nebraska, with a diameter of about 4 km (2.5 mi). Tornadoes that form ahead of an advancing cold front are often steered by southwesterly winds and, therefore, tend to move from the southwest toward the northeast at speeds usually between 20 and 40 knots. However, some have been clocked at speeds greater than 70 knots. Most tornadoes last only a few minutes and have an average path length of about 7 km (4 mi). There are cases

where they have reportedly traveled for hundreds of kilometers and have existed for many hours, such as the one that lasted over 7 hours and cut a path 470 km (292 mi) long through portions of Illinois and Indiana on May 26, 1917.*

TORNADO LIFE CYCLE   Major tornadoes usually evolve through a series of stages. The first stage is the *dust-whirl stage,* where dust swirling upward from the surface marks the tornado's circulation on the ground and a short funnel often extends downward from the thunderstorm's base. Damage during this stage is normally light. The next stage, called the *organizing stage,* finds the tornado increasing in intensity with an overall downward extent of the funnel. During the tornado's *mature stage,* damage normally is most severe as the funnel reaches its greatest width and is almost vertical

---

*Actually, this situation may have been several tornadoes (a family) that were generated by a single supercell thunderstorm as it moved along.

● FIGURE 14.37 A tornado in its mature stage roars over the Great Plains.

© D. Lloyd/WeatherStock

(see ● Fig. 14.37). The *shrinking stage* is characterized by an overall decrease in the funnel's width, an increase in the funnel's tilt, and a narrowing of the damage swath at the surface, although the tornado may still be capable of intense and sometimes violent damage. The final stage, called the *decay stage,* usually finds the tornado stretched into the shape of a rope. Normally, the tornado becomes greatly contorted before it finally dissipates. Although these are the typical stages of a major tornado, minor tornadoes may evolve only through the organizing stage. Some even skip the mature stage and go directly into the decay stage. However, when a tornado reaches its mature stage, its circulation usually stays in contact with the ground until it dissipates.

TORNADO OUTBREAKS Each year, tornadoes take the lives of many people. The yearly average is less than 100, although over 100 may die in a single day. In recent years, an alarming statistic is that 45 percent of all fatalities occurred in mobile homes. The deadliest tornadoes are those that occur in *families;* that is, different tornadoes spawned by the same thunderstorm. (Some thunderstorms produce a sequence of several tornadoes over 2 or more hours and over distances of 100 km or more.) Tornado families often are the result of a single, long-lived supercell thunderstorm. When a large number of tornadoes (typically 6 or more) form over a particular region, this constitutes what is termed a **tornado outbreak.**

A particularly devastating outbreak occurred on May 3, 1999, when 78 tornadoes marched across parts of Texas, Kansas, and Oklahoma. One tornado, whose width at times reached one mile and whose wind speed was measured by Doppler radar at 276 knots (318 mi/hr), moved through the southwestern section of Oklahoma City. Within its 40-mile path, it damaged or destroyed thousands of homes, injured nearly 600 people, claimed 38 lives, and caused over $1 billion in property damage.

One of the most violent outbreaks ever recorded occurred on April 3 and 4, 1974. During a 16-hour period, 148 tornadoes cut through parts of 13 states, killing 307 people, injuring more than 6000, and causing an estimated $600 million in damage. Some of these tornadoes were among the most powerful ever witnessed. The combined path of all the tornadoes during this *super outbreak* amounted to 4181 km (2598 mi), well over half of the total path for an average year. The greatest loss of life attributed to tornadoes occurred during the tri-state outbreak of March 18, 1925, when an estimated 695 people died as at least 7 tornadoes traveled a total of 703 km (437 mi) across portions of Missouri, Illinois, and Indiana.

TORNADO OCCURRENCE Tornadoes occur in many parts of the world, but no country experiences more tornadoes than the United States, which, in recent years, averages more than 1000 annually and experienced a record 1722 tornadoes during 2004. Although tornadoes have occurred in every state, including Alaska and Hawaii, the greatest number occur in the tornado belt, or *tornado alley,* of the Central Plains, which stretches from central Texas to Nebraska* (see ● Fig. 14.38).

The Central Plains region is most susceptible to tornadoes because it often provides the proper atmospheric setting for the development of the severe thunderstorms that spawn tornadoes. Here (especially in spring) warm, humid surface air is overlain by cooler, drier air aloft, producing a conditionally unstable atmosphere. When a strong vertical wind shear exists (usually provided by a low-level jet and by the polar jet stream) and the surface air is forced upward, large supercell thunderstorms capable of spawning tornadoes may form. Therefore, tornado frequency is highest during the spring and lowest during the winter when the warm surface air is normally absent.

---

*Many of the tornadoes that form along the Gulf Coast are generated by thunderstorms embedded within the circulation of hurricanes.

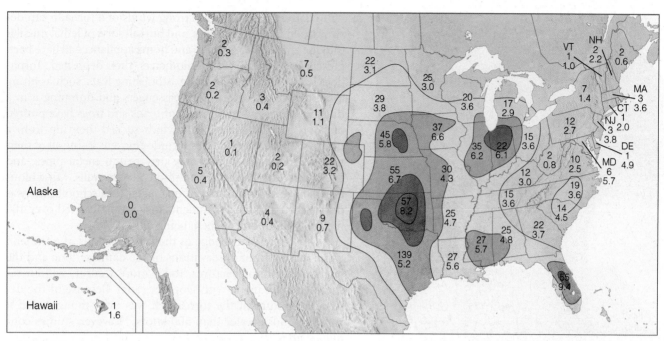

● FIGURE 14.38 Tornado incidence by state. The upper figure shows the average annual number of tornadoes observed in each state from 1953–2004. The lower figure is the average annual number of tornadoes per 10,000 square miles in each state during the same period. The darker the shading, the greater the frequency of tornadoes. (NOAA)

The frequency of tornadic activity shows a seasonal shift. For example, during the winter, tornadoes are most likely to form over the southern Gulf states when the polar-front jet is above this region, and the contrast between warm and cold air masses is greatest. In spring, humid Gulf air surges northward; contrasting air masses and the jet stream also move northward and tornadoes become more prevalent from the southern Atlantic states westward into the southern Great Plains. In summer, the contrast between air masses lessens, and the jet stream is normally near the Canadian border; hence, tornado activity tends to be concentrated from the northern plains eastward to New York State.

In ● Fig. 14.39 we can see that about three-fourths of all tornadoes in the United States develop from March to July. The month of May normally has the greatest number of tornadoes* (the average is about 6 per day) while the most violent tornadoes seem to occur in April when vertical wind shear tends to be present as well as when horizontal and vertical temperature and moisture contrasts are greatest. Although tornadoes have occurred at all times of the day and night, they are most frequent in the late afternoon (between 4:00 P.M. and 6:00 P.M.), when the surface air is most unstable; they are least frequent in the early morning before sunrise, when the atmosphere is most stable.

Although large, destructive tornadoes are most common in the Central Plains, they can develop anywhere if conditions are right. For example, a series of at least 36 tornadoes, more typical of those that form over the plains, marched through North and South Carolina on March 28, 1984, claiming

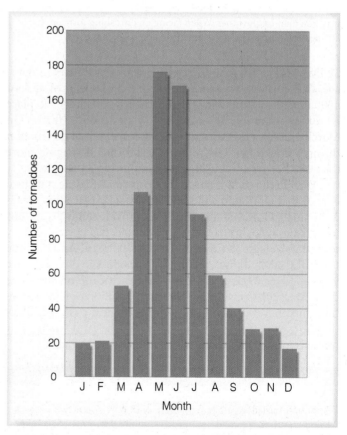

● FIGURE 14.39 Average number of tornadoes during each month in the United States. (NOAA)

---

*During May, 2003, a record 516 tornadoes touched down in the United States (an average of over 16 per day) — the most in any month ever.

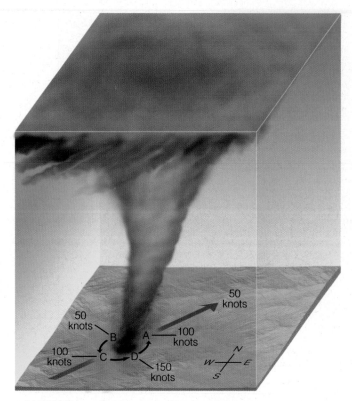

● **FIGURE 14.40** The total wind speed of a tornado is greater on one side than on the other. When facing an onrushing tornado, the strongest winds will be on your left side.

59 lives and causing hundreds of millions of dollars in damage. One tornado was enormous, with a diameter of at least 4000 m (2.5 mi) and winds that exceeded 200 knots. No place is totally immune to a tornado's destructive force. On March 1, 1983, a rare tornado cut a 5-km swath of destruction through downtown Los Angeles, California, damaging more than 100 homes and businesses and injuring 33 people.

Even in the central part of the United States, the statistical chance that a tornado will strike a particular place this year is quite small. However, tornadoes can provide many exceptions to statistics. Oklahoma City, for example, has been struck by tornadoes at least 34 times in the past 100 years. And the little town of Codell, Kansas, was hit by tornadoes in 3 consecutive years—1916, 1917, and 1918—and each time on the same date: May 20! Considering the many millions of tornadoes that must have formed during the geological past, it is likely that at least one actually moved across the land where your home is located, especially if it is in the Central Plains.

WEATHER WATCH

Although tornadoes are rare in Utah, with only about two per year being reported, a tornado rampaged through downtown Salt Lake City during August, 1999. While only on the ground for about 5 miles, the tornado damaged over 120 homes, injured a dozen people, produced one fatality, and caused over $50 million in damage.

TORNADO WINDS   The strong winds of a tornado can destroy buildings, uproot trees, and hurl all sorts of lethal missiles into the air. People, animals, and home appliances all have been picked up, carried several kilometers, then deposited. Tornadoes have accomplished some astonishing feats, such as lifting a railroad coach with its 117 passengers and dumping it in a ditch 25 meters away. Showers of toads and frogs have poured out of a cloud after tornadic winds sucked them up from a nearby pond. Other oddities include chickens losing all of their feathers, pieces of straw being driven into metal pipes, and frozen hot dogs being driven into concrete walls. Miraculous events have occurred, too. In one instance, a schoolhouse was demolished and the 85 students inside were carried over 100 meters without one of them being killed.

Our earlier knowledge of the furious winds of a tornado came mainly from observations of the damage done and the analysis of motion pictures. Today more accurate wind measurements are made with Doppler radar. Because of the destructive nature of the tornado, it was once thought that it packed winds greater than 500 knots. However, studies conducted after 1973 reveal that even the most powerful twisters seldom have winds exceeding 220 knots, and most tornadoes have winds of less than 125 knots. Nevertheless, being confronted with even a small tornado can be terrifying.

When a tornado is approaching from the southwest, its strongest winds are on its southeast side. We can see why in ● Fig. 14.40. The tornado is heading northeast at 50 knots. If its rotational speed is 100 knots, then its forward speed will add 50 knots to its southeastern side (position D) and subtract 50 knots from its northwestern side (position B). Hence, the most destructive and extreme winds will be on the tornado's southeastern side.

Many violent tornadoes (with winds exceeding 180 knots) contain smaller whirls that rotate within them. Such tornadoes are called *multi-vortex tornadoes* and the smaller whirls are called **suction vortices** (see ● Fig. 14.41). Suction vortices are only about 10 m (30 ft) in diameter, but they rotate very fast and can do a great deal of damage.

Seeking Shelter   The high winds of the tornado cause the most damage as walls of buildings buckle and collapse when blasted by the extreme wind force and by debris carried by the wind. Also, as high winds blow over a roof, lower air pressure forms above the roof. The greater air pressure inside the building then lifts the roof just high enough for the strong winds to carry it away. A similar effect occurs when the tornado's intense low-pressure center passes overhead. Because the pressure in the center of a tornado may be more than 100 mb (3 in.) lower than that of its surroundings, there is a momentary drop in outside pressure when the tornado is above the structure. It was once thought that opening windows and allowing inside and outside pressures to equalize would minimize the chances of the building exploding. However, it is now known that opening windows during a tornado actually increases the pressure on the opposite wall and *increases* the chances that the building will collapse. (The win-

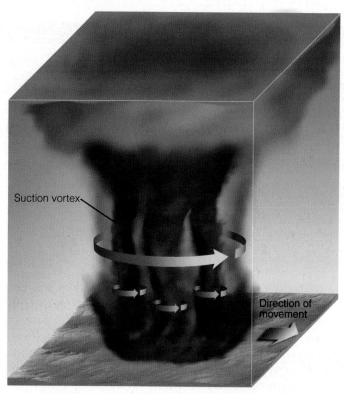

● FIGURE 14.41 A powerful multi-vortex tornado with three suction vortices.

dows are usually shattered by flying debris anyway.) So stay away from windows. Damage from tornadoes may also be inflicted on people and structures by flying debris. Hence, the wisest course to take when confronted with an approaching tornado is to *seek shelter immediately.*

At home, take shelter in a basement. In a large building without a basement, the safest place is usually in a small room, such as a bathroom, closet, or interior hallway, preferably on the lowest floor and near the middle of the edifice. Pull a mattress around you as the handles on the side make it easy to hang onto. Wear a bike or football helmet to protect your head from flying debris. At school, move to the hallway and lie flat with your head covered. In a mobile home, leave immediately and seek substantial shelter. If none exists, lie flat on the ground in a depression or ravine. Don't try to outrun an oncoming tornado in a car or truck, as tornadoes often cover erratic paths with speeds sometimes exceeding 70 knots (80 mi/hr). Stop your car and let the tornado go by or turn around on the road's shoulder and drive in the opposite direction. And do not take shelter under a freeway overpass, as the tornado's winds are actually funneled (strengthened) by the overpass structure. If caught outdoors in an open field, look for a ditch, streambed, or ravine, and lie flat with your head covered.

When tornadoes are likely to form during the next few hours, a **tornado watch** is issued by the Storm Prediction Center in Norman, Oklahoma, to alert the public that tornadoes may develop within a specific area during a certain time period. Many communities have trained volunteer spotters, who look for tornadoes after the watch is issued. (If a tornado

is spotted in the watch area, keep abreast of its movement by listening to the NOAA Weather Radio.) Once a tornado is spotted—either visually or on a radar screen—a **tornado warning** is issued by the local National Weather Service Office.* In some communities, sirens are sounded to alert people of the approaching storm. Radio and television stations interrupt regular programming to broadcast the warning. Although not completely effective, this warning system is apparently saving many lives. Despite the large increase in population in the tornado belt during the past 30 years, tornado-related deaths have actually shown a decrease (see ▼ Table 14.1).

The Fujita Scale   In the 1960s, the late Dr. T. Theodore Fujita, a noted authority on tornadoes at the University of Chicago, proposed a scale (called the **Fujita scale**) for classifying tornadoes according to their rotational wind speed. The tornado winds are estimated based on the damage caused by the storm. However, classifying a tornado based solely on the damage it causes is rather subjective. But the scale became widely used and is presented in ▼ Table 14.2.

The original Fujita scale, implemented in 1971, was based mainly on tornado damage incurred by a frame house. Because there are many types of structures susceptible to tornado damage, a new scale came into effect in February, 2007. Called the *Enhanced Fujita Scale*, or simply the **EF Scale**, the new scale attempts to provide a wide range of criteria in estimating a tornado's winds by using a set of 28 damage indicators. These indicators include items such as small barns, mobile homes, schools, and trees. Each item is then examined for the degree of damage it sustained. The combination of the damage indicators along with the degree of damage provides a range of probable wind speeds and an EF rating for the tornado. The wind estimates for the EF scale are given in ▼ Table 14.3.

---

*In October, 2007, the National Weather Service launched a new, more specific tornado warning system called *Storm Based Warnings.* The new system provides more precise information on where a tornado is located and where it is heading.

▼ TABLE 14.1   Average Annual Number of Tornadoes and Tornado Deaths by Decade*

DECADE	TORNADOES/YEAR	DEATHS/YEAR
1950–59	480	148
1960–69	681	94
1970–79	858	100
1980–89	819	52
1990–99	1,220	56
2000–07	1,319†	52

*Slightly less than a ten-year period.

†More tornadoes are being reported as populations increase and tornado-spotting technology improves.

▽ TABLE 14.2   Fujita Scale for Damaging Winds

SCALE	CATEGORY	MI/HR	KNOTS	EXPECTED DAMAGE
F0	Weak	40–72	35–62	Light: tree branches broken, sign boards damaged
F1		73–112	63–97	Moderate: trees snapped, windows broken
F2	Strong	113–157	98–136	Considerable: large trees uprooted, weak structures destroyed
F3		158–206	137–179	Severe: trees leveled, cars overturned, walls removed from buildings
F4	Violent	207–260	180–226	Devastating: frame houses destroyed
F5*		261–318	227–276	Incredible: structures the size of autos moved over 100 meters, steel-reinforced structures highly damaged

*The scale continues up to a theoretical F12. Very few (if any) tornadoes have wind speeds in excess of 318 mi/hr.

▽ TABLE 14.3   Modified (EF) Fujita Scale for Damaging Winds

EF SCALE	MI/HR*	KNOTS
EF0	65–85	56–74
EF1	86–110	75–95
EF2	111–135	96–117
EF3	136–165	118–143
EF4	166–200	144–174
EF5	>200	>174

*The wind speed is a 3-second gust estimated at the point of damage, based on a judgment of damage indicators.

Statistics reveal that the majority of tornadoes are relatively weak, with wind speeds less than about 110 mi/hr. Only a few percent each year are classified as violent, with perhaps one or two EF5 tornadoes reported annually (although several years may pass without the United States experiencing an EF5). However, it is the violent tornadoes that account for the majority of tornado-related deaths. As an example, a powerful EF5 tornado marched through the southern part of Andover, Kansas, on the evening of April 26, 1991. The tornado, which stayed on the ground for nearly 110 km (68 mi), destroyed more than 100 homes and businesses, injured several hundred people, and out of the 39 tornado fatalities in 1991, this violent tornado alone took the lives of 17. A powerful EF5 tornado moving through Hesston, Kansas, is shown in ● Fig. 14.42. And some of the extreme damage caused by the devastating EF5 tornado that roared through Oklahoma on May 3, 1999, is shown in ● Fig. 14.43.

● FIGURE 14.42 A devastating EF5 tornado about 200 meters wide plows through Hesston, Kansas, on March 13, 1990, leaving almost 300 people homeless and 13 injured.

© Wade Balzer/WeatherStock

● FIGURE 14.43 Total destruction caused by an EF5 tornado that devastated parts of Oklahoma on May 3, 1999.

One important reason for the number of deaths and extensive damage being caused by violent tornadoes is that, as the wind speed doubles, the force of the wind exerted on an object increases by a factor of four. Hence, the 200 mi/hr winds of an EF4 tornado exert four times as much force on a building as do the 100 mi/hr winds of an EF1.

Tornado Formation

Although not everything is known about the formation of a tornado, we do know that tornadoes tend to form with intense thunderstorms and that a conditionally unstable atmosphere is essential for their development. Most often they form with supercell thunderstorms in an environment with

WEATHER WATCH

It may be almost impossible to survive the powerful winds of a violent tornado if you are inside the wrong type of structure, such as a mobile home. During the May 3, 1999, tornado outbreak many people who abandoned their unprotected homes in favor of muddy ditches survived largely because the ditches were below ground level and out of the path of wind-blown objects. Many who stayed in the confines of their inadequate homes perished when tornado winds blew their homes away, leaving only the foundations.

strong vertical wind shear. The rotating air of the tornado may begin within a thunderstorm and work its way downward, or it may begin at the surface and work its way upward. First, we will examine tornadoes that form with supercells; then we will examine nonsupercell tornadoes.

SUPERCELL TORNADOES   Tornadoes that form with supercell thunderstorms are called **supercell tornadoes**. Earlier, we learned that a supercell is a thunderstorm that has a single rotating updraft that can exist for hours. Recall also that supercells form in a region of strong, vertical wind shear that causes the updraft inside the storm to rotate.

In ● Fig. 14.44a notice that there is wind direction shear, as the surface winds are southerly and a kilometer or so above the surface they are northerly. There is also wind speed shear as the wind speed increases rapidly with height. This wind shear causes the air near the surface to rotate about a horizontal axis much like a pencil rotates around its long axis. Such horizontal tubes of spinning air are called *vortex tubes*. (These spirally vortex tubes also form when a southerly low-level jet exists just above southerly surface winds.) If the strong updraft of a developing thunderstorm should tilt the

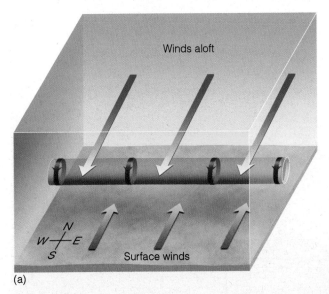

(a)

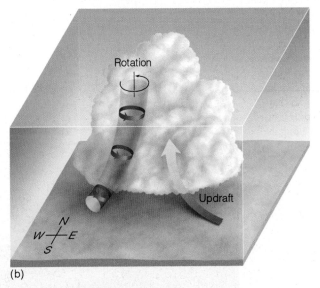

(b)

● FIGURE 14.44 (a) A spinning vortex tube created by wind shear. (b) The strong updraft in the developing thunderstorm carries the vortex tube into the thunderstorm producing a rotating air column that is oriented in the vertical plane.

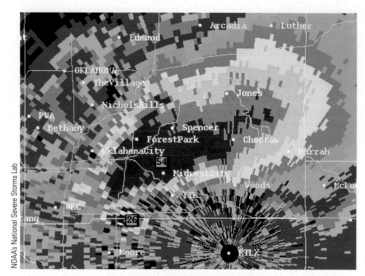

NOAA's National Severe Storms Lab

● FIGURE 14.45 A tornado-spawning supercell thunderstorm over Oklahoma City on May 3, 1999, shows a hook echo in its rainfall pattern on a Doppler radar screen. The colors red and orange represent the heaviest precipitation.

rotating tube upward and draw it into the storm, as illustrated in Fig. 14.44b, the tilted rotating tube then becomes a rotating air column inside the storm. The rising, spinning air is now part of the storm's structure called the *mesocyclone*— an area of lower pressure (a small cyclone) perhaps 5 to 10 kilometers across. The rotation of the updraft lowers the pressure in the mid-levels of the thunderstorm, which acts to increase the strength of the updraft.*

*You can obtain an idea of what might be taking place in the supercell by stirring a cup of coffee or tea with a spoon and watching the low pressure form in the middle of the beverage.

As we learned earlier in the chapter, the updraft is so strong in a supercell (sometimes 90 knots) that precipitation cannot fall through it. Southwesterly winds aloft usually blow the precipitation northeastward. If the mesocyclone persists, it can circulate some of the precipitation counterclockwise around the updraft. This swirling precipitation shows up on the radar screen, whereas the area inside the mesocyclone (nearly void of precipitation at lower levels) does not. The region inside the supercell where radar is unable to detect precipitation is known as the *bounded weak echo region (BWER)*. Meanwhile, as the precipitation is drawn into a cyclonic spiral around the mesocyclone, the rotating precipitation may, on the Doppler radar screen, unveil itself in the shape of a hook, called a **hook echo,** as shown in ● Fig. 14.45.

At this point in the storm's development, the updraft, the counterclockwise swirling precipitation, and the surrounding air may all interact to produce the *rear-flank downdraft* (to the south of the updraft), as shown in ● Fig. 14.46. The strength of the downdraft is driven by the amount of precipitation-induced cooling in the upper levels of the storm. The rear-flank downdraft appears to play an important role in producing tornadoes in classic supercells.

When the rear-flank downdraft strikes the ground, it may (under favorable shear conditions) interact with the forward-flank downdraft beneath the mesocyclone to initiate the formation of a tornado. At the surface, the cool air of the rear-flank downdraft (and the forward-flank downdraft) sweeps around the center of the mesocyclone, effectively cutting off the rising air from the warmer surrounding air. The lower half of the updraft now rises more slowly. The rising updraft, which we can imagine as a column of air, now shrinks horizontally and stretches vertically. This *vertical*

● FIGURE 14.46 A classic tornadic supercell thunderstorm showing updrafts and downdrafts, along with surface air flowing counterclockwise and in toward the tornado. The flanking line is a line of cumulus clouds that form as surface air is lifted into the storm along the gust front.

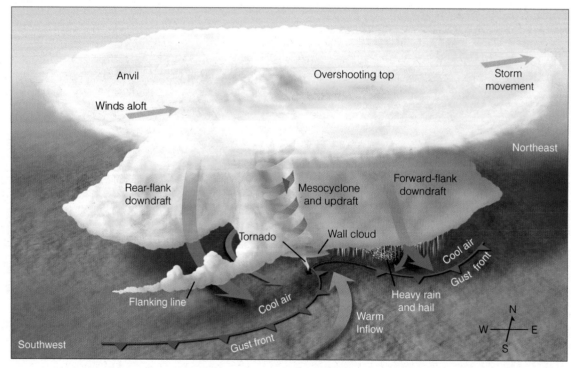

stretching of the spinning column of air causes the rising, spinning air to spin faster.* If this stretching process continues, the rapidly rotating air column may shrink into a narrow column of rapidly rotating air—a tornado vortex.

As air rushes upward and spins around the low-pressure core of the vortex, the air expands, cools, and, if sufficiently moist, condenses into a visible cloud—the *funnel cloud.* As the air beneath the funnel cloud is drawn into its core, the air cools rapidly and condenses, and the funnel cloud descends toward the surface. Upon reaching the ground, the tornado's circulation usually picks up dirt and debris, making it appear both dark and ominous. While the air along the outside of the funnel is spiraling upward, Doppler radar reveals that, within the core of violent tornadoes, the air is descending toward the extreme low pressure at the ground (which may be 100 mb lower than that of the surrounding air). As the air descends, it warms, causing the cloud droplets to evaporate. This process leaves the core free of clouds. Supercell tornadoes usually develop near the right rear sector of the storm, on the southwestern side of a northeastward-moving storm, as shown in Fig. 14.46.

Not all supercells produce tornadoes; in fact, perhaps less than 15 percent do. However, recent studies reveal that supercells are more likely to produce tornadoes when they interact with a pre-existing boundary, such as an old gust front (outflow boundary) that supplies the surface air with horizontal spin that can be tilted and lifted into the storm by its updraft. Many atmospheric situations may suppress tornado formation. For example, if the precipitation in the cloud is swept too far away from the updraft, or if too much precipitation wraps around the mesocyclone, the necessary interactions that produce the rear flank downdraft are disrupted, and a tornado is not likely to form. Moreover, tornadoes are not likely to form if the supercell is fed warm, moist air that is elevated above a deep layer of cooler surface air.

As we have seen, the first sign that a supercell is about to give birth to a tornado is the sight of *rotating clouds* at the base of the storm.† If the area of rotating clouds lowers, it becomes the *wall cloud.* Notice in Fig. 14.46 that the tornado extends from within the wall cloud to the earth's surface. Sometimes the air is so dry that the swirling, rotating wind remains invisible until it reaches the ground and begins to pick up dust. Unfortunately, people have mistaken these "invisible tornadoes" for dust devils, only to find out (often too late) that they were not. Occasionally, the funnel cannot be seen due to falling rain, clouds of dust, or darkness. Even when not clearly visible, many tornadoes have a distinctive roar that can be heard for several kilometers. This sound, which has been described as "a roar like a thousand freight trains," appears to be loudest when the tornado is touching the surface. However, not all tornadoes make this sound and, when these storms strike, they become silent killers.

Certainly, the likelihood of a thunderstorm producing a tornado increases when the storm becomes a supercell, but not all supercells produce tornadoes. And not all tornadoes come from rotating thunderstorms (supercells).

NONSUPERCELL TORNADOES Tornadoes that do not occur in association with a pre-existing wall cloud (or a mid-level mesocyclone) of a supercell are called **nonsupercell tornadoes**. These tornadoes may occur with intense multicell storms as well as with ordinary cell thunderstorms, even relatively weak ones. Some nonsupercell tornadoes extend from the base of a thunderstorm whereas others may begin on the ground and build upwards in the absence of a condensation funnel.

Nonsupercell tornadoes may form along a gust front where the cool downdraft of the thunderstorm forces warm, humid air upwards. Tornadoes that form along a gust front are commonly called **gustnadoes.** These relatively weak tornadoes normally are short-lived and rarely inflict significant damage. Gustnadoes are often seen as a rotating cloud of dust or debris rising above the surface.

Occasionally, rather weak, short-lived tornadoes will occur with rapidly building cumulus congestus clouds. Tornadoes such as these commonly form over east-central Colorado. Because they look similar to waterspouts that form over water, they are sometimes called **landspouts***(see ●Fig. 14.47).

● Figure 14.48 illustrates how a landspout can form. Suppose, for example, that the winds at the surface converge along a boundary, as illustrated in Fig. 14.48a. (The wind may converge due to topographic irregularities or any number of other factors, including temperature and moisture variations.) Notice that along the boundary, the air is rising, condensing, and forming into a cumulus congestus cloud. Notice also that along the surface at the boundary there is horizontal rotation (spin) created by the wind blowing in opposite directions along the boundary. If the developing cloud should move over the region of rotating air, the spinning air may be drawn up into the cloud by the storm's updraft. As the spinning, rising air shrinks in diameter, it produces a tornado-like structure, a *landspout* (see Fig. 14.47). Landspouts usually dissipate when rain falls through the cloud and destroys the updraft. Tornadoes may form in this manner along many types of converging wind boundaries, including sea breezes and gust fronts. Nonsupercell tornadoes and funnel clouds may also form with thunderstorms when cold air aloft (associated with an upper-level trough) moves over a region. Common along the west coast of North America, these short-lived tornadoes are sometimes called *cold-air funnels.*

---

*As the rotating air column stretches vertically into a narrow column, its rotational speed increases. You may recall from Chapter 10, on p. 270, that this situation is called the *conservation of angular momentum.*

†Occasionally, people will call a sky dotted with mammatus clouds "a tornado sky." Mammatus clouds may appear with both severe and nonsevere thunderstorms as well as with a variety of other cloud types (see Chapter 5). Mammatus clouds are not funnel clouds, do not rotate, and their appearance has no relationship to tornadoes.

---

*Landspouts occasionally form on the backside of a squall line where southerly winds ahead of a cold front and northwesterly winds behind it create swirling eddies that can be drawn into thunderstorms by their strong updrafts.

● FIGURE 14.47 A well-developed landspout moves over eastern Colorado.

Severe Weather and Doppler Radar

Most of our knowledge about what goes on inside a tornado-generating thunderstorm has been gathered through the use of *Doppler radar*. Remember from Chapter 7 that a radar transmitter sends out microwave pulses and that, when this

energy strikes an object, a small fraction is scattered back to the antenna. Precipitation particles are large enough to bounce microwaves back to the antenna. Consequently, as we saw earlier, the colorful area on the radar screen in Fig. 14.45, p. 402, represents precipitation intensity inside a supercell thunderstorm.

Doppler radar can do more than measure rainfall intensity, it can actually measure the speed at which precipitation is moving horizontally toward or away from the radar antenna. Because precipitation particles are carried by the wind, Doppler radar can peer into a severe storm and reveal its winds.

Doppler radar works on the principle that, as precipitation moves toward or away from the antenna, the returning radar pulse will change in frequency. A similar change occurs when the high-pitched sound (high frequency) of an approaching noise source, such as a siren or train whistle, becomes lower in pitch (lower frequency) after it passes by the person hearing it. This change in frequency in sound waves or microwaves is called the *Doppler shift* and this, of course, is where the Doppler radar gets its name.

A single Doppler radar cannot detect winds that blow parallel to the antenna. Consequently, two or more units probing the same thunderstorm are needed to give a complete three-dimensional picture of the winds within the storm. To help distinguish the storm's air motions, wind velocities can be displayed in color. Color contouring the wind field gives a good picture of the storm (see ● Fig. 14.49).

Even a single Doppler radar can uncover many of the features of a severe thunderstorm. For example, studies conducted in the 1970s revealed, for the first time, the existence of the swirling winds of the mesocyclone inside a supercell storm. Mesocyclones have a distinct image (signature) on the radar display. Tornadoes also have a distinct signature on the radar screen, known as the *tornado vortex signature (TVS),* which shows up as a region of rapidly (or abruptly) changing wind directions within the mesocyclone. (Look at Fig. 14.49.)

Unfortunately, the resolution of the Doppler radar is not high enough to measure actual wind speeds of most tornadoes, whose diameters are only a few hundred meters or less. However, a new and experimental Doppler system—called

● FIGURE 14.48
(a) Along the boundary of converging winds, the air rises and condenses into a cumulus congestus cloud. At the surface the converging winds along the boundary create a region of counterclockwise spin. (b) As the cloud moves over the area of rotation, the updraft draws the spinning air up into the cloud producing a nonsupercell tornado, or landspout. (Modified after Wakimoto and Wilson)

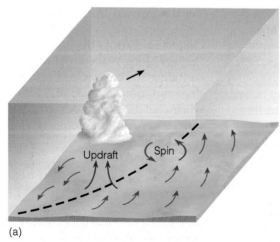

(a)

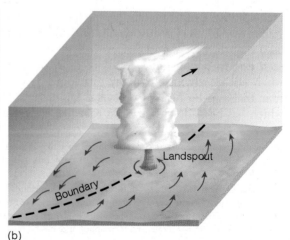

(b)

weather and possible flash flooding. In addition, the algorithms give advanced and improved warning of an approaching tornado. More reliable warnings, of course, will cut down on the number of false alarms.

Because the Doppler radar shows horizontal air motion within a storm, it can help to identify the magnitude of other severe weather phenomena, such as gust fronts, derechoes, microbursts, and wind shears that are dangerous to aircraft. Certainly, as more and more information from Doppler radar becomes available, our understanding of the processes that generate severe thunderstorms and tornadoes will be enhanced, and hopefully there will be an even better tornado and severe storm warning system, resulting in fewer deaths and injuries.

In an attempt to unravel some of the mysteries of the tornado, several studies are under way. In one study, scientists using an armada of observational vehicles, aircraft, and state-of-the-art equipment, including Doppler radar, pursued tornado-generating thunderstorms over portions of the Central Plains during the spring and summer. These observations are providing information on the inner workings of severe thun-

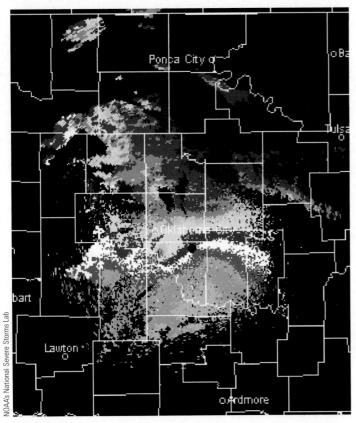

● FIGURE 14.49 Doppler radar display of winds associated with the supercell storm that moved through parts of Oklahoma City during the afternoon of May 3, 1999. The close packing of the horizontal winds blowing toward the radar (green and blue shades), and those blowing away from the radar (yellow and red shades), indicate strong cyclonic rotation and the presence of a tornado.

Doppler lidar—uses a light beam (instead of microwaves) to measure the change in frequency of falling precipitation, cloud particles, and dust. Because it uses a shorter wavelength of radiation, it has a narrower beam and a higher resolution than does Doppler radar. In an attempt to obtain tornado wind information at fairly close range (less than 10 km), smaller portable Doppler radar units (Doppler on wheels) are peering into tornado-generating storms (see ● Fig. 14.50).

The network of more than 150 Doppler radar units deployed at selected weather stations within the continental United States is referred to as **NEXRAD** (an acronym for *NEX*t Generation Weather *RAD*ar). The NEXRAD system consists of the WSR-88D* Doppler radar and a set of computers that perform a variety of functions.

The computers take in data, display them on a monitor, and run computer programs called *algorithms,* which, in conjunction with other meteorological data, detect severe weather phenomena, such as storm cells, hail, mesocyclones, and tornadoes. Algorithms provide a great deal of information to the forecasters that allows them to make better decisions as to which thunderstorms are most likely to produce severe

*The name WSR-88D stands for *Weather Surveillance Radar, 1988 Doppler.*

● FIGURE 14.50 Graduate students from the University of Oklahoma use a portable Doppler radar to probe a tornado near Hodges, Oklahoma.

© G. Kaufman

ACTIVE FIGURE 14.51 A powerful waterspout moves across Lake Tahoe, California. Visit the Meteorology Resource Center to view this and other active figures at academic.cengage.com/login

derstorms. At the same time, laboratory models of tornadoes in chambers (called *vortex chambers*), along with mathematical computer models, are offering new insights into the formation and development of these fascinating storms.

Waterspouts

A **waterspout** is a rotating column of air that is connected to a cumuliform cloud over a large body of water. The waterspout may be a tornado that formed over land and then traveled over water. In such a case, the waterspout is sometimes referred to as a *tornadic waterspout.* Such tornadoes can inflict major damage to ocean-going vessels, especially when the tornadoes are of the supercell variety. Waterspouts not associated with supercells that form over water, especially above warm, tropical coastal waters (such as in the vicinity of the Florida Keys, where almost 100 occur each month during the summer), are often referred to as *"fair weather" waterspouts.** These waterspouts are generally much smaller than an average tornado, as they have diameters usually between 3 and 100 meters. Fair weather waterspouts are also less intense, as their rotating winds are typically less than 45 knots. In addition, they tend to move more slowly than tornadoes and they only last for about 10 to 15 minutes, although some have existed for up to one hour.

Fair weather waterspouts tend to form in much the same way that landspouts do—when the air is conditionally unstable and cumulus clouds are developing. Some form with small thunderstorms, but most form with developing cumulus congestus clouds whose tops are frequently no higher than 3600 m (12,000 ft) and do not extend to the freezing level. Apparently, the warm, humid air near the water helps to create atmospheric instability, and the updraft beneath the resulting cloud helps initiate uplift of the surface air. Studies even suggest that gust fronts and converging sea breezes may play a role in the formation of some of the waterspouts that form over the Florida Keys.

The waterspout funnel is similar to the tornado funnel in that both are clouds of condensed water vapor with converging winds that rise about a central core. Contrary to popular belief, the waterspout does not draw water up into its core; however, swirling spray may be lifted several meters when the waterspout funnel touches the water. Apparently, the most destructive waterspouts are those that begin as tornadoes over land, then move over water. A photograph of a particularly well-developed and intense waterspout is shown in ●Fig. 14.51.

---

*"Fair weather" waterspouts may form over any large body of warm water. Hence, they occur frequently over the Great Lakes in summer.

SUMMARY

In this chapter, we examined thunderstorms and the atmospheric conditions that produce them. Thunderstorms are convective storms that produce lightning and thunder. Lightning is a discharge of electricity that occurs in mature thunderstorms. The lightning stroke momentarily heats the air to an incredibly high temperature. The rapidly expanding air produces a sound called thunder.

The ingredients for the isolated ordinary cell thunderstorm are humid surface air, plenty of sunlight to heat the ground, a conditionally unstable atmosphere, a "trigger" to start the air rising, and weak vertical wind shear. When these conditions prevail, and the air begins to rise, small cumulus clouds may grow into towering clouds and thunderstorms within 30 minutes.

When conditions are ripe for thunderstorm development, and moderate or strong vertical wind shear exists, the updraft in the thunderstorm may tilt and ride up and over the downdraft. As the forward edge of the downdraft (the gust front) pushes outward along the ground, the air is lifted and new cells form, producing a multicell thunderstorm—a storm with cells in various stages of development. Some multicell storms form as a complex of thunderstorms, such as the squall line (which forms as a line of thunderstorms either along or out ahead of an advancing cold front), and the Mesoscale Convective Complex (which forms as a cluster of storms). When convection in the multicell storm is strong, it may produce severe weather, such as strong damaging surface winds, hail, and flooding.

Supercell thunderstorms are large, intense thunderstorms with a single rotating updraft. The updraft and the downdraft in a supercell are nearly in balance, so that the storm may exist for many hours. Supercells are capable of producing severe weather, including strong damaging tornadoes.

Tornadoes are rapidly rotating columns of air with a circulation that reaches the ground. The rotating air of the tornado may begin within the thunderstorm or it may begin at the surface and extend upwards. Tornadoes can form with supercells, as well as with less intense thunderstorms. Tornadoes that do not form with supercells are the landspout and the gustnado. Most tornadoes are less than a few hundred meters wide with wind speeds less than 100 knots, although violent tornadoes may have wind speeds that exceed 250 knots. A violent tornado may actually have smaller whirls (suction vortices) rotating within it. With the aid of Doppler radar, scientists are probing tornado-spawning thunderstorms, hoping to better predict tornadoes and to better understand where, when, and how they form.

A normally small and less destructive cousin of the tornado is the "fair weather" waterspout that commonly forms above the warm waters of the Florida Keys and the Great Lakes in summer.

KEY TERMS

The following terms are listed (with page numbers) in the order they appear in the text. Define each. Doing so will aid you in reviewing the material covered in this chapter.

QUESTIONS FOR REVIEW

1. What is a thunderstorm?

2. What atmospheric conditions are necessary for the development of ordinary cell thunderstorms?

3. Describe the stages of development of an ordinary cell (air mass) thunderstorm.

4. How do downdrafts form in thunderstorms?

5. Why do ordinary cell thunderstorms most frequently form in the afternoon?

6. Explain why ordinary cell thunderstorms tend to dissipate much sooner than multicell storms.

7. How does the National Weather Service define a severe thunderstorm?

8. What is necessary for a multicell thunderstorm to form?

9. (a) How do gust fronts form?
   (b) What type of weather does a gust front bring when it passes?

10. (a) Describe how a microburst forms.
    (b) Why is the term *wind shear* often used in conjunction with a microburst?

11. How do derechoes form?

12. How does a squall line differ from a Mesoscale Convective Complex (MCC)?

13. Give a possible explanation for the generation of a prefrontal squall-line thunderstorm.

14. How does the cell in an ordinary cell thunderstorm differ from the cell in a supercell thunderstorm?

15. Describe the atmospheric conditions at the surface and aloft that are necessary for the development of most supercell thunderstorms. (Include in your answer the role that the low-level jet plays in the rotating updraft.)

16. What is the difference between an HP supercell and an LP supercell?

17. When thunderstorms are *training*, what are they doing?

18. In what region in the United States do dryline thunderstorms most frequently form? Why there?

19. Where does the highest frequency of thunderstorms occur in the United States? Why there?

20. Why is large hail more common in Kansas than in Florida?

21. Describe one process by which thunderstorms become electrified.

22. Explain how a cloud-to-ground lightning stroke develops.

23. Why is it unwise to seek shelter under an isolated tree during a thunderstorm? If caught out in the open, what should you do?

24. How does negative cloud-to-ground lightning differ from positive cloud-to-ground lightning?

25. How is thunder produced?

26. What is the primary difference between a tornado and a funnel cloud?

27. Give some average statistics about tornado size, winds, and direction of movement.

28. Why should you *not* open windows when a tornado is approaching?

29. Why is the central part of the United States more susceptible to tornadoes than any other region of the world?

30. Explain both how and why there is a shift in tornado activity from winter to summer.

31. How does a tornado *watch* differ from a tornado *warning*?

32. If you are in a single-story home (without a basement) during a tornado warning, what should you do?

33. Supercell thunderstorms that produce tornadoes form in a region of strong wind shear. Explain how the wind changes in speed and direction to produce this shear.

34. Explain how a nonsupercell tornado, such as a landspout, might form.

35. Describe how Doppler radar measures the winds inside a severe thunderstorm.

36. How has Doppler radar helped in the prediction of severe weather?

37. What atmospheric conditions lead to the formation of "fair weather" waterspouts?

QUESTIONS FOR THOUGHT

1. Why does the bottom half of a dissipating thunderstorm usually "disappear" before the top?

2. Sinking air warms, yet the downdrafts in a thunderstorm are usually cold. Why?

3. Explain why squall-line thunderstorms often form ahead of advancing cold fronts but seldom behind them.

4. A forecaster may say that he or she looks for "right-moving" thunderstorms when predicting severe weather. What does this mean?

5. Why is the old adage "lightning never strikes twice in the same place" wrong?

6. Suppose while you are on a high mountain ridge a thundercloud passes overhead. What would be the wisest thing to do—stand upright? lie down? or crouch? Explain.

7. If you are confronted in an open field by a large tornado and there is no way that you could outrun it, probably the only thing that you could do would be to run and lie down in a depression. If given the choice, would you run toward your right or left as the tornado approaches? Explain your reasoning.

8. Tornadoes apparently form in the region of a strong updraft rather than in a downdraft, yet they descend from the base of a cloud. Why?

9. Why are left-moving supercell thunderstorms uncommon in the Northern Hemisphere, yet are very common in the Southern Hemisphere?

PROBLEMS AND EXERCISES

1. On a map of the United States, place the surface weather conditions as well as weather conditions aloft (jet stream and so on) that are necessary for the formation of most supercell thunderstorms.

2. On the surface weather map in ●Fig. 14.52, at which number would you most likely observe a line of thunderstorms forming? Explain why you chose that location.

3. A multi-vortex tornado with a rotational wind speed of 125 knots is moving from southwest to northeast at 30 knots. Assume the suction vortices within this tornado have rotational winds of 100 knots:

   (a) What is the maximum wind speed of this multi-vortex tornado?

   (b) If you are facing the approaching tornado, on which side (northeast, northwest, southwest, or southeast) would the strongest winds be found? the weakest winds? Explain both of your answers.

   (c) According to Table 14.2 and Table 14.3, p. 400, how would this tornado be classified on the old Fujita scale and on the new EF scale?

4. If you see lightning and 10 seconds later you hear thunder, how far away is the lightning stroke?

5. Suppose several of your friends went on a storm-chasing adventure in the central United States. To help guide their chase, you stay behind, with an Internet-connected computer and a cellular phone. Which current weather and forecast maps would you use to guide their storm chase? Explain why you choose those maps.

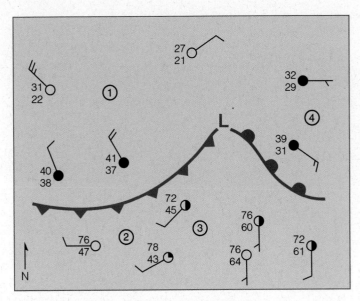

● FIGURE 14.52 Surface weather map for Exercise 2.

Hurricane Katrina over the Gulf of Mexico on August 28, 2005. With sustained winds of 152 knots (175 mi/hr) and a central pressure near 907 mb (26.78 in.), this large and powerful Category 5 hurricane takes aim on Louisiana and Mississippi.

NOAA

CHAPTER 15

Hurricanes

On September 18, 1926, as a hurricane approached Miami, Florida, people braced themselves for the devastating high winds and storm surge. Just before dawn the hurricane struck with full force—torrential rains, flooding, and easterly winds that gusted to over 100 miles per hour. Then, all of a sudden, it grew calm and a beautiful sunrise appeared. People wandered outside to inspect their property for damage. Some headed for work, and scores of adventurous young people crossed the long causeway to Miami Beach for the thrill of swimming in the huge surf. But the lull lasted for less than an hour. From the south, ominous black clouds quickly moved overhead. In what seemed like an instant, hurricane force winds from the west were pounding the area and pushing water from Biscayne Bay over the causeway. Many astonished bathers, unable to swim against the great surge of water, were swept to their deaths. Hundreds more drowned as Miami Beach virtually disappeared under the rising wind-driven tide. It is estimated that if a hurricane of this Category 4 magnitude were to hit Miami today, it would cause $87 billion in damages.

CONTENTS

Born over warm tropical waters and nurtured by a rich supply of water vapor, the *hurricane* can indeed grow into a ferocious storm that generates enormous waves, heavy rains, and winds that may exceed 150 knots. What exactly are hurricanes? How do they form? And why do they strike the east coast of the United States more frequently than the west coast? These are some of the questions we will consider in this chapter.

Tropical Weather

In the broad belt around the earth known as the tropics — the region between $23\frac{1}{2}°$ north and south of the equator — the weather is much different from that of the middle latitudes. In the tropics, the noon sun is always high in the sky, and so diurnal and seasonal changes in temperature are small. The daily heating of the surface and high humidity favor the development of cumulus clouds and afternoon thunderstorms. Most of these are individual thunderstorms that are not severe. Sometimes, however, they group together into loosely organized systems called *non-squall clusters*. On other occasions, the thunderstorms will align into a row of vigorous convective cells known as a *tropical squall cluster*, or *squall line*. The passage of a squall line is usually noted by a sudden wind gust followed immediately by a heavy downpour that may produce 3 cm (more than 1 in.) of rainfall in about 30 minutes. This deluge is then followed by several hours of relatively steady rainfall. Many of these tropical squall lines

are similar to the middle-latitude ordinary squall lines described in Chapter 14 on p. 379.

As it is warm all year long in the tropics, the weather is not characterized by four seasons which, for the most part, are determined by temperature variations. Rather, most of the tropics are marked by seasonal differences in precipitation. The greatest cloudiness and precipitation occur during the high-sun period, when the intertropical convergence zone moves into the region. Even during the dry season, precipitation can be irregular, as periods of heavy rain, lasting for several days, may follow an extreme dry spell.

The winds in the tropics generally blow from the east, northeast, or southeast. Because the variation of sea-level pressure is normally quite small, drawing isobars on a weather map provides little useful information. Instead of isobars, **streamlines** that depict wind flow are drawn. Streamlines are useful because they show where surface air converges and diverges. Occasionally, the streamlines will be disturbed by a weak trough of low pressure called a **tropical wave**, or **easterly wave**, because it tends to move from east to west. (see ● Fig. 15.1).

Tropical waves have wavelengths on the order of 2500 km (1550 mi) and travel from east to west at speeds between 10 and 20 knots. Look at Fig. 15.1 and observe that, on the western side of the trough (heavy dashed green line), where easterly and northeasterly surface winds diverge, sinking air produces generally fair weather. On its eastern side, southeasterly surface winds converge. The converging air rises, cools, and often condenses into showers and thunderstorms. Consequently, the main area of showers forms *behind* the trough. Occasionally, an easterly wave will intensify and grow into a hurricane.

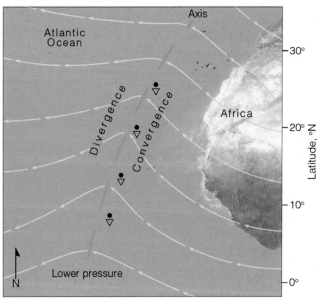

● FIGURE 15.1 A tropical wave (also called an easterly wave) moving off the coast of Africa over the Atlantic. The wave is shown by the bending of streamlines — lines that show wind flow patterns. (The heavy dashed green line is the axis of the trough.) The wave moves slowly westward, bringing fair weather on its western side and rain showers on its eastern side.

Anatomy of a Hurricane

A **hurricane** is an intense storm of tropical origin, with sustained winds exceeding 64 knots (74 mi/hr), which forms over the warm northern Atlantic and eastern North Pacific oceans. This same type of storm is given different names in different regions of the world. In the western North Pacific, it is called a **typhoon,** in India a *cyclone* and in Australia a *tropical cyclone.* By international agreement, **tropical cyclone** is the general term for all hurricane-type storms that originate over tropical waters. For simplicity, we will refer to all of these storms as hurricanes.

● Figure 15.2 is a photo of Hurricane Elena situated over the Gulf of Mexico. The storm is approximately 500 km

NOAA

● FIGURE 15.2 Hurricane Elena over the Gulf of Mexico about 130 km (80 mi) southwest of Apalachicola, Florida, as photographed from the space shuttle *Discovery* during September, 1985. Because this storm is situated north of the equator, surface winds are blowing counterclockwise about its center (eye). The central pressure of the storm is 955 mb, with sustained winds of 105 knots near its eye.

(310 mi) in diameter, which is about average for hurricanes. The area of broken clouds at the center is its **eye**. Elena's eye is almost 40 km (25 mi) wide. Within the eye, winds are light and clouds are mainly broken. The surface air pressure is very low, nearly 955 mb (28.20 in.).* Notice that the clouds align themselves into spiraling bands (called *spiral rain bands*) that swirl in toward the storm's center, where they wrap themselves around the eye. Surface winds increase in speed as they blow counterclockwise and inward toward this center. (In the Southern Hemisphere, the winds blow clockwise around the center.) Adjacent to the eye is the **eyewall**, a ring of intense thunderstorms that whirl around the storm's center and may extend upward to almost 18 km (59,000 ft) above sea level. Within the eyewall, we find the heaviest precipitation and the strongest winds, which, in this storm, are 105 knots, with peak gusts of 120 knots.

If we were to venture from west to east (left to right) at the surface through the storm in Fig. 15.2, what might we experience? As we approach the hurricane, the sky becomes overcast with cirrostratus clouds; barometric pressure drops slowly at first, then more rapidly as we move closer to the center. Winds blow from the north and northwest with ever-increasing speed as we near the eye. The high winds, which generate huge waves over 10 m (33 ft) high, are accompanied by heavy rainshowers. As we move into the eye, the winds slacken, rainfall ceases, and the sky brightens, as middle and high clouds appear overhead. The atmospheric pressure is

now at its lowest point (965 mb), some 50 mb lower than the pressure measured on the outskirts of the storm. The brief respite ends as we enter the eastern region of the eyewall. Here, we are greeted by heavy rain and strong southerly winds. As we move away from the eyewall, the pressure rises, the winds diminish, the heavy rain lets up, and eventually the sky begins to clear.

This brief, imaginary venture raises many unanswered questions. Why, for example, is the surface pressure lowest at the center of the storm? And why is the weather clear almost immediately outside the storm area? To help us answer such questions, we need to look at a vertical view, a profile of the hurricane along a slice that runs through its center. A model that describes such a profile is given in ● Fig. 15.3.

The model shows that the hurricane is composed of an organized mass of thunderstorms* that are an integral part of the storm's circulation. Near the surface, moist tropical air flows in toward the hurricane's center. Adjacent to the eye, this air rises and condenses into huge cumulonimbus clouds that produce heavy rainfall, as much as 25 cm (10 in.) per hour. Near the top of the clouds, the relatively dry air, having lost much of its moisture, begins to flow outward away from the center. This diverging air aloft actually produces a clockwise (*anticyclonic* in the Northern Hemisphere) flow of air several hundred kilometers from the eye. As this outflow reaches the storm's periphery, it begins to sink and warm, inducing clear skies. In the vigorous convective clouds of the eyewall, the air warms due to the release of large quantities of

---

*An extreme low pressure of 870 mb (25.70 in.) was recorded in Typhoon Tip while over the tropical Pacific ocean during October, 1979, and Hurricane Wilma had a pressure reading of 882 mb (26.04 in.) while over the Gulf of Mexico during October, 2005.

---

*These huge convective cumulonimbus clouds have suprisingly little lightning (and, hence, thunder) associated with them. Even so, for simplicity we will refer to these clouds as thunderstorms throughout this chapter.

ACTIVE FIGURE 15.3 A model that shows a vertical view of air motions and clouds in a typical hurricane in the Northern Hemisphere. The diagram is exaggerated in the vertical. Visit the Meteorology Resource Center to view this and other active figures at academic.cengage.com/login

latent heat. This produces slightly higher pressures aloft, which initiate downward air motion within the eye. As the air descends, it warms by compression. This process helps to account for the warm air and the absence of convective clouds in the eye of the storm (see ● Fig. 15.4).

As surface air rushes in toward the region of much lower surface pressure, it should expand and cool, and we might expect to observe cooler air around the eye, with warmer air farther away. But, apparently, so much heat is added to the air from the warm ocean surface that the surface air temperature remains fairly uniform throughout the hurricane.

● Figure 15.5 is a three-dimensional radar composite of Hurricane Katrina as it passes over the central area of the Gulf of Mexico. Compare Katrina's features with those of typical hurricanes illustrated in Fig. 15.2 and Fig. 15.3. Notice that the strongest radar echoes (heaviest rain) near the surface are located in the eyewall, adjacent to the eye. ● Figure 15.6 is an example of how surface winds blow around a hurricane in the Northern Hemisphere.

Hurricane Formation and Dissipation

We are now left with an important question: Where and how do hurricanes form? While there is no widespread agreement on how hurricanes actually form, it is known that certain necessary ingredients are required before a weak tropical disturbance will develop into a full-fledged hurricane.

THE RIGHT ENVIRONMENT Hurricanes form over tropical waters where the winds are light, the humidity is high in a deep layer extending up through the troposphere, and the surface water temperature is warm, typically 26.5°C (80°F) or greater, over a vast area* (see ● Fig. 15.7). These conditions usually prevail over the tropical and subtropical North

---

*It was once thought that for hurricane formation, the ocean must be sufficiently warm through a depth of about 200 meters. It is now known that hurricanes can form in the eastern North Pacific when the warm layer of ocean water is only about 20 m (65 ft) deep.

● FIGURE 15.4 The cloud mass is Hurricane Katrina's eyewall, and the clear area is Katrina's eye photographed inside the eye on August 28, 2005, from a NOAA reconnaissance (hurricane hunter) aircraft. For a satellite image of the storm on this day, look at the opening photo on p. 410.

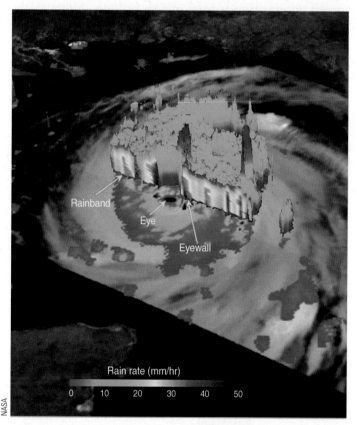

● FIGURE 15.5 A three-dimensional TRMM satellite view of Hurricane Katrina passing over the central Gulf of Mexico on August 28, 2005. The cutaway view shows concentric bands of heavy rain (red areas inside the clouds) encircling the eye. Notice that the heaviest rain (largest red area) occurs in the eyewall. The isolated tall cloud tower (in red) in the northern section of the eyewall indicates a cloud top of 16 km (52,000 ft) above the ocean surface. Such tall clouds in the eyewall often indicate that the storm is intensifying.

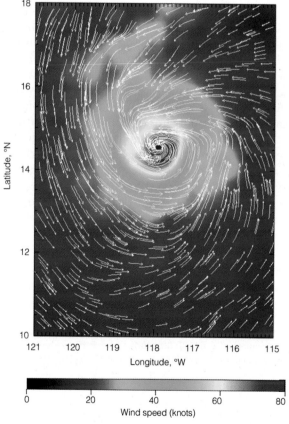

● FIGURE 15.6 Arrows show surface winds spinning counterclockwise around Hurricane Dora situated over the eastern tropical Pacific during August, 1999. Colors indicate surface wind speeds. Notice that winds of 80 knots (92 mi/hr) are encircling the eye (the dark dot in the center). Wind speed and direction obtained from QuikSCAT satellite. (NASA/JPL)

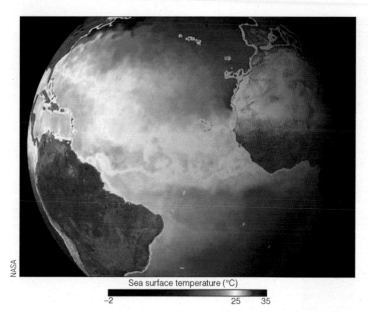

NASA

• FIGURE 15.7 Hurricanes form over warm, tropical waters. This image shows where sea surface temperatures in the tropical Atlantic exceed 28°C (82°F)—warm enough for tropical storm development—during May, 2002.

Sea surface temperature (°C)

−2   25   35

Atlantic and North Pacific oceans during the summer and early fall; hence, the hurricane season normally runs from June through November. • Fig. 15.8 shows the number of tropical storms and hurricanes that formed over the tropical Atlantic during the past 100 years. Notice that hurricane activity picks up in August, peaks in September, then drops off rapidly.

For a mass of unorganized thunderstorms to develop into a hurricane, the surface winds must converge. In the Northern Hemisphere, converging air spins counterclockwise about an area of surface low pressure. Because this type of rotation will not develop on the equator where the Coriolis force is zero (see Chapter 8), hurricanes form in tropical regions, usually between 5° and 20° latitude. (In fact, about

two-thirds of all tropical cyclones form between 10° and 20° of the equator.)

Hurricanes do not form spontaneously, but require some kind of "trigger" to start the air converging. We know, for example, from Chapter 10 that surface winds converge along the intertropical convergence zone (ITCZ). Occasionally, when a wave forms along the ITCZ, an area of low pressure develops, convection becomes organized, and the system grows into a hurricane. Weak convergence also occurs on the eastern side of a tropical wave, where hurricanes have been known to form. In fact, many if not most Atlantic hurricanes can be traced to tropical waves that form over Africa. However, only a small fraction of all of the tropical disturbances that form over the course of a year ever grow into hurricanes. Studies suggest that major Atlantic hurricanes are more numerous when the western part of Africa is relatively wet. Apparently, during the wet years, tropical waves are stronger, better organized, and more likely to develop into strong Atlantic hurricanes.

Convergence of surface winds may also occur along a pre-existing atmospheric disturbance, such as a front that has moved into the tropics from middle latitudes. Although the temperature contrast between the air on both sides of the front is gone, converging winds may still be present so that thunderstorms are able to organize.

Even when all of the surface conditions appear near perfect for the formation of a hurricane (for example, warm water, humid air, converging winds, and so forth), the storm may not develop if the weather conditions aloft are not just right. For instance, in the region of the trade winds, and especially near latitude 20°, the air is often sinking in association with the subtropical high. The sinking air warms and creates an inversion, known as the **trade wind inversion.** When the inversion is strong, it can inhibit the formation of intense thunderstorms and hurricanes. Also, hurricanes do not form where the upper-level winds are strong, creating strong wind shear. Strong wind shear tends to disrupt the

• FIGURE 15.8 The total number of hurricanes and tropical storms (red shade) and hurricanes only (yellow shade) that have formed during the past 100 years in the Atlantic Basin—the Atlantic Ocean, the Caribbean Sea, and the Gulf of Mexico. (NOAA)

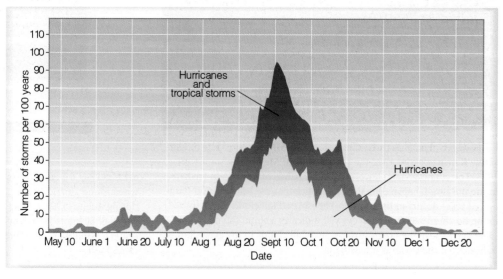

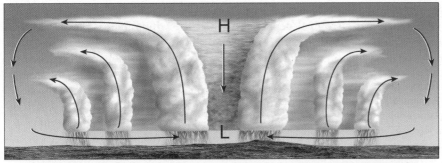

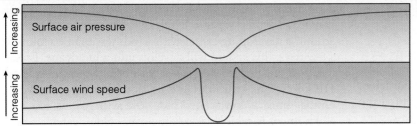

● FIGURE 15.9 The top diagram shows an intensifying tropical cyclone. As latent heat is released inside the clouds, the warming of the air aloft creates an area of high pressure, which induces air to move outward, away from the high. The warming of the air lowers the air density, which in turn lowers the surface air pressure. As surface winds rush in toward the surface low, they extract sensible heat, latent heat, and moisture from the warm ocean. As the warm, moist air flows in toward the center of the storm, it is swept upward into the clouds of the eyewall. As warming continues, surface pressure lowers even more, the storm intensifies, and the winds blow even faster. This situation increases the transfer of heat and moisture from the ocean surface. The middle diagram illustrates how the air pressure drops rapidly as you approach the eye of the storm. The lower diagram shows how surface winds normally reach maximum strength in the region of the eyewall.

organized pattern of convection and disperses heat and moisture, which are necessary for the growth of the storm.

The situation of strong winds aloft typically occurs over the tropical Atlantic during a major El Niño event. As a consequence, during El Niño there are usually fewer Atlantic hurricanes than normal. However, the warmer water of El Niño in the northern tropical Pacific favors the development of hurricanes in that region. During the cold water episode in the tropical Pacific (known as La Niña), winds aloft over the tropical Atlantic usually weaken and become easterly—a condition that favors hurricane development.*

THE DEVELOPING STORM The energy for a hurricane comes from the direct transfer of sensible heat and latent heat from the warm ocean surface. For a hurricane to form, a cluster of thunderstorms must become organized around a central area of surface low pressure. But it is not totally clear how this process occurs. One theory proposed that a hurricane forms in the following manner: Suppose, for example, that the trade wind inversion is weak and that thunderstorms start to organize along the ITCZ, or along a tropical wave. In the deep, moist conditionally unstable environment, a huge amount of latent heat is released inside the clouds during condensation. The process warms the air aloft, causing the temperature near the cluster of thunderstorms to be much higher than the air temperature at the same level farther away. This warming of the air aloft causes a region of higher pressure to form in the upper troposphere (see ● Fig. 15.9). This situation causes a horizontal pressure gradient aloft that induces the air aloft to move outward, away from the region of higher pressure in the anvils of the cumulonimbus clouds. This diverging air aloft, coupled with warming of the vertical air column, causes the surface pressure to drop and a small area of surface low pressure to form. The air now begins to

spin counterclockwise (Northern Hemisphere) and in toward the region of surface low pressure. As the air moves inward, its speed increases, just as ice skaters spin faster as their arms are brought in close to their bodies (the conservation of angular momentum).

As the air moves over the warm water, small swirling eddies transfer heat energy from the ocean surface into the overlying air. The warmer the water and the greater the wind speed, the greater the transfer of sensible and latent heat into the air above. As the air sweeps in toward the center of lower pressure, the rate of heat transfer increases because the wind speed increases. Similarly, the higher wind speed causes greater evaporation rates, and the overlying air becomes nearly saturated. The turbulent eddies then transfer the warm, moist air upward, where the water vapor condenses to fuel new thunderstorms. As the surface air pressure lowers, wind speeds increase, more evaporation occurs at the ocean surface, and thunderstorms become more organized. At the top of the thunderstorms, heat is lost by the clouds radiating infrared energy to space.

The driving force behind a hurricane is similar to that of a heat engine. In a heat engine, heat is taken in at a high temperature, converted into work, then ejected at a low temperature. In a hurricane, heat is taken in near the warm ocean surface, converted to kinetic energy (energy of motion or wind), and lost at its top through radiational cooling.

In a heat engine, the amount of work done is proportional to the difference in temperature between its input and output region. The maximum strength a hurricane can achieve is proportional to the difference in air temperature between the tropopause and the surface, and to the potential for evaporation from the sea surface. As a consequence, the warmer the ocean surface, the lower the minimum pressure of the storm, and the higher its winds. Because there is a limit to how intense the storm can become, peak wind gusts seldom exceed 200 knots.

---

*El Niño and La Niña are covered in Chapter 10 beginning on p. 276.

THE STORM DIES OUT If the hurricane remains over warm water, it may survive for a long time. For example, Hurricane Tina (1992) traveled for thousands of kilometers over deep, warm, tropical waters and maintained hurricane force winds for 24 days, making it one of the longest-lasting North Pacific hurricanes on record. However, most hurricanes last for less than a week.

Hurricanes weaken rapidly when they travel over colder water and lose their heat source. Studies show that if the water beneath the eyewall of the storm (the region of thunderstorms adjacent to the eye) cools by 2.5°C (4.5°F), the storm's energy source is cut off, and the storm will dissipate. Even a small drop in water temperature beneath the eyewall will noticeably weaken the storm. A hurricane can also weaken if the layer of warm water beneath the storm is shallow. In this situation, the strong winds of the storm generate powerful waves that produce turbulence in the ocean water under the storm. Such turbulence creates currents that bring to the surface cooler water from below. If the storm is moving slowly, it is more likely to lose intensity, as the eyewall will remain over the cooler water for a longer period.

Hurricanes also dissipate rapidly when they move over a large landmass. Here, they not only lose their energy source but friction with the land surface causes surface winds to decrease and blow more directly into the storm, an effect that causes the hurricane's central pressure to rise. And a hurricane, or any tropical system for that matter, will rapidly dissipate should it move into a region of strong vertical wind shear.

Our understanding of hurricane behavior is far from complete. However, with the aid of computer model simulations and research projects such as *RAINEX** (*Rain*band and Intensity Change *Ex*periment), scientists are gaining new insight into how tropical cyclones form, intensify, and ultimately die.

HURRICANE STAGES OF DEVELOPMENT Hurricanes go through a set of stages from birth to death. Initially, a *tropical disturbance* shows up as a mass of thunderstorms with only

*The *RAINEX* project consisted of reconnaissance aircraft flying into several hurricanes during the hurricane season of 2005. Equipped with sophisticated scientific instruments, including advanced Dopplar radar, the mission obtained high resolution data on each storm's structure, cloud configuration, and winds.

slight wind circulation. The tropical disturbance becomes a **tropical depression** when the winds increase to between 20 and 34 knots and several closed isobars appear about its center on a surface weather map. When the isobars are packed together and the winds are between 35 and 64 knots, the tropical depression becomes a **tropical storm.** (At this point, the storm gets a name.) The tropical storm is classified as a *hurricane* only when its winds exceed 64 knots (74 miles per hour).

• Figure 15.10 shows four tropical systems in various stages of development. Moving from east to west, we see a weak tropical disturbance (a tropical wave) crossing over Panama. Farther west, a tropical depression is organizing around a developing center with winds less than 25 knots. In a few days, this system will develop into a hurricane. Farther west is a full-fledged hurricane with peak winds in excess of 110 knots. The swirling band of clouds to the northwest is Emilia; once a hurricane (but now with winds less than 40 knots), it is rapidly weakening over colder water.

BRIEF REVIEW

Before reading the next several sections, here is a review of some of the important points about hurricanes.

- Hurricanes are tropical cyclones, comprised of an organized mass of thunderstorms.

- Hurricanes have peak winds about a central core (eye) that exceed 64 knots (74 mi/hr).

- The strongest winds and the heaviest rainfall normally occur in the eyewall—a ring of intense thunderstorms that surround the eye.

- Hurricanes form over warm tropical waters, where light surface winds converge, the humidity is high in a deep layer, and the winds aloft are weak.

- For a mass of thunderstorms to organize into a hurricane there must be some mechanism that triggers the formation, such as converging surface winds along the ITCZ, a pre-existing atmospheric disturbance, such as a weak front from the middle latitudes, or a tropical wave.

- Hurricanes derive their energy from the warm, tropical oceans and by evaporating water from the ocean's surface. Heat energy is converted to wind energy when the water vapor condenses and latent heat is released inside deep convective clouds.

- When hurricanes lose their source of warm water (either by moving over colder water or over a large landmass), they dissipate rapidly.

Up to this point, it is probably apparent that tropical cyclones called hurricanes are similar to middle-latitude cyclones in that, at the surface, both have central cores of low pressure and winds that spiral counterclockwise (in the Northern Hemisphere) about their respective centers. However, there are many differences between the two systems, which are described in the Focus section on p. 419.

FOCUS ON A SPECIAL TOPIC

How Do Hurricanes Compare with Middle-Latitude Storms?

By now, it should be apparent that a hurricane is much different from the mid-latitude cyclone that we discussed in Chapter 12. A hurricane derives its energy from the warm water and the latent heat of condensation, whereas the mid-latitude storm derives its energy from horizontal temperature contrasts. The vertical structure of a hurricane is such that its central column of air is warm from the surface upward; consequently, hurricanes are called *warm-core lows*. A hurricane weakens with height, and the area of low pressure at the surface may actually become an area of high pressure above 12 km (40,000 ft). Mid-latitude cyclones, on the other hand, are *cold-core lows* that usually intensify with increasing height, with a cold upper-level low or trough often existing above, or to the west of the surface low.

A hurricane usually contains an eye where the air is sinking, while mid-latitude cyclones are characterized by centers of rising air. Hurricane winds are strongest near the surface, whereas the strongest winds of the mid-latitude cyclone are found aloft in the jet stream.

Further contrasts can be seen on a surface weather map. Figure 1 shows Hurricane Rita over the Gulf of Mexico and a mid-latitude storm north of New England. Around the hurricane, the isobars are more circular, the pressure gradient is much steeper, and the winds are stronger. The hurricane has no fronts and is smaller (although Rita is a large Category 5 hurricane). There are similarities between the two systems: Both are areas of surface low pressure, with winds moving counterclockwise about their respective centers.

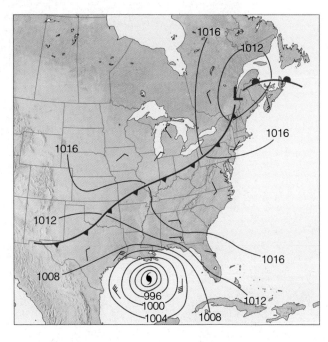

● FIGURE I Surface weather map for the morning of September 23, 2005, showing Hurricane Rita over the Gulf of Mexico and a middle-latitude storm system north of New England.

It is interesting to note that some northeasters (winter storms that move northeastward along the coastline of North America, bringing with them heavy precipitation, high surf, and strong winds) may actually possess some of the characteristics of a hurricane. For example, a particularly powerful northeaster during January, 1989, was observed to have a cloud-free eye, with surface winds in excess of 85 knots spinning about a warm inner core. Moreover, some *polar lows* — lows that develop over polar waters during winter — may exhibit many of the observed characteristics of a hurricane, such as a symmetric band of thunderstorms spiraling inward around a cloud-free eye, a

warm-core area of low pressure, and strong winds near the storm's center. In fact, when surface winds within these polar storms reach 58 knots, they are sometimes referred to as *Arctic hurricanes*.

Even though hurricanes weaken rapidly as they move inland, their circulation may draw in air with contrasting properties. If the hurricane links with an upper-level trough, it may actually become a mid-latitude cyclone. Swept eastward by upper-level winds, the remnants of an Atlantic hurricane can become an intense mid-latitude autumn storm in Europe.

HURRICANE MOVEMENT ● Figure 15.11 shows where most hurricanes are born and the general direction in which they move. Notice that hurricanes that form over the warm, tropical North Pacific and North Atlantic are steered by easterly winds and move west or northwestward at about 10 knots for a week or so. Gradually, they swing poleward around the subtropical high, and when they move far enough north, they become caught in the westerly flow, which curves them to the north or northeast. In the middle latitudes, the hurricane's forward speed normally increases, sometimes to

more than 50 knots. The actual path of a hurricane (which appears to be determined by the structure of the storm and the storm's interaction with the environment) may vary considerably. Some take erratic paths and make odd turns that occasionally catch weather forecasters by surprise (see ● Fig. 15.12). There have been many instances where a storm heading directly for land suddenly veered away and spared the region from almost certain disaster. As a case in point, Hurricane Elena, with peak winds of 90 knots, moved northwestward into the Gulf of Mexico on August 29, 1985. It then

● FIGURE 15.10 Visible satellite image showing four tropical systems, each in a different stage of its life cycle.

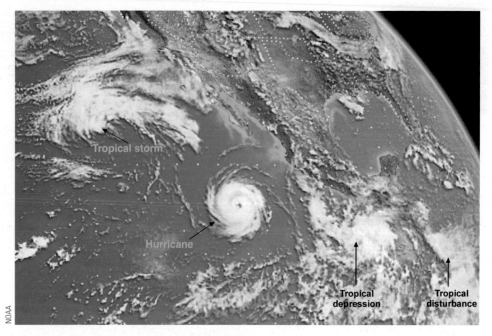

● FIGURE 15.11 Regions where tropical storms form (orange shading), the names given to storms, and the typical paths they take (red arrows).

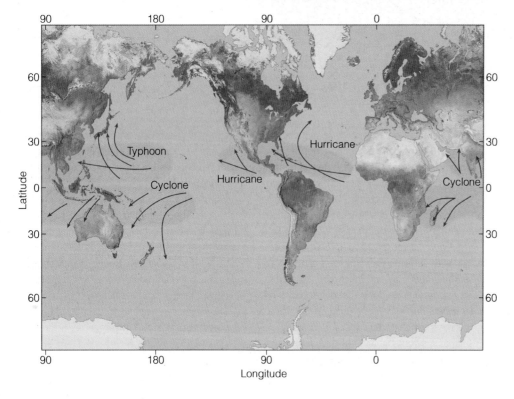

veered eastward toward the west coast of Florida. After stalling offshore, it headed northwest. After weakening, it then moved onshore near Biloxi, Mississippi, on the morning of September 2.

Look at Fig. 15.11 and notice there that it appears as if hurricanes do not form over the South Atlantic and the eastern South Pacific—directly east and west of South America. Cooler water, vertical wind shear, and the unfavorable position of the ITCZ discourages hurricanes from developing in these regions. Then, guess what? For the first time since satel-

lites began observing the south Atlantic, a tropical cyclone formed off the coast of Brazil during March, 2004 (see ● Fig. 15.13). So rare are tropical cyclones in this region that no government agency has an effective warning system for them, which is why the tropical cyclone was not given a name.

As we saw in an earlier section, many hurricanes form off the coast of Mexico over the North Pacific. In fact, this area usually spawns about nine hurricanes each year, which is slightly more than the yearly average of six storms born over the tropical North Atlantic. We can see in Fig. 15.11 that east-

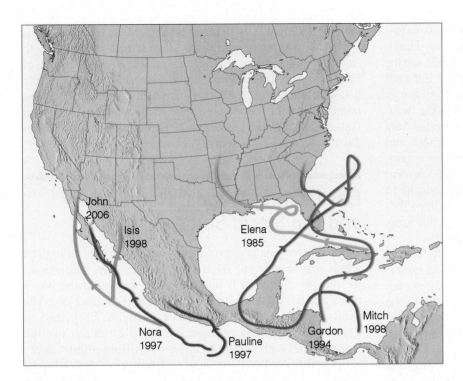

● FIGURE 15.12 Some erratic paths taken by hurricanes.

ern North Pacific hurricanes normally move westward, away from the coast, and so little is heard about them. When one does move northwestward, it normally weakens rapidly over the cool water of the North Pacific. Occasionally, however, one will curve northward or even northeastward and slam into Mexico, causing destructive flooding. Hurricane Tico left 25,000 people homeless and caused an estimated $66 million in property damage after passing over Mazatlán, Mexico, in October, 1983. The remains of Tico even produced record rains and flooding in Texas and Oklahoma. Even less frequently, a hurricane will stray far enough north to bring

summer rains to southern California and Arizona, as did the remains of Hurricane Nora during September, 1997. (Nora's path is shown in Fig. 15.12.) The only hurricane on record to reach the west coast of the United States with sustained hurricane-force winds did so in October, 1858, when a hurricane slammed into the extreme southern part of California near San Diego.

The Hawaiian Islands, which are situated in the central North Pacific between about 20° and 23°N, appear to be in the direct path of many eastern Pacific hurricanes and tropical storms. By the time most of these storms reach the islands,

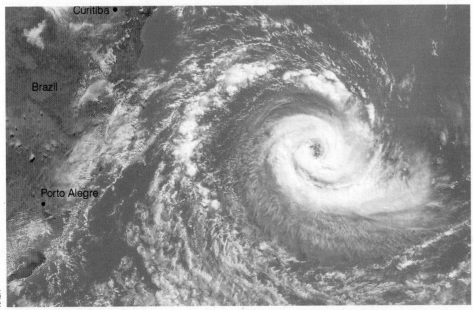

● FIGURE 15.13 An extremely rare tropical cyclone (with no name) near 28°S latitude spins clockwise over the south Atlantic off the coast of Brazil during March, 2004. Due to cool water and vertical wind shear, storms rarely form in this region of the Atlantic Ocean. In fact, this is the only tropical storm ever officially reported there.

however, they have weakened considerably, and pass harmlessly to the south or northeast. The exceptions were Hurricane Iwa during November, 1982, and Hurricane Iniki during September, 1992. Iwa lashed part of Hawaii with 100-knot winds and huge surf, causing an estimated $312 million in damages. Iniki, the worst hurricane to hit Hawaii in the twentieth century, battered the island of Kauai with torrential rain, sustained winds of 114 knots that gusted to 140 knots, and 20-foot waves that crashed over coastal highways. Major damage was sustained by most of the hotels and about 50 percent of the homes on the island. Iniki (the costliest hurricane in Hawaiian history with damage estimates of $1.8 billion) flattened sugarcane fields, destroyed the macadamia nut crop, injured about 100 people, and caused at least 7 deaths.

Hurricanes that form over the tropical North Atlantic also move westward or northwestward on a collision course with Central or North America. Most hurricanes, however, swing away from land and move northward, parallel to the coastline of the United States. A few storms, perhaps three per year, move inland, bringing with them high winds, huge waves, and torrential rain that may last for days.

A hurricane moving northward over the Atlantic will normally survive as a hurricane for a much longer time than will its counterpart at the same latitude over the eastern Pacific. The reason for this situation is that an Atlantic hurricane moving northward will usually stay over warmer water, whereas an eastern Pacific hurricane heading north will quickly move over much cooler water and, with its energy source cut off, will rapidly weaken.

Naming Hurricanes and Tropical Storms

In an earlier section, we learned that hurricanes are given a name when they reach tropical storm strength. Before hurricanes and tropical storms were assigned names, they were identified according to their latitude and longitude. This method was confusing, especially when two or more storms were present over the same ocean. To reduce the confusion, hurricanes were identified by letters of the alphabet. During World War II, names like Able and Baker were used. (These names correspond to the radio code words associated with each letter of the alphabet). This method also seemed cumbersome so, beginning in 1953, the National Weather Service began using female names to identify hurricanes. The list of names for each year was in alphabetical order, so that the names of the season's first storm began with the letter *A*, the second with *B*, and so on.

From 1953 to 1977, only female names were used. However, beginning in 1978, tropical storms in the eastern Pacific were alternately assigned female and male names, but not just English names, as Spanish and French ones were used too. This practice began for North Atlantic hurricanes in 1979. If a storm causes great damage and becomes infamous as a Cat-

egory 3 or higher, its name is retired for at least ten years. ▼ Table 15.1 gives the proposed list of names for both North Atlantic and eastern Pacific hurricanes. If the number of named storms in any year should exceed the names on the list, as occurred in 2005, then tropical storms are assigned names from the Greek alphabet, such as Alpha, Beta, and Gamma. In fact, the last of the 27 named tropical systems in 2005 was Zeta, which actually formed during January, 2006.

Devastating Winds, Flooding, and the Storm Surge

When a hurricane is approaching from the south, its highest winds are usually on its eastern (right) side. The reason for this phenomenon is that the winds that push the storm along add to the winds on the east side and subtract from the winds on the west (left) side. The hurricane illustrated in ●Fig. 15.14 is moving northward along the east coast of the United States with winds of 100 knots swirling counterclockwise about its center. Because the storm is moving northward at about 25 knots, sustained winds on its eastern side are about 125 knots, while on its western side, winds are only 75 knots.

Even though the hurricane in Fig. 15.14 is moving northward, there is a net transport of water directed eastward toward the coast. To understand this behavior, recall from Chapter 10 that as the wind blows over open water, the water beneath is set

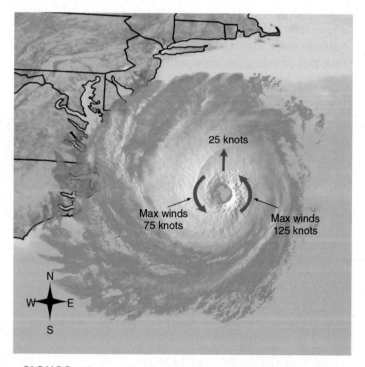

● FIGURE 15.14 A hurricane moving northward will have higher sustained winds on its eastern side than on its western side. If the hurricane moves from east to west, highest sustained winds will be on its northern side.

▼ TABLE 15.1   Names of Hurricanes and Tropical Storms

NORTH ATLANTIC HURRICANE NAMES				EASTERN NORTH PACIFIC HURRICANE NAMES			
2008	2009	2010	2011	2008	2009	2010	2011
Arthur	Ana	Alex	Arlene	Alma	Andres	Agatha	Adrian
Bertha	Bill	Bonnie	Bret	Boris	Blanca	Blas	Beatriz
Cristobal	Claudette	Colin	Cindy	Cristina	Carlos	Celia	Calvin
Dolly	Danny	Danielle	Don	Douglas	Dolores	Darby	Dora
Edouard	Erika	Earl	Emily	Elida	Enrique	Estelle	Eugene
Fay	Fred	Fiona	Franklin	Fausto	Felicia	Frank	Fernanda
Gustav	Grace	Gaston	Gert	Genevieve	Guillermo	Georgette	Greg
Hanna	Henri	Hermine	Harvey	Heman	Hilda	Howard	Hilary
Ike	Ida	Igor	Irene	Iselle	Ignacio	Isis	Irwin
Josephine	Joaquin	Julia	Jose	Julio	Jimena	Javier	Jova
Kyle	Kate	Karl	Katia	Karina	Kevin	Kay	Kenneth
Laura	Larry	Lisa	Lee	Lowell	Linda	Lester	Lidia
Marco	Mindy	Matthew	Maria	Marie	Marty	Madeline	Max
Nana	Nicholas	Nicole	Nate	Norbert	Nora	Newton	Norma
Omar	Odette	Otto	Ophelia	Odile	Olaf	Orlene	Otis
Paloma	Peter	Paula	Philippe	Polo	Patricia	Paine	Pilar
Rene	Rose	Richard	Rina	Rachel	Rick	Roslyn	Ramon
Sally	Sam	Shary	Sean	Simon	Sandra	Seymour	Selma
Teddy	Teresa	Tomas	Tammy	Trudy	Terry	Tina	Todd
Vicky	Victor	Virginie	Vince	Vance	Vivian	Virgil	Vicente
Wilfred	Wanda	Walter	Whitney	Winnie	Waldo	Winifred	Willa
				Xavier	Xina	Xavier	Xavier
				Yolanda	York	Yolanda	Yolanda
				Zeke	Zelda	Zeke	Zeke

in motion. If we imagine the top layer of water to be broken into a series of layers, then we find each layer moving to the *right* of the layer above (Northern Hemisphere). This type of movement (bending) of water with depth (called the *Ekman Spiral*) causes a net transport of water (known as *Ekman transport*) to the right of the surface wind in the Northern Hemisphere. Hence, the north wind on the hurricane's left (western) side causes a net transport of water toward the shore. Here, the water piles up and rapidly inundates the region.

The high winds of a hurricane also generate large waves, sometimes 10 to 15 m (33 to 49 ft) high. These waves move outward, away from the storm, in the form of *swells* that carry the storm's energy to distant beaches. Consequently, the effects of the storm may be felt days before the hurricane arrives.

Although the hurricane's high winds inflict a great deal of damage, it is the huge waves, high seas, and *flooding* that normally cause most of the destruction. The flooding is also responsible for the loss of many lives. In fact, the majority of hurricane-related deaths during the past century has been due to flooding. The flooding is due, in part, to winds pushing water onto the shore and to the heavy rains, which may exceed 63 cm (25 in.) in 24 hours.* Flooding is also aided by

---

*Hurricanes may sometimes have a beneficial aspect, in the sense that they can provide much needed rainfall in drought-stricken areas.

• FIGURE 15.15 When a storm surge moves in at high tide it can inundate and destroy a wide swath of coastal lowlands.

the low pressure of the storm. The region of low pressure allows the ocean level to rise (perhaps half a meter), much like a soft drink rises up a straw as air is withdrawn. (A drop of one millibar in air pressure produces a rise in ocean levels of one centimeter.) The combined effect of high water (which is usually well above the high-tide level), high winds, and the net Ekman transport toward the coast, produces the **storm surge** — an abnormal rise of several meters in the ocean level — which inundates low-lying areas and turns beachfront homes into piles of splinters (see • Fig. 15.15). The storm surge is particularly damaging when it coincides with normal high tides. Flooding, however, is not just associated with hurricanes, as destructive floods can occur with tropical storms that do not reach hurricane strength. An example of such a storm, tropical storm Allison, which inundated parts of Texas during June, 2001, is given in the Focus section on p. 425.

In an effort to estimate the possible damage a hurricane's sustained winds and storm surge could do to a coastal area, the

Saffir-Simpson scale was developed (see ▼ Table 15.2). The scale numbers (which range from 1 to 5) are based on actual conditions at some time during the life of the storm. As the hurricane intensifies or weakens, the category, or scale number, is reassessed accordingly. Major hurricanes are classified as Category 3 and above. In the western Pacific, a typhoon with sustained winds of at least 130 knots (150 mi/hr) — at the upper end of the wind speed range in Category 4 on the Saffir-Simpson scale — is called a **super typhoon**. • Figure 15.16 illustrates how the storm surge changes along the coast as hurricanes with increasing intensity make landfall. *

• Figure 15.17 shows the number of hurricanes that have made landfall along the coastline of the United States from 1900 through 2007. Out of a total of 181 hurricanes striking the American coastline, 73 (40 percent) were major hurricanes —

---

*Landfall is the position along the coast where the center of a hurricane passes from ocean to land.

Normal high tide   Category 1 [4-foot rise]   Category 3 [12-foot rise]   Category 5 [20-foot rise]

• FIGURE 15.16 The changing of the ocean level as different category hurricanes make landfall along the coast. The water typically rises about 4 feet with a Category 1 hurricane, but may rise to 22 feet (or more) with a Category 5 storm.

FOCUS ON A SPECIAL TOPIC

A Tropical Storm Named Allison

In late May, 2001, Allison began as a tropical wave that moved westward across the Atlantic. The wave continued its westward journey, and by the first of June it had moved across Central America and out over the Pacific Ocean. Here, it organized into a band of thunderstorms and a tropical depression. Upper-level winds guided the depression northward over Central America, then out over the Gulf of Mexico, where the warm water fueled the circulation; and just east of Galveston, Texas, the depression became tropical storm Allison. Packing winds of 53 knots, Allison made landfall over the east end of Galveston Island on June 5. It drifted inland and weakened (see Fig. 2).

On the eastern side of the storm, heavy rain fell over parts of Texas and Louisiana. Some areas of southeast Texas received as much as 25 cm (10 in.) of rain in less than five hours. Homes, streets, and highways flooded as heavy rain continued to pound the area. But the worst was yet to come.

On June 7, as the upper-level winds began to change, the remnants of Allison drifted southwestward toward Houston. Heavy rain fell over southeast Texas and Louisiana, where several tornadoes touched down. Over the Houston area, more than 50 cm (20 in.) of rain fell within a 12-hour period, submerging a vast part of the city. In six days the Port of Houston received a staggering 94 cm (which is over 3 ft) of rain.

The center of circulation drifted southward, moving off the Texas coast and out over the Gulf of Mexico on the evening of June 9. The flow aloft then guided the storm northeastward, where the storm made landfall again, but this time in southeastern Louisiana. Heavy rain continued to pound Louisiana, creating one of the worst floods on record—a station in

● FIGURE 2 Visible satellite image showing the remains of tropical storm Allison centered over Texas on the morning of June 6, 2001. Heavy rain is falling from the thick clouds over Louisiana and eastern Texas.

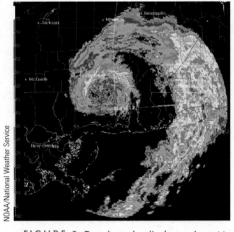

● FIGURE 3 Doppler radar display on June 11, 2001, showing bands of heavy rain swirling counterclockwise into the center of once tropical storm Allison. The center of the storm, which is over Mississippi, has actually deepened and formed somewhat of an eye.

southern Louisiana reported a rainfall total of 76 cm (30 in.). On June 11, a zone of maximum winds aloft (a jet streak) associated with the subtropical jet stream enhanced the outflow above the surface storm, and the remains of tropical storm Allison actually began to intensify over land. As the storm entered Mississippi, its central pressure lowered, wind gusts reached 52 knots, and the center of circulation developed a weak-looking eye (see Fig. 3). As the system trekked eastward, it weakened and lost its eye, but continued to dump heavy rain over the southern Gulf States. Eventually, on June 14, the storm reached the Carolina coast.

Unfortunately, the storm slowed, then turned northward over North Carolina. Flooding became a major problem—Doppler radar estimated that up to 53 cm (21 in.) of rain had fallen over parts of the state. Severe weather broke out in Georgia and in the Carolinas, where some areas reported hail and downed trees due to gusty winds. The storm moved northeastward, parallel to the coast. A cold front moving in from the west eventually hooked up with the moisture from Allison. This situation caused heavy rain to fall over the mid-Atlantic states and southern New England. The storm finally accelerated to the northeast, away from the coast on June 18.

Allison, which never developed hurricane strength winds, claimed the lives of 43 people, whose deaths were mainly due to flooding. The total damage from the storm totaled in the billions of dollars, with the Houston area alone sustaining over $2 billion in damage. If all the rain that fell from Allison could be placed in Texas, it would cover two-thirds of the state with water a foot deep.

Category 3 or higher. Hence, along the Gulf and Atlantic coasts, on the average, about five hurricanes make landfall every three years, two of which are major hurricanes with winds in excess of 95 knots (110 mi/hr) and a storm surge exceeding 2.5 m (8 ft).

Although the high winds of a hurricane can devastate a region, considerable damage may also occur from hurricane-spawned tornadoes. About one-fourth of the hurricanes that strike the United States produce tornadoes. In fact, in 2004 six tropical systems produced just over 300 tornadoes in the

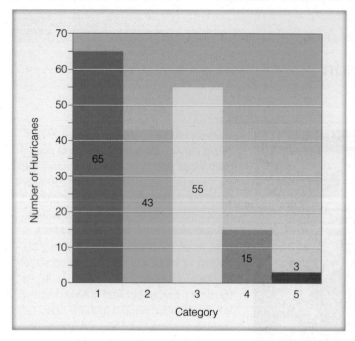

● FIGURE 15.17 The number of hurricanes (by each category) that made landfall along the coastline of the United States from 1900 through 2007. All of the hurricanes struck the Gulf or Atlantic coasts. Categories 3, 4, and 5 are considered major hurricanes.

southern and eastern United States. The exact mechanism by which these tornadoes form is not totally clear; however, studies suggest that surface topography may play a role by initiating the convergence (and hence, rising) of surface air. Moreover, tornadoes tend to form in the right front quadrant of an advancing hurricane,* where vertical wind speed shear is greatest. Studies also suggest that swathlike areas of extreme damage once attributed to tornadoes may actually be due to strong downdrafts (microbursts) associated with the large, intense thunderstorms around the eyewall.

In examining the extensive damage wrought by Hurricane Andrew during August, 1992, researchers theorized that the areas of most severe damage might have been caused by small whirling eddies perhaps 30 to 100 meters in diameter that occur in narrow bands. Many scientists today believe those rapidly rotating eddies were, in fact, small tornadoes. Lasting for about 10 seconds, the vortices appeared to have formed in a region of strong wind speed shear in the hurricane's eyewall, where the air was rapidly rising. As intense updrafts stretched the vortices vertically, they shrank horizontally, which induced them to spin faster, perhaps as fast as 70 knots. When the rotational winds of a vortice are added to the hurricane's steady wind, the total wind speed over a relatively small area may in-

*In the northeast quadrant of the hurricane shown in Fig. 15.14, on p. 422.

▼ TABLE 15.2 Saffir-Simpson Hurricane Damage-Potential Scale

SCALE NUMBER (CATEGORY)	CENTRAL PRESSURE		WINDS		STORM SURGE		DAMAGE
	mb	in.	mi/hr	knots	ft	m	
1	≥980*	≥28.94	74–95	64–82	4–5	~1.5	Damage mainly to trees, shrubbery, and unanchored mobile homes
2	965–979	28.50–28.91	96–110	83–95	6–8	~2.0–2.5	Some trees blown down; major damage to exposed mobile homes; some damage to roofs of buildings
3	945–964	27.91–28.47	111–130	96–113	9–12	~2.5–4.0	Foliage removed from trees; large trees blown down; mobile homes destroyed; some structural damage to small buildings
4	920–944	27.17–27.88	131–155	114–135	13–18	~4.0–5.5	All signs blown down; extensive damage to roofs, windows, and doors; complete destruction of mobile homes; flooding inland as far as 10 km (6 mi); major damage to lower floors of structures near shore
5	<920	<27.17	>155	>135	>18	>5.5	Severe damage to windows and doors; extensive damage to roofs of homes and industrial buildings; small buildings overturned and blown away; major damage to lower floors of all structures less than 4.5 m (15 ft) above sea level within 500 m of shore

*Symbol > means "greater than"; < means "less than"; ≥ means "equal to or greater than"; ~ means "approximately equal to."

crease substantially. In the case of Hurricane Andrew, isolated wind speeds may have reached 174 knots (200 mi/hr) over narrow stretches of south Florida.

Up until 2005, the annual death toll from hurricanes in the United States, over a span of about 30 years, averaged less than 50 persons.* Most of these fatalities were due to flooding. This relatively low total was due in part to the advanced warning provided by the National Weather Service and the fact that only a few really intense storms had made landfall during this time. But the hurricane death toll in the United States rose dramatically in 2005 when Hurricane Katrina slammed into Mississippi and Louisiana.

As Hurricane Katrina moved toward the coast, evacuation orders were given to residents living in low-lying areas, including the city of New Orleans. Many thousands of people moved to higher ground but, unfortunately, many people either refused to leave their homes or had no means of leaving, and were forced to ride out the storm. Tragically, more than 1300 people died either from Katrina's huge storm surge and high winds that demolished countless buildings, or from the flooding in New Orleans, when several levees broke and parts of the city were inundated with water over 20 feet deep.

---

*In other countries, the annual death toll was considerably higher. Estimates are that more than 3000 people died in Haiti from flooding and mud slides when Hurricane Jeanne moved through the Caribbean during September, 2004.

WEATHER WATCH

Are storm surges and tsunamis the same? Although there are similarities between the two, they are actually quite different. A storm surge is an onshore surge of ocean water caused primarily by the winds of a storm (most often a tropical cyclone) pushing sea water onto the coast. Tsunamis are waves generated by disturbances on the ocean floor caused most commonly by earthquakes. As the tsunami wave (or series of waves) moves into shallow water, it builds in height and rushes onto the land (sometimes unexpectedly), swooping up everything in its path, including cars, buildings, and people.

As the population density countinues to increase in vulnerable coastal areas, the potential for another hurricane-caused disaster increases also.

Some Notable Hurricanes

CAMILLE, 1969   Hurricane Camille (1969) stands out as one of the most intense hurricanes to reach the coastline of the United States during the twentieth century (see ▼ Table 15.3). With a central pressure of 909 mb, tempestuous winds reaching 160 knots (184 mi/hr) and a storm surge more than 7 m

▼ TABLE 15.3   The Thirteen Most Intense Hurricanes (at Landfall) to Strike the United States from 1900 through 2007

RANK	HURRICANE (MADE LANDFALL)	YEAR	CENTRAL PRESSURE (MILLIBARS/INCHES)	CATEGORY	DEATH TOLL
1	Florida (Keys)	1935	892/26.35	5	408
2	Camille (Mississippi)	1969	909/26.85	5	256
3	Andrew (South Florida)	1992	922/27.23	5	53
4	Katrina (Louisiana)	2005	920/27.17	3*	>1300
5	Florida (Keys)/South Texas	1919	927/27.37	4	>600†
6	Florida (Lake Okeechobee)	1928	929/27.43	4	1836
7	Donna (Long Island, NY)	1960	930/27.46	4	50
8	Texas (Galveston)	1900	931/27.49	4	>6000
9	Louisiana (Grand Isle)	1909	931/27.49	4	350
10	Louisiana (New Orleans)	1915	931/27.49	4	275
11	Carla (South Texas)	1961	931/27.49	4	46
12	Hugo (South Carolina)	1989	934/27.58	4	49
13	Florida (Miami)	1926	935/27.61	4	243

*Although the central pressure in Katrina's eye was quite low, Katrina's maximum sustained winds of 110 knots at landfall made it a Category 3 storm.

†More than 500 of this total were lost at sea on ships. (The > symbol means "greater than.")

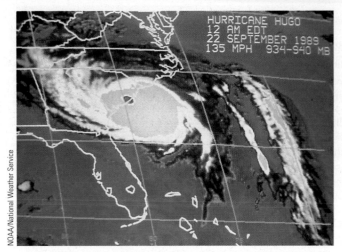

● FIGURE 15.18 A color-enhanced infrared satellite image of Hurricane Hugo with its eye over the coast near Charleston, South Carolina.

(23 ft) above the normal high-tide level, Camille, as a Category 5 storm, unleashed its fury on Mississippi, destroying thousands of buildings. During its rampage, it caused an estimated $1.5 billion in property damage and took more than 200 lives.

HUGO, 1989   During September, 1989, Hurricane Hugo, born as a cluster of thunderstorms, became a tropical depression off the coast of Africa, southeast of the Cape Verde Islands. The storm grew in intensity, tracked westward for several days, then turned northwestward, striking the island of St. Croix with sustained winds of 125 knots. After passing over the eastern tip of Puerto Rico, this large, powerful hurricane took aim at the coastline of South Carolina. With maximum winds estimated at about 120 knots (138 mi/hr), and a central pressure near 934 mb, Hugo made landfall as a Category 4 hurricane near Charleston, South Carolina, about midnight on September 21 (see ● Fig. 15.18). The high winds and storm surge, which ranged between 2.5 and 6 m (8 and 20 ft), hurled a thundering wall of water against the shore. This knocked out power, flooded streets, and caused widespread destruction to coastal communities. The total damage in the United States attributed to Hugo was over $7 billion, with a death toll of 21 in the United States and 49 overall.

ANDREW, 1992   Another devastating hurricane during the twentieth century was Hurricane Andrew. On August 21, 1992, as tropical storm Andrew churned westward across the Atlantic it began to weaken, prompting some forecasters to surmise that this tropical storm would never grow to hurricane strength. But Andrew moved into a region favorable for hurricane development. Even though it was outside the tropics near latitude 25°N, warm surface water and weak winds aloft allowed Andrew to intensify rapidly. And in just two days Andrew's winds increased from 45 knots to 122 knots, turning an average tropical storm into one of the most intense hurricanes to strike Florida in the past 105 years (see Table 15.3).

With winds of at least 130 knots (155 mi/hr) and a powerful storm surge, Andrew made landfall south of Miami on the morning of August 24 (see ● Fig. 15.19). The eye of the

● FIGURE 15.19 Color radar image of Hurricane Andrew as it moves onshore over south Florida on the morning of August 24, 1992. The National Hurricane Center (NHC) is located about 30 km (19 mi) from the center of the eye.

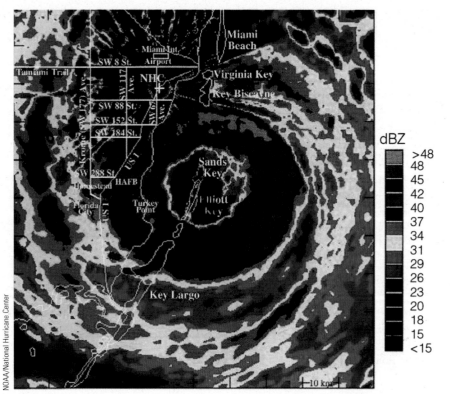

● FIGURE 15.20 A community in Homestead, Florida, devastated by Hurricane Andrew during August, 1992.

storm moved over Homestead, Florida. Andrew's fierce winds completely devastated the area (see ● Fig. 15.20), as 50,000 homes were destroyed, trees were leveled, and steel-reinforced tie beams weighing tons were torn free of townhouses and hurled as far as several blocks. Swaths of severe damage led scientists to postulate that peak winds may have approached 174 knots (200 mi/hr). Such winds may have occurred with small tornadoes, which added substantially to the storm's wind speed. In an instant, a wind gust of 142 knots (164 mi/hr) blew down a radar dome and inactivated several satellite dishes on the roof of the National Hurricane Center in Coral Gables. Observations reveal that some of Andrew's destruction may have been caused by microbursts in the severe thunderstorms of the eyewall. The hurricane roared westward across southern Florida, weakened slightly, then regained strength over the warm Gulf of Mexico. Surging northwestward, Andrew slammed into Louisiana as a Category 3 with 120-knot winds on the evening of August 25.

All told, Hurricane Andrew was one of the costliest natural disasters ever to hit the United States. It destroyed or damaged over 200,000 homes and businesses, left more than 160,000 people homeless, caused over $30 billion in damages, and took 53 lives, including 41 in Florida.

IVAN, 2004 Hurricane Ivan was an interesting but costly hurricane. It moved onshore just west of Gulf Shores, Alabama, on September 15, 2004 (see ● Fig. 15.21) as a strong Category 3 hurricane with winds 105 knots and a storm surge of about 5 meters (16 feet). The strongest winds and greatest damage occurred over an area near the border between Ala-

bama and Florida (see ● Fig. 15.22). As Ivan moved inland, it weakened and eventually linked up with a mid-latitude low. The remains of Ivan then split from the low and drifted southward, eventually ending up in the Gulf of Mexico, where it regained tropical storm strength. It made landfall for the second time along the Gulf Coast, but this time as a

● FIGURE 15.21 Visible satellite image of Hurricane Ivan as it makes landfall near Gulf Shores, Alabama, on September 15, 2004. Ivan is a major hurricane with winds of 105 knots (121 mi/hr) and a surface air pressure of 945 mb (27.91 in.).

(a)

(b)

● FIGURE 15.22 Beach homes along the Gulf Coast at Orange Beach, Alabama (a) before, and (b) after Hurricane Ivan made landfall during September, 2004. (Red arrows are for reference.)

tropical depression. All told, Ivan took 26 lives in the United States, produced a record 123 tornadoes over the southern and eastern states, and caused an estimated $14 billion in damages. (Ivan was one of five hurricanes to make landfall in the United States during 2004. Out of the five hurricanes that hit the United States, four impacted the state of Florida. More information on the record-setting Atlantic hurricane seasons of 2004 and 2005 is given in the Focus section on p. 431.)

KATRINA, 2005 Hurricane Katrina was the most costly hurricane to ever hit the United States. Forming over warm tropical water south of Nassau in the Bahamas, Katrina became a tropical storm on August 24, 2005, and a Category 1 hurricane just before making landfall in south Florida on August 25. (Katrina's path is given in Fig. 4, on p. 431.) It moved southwestward across Florida and out over the eastern Gulf of Mexico. As Katrina moved westward, it passed over a deep band of warm water called the *Loop current* that allowed Katrina to rapidly intensify. Within 12 hours, the hurricane increased from a Category 3 to a Category 5 storm with winds of 175 mi/hr and a central pressure of 902 mb (see the opening photo of Katrina on p. 410.)

Over the Gulf of Mexico, Katrina gradually turned northward toward Mississippi and Louisiana. As the powerful Category 5 hurricane moved slowly toward the coast, its rainbands near the center of the storm began to converge toward the storm's eye. This process cut off moisture to the eyewall. As the old eyewall dissipated, a new one formed farther away in a phenomenon meteorologists call **eyewall replacement**. The replacement of the eyewall weakened the storm such that Katrina made landfall near Buras, Louisiana, on August 29 (see ● Fig 15.23) as a strong Category 3 hurri-

cane with sustained winds of 110 knots (127 mi/hr), a central pressure of 920 mb, and a storm surge between 6 m and 9 m (20 and 30 ft).

Katrina's strong winds and high storm surge on its eastern side devastated southern Mississippi, with Biloxi, Gulf-

New Orleans ──→ X ──── Eye

● FIGURE 15.23 Hurricane Katrina just after making landfall along the Mississippi/Louisiana coast on the morning of August 29, 2005. Shown here, the storm is moving north with its eye due east of New Orleans. At landfall, Katrina had sustained winds of 110 knots, a central pressure of 920 mb (27.17 in.), and a storm surge over 20 feet.

The Record-Setting Atlantic Hurricane Seasons of 2004 and 2005

Both 2004 and 2005 were active years for hurricane development over the tropical North Atlantic. During 2004, nine storms became full-fledged hurricanes. Out of the five hurricanes that made landfall in the United States, three (Charley, Frances, and Jeanne) plowed through Florida, and one (Ivan) came onshore just west of the Florida panhandle (see Fig. 4), making this the first time since record-keeping began in 1861 that four hurricanes have impacted the state of Florida in one year. Total damage in the United States from the five hurricanes exceeded $40 billion.

Then, in 2005, a record twenty-seven named storms developed (the most in a single season), of which fifteen (another record) reached hurricane strength. The 2005 Atlantic hurricane season also had four hurricanes (Emily, Katrina, Rita, and Wilma) reach Category 5 intensity for the first time since reliable record-keeping began. And hurricane Wilma had the lowest central pressure ever measured in an Atlantic hurricane—882 mb (26.04 in.). Out of five hurricanes that made landfall in the United States, three (Dennis, Katrina, and Wilma) made landfall in hurricane-wary Florida and one (Ophelia) skirted northward along Florida's east

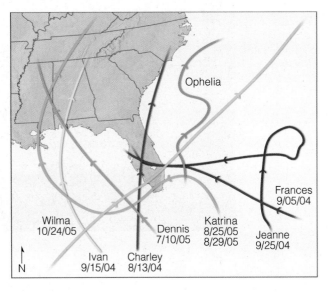

• FIGURE 4 The paths of eight hurricanes that impacted Florida during 2004 and 2005. Notice that in 2004 hurricanes Frances and Jeanne made landfall at just about the same spot along Florida's southeast coast. The date under the hurricane's name indicates the date the hurricane made landfall.

coast, giving Florida the dubious distinction of being the only state on record to experience eight hurricanes during the span of sixteen months (Fig. 4). Total damage in the United States from the five hurricanes that made landfall exceeded $100 billion.

Apparently, in 2004 and in 2005, very warm ocean water and weak vertical wind shear provided favorable conditions for hurricane development. In previous years, winds associated with a persistent upper-level trough over the eastern United States steered many tropical systems away from the coast before they could make landfall. However, in 2004 and in 2005, an area of high pressure replaced the trough, and winds tended to steer tropical cyclones on a more westerly track, toward the coastline of North America.

port, and Pass Christian being particularly hard hit (see •Fig. 15.24). The winds demolished all but the strongest structures, and the huge storm surge scoured areas up to 6 km (10 mi) inland.

New Orleans and the surrounding parishes actually escaped the brunt of Katrina's winds, as the eye passed just to the east of the city (Fig. 15.23). However, the combination of high winds, large waves, and a huge storm surge caused disastrous breaches in the levee system that protects New Orleans from the Mississippi River, the Gulf of Mexico, and Lake Pontchartrain. When the levees gave way, water up to 20 feet deep invaded a large part of the city, tragically before thousands of people could escape (see •Fig. 15.25). Less than a month later, powerful Hurricane Rita with sustained winds of 152 knots (175 mi/hr) moved over the Gulf of Mexico, south of New Orleans. Strong, tropical storm–force easterly winds, along with another storm surge, caused some of the repaired levees to break again, flooding parts of the city that just days earlier had been pumped dry. The death toll due to Hurri-

cane Katrina climbed to more than 1200, and the devastation wrought by the storm totaled more than $75 billion. Although Katrina may well be the most expensive hurricane on record, tragically it is not the deadliest.

OTHER DEVASTATING HURRICANES Before the era of satellites and radar, catastrophic losses of life had occurred. In 1900, more than 6000 people lost their lives when a hurricane slammed into Galveston, Texas, with a huge storm surge. (Look back at Table 15.3) Most of the deaths occurred in the low-lying coastal regions as flood waters pushed inland. In October, 1893, nearly 2000 people perished on the Gulf Coast of Louisiana as a giant storm surge swept that region. Spectacular losses are not confined to the Gulf Coast. Nearly 1000 people lost their lives in Charleston, South Carolina, during August of the same year. But these statistics are small compared to the more than 300,000 lives taken as a killer tropical cyclone and storm surge ravaged the coast of Bangladesh with flood waters in 1970. In April, 1991, a similar cyclone

● FIGURE 15.24 High winds and huge waves crash against a boat washed onto Highway 90 in Gulfport, Mississippi, as Hurricane Katrina makes landfall on the morning of August 29, 2005.

© John Bazemore/ AP Wide World

©Vincent Laforet/The *New York Times*

● FIGURE 15.25 Flood waters inundate New Orleans, Louisiana, during August, 2005, after the winds and storm surge from Hurricane Katrina caused several levee breaks.

devastated the area with reported winds of 127 knots and a storm surge of 7 m (23 ft). In all, the storm destroyed 1.4 million houses and killed 140,000 people and 1 million cattle. And again in November, 2007, Tropical Cyclone Sidr, a Category 4 storm with winds of 135 knots, moved into the region (see • Fig. 15.26), killing thousands of people, damaging or destroying over one million houses, and flooding more than two million acres. Estimates are Cyclone Sidr adversely affected more than 8.5 million people. Unfortunately, the potential for a repeat of this type of disaster remains high in Bangladesh, as many people live along the relatively low, wide flood plain that slopes outward to the bay. And, historically, this region is in a path frequently taken by tropical cyclones.

During late October, 1998, Hurricane Mitch became the most deadly hurricane to strike the Western Hemisphere since the Great Hurricane of 1780, which claimed approximately 22,000 lives in the eastern Caribbean. Mitch's high winds, huge waves (estimated maximum height 44 ft), and torrential rains destroyed vast regions of coastal Central America (for Mitch's path, see Fig.15.12 p. 421). In the mountainous regions of Honduras and Nicaragua, rainfall totals from the storm may have reached 190 cm (75 in.). The heavy rains produced floods and deep mud slides that swept away entire villages, including the inhabitants. Mitch caused over $5 billion in damages, destroyed hundreds of thousands of homes, and killed over 11,000 people. More than 3 million people were left homeless or were otherwise severely affected by this deadly storm.

Are major hurricanes on the increase worldwide? Will the intensity of hurricanes increase as the world warms? These questions are addressed in the Focus section on p. 434.

Hurricane Watches, Warnings, and Forecasts

With the aid of ship reports, satellites, radar, buoys, and reconnaissance aircraft, the location and intensity of hurricanes are pinpointed and their movements carefully monitored. When a hurricane poses a direct threat to an area, a **hurricane watch** is issued, typically 24 to 48 hours before the storm arrives, by the National Hurricane Center in Miami, Florida, or by the Pacific Hurricane Center in Honolulu, Hawaii. When it appears that the storm will strike an area, a **hurricane warning** is issued (see • Fig. 15.27). Along the east coast of North America, the warning is accompanied by a probability. The probability gives the percent chance of the hurricane's center passing within 105 km (65 mi) of a particular community. The warning is designed to give residents ample time to secure property and, if necessary, to evacuate the area.

Because hurricane-force winds can extend a considerable distance on either side of where the storm is expected to make landfall, a hurricane warning is issued for a rather large

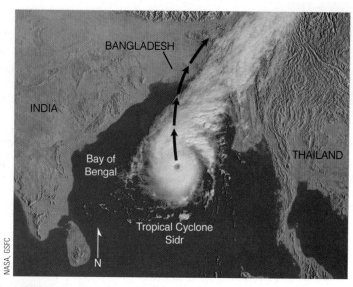

• FIGURE 15.26 Satellite image of Tropical Cyclone Sidr on November 14, 2007, as it moves northward over the Bay of Bengal toward Bangladesh. With winds of 135 knots and a central pressure estimated at 937 mb (27.67 in.), this strong Category 4 storm caused widespread destructio and the loss of many livestock and human lives when it made landfall on November 15, 2007. (The red dashed arrow shows Sidr's path.)

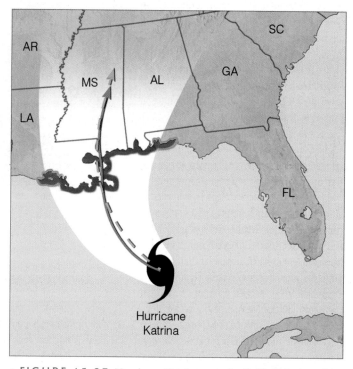

• FIGURE 15.27 Hurricane Katrina over the Gulf of Mexico with sustained winds of 145 mi/hr on August 28, 2005, at 1 A.M. CDT. The current movement of the storm is west-northwest at 8 mi/hr. The dashed orange line shows the hurricane's projected path; the solid purple line, the hurricane's actual path. Areas under a hurricane warning are in red. Those areas under a hurricane watch are in pink, while those areas under a tropical storm warning are in blue.

FOCUS ON AN ENVIRONMENTAL ISSUE

Hurricanes in a Warmer World

In the Focus section on p. 431, we saw that 2005 was a record year for Atlantic hurricanes, with 27 named storms, 15 hurricanes, and 5 storms reaching Category 5 status on the Saffir-Simpson scale. Was the record hurricane year 2005 related to global warming?

We know that hurricanes are fueled by warm tropical water—the warmer the water, the more fuel available to drive the storm. A mere 0.6°C (1°F) increase in sea-surface temperature will increase the maximum winds of a hurricane by about 5 knots, everything else being equal. During May, 2005, just before the hurricane season got underway, the surface water temperature over the tropical North Atlantic was considerably warmer than normal (see Fig. 5). Moreover, studies conducted by the National Center for Atmospheric Research (NCAR) in Boulder, Colorado, found that between June and October, 2005, sea-surface temperatures in the tropical Atlantic were about 0.9°C (1.6°F) warmer than the long-time (1901–1970) average for that region. The study concluded that about half of the warming (about 0.4°C) was due to global warming caused by increasing concentrations of greenhouse gases in the atmosphere. These findings suggest that global warming (described more completely in Chapter 16) may have had an effect on the intensity of some storms and on the number of tropical storms that formed, since the ocean surface remained warmer than normal well past October. Interestingly, during the hurricane season of 2007, only one weak hurricane (Humberto) made landfall in the United States, but two storms (Hurricanes Dean and Felix) reached Category 5 status over the warm Gulf of Mexico before weakening and making landfall south of the United States.

Climate models predict that, as the world warms, sea-surface temperatures in the tropics will rise by about 2°C (3.6°F) by the end of this century. Should these projections prove correct, a hurricane forming in today's atmosphere with maximum sustained winds of 125 knots (a Category 4 storm) could, in the warmer world, have maximum sustained winds of 140 knots (a Cate-

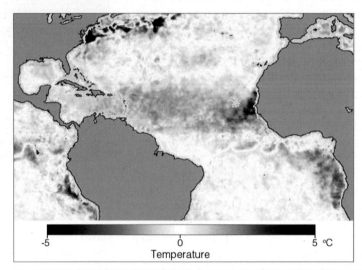

Sea-surface temperature departures from the twelve-year average (1985–1997) on May 30, 2005. Notice that the darker the red, the warmer the surface water. (NOAA)

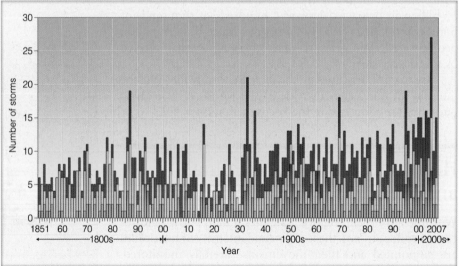

● FIGURE 6 The total number of tropical storms and hurricanes (red bars), hurricanes only (yellow bars), and Category 3 hurricanes or greater (green bars) in the Atlantic basin for the period 1851 through 2007. (NOAA)

gory 5 storm). As sea-surface temperatures rise, will hurricanes become more frequent? Presently, there is no clear answer to this question as some climate models predict more hurricanes, whereas others predict fewer storms, Since the earth's surface is gradually warming, are today's hurricanes more intense than those of the past? Several studies suggest that the frequency of major hurricanes (Category 3 and above) has been increasing (see Fig. 6). The problem with these

studies is that reliable records of tropical cyclones have only been available since the 1970s, when observations from satellites became more extensive. Sophisticated instruments today allow scientists to peer into hurricanes and examine their structure and winds with much greater clarity than in the past. Any trend in hurricane frequency or intensity will likely become clearer when more reliable information on past tropical cyclone activity becomes available.

coastal area, usually about 550 km (342 mi) in length. Since the average swath of hurricane damage is normally about one-third this length, much of the area is "over-warned." As a consequence, many people in a warning area feel that they are needlessly forced to evacuate. The evacuation order is given by local authorities* and typically only for those low-lying coastal areas directly affected by the storm surge. People at higher elevations or farther from the coast are not usually requested to leave, in part because of the added traffic problems this would create. This issue has engendered some controversy in the wake of Hurricane Andrew, since its winds were so devastating over inland south Florida during August, 1992. The time it takes to complete an evacuation puts a special emphasis on the timing and accuracy of the warning.

As Hurricane Katrina (see in Fig. 15.27) approaches land, will it intensify, maintain its strength, or weaken? Also, will it continue to move in the same direction and at the same speed? Such questions have challenged forecasters for some time. To forecast the intensity and movement of a hurricane, meteorologists use numerical weather prediction models, similar to those described in Chapter 13 on p. 344.

Information from satellites, buoys, and reconnaissance aircraft (that deploy dropsondes† into the eye of the storm) are fed into the models. The models then forecast the intensity and movement of the storm. There are a variety of forecast models, each one treating some aspect of the atmosphere (such as evaporation of water from the ocean's surface) in a slightly different manner. Often, the models do not agree on where the storm will move and on how strong it will be.

The problem of different models forecasting different paths for the same hurricane has been addressed by using the method of *ensemble forecasting.* You may recall from Chapter 13, on p. 348 that an ensemble forecast is based on running several forecast models (or different simulations of the same model), each beginning with slightly different weather information. If the forecast models (or different versions of the same model) all agree that a hurricane will move in a particular direction, the forecaster will have confidence in making a forecast of the storm's movement. If, on the other hand, the models do not agree, then the forecaster will have to decide which model (or models) is most likely correct in forecasting the hurricane's track.

The use of ensemble forecasting along with better forecast models has helped raise the level of skill in forecasting hurricane paths. For example, in the 1970s, the projected position of a hurricane three days into the future was off by an average of 708 km (440 mi). Today, the average error for the same forecast period has dropped to 278 km (173 mi). Unfortunately, the forecasting of hurricane intensity has shown little improvement since the early 1990s.

To help predict hurricane intensity, forecasters have been using statistical models that compare the behavior of the present storm with that of similar tropical storms in the past. The results using these models have not been encouraging. Another more recent model uses the depth of warm ocean water in front of the storm's path to predict the storm's intensity. Recall from an earlier discussion (p. 418) that if the reservoir of warm water ahead of the storm is relatively shallow, ocean waves generated by the hurricane's wind will turbulently bring deeper, cooler water to the surface. The cooler water will cut off the storm's energy source, and the hurricane will weaken. On the other hand, should a deep layer of warm water exist ahead of the hurricane, cooler water will not be brought to the surface, and the storm will either maintain its strength or intensify, as long as other factors remain the same. So, knowing the depth of warm surface water ahead of the storm is important in predicting whether a hurricane will intensify or weaken.* Moreover, as new hurricane-prediction models with greater resolution are implemented, and as our understanding of the nature of hurricanes increases, forecasting hurricane intensification and movement should improve.

Modifying Hurricanes

Because of the potential destruction and loss of lives that hurricanes can inflict, attempts have been made to reduce their winds by seeding them with silver iodide. The idea is to seed the clouds just outside the eyewall with just enough artificial ice nuclei so that the latent heat given off will stimulate cloud growth in this area of the storm. These clouds, which grow at the expense of the eyewall thunderstorms, actually form a new eyewall farther away from the hurricane's center. As the storm center widens, it pressure gradient should weaken, which may cause its spiraling winds to decrease in speed.

During project STORMFURY, a joint effort of the National Oceanic and Atmospheric Administration (NOAA) and the U.S. Navy, several hurricanes were seeded by aircraft. In 1963, shortly after Hurricane Beulah was seeded with silver iodide, surface pressure in the eye began to rise and the region of maximum winds moved away from the storm's center. Even more encouraging results were obtained from the multiple seeding of Hurricane Debbie in 1969. After one day of seeding, Debbie showed a 30 percent reduction in maximum winds. But many hurricanes that are not seeded show this type of behavior. So, the question remains: Would the winds have lowered naturally had the storm not been seeded? Several studies even cast doubt upon the theoretical basis for this kind of hurricane modification because hurricanes appear to contain too little supercooled water and too much natural ice. Consequently, there are many uncertainties about the effectiveness of seeding hurricanes in an attempt to reduce their winds, and all endeavors to modify hurricanes have been discontinued since the 1970s.

Other ideas have been proposed to weaken the winds of a hurricane. One idea is to place some form of oil (monomo-lecular film) on the water to retard the rate of evaporation and hence cut down on the release of latent heat inside the clouds. Some sailors, even in ancient times, would dump oil into the sea during stormy weather, claiming it reduced the winds around the ship. At this point, it's interesting to note that a recent mathematical study suggests that ocean spray has an effect on the winds of a hurricane. Apparently, the tiny spray reduces the friction between the wind and the sea surface. Consequently, with the same pressure gradient, the more ocean spray, the higher the winds. If this idea proves correct, limiting ocean spray from entering the air above may reduce the storm's winds. Perhaps the ancient sailors knew what they were doing after all.

SUMMARY

Hurricanes are tropical cyclones with winds that exceed 64 knots (74 mi/hr) and blow counterclockwise about their centers in the Northern Hemisphere. A hurricane consists of a mass of organized thunderstorms that spiral in toward the extreme low pressure of the storm's eye. The most intense thunderstorms, the heaviest rain, and the highest winds occur outside the eye, in the region known as the eyewall. In the eye itself, the air is warm, winds are light, and skies may be broken or overcast.

Hurricanes (and all tropical cyclones) are born over warm tropical waters where the air is humid, surface winds converge, and thunderstorms become organized in a region of weak upper-level winds. Surface convergence may occur along the ITCZ, on the eastern side of a tropical wave, or along a front that has moved into the tropics from higher latitudes. If the disturbance becomes more organized, it becomes a tropical depression. If central pressures drop and surface winds increase, the depression becomes a tropical storm. At this point, the storm is given a name. Some tropical storms continue to intensify into full-fledged hurricanes, as long as they remain over warm water and are not disrupted by strong vertical wind shear.

The energy source that drives the hurricane comes primarily from the warm tropical oceans and from the release of latent heat. A hurricane is like a heat engine in that energy for the storm's growth is taken in at the surface in the form of sensible and latent heat, converted to kinetic energy in the form of winds, then lost at the cloud tops through radiational cooling.

The easterly winds in the tropics usually steer hurricanes westward. In the Northern Hemisphere, most storms then gradually swing northwestward around the subtropical high. If the storm moves into middle latitudes, the prevailing westerlies steer it northeastward. Because hurricanes derive their energy from the warm surface water and from the latent heat of condensation, they tend to dissipate rapidly when they move over cold water or over a large mass of land, where surface friction causes their winds to decrease and flow into their centers.

Although the high winds of a hurricane can inflict a great deal of damage, it is usually the huge waves and the flooding associated with the storm surge that cause the most destruction and loss of life. The Saffir-Simpson hurricane scale was developed to estimate the potential destruction that a hurricane can cause.

KEY TERMS

The following terms are listed (with page numbers) in the order they appear in the text. Define each. Doing so will aid you in reviewing the material covered in this chapter.

streamlines, 412	tropical depression, 418
tropical wave (easterly wave), 412	tropical storm, 418
	storm surge, 424
hurricane, 412	Saffir-Simpson scale, 424
typhoon, 412	super typhoon, 424
tropical cyclone, 412	eyewall replacement, 430
eye (of hurricane), 413	hurricane watch, 433
eyewall, 413	hurricane warning, 433
trade wind inversion, 416	

QUESTIONS FOR REVIEW

1. What is a tropical (easterly) wave? How do these waves generally move in the Northern Hemisphere? Are showers found on the eastern or western side of the wave?
2. Why are streamlines, rather than isobars, used on surface weather maps in the tropics?
3. What is the name given to a hurricane-like storm that forms over the western North Pacific Ocean?
4. Describe the horizontal and vertical structure of a hurricane.
5. Why are skies often clear or partly cloudy in a hurricane's eye?
6. What conditions at the surface and aloft are necessary for hurricane development?
7. List three "triggers" that help in the initial stage of hurricane development.
8. (a) Hurricanes are sometimes described as a heat engine. What is the "fuel" that drives the hurricane?

(b) What determines the maximum strength (the highest winds) that the storm can achieve?

9. Would it be possible for a hurricane to form over land? Explain.

10. If a hurricane is moving westward at 10 knots, will the strongest winds be on its northern or southern side? Explain. If the same hurricane turns northward, will the strongest winds be on its eastern or western side?

11. What factors tend to weaken hurricanes?

12. Distinguish among a tropical depression, a tropical storm, and a hurricane.

13. In what ways is a hurricane different from a mid-latitude cyclone? In what ways are these two systems similar?

14. Why do most hurricanes move westward over tropical waters?

15. If the high winds of a hurricane are not responsible for inflicting the most damage, what is?

16. Most hurricane-related deaths are due to what?

17. Explain how a storm surge forms. How does it inflict damage in hurricane-prone areas?

18. Hurricanes are given names when the storm is in what stage of development?

19. When Hurricane Andrew moved over south Florida during August, 1992, what was it that caused the relatively small areas of extreme damage?

20. As Hurricane Katrina moved toward the Louisiana coast, it underwent eyewall replacement. What actually happened to the eyewall during this process?

21. How do meteorologists forecast the intensity and paths of hurricanes?

22. How does a hurricane watch differ from a hurricane warning?

23. Why have hurricanes been seeded with silver iodide?

24. Give two reasons why hurricanes are more likely to strike New Jersey than Oregon.

QUESTIONS FOR THOUGHT

1. Why are North Atlantic hurricanes more apt to form in October than in May?

2. Would it be possible for a hurricane to form in the tropical North Atlantic or North Pacific during December? Explain.

3. Would the winds of a hurricane decrease more quickly as the storm moves over cooler water or over warmer land? Explain.

4. Explain why the ocean surface water temperature is usually cooler after the passage of a hurricane. (Hint: The answer is not because the hurricane extracts heat from the water.)

5. Suppose, in the North Atlantic, an eastward-moving ocean vessel is directly in the path of a westward-moving hurricane. What would be the ship's wisest course—to veer to the north of the storm or to the south of the storm? Explain.

6. Suppose this year five tropical storms develop into full-fledged hurricanes over the North Atlantic Ocean. Would the name of the third hurricane begin with the letter *C*? Explain.

7. You are in Darwin, Australia (on the north shore), and a hurricane approaches from the north. Where would the highest storm surge be, to the east or west? Explain.

8. Occasionally when a hurricane moves inland, it will encounter a mountain range. Describe what will happen when this occurs. What will happen to the hurricane's intensity? What will cause the most damage (winds, storm surge, flooding)? Why?

9. Give several reasons how a hurricane that once began to weaken can strengthen.

10. Suppose a tropical storm in the Gulf of Mexico moves westward across Central America and out over the Pacific. If the storm maintains tropical storm strength the entire time, do you feel it should be given a new name over the Pacific? Explain your reasoning.

PROBLEMS AND EXERCISES

1. A hurricane just off the coast of northern Florida is moving northeastward, parallel to the eastern seaboard. Suppose that you live in North Carolina along the coast:
   (a) How will the surface winds in your area change direction as the hurricane's center passes due *east* of you? Illustrate your answer by making a sketch of the hurricane's movement and the wind flow around it.
   (b) If the hurricane passes east of you, the strongest winds would most likely be blowing from which direction? Explain your answer. (Assume that the storm does not weaken as it moves northeastward.)
   (c) The lowest sea-level pressure would most likely occur with which wind direction? Explain.

2. Use the Saffir-Simpson hurricane scale (see Table 15.2, p. 426) and the text material to determine the category of Hurricane Elena in Fig. 15.2, p. 413.

Visit the
Meteorology Resource Center
at
academic.cengage.com/login
for more assets, including questions for exploration, animations, videos, and more.

The warming in the Arctic has caused sea ice to thin by between 15 and 40 percent over the past 30 years. If the warming in this region continues as predicted, sea ice during the summer months may be totally absent by the middle of this century.

CHAPTER 16

The Earth's Changing Climate

A change in our climate however is taking place very sensibly. Both heats and colds are becoming much more moderate within the memory even of the middle-aged. Snows are less frequent and less deep. They do not often lie, below the mountains, more than one, two, or three days, and very rarely a week. They are remembered to have been formerly frequent, deep, and of long continuance. The elderly inform me the earth used to be covered with snow about three months in every year. The rivers, which then seldom failed to freeze over in the course of the winter, scarcely ever do now. This change has produced an unfortunate fluctuation between heat and cold, in the spring of the year, which is very fatal to fruits. In an interval of twenty-eight years, there was no instance of fruit killed by the frost in the neighborhood of Monticello. The accumulated snows of the winter remaining to be dissolved all together in the spring, produced those overflowings of our rivers, so frequent then, and so rare now.

Thomas Jefferson, *Notes on the State of Virginia*, 1781

CONTENTS

The climate is always changing. Evidence shows that climate has changed in the past, and nothing suggests that it will not continue to change. As the urban environment grows, its climate differs from that of the region around it. Sometimes the difference is striking, as when city nights are warmer than the nights of the outlying rural areas. Other times, the difference is subtle, as when a layer of smoke and haze covers the city or when the climate of a relatively small area—the microclimate—becomes modified by the light and warmth of a city street lamp (see • Fig. 16.1). Climate change, in the form of a persistent drought or a delay in the annual monsoon rains, can adversely affect the lives of millions. Even small changes can have an adverse effect when averaged over many years, as when grasslands once used for grazing gradually become uninhabited deserts.

Climate change is taking place right now as the world is warming at an alarming rate. Consequently, in the Northern Hemisphere, polar sea ice in winter does not extend as far south as it once did, Greenland ice is melting rapidly, and sea level is rising worldwide. The main cause of this **global warming** appears to be human (anthropogenic) activities. We will, therefore, first look at the evidence for climate change in the past; then, we will investigate the causes of climate change due to both natural variations and human intervention.

© C. Donald Ahrens

• FIGURE 16.1 Altering the microclimate. Notice in the picture that the leaves are still on the tree near the streetlight. Apparently, this sodium vapor lamp emits enough warmth and light during the night to trick the leaves into behaving as if it were September rather than the middle of November.

Reconstructing Past Climates

Not only is the earth's climate always changing, but a mere 18,000 years ago the earth was in the grip of a cold spell, with *alpine glaciers* extending their icy fingers down river valleys and huge ice sheets *(continental glaciers)* covering vast areas of North America and Europe (see • Fig. 16.2). The ice at that time measured several kilometers thick and extended as far south as New York and the Ohio River Valley. Perhaps the glaciers advanced 10 times during the last 2.5 million years, only to retreat. In the warmer periods, between glacier advances, average global temperatures were slightly higher than at present. Hence, some scientists feel that we are still in an ice age, but in the comparatively warmer part of it.

Presently, glaciers cover less than 10 percent of the earth's land surface. The total volume of ice over the face of the earth amounts to about 25 million cubic kilometers. Most of this ice is in the Greenland and Antarctic ice sheets, and its accumulation over time has allowed scientists to measure past climatic changes. If global temperatures were to rise enough so that all of this ice melted, the level of the ocean would rise about 65 m (213 ft) (see • Fig. 16.3). Imagine the catastrophic results: Many major cities (such as New York, Tokyo, and London) would be inundated. Even a rise in global temperature of several degrees Celsius might be enough to raise sea level by a meter or more, flooding coastal lowlands.

The study of the geological evidence left behind by advancing and retreating glaciers is one factor suggesting that global climate has undergone slow but continuous changes. To reconstruct past climates, scientists must examine and then carefully piece together all the available evidence. Unfortunately, the evidence only gives a general understanding of what past climates were like. For example, fossil pollen of a tundra plant collected in a layer of sediment in New England and dated to be 12,000 years old suggests that the climate of that region was much colder than it is today.

Other evidence of global climatic change comes from core samples taken from ocean floor sediments and ice from Greenland and Antarctica. A multiuniversity research project known as CLIMAP (Climate: long-range investigation mapping and prediction) studied the past million years of global climate. Thousands of meters of ocean sediment obtained with a hollow-centered drill were analyzed. This sediment contained the remains of calcium carbonate shells of organisms that once lived near the surface. Because certain organisms can only live within a narrow range of temperature, the distribution and type of organisms within the sediment indicate the temperature of the surface water.

In addition, the oxygen-isotope* ratio of these shells provided information about the sequence of glacier advances. For example, most of the oxygen in sea water is composed of 8 protons and 8 neutrons in its nucleus, giving it an atomic

---

*Isotopes are atoms whose nuclei have the same number of protons but different numbers of neutrons.

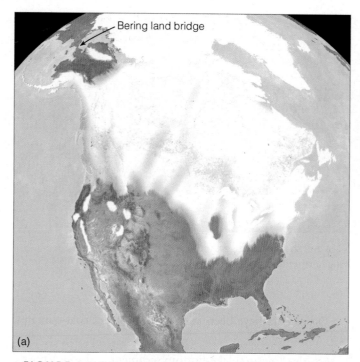

● FIGURE 16.2 Extent of glaciation about 18,000 years ago over (a) North America and over (b) western Europe.

weight of 16. However, about one out of every thousand oxygen atoms contains an extra 2 neutrons, giving it an atomic weight of 18. When ocean water evaporates, the heavy oxygen 18 tends to be left behind. Consequently, during periods of glacier advance, the oceans, which contain less water, have a higher concentration of oxygen 18. Since the shells of marine organisms are constructed from the oxygen atoms existing in ocean water, determining the ratio of oxygen 18 to oxygen 16 within these shells yields information about how the climate may have varied in the past. A higher ratio of oxygen 18 to oxygen 16 in the sediment record suggests a colder climate, whereas a lower ratio suggests a warmer climate. Using data such as these, the CLIMAP project was able to reconstruct the

earth's surface ocean temperature for various times during the past (see ● Fig. 16.4).

Vertical ice cores extracted from ice sheets in Antarctica and Greenland provide additional information on past temperature patterns. Glaciers form over land where temperatures are sufficiently low so that, during the course of a year, more snow falls than will melt. Successive snow accumulations over many years compact the snow, which slowly recrystallizes into ice. Since ice is composed of hydrogen and oxygen, examining the oxygen-isotope ratio in ancient cores provides a past record of temperature trends. Generally, the colder the air when the snow fell, the richer the concentration of oxygen 16 in the core. Moreover, bubbles of ancient air

● FIGURE 16.3 If all the ice locked up in glaciers and ice sheets were to melt, estimates are that this coastal area of south Florida would be under 65 m (213 ft) of water. Even a relatively small one-meter rise in sea level would threaten half of the world's population with rising seas. In fact, latest research suggests sea level will rise one meter or more by the end of this century due to the rapid melting of ice in Greenland and Antarctica.

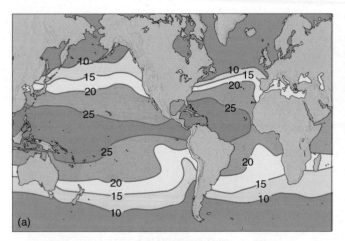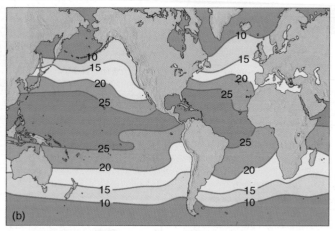

● FIGURE 16.4 (a) Sea surface isotherms (°C) during August 18,000 years ago and (b) during August today. Apparently, during the Ice Age (diagram a) the Gulf Stream shifted to a more easterly direction, depriving northern Europe of its warmth and causing a rapid north-to-south ocean surface temperature gradient.

trapped in the ice can be analyzed to determine the past composition of the atmosphere (see Fig. 16.14, p. 451).

Ice cores also record the causes of climate changes. One such cause is deduced from layers of sulfuric acid in the ice. The sulfuric acid originally came from large volcanic explosions that injected huge quantities of sulfur into the stratosphere. The resulting sulfate aerosols eventually fell to the earth in polar regions as acid snow, which was preserved in the ice sheets. The Greenland ice cores also provide a continuous record of sulfur from human sources. Moreover, ice cores at both poles are being analyzed for many chemicals that provide records of biological and physical changes in the climate system, such as a beryllium isotope (^{10}Be) that indicates solar activity. Various types of dust collected in the cores indicate whether the climate was arid or wet.

Scientists are even using the calcium carbonate material that forms into tiny stones (*otoliths*) in the inner ears of fish to reconstruct past temperatures of the Great Lakes region. As the otoliths grow, they extract oxygen from the lake water. The oxygen-isotope ratio then provides scientists with information on changes in water temperature over the life of the fish, whether it died last week or 10,000 years ago.

Still other evidence of climatic change comes from the study of annual growth rings of trees, called **dendrochronology.** As a tree grows, it produces a layer of wood cells under its bark. Each year's growth appears as a ring. The changes in thickness of the rings indicate climatic changes that may have taken place from one year to the next. The density of late growth tree rings is an even better indication of changes in climate. The presence of frost rings during particularly cold periods and the chemistry of the wood itself provide additional information about a changing climate. Tree rings are only useful in regions that experience an annual cycle and in trees that are stressed by temperature or moisture during their growing season. The growth of tree rings has been correlated with precipitation and temperature patterns for hundreds of years into the past in various regions of the world.

Other data have been used to reconstruct past climates, such as:

1. records of natural lake-bottom sediment and soil deposits
2. the study of pollen in deep ice caves, soil deposits, and sea sediments
3. certain geologic evidence (ancient coal beds, sand dunes, and fossils), and the change in the water level of closed basin lakes
4. documents concerning droughts, floods, crop yields, rain, snow, and dates of lakes freezing
5. the study of oxygen-isotope ratios of corals
6. dating calcium carbonate layers of stalactites in caves
7. borehole temperature profiles, which can be inverted to give records of past temperature change at the surface
8. deuterium (heavy hydrogen) ratios in ice cores, which indicate temperature changes

Even with all of this knowledge, our picture of past climates is still incomplete. With this shortcoming in mind, we will examine what the information gained about past climates does reveal.

CLIMATE THROUGHOUT THE AGES Throughout much of the earth's history, the global climate was probably between 8°C and 15°C, warmer than it is today. During most of this time, the polar regions were free of ice. These comparatively warm conditions, however, were interrupted by several periods of glaciation. Geologic evidence suggests that one glacial period occurred about 700 million years ago (m.y.a.) and another about 300 m.y.a. The most recent one — the *Pleistocene epoch* or, simply, the **Ice Age** — began about 2.5 m.y.a. Let's summarize the climatic conditions that led up to the Pleistocene.

About 65 m.y.a., the earth was warmer than it is now; polar ice caps did not exist. Beginning about 55 m.y.a., the earth entered a long cooling trend. After millions of years, polar ice appeared. As average temperatures continued to lower, the ice grew thicker, and by about 10 m.y.a. a deep blanket of ice covered the Antarctic. Meanwhile, snow and ice began to accumulate in high mountain valleys of the Northern Hemisphere, and alpine glaciers soon appeared.

About 2.5 m.y.a., continental glaciers appeared in the Northern Hemisphere, marking the beginning of the Pleistocene epoch. The Pleistocene, however, was not a period of continuous glaciation but a time when glaciers alternately advanced and retreated (melted back) over large portions of North America and Europe. Between the glacial advances were warmer periods called **interglacial periods,** which lasted for 10,000 years or more.

It was once thought that interglacial periods represented a more stable type of climate, unlike the large climate variations experienced during the colder part of the Ice Age. But analysis of Greenland ice cores suggests that the warm *Eemian interglacial period* (which lasted 19,000 years from 133,000 to 114,000 years ago) may have consisted of two major cold spells, lasting 2000 years and 6000 years. Some scientists speculate, however, that the cooling may represent a shift in the location of Greenland's moisture supply (which would change the oxygen-isotope ratio) rather than actual coolings, as these cold spells do not show up in the Antarctic ice core record.

The most recent North American glaciers reached their maximum thickness and extent about 18,000–22,000 years ago (y.a.). At that time, average temperatures in Greenland were about 10°C (18°F) lower than at present and tropical average temperatures were about 4°C (7°F) lower than they are today. Because a great deal of water was in the form of ice over land, sea level was perhaps 120 m (395 ft) lower than it is now. The lower sea level exposed vast areas of land, such as the *Bering land bridge* (a strip of land that connected Siberia to Alaska as shown in Fig. 16.2a), which allowed human and animal migration from Asia to North America.

The ice began to retreat about 14,000 y.a. as surface temperatures slowly rose, producing a warm spell called the *Bölling-Alleröd* period (see ●Fig. 16.5). Then, about 12,700 y.a., the average temperature suddenly dropped and northeastern North America and northern Europe reverted back to glacial conditions. About 1000 years later, the cold spell (known as the **Younger Dryas***) ended abruptly and temperatures rose rapidly in many areas. Beginning about 8000 y.a. the mean temperature dropped by as much as 2°C over central Europe. During this cold period, which was not experienced worldwide, the European alpine timberline fell about 200 m (600 ft). The cold period ended, temperatures began to rise, and by about 6000 y.a. the continental ice sheets over North America were gone. This warm spell during the current interglacial period, or *Holocene epoch,* is sometimes called the **mid-Holocene maximum,** and because this warm period favored the development of plants, it is also known as the *climatic optimum.* About 5000 y.a, a cooling trend set in, during which extensive alpine glaciers returned, but not continental glaciers.

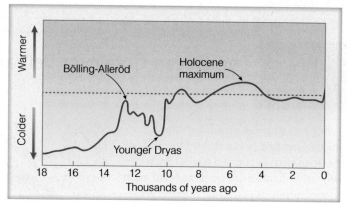

●**FIGURE 16.5** Relative air temperature variations (warmer and cooler periods) during the past 18,000 years. These data, which represent temperature records compiled from a variety of sources, only give an approximation of temperature changes. Some regions of the world experienced a cooling and other regions a warming that either preceded or lagged behind the temperature variations shown in the diagram.

It is interesting to note that ice core data from Greenland reveal that rapid shifts in climate (from ice age conditions to a much warmer state) took place in as little as three years over central Greenland around the end of the Younger Dryas. The data also reveal that similar rapid shifts in climate occurred several times toward the end of the Ice Age. What could cause such rapid changes in temperature? One possible explanation is given in the Focus section on p. 444.

CLIMATE DURING THE PAST 1000 YEARS ●Figure 16.6 shows how the average surface air temperature changed in the Northern Hemisphere during the last 1000 years. The data needed to reconstruct the temperature profile in Fig. 16.6 comes from a variety of sources, including tree rings, corals, ice cores, historical records, and thermometers. Notice that about 1000 y.a., the Northern Hemisphere was slightly cooler than average (where average represents the average temperature from 1961 to 1990). However, certain regions in the Northern Hemisphere were warmer than others. For example, during this time vineyards flourished and wine was produced in England, indicating warm, dry summers and the absence of cold springs. It was during the early part of the millennium that Vikings colonized Iceland and Greenland and traveled to North America.*

Notice in Fig. 16.6 that the temperature curve shows a relatively warm period during the 11th to the 14th centuries—relatively warm, but still cooler than the 20th century. During this time, the relatively mild climate of Western Europe began to show large variations. For several hundred years the climate grew stormy. Both great floods and great droughts occurred. Extremely cold winters were followed by relatively warm ones. During the cold spells, the English vineyards and the Viking settlements suffered. Europe experienced several famines during the 1300s.

*This exceptionally cold spell is named after the *Dryas,* an arctic flower.

*This relatively warm, tranquil period of several hundred years over western Europe is sometimes referred to in that region as the *Medieval Climatic Optimum.*

FOCUS ON A SPECIAL TOPIC

The Ocean Conveyor Belt and Climate Change

During the last glacial period, the climate around Greenland (and probably other areas of the world, such as northern Europe) underwent shifts, from ice-age temperatures to much warmer conditions in a matter of years. What could bring about such large fluctuations in temperature over such a short period of time? It now appears that a vast circulation of ocean water, known as the *conveyor belt*, plays a major role in the climate picture.

Figure 1 illustrates the movement of the ocean conveyor belt, or *thermohaline circulation*.* The conveyor-like circulation begins in the north Atlantic near Greenland and Iceland, where salty surface water is cooled through contact with cold Arctic air masses. The cold, dense water sinks and flows southward through the deep Atlantic Ocean, around Africa, and into the Indian and Pacific Oceans. In the North Atlantic, the sinking of cold water draws warm water northward from lower latitudes. As this water flows northward, evaporation increases the water's salinity (dissolved salt content) and density. When this salty, dense water reaches the far regions of the North Atlantic, it gradually sinks to great depths. This warm part of the conveyor delivers an incredible amount of tropical heat to the northern Atlantic. During the winter, this heat is transferred to the overlying atmosphere, and evaporation moistens the air. Strong westerly winds then carry this warmth and moisture into northern and western Europe, where it causes winters to be much warmer and wetter than one would normally expect for this latitude.

Ocean sediment records along with ice-core records from Greenland suggest that the giant conveyor belt has switched on and off during the last glacial period. Such events have apparently coincided with rapid changes in climate. For example, when the conveyor belt is strong, winters in northern Europe tend to be

*Thermohaline circulations are ocean circulations produced by differences in temperature and/or salinity. Changes in ocean water temperature or salinity create changes in water density.

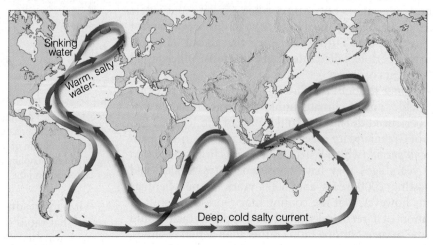

● FIGURE 1 The ocean conveyor belt. In the North Atlantic, cold, salty water sinks, drawing warm water northward from lower latitudes. The warm water provides warmth and moisture for the air above, which is then swept into northern Europe by westerly winds that keep the climate of that region milder than one would normally expect. When the conveyor belt stops, winters apparently turn much colder over northern Europe.

wet and relatively mild. However, when the conveyor belt is weak or stops altogether, winters in northern Europe appear to turn much colder. This switching from a period of milder winters to one of severe cold shows up many times in the climate record. One such event — the Younger Dryas — illustrates how quickly climate can change and how western and northern Europe's climate can cool within a matter of decades, then quickly return back to milder conditions.

Apparently, one mechanism that can switch the conveyor belt off is a massive influx of freshwater. For example, about 11,000 years ago during the Younger Dryas event, freshwater from a huge glacial lake began to flow down the St. Lawrence River and into the North Atlantic. This massive inflow of freshwater reduced the salinity (and, hence, density) of the surface water to the point that it stopped sinking. The conveyor shut down for about 1000 years during which time severe cold engulfed much of northern Europe. The conveyor belt started up again when freshwater began to drain down the Mississippi rather than into the North Atlantic. It was during this

time that milder conditions returned to northern Europe.

Will increasing levels of CO_2 have an effect on the conveyor belt? Some climate models predict that as CO_2 levels increase, more precipitation will fall over the North Atlantic. This situation reduces the density of the sea water and slows down the conveyor belt. In fact, if CO_2 levels double (from its current value), computer models predict that the conveyor belt will slow and that Europe will not warm as much as the rest of the world.

Winters were so cold over North America during the 1700s that soldiers in the Revolutionary War were able to drag cannons across the frozen Upper New York Bay from Staten Island to Manhattan.

Again look at Fig. 16.6 and observe that the Northern Hemisphere experienced a slight cooling during the 15th to 19th centuries. This cooling was significant enough in certain areas to allow alpine glaciers to increase in size and advance down river canyons. In many areas in Europe, winters were long and severe; summers, short and wet. The vineyards in England vanished, and farming became impossible in the more northern latitudes. Cut off from the rest of the world by an advancing ice pack, the Viking colony in Greenland perished.*

There is no evidence that this cold spell existed worldwide. However, over Europe, this cold period has come to be known as the **Little Ice Age.** During these colder times, one particular year stands out: 1816. In Europe that year, bad weather contributed to a poor wheat crop, and famine spread across the land. In Northern America, unusual blasts of cold arctic air moved through Canada and the northeastern United States between May and September. The cold spells brought heavy snow in June and killing frosts in July and August. In the warmer days that followed each cold snap, farmers replanted, only to have another cold outbreak damage the planting. The year 1816 has come to be known as "the year without a summer" or "eighteen hundred and froze-to-death." The unusually cold summer was followed by a bitterly cold winter.

TEMPERATURE TREND DURING THE PAST 100-PLUS YEARS   In the early 1900s, the average global surface temperature began to rise (see ● Fig. 16.7). Notice that, from about 1900 to 1945, the average temperature rose nearly 0.5°C. Following the warmer period, the earth began to cool slightly over the next 25 years or so. In the late 1960s and 1970s, the cooling trend ended over most of the Northern Hemisphere. In the mid-1970s, a warming trend set in that continued into the twenty-first century. In fact, over the Northern Hemisphere, the decade of the 1990s was the warmest of the 20th century, with 1998 and 2005 being the warmest years in over 1000 years.† It appears that the increase in average temperature experienced over the Northern Hemisphere during the 20th century is likely to have been the largest increase in temperature of any century during the past 1000 years.

The average warming experienced over the globe, however, has not been uniform. The greatest warming has occurred in the arctic and over the mid-latitude continents in winter and spring, whereas a few areas have not warmed in

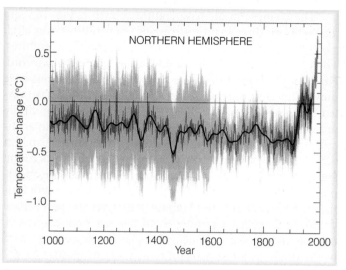

● FIGURE 16.6   The average temperature variations over the Northern Hemisphere for the last 1000 years relative to the 1961 to 1990 average (zero line). Yearly temperature data from tree rings, corals, ice cores, and historical records are shown in blue. Yearly temperature data from thermometers are in red. The black line represents a smoothing of the data. (The gray shading represents a statistical 95 percent confidence range in the annual temperature data, after Mann, et al., 1999.) (*Source:* From Climate Change 2001: The Scientific Basis, 2001, by J.T. Houghton, et al. Copyright © 2001 Cambridge University Press. Reprinted with permission of the Intergovernmental Panel on Climate Change.)

recent decades, such as areas of the oceans in the Southern Hemisphere and parts of Antarctica. The United States has experienced less warming than the rest of the world. Moreover, most of the warming has occurred at night—a situation that has lengthened the frost-free seasons in many mid- and high-latitude regions, although, in recent decades, the warming has been equally distributed between day and night.

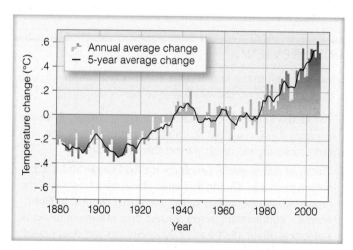

● FIGURE 16.7   The red and blue bars represent the annual average temperature variations over the globe (land and sea) from 1880 through 2006. Temperature changes are compared to the average surface temperature from 1951–1980. The dark solid line shows the five-year average temperature change. (NASA)

---

*Although climate change played a role in the demise of the Viking colony in northern Greenland, it was also their inability to adapt to the climate and to learn hunting and farming techniques from the Eskimos that led to their downfall.

†The exceptionally warm year of 1998 happened to coincide with a major El Niño warming of the tropical Pacific Ocean, whereas the warm year of 2005 did not.

The changes in air temperature shown in Fig. 16.7 are derived from three main sources: air temperatures over land, air temperatures over ocean, and sea surface temperatures. There are, however, uncertainties in the temperature record. For example, during this time period recording stations have moved, and techniques for measuring temperature have varied. Also, marine observing stations are scarce. In addition, urbanization (especially in developed nations) tends to artificially raise average temperatures as cities grow (the urban heat island effect). When urban warming is taken into account, and improved sea-surface temperature information is incorporated into the data, the warming during the twentieth century measures about 0.6°C (about 1°F). Over the past several decades this global warming trend has not only continued, but has increased to about 2.0°C (3.6°F) per century, with twelve of the warmest years on record occurring since 1995.

A global increase in temperature of 0.6°C may seem small, but global temperatures probably have not varied by more than 2°C during the past 10,000 years. Consequently, an increase of 0.6°C becomes significant when compared with temperature changes over thousands of years.

Up to this point we have examined the temperature record of the earth's surface and observed that the earth has been in a warming trend for more than 100 years. The main question regarding this global warming is whether the warming trend is due to natural variations in the climate system, or whether it is due to human activities. Or is it due to a combination of the two? As we will see later in this chapter, climate scientists believe that most of the recent warming is due to an enhanced greenhouse effect caused by increasing levels of greenhouse gases, such as CO_2.* If increasing levels of CO_2 are at least partly responsible for the warming, why has the United States, which produces copious amounts of CO_2, warmed less than the rest of the world? And what caused the exceptionally cold year of 1816—the year without a summer? These are among the questions we will address in the following sections.

BRIEF REVIEW

Before going on to the next section, here is a brief review of some of the facts and concepts we covered so far:

- The earth's climate is constantly undergoing change. Evidence suggests that throughout much of the earth's history, the earth's climate was much warmer than it is today.

- The most recent glacial period (or Ice Age) began about 2.5 million years ago. During this time, glacial advances were interrupted by warmer periods called *interglacial periods.* In North America, continental glaciers reached their maximum thickness and extent about 18,000 to 22,000 years ago and disappeared completely from North America by about 6000 years ago.

---

*The earth's atmospheric greenhouse effect is due mainly to the absorption and emission of infrared radiation by gases, such as water vapor, CO_2, methane, nitrous oxide, and chlorofluorocarbons. Refer back to Chapter 2 for additional information on this topic.

- The Younger Dryas event represents a time about 12,000 years ago when northeastern North America and northern Europe reverted back to glacier conditions.

- During the 20th century, the earth's surface temperature increased by about 0.6°C. This global warming has not only continued, but over the last several decades has increased to about 2°C per century (0.2°C/decade).

Possible Causes of Climate Change

Why does the earth's climate change? There are three "external" causes of climate change. They are:

1. changes in incoming solar radiation
2. changes in the composition of the atmosphere
3. changes in the earth's surface

Natural phenomena can cause climate to change by all three mechanisms, whereas human activities can change climate by both the second and third mechanisms. In addition to these external causes, there are "internal" causes of climate change, such as changes in the circulation patterns of the ocean and atmosphere, which redistribute energy within the climate system.

Part of the complexity of the climate system is the intricate interrelationship of the elements involved. For example, if temperature changes, many other elements may be altered as well. The interactions among the atmosphere, the oceans, and the ice are extremely complex and the number of possible interactions among these systems is enormous. No climatic element within the system is isolated from the others, which is why the complete picture of the earth's changing climate is not totally understood. With this in mind, we will first investigate how feedback systems work; then we will consider some of the current theories of climatic change.

CLIMATE CHANGE: FEEDBACK MECHANISMS   In Chapter 2, we learned that the earth-atmosphere system is in a delicate balance between incoming and outgoing energy. If this balance is upset, even slightly, global climate can undergo a series of complicated changes.

Let's assume that the earth-atmosphere system has been disturbed to the point that the earth has entered a slow warming trend. Over the years the temperature slowly rises, and water from the oceans rapidly evaporates into the warmer air. The increased quantity of water vapor absorbs more of the earth's infrared energy, thus strengthening the atmospheric greenhouse effect.

This strengthening of the greenhouse effect raises the air temperature even more, which, in turn, allows more water vapor to evaporate into the atmosphere. The greenhouse effect becomes even stronger, and the air temperature rises even more. This situation is known as the **water vapor–greenhouse feedback.** It represents a **positive feedback mechanism** because the initial increase in temperature is reinforced by the other processes. If this feedback were left unchecked, the

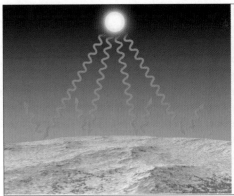
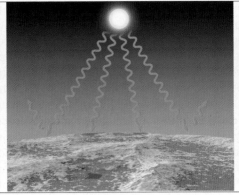
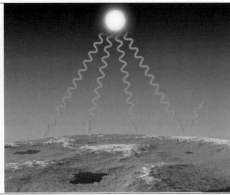

(a)
• High surface albedo
• Low absorption of sunlight
• Gradual surface warming

(b)
• Lower surface albedo
• Higher absorption of sunlight
• Surface warming increases

(c)
• Very low surface albedo
• Much higher absorption of sunlight
• Surface warming enhanced

● FIGURE 16.8 On a warming planet, the snow-albedo positive feedback would enhance the warming. (a) In polar regions snow reflects much of the sun's energy back to space. (b) If the air temperature were to gradually increase, some of the snow would melt, less sunlight would be reflected, and more sunlight would reach the ground, warming it more quickly. (c) The warm surface would enhance the snow melt which, in turn, would accelerate the rise in temperature.

earth's temperature would increase until the oceans evaporated away. Such a chain reaction is called a *runaway greenhouse effect.* This water vapor–greenhouse positive feedback mechanism works in the case of a cooling planet also. For instance, if the earth's climate system was cooling, this positive feedback mechanism would amplify the cooling.

Another positive feedback mechanism is the **snow-albedo feedback,** in which an increase in global surface air temperature might cause snow and ice to melt in polar latitudes. This melting would reduce the albedo (reflectivity) of the surface, allowing more solar energy to reach the surface, which would further raise the temperature (see ● Fig. 16.8).

All feedback mechanisms work simultaneously and in both directions. Consequently, the snow-albedo feedback produces a positive feedback on a cooling planet as well. Suppose, for example, the earth were in a slow cooling trend. Lower temperatures might allow for a greater snow cover in middle and high latitudes, which would increase the albedo of the surface so that much of the incoming sunlight would be reflected back to space. Lower temperatures might further increase the snow cover, causing the air temperature to lower even more. If left unchecked, the snow-albedo positive feedback would produce a *runaway ice age,* which is highly unlikely on earth because other feedback mechanisms in the atmospheric system would be working to moderate the magnitude of the cooling.

To counteract the positive feedback mechanisms there are **negative feedback mechanisms** — those that tend to weaken the interactions among the variables rather than reinforce them. For example, a warming planet emits more infrared radiation.* If the earth climate system were in a runaway

greenhouse effect the increase in radiant energy from the surface would greatly slow the rise in temperature and help to stabilize the climate. The increase in radiant energy from the surface as the planet warms is the strongest negative feedback in the climate system, and greatly lowers the possibility of a runaway greenhouse effect. Consequently, there is no evidence that a runaway greenhouse effect ever occurred on earth, and it is not very likely that it will occur in the future.

In summary, the earth-atmosphere system has a number of checks and balances called *feedback mechanisms* that help it counteract tendencies of climate change. Although we do not worry about a runaway greenhouse effect or an ice-covered earth anytime in the future, there is concern that large positive feedback mechanisms may be working in the climate system to produce accelerated melting of ice in polar regions, especially in Greenland.

CLIMATE CHANGE: PLATE TECTONICS AND MOUNTAIN BUILDING   Earlier, we saw that one of the external causes of climate change is a change in the surface of the earth. During the geologic past, the earth's surface has undergone extensive modifications. One involves the slow shifting of the continents and the ocean floors. This motion is explained in the widely accepted **theory of plate tectonics.** According to this theory, the earth's outer shell is composed of huge plates that fit together like pieces of a jigsaw puzzle. The plates, which slide over a partially molten zone below them, move in relation to one another. Continents are embedded in the plates and move along like luggage riding piggyback on a conveyor belt. The rate of motion is extremely slow, only a few centimeters per year.

Besides providing insights into many geological processes, plate tectonics also helps to explain past climates. For example, we find glacial features near sea level in Africa today, suggesting that the area underwent a period of glaciation hundreds of millions of years ago. Were temperatures at low

_____

*Recall from Chapter 2, p. 38, that the outgoing infrared radiation from the surface increases at a rate proportional to the fourth power of the surface's absolute temperature. This relationship is called the Stefan-Boltzmann law. In effect, doubling the absolute temperature of the earth's surface would result in 16 times more energy emitted.

elevations near the equator ever cold enough to produce ice sheets? Probably not. The ice sheets formed when this land mass was located at a much higher latitude. Over the many millions of years since then, the land has slowly moved to its present position. Along the same line, we can see how the fossil remains of tropical vegetation can be found under layers of ice in polar regions today.

According to plate tectonics, the now existing continents were at one time joined together in a single huge continent, which broke apart. Its pieces slowly moved across the face of the earth, thus changing the distribution of continents and ocean basins, as illustrated in ● Fig. 16.9. Some scientists feel that, when landmasses are concentrated in middle and high latitudes (as they are today), ice sheets are more likely to form. During these times, there is a greater likelihood that more sunlight will be reflected back into space from the snow that falls over the continent in winter. Less sunlight absorbed by the surface lowers the air temperature, which allows for a greater snow cover, and, over thousands of years, the formation of continental glaciers.*

The various arrangements of the continents may also influence the path of ocean currents, which, in turn, could not only alter the transport of heat from low to high latitudes but could also change both the global wind system and the climate in middle and high latitudes. As an example, suppose that plate movement "pinches off" a rather large body of high-latitude ocean water such that the transport of warm water into the region is cut off. In winter, the surface water would eventually freeze over with ice. This freezing would, in turn, reduce the amount of sensible and latent heat given up to the atmosphere. Furthermore, the ice allows snow to accumulate on top of it, thereby setting up conditions that could lead to even lower temperatures.

There are other mechanisms by which tectonic processes† may influence climate. In ● Fig. 16.10, notice that the

*The amplified cooling that takes place over the snow-covered land is the snow-albedo feedback mentioned earlier.

†Tectonic processes are large-scale processes that deform the earth's crust.

formation of oceanic plates (plates that lie beneath the ocean) begins at a *ridge*, where dense, molten material from inside the earth wells up to the surface, forming new sea floor material as it hardens. Spreading (on the order of several centimeters a year) takes place at the ridge center, where two oceanic plates move away from one another. When an oceanic plate encounters a lighter continental plate, it responds by diving under it, in a process called *subduction*. Heat and pressure then melt a portion of the subducting rock, which usually consists of volcanic rock and calcium-rich ocean sediment. The molten rock may then gradually work its way to the surface, producing volcanic eruptions that spew water vapor, carbon dioxide, and minor amounts of other gases into the atmosphere. The release of these gases (called *degassing*) usually takes place at other locations as well (for instance, at ridges where new crustal rock is forming).

Some scientists speculate that climatic change, taking place over millions of years, might be related to the rate at which the plates move and, hence, related to the amount of CO_2 in the air. For example, during times of rapid spreading, an increase in volcanic activity vents large quantities of CO_2 into the atmosphere, which enhances the atmospheric greenhouse effect, causing global temperatures to rise.

Millions of years later, when spreading rates decrease, less volcanic activity means less CO_2 is spewed into the atmosphere. A reduction in CO_2 levels weakens the greenhouse effect, which, in turn, causes global temperatures to drop. The accumulation of ice and snow over portions of the continents may promote additional cooling by reflecting more sunlight back to space.

A chain of volcanic mountains forming above a subduction zone may disrupt the airflow over them. By the same token, mountain building that occurs when two continental plates collide (like that which presumably formed the Himalayan mountains and Tibetan highlands) can have a marked influence on global circulation patterns and, hence, on the climate of an entire hemisphere.

Up to now, we have examined how climatic variations can take place over millions of years due to the movement of

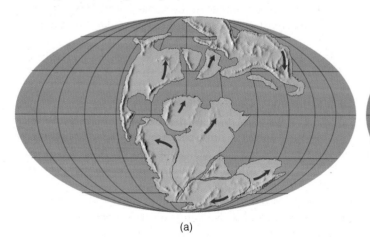
(a)

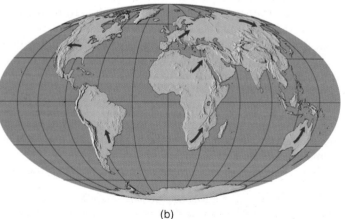
(b)

● FIGURE 16.9 Geographical distribution of (a) landmasses about 150 million years ago, and (b) today. Arrows show the relative direction of continental movement.

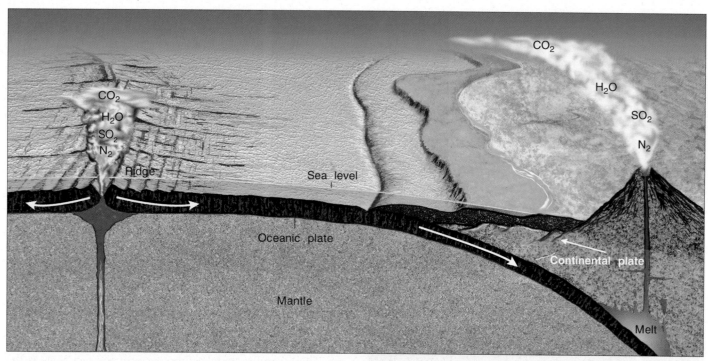

• FIGURE 16.10 The earth is composed of a series of moving plates. The rate at which plates move (spread) may influence global climate. During times of rapid spreading, increased volcanic activity may promote global warming by enriching the CO_2 content of the atmosphere.

continents and the associated restructuring of landmasses. We will now turn our attention to variations in the earth's orbit that may account for climatic fluctuations that take place on a time scale of tens of thousands of years.

CLIMATE CHANGE: VARIATIONS IN THE EARTH'S ORBIT   Another external cause of climate change involves a change in the amount of solar radiation that reaches the earth. A theory ascribing climatic changes to variations in the earth's orbit is the **Milankovitch theory,** named for the astronomer Milutin Milankovitch, who first proposed the idea in the 1930s. The basic premise of this theory is that, as the earth travels through space, three separate cyclic movements combine to produce variations in the amount of solar energy that falls on the earth.

The first cycle deals with changes in the shape (**eccentricity**) of the earth's orbit as the earth revolves about the sun. Notice in •Fig. 16.11 that the earth's orbit changes from being elliptical (dashed line) to being nearly circular (solid line). To go from circular to elliptical and back again takes about 100,000 years. The greater the eccentricity of the orbit (that is, the more elliptical the orbit), the greater the variation in solar energy received by the earth between its closest and farthest approach to the sun.

Presently, we are in a period of low eccentricity, which means that our annual orbit around the sun is more circular. Moreover, the earth is closer to the sun in January and farther away in July (see Chapter 3, p. 58). The difference in distance (which only amounts to about 3 percent) is responsible for a

nearly 7 percent increase in the solar energy received at the top of the atmosphere from July to January. When the difference in distance is 9 percent (a highly elliptical orbit), the difference in solar energy received between July and January will be on the order of 20 percent. In addition, the more eccentric orbit will change the length of seasons in each hemi-

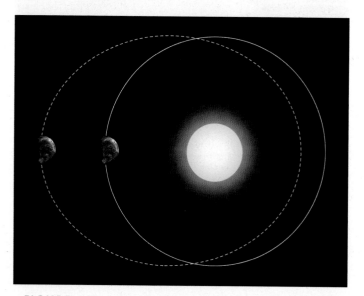

• FIGURE 16.11 For the earth's orbit to stretch from nearly a circular (solid line) to an elliptical orbit (dashed line) and back again takes nearly 100,000 years. (Diagram is highly exaggerated and is not to scale.)

• FIGURE 16.12 (a) Like a spinning top, the earth's axis of rotation slowly moves and traces out the path of a cone in space. (b) Presently the earth is closer to the sun in January, when the Northern Hemisphere experiences winter. (c) In about 11,000 years, due to precession, the earth will be closer to the sun in July, when the Northern Hemisphere experiences summer.

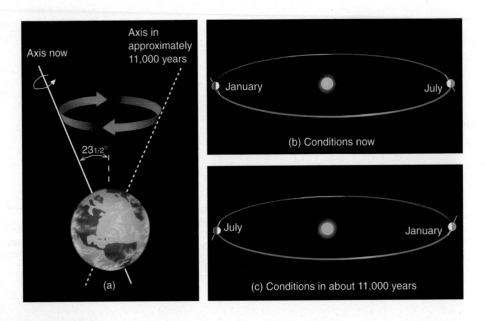

(b) Conditions now

(c) Conditions in about 11,000 years

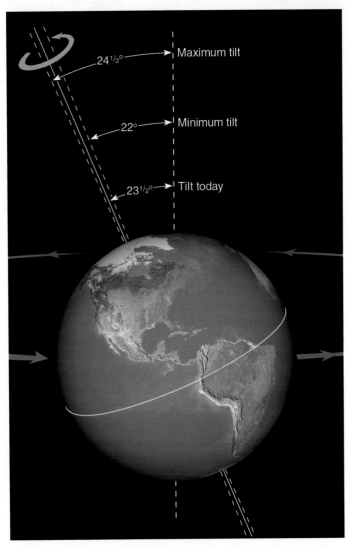

• FIGURE 16.13 The earth currently revolves around the sun while tilted on its axis by an angle of $23\frac{1}{2}°$. During a period of 41,000 years, this angle of tilt ranges from about 22° to $24\frac{1}{2}°$.

sphere by changing the length of time between the vernal and autumnal equinoxes.*

The second cycle takes into account the fact that, as the earth rotates on its axis, it wobbles like a spinning top. This wobble, known as the **precession** of the earth's axis, occurs in a cycle of about 23,000 years. Presently, the earth is closer to the sun in January and farther away in July. Due to precession, the reverse will be true in about 11,000 years (see • Fig. 16.12). In about 23,000 years we will be back to where we are today. This means, of course, that if everything else remains the same, 11,000 years from now seasonal variations in the Northern Hemisphere should be greater than at present. The opposite would be true for the Southern Hemisphere.

The third cycle takes about 41,000 years to complete and relates to the changes in tilt (**obliquity**) with respect to the earth's orbit. Presently, the earth's orbital tilt is $23\frac{1}{2}°$, but during the 41,000-year cycle the tilt varies from about 22° to $24\frac{1}{2}°$ (see • Fig. 16.13). The smaller the tilt, the less seasonal variation there is between summer and winter in middle and high latitudes; thus, winters tend to be milder and summers cooler.

Ice sheets over high latitudes of the Northern Hemisphere are more likely to form when less solar radiation reaches the surface in summer. Less sunlight promotes lower summer temperatures. During the cooler summer, snow from the previous winter may not totally melt. The accumulation of snow over many years increases the albedo of the surface. Less sunlight reaches the surface, summer temperatures continue to fall, more snow accumulates, and continental ice sheets gradually form. At this point, it is interesting to note that when all of the Milankovich cycles are taken into

*Although rather large percentage changes in solar energy can occur between summer and winter, the globally and annually averaged change in solar energy received by the earth (due to orbital changes) hardly varies at all. It is the distribution of incoming solar energy that changes, not the totals.

account, the present trend should be toward *cooler summers* over high latitudes of the Northern Hemisphere.

In summary, the Milankovitch cycles that combine to produce variations in solar radiation received at the earth's surface include:

1. changes in the shape *(eccentricity)* of the earth's orbit about the sun
2. *precession* of the earth's axis of rotation, or wobbling
3. changes in the tilt *(obliquity)* of the earth's axis

In the 1970s, scientists of the CLIMAP project found strong evidence in deep-ocean sediments that variations in climate during the past several hundred thousand years were closely associated with the Milankovitch cycles. More recent studies have strengthened this premise. For example, studies conclude that during the past 800,000 years, ice sheets have peaked about every 100,000 years. This conclusion corresponds naturally to variations in the earth's eccentricity. Superimposed on this situation are smaller ice advances that show up at intervals of about 41,000 years and 23,000 years. It appears, then, that eccentricity is the *forcing factor*—the external cause—for the frequency of glaciation, as it appears to control the severity of the climatic variation.

But orbital changes alone are probably not totally responsible for ice buildup and retreat. Evidence (from trapped air bubbles in the ice sheets of Greenland and Antarctica representing thousands of years of snow accumulation) reveals that CO_2 levels were about 30 percent lower during colder glacial periods than during warmer interglacial periods. Analysis of

air bubbles in Antarctic ice cores reveals that methane follows a pattern similar to that of CO_2 (see Fig. 16.14). This knowledge suggests that lower atmospheric CO_2 levels may have had the effect of amplifying the cooling initiated by the orbital changes. Likewise, increasing CO_2 levels at the end of the glacial period may have accounted for the rapid melting of the ice sheets.*

The latest research shows that temperature changes thousands of years ago actually *preceded* the CO_2 changes. This observation indicates that CO_2 is a positive feedback in the climate system, where higher temperatures lead to higher CO_2 levels and lower temperatures to lower CO_2 levels. Consequently, CO_2 is an internal, natural part of the earth's climate system.

Just why atmospheric CO_2 levels have varied as glaciers expanded and contracted is not clear, but it appears to be due to changes in biological activity taking place in the oceans. Perhaps, also, changing levels of CO_2 indicate a shift in ocean circulation patterns. Such shifts, brought on by changes in precipitation and evaporation rates, may alter the distribution of heat energy around the world. Alteration wrought in this manner could, in turn, affect the global circulation of winds, which may explain why alpine glaciers in the Southern Hemisphere expanded and contracted in tune with Northern Hemisphere glaciers during the last ice age, even though the Southern Hemisphere (according to the Milankovitch cycles) was not in an orbital position for glaciation.

---

*It is interesting to note that during peak CO_2 levels, its concentration of about 325 ppm was still lower than its concentration of 385 ppm in today's atmosphere.

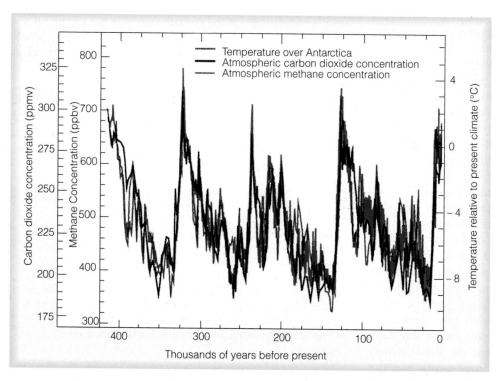

● FIGURE 16.14 Variations of temperature (red line, °C), carbon dioxide (black line, ppmv), and methane (blue line, ppbv). Concentrations of gases are derived from air bubbles trapped within the ice sheets of Antarctica and extracted from ice cores. Temperatures are derived from the analysis of oxygen isotopes. (Note: ppmv represents parts per million by volume, and ppbv represents parts per billion by volume.) (*Source:* From Climate Change 2001: The Scientific Basis, 2001, by J.T. Houghton, et al. Copyright © 2001 Cambridge University Press. Reprinted with permission of the Intergovernmental Panel on Climate Change.)

Still other factors may work in conjunction with the earth's orbital changes to explain the temperature variations between glacial and interglacial periods. Some of these are:

1. the amount of dust and other aerosols in the atmosphere
2. the reflectivity of the ice sheets
3. the concentration of other greenhouse gases
4. the changing characteristics of clouds
5. the rebounding of land, having been depressed by ice

Hence, the Milankovitch cycles, in association with other natural factors, may explain the advance and retreat of ice over periods of 10,000 to 100,000 years. But what caused the Ice Age to begin in the first place? And why have periods of glaciation been so infrequent during geologic time? The Milankovitch theory does not attempt to answer these questions.

CLIMATE CHANGE: ATMOSPHERIC PARTICLES Microscopic liquid and solid particles (aerosols) that enter the atmosphere from both human-induced (anthropogenic) and natural sources can have an effect on climate. The effect these particles have on the climate is exceedingly complex and depends upon a number of factors, such as the particle's size, shape, color, chemical composition, and vertical distribution above the surface. In this section, we will first examine aerosols in the lower atmosphere. Then we will examine the effect that volcanic aerosols in the stratosphere have on climate.

Aerosols in the Troposphere Aerosols enter the lower atmosphere in a variety of ways—from factory and auto emissions, agricultural burning, wildland fires, and dust storms. Many aerosols are not injected directly into the atmosphere, but form when gases convert to particles. Some particles (such as soil dust and sulfate particles) mainly reflect and scatter incoming sunlight, while others (such as smoky soot) readily absorb sunlight, which warms the air around them. Many aerosols that reduce the amount of sunlight reaching the earth's surface tend to cause net cooling of the surface air during the day. Certain aerosols also selectively absorb and emit infrared energy back to the surface, producing a net warming of the surface air at night. However, the overall net effect of human-induced (anthropogenic) aerosols on climate is to *cool the surface.*

In recent years, the effect of highly reflective **sulfate aerosols** on climate has been extensively researched. In the lower atmosphere, the majority of these particles comes from the combustion of sulfur-containing fossil fuels, but emissions from smoldering volcanoes can also be a significant source of tropospheric sulfate aerosols. Sulfur pollution, which has more than doubled globally since preindustrial times, enters the atmosphere mainly as sulfur dioxide gas. There, it transforms into tiny sulfate droplets or particles. Since these aerosols usually remain in the atmosphere for only a few days, they do not have time to spread around the globe. Hence, they are not well mixed and their effect is felt mostly over the Northern Hemisphere, especially over polluted regions. Over the oceans, a major source of sulfate aerosols comes from tiny drifting aquatic plants—phytoplankton—that produce *dimethylsulphide* (DMS). The DMS slowly diffuses into the atmosphere where it oxidizes to form sulfur dioxide, which in turn converts to sulfate aerosols.

Sulfate aerosols not only scatter incoming sunlight back to space, but they also serve as cloud condensation nuclei. Consequently, they have the potential for altering the physical characteristics of clouds. For example, if the number of sulfate aerosols and, hence, condensation nuclei inside a cloud should increase, the cloud would have to share its available moisture with the added nuclei, a situation that should produce many more (but smaller) cloud droplets. The greater number of droplets would reflect more sunlight and have the effect of brightening the cloud and reducing the amount of sunlight that reaches the surface.

In summary, sulfate aerosols reflect incoming sunlight, which tends to lower the earth's surface temperature during the day. Sulfate aerosols may also modify clouds by increasing their reflectivity. Because sulfate pollution has increased significantly over industrialized areas of eastern Europe, northeastern North America and China, the cooling effect brought on by these particles may explain:

1. why the Northern Hemisphere has warmed less than the Southern Hemisphere during the past several decades
2. why the United States has experienced less warming than the rest of the world
3. why up until the last few decades most of the global warming has occurred at night and not during the day, especially over polluted areas

Research is still being done, and the overall effect of tropospheric aerosols on the climate system is not totally understood. Information regarding the possible effect on climate from particles injected into the atmosphere during nuclear war is given in the Focus section on p. 454.

Volcanic Eruptions and Aerosols in the Stratosphere Volcanic eruptions can have a definitive impact on climate. During volcanic eruptions, fine particles of ash and dust (as well as gases) can be ejected into the stratosphere (see

---

WEATHER WATCH

Could atmospheric particles and a nuclear winter-type event have contributed to the demise of the dinosaurs? One theory proposes that about 65 million years ago a giant meteorite slammed into the earth with such impact that it sent billions of tons of dust and debris into the upper atmosphere. These particles greatly reduced the amount of sunlight reaching the earth's surface, causing cold, dark, and dismal conditions, as well as a disruption in the food chain, which may have adversely affected large plant-eating dinosaurs.

● FIGURE 16.15 Large volcanic eruptions rich in sulfur can affect climate. As sulfur gases in the stratosphere transform into tiny reflective sulfuric acid particles, they prevent a portion of the sun's energy from reaching the surface. Here, the Philippine volcano Mount Pinatubo erupts during June, 1991.

● Fig. 16.15). Scientists agree that the volcanic eruptions having the greatest impact on climate are those rich in sulfur gases. These gases, over a period of about two months, combine with water vapor in the presence of sunlight to produce tiny, reflective sulfuric acid particles that grow in size, forming a dense layer of haze. The haze may reside in the stratosphere for several years, absorbing and reflecting back to space a portion of the sun's incoming energy. The absorption of the sun's energy along with the absorption of infrared energy from the earth warms the stratosphere. The reflection of incoming sunlight by the haze tends to cool the air at the earth's surface, especially in the hemisphere where the eruption occurs.

Two of the largest volcanic eruptions of the 20th century in terms of their sulfur-rich veil, were that of El Chichón in Mexico during April, 1982, and Mount Pinatubo in the Philippines during June, 1991.* Mount Pinatubo ejected an estimated 20 million tons of sulfur dioxide (more than twice that of El Chichón) that gradually worked its way around the globe (see ● Fig. 16.16). For major eruptions such as this one, mathematical models predict that average hemispheric temperatures can drop by about 0.2° to 0.5°C or more for one to three years after the eruption. Model predictions agreed with temperature changes brought on by the Pinatubo eruption, as in early 1992 the mean global surface temperature had decreased by about 0.5°C (see ● Fig. 16.17). The cooling might even have been greater had the eruption not coincided with a major El Niño event that began in 1990 and lasted until early 1995 (see

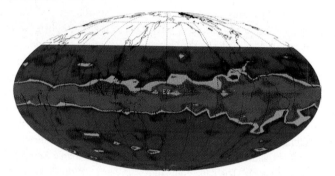

● FIGURE 16.16 Sulfur dioxide plume (dark red and green areas) from the eruption of Mount Pinatubo as measured by the Upper Atmosphere Research Satellite on September 21, 1991. Only three months after the eruption, the plume girdles the equator in the stratosphere at an altitude near 25 km. (NASA)

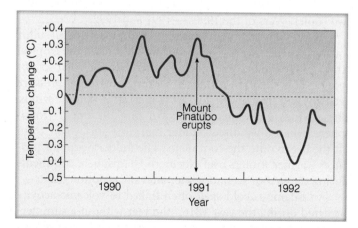

● FIGURE 16.17 Changes in average global air temperature from 1990 to 1992. After the eruption of Mount Pinatubo in June, 1991, the average global temperature by July, 1992, decreased by almost 0.5°C (0.9°F) from the 1981 to 1990 average (dashed line). (Data courtesy of John Christy, University of Alabama, Huntsville, and R. Spencer, NASA Marshall Space Flight Center.)

*The eruption of Mount Pinatubo in 1991 was many times greater than that of Mount St. Helens in the Pacific Northwest in 1980. In fact, the largest eruption of Mount St. Helens was a lateral explosion that pulverized a portion of the volcano's north slope. The ensuing dust and ash (and very little sulfur) had virtually no effect on global climate as the volcanic material was confined mostly to the lower atmosphere and fell out quite rapidly over a large area of the northwestern United States.

Nuclear Winter—Climate Change Induced by Nuclear War

A number of studies indicate that a nuclear war would drastically modify the earth's climate, instigating climate change unprecedented in recorded human history.

Researchers assume that a nuclear war would raise an enormous pall of thick, sooty smoke from massive fires that would burn for days, even weeks, following an attack. The smoke would drift higher into the atmosphere, where it would be caught in the upper-level westerlies and circle the middle latitudes of the Northern Hemisphere. Unlike soil dust, which mainly scatters and reflects incoming solar radiation, soot particles readily absorb sunlight. Hence, for months, or perhaps years, after the war, sunlight would virtually be unable to penetrate the smoke layer, bringing darkness or, at best, twilight at midday.

Such reduction in solar energy would cause surface air temperatures over landmasses to drop below freezing, even during the summer, resulting in extensive damage to plants and crops and the death of millions (or possibly billions) of people. The dark, cold, and gloomy conditions that would be brought on by nuclear war are often referred to as *nuclear winter*.

As the lower troposphere cools, the solar energy absorbed by the smoke particles in the upper troposphere would cause this region to warm. The end result would be a strong, stable temperature inversion extending from the surface up into the higher atmosphere. A strong inversion would lead to a number of adverse effects, such as suppressing convection, altering precipitation processes, and causing major changes in the general wind patterns.

The heating of the upper part of the smoke cloud would cause it to rise upward into the stratosphere, where it would then drift around the world. Thus, about one-third of the smoke would remain in the atmosphere for up to a decade. The other two-thirds would be washed out in a month or so by precipitation. This smoke lofting, combined with persisting sea ice formed by the initial cooling, would produce climatic change that would remain for more than a decade.

Virtually all research on nuclear winter, including models and analog studies, confirms this gloomy scenario. Observations of forest fires show lower temperatures under the smoke, confirming part of the theory. A three-year study involving more than 300 scientists from more than 30 countries conducted by the Scientific Committee On Problems of the Environment (SCOPE) of the International Council of Scientific Unions has detailed the climatic, environmental, and agricultural effects of nuclear winter. The implications of nuclear winter are clear: A nuclear war would drastically alter global climate and would devastate our living environment.

Even with improved global superpower relations, and the end of the Cold War, the danger of nuclear winter remains a possibility. Presently, the current global nuclear arsenal is more than that needed to produce the effects of a nuclear winter. As other nations develop nuclear capability, the potential for nuclear winter remains with us. It will not disappear until the global nuclear weapons arsenal numbers in the hundreds, not in the thousands.

Chapter 10, p. 276, for more information on El Niño). In spite of the El Niño, the eruption of Mount Pinatubo produced the two coolest years of the 1990s—1991 and 1992.

As previously noted, volcanic eruptions rich in sulfur warm the lower stratosphere. During the winter, the tropical stratosphere can become much warmer than the polar stratosphere. This situation produces a strong horizontal pressure gradient and strong west-to-east (zonal) stratospheric winds. These winds work their way down into the upper troposphere, where they direct milder maritime surface air from off the ocean onto the continents. The milder ocean air produces warmer winters over Northern Hemisphere continents during the first or second winter after the eruption occurs.

An infamous cold spell often linked to volcanic activity occurred during the year 1816, "the year without a summer" mentioned earlier. Apparently, a rather stable longwave pattern in the atmosphere produced unseasonably cold summer weather over eastern North America and western Europe. The cold weather followed the massive eruption in 1815 of Mount Tambora in Indonesia. In addition, a smaller volcanic erup-

tion occurred in 1809, from which the climate system may not have fully recovered when Tambora erupted in 1816.

In an attempt to correlate sulfur-rich volcanic eruptions with long-term trends in global climate, scientists are measuring the acidity of annual ice layers in Greenland and Antarctica. Generally, the greater the concentration of sulfuric acid particles in the atmosphere, the greater the acidity of the ice layer. Relatively acidic ice has been uncovered from about A.D. 1350 to about 1700, a time that corresponds to a cooling trend over Europe referred to as the *Little Ice Age*. Such findings suggest that sulfur-rich volcanic eruptions may have played an important role in triggering this com-

WEATHER WATCH

The year without a summer (1816) even had its effect on literature. Inspired (or perhaps dismayed) by the cold, gloomy, summer weather along the shores of Lake Geneva, Mary Shelley wrote the novel *Frankenstein*.

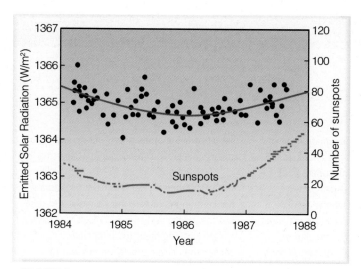

● FIGURE 16.18 Changes in solar energy output (upper curve) in watts per square meter as measured by the *Earth Radiation Budget Satellite*. Bottom curve represents the yearly average number of sunspots. As sunspot activity increases from minimum to maximum, the sun's energy output increases by about 0.1 percent. (From V. Ramanathan, B. R. Barstrom, and E. F. Harrison, "Climate and earth's radiation budget," *Physics Today*, May, 1988, Fig. 5.)

paratively cool period and, perhaps, other cool periods during the geologic past. Moreover, recent core samples taken from the northern Pacific Ocean reveal that volcanic eruptions in the northern Pacific were at least 10 times larger 2.6 million years ago (a time when Northern Hemisphere glaciation began) than previous volcanic events recorded elsewhere in the sediment.

CLIMATE CHANGE: VARIATIONS IN SOLAR OUTPUT

Measurements made by sophisticated radiometers aboard satellites suggest that the sun's energy output (called *brightness*) may vary slightly—by a fraction of one percent—with sunspot activity.

Sunspots are huge magnetic storms on the sun that show up as cooler (darker) regions on the sun's surface. They occur in cycles, with the number and size reaching a maximum approximately every 11 years. During periods of maximum sunspots, the sun emits more energy (about 0.1 percent more) than during periods of sunspot minimums (see ● Fig. 16.18). Evidently, the greater number of bright areas (*faculae*) around the sunspots radiate more energy, which offsets the effect of the dark spots.

It appears that the 11-year sunspot cycle has not always prevailed. Apparently, between 1645 and 1715, during the period known as the **Maunder minimum,*** there were few, if any, sunspots. It is interesting to note that the minimum occurred during the "Little Ice Age," a cool spell in the temperature record experienced mainly over Europe. Some scientists

---

*This period is named after E. W. Maunder, the British solar astronomer who first discovered the low sunspot period sometime in the late 1880s.

suggest that a reduction in the sun's energy output was, in part, responsible for this cold spell.

Fluctuations in solar output may account for climatic changes over time scales of decades and centuries. Many theories have been proposed linking solar variations to climate change, but none have been proven. However, instruments aboard satellites and solar telescopes on the earth are monitoring the sun to observe how its energy output may vary. To date, these measurements show that solar output has only changed a fraction of one percent over several decades. Because many years of data are needed, it may be some time before we fully understand the relationship between solar activity and climate change on earth.

BRIEF REVIEW

Before going on to the next section, here is a brief review of some of the facts and concepts we covered so far:

- The external causes of climate change include: (1) changes in incoming solar radiation; (2) changes in the composition of the atmosphere; (3) changes in the surface of the earth.
- The shifting of continents, along with volcanic activity and mountain building, are possible causes of climate change.
- The Milankovitch theory (in association with other natural forces) proposes that altering glacial and interglacial episodes during the past 2.5 million years are the result of small variations in the tilt of the earth's axis and in the geometry of the earth's orbit around the sun.
- Trapped air bubbles in the ice sheets of Greenland and Antarctica reveal that CO_2 levels and methane levels were lower during colder glacial periods and higher during warmer interglacial periods. But even when the levels were higher, they still were much lower than they are today.
- Sulfate aerosols in the troposphere reflect incoming sunlight, which tends to lower the earth's surface temperature during the day. Sulfate aerosols may also modify clouds by increasing the cloud's reflectivity.
- Volcanic eruptions rich in sulfur may be responsible for cooler periods that span years and decades in the geologic past.
- Fluctuation in solar output (brightness) may account for climatic changes over time scales of decades and centuries.

In previous sections, we saw how increasing levels of CO_2 may have contributed to changes in global climate spanning thousands and even millions of years. Today, we are undertaking a global scientific experiment by injecting vast quantities of greenhouse gases into our atmosphere without fully understanding the long-term consequences. The next section describes how CO_2 and other trace gases appear to be enhancing the earth's greenhouse effect, producing global warming.

Climate Models

Climate models that simulate the physical processes of the atmosphere (and the oceans) are called *General Circulation Models*, or *GCMs* for short. When an atmospheric component of a GCM is linked to an ocean component, the model is said to be "coupled" and the model is called an *Atmosphere–Ocean General Circulation Model*, or *AOGCM*. General circulation models use mathematics and the laws of physics to describe the general behavior of the atmosphere. To reduce some of the atmosphere's complexities, the models make simplified assumptions about the atmosphere and describe the atmosphere in more simplified physical terms. They also reduce many of the small-scale atmospheric processes (such as those due

to clouds) into a single approximation, or parameter, which is known as *parameterization*. The GCMs represent the atmosphere by dividing it up into grid squares, usually several hundred kilometers on a side. General circulation models simulate the behavior of the real atmosphere and describe the major circulation features as well as the seasonal and latitudinal temperature patterns.

A General Circulation Model is first run for a few decades to make sure that the model simulates the real atmosphere. Then, to see how some variables (such as increasing levels of CO_2) might influence the atmosphere, the model is repeatedly run with increasing concentrations of CO_2. In this manner, the GCMs

reveal how the atmosphere and its circulation might change with time, due to increasing levels of greenhouse gases. When the models are run with different scenarios (that is, varying concentrations of greenhouse gases and different forcing agents), the end result is usually a variation in the predicted temperature (such as those temperature projections shown in Fig. 16.19 on p. 457 and in Fig. 16.20 on p. 459).

Although General Circulation Models are not perfect (in that they do not take into account *all* natural factors that affect climate), today's models are extremely sophisticated, and serve as the most reliable tools available for estimating climate change.

Global Warming

We know from Chapter 2 that CO_2 is a greenhouse gas that strongly absorbs infrared radiation and plays a major role in warming the lower atmosphere. We also know that CO_2 has been increasing steadily in the atmosphere, primarily due to the burning of fossil fuel (see Fig. 1.5, p. 8). However, deforestation is also adding to this increase. In 2007, the annual average of CO_2 was about 385 ppm, and present estimates are that if CO_2 levels continue to increase at the same rate that they have been (about 1.9 ppm per year), atmospheric concentrations will rise to between 540 and 970 ppm by the end of this century. To complicate the picture, trace gases such as methane (CH_4), nitrous oxide (N_2O), and chlorofluorocarbons (CFCs), all readily absorb infrared radiation.* Collectively, these gases are approaching CO_2 in their ability to enhance the atmospheric greenhouse effect.

Numerical climate models (mathematical models that simulate climate) predict that by the end of this century increasing concentrations of greenhouse gases could result in an additional warming of about 3°C. The newest, most sophisticated models take into account a number of important relationships, including the interactions between the oceans and the atmosphere, the processes by which CO_2 is removed

from the atmosphere, and the cooling effect produced by sulfate aerosols in the lower atmosphere. The models also predict that as the air warms, additional water vapor will evaporate from the oceans into the air. The added water vapor (which is the most abundant greenhouse gas) will produce a positive feedback on the climate system by enhancing the atmospheric greenhouse effect and accelerating the temperature rise. (This is the *water vapor–greenhouse feedback* described on p. 446.) Without this feedback produced by the added water vapor, the models predict that the warming will be much less. (The models that simulate global climate and predict temperature changes are called *general circulation models*. More information on these models is given in the Focus section above.)

RECENT GLOBAL WARMING: PERSPECTIVE Since the beginning of the 20th century, the average global surface air temperature has risen by more than 0.8°C. Is this warming due to increasing greenhouse gases and an enhanced greenhouse effect? Before we can address this question, we need to review a few concepts we learned in Chapter 2.

Radiative Forcing Agents We know from Chapter 2 that our world without water vapor, CO_2, and other greenhouse gases would be a colder world—about 33°C (59°F) colder than at present. With an average surface temperature of about −18°C (0°F), much of the planet would be uninhabitable. In Chapter 2, we also learned that when the rate of the incoming solar en-

---

*Refer back to Chapter 1 and to Table 1.1, p. 5 for additional information on the concentration of these gases.

ergy balances the rate of outgoing infrared energy from the earth's surface and atmosphere, the earth-atmosphere system is in a state of *radiative equilibrium.* Increasing concentrations of greenhouse gases can disturb this equilibrium and are, therefore, referred to as **radiative forcing agents.** The **radiative forcing*** provided by extra CO_2 and other greenhouse gases increased by about 3 W/m^2 over the past several hundred years, with CO_2 contributing about 60 percent of the increase. So it is very likely that part of the warming during the last century is due to increasing levels of greenhouse gases. But what part does natural climate variability play in global warming? And with levels of CO_2 increasing by more than 25 percent since the early 1900s, why has the observed increase in global temperature been relatively small?

We know that the climate may change due to natural events. For example, changes in the sun's energy output (called *solar irradiance*) and volcanic eruptions rich in sulfur are two major natural radiative forcing agents. Studies show that since the middle 1700s, changes in the sun's energy output may have contributed a small positive forcing (about 0.12 W/m^2) on the climate system, most of which occurred during the first half of the 20th century. On the other hand, volcanic eruptions that inject sulfur-rich particles into the stratosphere produce a negative forcing, which lasts for a few years after the eruption. Because several major eruptions occurred between 1880 and 1920, as well as between 1960 and 1991, the combined change in radiative forcing due to both volcanic activity and solar activity over the past 25 to 45 years appears to be negative, which means that the net effect is that of cooling the earth's surface. Did this cooling in combination with the cooling produced by sulfur-rich aerosols in the lower troposphere reduce the overall warming of the earth's surface during the last century? The use of climate models can help answer this question. (Before going on to the next section, you may want to look at the Focus section on radiative forcing and climate change on p. 458.)

Climate Models and Recent Temperature Trends

The earth's average surface temperature increased by about 0.6°C from 1900 to 2000. How does this observed temperature change over the last century compare with temperature changes derived from climate models using different forcing agents? Before we look at what climate models reveal, it is important to realize that the interactions between the earth and its atmosphere are so complex that it is difficult to unequivocally *prove* that the earth's present warming trend is due entirely to increasing concentrations of greenhouse gases. The problem is that any human-induced signal of climate change is superimposed on a background of natural climatic variations ("noise"), such as the El Niño-Southern Oscillation (ENSO) phenome-

non (discussed in Chapter 10). Moreover, in the temperature observations, it is difficult to separate a signal from the noise of natural climate variability. However, today's more sophisticated climate models are much better at filtering out this noise while at the same time taking into account those forcing agents that are both natural and human-induced.

● Figure 16.19 shows the predicted changes in surface air temperature from 1860 to 2000 made by different climate models using various scenarios (different forcing agents). The gray line presents the actual changes in surface air temperature from 1860 to 2000. Notice that when only increasing levels of greenhouse gases are plugged into the model (yellow line), the model shows a surface temperature increase in excess of 1°C. When greenhouse gases and sulfate aerosols are both added to the model (blue line), the increase in surface temperature is much less; in fact, it is less than the temperature increases observed during the last century. However, when greenhouse gases, sulfate aerosols, and changes in solar radiation are *all* added to the model (red line), the projected temperature change and the observed temperature change closely match.

It is this match in projected and observed temperature trends that helps to explain why a global warming of 0.6°C measured during the last century was less than the warming projected by climate models that took into account only in-

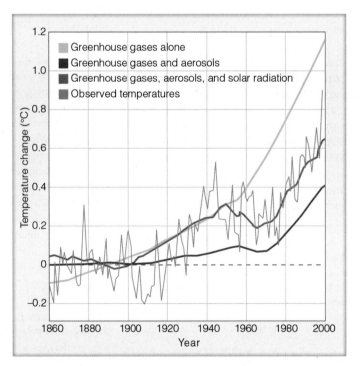

● FIGURE 16.19 Projected surface air temperature changes from different climate models. Model input from greenhouse gases only is shown in yellow; input from greenhouse gases plus sulfate aerosols is shown in blue; input from greenhouse gases, sulfate aerosols, and solar energy changes is shown in red. The gray line shows observed surface temperature change. The dashed line is the 1880 to 1999 mean temperature. (Redrawn from "The Science of Climate Change" by Tom M. L. Wigley, published by the Pew Center of Global Climate Change.)

---

*Radiative forcing is interpreted as an increase (positive) or a decrease (negative) in net radiant energy observed over an area at the tropopause. All factors being equal, an *increase in radiative forcing* may induce surface *warming,* whereas a *decrease* may induce surface *cooling.*

FOCUS ON AN ADVANCED TOPIC

Radiative Forcing — The Ins and Outs

To examine radiative forcing from a slightly different perspective, we need to remember a few important concepts from Chapter 2. First, recall that all objects emit radiation, and that the hotter the object, the more radiant energy it emits. Also recall that this relationship between temperature and radiation (called the Stefan-Boltzmann law) is written as

$$E = \sigma T^4$$

where E is the energy being emitted by each square meter of surface area of the object, T is the object's absolute temperature in Kelvins, and σ is the Stefan-Boltzmann constant. The constant σ is 5.67×10^{-8} with units of Watts per square meter per Kelvin to the fourth power, or W/m^2K^4. The units for energy (called the *flux radiance* or *emittance*) are watts per square meter (W/m^2).

When incoming energy from the sun equals outgoing infrared energy from the earth's surface, the earth is in a state of radiative equilibrium. The earth's radiative equilibrium temperature is about $-18°C$ ($0°F$).* Remember that due to the earth's greenhouse effect this temperature ($-18°C$) is about $33°C$ ($59°F$) lower than the earth's observed average temperature. To better understand radiative forcing, let's see how much energy the earth would be emitting while in radiative equilibrium.

Remember that to obtain this information we have to convert °C to Kelvins; otherwise at a temperature of $-18°C$ the earth would be emitting negative amounts of energy, which is meaningless, consequently

$$K = °C + 273$$
$$K = -18 + 273$$
$$K = 255.$$

If we plug this temperature (255 K) into the Stefan-Boltzmann equation we obtain

$$E = \sigma T^4$$

$$E = (5.67 \times 10^{-8} \left(\frac{W}{m^2K^4}\right))(255\ K)^4$$

$$E = 240 \left(\frac{W}{m^2}\right).$$

Thus, the earth in radiative equilibrium (and behaving as a blackbody) would be emitting 240 watts of infrared energy over each square meter of surface area.

Over the earth as a whole, outgoing infrared energy equals incoming solar energy (see Fig. 2). Consequently, without an atmospheric greenhouse effect, the earth's surface would (on average) emit $240\ W/m^2$ upward and, at the same time, receive $240\ W/m^2$ from the sun. But due to a greenhouse effect this equilibrium of $240\ W/m^2$ is only achieved at the *top* of the atmosphere. Moreover, as greenhouse gases slowly increase in concentration they alter this balance by gradually absorbing more and more of the earth's infrared radiation, thereby preventing this energy from escaping into space. So without any changes in the climate system, outgoing energy would gradually drop below a value of $240\ W/m^2$.

Climate models predict that, as long as everything else remains the same, a sudden doubling of the current levels of atmospheric CO_2 (about 385 ppm) would result in a net radiation reduction (imbalance) of about $4\ W/m^2$ at the top of the atmosphere. To restore this imbalance, the earth's surface and lower atmosphere must warm by about 1.2 Celsius degrees so that more infrared energy is directed upward.* Therefore, as levels of CO_2 and other greenhouse gases increase, they alter the amount of infrared energy lost to space and, in effect, *force* the atmosphere to respond by increasing the surface air temperature.

Any change in average net radiation that occurs at the top of the atmosphere (actually the top of the troposphere) which is due to

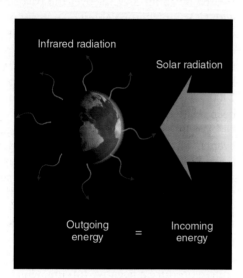

● FIGURE 2 In radiative equilibrium, over the earth as a whole, incoming solar energy equals outgoing infrared energy.

some change in the climate system (such as increasing levels of CO_2) is called *radiative forcing*. Therefore, greenhouse gases (which are increasing in concentration) are referred to as *radiative forcing agents*.*

As levels of greenhouse gases increase, they alter the infrared radiation leaving the atmosphere. This process enhances the atmospheric greenhouse effect, which causes the surface air temperature to rise. As the surface air warms, more evaporation occurs from the oceans, and the water-vapor content of the atmosphere increases. The added water vapor enhances the temperature rise, producing a positive feedback on the climate system. Most climate models show that doubling the concentration of CO_2 and allowing the atmospheric water-vapor content to rise will result in an increase in average surface air temperature of about $3°C$ — a much larger rise in temperature than produced by CO_2 alone.

---

*This temperature is calculated at an average distance from the sun with no atmospheric greenhouse effect.

*Remember that a small increase in temperature results in a great deal more energy emitted, as $E \sim T^4$.

*Additional examples of radiative forcing agents include changes in solar output and land surface modifications such as extensive deforestation.

creasing levels of greenhouse gases. So without negative forcing agents acting on the climate system, global warming during the last century would have likely been greater than observed.

It is climate studies using computer models such as these that have led scientists to conclude that most of the warming during the latter decades of the 20th century is very likely due to increasing levels of greenhouse gases. In fact, the Intergovernmental Panel on Climate Change (IPCC), a committee of over 2000 leading earth scientists, considered the issues of climate change in a report published in 1990 and updated in 1992, in 1995, in 2001, and again in 2007. The 2007 Fourth Assessment report of the IPCC states that:

Most of the observed increase in globally averaged temperatures since the mid-20th century is very likely* due to the observed increase in anthropogenic greenhouse gas concentrations.

FUTURE GLOBAL WARMING: PROJECTIONS   Today's climate models project that, due to increasing levels of greenhouse gases, the surface air temperature will increase substantially by the end of this century (see • Fig. 16.20). Notice, however, that the climate models do not all project the same amount of warming. Each model uses a different scenario describing how greenhouse gas emissions will change with time and how society will utilize energy in the future (see ▼ Table 16.1).

The IPCC in its 2007 report concluded that doubling the concentration of CO_2 would likely produce surface warming in the range of 2°C to 4.5°C, with the best estimate being 3°C. If, during this century, the surface temperature should in-

*In the report "very likely" means a greater than 90 percent probability.

crease by 2°C, the warming would be three times greater than that experienced during the 20th century. An increase of 4.5°C would have potentially devastating effects worldwide. Consequently, it is likely that the warming over this, the 21st century, will be much larger than the warming experienced during the 20th century, and probably greater than any warming during the past 10,000 years.

Uncertainties About Greenhouse Gases   There are, however, uncertainties in predicting the climate of the future. At this point in time, it is unclear how water and land will ultimately affect rising levels of CO_2. Currently, the oceans and the vegetation on land absorb about half of the CO_2 emitted by human sources. As a result, both oceans and landmasses play a major role in the climate system, yet the exact effect they will have on rising levels of CO_2 and global warming is not totally clear. For instance, the microscopic plants (phytoplankton) dwelling in the oceans extract CO_2 from the atmosphere during photosynthesis and store some of it below the oceans' surface, where they die. Will a warming earth trigger a large blooming of these microscopic plants, in effect reducing the rate at which atmospheric CO_2 is increasing?

Current models show that warming the earth tends to *reduce* both ocean and land intake of CO_2. Therefore, if levels of anthropogenic CO_2 emissions continue to increase at their present rate, more CO_2 should remain in the atmosphere to further enhance global warming. An example of how rising temperatures can play a role in altering the way landmasses absorb and emit CO_2 is found in the Alaskan tundra. There, temperatures in recent years have risen to the point where more frozen soil melts in summer than it used to. Accordingly,

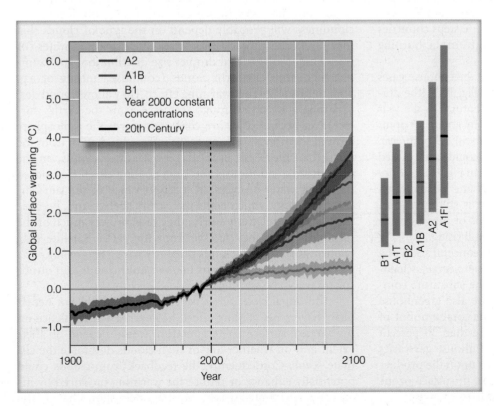

• FIGURE 16.20 Global average projected surface air temperature changes (°C) above the 1980–1999 average (dark purple zero line) for the years 2000 to 2100. Temperature changes inside the graph and to the right of the graph are based on multi-climate models with different scenarios. Each scenario describes how the average temperature will change based on different concentrations of greenhouse gases and various forcing agents. The black line shows global temperature change during the 20th century. The orange line shows projected temperature change where greenhouse gas concentrations are held constant at the year 2000 level. The vertical gray bars on the right side of the figure indicate the likely range of temperature change for each scenario. The thick solid bar within each gray bar gives the best estimate for temperature change for each scenario. (*Source:* Climate Change 2007, *The Physical Science Basis,* by the Working Group 1 contribution to the Fourth Assessment Report to the IPCC © 2007. Reprinted by permission of the Intergovernmental Panel on Climate Change.

▼ TABLE 16.1 The Projected Average Surface Air Temperature Ranges and Best Temperature Estimates for the Decade 2090–2099, Using Six Scenarios*

NAME OF SCENARIO	LIKELY TEMPERATURE RANGE, °C	ESTIMATED TEMPERATURE CHANGE, °C	SCENARIO DESCRIPTION
B1	1.1–2.9	1.8	Energy production, technology, and economy all focus on increased efficiency and minimal resource. Growth rate in high. Energy consumption is very low.
A1T	1.4–3.8	2.4	Energy produced using mostly *non-fossil* sources. Economic and technological growth is rapid. Energy consumption is high.
B2	1.4–3.8	2.4	Energy produced by the most effective means available. Economic and technological development are slow. Energy consumption is moderate.
A1B	1.7–4.4	2.8	Energy produced using a balance of *fossil fuels* and *non-fossil sources*. Economic and technological growth is rapid. Energy consumption is high.
A2	2.0–5.4	3.4	Energy is produced by the simplest means available. Global economic and technological growth is slow. Energy consumption is high.
A1FI	2.4–6.4	4.0	Energy produced using mostly *fossil fuels*. Economic and technological growth is rapid. Energy consumption is high.

*Temperature changes are relative to the average surface air temperature for the period 1980–1999.

during the warmer months, deep layers of exposed decaying peat moss release CO_2 into the atmosphere. Until recently, this region absorbed more CO_2 than it released. Now, however, much of the tundra acts as a producing source of CO_2.

At present, deforestation accounts for about one-fifth of the observed increase in atmospheric CO_2. Hence, changes in land use could influence levels of CO_2 concentrations, especially if the practice of deforestation is replaced by reforestation. Furthermore, it is unknown what future steps countries will take in limiting the emissions of CO_2 from the burning of fossil fuels.

Currently it is not known how quickly greenhouse gases will increase in the future. We can see in ●Fig. 16.21 the dramatic rise in CO_2 levels during the 20th century. In the year 1990, carbon dioxide levels were increasing by about 1.5 ppm/year, whereas today they are increasing by about 1.9 ppm/year. If this trend continues, CO_2 concentrations could easily exceed 550 ppm by the end of this, the 21st century. In Fig. 16.21 notice that the atmospheric concentration of methane has increased dramatically over the last 250 years, and it is still increasing. Also notice that atmospheric concentrations of nitrous oxide have risen quickly, and its concentration is still rising.

Since the mid-1990s, the atmospheric concentration of a group of greenhouse gases called *chlorofluorocarbons* (halocarbons) has been decreasing. However, the substitute compounds for chlorofluorocarbons, which are also greenhouse gases, have been increasing. Moreover, the total amount of surface ozone probably increased by more than 30 percent since 1750. The concentration of this greenhouse gas varies greatly from region to region, and depends upon the production of photochemical smog. The increase in surface ozone has probably led to a small increase in radiative forcing.

The Question of Clouds As the atmosphere warms and more water vapor is added to the air, global cloudiness might increase as well. How, then, would clouds—which come in a variety of shapes and sizes and form at different altitudes—affect the climate system? Clouds reflect incoming sunlight back to space, a process that tends to cool the climate, but clouds also emit infrared radiation to the earth, which tends to warm it. Just how the climate will respond to changes in cloudiness will probably depend on the type of clouds that form and their physical properties, such as liquid water (or ice) content, depth, and droplet size distribution. For example, high, thin cirriform clouds (composed mostly of ice) tend to promote a net warming effect: They allow a good deal of sunlight to pass through (which warms the earth's surface), yet because they are cold, they warm the atmosphere around them by absorbing more infrared radiation from the earth than they emit upward. Low stratified clouds, on the other hand, tend to promote a net cooling effect. Composed mostly of water droplets, they reflect much of the sun's incoming energy, which cools the earth's surface and, because their tops are relatively warm, they radiate away much of the infrared energy they receive from the earth. Satellite data confirm that, overall, clouds presently have a *net cooling effect* on our planet, which means that, without clouds, our atmosphere would be warmer.

Additional clouds in a warmer world would not necessarily have a net cooling effect, however. Their influence on the average surface air temperature would depend on their extent and on whether low or high clouds dominate the climate scene. Consequently, the feedback from clouds could potentially enhance or reduce the warming produced by increasing greenhouse gases. Most models show that as the surface air warms, there will be more convection, more convective-type clouds, and an increase in cirrus clouds. This situation would tend to provide a small positive feedback on the climate system, and the effect of clouds on cooling the earth would be diminished.*

CONSEQUENCES OF GLOBAL WARMING: THE POSSIBILITIES   If the world continues to warm as predicted by climate models, where will most of the warming take place? Climate models predict that land areas will warm more rapidly than the global average, particularly in the northern high latitudes in winter (see ● Fig. 16.22a). We can see in Fig. 16.22b that the greatest surface warming for the period 2001 to 2006 occurred over landmasses in the high latitudes of the Northern Hemisphere. These observations of global average temperature change suggest that climate models are on target with their warming projections.

As high-latitude regions of the Northern Hemisphere continue to warm, modification of the land may actually enhance the warming. For example, the dark green boreal forests of the high latitudes absorb up to three times as much solar energy as does the snow-covered tundra. Consequently, the winter temperatures in subarctic regions are, on the average, much higher than they would be without trees. If warming allows the boreal forests to expand into the tundra, the forests may accelerate the warming in that region. As the temperature rises, organic matter in the soil should decompose at a faster rate, adding more CO_2 to the air, which might accelerate the warming even more. Trees that grow in a climate zone defined by temperature may become especially hard hit as rising temperatures place them in an inhospitable environment. In a weakened state, they may become more susceptible to insects and disease.

As the world warms, total rainfall must increase to balance the increase in evaporation. But precipitation will not be evenly distributed as some areas will get more precipitation, and others less (see ● Fig. 16.23)† Notice in Fig. 16.23a that the models project an increase in winter precipitation over high latitudes of the Northern Hemisphere and a decrease in precipitation over areas of the subtropics. A decrease in precipitation in this region could have an adverse effect by placing added stress on agriculture. Some models even suggest that changes in global patterns of precipitation might cause more extreme rainfall events, such as floods and severe drought. In fact, it is interesting to note that during the warming of the 20th century, there appears to have been an increase in precipitation by as much as 10 percent over the middle- and high-latitude land areas of the Northern Hemisphere. In contrast, it appears that over subtropical land ar-

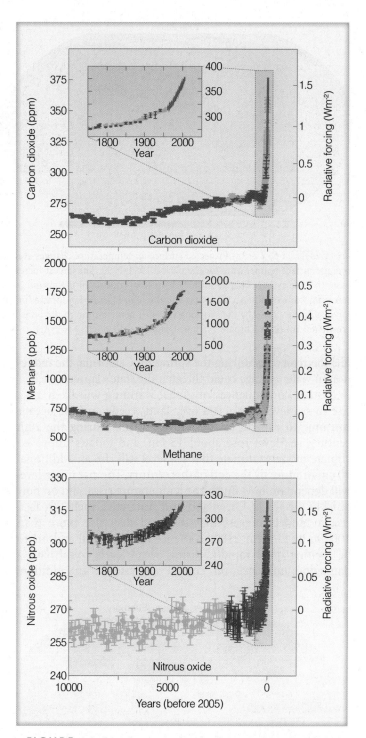

● FIGURE 16.21 Changes in the greenhouse gases carbon dioxide, methane, and nitrous oxide indicated from ice core and modern data. (*Source:* Climate Change 2007, *The Physical Science Basis,* by the Working Group 1 contribution to the Fourth Assessment Report to the IPCC © 2007. Reprinted by permission of the Intergovernmental Panel on Climate Change.

eas, a decrease in precipitation has occurred. It also appears that there has been an increase in the frequency of heavy precipitation events during the last 50 years or so.

In mountainous regions of western North America, where much of the precipitation falls in winter, precipitation might fall mainly as rain, causing a decrease in snow-melt runoff that

---

*In addition to the amount of distribution of clouds, the way in which climate models calculate the optical properties of a cloud (such as albedo) can have a large influence on the model's calculations. Also, there is much uncertainty as to how clouds will interact with aerosols, and what the net effect will be.

†As you look at Fig. 16.23, keep in mind that the stippled areas represent regions where more than 90 percent of the models agree about whether precipitation will increase or decrease, and white areas represent those regions where less than 66 percent of the models agree about how precipitation will change.

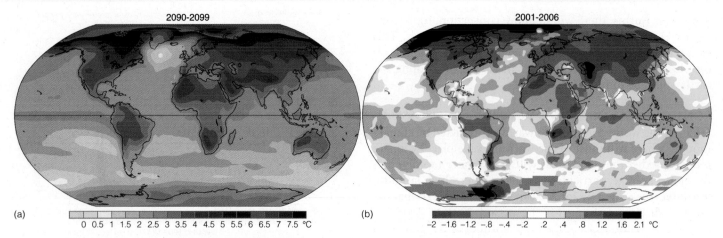

● FIGURE 16.22 (a) Projected surface air temperature changes averaged for the decade 2090–2099 (using the A1B scenario) compared to the average surface temperature for the period 1980–1999. The largest increase in air temperature is projected to be over landmasses and in the Arctic region. (b) The average change in surface air temperature for the period 2001–2006 compared to the average for the years 1951–1980. The greatest warming was over the Arctic region and the high-latitude landmasses of the Northern Hemisphere. (Diagram [a] (*Source:* Climate Change 2007, *The Physical Science Basis* by the Working Group 1 contribution to the Fourth Assessment Report to the IPCC. Reprinted by permission of the Intergovernmental Panel on Climate Change. Diagram [b] Courtesy NASA.)

fills the reservoirs during the spring. In California, the reduction in water storage could threaten the state's agriculture.

Other consequences of global warming will likely be a rise in sea level as glaciers over land recede and the oceans continue to expand as they slowly warm. During the 20th century, sea level rose about 15 cm, and today's improved climate models estimate that sea level will rise an additional 30 cm or more by the end of this century. The rise in sea level will depend on how much the temperature rises, and on how quickly the ice in Greenland and Antarctica melts. In fact, recent models suggest that sea level may rise more than 100 cm by the year 2100, as the ice in Greenland appears to be melting quite rapidly. Rising ocean levels could have a damaging influence on coastal ecosystems. In addition,

coastal groundwater supplies might become contaminated with saltwater. And as we saw in Chapter 15, as sea surface temperatures increase (other factors being equal) the intensity of hurricanes will likely increase as well.*

In polar regions, as elsewhere around the globe, rising temperatures produce complex interactions among temperature, precipitation, and wind patterns. Hence, in polar areas more snow might actually fall in the warmer (but still cold) air, causing snow to build up or, at least, stabilize over the continent of Antarctica. Over Greenland, which is experiencing rapid melting of ice and snow, any increase in precipita-

---

*For more information on hurricanes and global warming, read the Focus section "Hurricanes in a Warmer World" on p. 434.

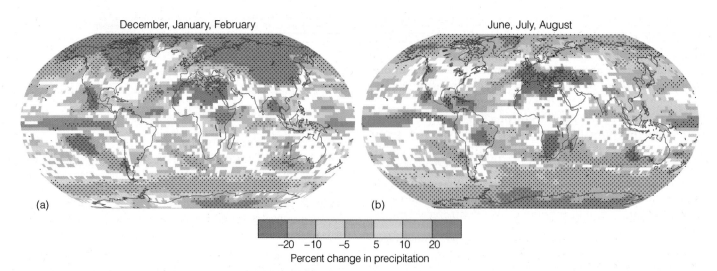

● FIGURE 16.23 Projected relative changes in precipitation (in percent) for the last decade of this century (2090–2099) compared to the average for the period 1980–1999. Values are multimodel averages for (a) December through February, and (b) June through August. The stippled areas represent regions where more than 90 percent of the models agree as to whether precipitation will increase or decrease; white regions show where less than 66 percent of the models agree about how precipitation will change. (*Source:* Climate Change 2007, *The Physical Science Basis,* by the Working Group 1 contribution to the Fourth Assessment Report to the IPCC. Reprinted by permission of the Intergovernmental Panel on Climate Change.

tion will likely be offset by rapid melting, and so the ice sheet is expected to continue to shrink. Presently, in the Arctic, warming has caused sea ice* to shrink and thin. During 2005, the extent of Arctic sea ice was at a record minimum for every month except May. If the warming in this region continues at its present rate, polar sea ice in summer may be totally absent by the middle of this century (see ● Fig. 16.24).

Increasing levels of CO_2 in a warmer world might have additional consequences. For example, higher levels of CO_2 might act as a "fertilizer" for some plants, accelerating their growth. Increased plant growth consumes more CO_2, which might retard the increasing rate of CO_2 in the environment. On the other hand, the increased plant growth might force some insects to eat more, resulting in a net loss in vegetation. It is possible that a major increase in CO_2 might upset the balance of nature, with some plant species becoming so dominant that others are eliminated. In tropical areas, where many developing nations are located, the warming may actually decrease crop yield, whereas in cold climates, where crops are now grown only marginally, the warming effect may actually increase crop yields. In a warmer world, higher latitudes might benefit from a longer growing season, and extremely cold winters might become less numerous with fewer bitter cold spells.

Following are some conclusions about global warming and its future impact on our climate system summarized from the 2007 Fourth Assessment Report of the Intergovernmental Panel on Climate Change (IPCC):

● The primary source of the increased atmospheric concentration of carbon dioxide since the pre-industrial period results from fossil fuel use, with land-use change providing another significant but smaller contribution. The at-

---

*Sea ice is formed by the freezing of sea water.

WEATHER WATCH

In our warmer world, many freshwater lakes in northern latitudes are freezing later in the fall and thawing earlier in the spring than they did in years past. Wisconsin's Lake Mendota, for example, now averages about 40 fewer days with ice than it did 150 years ago.

mospheric concentration of carbon dioxide in 2005 exceeds by far the natural range over the last 650,000 years (180 to 300 ppm) as determined from ice cores.

● Average Northern Hemisphere temperatures during the second half of the 20th century were *very likely* higher than during any other 50-year period in the last 500 years and *likely* the highest in at least the past 1,300 years.

● Temperatures of the most extreme hot nights, cold nights, and cold days are *likely* to have increased due to anthropogenic forcing. It is *more likely than not* that anthropogenic forcing has increased the risk of heat waves.

● Since IPCC's first report in 1990, assessed projections have suggested global average temperature increases between about 0.15°C and 0.3°C per decade for 1990 to 2005. This can now be compared with observed values of about 0.2°C per decade, strengthening confidence in near-term projections.

● Widespread changes in extreme temperatures have been observed over the last 50 years. Cold days, cold nights, and frost have become less frequent, whereas hot days, hot nights and heat waves have become more frequent.

● The average atmospheric water vapor content has increased since at least the 1980s over land and ocean as well as in the upper troposphere. The increase is broadly consistent with the extra water vapor that warmer air can hold.

● FIGURE 16.24 The extent of Arctic sea ice in (a) March, 2005, when the ice cover was at or near its maximum and in (b) September, 2005, when the ice cover was near or at its minimum. The orange line represents the (a) median maximum and (b) median minimum extent of the ice cover for the period 1979–2000.

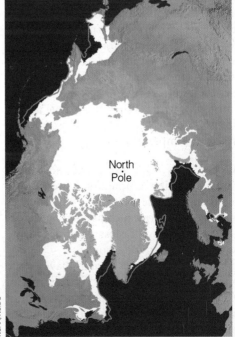

North Pole

North Pole

NOAA/NSIDG

(a) March 2005   (b) September 2005

- Observations since 1961 show that the average temperature of the global ocean has increased to depths of at least 3000 m and that the ocean has been absorbing more than 80% of the heat added to the climate system. Such warming causes seawater to expand, contributing to sea level rise.

- Average arctic temperatures increased at almost twice the global average rate in the past 100 years. Arctic temperatures have high decadal variability, and a warm period was also observed from 1925 to 1945.

- Mid-latitude westerly winds have strengthened in both hemispheres since the 1960s.

- Extratropical storm tracks are projected to move poleward, with consequent changes in wind, precipitation and temperature patterns, continuing the broad pattern of observed trends over the last half-century.

- Based on a range of models, it is *likely* that future tropical cyclones (typhoons and hurricanes) will become more intense, with larger peak wind speeds and more heavy precipitation associated with ongoing increases of tropical sea surface temperatures.

- New analyses of balloon-borne and satellite measurements of lower- and mid-tropospheric temperature show warming rates that are similar to those of the surface temperature record and are consistent within their respective uncertainties, largely reconciling a discrepancy noted in the Third Assessment Report.

- Global average sea level rose at an average rate of 1.8 [1.3 to 2.3] mm per year over 1961 to 2003. The rate was faster over 1993 to 2003: about 3.1 [2.4 to 3.8] mm per year.

- More intense and longer droughts have been observed over wider areas since the 1970s, particularly in the tropics and subtropics.

- Mountain glaciers and snow cover have declined on average in both hemispheres. Widespread decreases in glaciers and ice caps have contributed to sea level rise (ice caps do not include contributions from the Greenland and Antarctic Ice Sheets).

- Both past and future anthropogenic carbon dioxide emissions will continue to contribute to warming and sea level rise for more than a millennium, due to the time scales required for removal of this gas from the atmosphere.

- The observed widespread warming of the atmosphere and ocean, together with ice mass loss, support the conclusion that it is *extremely unlikely* that global climate change of the past 50 years can be explained without external forcing, and *very likely* that it is not due to known natural causes alone.

GLOBAL WARMING: LAND USE CHANGES All climate models predict that, as humanity continues to spew greenhouse gases into the air, the climate will change and the earth's surface will warm. But are humans changing the climate by other activities as well? Modification of the earth's surface taking place right now could potentially be influencing the immediate climate of certain regions. For example, studies show that about half the rainfall in the Amazon River Basin is returned to the atmosphere through evaporation and through transpiration from the leaves of trees. Consequently, clearing large areas of tropical rain forests in South America to create open areas for farms and cattle ranges will most likely cause a decrease in evaporative cooling. This decrease, in turn, could lead to a warming in that area of at least several degrees Celsius. In turn, the reflectivity of the deforested area will change. Similar changes in albedo result from the overgrazing and excessive cultivation of grasslands in semi-arid regions, causing an increase in desert conditions (a process known as **desertification**).

Currently, billions of acres of the world's range and cropland, along with the welfare of millions of people, are affected by desertification. Annually, millions of acres are reduced to a state of near or complete uselessness. The main cause is overgrazing, although overcultivation, poor irrigation practices, and deforestation also play a role. The effect this will have on climate, as surface albedos increase and more dust is swept into the air, is uncertain. (For a look at how a modified surface influences the inhabitants of a region in Africa, read the Focus section on p. 465.)

It is interesting to note that some scientists feel that humans may have been altering climate way before modern civilizations came along. For example, retired Professor William Ruddiman of the University of Virginia suggests that humans have been influencing climate change for the past 8000 years. Although some climate scientists vehemently oppose his ideas, Ruddiman speculates that without preindustrial farming, which produces methane and some carbon dioxide, we would have entered a naturally occurring ice age. He even suggests that the Little Ice Age of the 15th through the 19th centuries in Europe was human-induced because plagues, which killed millions of people, caused a reduction in farming. The reasoning behind this idea goes something like this: As forests are cleared for farming, levels of CO_2 and methane increase, producing a strong greenhouse effect and a rise in surface air temperature. When catastrophic plagues strike — the bubonic plague, for instance — high mortality rates cause farms to be abandoned. As forests begin to take over the untended land, levels of CO_2 and methane drop, causing a reduction in the greenhouse effect and a corresponding drop in air temperature. When the plague abates, the farms return, forests are cleared, levels of greenhouse gases go up, and surface air temperatures rise.

GLOBAL WARMING: EFFORTS TO CURB The most obvious way to curb global warming is to reduce greenhouse gas emissions by reducing the use of fossil fuels, such as oil and coal. Using alternative energy such as solar collectors and wind power — the world's two fastest growing energy sources — could also help with this endeavor.

In an attempt to mitigate the impact humans have on the climate system, representatives from 160 countries met at Kyoto, Japan, in 1997 to work out a formal agreement to limit greenhouse gas emissions in industrialized nations. The international agreement — called the *Kyoto Protocol* — was adopted in 1997, and was put into force in February, 2005.

FOCUS ON A SPECIAL TOPIC

The Sahel—An Example of Climatic Variability and Human Existence

The Sahel is in North Africa, located between about 14° and 18°N latitude (see Fig. 3). Bounded on the north by the dry Sahara and on the south by the grasslands of the Sudan, the Sahel is a semi-arid region of variable rainfall. Precipitation totals may exceed 50 cm (20 in.) in the southern portion while in the north, rainfall is scanty. Yearly rainfall amounts are also variable as a year with adequate rainfall can be followed by a dry one.

During the winter, the Sahel is dry, but, as summer approaches, the Intertropical Convergence Zone (ITCZ) with its rain usually moves into the region. The inhabitants of the Sahel are mostly nomadic people who migrate to find grazing land for their cattle and goats. In the early and middle 1960s, adequate rainfall led to improved pasturelands; herds grew larger and so did the population. However, in 1968, the annual rains did not reach as far north as usual, marking the beginning of a series of dry years and a severe drought.

The decrease in rainfall, along with overgrazing, turned thousands of square kilometers of pasture into barren wasteland. By 1973, when the severe drought reached its climax, rainfall totals were 50 percent of the long-term average, and perhaps 50 percent of the cattle and goats had died. The Sahara Desert had migrated southward into the northern fringes of the region, and a great famine had taken the lives of more than 100,000 people.

Although low rainfall years have been followed by wetter ones, relatively dry conditions have persisted over the region for the past 40 years or so. The overall dryness of the region has caused many of the larger, shallow lakes (such as Lake Chad) to shrink in size. The wetter years of the 1950s and 1960s appear to be due to the northward displacement of the ITCZ. The drier years, however, appear to be more related to the intensity of rain that falls during the so-called

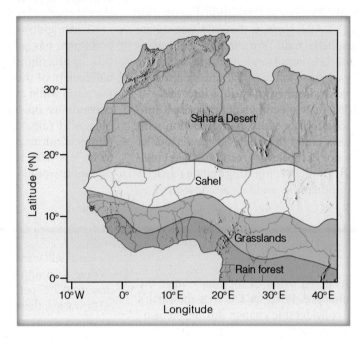

● FIGURE 3
The semi-arid Sahel of North Africa is bounded by the Sahara Desert to the north and grasslands to the south.

rainy season. But what causes the lack of intense rain? Some scientists feel that this situation is due to a *biogeophysical feedback mechanism* wherein less rainfall and reduced vegetation cover modify the surface and promote a positive feedback relationship: Surface changes act to reduce convective activity, which in turn promotes or reinforces the dry conditions. As an example, when the vegetation is removed from the surface (perhaps through overgrazing or excessive cultivation), the surface albedo (reflectivity) increases, and the surface temperature drops. But studies show that less vegetation cover does not always result in a higher albedo.

Since the mid-1970s the Sahara Desert has not progressively migrated southward into the Sahel. In fact, during dry years, the desert does migrate southward, but in wet years, it retreats. By the same token, vegetation cover throughout the Sahel is more extensive during the wetter years. Consequently, desertification is not pres-

ently overtaking the Sahel, nor is the albedo of the region showing much year-to-year change.

So the question remains: Why did the Sahel experience such devastating drought during the 1970s and 1980s? Recent studies suggest that the dry periods were due to a cooler North Atlantic Ocean. The cooler ocean appears to be the result of sulfate aerosols that enhance the formation of highly reflective clouds above the water. The increase in cloud reflectivity cooled the ocean surface, which in turn influenced the circulation of the atmosphere in such a way that the ITCZ did not, on average, move as far north. The sulfate pollution* apparently originated over North America, suggesting that human activities on one continent could potentially cause climate variability on another, with the end result being a disastrous famine.

*Recent studies show a correlation between sulfate particles ejected into the stratosphere from volcanic eruptions and past dry spells in the Sahel.

The Protocol sets mandatory targets for reducing greenhouse gas emissions in countries that adopt the plan. Although the percent by which each country reduces its emissions varies, the overall goal is to reduce greenhouse gas emissions in developed countries by at least 5 percent below existing 1990 levels during the 5-year period of 2008 through 2012.

The agreement gives countries flexibility in meeting their emission-reduction goals. For example, a country that plants forests can receive "credit" for reducing greenhouse gases, because trees act as a "sink" and remove CO_2 from the atmosphere. Other types of "credits" may be given to industrialized countries that establish emission-reducing projects in developing countries. Although the plan has gained worldwide acceptance, the United States has not signed the Protocol as of this writing. However, many large states such as California have implemented climate change policies. California's aggressive plan (adopted in 2006) sets targets for reducing greenhouse gas emissions to 1990 levels by the year 2020.

A study headed by Tom Wigley at the National Center for Atmospheric Research (NCAR), suggests that injecting sulfate aerosols into the stratosphere could slow down global warming. Using computer models to simulate climate, tons of sulfate aerosols—on the order of those lofted by Mount Pinatubo in 1991—were put into the stratosphere at various intervals.

The study concluded that injecting sulfate aerosols into the stratosphere every one to four years in conjunction with reducing greenhouse gases could provide a "grace period" of up to 20 years before major cutbacks in greenhouse gas emissions would be required. Of course, injecting sulfate particles into the stratosphere might have additional consequences, such as changing the temperature of the upper atmosphere and affecting the fragile ozone layer. Although the idea of injecting the stratosphere with sulfate particles has not been given much credence by scientists, the idea of reducing the impact of climate change through global scale technological fixes (called *geoengineering*) is intriguing. The science of geoengineering is fairly new, and typically poses costly technological challenges for the scientific community.

Cutting down on the emissions of greenhouse gases and pollutants has several potentially positive benefits. A reduction in greenhouse gas emissions could slow down the enhancement of the earth's greenhouse effect and reduce global warming while at the same time it would reduce a country's dependence on oil. A reduction in air pollutants might reduce acid rain, diminish haze, and slow the production of photochemical smog. Even if the greenhouse warming proves to be less than what modern climate models project, these measures would certainly benefit humanity.

SUMMARY

In this chapter, we considered some of the many ways the earth's climate can be changed. First, we saw that the earth's climate has undergone considerable change during the geologic past. Some of the evidence for a changing climate comes from tree rings (dendrochronology), chemical analysis of oxygen isotopes in ice cores and fossil shells, and geologic evidence left behind by advancing and retreating glaciers. The evidence from these suggests that, throughout much of the geologic past (long before humanity arrived on the scene), the earth was warmer than it is today. There were cooler periods, however, during which glaciers advanced over large sections of North America and Europe.

We examined some of the possible causes of climate change, noting that the problem is extremely complex, as a change in one variable in the climate system almost immediately changes other variables. One theory suggests that the shifting of the continents, along with volcanic activity and mountain building, may account for variations in climate that take place over millions of years.

The Milankovitch theory proposes that alternating glacial and interglacial episodes during the past 2.5 million years are the result of small variations in the tilt of the earth's axis and in the geometry of the earth's orbit around the sun. Another theory suggests that certain cooler periods in the geologic past may have been caused by volcanic eruptions rich in sulfur. Still another theory postulates that climatic variations on earth might be due to variations in the sun's energy output.

We looked at temperature trends over the past 100 years or so and found that, over this span of time, the earth has warmed by more than 0.8°C. It is very likely that most of the warming during the last 50 years is due to increasing concentrations of greenhouse gases. Sophisticated climate models project that, as levels of CO_2 and other greenhouse gases continue to increase, the earth will warm substantially by the end of this century. The average warming over the next several decades will likely be close to 0.2°C per decade. The models also predict that, as the earth warms, there will be a global increase in atmospheric water vapor, an increase in global precipitation, a more rapid melting of sea ice, and a rise in sea level.

KEY TERMS

The following terms are listed (with page numbers) in the order they appear in the text. Define each. Doing so will aid you in reviewing the material covered in this chapter.

global warming, 440
dendrochronology, 442
Ice Age, 442
interglacial period, 443
Younger Dryas (event), 443
mid-Holocene maximum, 443
Little Ice Age, 445
water vapor–greenhouse feedback, 446
positive feedback mechanism, 446
snow-albedo feedback, 447

negative feedback mechanism, 447
theory of plate tectonics, 447
Milankovitch theory, 449
eccentricity, 449
precession, 450
obliquity, 450
sulfate aerosols, 452
Maunder minimum, 455
radiative forcing agents, 457
radiative forcing, 457
desertification, 464

QUESTIONS FOR REVIEW

1. What methods do scientists use to determine climate conditions that have occurred in the past?
2. Explain how the changing climate influenced the formation of the Bering land bridge.
3. How does today's average global temperature compare with the average temperature during most of the past 1000 years?
4. What is the Younger Dryas episode? When did it occur?

5. How does a positive feedback mechanism differ from a negative feedback mechanism? Is the water vapor–greenhouse feedback considered positive or negative? Explain.

6. How does the theory of plate tectonics explain climate change over periods of millions of years?

7. Describe the Milankovitch theory of climatic change by explaining how each of the three cycles alters the amount of solar energy reaching the earth.

8. Given the analysis of air bubbles trapped in polar ice during the past 160,000 years, were CO_2 levels generally higher or lower during colder glacial periods? Were methane levels higher or lower at this time?

9. How do sulfate aerosols in the lower atmosphere affect surface air temperatures during the day?

10. Describe the scenario of nuclear winter.

11. Volcanic eruptions rich in sulfur warm the stratosphere. Do they tend to warm or cool the earth's surface? Explain.

12. Explain how variations in the sun's energy output might influence global climate.

13. Climate models predict that increasing levels of CO_2 will cause the mean global surface temperature to rise significantly by the year 2100. What other greenhouse gas *must* also increase in concentration in order for this condition to occur?

14. Describe some of the natural radiative forcing agents and their effect on climate.

15. (a) Describe how clouds influence the climate system.
    (b) Which clouds would tend to promote surface cooling: high clouds or low clouds?

16. In Fig. 16.19, p. 457, explain why the actual rise in surface air temperature (gray line) is much less than the projected rise in air temperature due to increasing levels of greenhouse gases (yellow line).

17. Why do climate scientists now believe that most of the warming experienced during the last 50 years was due to increasing levels of greenhouse gases?

18. List some of the consequences that global warming might have on the atmosphere and its inhabitants.

19. Is CO_2 the only greenhouse gas we should be concerned with for climate change? If not, what are the other gases?

20. Explain how the ocean's conveyor belt circulation works. How does the conveyor belt appear to influence the climate of northern Europe? (Hint: the answer is found in the Focus section on p. 444.)

QUESTIONS FOR THOUGHT

1. Ice cores extracted from Greenland and Antarctica have yielded valuable information on climate changes during the past few hundred thousands of years. What do you feel might be some of the limitations in using ice core information to evaluate past climate changes?

2. When glaciation was at a maximum (about 18,000 years ago), was global precipitation greater or less than at present? Explain your reasoning.

3. Consider the following climate change scenario. Warming global temperatures increase saturation vapor pressures over the ocean. As more water evaporates, increasing quantities of water vapor build up in the troposphere. More clouds form as the water vapor condenses. The clouds increase the albedo, resulting in decreased amounts of solar radiation reaching the earth's surface. Is this scenario plausible? What type(s) of feedback(s) is/are involved?

4. Explain why periods of glacial advance in the higher latitudes tend to occur with warmer winters and cooler summers.

5. Explain two different ways that an increase in sulfate particles might lower surface air temperatures.

6. Are ice ages in the Northern Hemisphere more likely when:
   (a) the tilt of the earth is at a maximum or a minimum?
   (b) the sun is closest to the earth during summer in the Northern Hemisphere, or during winter?
   Explain your reasoning for both (a) and (b).

7. Most climate models show that the poles will warm faster than the tropics. What effect will this have on winter storms in mid-latitudes?

8. The oceans are a major sink (absorber) of CO_2. One hypothesis states that as warming increases, less CO_2 will be dissolved in the oceans. Would you expect the earth to cool or to warm further? Why?

PROBLEMS AND EXERCISES

1. If the annual precipitation near Hudson Bay (latitude 55°N) is 38 cm (15 in.) per year, calculate how long it would take snow falling on this region to reach a thickness of 3000 m (about 10,000 ft). (Assume that all the precipitation falls as snow, that there is no melting during the summer, and that the annual precipitation remains constant. To account for compaction of the snow, use a water equivalent of 5 to 1.)

2. On a warming planet, the snow-albedo feedback produces a positive feedback. Make a diagram (or several diagrams) to illustrate this phenomenon. Now, with another diagram, show that the snow-albedo feedback produces a positive feedback on a cooling planet.

Visit the
Meteorology Resource Center
at
academic.cengage.com/login
for more assets, including questions for exploration, animations, videos, and more.

In mountainous regions, a variety of climate types can exist within a relatively short distance. Here, in Colorado, aspen change color in a continental-type climate, while the high peaks with a polar climate experience perpetual snow.
© Carr Clifton/Minden Pictures

CHAPTER 17

Global Climate

The climate is unbearable . . . At noon today the highest temperature measured was −33°C. We really feel that it is late in the season. The days are growing shorter, the sun is low and gives no warmth, katabatic winds blow continuously from the south with gales and drifting snow. The inner walls of the tent are like glazed parchment with several millimeters thick ice-armour . . . Every night several centimeters of frost accumulate on the walls, and each time you inadvertently touch the tent cloth a shower of ice crystals falls down on your face and melts. In the night huge patches of frost from my breath spread around the opening of my sleeping bag and melt in the morning. The shoulder part of the sleeping bag facing the tent-side is permeated with frost and ice, and crackles when I roll up the bag . . . For several weeks now my fingers have been permanently tender with numb fingertips and blistering at the nails after repeated frostbites. All food is frozen to ice and it takes ages to thaw out everything before being able to eat. At the depot we could not cut the ham, but had to chop it in pieces with a spade. Then we threw ourselves hungrily at the chunks and chewed with the ice crackling between our teeth. You have to be careful with what you put in your mouth. The other day I put a piece of chocolate from an outer pocket directly in my mouth and promptly got frostbite with blistering of the palate.

Ove Wilson (Quoted in David M. Gates, *Man and His Environment*)

CONTENTS

Our opening comes from a report by Norwegian scientists on their encounter with one of nature's cruelest climates—that of Antarctica. Their experience illustrates the profound effect that climate can have on even ordinary events, such as eating a piece of chocolate. Though we may not always think about it, climate profoundly affects nearly everything in the middle latitudes, too. For instance, it influences our housing, clothing, the shape of landscapes, agriculture, how we feel and live, and even where we reside, as most people will choose to live on a sunny hillside rather than in a cold, dark, and foggy river basin. Entire civilizations have flourished in favorable climates and have moved away from, or perished in, unfavorable ones. We learned early in this text that *climate* is the average of the day-to-day weather over a long duration. But the concept of climate is much larger than this, for it encompasses, among other things, the daily and seasonal extremes of weather within specified areas.

When we speak of climate, then, we must be careful to specify the spatial location we are talking about. For example, the Chamber of Commerce of a rural town may boast that its community has mild winters with air temperatures seldom below freezing. This may be true several meters above the ground in an instrument shelter, but near the ground the temperature may drop below freezing on many winter nights. This small climatic region near or on the ground is referred to as a **microclimate.** Because a much greater extreme in daily air temperatures exists near the ground than several meters above, the microclimate for small plants is far more harsh than the thermometer in an instrument shelter would indicate.

When we examine the climate of a small area of the earth's surface, we are looking at the **mesoclimate.** The size of the area may range from a few acres to several square kilometers. Mesoclimate includes regions such as forests, valleys, beaches, and towns. The climate of a much larger area, such as a state or a country, is called **macroclimate.** The climate extending over the entire earth is often referred to as **global climate.**

In this chapter, we will concentrate on the larger scales of climate. We will begin with the factors that regulate global climate; then we will discuss how climates are classified. Finally, we will examine the different types of climate.

A World with Many Climates

The world is rich in climatic types. From the teeming tropical jungles to the frigid polar "wastelands," there seems to be an almost endless variety of climatic regions. The factors that produce the climate in any given place—the **climatic controls**—are the same that produce our day-to-day weather. Briefly, the controls are the:

1. intensity of sunshine and its variation with latitude
2. distribution of land and water
3. ocean currents

WEATHER WATCH

Even "summers" in the Antarctic can be brutal. In 1912, during the Antarctic summer, Robert Scott of Great Britain not only lost the race to the South Pole to Norway's Roald Amundsen, but perished in a blizzard trying to return. Temperature data taken by Scott and his crew showed that the summer of 1912 was unusually cold, with air temperatures remaining below −34°C −30°F for nearly a month. These exceptionally low temperatures eroded the men's health and created an increase in frictional drag on the sleds the men were pulling. Just before Scott's death, he wrote in his journal that "no one in the world would have expected the temperatures and surfaces which we encountered at this time of year."

4. prevailing winds
5. positions of high- and low-pressure areas
6. mountain barriers
7. altitude

We can ascertain the effect these controls have on climate by observing the global patterns of two weather elements—temperature and precipitation.

GLOBAL TEMPERATURES • Figure 17.1 shows mean annual temperatures for the world. To eliminate the distorting effect of topography, the temperatures are corrected to sea level.* Notice that in both hemispheres the isotherms are oriented east-west, reflecting the fact that locations at the same latitude receive nearly the same amount of solar energy. In addition, the annual solar heat that each latitude receives decreases from low to high latitude; hence, annual temperatures tend to decrease from equatorial toward polar regions.†

The bending of the isotherms along the coastal margins is due in part to the unequal heating and cooling properties of land and water, and to ocean currents and upwelling. For example, along the west coast of North and South America, ocean currents transport cool water equatorward. In addition to this, the wind in both regions blows toward the equator, parallel to the coast. This situation favors upwelling of cold water (see Chapter 10), which cools the coastal margins. In the area of the eastern North Atlantic Ocean (north of 40°N), the poleward bending of the isotherms is due to the Gulf Stream and the North Atlantic Drift, which carry warm water northward.

The fact that landmasses heat up and cool off more quickly than do large bodies of water means that variations in temperature between summer and winter will be far greater over continental interiors than along the west coastal margins of continents. By the same token, the climates of interior continental regions will be more extreme, as they have (on the

---

*This correction is made by adding to each station above sea level an amount of temperature that would correspond to the normal (standard) temperature lapse rate of 6.5°C per 1000 m (3.6°F per 1000 ft).

†Average global temperatures for January and July are given in Figs. 3.20 and 3.21, respectively, on p. 74.

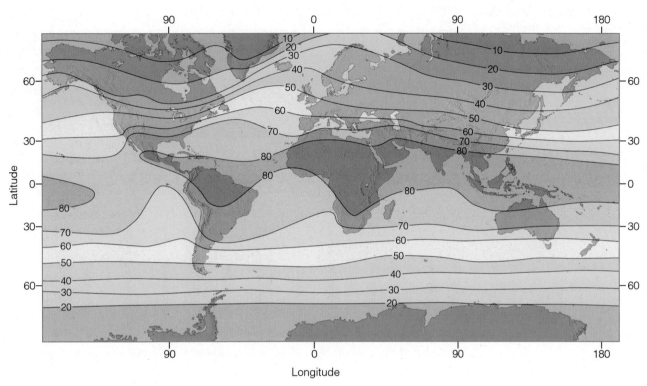

● FIGURE 17.1 Average annual sea-level temperatures throughout the world (°F).

average) higher summer temperatures and lower winter temperatures than their west-coast counterparts. In fact, west-coast climates are typically quite mild for their latitude.

The highest mean temperatures do not occur in the tropics, but rather in the subtropical deserts of the Northern Hemisphere. Here, the subsiding air associated with the subtropical anticyclones produces generally clear skies and low humidity. In summer, the high sun beating down upon a relatively barren landscape produces scorching heat.

The lowest mean temperatures occur over large landmasses at high latitudes. The coldest areas in the Northern Hemisphere are found in the interior of Siberia and Greenland, whereas the coldest area of the world is the Antarctic. During part of the year, the sun is below the horizon; when it is above the horizon, it is low in the sky and its rays do not effectively warm the surface. Consequently, the land remains snow- and ice-covered year-round. The snow and ice reflect perhaps 80 percent of the sunlight that reaches the surface. Much of the unreflected solar energy is used to transform the ice and snow into water vapor. The relatively dry air and the Antarctic's high elevation permit rapid radiational cooling during the dark winter months, producing extremely cold surface air. The extremely cold Antarctic helps to explain why, overall, the Southern Hemisphere is cooler than the Northern Hemisphere. Other contributing factors for a cooler Southern Hemisphere include the fact that polar regions of the Southern Hemisphere reflect more incoming sunlight, and the fact that less land area is found in tropical and subtropical areas of the Southern Hemisphere.

GLOBAL PRECIPITATION Appendix G shows the worldwide general pattern of annual precipitation, which varies from place to place. There are, however, certain regions that stand out as being wet or dry. For example, equatorial regions are typically wet, while the subtropics and the polar regions are relatively dry. The global distribution of precipitation is closely tied to the general circulation of winds in the atmosphere (Chapter 10) and to the distribution of mountain ranges and high plateaus.

● Figure 17.2 shows in simplified form how the general circulation influences the north-to-south distribution of precipitation to be expected on a uniformly water-covered earth. Precipitation is most abundant where the air rises; least abundant where it sinks. Hence, one expects a great deal of precipitation in the tropics and along the polar front, and little near subtropical highs and at the poles. Let's look at this in more detail.

In tropical regions, the trade winds converge along the Intertropical Convergence Zone (ITCZ), producing rising air, towering clouds, and heavy precipitation all year long. Poleward of the equator, near latitude 30°, the sinking air of the subtropical highs produces a "dry belt" around the globe. The Sahara Desert of North Africa is in this region. Here, annual rainfall is exceedingly light and varies considerably from year to year. Because the major wind belts and pressure systems shift with the season—northward in July and southward in January—the area between the rainy tropics and the dry subtropics is influenced by both the ITCZ and the subtropical highs.

• FIGURE 17.2 A vertical cross section along a line running north to south illustrates the main global regions of rising and sinking air and how each region influences precipitation.

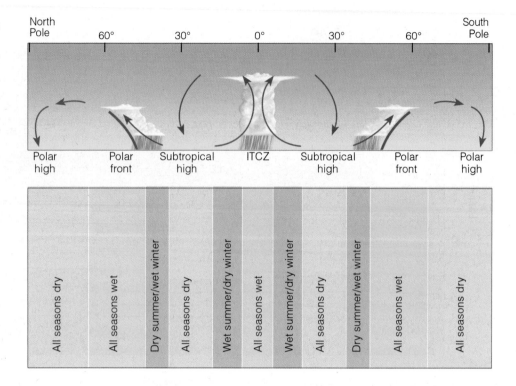

In the cold air of the polar regions there is little moisture, so there is little precipitation. Winter storms drop light, powdery snow that remains on the ground for a long time because of the low evaporation rates. In summer, a ridge of high pressure tends to block storm systems that would otherwise travel into the area; hence, precipitation in polar regions is meager in all seasons.

There are exceptions to this idealized pattern. For example, in middle latitudes the migrating position of the subtropical anticyclones also has an effect on the west-to-east distribution of precipitation. The sinking air associated with these systems is more strongly developed on their eastern side. Hence, the air along the eastern side of an anticyclone tends to be more stable; it is also drier, as cooler air moves equatorward because of the circulating winds around these systems. In addition, along coastlines, cold upwelling water cools the surface air even more, adding to the air's stability. Consequently, in summer, when the Pacific high moves to a position centered off the California coast, a strong, stable subsidence inversion forms above coastal regions. With the strong inversion and the fact that the anticyclone tends to steer storms to the north, central and southern California areas experience little, if any, rainfall during the summer months.

On the western side of subtropical highs, the air is less stable and more moist, as warmer air moves poleward. In summer, over the North Atlantic, the Bermuda high pumps moist tropical air northward from the Gulf of Mexico into the eastern two-thirds of the United States. The humid air is conditionally unstable to begin with, and by the time it moves over the heated ground, it becomes even more unstable. If conditions are right, the moist air will rise and condense into cumulus clouds, which may build into towering thunderstorms.

In winter, the subtropical North Pacific high moves south, allowing storms traveling across the ocean to penetrate the western states, bringing much needed rainfall to California after a long, dry summer. The Bermuda high also moves south in winter. Across much of the United States, intense winter storms develop and travel eastward, frequently dumping heavy precipitation as they go. Usually, however, the heaviest precipitation is concentrated in the eastern states, as moisture from the Gulf of Mexico moves northward ahead of these systems. Therefore, cities on the plains typically receive more rainfall in summer and those on the west coast have maximum precipitation in winter, whereas cities in the midwest and east usually have abundant precipitation all year long. • Figure 17.3 shows the average annual precipitation across the United States as well as the contrast in seasonal precipitation among a west coast city (San Francisco), a central plains city (Kansas City), and an eastern city (Baltimore).

Mountain ranges disrupt the idealized pattern of global precipitation (1) by promoting convection (because their slopes are warmer than the surrounding air) and (2) by forcing air to rise along their windward slopes (*orographic uplift*).

WEATHER WATCH

One of the wettest weather stations in the continental United States is located at Wynoochee Oxbow, Washington. Situated on the Olympic Peninsula, this station receives an average rainfall of 366 cm (144 in.)—a total 86 times greater than the average 4.3 cm (1.7 in.) for Death Valley, California.

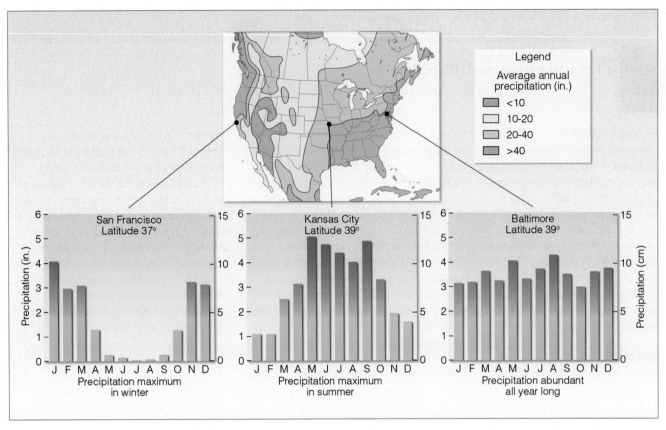

● FIGURE 17.3 Average annual precipitation across the United States along with variation in annual precipitation for three Northern Hemisphere cities.

Consequently, the windward side of mountains tends to be "wet." As air descends and warms along the leeward side, there is less likelihood of clouds and precipitation. Thus, the leeward (downwind) side of mountains tends to be "dry." As Chapter 6 points out, a region on the leeward side of a mountain where precipitation is noticeably less is called a *rain shadow*.

A good example of the rain shadow effect occurs in the northwestern part of Washington State. Situated on the western side at the base of the Olympic Mountains, the Hoh River Valley annually receives an average 380 cm (150 in.) of precipitation (see ● Fig. 17.4). On the eastern (leeward) side of this range, only about 100 km (62 mi) from the Hoh rain forest, the mean annual precipitation is less than 43 cm (17 in.), and irrigation is necessary to grow certain crops. ● Figure 17.5 shows a classic example of how topography produces several rain shadow effects. (Additional information on precipitation extremes is given in the Focus section on pp. 474–475.)

BRIEF REVIEW

Before going on to the section on climate classification, here is a brief review of some of the facts we have covered so far:

● The climate controls are the factors that govern the climate of any given region.

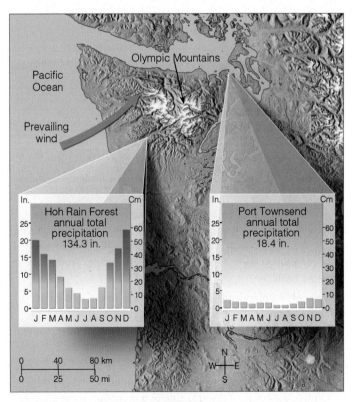

● FIGURE 17.4 The effect of the Olympic Mountains in Washington State on average annual precipitation.

Precipitation Extremes

Most of the "rainiest" places in the world are located on the windward side of mountains. For example, Mount Waialeale on the island of Kauai, Hawaii, has the greatest annual average rainfall on record: 1168 cm (460 in.). Cherrapunji, on the crest of the southern slopes of the Khasi Hills in northeastern India, receives an average of 1080 cm (425 in.) of rainfall each year, the majority of which falls during the summer monsoon, between April and October. Cherrapunji, which holds the greatest twelve-month rainfall total of 2647 cm (1042 in.), once received 380 cm (150 in.) of rain in just five days.

Record rainfall amounts are often associated with tropical storms. On the island of La Réunion (about 650 km east of Madagascar in the Indian Ocean), a tropical cyclone dumped 135 cm (53 in.) of rain on Belouve in twelve hours. Heavy rains of short duration often occur with severe thunderstorms that move slowly or stall over a region. On July 4, 1956, 3 cm (1.2 in.) of rain fell from a thunderstorm on Unionville, Maryland, in one minute.

Snowfalls tend to be heavier where cool, moist air rises along the windward slopes of mountains. One of the snowiest places in North America is located at the Paradise Ranger Station in Mt. Rainier National Park, Washington. Situated at an elevation of 1646 m (5400 ft) above sea level, this station receives an average 1758 cm (692 in.) of snow annually.

However, a record annual snowfall amount of 2896 cm (1140 in.) was recorded at Mt. Baker ski area during the winter of 1998–1999.

As we noted earlier, the driest regions of the world lie in the frigid polar region, the leeward side of mountains, and in the belt of subtropical high pressure, between 15° and 30° latitude. Arica in northern Chile holds the world record for lowest annual rainfall, 0.08 cm (0.03 in.). In the United States, Death Valley, California, averages only 4.5 cm (1.78 in.) of precipitation annually. Figure 1 gives additional information on world precipitation records.

KEY TO MAP

1	World's greatest annual average rainfall	1168 cm (460 in.)	Mt. Waialeale, Hawaii
2	Greatest 1-month rainfall total	930 cm (366 in.)	Cherrapunji, India, July, 1861
3	Greatest 12-hour rainfall total	135 cm (53 in.)	Belouve, La Réunion Island, February 28, 1964
4	Greatest 24-hour rainfall total in United States	109 cm (43 in.)	Alvin, Texas, July 25, 1979
5	Greatest 42-minute rainfall total	30 cm (12 in.)	Holt, Missouri, June 22, 1947
6	Greatest 1-minute rainfall total in United States	3 cm (1.2 in.)	Unionville, MD, July 4, 1956
7	Lowest annual average rainfall in Northern Hemisphere	3 cm (1.2 in.)	Bataques, Mexico
8	Lowest annual average rainfall in the world	0.08 cm (0.03 in.)	Arica, Chile
9	Greatest annual snowfall in United States	2896 cm (1140 in.)	Mt. Baker ski area, WA, 1998
10	Greatest snowfall in 1 month	991 cm (390 in.)	Tamarack, CA, January, 1911
11	Greatest snowfall in 24 hours	193 cm (76 in.)	Silverlake, Boulder, CO, April 14–15, 1921
12	Longest period without measurable precipitation in U.S. (993 days)	0.0 cm (0.0 in.)	Bagdad, CA, August, 1909, to May, 1912

- The hottest places on earth tend to occur in the subtropical deserts of the Northern Hemisphere, where clear skies and sinking air, coupled with low humidity and a high summer sun beating down upon a relatively barren landscape, produce extreme heat.

- The coldest places on earth tend to occur in the interior of high-latitude landmasses. The coldest areas of the Northern Hemisphere are found in the interior of Siberia and Greenland, whereas the coldest area of the world is the Antarctic.

- The wettest places in the world tend to be located on the windward side of mountains where warm, humid air rises upslope. On the downwind (leeward) side of a mountain there often exists a "dry" region, known as a *rain shadow*.

Climatic Classification

The climatic controls interact to produce such a wide array of different climates that no two places experience exactly the same climate. However, the similarity of climates within a given area allows us to divide the earth into climatic regions.

THE ANCIENT GREEKS   By considering temperature and worldwide sunshine distribution, the ancient Greeks categorized the world into three climatic regions:

1. A low-latitude *tropical* (or *torrid*) *zone;* bounded by the northern and southern limit of the sun's vertical rays

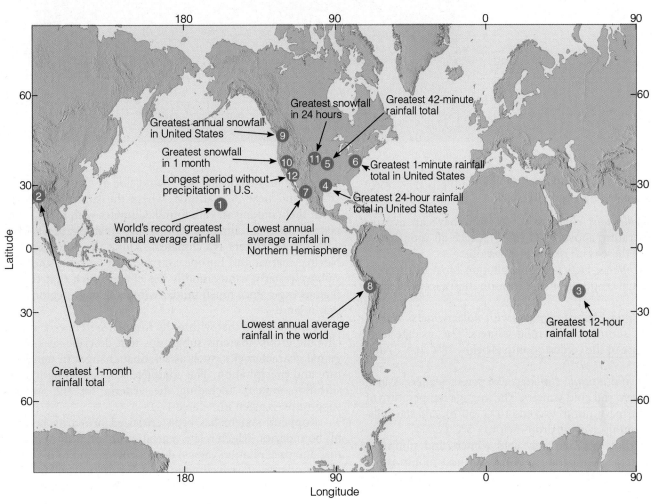

● FIGURE I   Some precipitation records throughout the world.

(23½°N and 23½°S); here, the noon sun is always high, day and night are of nearly equal length, and it is warm year-round.

2. A high-latitude *polar* (or *frigid*) *zone*; bounded by the Arctic or Antarctic Circle; cold all year long due to long periods of winter darkness and a low summer sun.

3. A middle-latitude *temperate zone*; sandwiched between the other two zones; has distinct summer and winter, so exhibits characteristics of both extremes.

Such a sunlight, or temperature-based, climatic scheme is, of course, far too simplistic. It excludes precipitation, so there is no way to differentiate between wet and dry regions.

The best classification of climates would take into account as many meteorological factors as can possibly be obtained.

THE KÖPPEN SYSTEM   A widely used classification of world climates based on the annual and monthly averages of temperature and precipitation was devised by the famous German scientist Waldimir Köppen (1846–1940). Initially published in 1918, the original **Köppen classification system** has since been modified and refined. Faced with the lack of adequate observing stations throughout the world, Köppen related the distribution and type of native vegetation to the various climates. In this way, climatic boundaries could be approximated where no climatological data were available.

● FIGURE 17.5 The effect of topography on average annual precipitation along a line running from the Pacific Ocean through central California into western Nevada.

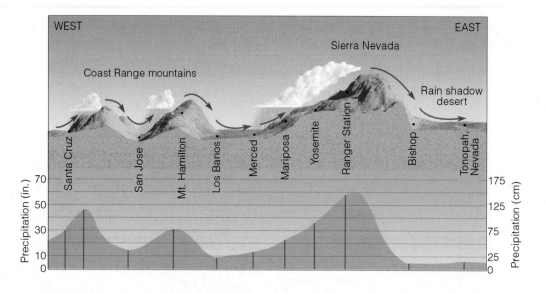

Köppen's scheme employs five major climatic types; each type is designated by a capital letter:

A *Tropical moist climates:* All months have an average temperature above 18°C (64°F). Since all months are warm, there is no real winter season.

B *Dry climates:* Deficient precipitation most of the year. Potential evaporation and transpiration exceed precipitation.

C *Moist mid-latitude climates with mild winters:* Warm-to-hot summers with mild winters. The average temperature of the coldest month is below 18°C (64°F) and above −3°C (27°F).

D *Moist mid-latitude climates with severe winters:* Warm summers and cold winters. The average temperature of the warmest month exceeds 10°C (50°F), and the coldest monthly average drops below −3°C (27°F).

E *Polar climates:* Extremely cold winters and summers. The average temperature of the warmest month is below 10°C (50°F). Since all months are cold, there is no real summer season.

In mountainous country, where rapid changes in elevation bring about sharp changes in climatic type, delineating the climatic regions is impossible. These regions are designated by the letter *H,* for highland climates.

● Figure 17.6 gives a simplified overview of the major climate types throughout the world, according to Köppen's system. Superimposed on the map are some of the climatic controls. These include the average annual positions of the semi-permanent high- and low-pressure areas, the average position of the intertropical convergence zone in January and July, the major mountain ranges and deserts of the world, and some of the major ocean currents. Notice how the climate controls impact the climate in different regions of the world. As we would expect, due to changes in the intensity and amount of solar energy, polar climates are found at high latitudes and tropical climates at low latitudes. Dry climates

tend to be located on the downwind side of major mountain chains and near 30° latitude, where the subtropical highs (with their sinking air) are found. Climates with more moderate winters (C climates) tend to be equatorward of those with severe winters (D climates). Along the west coast of North America and Europe, warm ocean currents and prevailing westerly winds modify the climate such that coastal regions experience much milder winters than do regions farther inland.

Keep in mind that within the Köppen system each major climatic group contains subgroups that describe special regional characteristics, such as seasonal changes in temperature and precipitation. The complete Köppen climatic classification system, including the criteria for the various subgroups, is given in ▼ Table 17.1.

Köppen's system has been criticized primarily because his boundaries (which relate vegetation to monthly temperature and precipitation values) do not correspond to the natural boundaries of each climatic zone. In addition, the Köppen system implies that there is a sharp boundary between climatic zones, when in reality there is a gradual transition.

The Köppen system has been revised several times, most notably by the German climatologist Rudolf Geiger, who worked with Köppen on amending the climatic boundaries of certain regions. A popular modification of the Köppen system was developed by the American climatologist Glenn T. Trewartha, who redefined some of the climatic types and altered the climatic world map by putting more emphasis on the lengths of growing seasons and average summer temperatures.

THORNTHWAITE'S SYSTEM   To correct some of the Köppen deficiencies, the American climatologist C. Warren Thornthwaite (1899–1963) devised a new classification system in the early 1930s. Both systems utilized temperature and precipitation measurements and both related natural vegetation to climate. However, to emphasize the importance of precipitation (P) and evaporation (E) on plant growth, Thornthwaite developed a *P/E ratio,* which is essentially monthly precipitation divided by monthly evaporation. The annual sum of the P/E ratios gives the **P/E index.** Using this index, the Thornthwaite system defines five major humidity provinces and their characteristic vegetations: rain forest, forest, grassland, steppe, and desert.

To better describe the moisture available for plant growth, Thornthwaite proposed a new classification system in 1948 and slightly revised it in 1955. His new scheme emphasized the concept of *potential evapotranspiration** (PE), which is the amount of moisture that would be lost from the soil and vegetation if the moisture were available.

Thornthwaite incorporated potential evapotranspiration into a moisture index that depends essentially on the differ-

Evapotranspiration refers to the evaporation from soil and transpiration of plants.

ences between precipitation and PE. The index is high in moist climates and negative in arid climates. An index of 0 marks the boundary between wet and dry climates.

The Global Pattern of Climate

Figure 17.7 gives a more detailed view of how the major climatic regions and subregions of the world are distributed based mainly on the work of Köppen. (The major climatic types along with their subdivisions are given in Table 17.1.) We will first examine humid tropical climates in low latitudes and then we'll look at middle latitude and polar climates. Bear in mind that each climatic region has many subregions of local climatic differences wrought by such factors as topography, elevation, and large bodies of water. Remember, too, that boundaries of climatic regions represent gradual

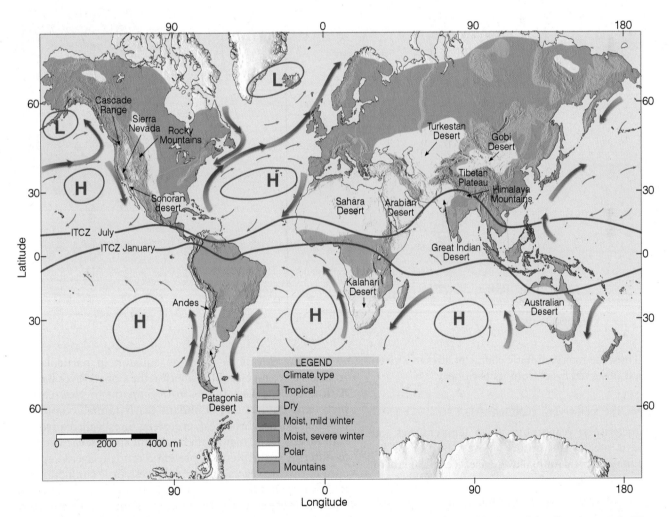

● FIGURE 17.6 A simplified overview of the major climate types according to Köppen, along with some of the climatic controls. The large Hs and Ls on the map represent the average position of the semi-permanent high- and low-pressure areas. The solid red lines show the average position of the intertropical covergence zone (ITCZ) in January and July. The ocean currents in red are warm, whereas those in blue are cold. The major mountain ranges and deserts of the world also are included.

● FIGURE 17.7 Worldwide distribution of climatic regions (after Köppen).

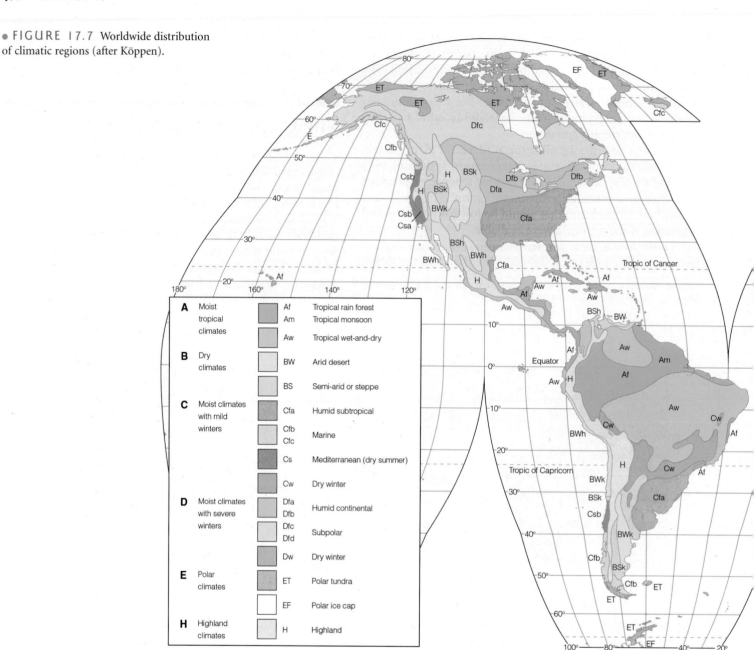

transitions. Thus, the major climatic characteristics of a given region are best observed away from its periphery.

TROPICAL MOIST CLIMATES (GROUP A)

General characteristics: year-round warm temperatures (all months have a mean temperature above 18°C, or 64°F); abundant rainfall (typical annual average exceeds 150 cm, or 59 in.).

Extent: northward and southward from the equator to about latitude 15° to 25°.

Major types (based on seasonal distribution of rainfall): *tropical wet* (Af), *tropical monsoon* (Am), and *tropical wet and dry* (Aw).

At low elevations near the equator, in particular the Amazon lowland of South America, the Congo River Basin of Africa, and the East Indies from Sumatra to New Guinea, high temperatures and abundant yearly rainfall combine to produce a dense, broadleaf, evergreen forest called a **tropical rain forest.** Here, many different plant species, each adapted to differing light intensity, present a crudely layered appearance of diverse vegetation. In the forest, little sunlight is able to penetrate to the ground through the thick crown cover. As a result, little plant growth is found on the forest floor. However, at the edge of the forest, or where a clearing has been made, abundant sunlight allows for the growth of tangled shrubs and vines, producing an almost impenetrable *jungle* (see ● Fig. 17.8).

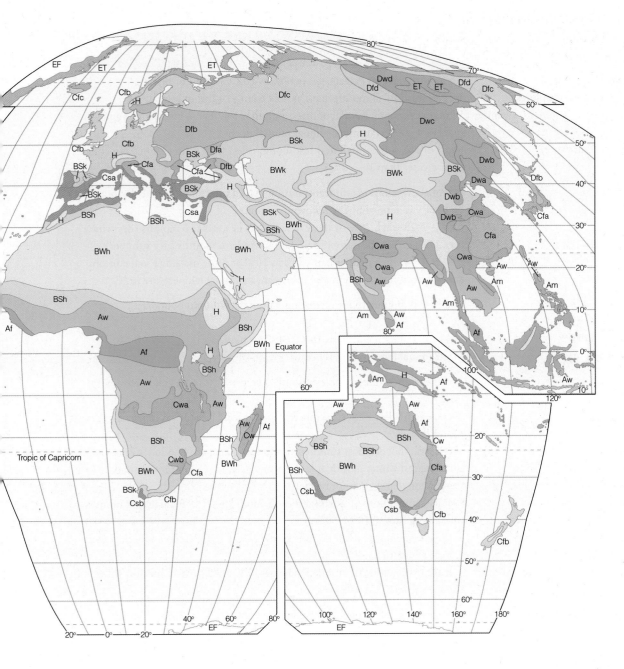

Within the **tropical wet climate*** (Af), seasonal temperature variations are small (normally less than 3°C) because the noon sun is always high and the number of daylight hours is relatively constant. However, there is a greater variation in temperature between day (average high about 32°C) and night (average low about 22°C) than there is between the warmest and coolest months. This is why people remark that winter comes to the tropics at night. The weather here is monotonous and sultry. There is little change in temperature from one day to the next. Furthermore, almost every day, towering cumulus clouds form and produce heavy, localized showers by early afternoon. As evening approaches, the showers usually end and skies clear. Typical annual rainfall totals are greater than 150 cm (59 in.) and, in some cases, especially along the windward side of hills and mountains, the total may exceed 400 cm (157 in.).

The high humidity and cloud cover tend to keep maximum temperatures from reaching extremely high values. In fact, summer afternoon temperatures are normally higher in middle latitudes than here. Nighttime radiational cooling can produce saturation and, hence, a blanket of dew and—occasionally—fog covers the ground.

An example of a station with a tropical wet climate (Af) is Iquitos, Peru (see ● Fig. 17.9). Located near the equator (latitude 4°S), in the low basin of the upper Amazon River, Iquitos has an average annual temperature of 25°C (77°F),

*The tropical wet climate is also known as the *tropical rain forest climate*.

▼ TABLE 17.1 Köppen's Climatic Classification System

1ST	2ND	3RD	CLIMATIC CHARACTERISTICS	CRITERIA
A			Humid tropical	All months have an average temperature of 18°C (64°F) or higher
	f		Tropical wet (rain forest)	Wet all seasons; all months have at least 6 cm (2.4 in.) of rainfall
	w		Tropical wet and dry (savanna)	Winter dry season; rainfall in driest month is less than 6 cm (2.4 in.) and less than $10 - P/25$ (P is mean annual rainfall in cm)
	m		Tropical monsoon	Short dry season; rainfall in driest month is less than 6 cm (2.4 in.) but equal to or greater than $10 - P/25$
B			Dry	Potential evaporation and transpiration exceed precipitation. The dry/humid boundary is defined by the following formulas: $p = 2t + 28$ when 70% or more of rain falls in warmer 6 months (dry winter) $p = 2t$ when 70% or more of rain falls in cooler 6 months (dry summer) $p = 2t + 14$ when neither half year has 70% or more of rain (p is the mean annual precipitation in cm and t is the mean annual temperature in °C)*
	S		Semi-arid (steppe)	The BS/BW boundary is exactly one-half the dry/humid boundary
	W		Arid (desert)	
		h	Hot and dry	Mean annual temperature is 18°C (64°F) or higher
		k	Cool and dry	Mean annual temperature is below 18°C (64°F)
C			Moist with mild winters	Average temperature of coolest month is below 18°C (64°F) and above −3°C (27°F)
	w		Dry winters	Average rainfall of wettest summer month at least 10 times as much as in driest winter month
	s		Dry summers	Average rainfall of driest summer month less than 4 cm (1.6 in.); average rainfall of wettest winter month at least 3 times as much as in driest summer month
	f		Wet all seasons	Criteria for w and s cannot be met
		a	Summers long and hot	Average temperature of warmest month above 22°C (72°F); at least 4 months with average above 10°C (50°F)
		b	Summers long and cool	Average temperature of all months below 22°C (72°F); at least 4 months with average above 10°C (50°F)
		c	Summers short and cool	Average temperature of all months below 22°C (72°F); 1 to 3 months with average above 10°C (50°F)
D			Moist with cold winters	Average temperature of coldest month is −3°C (27°F) or below; average temperature of warmest month is greater than 10°C (50°F)
	w		Dry winters	Same as under Cw
	s		Dry summers	Same as under Cs
	f		Wet all seasons	Same as under Cf
		a	Summers long and hot	Same as under Cfa
		b	Summers long and cool	Same as under Cfb
		c	Summers short and cool	Same as under Cfc
		d	Summers short and cool; winters severe	Average temperature of coldest month is −38°C (−36°F) or below
E			Polar climates	Average temperature of warmest month is below 10°C (50°F)
	T		Tundra	Average temperature of warmest month is greater than 0°C (32°F) but less than 10°C (50°F)
	F		Ice cap	Average temperature of warmest month is 0°C (32°F) or below

*The dry/humid boundary is defined in English units as: $p = 0.44t - 3$ (dry winter); $p = 0.44t - 14$ (dry summer); and $p = 0.44t - 8.6$ (rainfall evenly distributed); where p is mean annual rainfall in inches and t is mean annual temperature in °F.

● FIGURE 17.8 Tropical rain forest near Iquitos, Peru. (Climatic information for this region is presented in Fig. 17.9.)

WEATHER WATCH

Hot and humid Belem, Brazil—a city situated near the equator with a tropical wet climate—had an all-time record high temperature of 98°F, exactly 2°F *less* than the highest temperature (100°F) ever measured in Prospect Creek, Alaska, a city with a subpolar climate.

with an annual temperature range of only 2.2°C (4°F). Notice also that the monthly rainfall totals vary more than do the monthly temperatures. This is due primarily to the migrating position of the Intertropical Convergence Zone (ITCZ) and its associated wind-flow patterns. Although monthly precipitation totals vary considerably, the average for each month exceeds 6 cm, and consequently no month is considered deficient of rainfall.

Take a minute and look again at Fig. 17.8. From the photo, one might think that the soil beneath the forest's canopy would be excellent for agriculture. Actually, this is not true. As heavy rain falls on the soil, the water works its way downward, removing nutrients in a process called *leaching*. Strangely enough, many of the nutrients needed to sustain the lush forest actually come from dead trees that decompose. The roots of the living trees absorb this matter before the rains leach it away. When the forests are cleared for agricultural purposes, or for the timber, what is left is a thick red soil

● FIGURE 17.9 Temperature and precipitation data for Iquitos, Peru, latitude 4°S. A station with a tropical wet climate (Af). (This type of diagram is called a *climograph*. It shows monthly mean temperatures with a solid red line and monthly mean precipitation with blue bar graphs.)

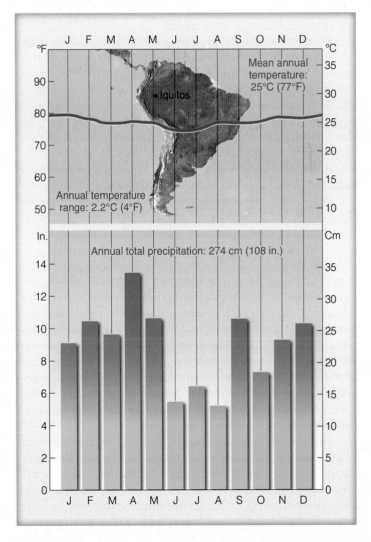

FIGURE 17.10 Baobob and acacia trees illustrate typical trees of the East African grassland savanna, a region with a tropical wet-and-dry climate (Aw).

© J. L. Medeiros

called **laterite.** When exposed to the intense sunlight of the tropics, the soil may harden into a bricklike consistency, making cultivation almost impossible.

Köppen classified tropical wet regions, where the monthly precipitation totals drop below 6 cm for perhaps one or two months, as **tropical monsoon climates** (Am). Here, yearly rainfall totals are similar to those of the tropical wet climate, usually exceeding 150 cm a year. Because the dry season is brief and copious rains fall throughout the rest of the year, there is sufficient soil moisture to maintain the tropical rain forest through the short dry period. Tropical monsoon climates can be seen in Fig. 17.7 along the coasts of Southeast Asia, India, and in northeastern South America.

Poleward of the tropical wet region, total annual rainfall diminishes, and there is a gradual transition from the tropical wet climate to the **tropical wet-and-dry climate** (Aw), where a distinct dry season prevails. Even though the annual precipitation usually exceeds 100 cm, the dry season, where the monthly rainfall is less than 6 cm (2.4 in.), lasts for more than two months. Because tropical rain forests cannot survive this "drought," the jungle gradually gives way to tall, coarse **savanna grass,** scattered with low, drought-resistant deciduous trees (see Fig. 17.10). The dry season occurs during the winter (low sun period), when the region is under the influence of the subtropical highs. In summer, the ITCZ moves poleward, bringing with it heavy precipitation, usually in the form of showers. Rainfall is enhanced by slow moving shallow lows that move through the region.

Tropical wet-and-dry climates not only receive less total rainfall than the tropical wet climates, but the rain that does occur is much less reliable, as the total rainfall often fluctuates widely from one year to the next. In the course of a single year, for example, destructive floods may be followed by serious droughts. As with tropical wet regions, the daily range of temperature usually exceeds the annual range, but the climate here is much less monotonous. There is a cool season in winter when the maximum temperature averages 30°C to 32°C (86°F to 90°F). At night, the low humidity and clear skies allow for rapid radiational cooling and, by early morning, minimum temperatures drop to 20°C (68°F) or below.

From Fig. 17.7, pp. 478–479, we can see that the principal areas having a tropical wet-and-dry climate (Aw) are those located in western Central America, in the region both north and south of the Amazon Basin (South America), in south-central and eastern Africa, in parts of India and Southeast Asia, and in northern Australia. In many areas (especially within India and Southeast Asia), the marked variation in precipitation is associated with the *monsoon*—the seasonal reversal of winds.

As we saw in Chapter 9, the monsoon circulation is due in part to differential heating between landmasses and oceans. During winter in the Northern Hemisphere, winds blow outward, away from a cold, shallow high-pressure area centered over continental Siberia. These downslope, relatively dry northeasterly winds from the interior provide India and Southeast Asia with generally fair weather and the dry season. In summer, the wind-flow pattern reverses as air flows into a developing thermal low over the continental interior. The humid air from the water rises and condenses, resulting in heavy rain and the wet season. (A more detailed look at the winter and summer monsoon is shown in Fig. 9.30 on p. 243.)

An example of a station with a tropical wet-and-dry climate (Aw) is given in Fig. 17.11. Located at latitude 11°N in west Africa, Timbo, Guinea, receives an annual average 163 cm (64 in.) of rainfall. Notice that the rainy season is during the summer when the ITCZ has migrated to its most northern position. Note also that practically no rain falls during the months of December, January, and February, when the region comes under the domination of the subtropical high-pressure area and its sinking air.

The monthly temperature patterns at Timbo are characteristic of most tropical wet-and-dry climates. As spring approaches, the noon sun is slightly higher, and the more intense sunshine produces greater surface heating and higher afternoon temperatures—usually above 32°C (90°F) and occasionally above 38°C (100°F)—creating hot, dry desertlike conditions. After this brief hot season, a persistent cloud cover and the evaporation of rain tends to lower the temperature during the summer. The warm, muggy weather of summer often resembles that of the tropical wet climate (Af). The rainy summer is followed by a warm, relatively dry period, with afternoon temperatures usually climbing above 30°C (86°F).

Poleward of the tropical wet-and-dry climate, the dry season becomes more severe. Clumps of trees are more isolated and the grasses dominate the landscape. When the potential annual water loss through evaporation and transpiration exceeds the annual water gain from precipitation, the climate is described as dry.

DRY CLIMATES (GROUP B)

General characteristics: deficient precipitation most of the year; potential evaporation and transpiration exceed precipitation.

Extent: the subtropical deserts extend from roughly 20° to 30° latitude in large continental regions of the middle latitudes, often surrounded by mountains.

Major types: arid (BW)—the "true desert"—and semi-arid (BS).

A quick glance at Fig. 17.7, pp. 478–479, reveals that, according to Köppen, the dry regions of the world occupy more land area (about 26 percent) than any other major climatic type. Within these dry regions, a deficiency of water exists. Here, the potential annual loss of water through evaporation is greater than the annual water gained through precipitation. Thus, classifying a climate as dry depends not only on precipitation totals but also on temperature, which greatly influences evaporation. For example, 35 cm (14 in.) of precipitation in a hot climate will support only sparse vegetation, while the same amount of precipitation in much colder northcentral Canada will support a conifer forest. In addition, a region with a low annual rainfall total is more likely to be classified as dry if the majority of precipitation is concentrated during the warm summer months, when evaporation rates are greater.

Precipitation in a dry climate is both meager and irregular. Typically, the lower the average annual rainfall, the greater its variability. For example, a station that reports an annual rainfall of 5 cm (2 in.) may actually measure no rainfall for two years; then, in a single downpour, it may receive 10 cm (4 in.).

The major dry regions of the world can be divided into two primary categories. The first includes the area of the

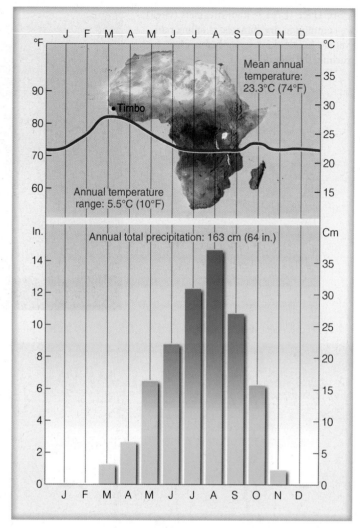

● **FIGURE 17.11** Climatic data for Timbo, Guinea, latitude 11°N. A station with a tropical wet-and-dry climate (Aw).

subtropics (between latitude 15° and 30°), where the sinking air of the subtropical anticyclones produces generally clear skies. The second is found in the continental areas of the middle latitudes. Here, far removed from a source of moisture, areas are deprived of precipitation. Dryness here is often accentuated by mountain ranges that produce a rain shadow effect.

Köppen divided dry climates into two types based on their degree of dryness: the *arid* (BW)* and the *semi-arid,* or *steppe* (BS). These two climatic types can be divided even further. For example, if the climate is hot and dry with a mean annual temperature above 18°C (64°F), it is either BWh or BSh (the *h* is for *heiss,* meaning "hot" in German). On the other hand, if the climate is cold (in winter, that is) and dry with a mean annual temperature below 18°C, then it is either BWk or BSk (where the *k* is for *kalt,* meaning "cold" in German).

The **arid climates** (BW) occupy about 12 percent of the world's land area. From Fig. 17.7, pp. 478–479, we can see

*The letter *W* is for *Wüste,* the German word for "desert."

● FIGURE 17.12 Rain streamers (virga) are common in dry climates, as falling rain evaporates into the drier air before ever reaching the ground.

● FIGURE 17.13 Creosote bushes and cacti are typical of the vegetation found in the arid southwestern American deserts (BWh).

that this climatic type is found along the west coast of South America and Africa and over much of the interior of Australia. Notice, also, that a swath of arid climate extends from northwest Africa all the way into central Asia. In North America, the arid climate extends from northern Mexico into the southern interior of the United States and northward along the leeward slopes of the Sierra Nevada. This region includes both the Sonoran and Mojave deserts and the Great Basin.

The southern desert region of North America is dry because it is dominated by the subtropical high most of the year, and winter storm systems tend to weaken before they move into the area. The northern region is in the rain shadow of the Sierra Nevada. These regions are deficient in precipitation all year long, with many stations receiving less than 13 cm (5 in.) annually. As noted earlier, the rain that does fall is spotty, often in the form of scattered summer afternoon showers. Some of these showers can be downpours that change a gentle gully into a raging torrent of water. More often than not, however, the rain evaporates into the dry air before ever reaching the ground, and the result is rain streamers (virga) dangling beneath the clouds (see ● Fig. 17.12).

Contrary to popular belief, few deserts are completely without vegetation. Although meager, the vegetation that does exist must depend on the infrequent rains. Thus, most of the native plants are **xerophytes**—those capable of surviving prolonged periods of drought (see ● Fig. 17.13). Such vegetation includes various forms of cacti and short-lived plants that spring up during the rainy periods.

In low-latitude deserts (BWh), intense sunlight produces scorching heat on the parched landscape. Here, air temperatures are as high as anywhere in the world. Maximum daytime readings during the summer can exceed 50°C (122°F), although 40°C to 45°C (104°F to 113°F) are more common. In the middle of the day, the relative humidity is usually between 5 and 25 percent. At night, the air's relatively low water-vapor content allows for rapid radiational cooling. Minimum temperatures often drop below 25°C (77°F). Thus, arid climates have large daily temperature ranges, often between 15°C and 25°C (27°F and 45°F) and occasionally higher.

During the winter, temperatures are more moderate, and minimums may, on occasion, drop below freezing. The variation in temperature from summer to winter produces large annual temperature ranges. We can see this in the climate record for Phoenix, Arizona (see ● Fig. 17.14), a city in the southwestern United States with a BWh climate. Notice that the average annual temperature in Phoenix is 22°C (72°F), and that the average temperature of the warmest month (July) reaches a sizzling 32°C (90°F). As we would expect, rainfall is meager in all months. There is, however, a slight maximum in July and August. This is due to the summer monsoon, when more humid, southerly winds are likely to sweep over the region and develop into afternoon showers and thunderstorms (see Fig. 9.32, p. 244).

In middle-latitude deserts (BWk), average annual temperatures are lower. Summers are typically warm to hot, with afternoon temperatures frequently reaching 40°C (104°F). Winters are usually extremely cold, with minimum temperatures sometimes dropping below −35°C (−31°F). Many of these deserts lie in the rain shadow of an extensive mountain chain, such as the Sierra Nevada and the Cascade mountains in North America, the Himalayan Mountains in Asia, and the Andes in South America. The meager precipitation that falls comes from an occasional summer shower or a passing mid-latitude cyclonic storm in winter.

Again, refer to Fig. 17.7 and notice that around the margins of the arid regions, where rainfall amounts are greater, the climate gradually changes into **semi-arid** (BS). This region is called **steppe** and typically has short bunch grass, scattered low bushes, trees, or sagebrush (see ● Fig. 17.15). In North America, this climatic region includes most of the Great Plains, the southern coastal sections of California, and the northern valleys of the Great Basin. As in the arid region, northern areas experience lower winter temperatures and more frequent snowfalls. Annual precipitation is generally between 20 and 40 cm (8 and 16 in.). The climatic record for Denver, Colorado (see ● Fig. 17.16), exemplifies the semi-arid (BSk) climate.

As average rainfall amounts increase, the climate gradually changes to one that is more humid. Hence, the semi-arid (steppe) climate marks the transition between the arid and the humid climatic regions. (Before reading about moist climates, you may wish to read the Focus section on p. 487 about deserts that experience drizzle but little rainfall.)

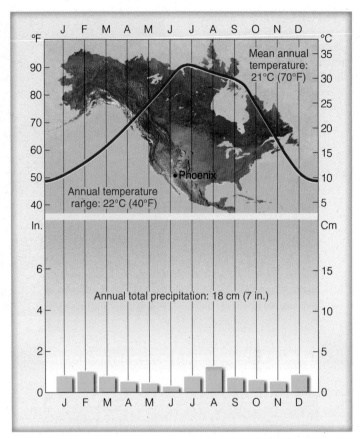

● FIGURE 17.14 Climatic data for Phoenix, Arizona, latitude 33.5°N. A station with an arid climate (BWh).

MOIST SUBTROPICAL MID-LATITUDE CLIMATES (GROUP C)

General characteristics: humid with mild winters (i.e., average temperature of the coldest month below 18°C, or 64°F, and above −3°C, or 27°F).

Extent: on the eastern and western regions of most continents, from about 25° to 40° latitude.

Major types: humid subtropical (Cfa), marine (Cfb), and dry-summer subtropical, or Mediterranean (Cs).

The Group C climates of the middle latitudes have distinct summer and winter seasons. Additionally, they have ample precipitation to keep them from being classified as dry. Although winters can be cold, and air temperatures can change appreciably from one day to the next, no month has a mean temperature below −3°C (27°F), for if it did, it would be classified as a D climate—one with severe winters.

The first C climate we will consider is the **humid subtropical climate** (Cfa).* Notice in 17.7, pp. 478–479, that Cfa

*In the Cfa climate, the "f" means that all seasons are wet and the "a" means that summers are long and hot. A more detailed explanation is given in Table 17.1 on p. 480.

● FIGURE 17.15 Cumulus clouds forming over the steppe grasslands of western North America, a region with a semi-arid climate (BS).

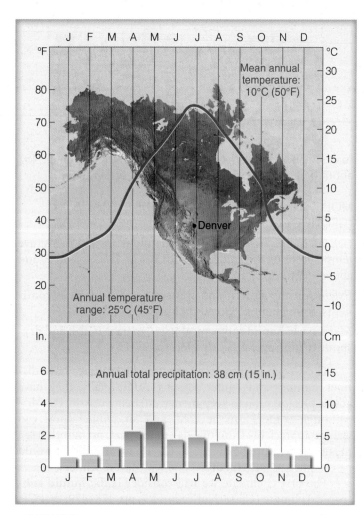

● FIGURE 17.16 Climatic data for Denver, Colorado, latitude 40°N. A station with a semi-arid climate (BSk).

climates are found principally along the east coasts of continents, roughly between 25° and 40° latitude. They dominate the southeastern section of the United States, as well as eastern China and southern Japan. In the Southern Hemisphere, they are found in southeastern South America and along the southeastern coasts of Africa and Australia.

A trademark of the humid subtropical climate is its hot, muggy summers. This sultry summer weather occurs because Cfa climates are located on the western side of subtropical highs, where maritime tropical air from lower latitudes is swept poleward into these regions. Generally, summer dew-point temperatures are high (often exceeding 23°C, or 73°F) and so is the relative humidity, even during the middle of the day. The high humidity combines with the high air temperature (usually above 32°C, or 90°F) to produce more oppressive conditions than are found in equatorial regions. Summer morning low temperatures often range between 21°C and 27°C (70°F and 81°F). Occasionally, a weak summer cool front will bring temporary relief from the sweltering conditions. However, devastating heat waves, sometimes lasting many weeks, can occur when an upper-level ridge moves over the area.

Winters tend to be relatively mild, especially in the lower latitudes, where air temperatures rarely dip much below freezing. Poleward regions experience winters that are colder and harsher. Here, frost, snow, and ice storms are more common, but heavy snowfalls are rare. Winter weather can be quite changeable, as almost summerlike conditions can give way to cold rain and wind in a matter of hours when a middle-latitude storm and its accompanying fronts pass through the region.

Humid subtropical climates experience adequate and fairly well-distributed precipitation throughout the year, with typical annual averages between 80 and 165 cm (31 and

FOCUS ON AN OBSERVATION

A Desert with Clouds and Drizzle

We already know that not all deserts are hot. By the same token, not all deserts are sunny. In fact, some coastal deserts experience considerable cloudiness, especially low stratus and fog.

Amazingly, these coastal deserts are some of the driest places on earth. They include the Atacama Desert of Chile and Peru, the coastal Sahara Desert of northwest Africa, the Namib Desert of southwestern Africa, and a portion of the Sonoran Desert in Baja, California (see Fig. 2). On the Atacama Desert, for example, some regions go without measurable rainfall for decades. And Arica, in northern Chile, has an annual rainfall of only 0.08 cm (0.03 in.).

The cause of this aridity is, in part, due to the fact that each region is adjacent to a large body of relatively cool water. Notice in Fig. 2 that these deserts are located along the western coastal margins of continents, where a subtropical high-pressure area causes prevailing winds to move cool water from higher latitudes along the coast. In addition, these winds help to accentuate the water's coldness by initiating *up-welling* — the rising of cold water from lower levels. The combination of these conditions tends to produce coastal water temperatures between 10°C and 15°C (50°F and 59°F), which is quite cool for such low latitudes. As surface air sweeps across the cold water, it is chilled to its dew point, often producing a blanket of fog and low clouds, from which drizzle falls. The drizzle, however, accounts for very little rainfall. In most regions, it is only enough to dampen the streets with a mere trace of precipitation.

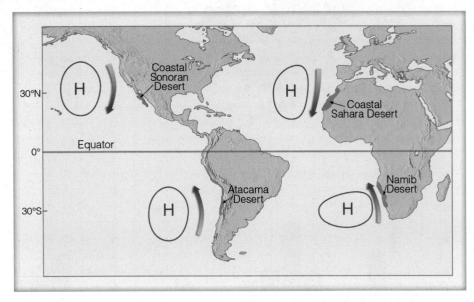

● FIGURE 2 Location of coastal deserts (dark orange shade) that experience frequent fog, drizzle, and low clouds. (Blue arrows indicate prevailing winds and the movement of cool ocean currents.)

As the cool stable air moves inland, it warms, and the water droplets evaporate. Hence, most of the cloudiness and drizzle is found along the immediate coast. Although the relative humidity of this air is high, the dew-point temperature is comparatively low (often near that of the coastal surface water). Inland, further warming causes the air to rise. However, a stable subsidence inversion, associated with the subtropical highs, inhibits vertical motions by capping the rising air, causing it to drift back toward the ocean, where it sinks, completing a rather strong sea breeze circulation. The posi-

tion of the subtropical highs, which tend to remain almost stationary, plays an additional role by preventing the Intertropical Convergence Zone with its rising, unstable air from entering the region.

And so we have a desert with clouds and drizzle — a desert that owes its existence, in part, to its proximity to rather cold ocean water and, in part, to the position and air motions of a subtropical high.

---

65 in.). In summer, when thunderstorms are common, much of the precipitation falls as afternoon showers. Tropical storms entering the United States and China can substantially add to their summer and autumn rainfall totals. Winter precipitation most often occurs with eastward-trekking middle-latitude cyclonic storms. In the southeastern United States, the abundant rainfall supports a thick pine forest that becomes mixed with oak at higher latitudes. The climate data for Mobile, Alabama, a city with a Cfa climate, is given in ● Fig. 17.17.

Glance back at Fig. 17.7, pp. 478–479, and observe that C climates extend poleward along the western side of most continents from about latitude 40° to 60°. These regions are dominated by prevailing winds from the ocean that moderate the climate, keeping winters considerably milder than stations located at the same latitude farther inland. In addition to this, summers are quite cool. When the summer season is both short and cool, the climate is designated as Cfc. Equatorward, where summers are longer (but still cool),

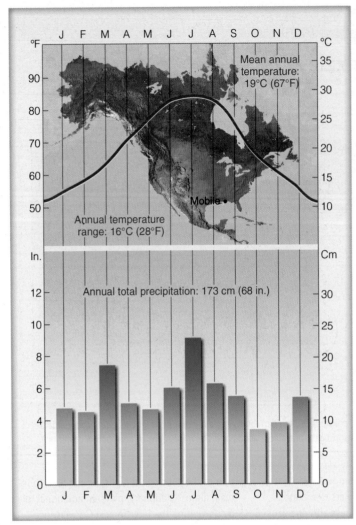

● FIGURE 17.17 Climatic data for Mobile, Alabama, latitude 30°N. A station with a humid subtropical climate (Cfa).

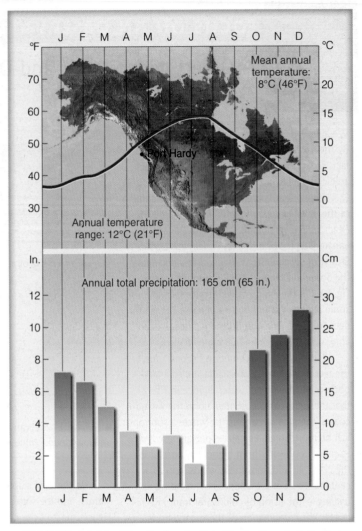

● FIGURE 17.18 Climatic data for Port Hardy, Canada, latitude 51°N. A station with a marine climate (Cfb).

the climate is classified as *west coast marine,* or simply **marine, Cfb.** *

Where mountains parallel the coastline, such as along the west coasts of North and South America, the marine influence is restricted to narrow belts. Unobstructed by high mountains, prevailing westerly winds pump ocean air over much of western Europe and thus provide this region with a marine climate (Cfb).

During much of the year, marine climates are characterized by low clouds, fog, and drizzle. The ocean's influence produces adequate precipitation in all months, with much of it falling as light or moderate rain associated with maritime polar air masses. Snow does fall, but frequently it turns to slush after only a day or so. In some locations, topography greatly enhances precipitation totals. For example, along the west coast of North America, coastal mountains not only

*In the Cfb climate, the "b" means that summers are cooler than in those regions experiencing a Cfa climate. The temperature criteria for the various sub-regions is given in Table 17.1, p. 480.

force air upward, enhancing precipitation, they also slow the storm's eastward progress, which enables the storm to drop more precipitation on the area.

Along the northwest coast of North America, rainfall amounts decrease in summer. This phenomenon is caused by the northward migration of the subtropical Pacific high, which is located southwest of this region. The summer decrease in rainfall can be seen by examining the climatic record of Port Hardy (see ● Fig. 17.18), a station situated along the coast of Canada's Vancouver Island. The data illustrate another important characteristic of marine climates: the low annual temperature range for such a high-latitude station. The ocean's influence keeps daily temperature ranges low as well. In this climate type, it rains on many days and when it is not raining, skies are usually overcast. The heavy rains produce a dense forest of Douglas fir.

Moving equatorward of marine climates, the influence of the subtropical highs becomes greater, and the summer dry period more pronounced. Gradually, the climate changes from marine to one of **dry-summer subtropical** (Cs), or **Mediterranean,** because it also borders the coastal areas of the Mediterranean Sea. (Here the lower case "s" stands for "summer dry.") Along the west coast of North America, Portland, Oregon—because it has rather dry summers—marks the transition between the marine climate and the dry-summer subtropical climate to the south.

The extreme summer aridity of the Mediterranean climate, which in California may exist for five months, is caused by the sinking air of the subtropical highs. In addition, these anticyclones divert summer storm systems poleward. During the winter, when the subtropical highs move equatorward, mid-latitude storms from the ocean frequent the region, bringing with them much needed rainfall. Consequently, Mediterranean climates are characterized by mild, wet winters, and mild-to-hot, dry summers.

Where surface winds parallel the coast, upwelling of cold water helps keep the water itself and the air above it cool all summer long. In these coastal areas, which are often shrouded in low clouds and fog, the climate is called *coastal Mediterranean* (Csb). Here, summer daytime maximum temperatures usually reach about 21°C (70°F), while overnight lows often drop below 15°C (59°F). Inland, away from the ocean's influence, summers are hot and winters are a little cooler than coastal areas. In this *interior Mediterranean climate* (Csa), summer afternoon temperatures usually climb above 34°C (93°F) and occasionally above 40°C (104°F).

● Figure 17.19 contrasts the coastal Mediterranean climate of San Francisco, California, with the interior Mediterranean

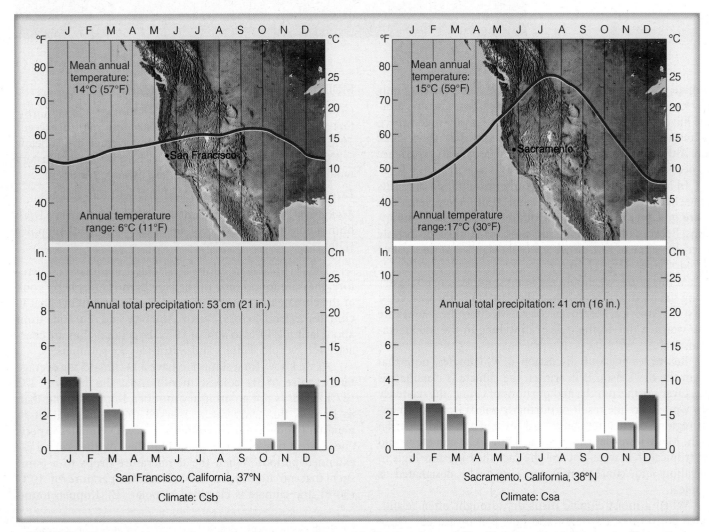

● FIGURE 17.19 Comparison of a coastal Mediterranean climate, Csb (San Francisco, at left), with an interior Mediterranean climate, Csa (Sacramento, at right).

● FIGURE 17.20 In the Mediterranean-type climates of North America, typical chaparral vegetation includes chamisa, manzanita, and foothill pine.

© C. Donald Ahrens

climate of Sacramento, California. While Sacramento is only 130 km (80 mi) inland from San Francisco, Sacramento's average July temperature is 9°C (16°F) higher. As we would expect, Sacramento's annual temperature range is considerably higher, too. Although Sacramento and San Francisco both experience an occasional frost, snow in these areas is a rarity.

In Mediterranean climates, yearly precipitation amounts range between 30 and 90 cm (12 and 35 in.). However, much more precipitation falls on surrounding hillsides and mountains. Because of the summer dryness, the land supports only a scrubby type of low-growing woody plants and trees called *chaparral* (see ● Fig. 17.20).

At this point, we should note that summers are not as dry along the Mediterranean Sea as they are along the west coast of North America. Moreover, coastal Mediterranean areas are also warmer, due to the lack of upwelling in the Mediterranean Sea.

Before leaving our discussion of C climates, note that when the dry season is in winter, the climate is classified as Cw. Over northern India and portions of China, the relatively dry winters are the result of northerly winds from continental regions circulating southward around the cold Siberian high. Many lower-latitude regions with a Cw climate would be tropical if it were not for the fact they are too high in elevation and, consequently, too cool to be designated as tropical.

When a moist climate turns dry, drought often results. What constitutes a drought and how is it measured? These are some of the questions addressed in the Focus section on pp. 492-493.

MOIST CONTINENTAL CLIMATES (GROUP D)

General characteristics: warm-to-cool summers and cold winters (i.e., average temperature of warmest month exceeds 10°C, or 50°F, and the coldest monthly average drops below −3°C, or 27°F); winters are severe with snowstorms, blustery winds, bitter cold; climate controlled by large continent.

Extent: north of moist subtropical mid-latitude climates.

Major types: humid continental with hot summers (Dfa), humid continental with cool summers (Dfb), and subpolar (Dfc).

The D climates are controlled by large landmasses. Therefore, they are found only in the Northern Hemisphere. Look at the climate map, Fig. 17.7, pp. 478–479, and notice that D climates extend across North America and Eurasia, from about latitude 40°N to almost 70°N. In general, they are characterized by cold winters and warm-to-cool summers.

As we know, for a station to have a D climate, the average temperature of its coldest month must dip below −3°C (27°F). This is not an arbitrary number. Köppen found that, in Europe, this temperature marked the southern limit of persistent snow cover in winter.* Hence, D climates experience a great deal of winter snow that stays on the ground for extended periods. When the temperature drops to a point such that no month has an average temperature of 10°C (50°F), the climate is classified as polar (E). Köppen found

*In North America, studies suggest that an average monthly temperature of 0°C (32°F) or below for the coldest month seems to correspond better to persistent winter snow cover.

that the average monthly temperature of 10°C tended to represent the minimum temperature required for tree growth. So no matter how cold it gets in a D climate (and winters can get extremely cold), there is enough summer warmth to support the growth of trees.

There are two basic types of D climates: the **humid continental** (Dfa and Dfb) and the **subpolar** (Dfc). Humid continental climates are observed from about latitude 40°N to 50°N (60°N in Europe). Here, precipitation is adequate and fairly evenly distributed throughout the year, although interior stations experience maximum precipitation in summer. Annual precipitation totals usually range from 50 to 100 cm (20 to 40 in.). Native vegetation in the wetter regions includes forests of spruce, fir, pine, and oak. In autumn, nature's pageantry unveils itself as the leaves of deciduous trees turn brilliant shades of red, orange, and yellow (see ● Fig. 17.21).

Humid continental climates are subdivided on the basis of summer temperatures. Where summers are long and hot,* the climate is described as *humid continental with hot summers* (Dfa). Here summers are often hot and humid, especially in the southern regions. Midday temperatures often exceed 32°C (90°F) and occasionally 40°C (104°F). Summer nights are usually warm and humid, as well. The frost-free season normally lasts from five to six months, long enough to grow a wide variety of crops. Winters tend to be windy, cold,

*"Hot" means that the average temperature of the warmest month is above 22°C (72°F) and at least four months have a monthly mean temperature above 10°C (50°F).

and snowy. Farther north, where summers are shorter and not as hot,* the climate is described as *humid continental with long cool summers* (Dfb). In Dfb climates, summers are not only cooler but much less humid. Temperatures may exceed 35°C (95°F) for a time, but extended hot spells lasting many weeks are rare. The frost-free season is shorter than in the Dfa climate, and normally lasts between three and five months. Winters are long, cold, and windy. It is not uncommon for temperatures to drop below −30°C (−22°F) and stay below −18°C (0°F) for days and sometimes weeks. Autumn is short, with winter often arriving right on the heels of summer. Spring, too, is short, as late spring snowstorms are common, especially in the more northern latitudes.

● Figure 17.22 compares the Dfa climate of Des Moines, Iowa, with the Dfb climate of Winnipeg, Canada. Notice that both cities experience a large annual temperature range. This is characteristic of climates located in the northern interior of continents. In fact, as we move poleward, the annual temperature range increases. In Des Moines, it is 31°C (56°F), while 950 km (590 mi) to the north in Winnipeg, it is 38°C (68°F). The summer precipitation maximum expected for these interior continental locations shows up well in Fig. 17.22. Most of the summer rain is in the form of convective showers, although an occasional weak frontal system can produce more widespread precipitation, as can a cluster of

*"Not as hot" means that the average temperature of the warmest month is below 22°C (72°F) and at least four months have a monthly mean temperature above 10°C (50°F).

● FIGURE 17.21
The leaves of deciduous trees burst into brilliant color during autumn over the countryside of Adirondack Park, a region with a humid continental climate.

FOCUS ON A SPECIAL TOPIC

When a Dry Spell Is Not a Drought, and a Drought Does Not Mean "Dry"

When a region's average precipitation drops dramatically for an extended period of time, drought may result. The word *drought* refers to a period of abnormally dry weather that produces a number of negative consequences, such as crop damage or an adverse impact on a community's water supply. Keep in mind that drought is more than a dry spell. In the dry, summer subtropical (Csa) climate of California's Central Valley, it may not rain from May through September. This dry spell is normal for this region and, therefore, would not be considered a drought. However, if this summer dry spell were to occur in the humid subtropical (Cfa) climate of the southeastern United States, the lack of rain could be disastrous for many aspects of the community, and a drought would ensue.

During the extensive dry summer, the Central Valley of California produces vast quantities of fruits, nuts, and vegetables. These crops can only grow with the aid of extensive irrigation. Large lakes and reservoirs in the nearby foothills provide summer water for the Valley's agriculture. The lakes are mainly filled by precipitation that falls over the mountains during the fall, winter, and spring. Should a prolonged dry period lasting several years occur over the mountains, lake levels would drop and irrigation in the Valley would diminish. This situation would place great hardship on the Valley's economy and agricultural production. So an extended winter dry spell in the mountains could potentially cause a water deficiency in the Valley. Without sufficient water resources to supply its agricultural needs, the Valley would (by definition) be in a drought.

In an attempt to measure drought severity, Wayne Palmer, a scientist with the National Weather Service, developed the *Palmer Drought Severity Index (PDSI)*. The index takes into account average temperature and precipitation values to define drought severity. The index is most effective in assessing long-term drought that lasts several months or more. Drought conditions are indicated by a set of numbers that range from 0 (normal) to −4

▼ TABLE 1   Palmer Drought Severity Index

VALUE	DROUGHT	VALUE	MOISTURE
−4.0 or less	Extreme	+4.0 or greater	Extremely Moist
−3.0 to −3.9	Severe	+3.0 to 3.9	Very Moist
−2.0 to −2.9	Moderate	+2.0 to 2.9	Unusually Moist
−1.9 to +1.9	Normal	−1.9 to +1.9	Normal

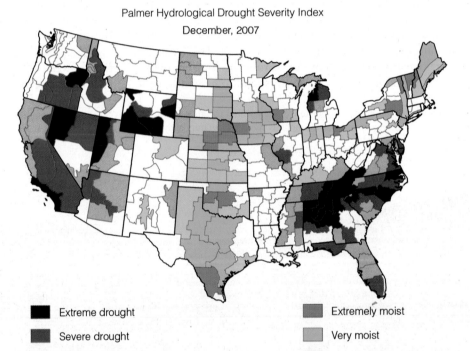

Palmer Hydrological Drought Severity Index
December, 2007

■ Extreme drought
■ Severe drought
■ Moderate drought

■ Extremely moist
■ Very moist
■ Moderately moist

● FIGURE 3 The Palmer Hydrological Drought Index for December, 2007, showing long-term drought conditions. The large black dot in Alabama is Birmingham. (National Climatic Data Center, NOAA)

---

thunderstorms—the Mesoscale Convective Complex described in Chapter 14. The weather in both climatic types can be quite changeable, especially in winter, when a brief warm spell is replaced by blustery winds and temperatures plummeting well below −30°C (−22°F).

When winters are severe and summers short and cool, with only one to three months having a mean temperature exceeding 10°C (50°F), the climate is described as *subpolar* (Dfc). From Fig. 17.7 we can see that, in North America, this climate occurs in a broad belt across Canada and Alaska; in

(extreme drought). (See ▽ Table 1.) The index also assesses wet conditions with numbers that range from +2 (unusually moist) to +4 (extremely moist). The *Palmer Hydrological Drought Index (PHDI)* expands the PDSI by taking into account additional water (hydrological) information, such as a region's groundwater reserves and reservoir levels.

Figure 3 shows the PHDI across the United States for December, 2007. Notice that several regions are in an extreme drought (dark red shade), including a large portion of the southeastern United States. A city in the heart of this drought is Birmingham, Alabama. The climograph for Birmingham is given in Fig. 4. Notice that the annual average precipitation for Birmingham is 140 cm (55 in.) and for the year 2007, Birmingham only recorded precipitation totaling 74 cm (29 in.), which is about 53 percent of the long-term average. Note also that the average temperature for 2007 was 18.5°C (65.3°F), nearly 1.9°C (3.5°F) higher than normal. If these extreme drought conditions were to persist in Birmingham for many years, would the climate ultimately be considered "dry"?

If we plug the precipitation and temperature values for 2007 into Table 17.1, p. 480, we would find that according to Köppen, Birmingham would have to experience annual average precipitation totals below 51 cm (20 in.) over many years to change from a humid subtropical (Cfa) climate to one considered dry. In fact, the value 51 cm (20 in.) is about 31 percent less than the total precipitation for Birmingham in 2007 and nearly 64 percent below Birmingham's annual average precipitation. Consequently, the extreme drought conditions that plagued Birmingham during 2007 if extended over many years would certainly have a devastating effect on the area, but this extreme drought would not change Birmingham's climate classification from humid to dry. However, look at Fig. 3 and determine which areas in the United States you feel *would* have the greatest likelihood of changing from a humid climate to a dry climate.

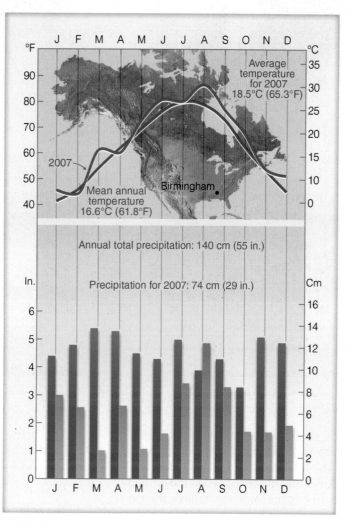

● FIGURE 4 Climatic data for Birmingham, Alabama. The shaded blue bars represent average monthly values of precipitation; those shaded orange are monthly precipitation values for 2007. The solid red line represents average monthly temperatures, whereas the purple line shows average monthly temperatures for 2007.

Eurasia, it stretches from Norway over much of Siberia. The exceedingly low temperatures of winter account for these areas being the primary source regions for continental polar and arctic air masses. Extremely cold winters coupled with cool summers produce large annual temperature ranges, as exemplified by the climate data in ●Fig. 17.23 for Fairbanks, Alaska.

Precipitation is comparatively light in the subpolar climates, especially in the interior regions, with most places receiving less than 50 cm (20 in.) annually. A good percentage of

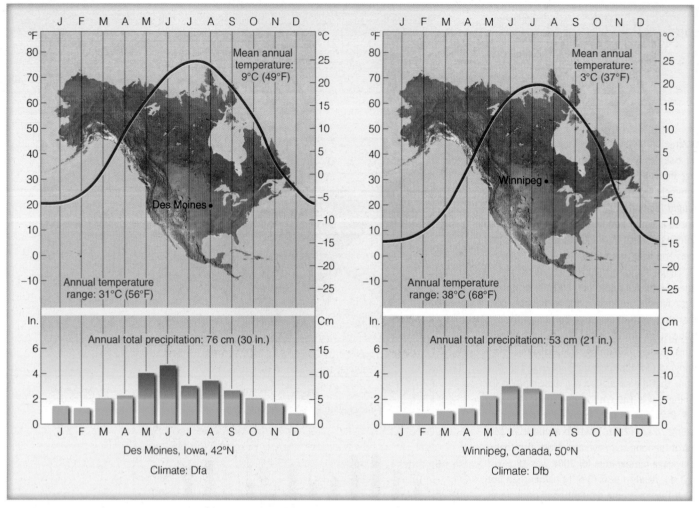

● FIGURE 17.22 Comparison of a humid continental hot summer climate, Dfa (Des Moines, at left), with a humid continental cool summer climate, Dfb (Winnipeg, at right).

the precipitation falls when weak cyclonic storms move through the region in summer. The total snowfall is usually not large but the cold air prevents melting, so snow stays on the ground for months at a time. Because of the low temperatures, there is a low annual rate of evaporation that ensures adequate moisture to support the boreal* forests of conifers and birches known as **taiga** (see ● Fig. 17.24). Hence, the subpolar climate is known also as a *boreal climate* and as a *taiga climate*.

In the taiga region of northern Siberia and Asia, where the average temperature of the coldest month drops to a frigid −38°C (−36°F) or below, the climate is designated Dfd. Where the winters are considered dry, the climate is designated Dwd.

POLAR CLIMATES (GROUP E)

General characteristics: year-round low temperatures (i.e., average temperature of the warmest month is below 10°C, or 50°F).

---
*The word *boreal* comes from the ancient Greek *Boreas,* meaning "wind from the north."

Extent: northern coastal areas of North America and Eurasia; Greenland and Antarctica.

Major types: polar tundra (ET) and polar ice caps (EF).

In the **polar tundra** (ET), the average temperature of the warmest month is below 10°C (50°F), but above freezing. (See ● Fig. 17.25, the climate data for Barrow, Alaska.) Here, the ground is permanently frozen to depths of hundreds of meters, a condition known as **permafrost.** Summer weather, however, is just warm enough to thaw out the upper meter or so of soil. Hence, during the summer, the tundra turns swampy and muddy. Annual precipitation on the tundra is meager, with most stations receiving less than 20 cm (8 in.). In lower latitudes, this would constitute a desert, but in the cold polar regions evaporation rates are very low and moisture remains adequate. Because of the extremely short growing season, *tundra vegetation* consists of mosses, lichens, dwarf trees, and scattered woody vegetation, fully grown and only several centimeters tall (see ● Fig. 17.26).

Even though summer days are long, the sun is never very high above the horizon. Additionally, some of the sunlight

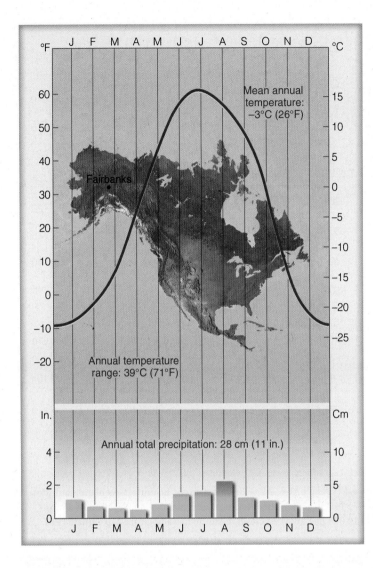

that reaches the surface is reflected by snow and ice, while some is used to melt the frozen soil. Consequently, in spite of the long hours of daylight, summers are quite cool. The cool summers and the extremely cold winters produce large annual temperature ranges.

When the average temperature for every month drops below freezing, plant growth is impossible, and the region is perpetually covered with snow and ice. This climatic type is known as **polar ice cap** (EF). It occupies the interior ice sheets of Greenland and Antarctica, where the depth of ice in some places measures thousands of meters. In this region, temperatures are never much above freezing, even during the middle of "summer." The coldest places in the world are located here. Precipitation is extremely meager, with many places receiving less than 10 cm (4 in.) annually. Most precipitation falls as snow during the "warmer" summer. Strong downslope katabatic winds frequently whip the snow about, adding to the climate's harshness. The data in ● Fig. 17.27 for Eismitte, Greenland, illustrate the severity of an EF climate.

HIGHLAND CLIMATES (GROUP H)   It is not necessary to visit the polar regions to experience a polar climate. Because temperature decreases with altitude, climatic changes experienced when climbing 300 m (1000 ft) in elevation are about equivalent in high latitudes to horizontal changes experienced when traveling 300 km (186 mi) northward. (This distance is equal to about 3° latitude.) Therefore, when ascending a high mountain, one can travel through many climatic regions in a relatively short distance.

● FIGURE 17.24 Coniferous forests (taiga) such as this occur where winter temperatures are low and precipitation is abundant.

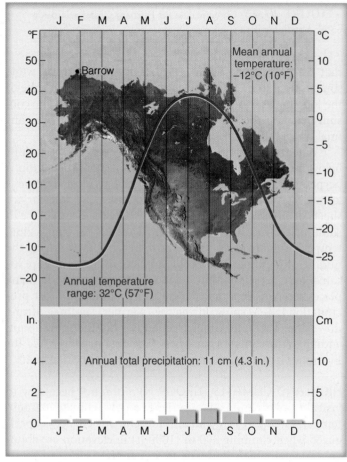

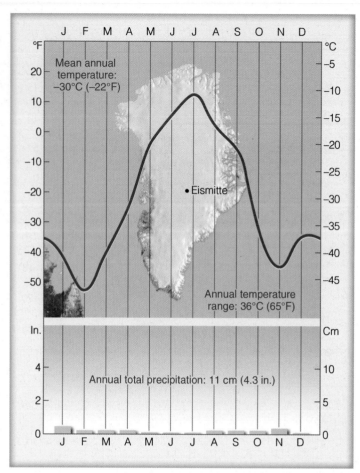

● FIGURE 17.25 Climatic data for Barrow, Alaska, latitude 71°N. A station with a polar tundra climate (ET).

● FIGURE 17.27 Climatic data for Eismitte, Greenland, latitude 71°N. Located in the interior of Greenland at an elevation of almost 10,000 feet above sea level. Eismitte has a polar ice cap climate (EF).

● FIGURE 17.26 Tundra vegetation in Alaska. This type of tundra is composed mostly of sedges and dwarfed wildflowers that bloom during the brief growing season.

© Michio Hoshino/Minden Pictures

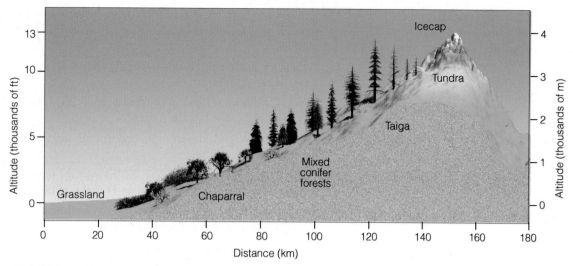

● FIGURE 17.28 Vertical view of changing vegetation and climate due to elevation in the central Sierra Nevada.

● Figure 17.28 shows how the climate and vegetation change along the western slopes of the central Sierra Nevada. (See Fig. 17.5, p. 476, for the precipitation patterns for this region.) Notice that, at the base of the mountains, the climate and vegetation represent semi-arid conditions, while in the foothills the climate becomes Mediterranean and the vegetation changes to chaparral. Higher up, thick fir and pine for-ests prevail. At still higher elevations, the climate is subpolar and the taiga gives way to dwarf trees and tundra vegetation. Near the summit there are permanent patches of ice and snow, with some small glaciers nestled in protected areas. Hence, in less than 13,000 vertical feet, the climate has changed from semi-arid to polar.

SUMMARY

In this chapter, we examined global temperature and precipitation patterns, as well as the various climatic regions throughout the world. Tropical climates are found in low latitudes, where the noon sun is always high, day and night are of nearly equal length, every month is warm, and no real winter season exists. Some of the rainiest places in the world exist in the tropics, especially where warm, humid air rises upslope along mountain ranges.

Dry climates prevail where potential evaporation and transpiration exceed precipitation. Some deserts, such as the Sahara, are mainly the result of sinking air associated with the subtropical highs, while others, due to the rain shadow effect, are found on the leeward side of mountains. Many deserts form in response to both of these effects.

Middle latitudes are characterized by a distinct winter and summer season. Winters tend to be milder in lower latitudes and more severe in higher latitudes. Along the east coast of some continents, summers tend to be hot and humid as moist air sweeps poleward around the subtropical highs. The air often rises and condenses into afternoon thunderstorms in this humid subtropical climate. The west coasts of many continents tend to be drier, especially in summer, as the combination of cool ocean water and sinking air of the sub-tropical highs, to a large degree, inhibit the formation of cumuliform clouds.

In the middle of large continents, such as North America and Eurasia, summers are usually wetter than winters. Winter temperatures are generally lower than those experienced in coastal regions. As one moves northward, summers become shorter and winters longer and colder. Polar climates prevail at high latitudes, where winters are severe and there is no real summer. When ascending a high mountain, one can travel through many climatic zones in a relatively short distance.

KEY TERMS

The following terms are listed (with page numbers) in the order they appear in the text. Define each. Doing so will aid you in reviewing the material covered in this chapter.

microclimate, 470
mesoclimate, 470
macroclimate, 470
global climate, 470
climatic controls, 470
Köppen classification
  system, 475

P/E index, 477
tropical rain forest, 478
tropical wet climate, 479
laterite, 482
tropical monsoon climate, 482
tropical wet-and-dry climate,
  482

QUESTIONS FOR REVIEW

1. What factors determine the global pattern of precipitation?

2. Explain why, in North America, precipitation typically is a maximum along the West Coast in winter, a maximum on the Central Plains in summer, and fairly evenly distributed between summer and winter along the East Coast.

3. Why are the lowest temperatures in polar regions observed in the interior of large landmasses?

4. What climatic information did Köppen use in classifying climates?

5. How did Köppen define tropical climate? How did he define a polar climate?

6. According to Köppen's climatic system (Fig. 17.7, pp. 478–479), what major climatic type is most abundant in each of the following areas:

    (a) in North America

    (b) in South America, and

    (c) throughout the world?

7. What is the primary factor that makes a dry climate "dry"?

8. In which climatic region would each of the following be observed: tropical rain forest, xerophytes, steppe, taiga, tundra, and savanna?

9. What are the controlling factors (the climatic controls) that produce the following climatic regions:

    (a) tropical wet and dry

    (b) Mediterranean

    (c) marine

    (d) humid subtropical

    (e) subpolar

    (f) polar ice cap

10. How do C-type climates differ from D-type climates?

11. Why are large annual temperature ranges characteristic of D-type climates?

12. Why are D climates found in the Northern Hemisphere but not in the Southern Hemisphere?

13. Explain why a tropical rain forest climate will support a tropical rain forest, while a tropical wet-and-dry climate will not.

14. Why are marine climates (Cs) usually found on the west coast of continents?

15. What is the primary distinction between a Cfa and a Dfa climate?

16. Explain how arid deserts can be found adjacent to oceans.

17. Why did Köppen use the 10°C (50°F) average temperature for July to distinguish between D and E climates?

18. What accounts for the existence of a BWk climate in the western Great Basin of North America?

19. Barrow, Alaska, receives a mere 11 cm (4.3 in.) of precipitation annually. Explain why its climate is not classified as arid or semi-arid.

20. Explain why subpolar climates are also known as boreal climates and taiga climates.

QUESTIONS FOR THOUGHT

1. Why do cities directly east of the Rockies (such as Denver, Colorado) receive much more precipitation than cities east of the Sierra Nevada (such as Reno and Lovelock, Nevada)?

2. What climatic controls affect the climate in your area?

3. Los Angeles, Seattle, and Boston are all coastal cities, yet Boston has a continental rather than a marine climate. Explain why.

4. Why are many structures in polar regions built on pilings?

5. Why are summer afternoon temperatures in a humid subtropical climate (Cfa) often higher than in a tropical wet climate (Af)?

6. Why are humid subtropical climates (Cfa) found in regions bounded by 20° and 40° (N or S) latitudes, and nowhere else?

7. In which of the following climate types is virga likely to occur most frequently: humid continental, arid desert, or polar tundra? Explain why.

8. As shown in Figure 17.19, p. 489, San Francisco and Sacramento, California, have similar mean annual temperatures but different annual temperature ranges. What factors control the annual temperature ranges at these two locations?

9. Why is there a contrast in climate types on either side of the Rocky Mountains, but not on either side of the Appalachian Mountains?

10. Over the past 100 years or so the earth has warmed by more than 0.7°C (1.3°F). If this warming should con-

tinue over the next 100 years, explain how this rise in temperature might influence the boundary between C and D climates. How would the warming influence the boundary between D and E climates?

PROBLEMS AND EXERCISES

1. Suppose a city has the mean annual precipitation and temperature given in ▼ Table 17.2. Based on Köppen's climatic types, how would this climate be classified? On a map of North America, approximately where would this city be located? What type of vegetation would you expect to see there? Answer these same questions for the data in ▼ Table 17.3.

2. Compare the following climate classifications for your area:
   (a) Ancient Greeks
   (b) Köppen system
   (c) Thornthwaite's system
   Which classification system is best for your area's mesoclimate? Macroclimate?

3. On a blank map of the world, roughly outline where Köppen's major climatic regions are located.

▼ TABLE 17.2

	JAN.	FEB.	MAR.	APR.	MAY	JUNE	JULY	AUG.	SEPT.	OCT.	NOV.	DEC.	YEAR
Temperature (°F)	40	42	50	60	68	77	80	79	73	62	49	42	60
Precipitation (in.)	4.9	4.2	5.3	3.7	3.8	3.2	4.0	3.3	2.7	2.5	3.4	4.1	45

▼ TABLE 17.3

	JAN.	FEB.	MAR.	APR.	MAY	JUNE	JULY	AUG.	SEPT.	OCT.	NOV.	DEC.	YEAR
Temperature (°F)	18	18	29	42	55	65	70	68	60	48	36	23	44
Precipitation (in.)	1.9	1.5	2.2	2.6	2.9	3.6	3.8	3.0	3.1	2.9	2.8	1.9	32

Visit the
Meteorology Resource Center
at
academic.cengage.com/login
for more assets, including questions for exploration, animations, videos, and more.

A thick layer of smoke and haze covers Santiago, Chile.
M.I.G./Baeza/Photo Researchers, Inc.

CHAPTER 18

Air Pollution

A ir pollution makes the earth a less pleasant place to live. It reduces the beauty of nature. This blight is particularly noticed in mountain areas. Views that once made the pulse beat faster because of the spectacular panorama of mountains and valleys are more often becoming shrouded in smoke. When once you almost always could see giant boulders sharply etched in the sky and the tapered arrowheads of spired pines, you now often see a fuzzy picture of brown and green. The polluted air acts like a translucent screen pulled down by an unhappy God.

Louis J. Battan, *The Unclean Sky*

☼ CONTENTS

Every deep breath fills our lungs mostly with gaseous nitrogen and oxygen. Also inhaled, in minute quantities, may be other gases and particles, some of which could be considered pollutants. These contaminants come from car exhaust, chimneys, forest fires, factories, power plants, and other sources related to human activities.

Virtually every large city has to contend in some way with air pollution, which clouds the sky, injures plants, and damages property. Some pollutants merely have a noxious odor, whereas others can cause severe health problems. The cost is high. In the United States, for example, outdoor air pollution takes its toll in health care and lost work productivity at an annual expense that runs into *billions* of dollars. Estimates are that, worldwide, nearly one billion people in urban environments are continuously being exposed to health hazards from air pollutants.

This chapter takes a look at this serious contemporary concern. We begin by briefly examining the history of problems in this area, and then go on to explore the types and sources of air pollution, as well as the weather that can produce an unhealthful accumulation of pollutants. Finally, we investigate how air pollution influences the urban environment and also how it brings about unwanted acid precipitation.

A Brief History of Air Pollution

Strictly speaking, air pollution is not a new problem. More than likely it began when humans invented fire whose smoke choked the inhabitants of poorly ventilated caves. In fact, very early accounts of air pollution characterized the phenomenon as "smoke problems," the major cause being people burning wood and coal to keep warm.

To alleviate the smoke problem in old England, King Edward I issued a proclamation in 1273 forbidding the use of sea coal, an impure form of coal that produced a great deal of soot and sulfur dioxide when burned. One person was reputedly executed for violating this decree. In spite of such restrictions, the use of coal grew as a heating fuel during the fifteenth and sixteenth centuries.

As industrialization increased, the smoke problem worsened. In 1661, the prominent scientist John Evelyn wrote an essay deploring London's filthy air. And by the 1850s, London had become notorious for its "pea soup" fog, a thick mixture of smoke and fog that hung over the city. These fogs could be dangerous. In 1873, one was responsible for as many as 700 deaths. Another in 1911 claimed the lives of 1150 Londoners. To describe this chronic atmospheric event, a physician, Harold Des Voeux, coined (around 1911) the word *smog*, meaning a combination of smoke and fog.

Little was done to control the burning of coal as time went by, primarily because it was extremely difficult to counter the basic attitude of the powerful industrialists: "Where there's muck, there's money." London's acute smog problem intensified. Then, during the first week of December, 1952, a major disaster struck. The winds died down over London and the fog and smoke became so thick that people walking along the street literally could not see where they were going. People wore masks over their mouths and found their way along the sidewalks by feeling the walls of buildings. This particular disastrous smog lasted 5 days and took nearly 4000 lives, prompting Parliament to pass a Clean Air Act in 1956. Additional air pollution incidents occurred in England during 1956, 1957, and 1962, but due to the strong legislative measures taken against air pollution, London's air today is much cleaner, and "pea soup" fogs are a thing of the past.

Air pollution episodes were by no means limited to Great Britain. During the winter of 1930, for instance, Belgium's highly industrialized Meuse Valley experienced an air pollution tragedy when smoke and other contaminants accumulated in a narrow steep-sided valley. The tremendous buildup of pollutants caused about 600 people to become ill, and ultimately 63 died. Not only did humans suffer, but cattle, birds, and rats fell victim to the deplorable conditions.

The industrial revolution brought air pollution to the United States, as homes and coal-burning industries belched smoke, soot, and other undesirable emissions into the air. Soon, large industrial cities, such as St. Louis and Pittsburgh (which became known as the "Smoky City"), began to feel the effects of the ever-increasing use of coal. As early as 1911, studies documented the irritating effect of smoke particles on the human respiratory system and the "depressing and devitalizing" effects of the constant darkness brought on by giant, black clouds of smoke. By 1940, the air over some cities had become so polluted that automobile headlights had to be turned on during the day.

The first major documented air pollution disaster in the United States occurred at Donora, Pennsylvania, during October, 1948, when industrial pollution became trapped in the Monongahela River Valley. During the ordeal, which lasted 5 days, more than 20 people died and thousands became ill.[*] Several times during the 1960s, air pollution levels became dangerously high over New York City. Meanwhile, on the West Coast, in cities such as Los Angeles, the ever-rising automobile population, coupled with the large petroleum processing plants, were instrumental in generating a different type of pollutant, photochemical smog—one that forms in sunny weather and irritates the eyes. Toward the end of World War II, Los Angeles had its first (of many) smog alerts.

Air pollution episodes in Los Angeles, New York, and other large American cities led to the establishment of much stronger emission standards for industry and automobiles. The Clean Air Act of 1970, for example, empowered the federal government to set emission standards that each state was required to enforce. The Clean Air Act was revised in 1977 and updated by Congress in 1990 to include even stricter emission requirements for autos and industry. The new version of the Act also includes incentives to encourage companies to lower emissions of those pollutants contributing to the current problem of acid rain. Moreover, amendments to the Act have identified 189 toxic air pollutants for regulation.

---

[*]Additional information about the Donora air pollution disaster is given in the Focus section on p. 521.

In 2001, the United States Supreme Court, in a unanimous ruling, made it clear that cost need not be taken into account when setting clean air standards.

Types and Sources of Air Pollutants

Air pollutants are airborne substances (either solids, liquids, or gases) that occur in concentrations high enough to threaten the health of people and animals, to harm vegetation and structures, or to toxify a given environment. Air pollutants come from both natural sources and human activities. Examples of natural sources include wind picking up dust and soot from the earth's surface and carrying it aloft, volcanoes belching tons of ash and dust into our atmosphere, and forest fires producing vast quantities of drifting smoke (see ●Fig. 18.1).

Human-induced pollution enters the atmosphere from both *fixed sources* and *mobile sources.* Fixed sources encompass industrial complexes, power plants, homes, office buildings, and so forth; mobile sources include motor vehicles, ships, and jet aircraft. Certain pollutants are called **primary air pollutants** because they enter the atmosphere directly—from smokestacks and tail pipes, for example. Other pollutants, known as **secondary air pollutants,** form only when a chemical reaction occurs between a primary pollutant and some other component of air, such as water vapor or another pollutant. ▼ Table 18.1 summarizes some of the sources of primary air pollutants.

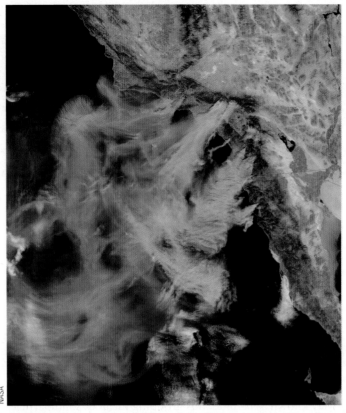

●FIGURE 18.1 Strong northeasterly Santa Ana winds on October 28, 2003, blew the smoke from massive wild fires across southern California out over the Pacific Ocean.

▼ TABLE 18.1   Some of the Sources of Primary Air Pollutants

	SOURCES		POLLUTANTS
Natural			
	Volcanic eruptions		Particles (dust, ash), gases (SO_2, CO_2)
	Forest fires		Smoke, unburned hydrocarbons, CO_2, nitrogen oxides, ash
	Dust storms		Suspended particulate matter
	Ocean waves		Salt particles
	Vegetation		Hydrocarbons (VOCs),* pollens
	Hot springs		Sulfurous gases
Human caused			
Industrial	Paper mills		Particulate matter, sulfur oxides
	Power Plants	Coal	Ash, sulfur oxides, nitrogen oxides
		Oil	Sulfur oxides, nitrogen oxides, CO
	Refineries		Hydrocarbons, sulfur oxides, CO
	Manufacturing	Sulfuric acid	SO_2, SO_3, and H_2SO_4
		Phosphate fertilizer	Particulate matter, gaseous fluoride
		Iron and steel mills	Metal oxides, smoke, fumes, dust, organic and inorganic gases
		Plastics	Gaseous resin
		Varnish/paint	Acrolein, sulfur compounds
Personal	Automobiles		CO, nitrogen oxides, hydrocarbons (VOCs), particulate matter
	Home furnaces/fireplaces		CO, particulate matter
	Open burning of refuse		CO, particulate matter

*VOCs are volatile organic compounds; they represent a class of organic compounds, most of which are hydrocarbons.

● FIGURE 18.2 (a) Estimates of emissions of the primary air pollutants in the United States on a per weight basis. (b) The primary sources for the pollutants. (Data courtesy of United States Environmental Protection Agency.)

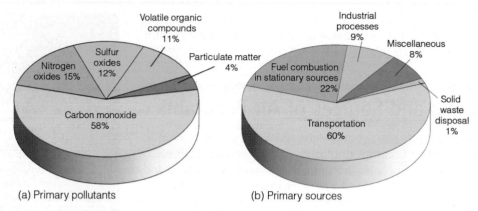

(a) Primary pollutants

(b) Primary sources

● Figure 18.2a shows that carbon monoxide is the most abundant primary air pollutant in the United States. The primary source for all pollutants is transportation (motor vehicles, and so on), with fuel combustion from stationary (fixed) sources coming in a distant second (see Fig. 18.2b). Although hundreds of pollutants are found in our atmosphere, most fall into five groups, which are summarized in the following section.

PRINCIPAL AIR POLLUTANTS The term **particulate matter** represents a group of solid particles and liquid droplets that are small enough to remain suspended in the air. Collectively known as *aerosols*, this grouping includes solid particles that may irritate people but are usually not poisonous, such as soot (tiny solid carbon particles), dust, smoke, and pollen. Some of the more dangerous substances include asbestos fibers and arsenic. Tiny liquid droplets of sulfuric acid, PCBs, oil, and various pesticides are also placed into this category.

Because it often dramatically reduces visibility in urban environments, particulate matter pollution is the most noticeable (see ● Fig. 18.3). Some particulate matter collected in cities includes iron, copper, nickel, and lead. This type of pollution can immediately influence the human respiratory system. Once inside the lungs, it can make breathing difficult, particularly for those suffering from chronic respiratory disorders. Lead particles especially are dangerous as they tend to fall out of the atmosphere and become absorbed into the body through ingestion of contaminated food and water supplies. Lead accumulates in bone and soft tissues, and in high concentrations can cause brain damage, convulsions, and death. Even at low doses, lead can be particularly dangerous to fetuses, infants, and children who, when exposed, may suffer central nervous system damage.

Particulate pollution not only adversely affects the lungs, but recent studies suggest that particulate matter can interfere with the normal rhythm of the human heart. Apparently, as this type of pollution increases, there is a subtle change in a person's heart rate. For a person with an existing cardiac problem, a change in heart rate can produce serious consequences. In fact, one study estimated that, each year, particulate pollution may be responsible for as many as 10,000 heart disease fatalities in the United States.

Of the nearly 7 million metric tons of particulate matter emitted over the United States each year, about 40 percent comes from industrial processes, with highway vehicles accounting for about 17 percent. One main problem is that

ACTIVE FIGURE 18.3 (a) Denver, Colorado, on a clear day, and (b) on a day when particulate matter and other pollutants greatly reduce visibility. Visit the Meteorology Resource Center to view this and other active figures at academic.cengage.com/login

particulate pollution may remain in the atmosphere for some time depending on the size and the amount of precipitation that occurs. For example, larger, heavier particles with diameters greater than about 10 μm* (0.01 mm) tend to settle to the ground in about a day or so after being emitted; whereas fine, lighter particles with diameters less than 1 μm (0.001 mm) can remain suspended in the lower atmosphere for several weeks.

Finer particles with diameters smaller than 10 μm are referred to as *PM-10.* These particles pose the greatest health risk, as they are small enough to penetrate the lung's natural defense mechanisms. Moreover, winds can carry these fine particles great distances before they finally reach the surface. In fact, suspended particles from sources in Europe and the former Soviet Union are believed responsible for the brownish cloud layer called *Arctic haze* that forms over the Arctic each spring. And strong winds over northern China can pick up dust particles and sweep them eastward, where they may settle on North America. This *Asian dust* can reduce visibility, produce spectacular sunrises and sunsets, and coat everything with a thin veneer of particles (see ● Fig. 18.4).

Studies show that particulate matter with diameters less than 2.5 μm, called *PM-2.5,* are especially dangerous. For one thing, they can penetrate farther into the lungs. Moreover, these tiny particles frequently consist of toxic or carcinogenic (cancer-causing) combustion products. Of recent concern are the PM-2.5 particles found in diesel soot. Relatively high amounts of these particles have been measured inside school buses with higher amounts observed downwind of traffic corridors and truck terminals.

Rain and snow remove many of these particles from the air; even the minute particles are removed by ice crystals and

---

*Recall that one micrometer (μm) is one-millionth of a meter. (The thickness of this page is about 100 micrometers.)

cloud droplets. In fact, numerical simulations of air pollution suggest that the predominant removal mechanism occurs when these particles act as nuclei for cloud droplets and ice crystals. Moreover, a long-lasting accumulation of suspended particles (especially those rich in sulfur) is not only aesthetically unappealing but has the potential for affecting the climate, as some particles reflect incoming sunlight, while others absorb outgoing infrared energy from the earth's surface.

Many of the suspended particles are hygroscopic, as water vapor readily condenses onto them. As a thin film of water forms on the particles, they grow in size. When they reach a diameter between 0.1 and 1.0 μm these *wet haze* particles effectively scatter incoming sunlight to give the sky a milky white appearance. The particles are usually sulfate or nitrate particulate matter from combustion processes, such as those produced by diesel engines and power plants. The hazy air mass may become quite thick, and on humid summer days it often becomes well defined, as illustrated in ● Figure 18.5.

Carbon monoxide (CO), a major pollutant of city air, is a colorless, odorless, poisonous gas that forms during the incomplete combustion of carbon-containing fuels. As we saw earlier, carbon monoxide is the most plentiful of the primary pollutants (Fig. 18.2a)

The Environmental Protection Agency (EPA) estimates that over 60 million metric tons of carbon monoxide enter the air annually over the United States alone—about half from highway vehicles. However, due to stricter air quality

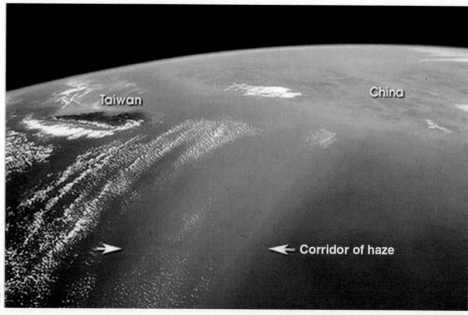

● FIGURE 18.4 A thick haze about 200 km wide and about 600 km long covers a portion of the East China Sea on March 4, 1996. The haze is probably a mixture of industrial air pollution, dust, and smoke.

● FIGURE 18.5 Cumulus clouds and a thunderstorm rise above the thick layer of haze that frequently covers the eastern half of the United States on humid summer days.

© C. Donald Ahrens

standards and the use of emission-control devices, carbon monoxide levels have decreased by about 40 percent since the early 1970s.

Fortunately, carbon monoxide is quickly removed from the atmosphere by microorganisms in the soil, for even in small amounts this gas is dangerous. Hence, it poses a serious problem in poorly ventilated areas, such as highway tunnels and underground parking garages. Because carbon monoxide cannot be seen or smelled, it can kill without warning. Here's how: Normally, your cells obtain oxygen through a blood pigment called *hemoglobin,* which picks up oxygen from the lungs, combines with it, and carries it throughout your body. Unfortunately, human hemoglobin prefers carbon monoxide to oxygen, so if there is too much carbon monoxide in the air you breathe, your brain will soon be starved of oxygen, and headache, fatigue, drowsiness, and even death may result.*

Sulfur dioxide (SO$_2$) is a colorless gas that comes primarily from the burning of sulfur-containing fossil fuels (such as coal and oil). Its primary source includes power plants, heating devices, smelters, petroleum refineries, and paper mills. However, it can enter the atmosphere naturally during volcanic eruptions and as sulfate particles from ocean spray.

Sulfur dioxide readily oxidizes to form the secondary pollutants *sulfur trioxide* (SO$_3$) and, in moist air, highly corrosive *sulfuric acid* (H$_2$SO$_4$). Winds can carry these particles great distances before they reach the earth as undesirable contaminants. When inhaled into the lungs, high concentrations of sulfur dioxide aggravate respiratory problems, such as asthma, bronchitis, and emphysema. Sulfur dioxide in large quantities can cause injury to certain plants, such as lettuce and spinach, sometimes producing bleached marks on their leaves and reducing their yield.

---

*Should you become trapped in your car during a snowstorm, and you have your engine and heater running to keep warm, roll down the window just a little. This action will allow the escape of any carbon monoxide that may have entered the car through leaks in the exhaust system.

Volatile organic compounds (VOCs) represent a class of organic compounds that are mainly **hydrocarbons**—individual organic compounds composed of hydrogen and carbon. At room temperature they occur as solids, liquids, and gases. Even though thousands of such compounds are known to exist, methane (which occurs naturally and poses no known dangers to health) is the most abundant. Other volatile organic compounds include benzene, formaldehyde, and some chlorofluorocarbons. The Environmental Protection Agency estimates that over 18 million metric tons of VOCs are emitted into the air over the United States each year, with about 34 percent of the total coming from vehicles used for transportation and about 50 percent from industrial processes.

Certain VOCs, such as benzene (an industrial solvent) and benzo-a-pyrene (a product of burning wood, tobacco, and barbecuing), are known to be carcinogens—cancer-causing agents. Although many VOCs are not intrinsically harmful, some will react with nitrogen oxides in the presence of sunlight to produce secondary pollutants, which are harmful to human health.

Nitrogen oxides are gases that form when some of the nitrogen in the air reacts with oxygen during the high-temperature combustion of fuel. The two primary nitrogen pollutants are **nitrogen dioxide (NO$_2$)** and **nitric oxide (NO),** which, together, are commonly referred to as NO$_x$—or simply, *oxides of nitrogen.*

Although both nitric oxide and nitrogen dioxide are produced by natural bacterial action, their concentration in urban environments is between 10 and 100 times greater than in nonurban areas. In moist air, nitrogen dioxide reacts with water vapor to form corrosive nitric acid (HNO$_3$), a substance that adds to the problem of acid rain, which we will address later.

The primary sources of nitrogen oxides are motor vehicles, power plants, and waste disposal systems. High concentrations are believed to contribute to heart and lung problems, as well as lowering the body's resistance to respiratory infections. Studies on test animals suggest that nitrogen oxides may encourage the spread of cancer. Moreover, nitrogen oxides are highly reactive gases that play a key role in producing ozone and other ingredients of photochemical smog.

OZONE IN THE TROPOSPHERE   As mentioned earlier, the word **smog** originally meant the combining of smoke and fog. Today, however, the word mainly refers to the type of smog that forms in large cities, such as Los Angeles. Because this type of smog forms when chemical reactions take place in the presence of sunlight (called *photochemical reactions*), it is termed **photochemical smog,** or *Los Angeles-type smog.* When the smog is composed of sulfurous smoke and foggy air, it is usually called *London-type smog.*

The main component of photochemical smog is the gas **ozone (O$_3$).** Ozone is a noxious substance with an unpleasant odor that irritates eyes and the mucous membranes of the respiratory system, aggravating chronic diseases, such as asthma and bronchitis. Even in healthy people, exposure to relatively low concentrations of ozone for six or seven hours during periods of moderate exercise can significantly reduce lung function. This situation often is accompanied by symptoms such as chest pain, nausea, coughing, and pulmonary congestion. Ozone also attacks rubber, retards tree growth, and damages crops. Each year, in the United States alone, ozone is responsible for crop yield losses of several billion dollars.

We will see later that ozone forms naturally in the stratosphere through the combining of molecular oxygen and atomic oxygen. There, *stratospheric ozone* provides a protective shield against the sun's harmful ultraviolet rays. However, near the surface, in polluted air, ozone—often referred to as *tropospheric* (or *ground-level) ozone*—is a secondary pollutant that is not emitted directly into the air. Rather, it forms from a complex series of chemical reactions involving other pollutants, such as nitrogen oxides and volatile organic compounds (hydrocarbons). Because sunlight is required to produce ozone, concentrations of tropospheric ozone are normally higher during the afternoons (see ●Fig. 18.6) and during the summer months, when sunlight is more intense.

In polluted air, ozone production occurs along the following lines. Sunlight (with wavelengths shorter than about 0.41 μm) dissociates nitrogen dioxide into nitric oxide and atomic oxygen, which may be expressed by

$$NO_2 + \text{solar radiation} \rightarrow NO + O.$$

The atomic oxygen then combines with molecular oxygen (in the presence of a third molecule, M), to form ozone, as

$$O_2 + O + M \rightarrow O_3 + M.$$

The ozone is then destroyed by combining with nitric oxide; thus

$$O_3 + NO \rightarrow NO_2 + O_2.$$

If sunlight is present, however, the newly formed nitrogen dioxide will break down into nitric oxide and atomic oxygen. The atomic oxygen then combines with molecular oxygen to

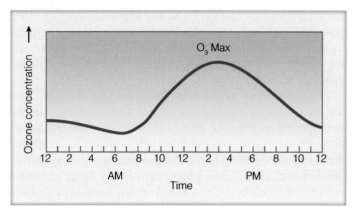

● FIGURE 18.6 Average hourly concentrations of ozone measured at six major cities over a two-year period.

form ozone again. Consequently, large concentrations of ozone can form in polluted air only if some of the nitric oxide reacts with other gases *without removing ozone in the process.* Under these conditions, certain hydrocarbons (emitted by autos and industrial sources) and the hydroxyl radical come into play.

The formation of the hydroxyl radical (OH) begins when ultraviolet radiation (at wavelengths of about 0.31 μm and below) dissociates some of the ozone into molecular oxygen and atomic oxygen; accordingly

$$O_3 + \text{UV radiation} \rightarrow O_2 + O.$$

The atomic oxygen formed is in an excited state, which means it can react with a variety of other molecules, including water vapor, to produce two hydroxyl radical molecules; thus

$$O + H_2O \rightarrow OH + OH.$$

The OH is called a "radical" because it contains an unpaired electron. This situation allows the OH molecule to react with many other atoms and molecules, including unburned or partially burned hydrocarbons (RH) released into the air by automobiles and industry, as

$$OH + RH \rightarrow R\cdot + H_2O.$$

The product R· represents an organic hydrocarbon that can have a complex molecular structure. The R· is then able to react with molecular oxygen to form RO$_2$·, a reactive molecule that removes nitric oxide by combining with it to form nitrogen dioxide, shown by the expression

$$RO_2\cdot + NO \rightarrow NO_2 + \text{other products.}$$

In this manner, nitric oxide can react with hydrocarbons to form nitrogen dioxide *without removing ozone.* Hence, the reactive hydrocarbons in polluted air allow ozone concentrations to increase by preventing nitric oxide from destroying the ozone as rapidly as it is formed.

The hydrocarbons (VOCs) also react with oxygen and nitrogen dioxide to produce other undesirable contaminants, such as *PAN* (peroxyacetyl nitrate)—a pollutant that irritates eyes and is extremely harmful to vegetation—and organic compounds. Ozone, PAN, and small amounts of other oxidating pollutants are the ingredients of photochemical smog. Instead of being specified individually, these pollutants are sometimes grouped under a single heading called *photochemical oxidants.**

In addition to the human-related sources discussed earlier, many hydrocarbons (VOCs) also occur naturally in the atmosphere, as they are given off by vegetation. Oxides of nitrogen drifting downwind from urban areas can react with these natural hydrocarbons and produce smog in relatively uninhabited areas. This phenomenon has been observed downwind of cities such as Los Angeles, London, and New York. Some regions have so much natural (background) hydrocarbon that it may be difficult to reduce ozone levels as much as desired.

In spite of vast efforts to control ozone levels in some major metropolitan areas, results have been generally disappointing because ozone, as we have seen, is a secondary pollutant that forms from chemical reactions involving other pollutants. Ozone production should decrease in most areas when emissions of *both* nitrogen oxides and hydrocarbons (VOCs) are reduced. However, the reduction of only one of these pollutants will not necessarily diminish ozone production because the oxides of nitrogen act as a catalyst for producing ozone in the presence of hydrocarbons (VOCs). Air pollution models suggest that, in the Los Angeles basin, substantially reducing the emissions of hydrocarbons (VOCs) should also reduce ozone production.

Up to now, we have concentrated on ozone in the troposphere, primarily in a polluted environment. The next section examines the formation and destruction of ozone in the upper atmosphere—in the stratosphere.

OZONE IN THE STRATOSPHERE

Recall from Chapter 1 that the stratosphere is a region of the atmosphere that lies above the troposphere between about 10 and 50 km (6 and 31 mi) above the earth's surface. The atmosphere is stable in the stratosphere, as there exists a strong temperature inversion—the air temperature increases rapidly with height (look back at Fig. 1.11, p. 13). The inversion is due, in part, to the gas ozone that absorbs ultraviolet radiation at wavelengths below about 0.3 μm.

In the stratosphere, above middle latitudes, ozone is most dense at an altitude near 25 km (see ● Fig. 18.7). Even at this altitude, its concentration is quite small, as there are only about 12 ozone molecules for every million air molecules (12 ppm).† Although thin, this layer of ozone is significant, for it shields earth's inhabitants from harmful amounts of ultraviolet solar radiation. This protection is fortunate because ultraviolet radiation at wavelengths below 0.3 μm has enough energy to cause skin cancer in humans. Also, UV radiation at

*An *oxidant* is a substance (such as ozone) whose oxygen combines chemically with another substance.

†With a concentration of ozone of only 12 parts per million in the stratosphere, the composition of air here is about the same as it is near the earth's surface—mainly 78 percent nitrogen and 21 percent oxygen.

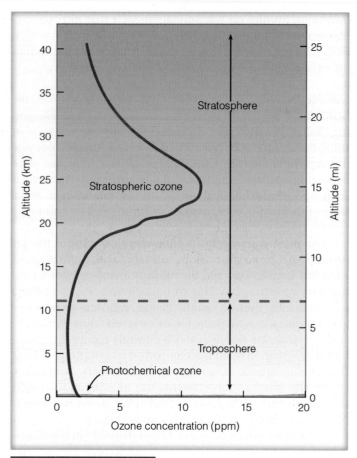

ACTIVE FIGURE 18.7 The average distribution of ozone above the earth's surface in the middle latitudes. Visit the Meteorology Resource Center to view this and other active figures at academic. cengage.com/login

0.26 μm can destroy acids in DNA (deoxyribonucleic acid), the substance that transmits the hereditary blueprint from one generation to the next.

If the concentration of stratospheric ozone decreases, the following are expected to occur:

● An increase in the number of cases of skin cancer.

● A sharp increase in eye cataracts and sunburning.

● Suppression of the human immune system.

● An adverse impact on crops and animals due to an increase in ultraviolet radiation.

● A reduction in the growth of ocean phytoplankton.

● A cooling of the stratosphere that could alter stratospheric wind patterns, possibly affecting the destruction of ozone.

Stratospheric Ozone: Production-Destruction

Ozone (O_3) forms naturally in the stratosphere by the combining of atomic oxygen (O) with molecular oxygen (O_2)—in the presence of another molecule. Although it forms mainly above 25 km, ozone gradually drifts downward by mixing processes, producing a peak concentration in middle latitudes near 25 km. (In polar regions, its maximum concentration is found at lower levels.) Ozone is broken down into molecular and atomic oxygen by absorbing ultraviolet (UV) radiation with wavelengths between 0.2 and 0.3 mm (see Fig. 18.8). Thus

$$O_3 + UV \rightarrow O_2 + O.$$

Ozone is destroyed by colliding with other atoms and molecules. For example, ozone and atomic oxygen combine to form two oxygen molecules, as

$$O_3 + O \rightarrow 2\,O_2.$$

Likewise, the combination of two ozone molecules destroys ozone, as

$$O_3 + O_3 \rightarrow 3\,O_2.$$

These equations also represent the net result of a number of complex chemical reactions that include trace gases of nitrogen, hydrogen, and chlorine. For example, two natural destructive gases for ozone are nitric oxide (NO) and nitrogen dioxide (NO$_2$), which, as we have seen, are collectively known as *oxides of nitrogen*. The origin of these gases begins at the earth's surface as soil bacteria produce N$_2$O (nitrous oxide). This gas gradually finds its way into the stratosphere where, above about 25 km, solar energy converts some of it into ozone-destroying oxides of nitrogen. In the stratosphere just a small amount of nitric oxide can destroy a large amount of ozone. The following sequence of chemical reactions will illustrate why. In step 1, the nitric oxide quickly combines with ozone to form nitrogen dioxide and molecular oxygen. Then, in step 2, the nitrogen dioxide combines with atomic oxygen to form nitric oxide and molecular oxygen. Thus

$$\begin{aligned}
&\textit{Step 1} \quad NO + O_3 \rightarrow NO_2 + O_2. \\
&\textit{Step 2} \quad NO_2 + O \rightarrow NO + O_2.
\end{aligned}$$

The nitric oxide (NO) released in step 2 is now ready to start destroying ozone again.

Stratospheric ozone is maintained by a delicate natural balance between production and destruction. Could this balance be upset?

Stratospheric Ozone: Upsetting the Balance

The concentration of ozone in the stratosphere may be changed by natural events. In the upper atmosphere, both cosmic rays* and solar particles can produce secondary electrons having sufficient energy to separate molecular nitrogen (N$_2$) into two nitrogen atoms (N). The nitrogen atoms combine with free atomic oxygen to form nitric oxide which, in turn, rapidly destroys ozone. Furthermore, large volcanic eruptions, such as the Philippine's Mt. Pinatubo in mid-June, 1991, can inject ozone-destroying chemicals into the stratosphere. And scientists, using measurements from satellites, discovered

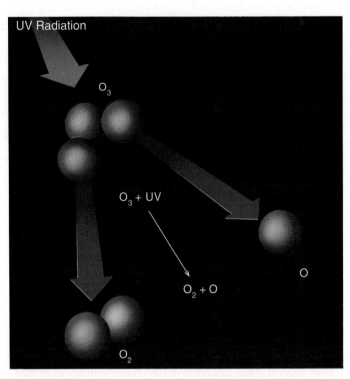

● FIGURE 18.8 An ozone molecule absorbing ultraviolet radiation can become molecular oxygen (O$_2$) and atomic oxygen (O).

that even changes in ultraviolet radiation from the sun can cause small variations in the amount of stratospheric ozone.

It is now apparent that human activities are also altering the amount of stratospheric ozone. This possibility was first brought to light in the early 1970s as Congress pondered over whether or not the United States should build a supersonic jet transport. One of the gases emitted from the engines of this aircraft is nitric oxide. Although the aircraft was designed to fly in the stratosphere below the level of maximum ozone, it was feared that the nitric oxide would eventually have an adverse effect on the ozone. This factor was one of many considered when Congress decided to halt the development of the United States' version of the supersonic transport.

More recently, concerns involve emissions of chemicals at the earth's surface, such as nitrous oxide emitted from nitrogen fertilizers (which may drift into the stratosphere, where it could destroy ozone) and *chlorofluorocarbons* (CFCs). Until the late 1970s, when the United States banned all nonessential uses of chlorofluorocarbons, they were the most widely used propellants in spray cans, such as deodorants and hairsprays. In the troposphere, these gases are quite safe, being nonflammable, nontoxic, and unable to chemically combine with other substances.* Hence, these gases slowly diffused upward without being destroyed. They apparently enter the stratosphere

1. near breaks in the tropopause, especially in the vicinity of jet streams

---

*Cosmic rays are high-energy atomic nuclei and atomic particles that travel through space at extremely high speeds. Most cosmic rays are believed to come from exploding stars (hypernovae), although some are produced in solar flares.

---

*Recall from Chapter 3 that CFCs do act as strong greenhouse gases in the troposphere.

2. in building thunderstorms, especially those that develop in the tropics along the Intertropical Convergence Zone and penetrate the lower stratosphere.

Once chlorofluorocarbon molecules reach the middle stratosphere, ultraviolet energy that is normally absorbed by ozone breaks them up, releasing atomic *chlorine* in the process. Note in the following sequence of reactions how chlorine destroys ozone rapidly. In step 1, atomic chlorine (Cl) combines with ozone, forming chlorine monoxide (ClO)—a new substance—and molecular oxygen (O_2). Almost immediately, the chlorine monoxide combines with free atomic oxygen (step 2) to produce chlorine atoms and molecular oxygen. The free chlorine atoms are now ready to combine with and destroy more ozone molecules. Estimates are that a single chlorine atom removes as many as 100,000 ozone molecules before it is taken out of action by combining with other substances.

$$\text{Step 1} \quad Cl + O_3 \rightarrow ClO + O_2.$$
$$\text{Step 2} \quad ClO + O \rightarrow + Cl\ ClO_2.$$

Fortunately, chlorine atoms do not exist in the stratosphere forever. They are removed as chlorine monoxide combines with nitrogen dioxide to form chlorine nitrate, $ClONO_2$ (step 3). In step 4, free chlorine atoms combine with methane to form hydrogen chloride (HCl) and a new substance, CH_3.

$$\text{Step 3} \quad ClO + NO_2 \rightarrow ClONO_2.$$
$$\text{Step 4} \quad CH_4 + Cl \rightarrow HCl + CH_3.$$

Since the average lifetime of a CFC molecule is about 50 to 100 years, any increase in the concentration of CFCs is long-lasting and a genuine threat to the concentration of ozone. Given this fact and the additional knowledge that CFCs contribute to the earth's greenhouse effect, an international agreement called the *Montreal Protocol* was signed in 1987. This agreement established a timetable for diminishing CFC emissions and the use of bromine compounds (halons), which destroy ozone at a rate 50 times greater than do chlorine compounds.[*]

During November, 1992, representatives of more than half the world's nations met in Copenhagen to update and revise the treaty. Provisions of the meeting called for a quicker phase-out of the previously targeted ozone-destroying chemicals and the establishment of a permanent fund to help Third World nations find the technology to develop ozone-friendly chemicals.[†] The phase-out appears to be working, as global concentrations of atmospheric chlorine and bromine have been decreasing (see ● Fig. 18.9).

Although the use of CFCs has decreased 96 percent between 1986 and 2005, there are still millions of kilograms in the troposphere that will continue to slowly diffuse upward. In a 1991 study, an international panel of over 80 scientists

---

[*]There are many chemical reactions that involve chlorine and bromine and the destruction of ozone in the stratosphere. The example given so far is just one of them.

[†]The phase-out of CFCs for domestic use has stimulated a thriving black market.

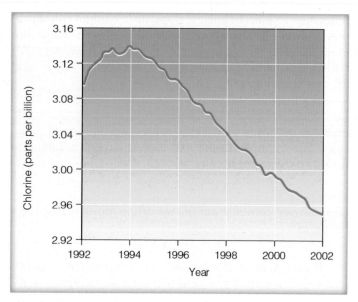

● FIGURE 18.9 Global average concentration of atmospheric chlorine. (Data from NOAA)

concluded that the ozone layer thinned by about 3 percent during the summer from 1979 to 1991 over heavily populated areas of the Northern Hemisphere. Moreover, recent studies show that stratospheric ozone concentrations above the United States presently are about 3 percent below normal is summer and about 5 percent below normal in winter.

Satellite measurements in 1992 and 1993 revealed ozone concentrations had dropped to record low levels over much of the globe. The decrease appears to stem from ozone-destroying chemicals and from the 1991 volcanic eruption of Mt. Pinatubo that sent tons of sulfur dioxide gas into the stratosphere, where it formed tiny droplets of sulfuric acid. These droplets not only enhance the ozone destructiveness of the chlorine chemicals but also alter the circulation of air in the stratosphere, making it more favorable for ozone depletion. During the mid-1990s, wintertime ozone levels dropped well below normal over much of the Northern Hemisphere. This decrease apparently was due to ozone-destroying pollution along with natural cold stratospheric weather patterns that favored ozone reduction.

Presently, there are two major substitutes for CFCs, *hydrochlorofluorocarbons* (HCFCs) and *hydrofluorocarbons* (HFCs). The HCFCs contain fewer chlorine atoms per molecule than CFCs and, therefore, pose much less danger to the ozone layer, whereas HFCs contain no chlorine. These gases may have to be phased out, however, as both are greenhouse gases that can enhance global warming.

At present, most scientists believe that ozone levels in the stratosphere throughout the world (except over Antarctica) will return to pre-1980 levels by the year 2050. Over Antarctica, ozone concentrations will likely remain low until about 2070. In fact, ozone concentrations over springtime Antarctica have been exceptionally low. This sharp drop in ozone is known as the **ozone hole**. (More information on the ozone hole is provided in the Focus section on p. 511.)

FOCUS ON AN ENVIRONMENTAL ISSUE

The Ozone Hole

In 1974, two chemists from the University of California at Irvine — F. Sherwood Rowland and Mario J. Molina — warned that increasing levels of CFCs would eventually deplete stratospheric ozone on a global scale. Their studies suggested that ozone depletion would occur gradually and would perhaps not be detectable for many years to come. It was surprising, then, when British researchers identified a year-to-year decline in stratospheric ozone over Antarctica. Their findings, corroborated later by satellites and balloon-borne instruments, showed that since the late 1970s ozone concentrations have diminished each year during the months of September and October. This decrease in stratospheric ozone over springtime Antarctica is known as the *ozone hole*. In years of severe depletion, such as in 2006, the ozone hole covers almost twice the area of the Antarctic continent (see Fig.1).

To understand the causes behind the ozone hole, scientists in 1986 organized the first *National Ozone Expedition*, NOZE-1, which set up a fully instrumented observing station near McMurdo Sound, Antarctica. During 1987, with the aid of instrumented aircraft, NOZE-2 got under way. The findings from these research programs helped scientists put together the pieces of the ozone puzzle.

The stratosphere above Antarctica has one of the world's highest ozone concentrations. Most of this ozone forms over the tropics and is brought to the Antarctic by stratospheric winds. During September and October (spring in the Southern Hemisphere), a belt of stratospheric winds called the *polar vortex* encircles the Antarctic region near 66°S latitude, essentially isolating the cold Antarctic stratospheric air from the warmer air of the middle latitudes. During the long dark Antarctic winter, temperatures inside the vortex can drop to −85°C (−121°F). This frigid air allows for the formation of *polar stratospheric clouds*. These ice clouds are critical in facilitating chemical interactions among nitrogen, hydrogen, and chlorine atoms, the end product of which is the destruction of ozone.

In 1986, the NOZE-1 study detected unusually high levels of chlorine compounds in the stratosphere, and, in 1987, the instrumented aircraft of NOZE-2 measured enormous increases in chlorine compounds when it entered the polar vortex. These findings, in conjunction with other chemical discoveries, allowed scientists to pinpoint *chlorine* from CFCs as the main cause of the ozone hole.

Even with a decline in ozone-destroying chemicals, the largest Antarctic ozone hole observed to date occurred during September, 2006 (Fig. 1). Apparently, these yearly variations in the size and depth of the ozone hole are mainly due to changes in polar stratospheric temperatures.

In the Northern Hemisphere's polar Arctic, airborne instruments and satellites during the late 1980s and 1990s measured high levels of ozone-destroying chlorine compounds in the stratosphere. By 1997, springtime ozone levels in the Arctic were about 40 percent below average (see Fig. 2). But observations could not detect an ozone hole like the one that forms over the Antarctic.

Apparently, several factors inhibit massive ozone loss in the Arctic. For one thing, in the stratosphere, the circulation of air over the Arctic differs from that over the Antarctic. Then, too, the Arctic stratosphere is normally too warm for

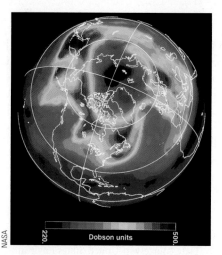

● FIGURE 2   Color image of total ozone amounts over the Northern Hemisphere for March 24, 1997. Notice that minimum ozone values (purple shades) appear over a region near the North Pole. The color scale on the bottom of the image shows total ozone values in Dobson units (DU). A Dobson unit is the physical thickness of the ozone layer if it were brought to the earth's surface (500 DU equals 5 mm).

the widespread development of clouds that help activate chlorine molecules. However, it appears that a very cold Arctic stratosphere, along with ozone-destroying chemicals were responsible for the low readings in 1997. Moreover, during January, 2000, more polar stratospheric clouds formed over the Arctic, and they lasted longer than during any previous year. This situation contributed to significant ozone loss.

Ozone depletion is not just confined to the stratospheric Arctic and Antarctic. For example, in March, 1995, satellite measurements revealed that the United States' ozone levels fell between 15 and 20 percent below the values observed during March, 1979.

We still have much to learn about stratospheric ozone and the processes that both form and destroy it. Presently, atmospheric studies are providing more information so that a more complete assessment of the ozone problem will become available in the future. Studies show that the ozone hole is still there — some years it is stronger, some years it is weaker.

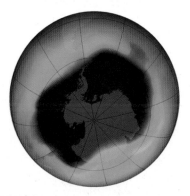

● FIGURE 1   Ozone distribution over the South Pole on September 29, 2006, as measured by ozone monitoring equipment on NASA's Aura Satellite. Notice that the lowest ozone concentration or ozone hole (purple shades) covers most of Antarctica. (NASA)

AIR POLLUTION: TRENDS AND PATTERNS Over the past decades, strides have been made in the United States to improve the quality of the air we breathe. ● Figure 18.10 shows the estimated emission trends over the United States for the primary pollutants. Notice that since the Clean Air Act of 1970, emissions of most pollutants have fallen off substantially, with lead showing the greatest reduction, primarily due to the gradual elimination of leaded gasoline.

Although the situation has improved, we can see from Fig. 18.10 that much more needs to be done, as large quantities of pollutants still spew into our air. In fact, many areas of the country do not conform to the standards for air quality set by the Clean Air Act of 1990. A large part of the problem of pollution control lies in the fact that even with stricter emission laws, increasing numbers of autos (estimates are that more than 198 million are on the road today) and other sources can overwhelm control efforts.

Clean air standards are established by the Environmental Protection Agency. *Primary ambient air quality standards* are set to protect human health, whereas *secondary standards* protect human welfare, as measured by the effects of air pollution on visibility, crops, and buildings. The National Ambient Air Quality Standards (NAAQS) currently in effect for six pollutants are given in ▼ Table 18.2. The table shows the concentration of each pollutant needed to exceed the NAAQS. Those

● FIGURE 18.10 Emission estimates of six pollutants in the United States from 1940 through 2003. (Data courtesy of United States Environmental Protection Agency.)

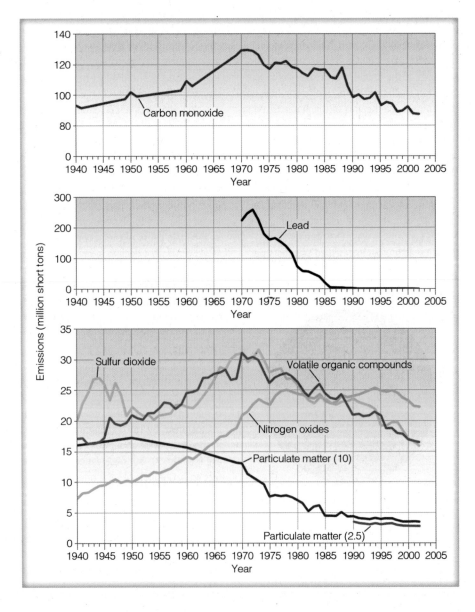

Mexico City lies in a broad basin surrounded by tall mountains. With 20 million inhabitants and 5.5 million vehicles traveling in and around the city daily, Mexico City exceeds the country's ground-level ozone standards about 284 days per year on average.

POLLUTANT	AVERAGING PERIOD	PRIMARY NAAQS	SECONDARY NAAQS
Ozone (O$_3$)	1-hour	0.12 ppm	0.12 ppm
	8-hour	0.08 ppm	0.08 ppm
Carbon Monoxide (CO)	1-hour	35 ppm	—
	8-hour	9 ppm	—
Sulfur Dioxide (SO$_2$)	3-hour	—	0.5 ppm
	24-hour	0.14 ppm	—
	Annual	0.030 ppm	—
Nitrogen Dioxide (NO$_2$)	Annual	0.053 ppm	0.053 ppm
Respirable Particulate Matter (10 μm or less) PM 10	24-hour	150 μg/m^3	150 μg/m^3
	Annual	50 μg/m^3	50 μg/m^3
Respirable Particulate Matter (2.5 μm or less) PM 2.5	24-hour	65 μg/m^3	65 μg/m^3
	Annual	15 μg/m^3	15 μg/m^3
Lead (Pb)	Calendar Quarter	1.5 μg/m^3	1.5 μg/m^3

areas that do not meet air quality standards are called *nonattainment areas.* Even with stronger emission laws, estimates are that more than 80 million Americans are breathing air that does not meet at least one of the standards (see ●Fig. 18.11).

To indicate the air quality in a particular region, the EPA developed the **air quality index (AQI)**.* The index includes the pollutants carbon monoxide, sulfur dioxide, nitrogen dioxide, particulate matter, and ozone. On any given day, the pollutant measuring the highest value is the one used in the index. The pollutant's measurement is then converted to a number that ranges from 0 to 500 (see ▼ Table 18.3). When the pollutant's value is the same as the primary ambient air quality standard, the pollutant is assigned an AQI number of 100. A pollutant is considered unhealthful when its AQI value exceeds 100. When the AQI value is between 51 and 100, the air quality is described as "moderate." Although these levels

*In June, 2000, the EPA updated the pollutant standard index (PSI) and renamed it the air quality index (AQI).

▼ TABLE 18.3   The Air Quality Index (AQI)

AQI VALUE	AIR QUALITY	GENERAL HEALTH EFFECTS	RECOMMENDED ACTIONS
0–50	Good	None	None
51–100	Moderate	There may be a moderate health concern for a very small number of individuals. People unusually sensitive to ozone may experience respiratory symptoms.	When O$_3$ AQI values are in this range, unusually sensitive people should consider limiting prolonged outdoor exposure.
101–150	Unhealthy for sensitive groups	Mild aggravation of symptoms in susceptible persons.	Active people with respiratory or heart disease should limit prolonged outdoor exertion.
151–200	Unhealthy	Aggravation of symptoms in susceptible persons, with irritation symptoms in the healthy population.	Active children and adults with respiratory or heart disease should avoid extended outdoor activities; everyone else, especially children, should limit prolonged outdoor exertion.
201–300	Very Unhealthy	Significant aggravation of symptoms and decreased exercise tolerance in persons with heart or lung disease, with widespread symptoms in the healthy population.	Active children and adults with existing heart or lung disease should avoid outdoor activities and exertion. Everyone else, especially children, should limit outdoor exertion.
301–500	Hazardous	Significant aggravation of symptoms. Premature onset of certain diseases. Premature death may occur in ill or elderly people. Healthy people may experience a decrease in exercise tolerance.	Everyone should avoid all outdoor exertion and minimize physical outdoor activities. Elderly and persons with existing heart or lung disease should stay indoors.

● FIGURE 18.11 The number of unhealthful days (by county) across the United States for any one of the five pollutants (CO, SO$_2$, NO$_2$, O$_3$, and particulate matter) during 2003. (Data courtesy of United States Environmental Protection Agency.)

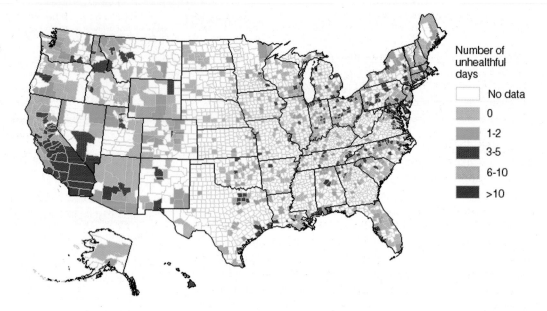

Number of unhealthful days

	No data
	0
	1–2
	3–5
	6–10
	>10

may not be harmful to humans during a 24-hour period, they may exceed long-term standards. Notice that the AQI is color-coded, with each color corresponding to an AQI level of health concern. The color green indicates "good" air quality; the color red, "unhealthy" air; and maroon, "hazardous" air quality. Table 18.3 also shows the health effects and the precautions that should be taken when the AQI value reaches a certain level.

Higher emission standards, along with cleaner fuels (such as natural gas), have made the air over our large cities cleaner today than it was years ago. In fact, total emissions of toxic chemicals spewed into the skies over the United States have

been declining steadily since the EPA began its inventory of these chemicals in 1987. But the control of ozone in polluted air is still a pervasive problem. Because ozone is a secondary pollutant, its formation is controlled by the concentrations of other pollutants, namely nitrogen oxides and hydrocarbons (VOCs). Moreover, weather conditions play a vital role in ozone formation, as ozone reaches its highest concentrations in hot sunny weather when surface winds are light and a stagnant high-pressure area covers the region. As a result of these factors, year-to-year ozone trends are quite variable, although the Los Angeles area has shown a steady decline in the number of unhealthful days due to ozone (see ● Fig. 18.12).

● FIGURE 18.12 The number of days ozone exceeded the 8-hour federal standard (0.08 ppm) and maximum 8-hour ozone concentration (ppm) for Los Angeles and surrounding areas in the South Coast air basin. (Courtesy of South Coast Air Pollution District.)

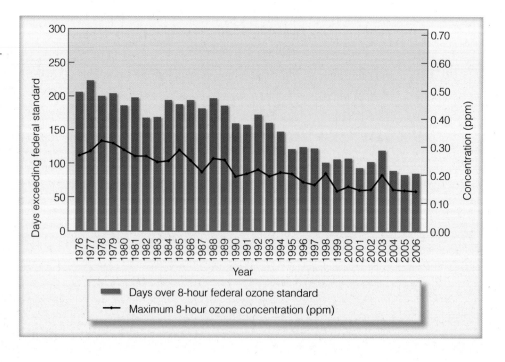

BRIEF REVIEW

Before going on to the next several sections, here is a brief review of some of the important points presented so far.

- Near the surface, primary air pollutants (such as particulate matter, CO, SO_2, NO, NO_2, and VOCs) enter the atmosphere directly, whereas secondary air pollutants (such as O_3) form when a chemical reaction takes place between a primary pollutant and some other component of air.

- The word "smog" (coined in London in the early 1900s) originally meant the combining of smoke and fog. Today, the word mainly refers to photochemical smog—pollutants that form in the presence of sunlight.

- Stratospheric ozone forms naturally in the stratosphere and provides a protective shield against the sun's harmful ultraviolet rays. Tropospheric (ground-level) ozone that forms in polluted air is a health hazard and is the primary ingredient of photochemical smog.

- Human-induced chemicals, such as chlorofluorocarbons (CFCs), have been altering the amount of ozone in the stratosphere by releasing chlorine, which rapidly destroys ozone.

- Even though the emissions of most pollutants have declined across the United States since 1970, millions of Americans are breathing air that does not meet air quality standards.

Up to now we have concentrated on outdoor air pollution. The Focus section on p. 516 looks at pollutants inside the home.

Factors That Affect Air Pollution

If you live in a region that periodically experiences photochemical smog, you may have noticed that these episodes often occur with clear skies, light winds, and generally warm sunny weather. Although this may be "typical" air pollution weather, it by no means represents the only weather conditions necessary to produce high concentrations of pollutants, as we will see in the following sections.

THE ROLE OF THE WIND   The wind speed plays a role in diluting pollution. When vast quantities of pollutants are spewed into the air, the wind speed determines how quickly the pollutants mix with the surrounding air and, of course, how fast they move away from their source. Strong winds tend to lower the concentration of pollutants by spreading them apart as they move downstream. Moreover, the stronger the wind, the more turbulent the air. Turbulent air produces swirling eddies that dilute the pollutants by mixing them with the cleaner surrounding air. Hence, when the wind dies down, pollutants are not readily dispersed, and they tend to become more concentrated (see ● Fig. 18.13).

THE ROLE OF STABILITY AND INVERSIONS   Recall from Chapter 6 that atmospheric stability determines the extent to which air will rise. Remember also that an unstable atmosphere favors vertical air currents, whereas a stable atmosphere strongly resists upward vertical motions. Consequently, smoke emitted into a stable atmosphere tends to spread horizontally, rather than mix vertically.

The stability of the atmosphere is determined by the way the air temperature changes with height (the lapse rate). When the measured air temperature decreases rapidly as we

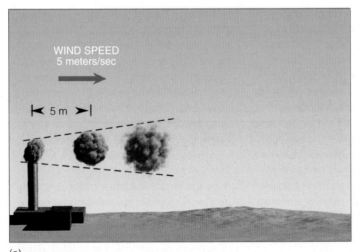
(a)

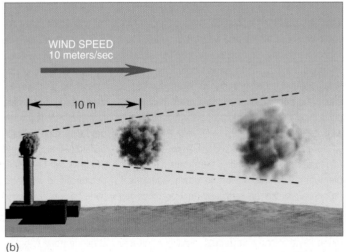
(b)

● FIGURE 18.13 If each chimney emits a puff of smoke every second, then where the wind speed is low (a), the smoke puffs are closer together and more concentrated. Where the wind speed is greater (b), the smoke puffs are farther apart and more diluted as turbulent eddies mix the smoke with the surrounding air.

FOCUS ON AN ENVIRONMENTAL ISSUE

Indoor Air Pollution

When people think of air pollution, most think of outside air where automobiles, factories, and power plants spew countless tons of contaminants into the air. But, surprisingly, the air we breathe inside our homes and other structures can be between 5 and 100 times more polluted than the air we breathe outdoors, even in the largest, most industrialized, cities! Since we spend so much time indoors, the quality of air we breathe there can have a far greater impact on our health and well-being than the air outdoors.

The Environmental Protection Agency has identified many sources of indoor air pollution, ranging from building materials, pressed wood products, furnishings, and home cleaning products, to pesticides, adhesives, and personal care products. In addition, heating sources (such as unvented kerosene heaters, wood stoves, and fireplaces) can release a variety of pollutants into a home. The pollution impact of any heating source depends upon several factors, such as how old the source is, the level of maintenance it receives, as well as its location and access to ventilation. For example, a gas cooking stove or heating stove with improper fittings and adjustments can cause a significant emission of carbon monoxide. New carpets and padding (as well as the adhesives used in their installation) can emit volatile organic compounds.

Some pollution sources, such as building materials and foam insulation, produce a constant stream of pollutants, whereas activities like tobacco smoking emit pollutants into the air on an intermittent basis. In certain instances, outside pollution is brought indoors. Some pollutants enter homes and other structures through cracks and holes in foundations and basements, as is the case with radon.

Radon is a colorless, odorless gas — a natural radioactive compound — that forms as the uranium in soil and rock breaks down. Radon is found everywhere on earth and only becomes a problem when it leaks out of the soil and becomes trapped inside homes and buildings. The radon gas that seeps in through cracks and other openings in a building can accumulate to levels that create a serious health threat. Studies by the EPA have shown that as many as 10 percent of the homes (about 8 million) in the United States

● FIGURE 3 With candles burning, a fire in the fireplace, and stain-resistant material on rugs and carpets, there are probably more air pollutants in this living room than there are in a similar size volume of air outdoors.

may have elevated levels of radon. The level of radon varies greatly from structure to structure and can only be measured by devices known as *radon detectors.* Inside a home, the radon decays into *polonium,* a solid substance that attaches itself to dust in the air. The tiny dust particles can be inhaled deep into the lungs, where they attach to lung tissue. As the polonium decays, it damages the lung tissue, sometimes producing mutated cells that may develop into lung cancer.

Another major chemical pollutant found inside our homes is *formaldehyde.* It is a colorless, pungent-smelling gas, used widely to manufacture building materials, insulation, and other household products. In most homes, the significant sources of formaldehyde are urethane-formaldehyde foam insulation and pressed wood products, such as particle board and plywood paneling. Emissions of formaldehyde from new products can be greatly affected by indoor temperatures and ventilation. Exposure to formaldehyde can cause watery eyes, burning sensations in the nose and throat, breathing difficulties, and nausea. It can also trigger attacks in individuals suffering from asthma.

Another polluter of our indoor air is *asbestos,* a mineral fiber once used in insulation and as a fire-retardant. In recent years, manufacturers have voluntarily reduced the use of asbestos, but

much asbestos still remains in furnace and pipe insulation, texturing materials, and floor tiles of older buildings. The most lethal fibers of asbestos are invisible. When inhaled, these tiny particles can accumulate and remain deep in the lungs for extended periods of time, where they damage tissue and potentially cause cancer or *asbestosis,* a permanent scarring of the lung that can be fatal.

Smoking tobacco indoors can also create an extremely dangerous health situation. Environmental tobacco smoke is a complex mixture of more than 4700 different compounds. These pollutants enter the body as particles and as gases, such as carbon monoxide and hydrogen cyanide. Exposure to tobacco smoke greatly increases the risk of developing lung cancer in both smokers and nonsmokers (studies show that the non-smoking spouses of smokers experience a 30 percent increase in the occurrence of lung cancer). The small children of smokers are also more likely to fall victim to such illnesses as bronchitis and pneumonia. Heart disease is closely linked with exposure to tobacco smoke, as is premature aging of the skin.

In some regions, a phenomenon known as *"sick building syndrome"* has become the focus of public concern. Several illnesses, such as *Legionnaire's disease* (which is spread through contaminated air conditioning systems), have been attributed to specific building problems, but other, less definite, illnesses have been traced to the indoor office environment as well. These complaints range from dry mucous membranes, sneezing, fatigue, and irritability to forgetfulness and nausea. Unfortunately, the nature of these symptoms and their random occurrence makes tracing the source difficult, if not impossible.

Indoor pollutants are responsible for a wide variety of health problems. Irritation of the eyes, nose, and throat, headaches, fatigue, and dizziness are but a few of the maladies attributed to indoor air pollutants. While these symptoms can be annoying, other life-threatening diseases can occur after prolonged exposure to many of the substances mentioned earlier, such as radon, asbestos, and other toxic compounds.

move up into the atmosphere, the atmosphere tends to be more unstable and pollutants tend to be mixed vertically as illustrated in ● Fig. 18.14a. If, however, the measured air temperature either decreases quite slowly as we ascend, or actually increases with height (remember that this is called an *inversion*), the atmosphere is stable. An inversion represents an extremely stable atmosphere where warm air lies above cool air (see Fig. 18.14b). Any air parcel that attempts to rise into the inversion will, at some point, be cooler and heavier (more dense) than the warmer air surrounding it. Hence, the inversion acts like a lid on vertical air motions.

The inversion depicted in Fig. 18.14b is called a **radiation** (or **surface**) **inversion**. This type of inversion typically forms during the night and early morning hours when the sky is clear and the winds are light. As we saw earlier in Chapter 3, radiation inversions also tend to be well developed during the long nights of winter.

In Figure 18.14b, notice that within the stable inversion, the smoke from the shorter stacks does not rise very high, but spreads out, contaminating the area around it. In the relatively unstable air above the inversion, smoke from the taller stack is able to rise and become dispersed. Since radiation inversions are often rather shallow, it should be apparent why taller chimneys have replaced many of the shorter ones. In fact, taller chimneys disperse pollutants better than shorter ones even in the absence of a surface inversion because the taller chimneys are able to mix pollutants throughout a greater volume of air. Although these taller stacks do improve the air quality in their immediate area, they may also contribute to the acid rain problem by allowing the pollutants to be swept great distances downwind.

As the sun rises and the surface warms, the radiation inversion normally weakens and disappears before noon. By afternoon, the atmosphere is sufficiently unstable so that, with adequate winds, pollutants are able to disperse vertically (Fig. 18.14b). The changing atmospheric stability, from stable in the early morning to conditionally unstable in the afternoon, can have a profound effect on the daily concentrations of pollution in certain regions. For example, on a busy city street corner, carbon monoxide levels can be considerably higher in the early morning than in the early afternoon (with the same flow of traffic). Changes in atmospheric stability can also cause smoke plumes from chimneys to change during the course of a day. (Some of these changes are described in the Focus section on p. 518.)

Radiation inversions normally last just a few hours, while **subsidence inversions** may persist for several days or longer. Subsidence inversions, therefore, are the ones commonly associated with major air pollution episodes. They form as the air above a deep anticyclone slowly sinks (subsides) and warms.*

A typical temperature profile of a subsidence inversion is shown in ● Fig. 18.15. Notice that in the relatively unstable air beneath the inversion, the pollutants are able to mix vertically

*Remember from Chapter 2 that sinking air always warms because it is being compressed by the surrounding air.

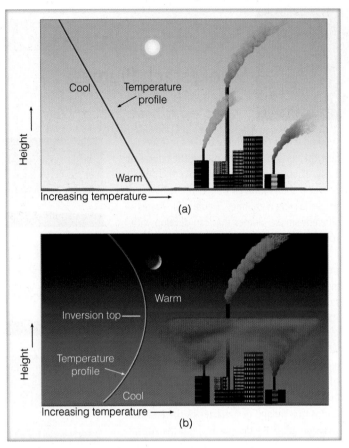

● **FIGURE 18.14** (a) During the afternoon, when the atmosphere is most unstable, pollutants rise, mix, and disperse downwind. (b) At night when a radiation inversion exists, pollutants from the shorter stacks are trapped within the inversion, while pollutants from the taller stack, above the inversion, are able to rise and disperse downwind.

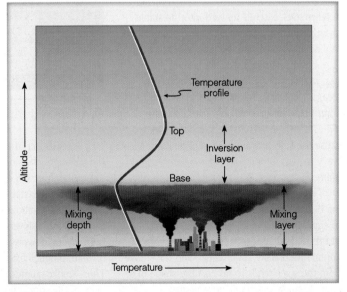

● **FIGURE 18.15** The inversion layer acts as a lid on the pollutants below. If the inversion lowers, the mixing depth decreases and the pollutants are concentrated within a smaller volume.

FOCUS ON AN OBSERVATION

Smokestack Plumes

We know that the stability of the air (especially near the surface) changes during the course of a day. These changes can influence the pollution near the ground as well as the behavior of smoke leaving a chimney. Figure 4 illustrates different smoke plumes that can develop with adequate wind, but different types of stability.

In Fig. 4a, it is early morning, the winds are light, and a radiation inversion extends from the surface to well above the height of the smokestack. In this stable environment, there is little up and down motion, so the smoke spreads horizontally rather than vertically. When viewed from above, the smoke plume resembles the shape of a fan. For this reason, it is referred to as a *fanning smoke plume*.

Later in the morning, the surface air warms quickly and destabilizes as the radiation inversion gradually disappears from the surface upward (Fig. 4b). However, the air above the chimney is still stable, as indicated by the presence of the inversion. Consequently, vertical motions are confined to the region near the surface. Hence, the smoke mixes downwind, increasing the concentration of pollution at the surface — sometimes to dangerously high levels. This effect is called *fumigation*. Here again, we can see why a taller smokestack is preferred. A taller stack extends upward into the stable layer, producing a fanning plume that does not mix downward toward the ground.

If daytime heating of the ground continues, the depth of atmospheric instability increases. Notice in Fig. 4c that the inversion has completely disappeared. Light-to-moderate winds combine with rising and sinking air to cause the smoke to move up and down in a wavy pattern, producing a *looping smoke plume*.

The continued rising of warm air and sinking of cool air can cause the temperature profile to equal that of the dry adiabatic rate (Fig. 4d). In this neutral atmosphere, vertical and horizontal motions are about equal, and the smoke from the stack tends to take on the shape of a cone, forming a *coning smoke plume*.

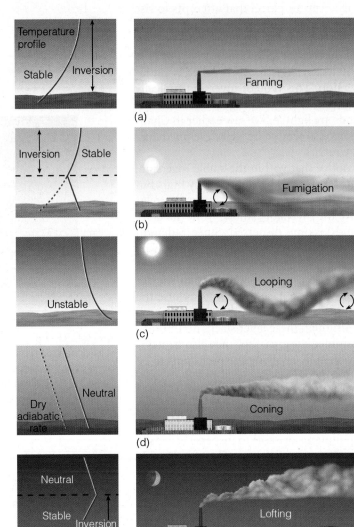

● FIGURE 4 As the vertical temperature profile changes during the course of a day (a through e), the pattern of smoke emitted from the stack changes as well.

After sunset, the ground cools rapidly and the radiation inversion reappears. When the top of the inversion extends upward to slightly above the stack, stable air is near the ground with neutral air above (Fig. 4e). Because the stable air in the inversion prevents the smoke from mixing downward, the smoke is carried upward, producing a *lofting smoke plume*. Thus, smoke plumes provide a clue to the stability of the atmosphere, and knowing the sta-

bility yields important information about the dispersion of pollutants.

Of course, other factors influence the dispersion of pollutants from a chimney, including the pollutants' temperature and exit velocity, wind speed and direction, and, as we saw in an earlier section, the chimney's height. Overall, taller chimneys, greater wind speeds, and higher exit velocities result in a lower concentration of pollutants.

up to the inversion base. The stable air of the inversion, however, inhibits vertical mixing and acts like a lid on the pollution below, preventing it from entering into the inversion.

In Fig. 18.15, the region of relatively unstable (well mixed) air that extends from the surface to the base of the inversion is referred to as the **mixing layer.** The vertical extent of the mixing layer is called the **mixing depth.** Observe that if the inversion rises, the mixing depth increases and the pollutants would be dispersed throughout a greater volume of air; if the inversion lowers, the mixing depth would decrease and the pollutants would become more concentrated, sometimes reaching unhealthy levels. Since the atmosphere tends to be most unstable in the afternoon and most stable in the early morning, we typically find the greatest mixing depth in the afternoon and the most shallow one (if one exists at all) in the early morning. Consequently, during the day, the top of the mixing layer may clearly be visible (see ● Fig. 18.16). Moreover, during take-off or landing on daylight flights out of large urban areas, the top of the mixing layer may sometimes be observed.

The position of the semipermanent Pacific high off the coast of California contributes greatly to the air pollution in that region. The Pacific high promotes subsiding air, which warms the air aloft. Surface winds around the high promote upwelling of ocean water. Upwelling—the rising of cold water from below—makes the surface water cool, which, in turn, cools the air above. Warm air aloft coupled with cool, surface (marine) air together produce a strong and persistent subsidence inversion—one that exists 80 to 90 percent of the time over the city of Los Angeles between June and October, the smoggy months. The pollutants trapped within

● FIGURE 18.16 A thick layer of polluted air is trapped in the valley. The top of the polluted air marks the base of a subsidence inversion.

the cool marine air are occasionally swept eastward by a sea breeze. This action carries smog from the coastal regions into the interior valleys producing a *smog front* (see ● Fig. 18.17).

THE ROLE OF TOPOGRAPHY The shape of the landscape (topography) plays an important part in trapping pollutants. We know from Chapter 3 that, at night, cold air tends to

● FIGURE 18.17 The leading edge of cool, marine air carries pollutants into Riverside, California, as an advancing smog front.

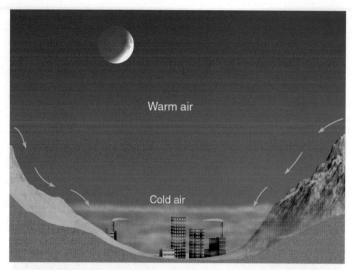

Warm air

Cold air

● FIGURE 18.18 At night, cold air and pollutants drain downhill and settle in low-lying valleys.

drain downhill, where it settles into low-lying basins and valleys. The cold air can have several effects: It can strengthen a pre-existing surface inversion, and it can carry pollutants downhill from the surrounding hillsides (see ● Fig. 18.18).

Valleys prone to pollution are those completely encased by mountains and hills. The surrounding mountains tend to block the prevailing wind. With light winds, and a shallow mixing layer, the poorly ventilated cold valley air can only slosh back and forth like a murky bowl of soup.

Air pollution concentrations in mountain valleys tend to be greatest during the colder months. During the warmer months, daytime heating can warm the sides of the valley to the point that upslope valley winds vent the pollutants up-

ward, like a chimney. Valleys susceptible to stagnant air exist in just about all mountainous regions.

The pollution problem in several large cities is, at least, partly due to topography. For example, the city of Los Angeles is surrounded on three sides by hills and mountains. Cool marine air from off the ocean moves inland and pushes against the hills, which tend to block the air's eastward progress. Unable to rise, the cool air settles in the basin, trapping pollutants from industry and millions of autos. Baked by sunlight, the pollutants become the infamous photochemical smog (see ● Fig. 18.19). By the same token, the "mile high" city of Denver, Colorado, sits in a broad shallow basin that frequently traps both cold air and pollutants.

SEVERE AIR POLLUTION POTENTIAL   The greatest potential for an episode of severe air pollution occurs when all of the factors mentioned in the previous sections come together simultaneously. Ingredients for a major buildup of atmospheric pollution are:

- many sources of air pollution (preferably clustered close together)
- a deep high-pressure area that becomes stationary over a region
- light surface winds that are unable to disperse the pollutants
- a strong subsidence inversion produced by the sinking of air aloft
- a shallow mixing layer with poor ventilation
- a valley where the pollutants can accumulate
- clear skies so that radiational cooling at night will produce a surface inversion, which can cause an even greater buildup of pollutants near the ground

● FIGURE 18.19 A thick layer of smog covers the city of Los Angeles.

© David R. Frazier/Photo Researchers, Inc.

Five Days in Donora—An Air Pollution Episode

On Tuesday morning, October 26, 1948, a cold surface anticyclone moved over the eastern half of the United States. There was nothing unusual about this high-pressure area; with a central pressure of only 1025 mb (30.27 in.), it was not exceptionally strong (see Fig. 5). Aloft, however, a large blocking-type ridge formed over the region, and the jet stream, which moves the surface pressure features along, was far to the west. Consequently, the surface anticyclone became entrenched over Pennsylvania and remained nearly stationary for five days.

The widely spaced isobars around the high-pressure system produced a weak pressure gradient and generally light winds throughout the area. These light winds, coupled with the gradual sinking of air from aloft, set the stage for a disastrous air pollution episode.

On Tuesday morning, radiation fog gradually settled over the moist ground in Donora, a small town nestled in the Monongahela Valley of western Pennsylvania. Because Donora rests on bottom land, surrounded by rolling hills, its residents were accustomed to fog, but not to what was to follow.

The strong radiational cooling that formed the fog, along with the sinking air of the anticyclone, combined to produce a strong temperature inversion. Light, downslope winds spread cool air and contaminants over Donora from the community's steel mill, zinc smelter, and sulfuric acid plant.

The fog with its burden of pollutants lingered into Wednesday. Cool drainage winds during the night strengthened the inversion and added more effluents to the already filthy air. The dense fog layer blocked sunlight from reaching the ground. With essentially no surface heating, the mixing depth lowered and the pollution became more concentrated. Unable to

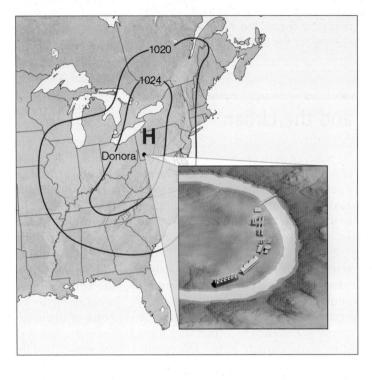

● **FIGURE 5**
Surface weather map that shows a stagnant anticyclone over the eastern United States on October 26, 1948. The insert map shows the town of Donora on the Monongahela River.

mix and disperse both horizontally and vertically, the dirty air became confined to a shallow, stagnant layer.

Meanwhile, the factories continued to belch impurities into the air (primarily sulfur dioxide and particulate matter) from stacks no higher than 40 m (130 ft) tall. The fog gradually thickened into a moist clot of smoke and water droplets. By Thursday, the visibility had decreased to the point where one could barely see across the street. At the same time, the air had a penetrating, almost sickening, smell of sulfur dioxide. At this point, a large percentage of the population became ill.

The episode reached a climax on Saturday, as 17 deaths were reported. As the death rate mounted, alarm swept through the town. An

emergency meeting was called between city officials and factory representatives to see what could be done to cut down on the emission of pollutants.

The light winds and unbreathable air persisted until, on Sunday, an approaching storm generated enough wind to vertically mix the air and disperse the pollutants. A welcome rain then cleaned the air further. All told, the episode had claimed the lives of 22 people. During the five-day period, about half of the area's 14,000 inhabitants experienced some ill effects from the pollution. Most of those affected were older people with a history of cardiac or respiratory disorders.

● and, for photochemical smog, adequate sunlight to produce secondary pollutants, such as ozone

Light winds and poor vertical mixing can produce a condition known as **atmospheric stagnation.** When this condition prevails for several days to a week or more, the buildup

of pollutants can lead to some of the worst air pollution disasters on record, such as the one in the valley city of Donora, Pennsylvania, where in 1948 seventeen people died within fourteen hours. (Additional information on the Donora disaster is found in the Focus section above.)

Air Pollution and the Urban Environment

For more than 100 years, it has been known that cities are generally warmer than surrounding rural areas. This region of city warmth, known as the **urban heat island,** can influence the concentration of air pollution. However, before we look at its influence, let's see how the heat island actually forms.

The urban heat island is due to industrial and urban development. In rural areas, a large part of the incoming solar energy is used to evaporate water from vegetation and soil. In cities, where less vegetation and exposed soil exists, the majority of the sun's energy is absorbed by urban structures and asphalt. Hence, during warm daylight hours, less evaporative cooling in cities allows surface temperatures to rise higher than in rural areas.*

At night, the solar energy (stored as vast quantities of heat in city buildings and roads) is slowly released into the city air. Additional city heat is given off at night (and during the day) by vehicles and factories, as well as by industrial and domestic heating and cooling units. The release of heat energy is retarded by the tall vertical city walls that do not allow infrared radiation to escape as readily as do the relatively level surfaces of the surrounding countryside. The slow release of heat tends to keep nighttime city temperatures higher than those of the faster cooling rural areas. Overall, the heat island is strongest

1. at night when compensating sunlight is absent
2. during the winter when nights are longer and there is more heat generated in the city
3. when the region is dominated by a high-pressure area with light winds, clear skies, and less humid air.

Over time, increasing urban heat islands affect climatological temperature records, producing artificial warming in climatic records taken in cities.

*The cause of the urban heat island is quite involved. Depending on the location, time of year, and time of day, any or all of the following differences between cities and their surroundings can be important: albedo (reflectivity of the surface), surface roughness, emissions of heat, emissions of moisture, and emissions of particles that affect net radiation and the growth of cloud droplets.

The constant outpouring of pollutants into the environment may influence the climate of a city. Certain particles reflect solar radiation, thereby reducing the sunlight that reaches the surface. Some particles serve as nuclei upon which water and ice form. Water vapor condenses onto these particles when the relative humidity is as low as 70 percent, forming haze that greatly reduces visibility. Moreover, the added nuclei increase the frequency of city fog.*

Studies suggest that precipitation may be greater in cities than in the surrounding countryside. This phenomenon may be due in part to the increased roughness of city terrain, brought on by large structures that cause surface air to slow and gradually converge. This piling-up of air over the city then slowly rises, much like toothpaste does when its tube is squeezed. At the same time, city heat warms the surface air, making it more unstable, which enhances rising air motions, which, in turn, aid in forming clouds and thunderstorms. This process helps explain why both clouds and storms tend to be more frequent over cities. ▼ Table 18.4 summarizes the environmental influence of cities by contrasting the urban environment with the rural.

On clear still nights when the heat island is pronounced, a small thermal low-pressure area forms over the city. Sometimes a light breeze—called a **country breeze**—blows from the countryside into the city. If there are major industrial areas along the city's outskirts, pollutants are carried into the heart of town, where they become even more concentrated. Such an event is especially likely if an inversion inhibits vertical mixing and dispersion (see ● Fig. 18.20).

*The impact that tiny liquid and solid particles, aerosols, may have on a larger scale is complex and depends upon many factors.

▼ TABLE 18.4 Contrast of the Urban and Rural Environment (Average Conditions)*

CONSTITUENTS	URBAN AREA (CONTRASTED TO RURAL AREA)
Mean pollution level	higher
Mean sunshine reaching the surface	lower
Mean temperature	higher
Mean relative humidity	lower
Mean visibility	lower
Mean wind speed	lower
Mean precipitation	higher
Mean amount of cloudiness	higher
Mean thunderstorm (frequency)	higher

*Values are omitted because they vary greatly depending upon city, size, type of industry, and season of the year.

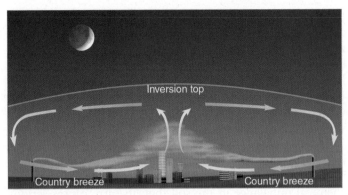

● FIGURE 18.20 On a clear, relatively calm night, a weak country breeze carries pollutants from the outskirts into the city, where they concentrate and rise due to the warmth of the city's urban heat island. This effect may produce a pollution (or dust) dome from the suburbs to the center of town.

Pollutants from urban areas may even affect the weather downwind from them. In a controversial study conducted at La Porte, Indiana — a city located about 50 km downwind of the industries of south Chicago — scientists observed that La Porte had experienced a notable increase in annual precipitation since 1925. Because this rise closely followed the increase in steel production, it was suggested that the phenomenon was due to the additional emission of particles or moisture (or both) by industries to the west of La Porte.

A study conducted in St. Louis, Missouri (the Metropolitan Meteorological Experiment, or *METROMEX*) indicated that the average annual precipitation downwind from this city increased by about 10 percent. These increases closely followed industrial development upwind. This study also demonstrated that precipitation amounts were significantly greater on weekdays (when pollution emissions were higher) than on weekends (when pollution emissions were lower). Corroborative findings have been reported for Paris, France, and for other cities as well. However, in areas with marginal humidity to support the formation of clouds and precipitation, studies suggest that the rate of precipitation may actually decrease as excess pollutant particles (nuclei) compete for the available moisture, similar to the effect of overseeding a cloud, discussed in Chapter 7. Moreover, studies using satellite data indicate that fine airborne particles, concentrated over an area, can greatly reduce precipitation.

Acid Deposition

Air pollution emitted from industrial areas, especially products of combustion, such as oxides of sulfur and nitrogen, can be carried many kilometers downwind. Either these particles and gases slowly settle to the ground in dry form *(dry deposition)* or they are removed from the air during the formation of cloud particles and then carried to the ground in rain and snow *(wet deposition)*. **Acid rain** and *acid precipitation* are common terms used to describe wet deposition,

while **acid deposition** encompasses both dry and wet acidic substances. How, then, do these substances become acidic?

Emissions of sulfur dioxide (SO_2) and oxides of nitrogen may settle on the local landscape, where they transform into acids as they interact with water, especially during the formation of dew or frost. The remaining airborne particles may transform into tiny dilute drops of sulfuric acid (H_2SO_4) and nitric acid (HNO_3) during a complex series of chemical reactions involving sunlight, water vapor, and other gases. These acid particles may then fall slowly to earth, or they may adhere to cloud droplets or to fog droplets, producing **acid fog.** They may even act as nuclei on which the cloud droplets begin to grow. When precipitation occurs in the cloud, it carries the acids to the ground. Because of this, precipitation is becoming increasingly acidic in many parts of the world, especially downwind of major industrial areas.

Airborne studies conducted during the middle 1980s revealed that high concentrations of pollutants that produce acid rain can be carried great distances from their sources. For example, scientists in one study discovered high concentrations of pollutants about 600 km off the east coast of North America. It is suspected that they came from industrial East Coast cities. Although most pollutants are washed from the atmosphere during storms, some may be swept over the Atlantic, reaching places like Bermuda and Ireland. Acid rain knows no national boundaries.

Although studies suggest that acid precipitation may be nearly worldwide in distribution, regions noticeably affected are eastern North America, central Europe, and Scandinavia. Sweden contends that most of the sulfur emissions responsible for its acid precipitation are coming from factories in England. In some places, acid precipitation occurs naturally, such as in northern Canada, where natural fires in exposed coal beds produce tremendous quantities of sulfur dioxide. By the same token, acid fog can form by natural means.

Precipitation is naturally somewhat acidic. The carbon dioxide occurring naturally in the air dissolves in precipitation, making it slightly acidic with a pH between 5.0 and 5.6. Consequently, precipitation is considered acidic when its pH is below about 5.0 (see ● Fig. 18.21). In the northeastern United States, where emissions of sulfur dioxide are primarily responsible for the acid precipitation, typical pH values range between 4.0 and 4.7 (see ● Fig. 18.22). But acid precipitation is not confined to the Northeast; the acidity of precipitation has increased in the southeastern states, too. Along the West Coast, the main cause of acid deposition appears to be the oxides of nitrogen released in automobile exhaust. In Los Angeles, acid fog is a more serious problem than acid rain, especially along the coast, where fog is most prevalent. The fog's pH is usually between 4.4 and 4.8, although pH values of 3.0 and below have been measured.

High concentrations of acid deposition can damage plants and water resources (freshwater ecosystems seem to be particularly sensitive to changes in acidity). Concern centers chiefly on areas where interactions with alkaline soil are unable to neutralize the acidic inputs. Studies indicate that

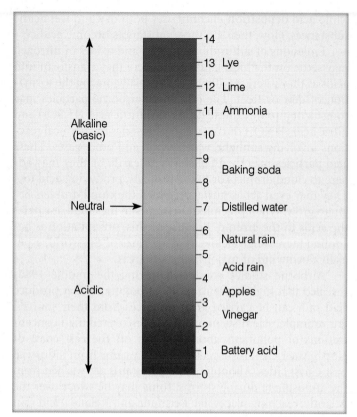

● FIGURE 18.21 The pH scale ranges from 0 to 14, with a value of 7 considered neutral. Values greater than 7 are alkaline and below 7 are acidic. The scale is logarithmic, which means that rain with pH 3 is 10 times more acidic than rain with pH 4 and 100 times more acidic than rain with pH 5.

thousands of lakes in the United States and Canada are so acidified that entire fish populations may have been adversely affected. In an attempt to reduce acidity, lime (calcium carbonate, $CaCO_3$) is being poured into some lakes. Natural alkaline soil particles can be swept into the air where they neutralize the acid. If it were not for airborne alkaline dust, China and the western United States would have a more severe acid rain problem than they do.

Many trees in Germany show signs of a blight that is due, in part, to acid deposition. Apparently, acidic particles raining down on the forest floor for decades have caused a chemical imbalance in the soil that, in turn, causes serious deficiencies in certain elements necessary for the trees' growth. The trees are thus weakened and become susceptible to insects and drought. The same type of processes may be affecting North American forests, but at a much slower pace, as many forests at higher elevations from southeastern Canada to South Carolina appear to be in serious decline (see ● Fig. 18.23). Moreover, acid precipitation is a problem in the mountainous West where high mountain lakes and forests seem to be most affected.

Also, acid deposition is eroding the foundations of structures in many cities throughout the world. In Rome, the acidity of rainfall is beginning to disfigure priceless outdoor fountain sculptures and statues. The estimated annual cost of this damage to building surfaces, monuments, and other structures is more than $2 billion.

Control of acid deposition is a difficult political problem because those affected by acid rain can be quite distant from those who cause it. Technology can control sulfur emissions

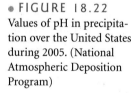

● FIGURE 18.22
Values of pH in precipitation over the United States during 2005. (National Atmospheric Deposition Program)

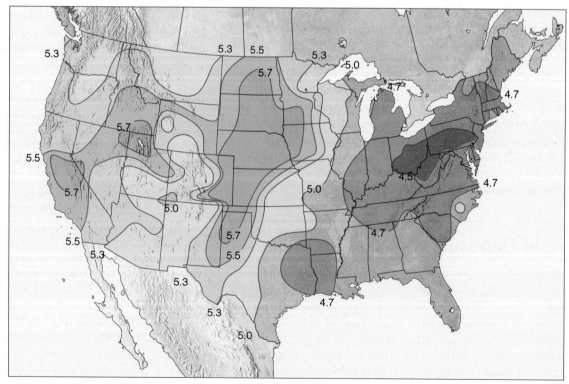

● FIGURE 18.23 The effects of acid fog in the Great Smoky Mountains of Tennessee.

(for example, stack scrubbers and fluidized bed combustion) and nitrogen emissions (catalytic converters on cars), but some people argue the cost is too high. If the United States turns more to coal-fired power plants, which are among the leading sources of sulfur oxide emissions, many scientists believe that the acid deposition problem will become more acute.

In an attempt to better understand acid deposition, the National Center for Atmospheric Research (NCAR) and the Environmental Protection Agency have been working to develop computer models that better describe the many physical and chemical processes contributing to acid deposition. To deal with the acid deposition problem, the Clean Air Act of 1990 imposed a reduction in the United States' emissions of sulfur dioxide and nitrogen dioxide. Canada has imposed pollution control standards and set a goal of reducing industrial air pollution by 50 percent.

SUMMARY

In this chapter, we found that air pollution has plagued humanity for centuries. Air pollution problems began when people tried to keep warm by burning wood and coal. These problems worsened during the industrial revolution as coal became the primary fuel for both homes and industry. Even though many American cities do not meet all of the air quality standards set by the federal Clean Air Act of 1990, the air over our large cities is cleaner today than it was years ago due to stricter emission standards and cleaner fuels.

We examined the types and sources of air pollution and found that primary air pollutants enter the atmosphere directly, whereas secondary pollutants form by chemical reactions that involve other pollutants. The secondary pollutant ozone is the main ingredient of photochemical smog—a smog that irritates the eyes and forms in the presence of sunlight. In polluted air near the surface, ozone forms during a series of chemical reactions involving nitrogen oxides and hydrocarbons (VOCs). In the stratosphere, ozone is a naturally occurring gas that protects us from the sun's harmful ultraviolet rays. We examined how ozone forms in the stratosphere and how it may be altered by natural means. We also learned that human-induced gases, such as chlorofluorocarbons, work their way into the stratosphere where they release chlorine that rapidly destroys ozone, especially in polar regions.

We looked at the air quality index and found that a number of areas across the United States still have days considered unhealthy by the standards set by the United States Environmental Protection Agency. We also looked at the main factors affecting air pollution and found that most air pollution episodes occur when the winds are light, skies are clear, the mixing layer is shallow, the atmosphere is stable, and a strong inversion exists. These conditions usually prevail when a high-pressure area stalls over a region.

We observed that, on the average, urban environments tend to be warmer and more polluted than the rural areas that surround them. We saw that pollution from industrial areas can modify environments downwind from them, as oxides of sulfur and nitrogen are swept into the air, where they may transform into acids that fall to the surface. Acid deposition, a serious problem in many regions of the world, knows no national boundaries—the pollution of one country becomes the acid rain of another.

KEY TERMS

The following terms are listed (with page numbers) in the order they appear in the text. Define each. Doing so will aid you in reviewing the material covered in this chapter.

QUESTIONS FOR REVIEW

1. What are some of the main sources of air pollution?
2. List a few of the substances that fall under the category of particulate matter.
3. How does PM-10 particulate matter differ from that called PM-2.5? Which poses the greatest risk to human health?
4. List two ways particulate matter is removed from the atmosphere.
5. Describe the primary sources and some of the health problems associated with each of the following pollutants:
   (a) carbon monoxide (CO)
   (b) sulfur dioxide (SO$_2$)
   (c) volatile organic compounds (VOCs)
   (d) nitrogen oxides
6. How does London-type smog differ from Los Angeles-type smog?
7. What is the main component of photochemical smog? What are some of the adverse health effects of photochemical smog? Why is the production of photochemical smog more prevalent during the summer and early fall than during the middle of winter?
8. In polluted air, (a) describe the role that NO$_2$ plays in the production of tropospheric (ground-level) ozone, and (b) the role that NO plays in the destruction of tropospheric (ground-level) ozone. What role do hydrocarbons (VOCs) play in the production of ozone and other photochemical oxidants?
9. Describe the main processes that account for stratospheric ozone production and destruction. What natural and human-produced substances could alter the concentration of ozone in the stratosphere?
10. Why is stratospheric ozone beneficial to life on earth, whereas tropospheric (ground-level) ozone is not?
11. (a) How are CFCs related to the destruction of stratospheric ozone?
    (b) If all of the ozone in the stratosphere were destroyed, what possible effects might this have on the earth's inhabitants?

12. Explain how scientists believe the Antarctic "ozone hole" forms.
13. Describe some of the sources of pollution found inside a home or building.
14. Why are high levels of radon inside a home so dangerous?
15. What are primary ambient air quality standards and secondary standards intended to do? What are nonattainment areas?
16. Why is a light wind, rather than a strong wind, more conducive to high concentrations of air pollution?
17. How does atmospheric stability influence the accumulation of air pollutants?
18. Why is it that polluted air and inversions seem to go hand in hand?
19. Give several reasons why taller smoke stacks are better than shorter ones at improving the air quality in their immediate area.
20. How does the mixing depth normally change during the course of a day? As the mixing depth changes, how does it affect the concentration of pollution near the surface?
21. How does topography influence the concentration of pollutants in cities such as Los Angeles and Denver? In mountainous terrain?
22. List the factors that can lead to a major buildup of atmospheric pollution.
23. What is an urban heat island? Is it more strongly developed at night or during the day? Explain.
24. What causes the "country breeze"? Why is it usually more developed at night than during the day? Would it be more easily developed in summer or winter? Explain.
25. How can pollution play a role in influencing the precipitation downwind of certain large industrial complexes?
26. Why is acid deposition considered a serious problem in many regions of the world? How does precipitation become acidic?

QUESTIONS FOR THOUGHT

1. (a) Suppose clouds of nitrogen dioxide drift slowly from a major industrial complex over a relatively unpopulated area. If the area is essentially "free" of hydrocarbons, would you expect high levels of tropospheric (ground-level) ozone to form? Explain.
   (b) Now suppose that the clouds of nitrogen dioxide drift slowly over an area that has a high concentration of hydrocarbons (VOCs), from both natural and industrial sources. Would you expect high levels of tropospheric (ground-level) ozone to form under these conditions? Explain your reasoning.
2. For least-polluted conditions, what would be the best time of day for a farmer to burn agricultural debris? Explain why you chose that time of day.
3. Why are most severe air pollution episodes associated with subsidence inversions rather than radiation inversions?

4. What surface and upper-air conditions lead to atmospheric stagnation?

5. Table 18.4, p. 522, shows that cloudiness is generally greater in urban areas than in rural areas. Since clouds reflect a great deal of incoming sunlight, they tend to keep daytime temperatures lower. Why then, during the day, are urban areas generally warmer than surrounding rural areas?

6. Acid snow can be a major problem. In the high mountains of the western United States, especially downwind of a major metropolitan or industrial area, explain why, for a high-mountain lake, acid snow can be a greater problem than acid rain, even when both have the same pH.

7. What atmospheric conditions would produce a fumigation-type plume downwind of a smokestack during the *middle* of the afternoon?

8. Why do we want to reduce high ozone concentrations at the earth's surface while, at the same time, we do not want to reduce ozone concentrations in the stratosphere?

9. Give a few reasons why, in industrial areas, nighttime pollution levels might be higher than daytime levels.

10. A large industrial smokestack located within an urban area emits vast quantities of sulfur dioxide and nitrogen dioxide. Following criticism from local residents that emissions from the stack are contributing to poor air quality in the area, the management raises the height of the stack from 10 m (33 ft) to 100 m (330 ft). Will this increase in stack height change any of the existing air quality problems? Will it create any new problems? Explain.

11. If the sulfuric acid and nitric acid in rainwater are capable of adversely affecting soil, trees, and fish, why doesn't this same acid adversely affect people when they walk in the rain?

PROBLEMS AND EXERCISES

1. Keep a log of the daily AQI readings in your area and note the pollutants listed in the index. Also, keep a record of the daily weather conditions, such as cloud cover, high temperature for the day, average wind direction and speed, etc. See if there is any relationship between these weather conditions and high AQI readings for certain pollutants.

2. Suppose the AQI reading for ozone is 300.
   (a) How would the air be described on this day?
   (b) What would be the general health effects, and what precautions should a person take under these conditions?

3. Suppose the air temperature at the surface is 30°C (86°F). Further suppose that the air temperature decreases at the dry adiabatic rate (10°C/km) up to the base of a strong subsidence inversion, situated about 2000 m above the surface. If the surface air temperature increases to 40°C (104°F), and the air temperature continues to decrease at the dry adiabatic rate up to the base of the inversion, will the mixing depth increase or decrease? Explain your answer with a diagram. Will pollutants found within the mixing layer be more concentrated or less concentrated? Explain.

4. Rain with a pH of 2 is how much more acidic than rain with a pH of 5?

5. Keep a log of daily AQI readings for one urban and one rural location for days on which both areas have similar weather conditions; note the weather conditions. Compare them. How do the weather conditions influence the AQI in both locations? Do the weather conditions contribute to greater differences in AQI readings between the two locations? Explain.

Visit the
Meteorology Resource Center
at
academic.cengage.com/login
for more assets, including questions for exploration, animations, videos, and more.

Sunlight bending through ice crystals in cirriform clouds produces bands of color called sundogs, or parhelia, on both sides of the sun on this cold winter day in Minnesota.

Photo © 2002 STAR TRIBUNE/Minneapolis-St. Paul

CHAPTER 19

Light, Color, and Atmospheric Optics

The sky is clear, the weather cold, and the year, 1818. Near Baffin Island in Canada, a ship with full sails enters unknown waters. On board are the English brothers James and John Ross, who are hoping to find the elusive "Northwest Passage," the waterway linking the Atlantic and Pacific oceans. On this morning, however, their hopes would be dashed, for directly in front of the vessel, blocking their path, is a huge towering mountain range. Disappointed, they turn back and report that the Northwest Passage does not exist. About seventy-five years later Admiral Perry met the same barrier and called it "Crocker land."

What type of treasures did this mountain conceal — gold, silver, precious gems? The curiosity of explorers from all over the world had been aroused. Speculation was the rule, until, in 1913, the American Museum of Natural History commissioned Donald MacMillan to lead an expedition to solve the mystery of Crocker land. At first, the journey was disappointing. Where Perry had seen mountains, MacMillan saw only vast stretches of open water. Finally, ahead of his ship was Crocker land, but it was more than two hundred miles farther west from where Perry had encountered it. MacMillan sailed on as far as possible. Then he dropped anchor and set out on foot with a small crew of men.

As the team moved toward the mountains, the mountains seemed to move away from them. If they stood still, the mountains stood still; if they started walking, the mountains receded again. Puzzled, they trekked onward over the glittering snow-fields until huge mountains surrounded them on three sides. At last the riches of Crocker land would be theirs. But in the next instant the sun disappeared below the horizon and, as if by magic, the mountains dissolved into the cold arctic twilight. Dumbfounded, the men looked around only to see ice in all directions — not a mountain was in sight. There they were, the victims of one of nature's greatest practical jokes, for Crocker land was a mirage.

 CONTENTS

The sky is full of visual events. Optical illusions (mirages) can appear as towering mountains or wet roadways. In clear weather, the sky can appear blue, while the horizon appears milky white. Sunrises and sunsets can fill the sky with brilliant shades of pink, red, orange, and purple. At night, the sky is black, except for the light from the stars, planets, and the moon. The moon's size and color seem to vary during the night, and the stars twinkle. To understand what we see in the sky, we will take a closer look at sunlight, examining how it interacts with the atmosphere to produce an array of atmospheric visuals.

White and Colors

We know from Chapter 2 that nearly half of the solar radiation that reaches the atmosphere is in the form of visible light. As sunlight enters the atmosphere, it is either absorbed, reflected, scattered, or transmitted on through. How objects at the surface respond to this energy depends on their general nature (color, density, composition) and the wavelength of light that strikes them. How do we see? Why do we see various colors? What kinds of visual effects do we observe because of the interaction between light and matter? In particular, what can we see when light interacts with our atmosphere?

We perceive light because radiant energy from the sun travels outward in the form of electromagnetic waves. When these waves reach the human eye, they stimulate antenna-like nerve endings in the retina. These antennae are of two types—*rods* and *cones*. The rods respond to all wavelengths of visible light and give us the ability to distinguish light from dark. If people possessed rod-type receptors only, then only black and white vision would be possible. The cones respond to specific wavelengths of visible light. Radiation with a wavelength between 0.4 and 0.7 micrometers (µm) strikes the cones, which immediately fire an impulse through the nervous system to the brain, and we perceive this impulse as the sensation of color. (Color blindness is caused by missing or malfunctioning cones.) Wavelengths of radiation shorter than 0.4 µm, or longer than 0.7 µm, do not stimulate color vision in humans.

White light is perceived when all visible wavelengths strike the cones of the eye with nearly equal intensity.* Because the sun radiates almost half of its energy as visible light, all visible wavelengths from the midday sun reach the cones, and the sun usually appears white. A star that is cooler than our sun radiates most of its energy at slightly longer wavelengths; therefore, it appears redder. On the other hand, a star much hotter than our sun radiates more energy at shorter wavelengths and thus appears bluer. A star whose temperature is about the same as the sun's appears white.

---

*Recall from Chapter 2 that visible white light is a combination of waves with different wavelengths. The wavelengths of visible light in decreasing order are: red (longest), orange, yellow, green, blue, and violet (shortest).

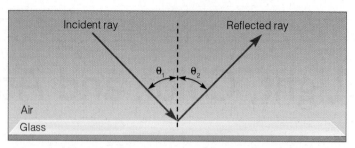

● FIGURE 19.1 For a ray of light striking a flat, smooth surface, the angle at which the incident ray strikes the surface (the angle of incidence, or θ_1) is equal to the angle at which the reflected ray leaves the surface (the angle of reflection, or θ_2). This phenomenon is called *Snell's law.*

Objects that are not hot enough to emit radiation at visible wavelengths can still have color. Everyday objects we see as red are those that absorb all visible radiation except red. The red light is reflected from the object to our eyes. Blue objects have blue light returning from them, since they absorb all visible wavelengths except blue. Some surfaces absorb all visible wavelengths and reflect no light at all. Since no radiation strikes the rods or cones, these surfaces appear black. Therefore, when we see colors, we know that light must be reaching our eyes.

White Clouds and Scattered Light

One exciting feature of the atmosphere can be experienced when we watch the underside of a puffy, growing cumulus cloud change color from white to dark gray or black. When we see this change happen, our first thought is usually, "It's going to rain." Why is the cloud initially white? Why does it change color? To answer these questions, let's investigate the concept of *scattering*.

When sunlight bounces off a surface at the same angle at which it strikes the surface, we say that the light is **reflected,** and call this phenomenon *reflection* (see ● Fig. 19.1). There are various constituents of the atmosphere, however, that tend to deflect solar radiation from its path and send it out in all directions. We know from Chapter 2 that radiation reflected in this way is said to be **scattered.*** (Scattered light is also called *diffuse light.*) During the scattering process, no energy is gained or lost and, therefore, no temperature changes occur. In the atmosphere, scattering is usually caused by small objects, such as air molecules, fine particles of dust, water molecules, and some pollutants. Just as the ball in a pinball machine bounces off the pins in many directions, so solar radiation is knocked about by small particles in the atmosphere.

Cloud droplets about 10 µm or so in diameter are large enough to effectively scatter all wavelengths of visible radiation more or less equally, a phenomenon we call *geometric*

---

*The concept of scattered light is illustrated in Fig. 2.15, p. 46.

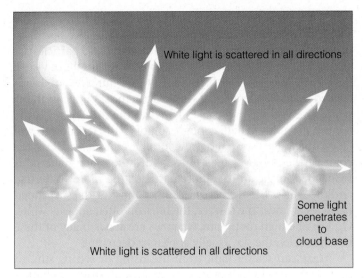

● FIGURE 19.2 Since tiny cloud droplets scatter visible light in all directions, light from many billions of droplets turns a cloud white.

scattering. Clouds, even small ones, are optically thick, meaning that they are able to scatter vast amounts of sunlight and there is very little chance sunlight will pass through unscattered. These same clouds are poor absorbers of sunlight. Hence, when we look at a cloud, it appears white because countless cloud droplets scatter all wavelengths of visible sunlight in all directions (see ● Fig. 19.2).

As a cloud grows larger and taller, more sunlight is reflected from it and less light can penetrate all the way through it (see ● Fig. 19.3). In fact, relatively little light penetrates a cloud whose thickness is 1000 m (3300 ft). Since little sunlight reaches the underside of the cloud, little light is scat-

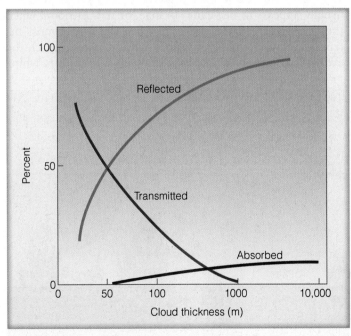

● FIGURE 19.3 Average percent of radiation reflected, absorbed, and transmitted by clouds of various thickness.

tered, and the cloud base appears dark. At the same time, if droplets near the cloud base grow larger, they become less effective scatterers and better absorbers. As a result, the meager amount of visible light that does reach this part of the cloud is absorbed rather than scattered, which makes the cloud appear even darker. These same cloud droplets may even grow large and heavy enough to fall to the earth as rain. From a casual observation of clouds, we know that dark, threatening ones frequently produce rain. Now, we know why they appear so dark.

Blue Skies and Hazy Days

The sky appears blue because light that stimulates the sensation of blue color is reaching the retina of the eye. How does this happen?

Individual air molecules are much smaller than cloud droplets—their diameters are small even when compared with the wavelength of visible light. Each air molecule of oxygen and nitrogen is a *selective scatterer* in that each scatters shorter waves of visible light much more effectively than longer waves. This selective scattering is also known as *Rayleigh scattering* (see ▼ Table 19.1).

As sunlight enters the atmosphere, the shorter visible wavelengths of violet, blue, and green are scattered more by atmospheric gases than are the longer wavelengths of yellow, orange, and especially red.* In fact, violet light is scattered about 16 times more than red light. Consequently, as we view the sky, the scattered waves of violet, blue, and green strike

*The reason for this fact is that the intensity of Rayleigh scattering varies as $1/\lambda^4$, where λ is the wavelength of radiation.

▼ TABLE 19.1   The Various Types of Scattering of Visible Light

TYPE OF PARTICLE	PARTICLE DIAMETER (MICROMETERS, μm)	TYPE OF SCATTERING	PHENOMENA
Air molecules	0.0001 to 0.001	Rayleigh	Blue sky, red sunsets
Aerosols (pollutants)	0.01 to 1.0	Mie	Brownish smog
Cloud droplets	10 to 100	Geometric	White clouds

● FIGURE 19.4 The sky appears blue because billions of air molecules selectively scatter the shorter wavelengths of visible light more effectively than the longer ones. This causes us to see blue light coming from all directions.

the eye from all directions. Because our eyes are more sensitive to blue light, these waves, viewed together, produce the sensation of blue coming from all around us (see ● Fig. 19.4). Therefore, when we look at the sky it appears blue (see ● Fig. 19.5). (Earth, by the way, is not the only planet with a colorful sky. On Mars, dust in the air turns the sky red at midday and purple at sunset.)

The selective scattering of blue light by air molecules and very small particles can make distant mountains appear blue, such as the Blue Ridge Mountains of Virginia and the Blue Mountains of Australia (see ● Fig. 19.6). In some places, a *blue haze* may cover the landscape, even in areas far removed from human contamination. Although its cause is still controversial, the blue haze appears to be the result of a particular process. Extremely tiny particles (hydrocarbons called *terpenes*) are released by vegetation to combine chemically with small amounts of ozone. This reaction produces tiny particles (about 0.2 μm in diameter) that selectively scatter blue light.

● FIGURE 19.5 Blue skies and white clouds. The selective scattering of blue light by air molecules produces the blue sky, while the scattering of all wavelengths of visible light by liquid cloud droplets produces the white clouds.

When small particles, such as fine dust and salt, become suspended in the atmosphere, the color of the sky begins to change from blue to milky white. Although these particles are small, they are large enough to scatter all wavelengths of visible light fairly evenly in all directions. When our eyes are bombarded by all wavelengths of visible light, the sky appears milky white, the visibility lowers, and we call the day "hazy." If the relative humidity is high enough, soluble particles (nuclei) will "pick up" water vapor and grow into haze particles. Thus, the color of the sky gives us a hint about how much material is suspended in the air: the more particles, the more scattering, and the whiter the sky becomes. Since most of the suspended particles are near the surface, the horizon often appears white. On top of a high mountain, when we are above many of these haze particles, the sky usually appears a deep blue.

Haze can scatter light from the rising or setting sun, so that we see bright lightbeams, or **crepuscular rays,** radiating across the sky. A similar effect occurs when the sun shines through a break in a layer of clouds (see ● Fig. 19.7). Dust, tiny water droplets, or haze in the air beneath the clouds scatter sunlight, making that region of the sky appear bright with rays. Because these rays seem to reach downward from clouds, some people will remark that the "sun is drawing up water." In England, this same phenomenon is referred to as "Jacob's ladder." No matter what these sunbeams are called, it is the scattering of sunlight by particles in the atmosphere that makes them visible.

Red Suns and Blue Moons

At midday, the sun seems a brilliant white, while at sunset it usually appears to be yellow, orange, or red. At noon, when the sun is high in the sky, light from the sun is most intense—all wavelengths of visible light are able to reach the eye with about equal intensity, and the sun appears white.

© C. Donald Ahrens

● FIGURE 19.6 The Blue Ridge Mountains in Virginia. The blue haze is caused by the scattering of blue light by extremely small particles—smaller than the wavelengths of visible light. Notice that the scattered blue light causes the most distant mountains to become almost indistinguishable from the sky.

(Looking directly at the sun, especially during this time of day, can cause irreparable damage to the eye. Normally, we get only glimpses or impressions of the sun out of the corner of our eye.)

Near sunrise or sunset, however, the rays coming directly from the sun strike the atmosphere at a low angle. They must pass through much more atmosphere than at any other time during the day. (When the sun is 4° above the horizon, sunlight must pass through an atmosphere more than 12 times thicker than when the sun is directly overhead.) By the time sunlight has penetrated this large amount of air, most of the shorter waves of visible light have been scattered away by the air molecules. Just about the only waves from a setting sun that make it on through the atmosphere on a fairly direct path are the yellow, orange, and red. Upon reaching the eye, these waves produce a bright yellow-orange sunset (see ● Fig. 19.8).

Bright, yellow-orange sunsets only occur when the atmosphere is fairly clean, as it would be after a recent rain. If the atmosphere contains many fine particles whose diameters are a little larger than air molecules, slightly longer (yellow) waves also would be scattered away. Only orange and red waves would penetrate through to the eye, and the sun would appear red-orange. When the atmosphere becomes loaded with particles, only the longest red wavelengths are able to penetrate the atmosphere, and we see a red sun.

Natural events may produce red sunrises and sunsets over the oceans. For example, the scattering characteristics of small suspended salt particles and water molecules are responsible for the brilliant red suns that can be observed from a beach (see ● Fig. 19.9). Volcanic eruptions rich in sulfur can produce red sunsets, too. Such red sunsets are actually produced by a highly reflective cloud of sulfuric acid droplets,

● FIGURE 19.7 The scattering of sunlight by dust and haze produces these white bands of crepuscular rays.

● FIGURE 19.8 Because of the selective scattering of radiant energy by a thick section of atmosphere, the sun at sunrise and sunset appears either yellow, orange, or red. The more particles in the atmosphere, the more scattering of sunlight, and the redder the sun appears.

formed from sulfur dioxide gas injected into the stratosphere during powerful eruptions, like that of the Mexican volcano El Chichón in 1982 and the Philippine volcano Mt. Pinatubo in 1991. These fine particles, moved by the winds aloft, circled the globe, producing beautiful sunrises and sunsets for months and even years after the eruptions. These same volcanic particles in the stratosphere can turn the sky red after sunset, as some of the red light from the setting sun bounces off the bottom of the particles back to the earth's surface. Generally, these volcanic red sunsets occur about an hour after the actual sunset (see ● Fig. 19.10).

● FIGURE 19.9 Red sunset near the coast of Iceland. The reflection of sunlight off the slightly rough water is producing a glitter path.

Occasionally, the atmosphere becomes so laden with dust, smoke, and pollutants that even red waves are unable to pierce the filthy air. An eerie effect then occurs. Because no visible waves enter the eye, the sun literally disappears before it reaches the horizon.

The scattering of light by large quantities of atmospheric particles can cause some rather unusual sights. If the dust, smoke particles, or pollutants are roughly uniform in size, they can selectively scatter the sun's rays. Even at noon, various colored suns have appeared: orange suns, green suns, and even blue suns. For blue suns to appear, the size of the suspended particles must be similar to the wavelength of visible light. When these particles are present they tend to scatter red light more than blue, which causes a bluing of the sun and a reddening of the sky. Although rare, the same phenomenon can happen to moonlight, making the moon appear blue; thus, the expression "once in a blue moon."

In summary, the scattering of light by small particles in the atmosphere causes many familiar effects: white clouds, blue skies, hazy skies, crepuscular rays, and colorful sunsets. In the absence of any scattering, we would simply see a white sun against a black sky—not an attractive alternative.

Twinkling, Twilight, and the Green Flash

Light that passes through a substance is said to be *transmitted.* Upon entering a denser substance, transmitted light slows in speed. If it enters the substance at an angle, the light's path also bends. This bending is called **refraction.** The amount of refraction depends primarily on two factors: the density of the material and the angle at which the light enters the material.

Refraction can be demonstrated in a darkened room by shining a flashlight into a beaker of water (see ● Fig. 19.11). If the light is held directly above the water so that the beam

● FIGURE 19.10 Bright red sky over California produced by the sulfur-rich particles from the volcano Mt. Pinatubo during September, 1992. The photo was taken about an hour after sunset.

strikes the surface of the water straight on, no bending occurs. But, if the light enters the water at some angle, it bends toward the *normal,* which is the dashed line in the diagram running perpendicular to the air-water boundary. (The normal is simply a line that intersects any surface at a right angle. We use it as a reference to see how much bending occurs as light enters and leaves various substances.) A small mirror on the bottom of the beaker reflects the light upward. This reflected light bends away from the normal as it re-enters the air. We can summarize these observations as follows: *Light that travels from a less-dense to a more-dense medium loses speed and bends toward the normal, while light that enters a less-dense medium increases in speed and bends away from the normal.*

The refraction of light within the atmosphere causes a variety of visual effects. At night, for example, the light from the stars that we see directly above us is not bent, but starlight that enters the earth's atmosphere at an angle is bent. In fact, a star whose light enters the atmosphere just above the horizon has more atmosphere to penetrate and is thus refracted the most. As we can see in ● Fig. 19.12, the bending is toward the normal as the light enters the more-dense atmosphere. By

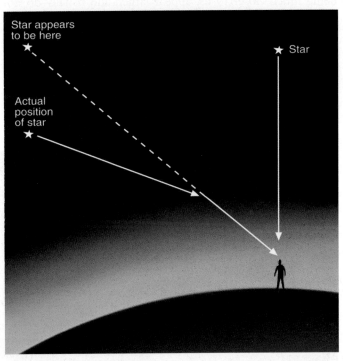

● FIGURE 19.12 Due to the bending of starlight by the atmosphere, stars not directly overhead appear to be higher than they really are.

● FIGURE 19.11 The behavior of light as it enters and leaves a more-dense substance, such as water.

▼ TABLE 19.2 The Amount of Atmospheric Refraction (Bending) in Minutes Viewed at Sea Level Under Standard Atmospheric Conditions (60 minutes equals 1°)

ELEVATION ABOVE HORIZON (DEGREES)	REFRACTION (MINUTES)
0°	35.0
5°	10.0
20°	2.6
40°	1.2
60°	0.6
90°	0.0

the time this "bent" starlight reaches our eyes, the star appears to be higher than it actually is because our eyes cannot detect that the light path is bent. We see light coming from a particular direction and interpret the star to be in that direction. So, the next time you take a midnight stroll, point to any star near the horizon and remember: This is where the star appears to be. To point to the star's true position, you would have to lower your arm just a bit (about one-half a degree, according to ▼ Table 19.2).

As starlight enters the atmosphere, it often passes through regions of differing air density. Each of these regions deflects and bends the tiny beam of starlight, constantly changing the apparent position of the star. This causes the star to appear to *twinkle* or flicker, a condition known as **scintillation**. Planets, being much closer to us, appear larger, and usually do not twinkle because their size is greater than the angle at which their light deviates as it penetrates the atmosphere. Planets sometimes twinkle, however, when they are near the horizon, where the bending of their light is greatest.

The refraction of light by the atmosphere has some other interesting consequences. For example, the atmosphere gradually bends the rays from a rising or setting sun or moon. Because light rays from the lower part of the sun (or moon) are bent more than those from the upper part, the sun appears to flatten out on the horizon, taking on an elliptical shape. (The sun in Fig. 19.14, p. 537, shows this effect.) Also, since light is bent most on the horizon, the sun and moon both appear to be higher than they really are. Consequently, they both rise about two minutes earlier and set about two minutes later than they would if there were no atmosphere (see ● Fig. 19.13).

You may have noticed that on clear days the sky is often bright for some time after the sun sets. The atmosphere refracts and scatters sunlight to our eyes, even though the sun itself has disappeared from our view. (Look back at Fig. 19.10.) **Twilight** is the name given to the time after sunset (and immediately before sunrise) when the sky remains illuminated and allows outdoor activities to continue without artificial lighting. (*Civil twilight* lasts from sunset until the sun is 6° below the horizon, while *astronomical twilight* lasts until the sky is completely dark and the astronomical observation of the faintest stars is possible.)

The length of twilight depends on season and latitude. During the summer in middle latitudes, twilight adds about 30 minutes of light to each morning and evening for outdoor activities. The duration of twilight increases with increasing latitude, especially in summer. At high latitudes during the summer, morning and evening twilight may converge, producing a *white night*—a nightlong twilight.

In general, without the atmosphere, there would be no refraction or scattering, and the sun would rise later and set earlier than it now does. Instead of twilight, darkness would arrive immediately when the sun disappears below the horizon. Imagine the number of sandlot baseball games that would be called because of instant darkness.

Occasionally, a flash of green light—called the **green flash**—may be seen near the upper rim of a rising or setting sun (see ● Fig. 19.14). Remember from our earlier discussion that, when the sun is near the horizon, its light must penetrate a thick section of atmosphere. This thick atmosphere refracts sunlight, with purple and blue light bending the most, and red light the least. Because of this bending, more

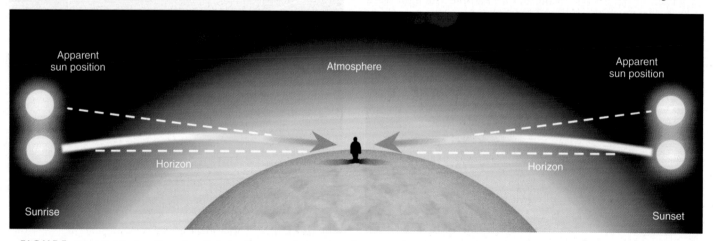

● FIGURE 19.13 The bending of sunlight by the atmosphere causes the sun to rise about two minutes earlier, and set about two minutes later, than it would otherwise.

● FIGURE 19.14 The very light green on the upper rim of the sun is the green flash. Also, observe how the atmosphere makes the sun appear to flatten on the horizon into an elliptical shape.

blue light should appear along the top of the sun. But because the atmosphere selectively scatters blue light, very little reaches us, and we see green light instead.

Usually, the green light is too faint to see with the human eye. However, under certain atmospheric conditions, such as when the surface air is very hot or when an upper-level inversion exists, the green light is magnified by the atmosphere. When this happens, a momentary flash of green light appears, often just before the sun disappears from view.

The flash usually lasts about a second, although in polar regions it can last longer. Here, the sun slowly changes in elevation and the flash may exist for many minutes. Members of Admiral Byrd's expedition in the south polar region reported seeing the green flash for 35 minutes in September as the sun slowly rose above the horizon, marking the end of the long winter.

BRIEF REVIEW

Up to this point, we have examined how light can interact with our atmosphere. Before going on, here is a review of some of the important concepts and facts we have covered:

- When light is scattered, it is sent in all directions — forward, sideways, and backward.
- White clouds, blue skies, hazy skies, crepuscular rays, and colorful sunsets are the result of sunlight being scattered.
- The bending of light as it travels through regions of differing density is called *refraction*.
- As light travels from a less-dense substance (such as outer space) and enters a more-dense substance at an angle (such as our atmosphere), the light bends downward, toward the normal. This effect causes stars, the moon, and the sun to appear just a tiny bit higher than they actually are.

The Mirage: Seeing Is Not Believing

In the atmosphere, when an object appears to be displaced from its true position, we call this phenomenon a **mirage.** A mirage is not a figment of the imagination — our minds are not playing tricks on us, but the atmosphere is.

Atmospheric mirages are created by light passing through and being bent by air layers of different densities. Such changes in air density are usually caused by sharp changes in air temperature. The greater the rate of temperature change, the greater the light rays are bent. For example, on a warm, sunny day, black road surfaces absorb a great deal of solar energy and become very hot. Air in contact with these hot surfaces warms by conduction and, because air is a poor thermal conductor, we find much cooler air only a few meters higher. On hot days, these road surfaces often appear wet (see ● Fig. 19.15). Such "puddles" disappear as we approach them, and advancing cars seem to swim in them. Yet, we know the road is dry. The apparent wet pavement above a road is the result of blue skylight refracting up into our eyes as it travels through air of different densities. A similar type of mirage occurs in deserts during the hot summer. Many thirsty travelers have been disappointed to find that what appeared to be a water hole was in actuality hot desert sand.

Sometimes, these "watery" surfaces appear to *shimmer.* The shimmering results as rising and sinking air near the ground constantly change the air density. As light moves through these regions, its path also changes, causing the shimmering effect.

When the air near the ground is much warmer than the air above, objects may not only appear to be lower than they really are, but also (often) inverted. These mirages are called **inferior** (lower) **mirages.** The tree in ● Fig. 19.16 certainly doesn't grow upside down. So why does it look that way? It appears to be inverted because light reflected from the top of

● FIGURE 19.15 The road in the photo appears wet because blue skylight is bending up into the camera as the light passes through air of different densities.

the tree moves outward in all directions. Rays that enter the hot, less-dense air above the sand are refracted upward, entering the eye from below. The brain is fooled into thinking that these rays came from below the ground, which makes the

tree appear upside down. Some light from the top of the tree travels directly toward the eye through air of nearly constant density and, therefore, bends very little. These rays reach the eye "straight-on," and the tree appears upright. Hence, off in the distance, we see a tree and its upside-down image beneath it. (Some of the trees in Fig. 19.15 show this effect.)

The atmosphere can play optical jokes on us in extremely cold areas, too. In polar regions, air next to a snow surface can be much colder than the air many meters above. Because the air in this cold layer is very dense, light from distant objects entering it bends toward the normal in such a way that the

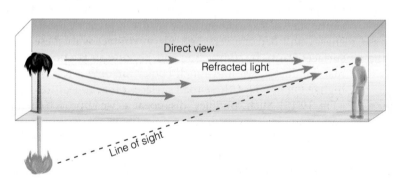

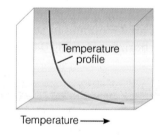

● FIGURE 19.16
Inferior mirage over hot desert sand.

● FIGURE 19.17
The formation of a superior mirage. When cold air lies close to the surface with warm air aloft, light from distant mountains is refracted toward the normal as it enters the cold air. This causes an observer on the ground to see mountains higher and closer than they really are.

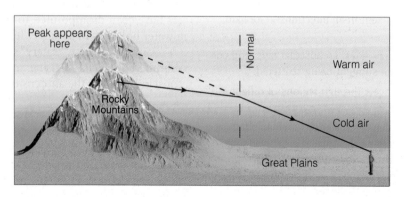

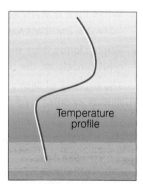

The *Fata Morgana*

A special type of superior mirage is the *Fata Morgana*, a mirage that transforms a fairly uniform horizon into one of vertical walls and columns with spires (see Fig. 1). According to legend, *Fata Morgana* (Italian for "fairy Morgan") was the half-sister of King Arthur. Morgan, who was said to live in a crystal palace beneath the water, had magical powers that could build fantastic castles out of thin air. Looking across the Straits of Messina (between Italy and Sicily), residents of Reggio, Italy, on occasion would see buildings, castles, and sometimes whole cities appear, only to vanish again in minutes. The *Fata Morgana* is observed where the air temperature increases with height above the surface, slowly at first, then more rapidly, then slowly again. Consequently, mirages like the *Fata Morgana* are frequently seen where warm air rests above a cold surface, such as above large bodies of water and in polar regions.

● FIGURE 1 The *Fata Morgana* mirage over water. The mirage is the result of refraction—light from small islands and ships is bent in such a way as to make them appear to rise vertically above the water.

objects can appear to be shifted upward. This phenomenon is called a **superior** (upward) **mirage.** ● Figure 19.17 shows the atmospheric conditions favorable for a superior mirage. (A special type of mirage, the ***Fata Morgana,*** is described in the Focus section above.)

Halos, Sundogs, and Sun Pillars

A ring of light encircling and extending outward from the sun or moon is called a **halo.** Such a display is produced when sunlight or moonlight is refracted as it passes through ice crystals. Hence, the presence of a halo indicates that *cirriform clouds* are present.

The most common type of halo is the 22° halo—a ring of light 22° from the sun or moon.* Such a halo forms when tiny suspended column-type ice crystals (with diameters less than 20 μm) become randomly oriented as air molecules constantly bump against them. The refraction of light rays through these

crystals forms a halo like the one shown in ● Fig. 19.18. Less common is the 46° halo, which forms in a similar fashion to the 22° halo (see ● Fig. 19.19). With the 46° halo, however, the

● FIGURE 19.18 A 22° halo around the sun, produced by the refraction of sunlight through ice crystals.

*Extend your arm and spread your fingers apart. An angle of 22° is about the distance from the tip of the thumb to the tip of the little finger.

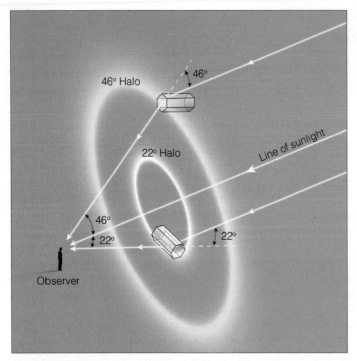

● FIGURE 19.19 The formation of a 22° and a 46° halo with column-type ice crystals.

● FIGURE 19.21 Refraction and dispersion of light through a glass prism.

light is refracted through column-type ice crystals that have diameters in a narrow range between about 15 and 25 μm.

Occasionally, a bright arc of light may be seen at the top of a 22° halo (see ● Fig. 19.20). Since the arc is tangent to the halo, it is called a **tangent arc.** Apparently, the arc forms as large six-sided (hexagonal), pencil-shaped ice crystals fall with their long axes horizontal to the ground. Refraction of sunlight through the ice crystals produces the bright arc of light. When the sun is on the horizon, the arc that forms at the top of the halo is called an *upper tangent arc.* When the

sun is above the horizon, a *lower tangent arc* may form on the lower part of the halo beneath the sun. The shape of the arcs changes greatly with the position of the sun.

A halo is usually seen as a bright, white ring, but there are refraction effects that can cause it to have color. To understand this, we must first examine refraction more closely.

When white light passes through a glass prism, it is refracted and split into a spectrum of visible colors (see ● Fig. 19.21). Each wavelength of light is slowed by the glass, but each is slowed a little differently. Because longer wavelengths (red) slow the least and shorter wavelengths (violet) slow the most, red light bends the least, and violet light bends the most. The breaking up of white light by "selective" refraction is called **dispersion.** As light passes through ice crystals, dispersion causes red light to be on the inside of the halo and blue light on the outside.

When hexagonal platelike ice crystals with diameters larger than about 30 μm are present in the air, they tend to fall slowly and orient themselves horizontally (see ● Fig. 19.22). (The horizontal orientation of these ice crystals prevents a ring halo.) In this position, the ice crystals act as small prisms, refracting and dispersing sunlight that passes through them. If the sun is near the horizon in such a configuration that it, ice crystals, and observer are all in the same horizontal plane, the observer will see a pair of brightly colored spots, one on either side of the sun. These colored spots are called **sundogs,** *mock suns,* or **parhelia**—meaning "with the sun" (see ● Fig. 19.23, and the opening photo on p. 528. The colors usually grade from red (bent least) on the inside closest to the sun to blue (bent more) on the outside.

Whereas sundogs, tangent arcs, and halos are caused by *refraction* of sunlight *through* ice crystals, **sun pillars** are caused by *reflection* of sunlight *off* ice crystals. Sun pillars ap-

● FIGURE 19.20 Halo with an upper tangent arc.

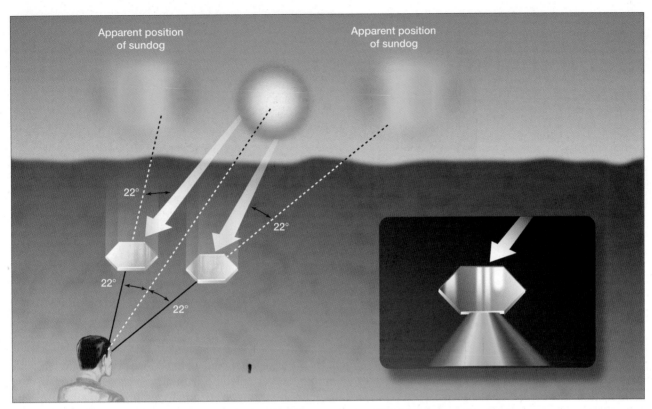

● FIGURE 19.22 Platelike ice crystals falling with their flat surfaces parallel to the earth produce sundogs.

● FIGURE 19.23 The bright areas on each side of the sun are sundogs.

● FIGURE 19.24 A brilliant red sun pillar extending upward above the sun, produced by the reflection of sunlight off ice crystals.

NASA

pear most often at sunrise or sunset as a vertical shaft of light extending upward or downward from the sun (see ● Fig. 19.24). Pillars may form as hexagonal platelike ice crystals fall with their flat bases oriented horizontally. As the tiny crystals fall in still air, they tilt from side to side like a falling leaf. This motion allows sunlight to reflect off the tipped surfaces of the crystals, producing a relatively bright area in the sky above or below the sun. Pillars may also form as sunlight

reflects off hexagonal pencil-shaped ice crystals that fall with their long axes oriented horizontally. As these crystals fall, they can rotate about their horizontal axes, producing many orientations that reflect sunlight. So, look for sun pillars when the sun is low on the horizon and cirriform (ice crystal) clouds are present. ● Figure 19.25 is a summary of some of the optical phenomena that form when cirriform clouds are present.

Rainbows

Now we come to one of the most spectacular light shows observed on the earth—the rainbow. **Rainbows** occur when rain is falling in one part of the sky, and the sun is shining in another. (Rainbows also may form by the sprays from waterfalls and water sprinklers.) To see the rainbow, we must face the falling rain with the sun at our backs. Look at ● Fig. 19.26 closely and note that, when we see a rainbow in the evening, we are facing east toward the rainshower. Behind us—in the west—it is clear. Because clouds tend to move from west to east in middle latitudes, the clear skies in the west suggest that the showers will give way to clearing. However, when we see a rainbow in the morning, we are facing west, toward the rainshower. It is a good bet that the clouds and showers will move toward us and it will rain soon. These observations explain why the following weather rhyme became popular:

> Rainbow in morning, sailors take warning,
> Rainbow at night, a sailor's delight.*

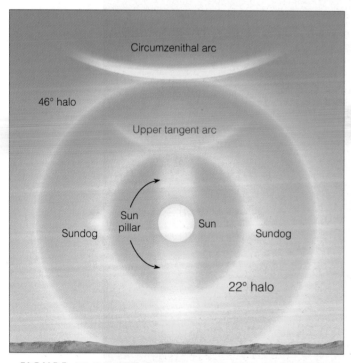

● FIGURE 19.25 Optical phenomena that form when cirriform ice crystal clouds are present. (A picture of the circumzenithal arc is in Fig. 2 on p. 545.)

---

*This rhyme is often used with the words "red sky" in the place of rainbow. The red sky makes sense when we consider that it is the result of red light from a rising or setting sun being reflected from the underside of clouds above us. In the morning, a red sky indicates that it is clear to the east and cloudy to the west. A red sky in the evening suggests the opposite.

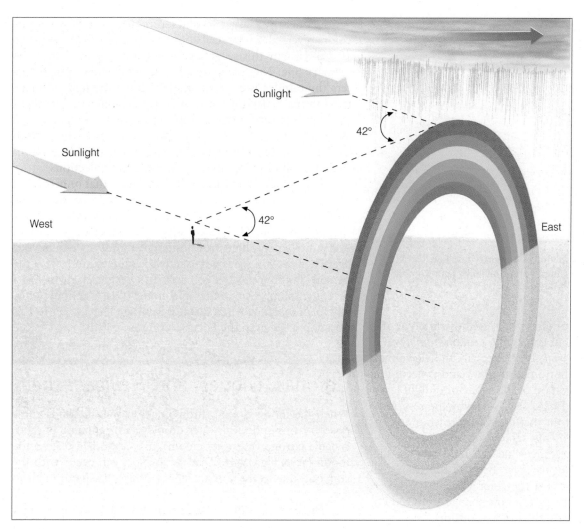

• FIGURE 19.26 When you observe a rainbow, the sun is always to your back. In middle latitudes, a rainbow in the evening indicates that clearing weather is ahead.

When we look at a rainbow we are looking at sunlight that has entered the falling drops, and, in effect, has been redirected back toward our eyes. Exactly how this process happens requires some discussion.

As sunlight enters a raindrop, it slows and bends, with violet light refracting the most and red light the least (see Fig. 19.27). Although most of this light passes right on through the drop and is not seen by us, some of it strikes the backside of the drop at such an angle that it is reflected within the drop. The angle at which this occurs is called the *critical angle.* For water, this angle is 48°. Light that strikes the back of a raindrop at an angle exceeding the critical angle bounces off the back of the drop and is *internally reflected* toward our eyes (see • Fig. 19.27a). Because each light ray bends differently from the rest, each ray emerges from the drop at a slightly different angle. For red light, the angle is 42° from the beam of sunlight; for violet light, it is 40° (see Fig. 19.27b). The light leaving the drop is, therefore, dispersed into a spectrum of colors from red to violet. Since we see only a single color from each drop, it takes myriads of raindrops (each refracting

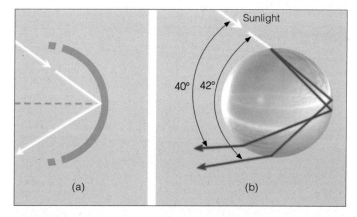

• FIGURE 19.27 Sunlight internally reflected and dispersed by a raindrop. (a) The light ray is internally reflected only when it strikes the backside of the drop at an angle greater than the critical angle for water. (b) Refraction of the light as it enters the drop causes the point of reflection (on the back of the drop) to be different for each color. Hence, the colors are separated from each other when the light emerges from the raindrop.

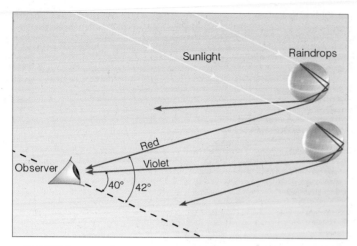

● FIGURE 19.28 The formation of a primary rainbow. The observer sees red light from the upper drop and violet light from the lower drop.

and reflecting light back to our eyes at slightly different angles) to produce the brilliant colors of a *primary rainbow*.

Figure 19.27b might lead us erroneously to believe that red light should be at the bottom of the bow and violet at the top. A more careful observation of the behavior of light leaving two drops (see ● Fig. 19.28) shows us why the reverse is true. When violet light from the *lower drop* reaches an observer's eye, red light from the same drop is incident elsewhere, toward the waist. Notice that red light reaches the observer's eye from the *higher drop*. Because the color red comes from higher drops and the color violet from lower drops, the colors of a primary rainbow change from red on the outside (top) to violet on the inside (bottom).

Frequently, a larger second (secondary) rainbow with its colors reversed can be seen above the primary bow (see ● Fig. 19.29). Usually this secondary bow is much fainter than the primary one. The *secondary bow* is caused when sunlight enters the raindrops at an angle that allows the light to make two internal reflections in each drop. Each reflection weakens the light intensity and makes the bow dimmer. ● Figure 19.30 shows that the color reversals—with red now at the bottom and violet on top—are due to the way the light emerges from each drop after going through two internal reflections.

As you look at a rainbow, keep in mind that only one ray of light is able to enter your eye from each drop. Every time you move, whether it be up, down, or sideways, the rainbow moves with you. The reason why this happens is that, with every movement, light from different raindrops enters your eye. The bow you see is not exactly the same rainbow that the person standing next to you sees. In effect, each of us has a personal rainbow to ponder and enjoy! (Can a rainbow actually form on a day when it is not raining? The answer to this question is given in the Focus section on p. 545.)

Coronas, Glories, and *Heiligenschein*

When the moon is seen through a thin veil of clouds composed of tiny spherical water droplets, a bright ring of light, called a **corona** (meaning crown), may appear to rest on the moon (see ● Fig. 19.31). The same effect can occur with the sun, but, due to the sun's brightness, it is usually difficult to see.

The corona is due to **diffraction**—the bending of light as it passes *around* objects. To understand the corona, imagine water waves moving around a small stone in a pond. As the waves spread around the stone, the trough of one wave may meet the crest of another wave. This situation results in

● FIGURE 19.29 A primary and a secondary rainbow.

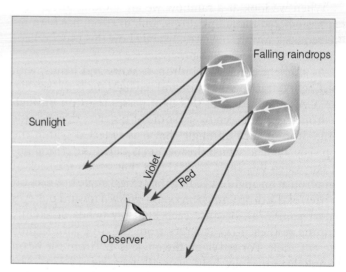

● FIGURE 19.30 Two internal reflections are responsible for the weaker, secondary rainbow. Notice that the eye sees violet light from the upper drop and red light from the lower drop.

FOCUS ON AN OBSERVATION

Can It Be a Rainbow If It Is Not Raining?

Up to this point, we have seen that the sky is full of atmospheric visuals. One that we closely examined was the rainbow. Look back at Fig. 19.29, p. 544, and then look at Fig. 2. Is the color display in Fig. 2 a rainbow? The colors definitely show a rainbow-like brilliance. For this reason, some people will call this phenomenon a rainbow. But remember, for a rainbow to form it must be raining, and on this day it is not.

The color display in Fig. 2 is due to the refraction of light through ice crystals. Earlier in this chapter we saw that the refraction of light produces a variety of visuals, such as halos, tangent arcs, and sundogs. (Refer back to Fig. 19.25, p. 542.) The color display in Fig. 2 is a type of refraction phenomena called a *circumzenithal arc*. The photograph was taken in the afternoon, during late winter, when the sun was about 30° or so above the western horizon and the sky was full of ice crystal (cirrus) clouds. The arc was almost directly overhead, at the zenith; hence, its name — "circumzenithal."

The circumzenithal arc forms about 45° above the sun as platelike ice crystals fall with their flat surfaces parallel to the ground. Remember that this is similar to the formation of a sundog (see Fig. 19.22, p. 541.), except that in the formation of a circumzenithal arc, sunlight enters the top of the crystal and exits one of its sides (see Fig. 3). This is why the short-lived circumzenithal arc can only form when the sun is lower than 32° above the horizon. When the sun is higher than this angle, the refracted light cannot be seen by the observer.

There is a wide variety of other refraction phenomena (too many to describe here) that may at first glance appear as a rainbow. Keep in mind, however, that rainbows only form when it is raining in one part of the sky and the sun is shining in another.

© C. Donald Ahrens

● FIGURE 2 Is this a rainbow? The photograph was taken looking almost straight up.

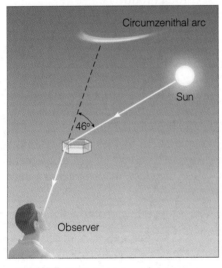

Circumzenithal arc

Sun

46°

Observer

● FIGURE 3 The formation of a circumzenithal arc.

the waves canceling each other, thus producing calm water. (This is known as *destructive interference.*) Where two crests come together (*constructive interference*), they produce a much larger wave. The same thing happens when light passes around tiny cloud droplets. Where light waves constructively interfere, we see bright light; where destructive interference occurs, we see darkness. Sometimes, the corona appears white, with alternating bands of light and dark. On other occasions, the rings have color (see ● Fig. 19.32).

The colors appear when the cloud droplets (or any kind of small particles, such as volcanic ash), are of uniform size. Because the amount of bending due to diffraction depends upon the wavelength of light, the shorter wavelength blue light appears on the inside of a ring, while the longer wavelength red light appears on the outside. These colors may repeat over and over, becoming fainter as each ring is farther from the moon or sun. Also, the smaller the cloud droplets,

WEATHER WATCH

During the summer, rainbows are occasionally seen after a thunderstorm. Because of this fact, the Shoshone Indians viewed the rainbow as a giant serpent that would sometimes rub its back on the icy sky and hurl pieces of ice (hail) to the ground.

● FIGURE 19.31 The corona around the moon results from the diffraction of light by tiny liquid cloud droplets of uniform size.

● FIGURE 19.33 Cloud iridescence.

the larger the ring diameter. Therefore, clouds that have recently formed (such as thin altostratus and altocumulus) are the best corona producers.

When different size droplets exist within a cloud, the corona becomes distorted and irregular. Sometimes the cloud exhibits patches of color, often pastel shades of pink, blue, or green. These bright areas produced by diffraction are called **iridescence** (see ● Fig. 19.33). Cloud iridescence is most often seen within 20° of the sun, and is often associated with clouds such as cirrocumulus and altocumulus.

Like the corona, the **glory** is also a diffraction phenomenon. When an aircraft flies above a cloud layer composed of water droplets less than 50 μm in diameter, a set of colored rings, called the *glory,* may appear around the shadow of the aircraft (see ● Fig. 19.34). The same effect can happen when you stand with your back to the sun and look into a cloud or fog bank, as a bright ring of light may be seen around the shadow of your head. In this case, the glory is called the *brocken bow,* after the Brocken Mountains in Germany, where it is particularly common.

For the glory and the brocken bow to occur, the sun must be to your back, so that sunlight can be returned to your eye from the water droplets. Sunlight that enters the small water droplet along its edge is refracted, then reflected off the backside of the droplet. The light then exits at the other side of the droplet, being refracted once again (see ● Fig. 19.35). However, in order for the light to be returned to your eyes, the light actually clings ever so slightly to the edge of the droplet—the light actually skims along the surface of the droplet as a *surface wave* for

● FIGURE 19.32 Corona around the sun photographed in Colorado. This type of corona, called *Bishop's ring,* is the result of diffraction of sunlight by tiny volcanic particles emitted from the volcano El Chichón in 1982.

● FIGURE 19.34 The series of rings surrounding the shadow of the aircraft is called the *glory.*

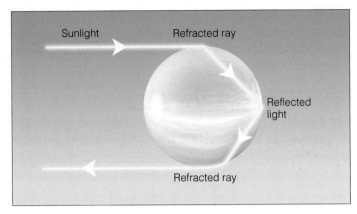

● FIGURE 19.35 Light that produces the glory follows this path in a water droplet.

a short distance. Diffraction of light coming from the edges of the droplets produces the ring of light we see as the glory and the brocken bow. The colorful rings may be due to the various angles at which different colors leave the droplet.

On a clear morning with dew on the grass, stand facing the dew with your back to the sun and observe that, around the shadow of your head, is a bright area—the **Heiligenschein** (German for halo). The *Heiligenschein* forms when sunlight, which falls on nearly spherical dew drops, is focused and reflected back toward the sun along nearly the same path that it took originally. (Light reflected in this manner is said to be *retroreflected*.) The light, however, does not travel along the exact path; it actually spreads out just enough to be seen as bright white light around the shadow of your head on a dew-covered lawn (see ● Fig. 19.36).

● FIGURE 19.36 The *Heiligenschein* is the ring of light around the shadow of the observer's head.

SUMMARY

The scattering of sunlight in the atmosphere can produce a variety of atmospheric visuals, from hazy days and blue skies to crepuscular rays and blue moons. Refraction (bending) of light by the atmosphere causes stars near the horizon to appear higher than they really are. It also causes the sun and moon to rise earlier and set later than they otherwise would. Under certain atmospheric conditions, the amplification of green light near the upper rim of a rising or setting sun produces the illusive green flash.

Mirages form when refraction of light displaces objects from their true positions. Inferior mirages cause objects to appear lower than they really are, while superior mirages displace objects upward.

Halos and sundogs form from the refraction of light through ice crystals. Sun pillars are the result of sunlight reflecting off gently falling ice crystals. The refraction, reflection, and dispersion of light in raindrops create a rainbow. To see a rainbow, the sun must be to your back, and rain must be falling in front of you. Diffraction of light produces coronas, glories,

and cloud iridescence. We can see the *Heiligenschein* on a clear morning when sunlight falls on nearly spherical dew drops.

KEY TERMS

The following terms are listed (with page numbers) in the order they appear in the text. Define each. Doing so will aid you in reviewing the material covered in this chapter.

reflected light, 530
scattered light, 530
crepuscular rays, 532
refraction (of light), 534
scintillation, 536
twilight, 536
green flash, 536
mirage, 537
inferior mirage, 537
superior mirage, 539
Fata Morgana, 539

halo, 539
tangent arc, 540
dispersion (of light), 540
sundog, parhelia, 540
sun pillars, 540
rainbow, 542
corona, 544
diffraction, 544
iridescence, 546
glory, 546
Heiligenschein, 547

QUESTIONS FOR REVIEW

1. (a) Why are cumulus clouds normally white?
   (b) Why do the undersides of building cumulus clouds frequently change color from white to dark gray or even black?
2. Explain why the sky is blue during the day and black at night.
3. What can make a setting (or rising) sun appear red?
4. Why do stars "twinkle"?
5. Explain why the horizon sky appears white on a hazy day.
6. How does light bend as it enters a more-dense substance at an angle? How does it bend upon leaving the more-dense substance? Make a sketch to illustrate your answer.
7. Since twilight occurs without the sun being visible, how does it tend to lengthen the day?
8. On a clear, dry, warm day, why do dark road surfaces frequently appear wet?
9. What atmospheric conditions are necessary for an inferior mirage? A superior mirage?
10. What process (refraction or scattering) produces crepuscular rays?
11. (a) Describe how a halo forms.
    (b) How is the formation of a halo different from that of a sundog?
12. At what time of day would you expect to observe the green flash?
13. Explain how sun pillars form.
14. What process (refraction, scattering, diffraction) is responsible for lengthening the day?
15. Why can a rainbow only be observed if the sun is toward the observer's back?
16. Why are secondary rainbows higher and much dimmer than primary rainbows? Explain your answer with the aid of a diagram.
17. Explain why this rhyme makes sense:
    Rainbow in morning, joggers take warning.
    Rainbow at night (evening), jogger's delight.
18. Suppose you look at the moon and see a bright ring of light that appears to rest on its surface.
    (a) Is this ring of light a halo or a corona?
    (b) What type of clouds (water or ice) must be present for this type of optical phenomenon to occur?
    (c) Is this ring of light produced mainly by refraction or diffraction?
19. Would you expect to see the glory when flying in an aircraft on a perfectly clear day? Explain.
20. Explain how light is able to reach your eyes when you see
    (a) a corona
    (b) a glory
    (c) the *Heiligenschein*
21. How would you distinguish a corona from a halo?
22. What process is primarily responsible for the formation of cloud iridescence—reflection, refraction, or diffraction of light?

QUESTIONS FOR THOUGHT

1. Explain why on a cloudless day the sky will usually appear milky white before it rains and a deep blue after it rains.
2. How long does twilight last on the moon? (Hint: The moon has no atmosphere.)
3. Why is it often difficult to see the road while driving on a foggy night with your high beam lights on?
4. What would be the color of the sky if air molecules scattered the longest wavelengths of visible light and passed the shorter wavelengths straight through? (Use a diagram to help explain your answer.)
5. Explain why the colors of the planets are not related to the temperatures of the planets, while the colors of the stars are related to the temperatures of the stars.
6. If there were no atmosphere surrounding the earth, what color would the sky be at sunrise? At sunset? What color would the sun be at noon? At sunrise? At sunset?
7. Why are rainbows seldom observed at noon?
8. On a cool, clear summer day, a blue haze often appears over the Great Smoky Mountains of Tennessee. Explain why the blue haze usually changes to a white haze as the relative humidity of the air increases.
9. During a lunar eclipse, the earth, sun, and moon are aligned as shown in ●Fig. 19.37. The earth blocks sunlight from directly reaching the moon's surface, yet the surface of the moon will often appear a pale red color during a lunar eclipse. How can you account for this phenomenon?

● FIGURE 19.37

10. Explain why smoke rising from a cigarette often appears blue, yet appears white when blown from the mouth.
11. During Ernest Shackleton's last expedition to Antarctica, on May 8, 1915, seven days after the sun had set for the winter, he saw the sun reappear. Explain how this event—called the *Novaya Zemlya effect*—can occur.
12. Could a superior mirage form over land on a hot, sunny day? Explain.
13. Explain why it is easier to get sunburned on a high mountain than in the valley below. (The answer is not that you are closer to the sun on top of the mountain.)
14. Why are stars more visible on a clear night when there is no moon than on a clear night with a full moon?
15. During the day, clouds are white, and the sky is blue. Why then, during a full moon, do cumulus clouds appear faintly white, while the sky does not appear blue?

PROBLEMS AND EXERCISES

1. Choose a 3-day period in which to observe the sky 5 times each day. Record in a notebook the number of times you see halos, crepuscular rays, coronas, cloud iridescence, sun dogs, rainbows, and other phenomena.

2. Make your own rainbow. On a sunny afternoon or morning, turn on the water sprinkler to create a spray of water drops. Stand as close to the spray as possible (without getting soaked) and observe that, as you move up, down, and sideways, the bow moves with you.
   (a) Explain why this happens.
   (b) Also, explain with the use of a diagram why the sun must be at your back in order to see the bow.

3. Take a large beaker or bottle and fill it with water. Add a small amount of nonfat powdered milk and stir until the water turns a faint milky white. Shine white light into the beaker, and, on the opposite side, hold a white piece of paper.
   (a) Explain why the milk has a blue cast to it and why the light shining on the paper appears ruddy.
   (b) What do you know about the size of the milk particles?
   (c) How does this demonstration relate to the color of the sky and the color of the sun—at sunrise and sunset?

Visit the
Meteorology Resource Center
at
academic.cengage.com/login
for more assets, including questions for exploration, animations, videos, and more.

Units, Conversions, Abbreviations, and Equations

LENGTH

1 kilometer (km)	=	1000 m
	=	3281 ft
	=	0.62 mi
1 mile (mi)	=	5280 ft
	=	1609 mi
	=	1.61 km
1 meter (m)	=	100 cm
	=	3.28 ft
	=	39.37 in.
1 foot (ft)	=	12 in.
	=	30.48 cm
	=	0.305 m
1 centimeter (cm)	=	0.39 in.
	=	0.01 m
	=	10 mm
1 inch (in.)	=	2.54 cm
	=	0.08 ft
1 millimeter (mm)	=	0.1 cm
	=	0.001 m
	=	0.039 in.
1 micrometer (μm)	=	0.0001 cm
	=	0.000001 m
1 degree latitude	=	111 km
	=	60 nautical mi
	=	69 statute mi

AREA

1 square centimeter (cm^2)	=	0.15 in.^{2}
1 square inch (in.^{2})	=	6.45 cm^2
1 square meter (m^2)	=	10.76 ft^2
1 square foot (ft^2)	=	0.09 m^2

VOLUME

1 cubic centimeter (cm^3)	=	0.06 in.^{3}
1 cubic inch (in.^{3})	=	16.39 cm^3
1 liter (l)	=	1000 cm^3
	=	0.264 gallon (gal) U.S.

SPEED

1 knot	=	1 nautical mi/hr
	=	1.15 statute mi/hr
	=	0.51 m/sec
	=	1.85 km/hr
1 mile per hour (mi/hr)	=	0.87 knot
	=	0.45 m/sec
	=	1.61 km/hr
1 kilometer per hour (km/hr)	=	0.54 knot
	=	0.62 mi/hr
	=	0.28 m/sec
1 meter per second (m/sec)	=	1.94 knots
	=	2.24 mi/hr
	=	3.60 km/hr

FORCE

1 dyne	=	1 gram centimeter per second per second
	=	2.2481×10^{-6} pound (lb)
1 newton (N)	=	1 kilogram meter per second per second
	=	10^5 dynes
	=	0.2248 lb

MASS

1 gram (g)	=	0.035 ounce
	=	0.002 lb
1 kilogram (kg)	=	1000 g
	=	2.2 lb

ENERGY

1 erg	=	1 dyne per cm
	=	2.388×10^{-8} cal
1 joule (J)	=	1 newton meter
	=	0.239 cal
	=	10^7 erg
1 calorie (cal)	=	4.186 J
	=	4.186×10^7 erg

PRESSURE

1 millibar (mb)	=	1000 dynes/cm^2
	=	0.75 millimeter of mercury (mm Hg)
	=	0.02953 inch of mercury (in. Hg)
	=	0.01450 pound per square inch (lb/in.^{2})
	=	100 pascals (Pa)
1 standard atmosphere	=	1013.25 mb
	=	760 mm Hg
	=	29.92 in. Hg
	=	14.7 lb/in.^{2}
1 inch of mercury	=	33.865 mb
1 millimeter of mercury	=	1.3332 mb
1 pascal	=	0.01 mb
	=	1 N/m^2
1 hectopascal (hPa)	=	1 mb
1 kilopascal (kPa)	=	10 mb

POWER

1 watt (W)	=	1 J/sec
	=	14.3353 cal/min
1 cal/min	=	0.06973 W
1 horse power (hp)	=	746 W

POWERS OF TEN

PREFIX

nano	one-billionth	=	10^{-9}	=	0.000000001
micro	one-millionth	=	10^{-6}	=	0.000001
milli	one-thousandth	=	10^{-3}	=	0.001
centi	one-hundredth	=	10^{-2}	=	0.01
deci	one-tenth	=	10^{-1}	=	0.1
hecto	one hundred	=	10^2	=	100
kilo	one thousand	=	10^3	=	1000
mega	one million	=	10^6	=	1,000,000
giga	one billion	=	10^9	=	1,000,000,000

TEMPERATURE

$°C = \frac{5}{9} \, (°F - 32)$

To convert degrees Fahrenheit (°F) to degrees Celsius (°C): Subtract 32 degrees from °F, then divide by 1.8.

To convert degrees Celsius (°C) to degrees Fahrenheit (°F): Multiply °C by 1.8, then add 32 degrees.

To convert degrees Celsius (°C) to Kelvins (K): Add 273 to Celsius temperature, as

$$K = °C + 273.$$

▼ TABLE A.I   Temperature Conversions

°F	°C	°F	°C	°F	°C	°F	°C	°F	°C	°F	°C	°F	°C	°F	°C
−40	−40	−20	−28.9	0	−17.8	20	−6.7	40	4.4	60	15.6	80	26.7	100	37.8
−39	−39.4	−19	−28.3	1	−17.2	21	−6.1	41	5.0	61	16.1	81	27.2	101	38.3
−38	−38.9	−18	−27.8	2	−16.7	22	−5.6	42	5.6	62	16.7	82	27.8	102	38.9
−37	−38.3	−17	−27.2	3	−16.1	23	−5.0	43	6.1	63	17.2	83	28.3	103	39.4
−36	−37.8	−16	−26.7	4	−15.6	24	−4.4	44	6.7	64	17.8	84	28.9	104	40.0
−35	−37.2	−15	−26.1	5	−15.0	25	−3.9	45	7.2	65	18.3	85	29.4	105	40.6
−34	−36.7	−14	−25.6	6	−14.4	26	−3.3	46	7.8	66	18.9	86	30.0	106	41.1
−33	−36.1	−13	−25.0	7	−13.9	27	−2.8	47	8.3	67	19.4	87	30.6	107	41.7
−32	−35.6	−12	−24.4	8	−13.3	28	−2.2	48	8.9	68	20.0	88	31.1	108	42.2
−31	−35.0	−11	−23.9	9	−12.8	29	−1.7	49	9.4	69	20.6	89	31.7	109	42.8
−30	−34.4	−10	−23.3	10	−12.2	30	−1.1	50	10.0	70	21.1	90	32.2	110	43.3
−29	−33.9	−9	−22.8	11	−11.7	31	−0.6	51	10.6	71	21.7	91	32.8	111	43.9
−28	−33.3	−8	−22.2	12	−11.1	32	0.0	52	11.1	72	22.2	92	33.3	112	44.4
−27	−32.8	−7	−21.7	13	−10.6	33	0.6	53	11.7	73	22.8	93	33.9	113	45.0
−26	−32.2	−6	−21.1	14	−10.0	34	1.1	54	12.2	74	23.3	94	34.4	114	45.6
−25	−31.7	−5	−20.6	15	−9.4	35	1.7	55	12.8	75	23.9	95	35.0	115	46.1
−24	−31.1	−4	−20.0	16	−8.9	36	2.2	56	13.3	76	24.4	96	35.6	116	46.7
−23	−30.6	−3	−19.4	17	−8.3	37	2.8	57	13.9	77	25.0	97	36.1	117	47.2
−22	−30.0	−2	−18.9	18	−7.8	38	3.3	58	14.4	78	25.6	98	36.7	118	47.8
−21	−29.4	−1	−18.3	19	−7.2	39	3.9	59	15.0	79	26.1	99	37.2	119	48.3

▼ TABLE A.2   SI Units* and Their Symbols

QUANTITY	NAME	UNITS	SYMBOL
length	meter	m	m
mass	kilogram	kg	kg
time	second	sec	sec
temperature	Kelvin	K	K
density	kilogram per cubic meter	kg/m^3	kg/m^3
speed	meter per second	m/sec	m/sec
force	newton	$m \cdot kg/sec^2$	N
pressure	pascal	N/m^2	Pa
energy	joule	$N \cdot m$	J
power	watt	J/sec	W

*SI stands for Système International, which is the international system of units and symbols.

▼ TABLE A.3   Some Useful Equations and Constants

NAME	EQUATION	CONSTANTS AND ABBREVIATIONS
Gas law (equation of state)	$p = \rho RT$	R = 287 J/kg · K (SI) or R = 2.87×10^6 erg/g · K p = pressure is N/m^2 (SI) ρ = density (kg/m^3) T = temperature (K)
Stefan-Boltzmann law	$E = \sigma T^4$	σ = 5.67×10^{-8} W/m^2 · K^4 (SI) or σ = 5.67×10^{-5} erg/cm^2 · K^4 · sec E = radiation emitted in W/m^2 (SI)
Wien's law	$\lambda_{max} = \dfrac{w}{T}$	w = 0.2897 μm K λ_{max} = wavelength (μm)
Solar constant		1376 W/m^2 (SI) 1.97 cal/cm^2/min
Geostrophic wind equation	$V_g = \dfrac{1}{2\Omega \sin\phi\rho} \dfrac{\Delta p}{d}$	V_g = geostrophic wind (m/sec) Ω = 7.29×10^{-5} radian*/sec ϕ = latitude d = distance (m) Δp = pressure difference (N/m^2)
Coriolis parameter	$f = 2\Omega \sin\phi$	g = force of gravity (9.8 m/sec^2)
Hydrostatic equation	$\dfrac{\Delta p}{\Delta z} = -\rho g$	Δz = change in height (m)

*2π radians equal 360°.

Weather Symbols and the Station Model

SIMPLIFIED SURFACE-STATION MODEL

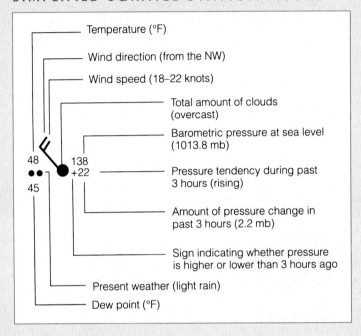

- Temperature (°F)
- Wind direction (from the NW)
- Wind speed (18–22 knots)
- Total amount of clouds (overcast)
- Barometric pressure at sea level (1013.8 mb)
- Pressure tendency during past 3 hours (rising)
- Amount of pressure change in past 3 hours (2.2 mb)
- Sign indicating whether pressure is higher or lower than 3 hours ago
- Present weather (light rain)
- Dew point (°F)

CLOUD COVERAGE

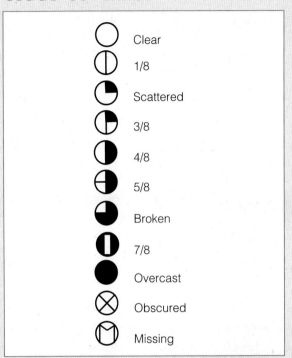

- Clear
- 1/8
- Scattered
- 3/8
- 4/8
- 5/8
- Broken
- 7/8
- Overcast
- Obscured
- Missing

UPPER-AIR MODEL (500 MB)

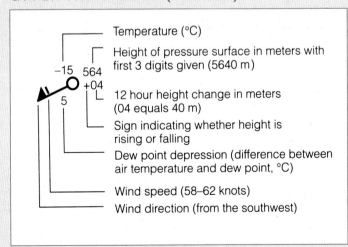

- Temperature (°C)
- Height of pressure surface in meters with first 3 digits given (5640 m)
- 12 hour height change in meters (04 equals 40 m)
- Sign indicating whether height is rising or falling
- Dew point depression (difference between air temperature and dew point, °C)
- Wind speed (58–62 knots)
- Wind direction (from the southwest)

COMMON WEATHER SYMBOLS

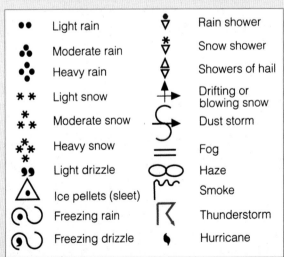

- Light rain
- Moderate rain
- Heavy rain
- Light snow
- Moderate snow
- Heavy snow
- Light drizzle
- Ice pellets (sleet)
- Freezing rain
- Freezing drizzle
- Rain shower
- Snow shower
- Showers of hail
- Drifting or blowing snow
- Dust storm
- Fog
- Haze
- Smoke
- Thunderstorm
- Hurricane

WIND ENTRIES

	MILES (STATUTE) PER HOUR	KNOTS	KILOMETERS PER HOUR
◎	Calm	Calm	Calm
	1–2	1–2	1–3
	3–8	3–7	4–13
	9–14	8–12	14–19
	15–20	13–17	20–32
	21–25	18–22	33–40
	26–31	23–27	41–50
	32–37	28–32	51–60
	38–43	33–37	61–69
	44–49	38–42	70–79
	50–54	43–47	80–87
	55–60	48–52	88–96
	61–66	53–57	97–106
	67–71	58–62	107–114
	72–77	63–67	115–124
	78–83	68–72	125–134
	84–89	73–77	135–143
	119–123	103–107	144–198

PRESSURE TENDENCY

∧	Rising, then falling	
⌐	Rising, then steady; or rising, then rising more slowly	Barometer now higher than 3 hours ago
/	Rising steadily or unsteadily	
✓	Falling or steady, then rising; or rising, then rising more quickly	
—	Steady, same as 3 hours ago	
∨	Falling, then rising, same or lower than 3 hours ago	
⌐	Falling, then steady; or falling, then falling more slowly	Barometer now lower than 3 hours ago
\	Falling steadily, or unsteadily	
∧	Steady or rising, then falling; or falling, then falling more quickly	

FRONT SYMBOLS

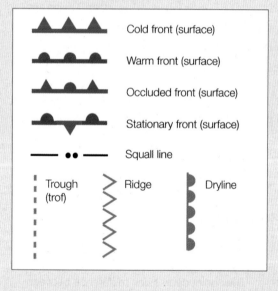

Cold front (surface)

Warm front (surface)

Occluded front (surface)

Stationary front (surface)

Squall line

Trough (trof)   Ridge   Dryline

Beaufort Wind Scale (Over Land)

▼ TABLE C.1 Estimating Wind Speed from Surface Observation

BEAUFORT NUMBER	DESCRIPTION	WIND SPEED MI/HR	KNOTS	KM/HR	OBSERVATIONS
0	Calm	0–1	0–1	0–2	Smoke rises vertically
1	Light air	1–3	1–3	2–6	Direction of wind shown by drifting smoke, but not by wind vanes
2	Slight breeze	4–7	4–6	7–11	Wind felt on face; leaves rustle; wind vanes moved by wind; flags stir
3	Gentle breeze	8–12	7–10	12–19	Leaves and small twigs move; wind will extend light flag
4	Moderate breeze	13–18	11–16	20–29	Wind raises dust and loose paper; small branches move; flags flap
5	Fresh breeze	19–24	17–21	30–39	Small trees with leaves begin to sway; flags ripple
6	Strong breeze	25–31	22–27	40–50	Large tree branches in motion; whistling heard in telegraph wires; umbrellas used with difficulty
7	High wind	32–38	28–33	51–61	Whole trees in motion; inconvenience felt walking against wind; flags extend
8	Gale	39–46	34–40	62–74	Wind breaks twigs off trees; walking is difficult
9	Strong gale	47–54	41–47	75–87	Slight structural damage occurs (signs and antennas blown down)
10	Whole gale	55–63	48–55	88–101	Trees uprooted; considerable damage occurs
11	Storm	64–74	56–64	102–119	Winds produce widespread damage
12	Hurricane	≥ 75	≥ 65	≥ 120	Winds produce extensive damage

APPENDIX D

Humidity and Dew-Point Tables (Psychrometric Tables)

To obtain the dew point (or relative humidity), simply read down the temperature column and then over to the wet-bulb depression. For example, in Table D.1, a temperature of 10°C with a wet-bulb depression of 3°C produces a dew-point temperature of 4°C. (Dew-point temperature and relative humidity readings are appropriate for pressures near 1000 mb.)

▼ TABLE D.1   Dew-Point Temperature (°C)

	WET-BULB DEPRESSION (DRY-BULB TEMPERATURE MINUS WET-BULB TEMPERATURE) (°C)																
	0.5	1.0	1.5	2.0	2.5	3.0	3.5	4.0	4.5	5.0	7.5	10.0	12.5	15.0	17.5	20.0
−20	−25	−33														
−17.5	−21	−27	−38													
−15	−19	−23	−28													
−12.5	−15	−18	−22	−29												
−10	−12	−14	−18	−21	−27	−36										
−7.5	−9	−11	−14	−17	−20	−26	−34									
−5	−7	−8	−10	−13	−16	−19	−24	−31								
−2.5	−4	−6	−7	−9	−11	−14	−17	−22	−28	−41						
0	−1	−3	−4	−6	−8	−10	−12	−15	−19	−24						
2.5	1	0	−1	−3	−4	−6	−8	−10	−13	−16						
5	4	3	2	0	−1	−3	−4	−6	−8	−10	−48					
7.5	6	6	4	3	2	1	−1	−2	−4	−6	−22					
10	9	8	7	6	5	4	2	1	0	−2	−13					
12.5	12	11	10	9	8	7	6	4	3	2	−7	−28				
15	14	13	12	12	11	10	9	8	7	5	−2	−14				
17.5	17	16	15	14	13	12	12	11	10	8	2	−7	−35			
20	19	18	18	17	16	15	14	14	13	12	6	−1	−15			
22.5	22	21	20	20	19	18	17	16	16	15	10	3	−6	−38		
25	24	24	23	22	21	21	20	19	18	18	13	7	0	−14		
27.5	27	26	26	25	24	23	23	22	21	20	16	11	5	−5	−32	
30	29	29	28	27	27	26	25	25	24	23	19	14	9	2	−11	
32.5	32	31	31	30	29	29	28	27	26	26	22	18	13	7	−2	
35	34	34	33	32	32	31	31	30	29	28	25	21	16	11	4	
37.5	37	36	36	35	34	34	33	32	32	31	28	24	20	15	9	0
40	39	39	38	38	37	36	36	35	34	34	30	27	23	18	13	6
42.5	42	41	41	40	40	39	38	38	37	36	33	30	26	22	17	11
45	44	44	43	43	42	42	41	40	40	39	36	33	29	25	21	15
47.5	47	46	46	45	45	44	44	43	42	42	39	35	32	28	24	19
50	49	49	48	48	47	47	46	45	45	44	41	38	35	31	28	23

AIR (DRY-BULB) TEMPERATURE (°C)

▼ TABLE D.2   Relative Humidity (Percent)

AIR (DRY-BULB) TEMPERATURE (°C)	WET-BULB DEPRESSION (DRY-BULB TEMPERATURE MINUS WET-BULB TEMPERATURE) (°C)																	
	0.5	1.0	1.5	2.0	2.5	3.0	3.5	4.0	4.5	5.0	7.5	10.0	12.5	15.0	17.5	20.0	22.5	25.0
−20	70	41	11															
−17.5	75	51	26	2														
−15	79	58	38	18														
−12.5	82	65	47	30	13													
−10	85	69	54	39	24	10												
−7.5	87	73	60	48	35	22	10											
−5	88	77	66	54	43	32	21	11	1									
−2.5	90	80	70	60	50	42	37	22	12	3								
0	91	82	73	65	56	47	39	31	23	15								
2.5	92	84	76	68	61	53	46	38	31	24								
5	93	86	78	71	65	58	51	45	38	32	1							
7.5	93	87	80	74	68	62	56	50	44	38	11							
10	94	88	82	76	71	65	60	54	49	44	19							
12.5	94	89	84	78	73	68	63	58	53	48	25	4						
15	95	90	85	80	75	70	66	61	57	52	31	12						
17.5	95	90	86	81	77	72	68	64	60	55	36	18	2					
20	95	91	87	82	78	74	70	66	62	58	40	24	8					
22.5	96	92	87	83	80	76	72	68	64	61	44	28	14	1				
25	96	92	88	84	81	77	73	70	66	63	47	32	19	7				
27.5	96	92	89	85	82	78	75	71	68	65	50	36	23	12	1			
30	96	93	89	86	82	79	76	73	70	67	52	39	27	16	6			
32.5	97	93	90	86	83	80	77	74	71	68	54	42	30	20	11	1		
35	97	93	90	87	84	81	78	75	72	69	56	44	33	23	14	6		
37.5	97	94	91	87	85	82	79	76	73	70	58	46	36	26	18	10	3	
40	97	94	91	88	85	82	79	77	74	72	59	48	38	29	21	13	6	
42.5	97	94	91	88	86	83	80	78	75	72	61	50	40	31	23	16	9	2
45	97	94	91	89	86	83	81	78	76	73	62	51	42	33	26	18	12	6
47.5	97	94	92	89	86	84	81	79	76	74	63	53	44	35	28	21	15	9
50	97	95	92	89	87	84	82	79	77	75	64	54	45	37	30	23	17	11

▼ TABLE D.3   Dew-Point Temperature (°F)

WET-BULB DEPRESSION (DRY-BULB TEMPERATURE MINUS WET-BULB TEMPERATURE) (°F)

AIR (DRY-BULB) TEMPERATURE (°F)

AIR (°F)	1	2	3	4	5	6	7	8	9	10	11	12	13	14	15	16	17	18	19	20	25	30	35	40
0	−7	−20																						
5	−1	−9	−24																					
10	5	−2	−10	−27																				
15	11	6	0	−9	−26																			
20	16	12	8	2	−7	−21																		
25	22	19	15	10	5	−3	−15	−51																
30	27	25	21	18	14	8	2	−7	−25															
35	33	30	28	25	21	17	13	7	0	−11														
40	38	35	33	30	28	25	21	18	13	7	−1	−14												
45	43	41	38	36	34	31	28	25	22	18	13	7	−1	−14										
50	48	46	44	42	40	37	34	32	29	26	22	18	13	8	0	−13								
55	53	51	50	48	45	43	41	38	36	33	30	27	24	20	15	9	1	−12						
60	58	57	55	53	51	49	47	45	43	40	38	35	32	29	25	21	17	11	4	−8				
65	63	62	60	59	57	55	53	51	49	47	45	42	40	37	34	31	27	24	19	14				
70	69	67	65	64	62	61	59	57	55	53	51	49	47	44	42	39	36	33	30	26	−11			
75	74	72	71	69	68	66	64	63	61	59	57	55	54	51	49	47	44	42	39	36	15			
80	79	77	76	74	73	72	70	68	67	65	63	62	60	58	56	54	52	50	47	44	28	−7		
85	84	82	81	80	78	77	75	74	72	71	69	68	66	64	62	61	59	57	54	52	39	19		
90	89	87	86	85	83	82	81	79	78	76	75	73	72	70	69	67	65	63	61	59	48	32		
95	94	93	91	90	89	87	86	85	83	81	80	79	78	76	74	73	71	70	68	66	56	43	24	
100	99	98	96	95	94	93	91	90	89	87	86	85	83	82	80	79	77	76	74	72	63	52	37	12
105	104	103	101	100	99	98	96	95	94	93	91	90	89	87	86	84	83	82	80	78	70	61	48	30
110	109	108	106	105	104	103	102	100	99	98	97	95	94	93	91	90	89	87	86	84	77	68	57	43
115	114	113	112	110	109	108	107	106	104	103	102	101	99	98	97	96	94	93	92	90	83	75	65	54
120	119	118	117	115	114	113	112	111	110	108	107	106	105	104	102	101	100	98	97	96	89	81	73	63

▼ TABLE D.4   Relative Humidity (Percent)

AIR (DRY-BULB) TEMPERATURE (°F)	WET-BULB DEPRESSION (DRY-BULB TEMPERATURE MINUS WET-BULB TEMPERATURE) (°F)																						
	1	2	3	4	5	6	7	8	9	10	11	12	13	14	15	16	17	18	19	20	25	30	35	40
0	67	31	1																					
5	73	46	20																					
10	78	56	34	13																				
15	82	64	46	29	11																			
20	85	70	55	40	26	12																		
25	87	74	62	49	37	25	13	1																
30	89	78	67	56	46	36	26	16	6															
35	91	81	72	63	54	45	36	27	19	10	2													
40	92	83	75	68	60	52	45	37	29	22	15	7												
45	93	86	78	71	64	57	51	44	38	31	25	18	12	6										
50	93	87	80	74	67	61	55	49	43	38	32	27	21	16	10	5								
55	94	88	82	76	70	65	59	54	49	43	38	33	28	23	19	14	9	5						
60	94	89	83	78	73	68	63	58	53	48	43	39	34	30	26	21	17	13	9	5				
65	95	90	85	80	75	70	66	61	56	52	48	44	39	35	31	27	24	20	16	12				
70	95	90	86	81	77	72	68	64	59	55	51	48	44	40	36	33	29	25	22	19	3			
75	96	91	86	82	78	74	70	66	62	58	54	51	47	44	40	37	34	30	27	24	7	1		
80	96	91	87	83	79	75	72	68	64	61	57	54	50	47	44	41	38	35	32	29	15	3		
85	96	92	88	84	80	76	73	69	66	62	59	56	52	49	46	43	41	38	35	32	20	8		
90	96	92	89	85	81	78	74	71	68	65	61	58	55	52	49	47	44	41	39	36	24	13		
95	96	93	89	86	82	79	75	72	69	66	63	60	57	54	51	49	46	43	41	38	27	17		
100	96	93	89	86	83	80	77	73	70	68	65	62	59	56	54	51	49	46	44	41	30	21	12	4
105	97	93	90	87	83	80	77	74	71	69	66	63	60	58	55	53	50	48	46	43	33	23	15	7
110	97	93	90	87	84	81	78	75	73	70	67	65	62	60	57	55	52	50	48	46	36	26	18	11
115	97	94	91	88	85	82	79	76	74	71	68	66	63	61	58	56	54	52	49	47	37	28	21	13
120	97	94	91	88	85	82	80	77	74	72	69	67	65	62	60	58	55	53	51	49	40	31	23	17

Instant Weather Forecast Chart

This chart is a guide to forecasting the weather. It is applicable to most of the United States, especially the eastern two-thirds. It works best during the fall, winter, and spring, when the weather systems are most active.

▼ TABLE E.1   Instant Weather Forecast Chart

SEA-LEVEL PRESSURE (MB)	PRESSURE TENDENCY	SURFACE WIND DIRECTION	SKY CONDITION	24-HR WEATHER FORECAST (SEE WEATHER FORECAST CODE)
1023 or higher (30.21 in.)	rising, steady, or falling	any direction	clear, high clouds, or Cu	1, 18 (in winter, 14)
1022 to 1016 (30.20 in. to 30.00 in.)	rising or steady	SW, W, NW, N	clear, high clouds or Cu	1, 18
	falling or steady	SW, S, SE	clear, high clouds	1, 3, 17, 5
	falling	SW, S, SE	middle or low clouds	6, 17
	falling	E, NE	middle or low clouds	6, 14
	falling or steady	E, NE	clear or high clouds	3, 5, 14
1015 to 1009 (29.99 in. to 29.80 in.)	rising	SW, W, NW, N	clear	1, 14
			overcast	2, 16
			precipitation	11, 2, 16
	falling	any direction	clear	3, 17 (dry climate summer, 1, 15)
	falling or steady	SW, S, SE	high clouds	3, 17, 5
	falling	SW, S, SE	middle or low clouds	7
	falling	E, NE	middle or low clouds	7, 12, 14
	falling	SE, E, NE	overcast, precipitation	9
	falling	S, SW	overcast, precipitation	10, 13
1008 or below (29.79 in.)	rising	SW, W, NW, N	clear	1, 12
	rising	SW, W, NW, N	overcast	2, 12, 16
	rising	SW, W, NW, N	overcast with precipitation	11, 12, 16
	rising	NE	overcast	4, 12, 13, 14
	rising	NE	overcast with precipitation	11, 12, 13, 14
	rising or steady	SW, S, SE	clear	3, 6, 8, 12, 15
	falling	SW, S, SE	overcast	7, 8, 12, 13
	falling	SW, S, SE	overcast with precipitation	8, 10, 12, 13, 16
	falling	N	overcast	4, 14
	falling or steady	E, NE	overcast	7, 12, 14
	falling	E, NE	overcast with precipitation	8, 9, 12, 13

Weather Forecast Code

1 = clear or scattered clouds
2 = clearing
3 = increasing clouds
4 = continued overcast
5 = precipitation possible within 24 hours
6 = precipitation possible within 12 hours

7 = precipitation possible within 8 hours
8 = possible period of heavy precipitation
9 = precipitation continuing
10 = precipitation ending within 12 hours
11 = precipitation ending within 6 hours
12 = windy

13 = possible wind shift to W, NW, or N
14 = continued cool or cold
15 = continued mild or warm
16 = turning colder
17 = slowly rising temperatures
18 = little temperature change

Changing GMT and UTC to Local Time

The system of time used in meteorology is Greenwich Mean Time (GMT), which is also known as Coordinated Universal Time (UTC), and as Zulu (Z) Time. This is the time measured on the prime meridian (0° longitude) in Greenwich, England. Because Eastern Standard Time (EST) is 5 hours slower than GMT, to convert from GMT to EST simply requires subtracting 5 hours from GMT. Conversely, to change EST to GMT entails adding 5 hours to EST. Figure F.1 shows how to convert to GMT in various time zones of North America. Since in meteorology the time is given on a 24-clock, Table F.1 shows the relationship between the familiar two 12-hour periods of A.M. and P.M. and the 24-hour system.

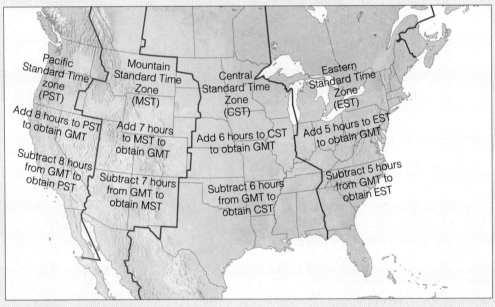

● FIGURE F.1 The time zones of North America.

▼ TABLE F.1   Conversion of A.M. and P.M. Time System to 24-Hour System

TIME	24-HR SYSTEM TIME	TIME	24-HR SYSTEM TIME	TIME	24-HR SYSTEM TIME	TIME	24-HR SYSTEM TIME
A.M.		**A.M.**		**P.M.**		**P.M.**	
12:00 (midnight)	0000	6:00	0600	12:00 (noon)	1200	6:00	1800
1:00	0100	7:00	0700	1:00	1300	7:00	1900
2:00	0200	8:00	0800	2:00	1400	8:00	2000
3:00	0300	9:00	0900	3:00	1500	9:00	2100
4:00	0400	10:00	1000	4:00	1600	10:00	2200
5:00	0500	11:00	1100	5:00	1700	11:00	2300

Average Annual Global Precipitation

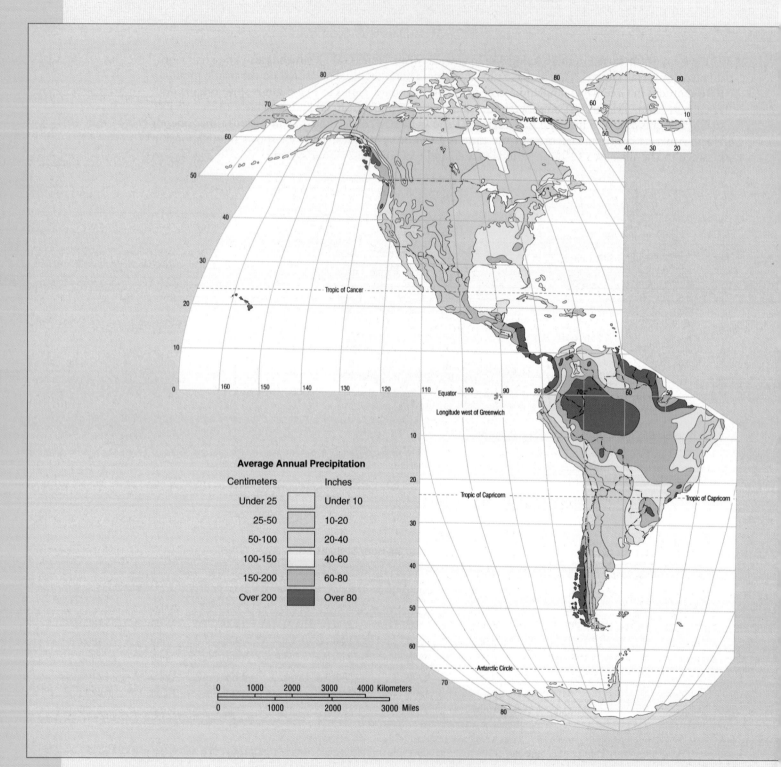

Average Annual Precipitation

Centimeters		Inches
Under 25		Under 10
25–50		10–20
50–100		20–40
100–150		40–60
150–200		60–80
Over 200		Over 80

● FIGURE G.1 World map of average annual precipitation.

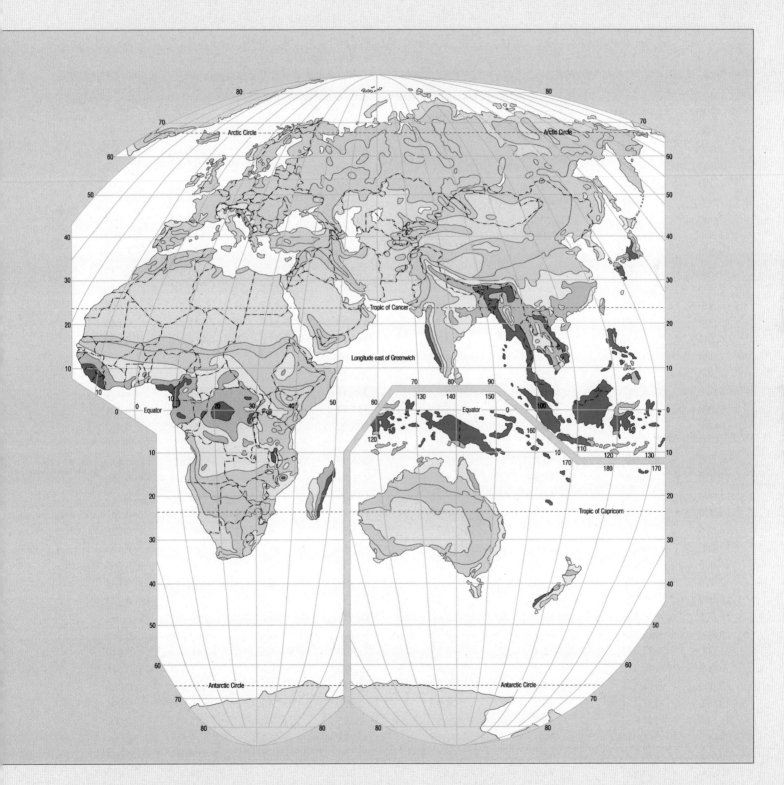

A Western Paragraphic Projection developed at Western Illinois University

APPENDIX H

Standard Atmosphere

ALTITUDE				PRESSURE	TEMPERATURE		DENSITY
METERS	FEET	KILOMETERS	MILES	MILLIBARS	°C	°F	kg/m³
0	0	0.0	0.0	1013.25	15.0	(59.0)	1.225
500	1,640	0.5	0.3	954.61	11.8	(53.2)	1.167
1,000	3,280	1.0	0.6	898.76	8.5	(47.3)	1.112
1,500	4,921	1.5	0.9	845.59	5.3	(41.5)	1.058
2,000	6,562	2.0	1.2	795.01	2.0	(35.6)	1.007
2,500	8,202	2.5	1.5	746.91	−1.2	(29.8)	0.957
3,000	9,842	3.0	1.9	701.21	−4.5	(23.9)	0.909
3,500	11,483	3.5	2.2	657.80	−7.7	(18.1)	0.863
4,000	13,123	4.0	2.5	616.60	−11.0	(12.2)	0.819
4,500	14,764	4.5	2.8	577.52	−14.2	(6.4)	0.777
5,000	16,404	5.0	3.1	540.48	−17.5	(0.5)	0.736
5,500	18,045	5.5	3.4	505.39	−20.7	(−5.3)	0.697
6,000	19,685	6.0	3.7	472.17	−24.0	(−11.2)	0.660
6,500	21,325	6.5	4.0	440.75	−27.2	(−17.0)	0.624
7,000	22,965	7.0	4.3	411.05	−30.4	(−22.7)	0.590
7,500	24,606	7.5	4.7	382.99	−33.7	(−28.7)	0.557
8,000	26,247	8.0	5.0	356.51	−36.9	(−34.4)	0.526
8,500	27,887	8.5	5.3	331.54	−40.2	(−40.4)	0.496
9,000	29,528	9.0	5.6	308.00	−43.4	(−46.1)	0.467
9,500	31,168	9.5	5.9	285.84	−46.6	(−51.9)	0.440
10,000	32,808	10.0	6.2	264.99	−49.9	(−57.8)	0.413
11,000	36,089	11.0	6.8	226.99	−56.4	(−69.5)	0.365
12,000	39,370	12.0	7.5	193.99	−56.5	(−69.7)	0.312
13,000	42,651	13.0	8.1	165.79	−56.5	(−69.7)	0.267
14,000	45,932	14.0	8.7	141.70	−56.5	(−69.7)	0.228
15,000	49,213	15.0	9.3	121.11	−56.5	(−69.7)	0.195
16,000	52,493	16.0	9.9	103.52	−56.5	(−69.7)	0.166
17,000	55,774	17.0	10.6	88.497	−56.5	(−69.7)	0.142
18,000	59,055	18.0	11.2	75.652	−56.5	(−69.7)	0.122
19,000	62,336	19.0	11.8	64.674	−56.5	(−69.7)	0.104
20,000	65,617	20.0	12.4	55.293	−56.5	(−69.7)	0.089
25,000	82,021	25.0	15.5	25.492	−51.6	(−60.9)	0.040
30,000	98,425	30.0	18.6	11.970	−46.6	(−51.9)	0.018
35,000	114,829	35.0	21.7	5.746	−36.6	(−33.9)	0.008
40,000	131,234	40.0	24.9	2.871	−22.8	(−9.0)	0.004
45,000	147,638	45.0	28.0	1.491	−9.0	(15.8)	0.002
50,000	164,042	50.0	31.1	0.798	−2.5	(27.5)	0.001
60,000	196,850	60.0	37.3	0.220	−26.1	(−15.0)	0.0003
70,000	229,659	70.0	43.5	0.052	−53.6	(−64.5)	0.00008
80,000	262,467	80.0	49.7	0.010	−74.5	(−102.1)	0.00002

Hurricane Tracking Chart

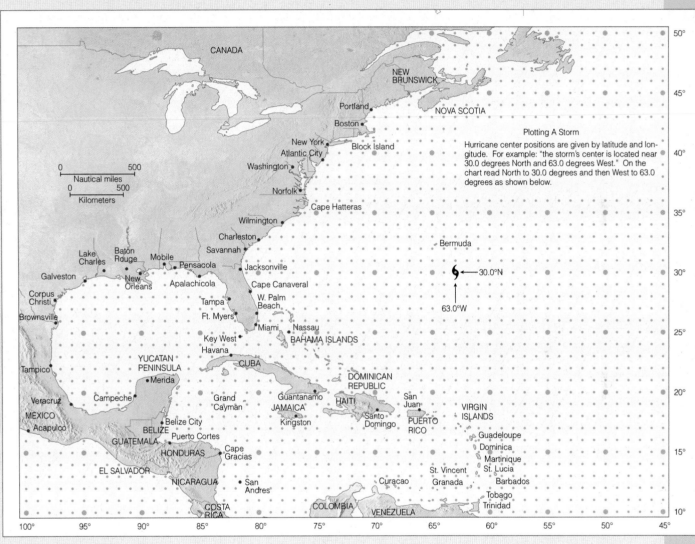

Plotting A Storm

Hurricane center positions are given by latitude and longitude. For example: "the storm's center is located near 30.0 degrees North and 63.0 degrees West." On the chart read North to 30.0 degrees and then West to 63.0 degrees as shown below.

FIGURE I.1

Adiabatic Chart

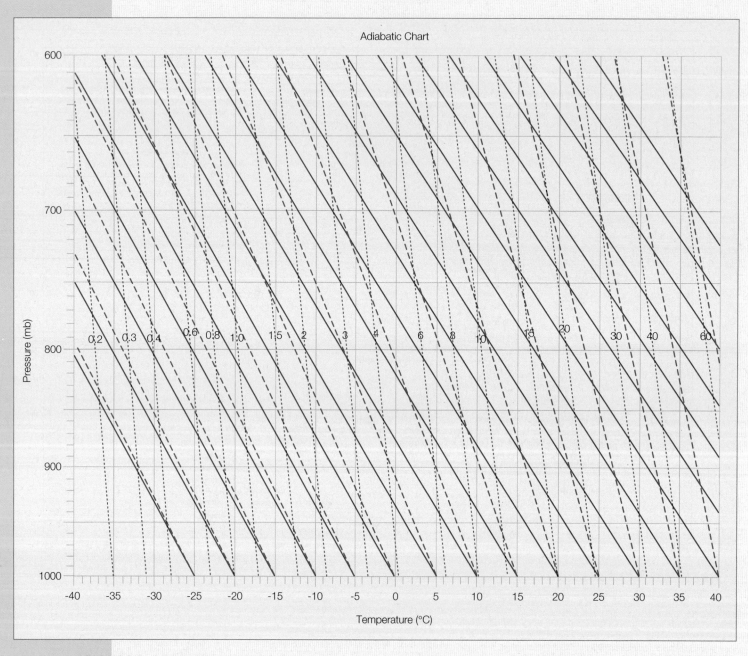

● FIGURE J.1 Adiabatic chart.

Absolute humidity The mass of water vapor in a given volume of air. It represents the density of water vapor in the air.

Absolute vorticity *See* Vorticity.

Absolute zero A temperature reading of −273°C, −460°F, or 0K. Theoretically, there is no molecular motion at this temperature.

Absolutely stable atmosphere An atmospheric condition that exists when the environmental lapse rate is less than the moist adiabatic rate. This results in a lifted parcel of air being colder than the air around it.

Absolutely unstable atmosphere An atmospheric condition that exists when the environmental lapse rate is greater than the dry adiabatic rate. This results in a lifted parcel of air being warmer than the air around it.

Accretion The growth of a precipitation particle by the collision of an ice crystal or snowflake with a supercooled liquid droplet that freezes upon impact.

Acid deposition The depositing of acidic particles (usually sulfuric acid and nitric acid) at the earth's surface. Acid deposition occurs in dry form (*dry deposition*) or wet form (*wet deposition*). Acid rain and acid precipitation often denote wet deposition. (*See* Acid rain.)

Acid fog *See* Acid rain.

Acid rain Cloud droplets or raindrops combining with gaseous pollutants, such as oxides of sulfur and nitrogen, to make falling rain (or snow) acidic—pH less than 5.0. If fog droplets combine with such pollutants it becomes *acid fog*.

Actual Vapor Pressure *See* Vapor pressure.

Adiabatic process A process that takes place without a transfer of heat between the system (such as an air parcel) and its surroundings. In an adiabatic process, compression always results in warming, and expansion results in cooling.

Advection The horizontal transfer of any atmospheric property by the wind.

Advection fog Occurs when warm, moist air moves over a cold surface and the air cools to below its dew point.

Advection-radiation fog Fog that forms as relatively warm moist air moves over a colder surface that cooled mainly by radiational cooling.

Aerosols Tiny suspended solid particles (dust, smoke, etc.) or liquid droplets that enter the atmosphere from either natural or human (anthropogenic) sources, such as the burning of fossil fuels. Sulfur-containing fossil fuels, such as coal, produce *sulfate aerosols*.

Aerovane A wind instrument that indicates or records both wind speed and wind direction. Also called a *skyvane*.

Aggregation The clustering together of ice crystals to form snowflakes.

Air density *See* Density.

Air glow A faint glow of light emitted by excited gases in the upper atmosphere. Air glow is much fainter than the aurora.

Air mass A large body of air that has similar horizontal temperature and moisture characteristics.

Air-mass thunderstorm *See* Ordinary thunderstorm.

Air-mass weather A persistent type of weather that may last for several days (up to a week or more). It occurs when an area comes under the influence of a particular air mass.

Air parcel *See* Parcel of air.

Air pollutants Solid, liquid, or gaseous airborne substances that occur in concentrations high enough to threaten the health of people and animals, to harm vegetation and structures, or to toxify a given environment.

Air pressure (atmospheric pressure) The pressure exerted by the mass of air above a given point, usually expressed in millibars (mb), inches of mercury (Hg) or in hectopascals (hPa).

Air Quality Index (AQI) An index of air quality that provides daily air pollution concentrations. Intervals on the scale relate to potential health effects.

Aitken nuclei *See* Condensation nuclei.

Albedo The percent of radiation returning from a surface compared to that which strikes it.

Aleutian low The subpolar low-pressure area that is centered near the Aleutian Islands on charts that show mean sea-level pressure.

Altimeter An instrument that indicates the altitude of an object above a fixed level. Pressure altimeters use an aneroid barometer with a scale graduated in altitude instead of pressure.

Altocumulus A middle cloud, usually white or gray. Often occurs in layers or patches with wavy, rounded masses or rolls.

Altocumulus castellanus An altocumulus cloud showing vertical development. Individual cloud elements have towerlike tops, often in the shape of tiny castles.

Altostratus A middle cloud composed of gray or bluish sheets or layers of uniform appearance. In the thinner regions, the sun or moon usually appears dimly visible.

Analogue forecasting method A forecast made by comparison of past large-scale synoptic weather patterns that resemble a given (usually current) situation in its essential characteristics.

Analysis The drawing and interpretation of the patterns of various weather elements on a surface or upper-air chart.

Anemometer An instrument designed to measure wind speed.

Aneroid barometer An instrument designed to measure atmospheric pressure. It contains no liquid.

Angular momentum The product of an object's mass, speed, and radial distance of rotation.

Annual range of temperature The difference between the warmest and coldest months at any given location.

Anticyclone An area of high atmospheric pressure around which the wind blows clockwise in the Northern Hemisphere and counterclockwise in the Southern Hemisphere. Also called a *high*.

Apparent temperature What the air temperature "feels like" for various combinations of air temperature and relative humidity.

Arctic front In northern latitudes, the semi-permanent front that separates deep cold arctic air from the more shallow, less-cold polar air.

Arctic Oscillation (AO) A reversal of atmospheric pressure over the Arctic that produces changes in the upper-level westerly winds over northern latitudes. These changes in upper-level winds influence winter weather patterns over North America, Greenland, and Europe.

Arcus cloud *See* Shelf cloud.

Arid climate An extremely dry climate—drier than the semi-arid climate. Often referred to as a "true desert" climate.

Asbestos A general name for the fibrous variety of silicate minerals that are incombustible and resist chemicals and heat. Once used for fireproofing, electrical insulation, building materials, chemical filters, and brake linings.

ASOS Acronym for *Automated Surface Observing Systems.* A system designed to provide continuous information of wind, temperature, pressure, cloud base height, and runway visibility at selected airports.

Atmosphere The envelope of gases that surround a planet and are held to it by the planet's gravitational attraction. The earth's atmosphere is mainly nitrogen and oxygen.

Atmospheric boundary layer The layer of air from the earth's surface usually up to about 1 km (3300 ft) where the wind is influenced by friction of the earth's surface and objects on it. Also called the *planetary boundary layer* and the *friction layer.*

Atmospheric greenhouse effect The warming of an atmosphere by its absorbing and emitting infrared radiation while allowing shortwave radiation to pass on through. The gases mainly responsible for the earth's atmospheric greenhouse effect are water vapor and carbon dioxide. Also called the *greenhouse effect.*

Atmospheric models Simulation of the atmosphere's behavior by mathematical equations or by physical models.

Atmospheric stagnation A condition of light winds and poor vertical mixing that can lead to a high concentration of pollutants. Air stagnations are most often associated with fair weather, an inversion, and the sinking air of a high-pressure area.

Atmospheric window The wavelength range between 8 and 11 μm in which little absorption of infrared radiation takes place.

Attenuation Any process in which the rate of flow of a beam of energy decreases (mainly due to absorption or scattering) with increasing distance from the energy source.

Aurora Glowing light display in the nighttime sky caused by excited gases in the upper atmosphere giving off light. In the Northern Hemisphere it is called the *aurora borealis* (northern lights); in the Southern Hemisphere, the *aurora australis* (southern lights).

Autumnal equinox The equinox at which the sun approaches the Southern Hemisphere and passes directly over the equator. Occurs around September 23.

AWIPS Acronym for *Advanced Weather Interactive Processing System.* New computerized system that integrates and processes data received at a weather forecasting office from NEXRAD, ASOS, and analysis and guidance products prepared by NMC.

Back-door cold front A cold front moving south or southwest along the Atlantic seaboard of the United States.

Backing wind A wind that changes direction in a counterclockwise sense (e.g., north to northwest to west).

Ball lightning A rare form of lightning that may consist of a reddish, luminous ball of electricity or charged air.

Banner cloud A cloud extending downwind from an isolated mountain peak, often on an otherwise cloud-free day.

Baroclinic (atmosphere) The state of the atmosphere where surfaces of constant pressure intersect surfaces of constant density. On an isobaric chart, isotherms cross the contour lines, and temperature advection exists.

Baroclinic instability A type of instability arising from a meridional (north to south) temperature gradient, a strong vertical wind speed shear, temperature advection, and divergence in the flow aloft. Many mid-latitude cyclones develop as a result of this instability.

Barograph A recording barometer.

Barometer An instrument that measures atmospheric pressure. The two most common barometers are the *mercury barometer* and the *aneroid barometer.*

Barotropic (atmosphere) A condition in the atmosphere where surfaces of constant density parallel surfaces of constant pressure.

Bead lightning Lightning that appears as a series of beads tied to a string.

Bergeron process *See* Ice-crystal process.

Bermuda high *See* Subtropical high.

Billow clouds Broad, nearly parallel lines of wavelike clouds oriented at right angles to the wind. Also called Kelvin–Helmholtz wave clouds.

Bimetallic thermometer A temperature-measuring device usually consisting of two dissimilar metals that expand and contract differentially as the temperature changes.

Blackbody A hypothetical object that absorbs all of the radiation that strikes it. It also emits radiation at a maximum rate for its given temperature.

Black ice A thin sheet of ice that appears relatively dark and may form as supercooled droplets, drizzle, or light rain come in contact with a road surface that is below freezing. Also, thin dark-appearing ice that forms on freshwater or saltwater ponds, or lakes.

Blizzard A severe weather condition characterized by low temperatures and strong winds (greater than 35 mi/hr) bearing a great amount of snow either falling or blowing. When these conditions continue after the falling snow has ended, it is termed a *ground blizzard.*

Boulder winds Fast-flowing, local downslope winds that may attain speeds of 100 knots or more. They are especially strong along the eastern foothills of the Rocky Mountains near Boulder, Colorado.

Boundary layer *See* Atmospheric boundary layer.

Bow echo A line of thunderstorms on a radar screen that appears in the shape of a bow. Bow echoes are often associated with damaging straight-line winds and small tornadoes.

Brocken bow A bright ring of light seen around the shadow of an observer's head as the observer peers into a cloud or fog bank. Formed by *diffraction* of light.

Buoyant force (buoyancy) The upward force exerted upon an air parcel (or any object) by virtue of the density (mainly temperature) difference between the parcel and that of the surrounding air.

Buys-Ballot's law A law describing the relationship between the wind direction and the pressure distribution. In the Northern Hemisphere, if you stand with your back to the surface wind, then turn clockwise about 30°, lower pressure will be to your left. In the Southern Hemisphere, stand with your back to the surface wind, then turn counterclockwise about 30°; lower pressure will be to your right.

California current   The ocean current that flows southward along the west coast of the United States from about Washington to Baja, California.

California norther   A strong, dry, northerly wind that blows in late spring, summer, and early fall in northern and central California. Its warmth and dryness are due to downslope compressional heating.

Cap cloud   See Pileus cloud.

Carbon dioxide   (CO_2) A colorless, odorless gas whose concentration is about 0.038 percent (385 ppm) in a volume of air near sea level. It is a selective absorber of infrared radiation and, consequently, it is important in the earth's atmospheric greenhouse effect. Solid CO_2 is called dry ice.

Carbon monoxide (CO)   A colorless, odorless, toxic gas that forms during the incomplete combustion of carbon-containing fuels.

Ceiling   The height of the lowest layer of clouds when the weather reports describe the sky as broken or overcast.

Ceiling balloon   A small balloon used to determine the height of the cloud base. The height is computed from the balloon's ascent rate and the time required for its disappearance into the cloud.

Ceilometer   An instrument that automatically records cloud height.

Celsius scale   A temperature scale where zero is assigned to the temperature where water freezes and 100 to the temperature where water boils (at sea level).

Centripetal acceleration   The inward-directed acceleration on a particle moving in a curved path.

Centripetal force   The radial force required to keep an object moving in a circular path. It is directed toward the center of that curved path.

Chaos   The property describing a system that exhibits erratic behavior in that very small changes in the initial state of the system rapidly lead to large and apparently unpredictable changes sometime in the future.

Chinook wall cloud   A bank of clouds over the Rocky Mountains that signifies the approach of a chinook.

Chinook wind   A warm, dry wind on the eastern side of the Rocky Mountains. In the Alps, the wind is called a Foehn.

Chlorofluorocarbons (CFCs)   Compounds consisting of methane (CH_4) or ethane (C_2H_6) with some or all of the hydrogen replaced by chlorine or fluorine. Used in fire extinguishers, as refrigerants, as solvents for cleaning electronic microcircuits, and as propellants. CFCs contribute to the atmospheric greenhouse effect and destroy ozone in the stratosphere.

Cirrocumulus   A high cloud that appears as a white patch of clouds without shadows. It consists of very small elements in the form of grains or ripples.

Cirrostratus   High, thin, sheetlike clouds, composed of ice crystals. They frequently cover the entire sky and often produce a halo.

Cirrus   A high cloud composed of ice crystals in the form of thin, white, featherlike clouds in patches, filaments, or narrow bands.

Clear air turbulence (CAT)   Turbulence encountered by aircraft flying through cloudless skies. Thermals, wind shear, and jet streams can each be a factor in producing CAT.

Clear ice   A layer of ice that appears transparent because of its homogeneous structure and small number and size of air pockets.

Climate   The accumulation of daily and seasonal weather events over a long period of time.

Climatic controls   The relatively permanent factors that govern the general nature of the climate of a region.

Climatic optimum   See Mid-Holocene maximum.

Climatological forecast   A weather forecast, usually a month or more in the future, which is based upon the climate of a region rather than upon current weather conditions.

Cloud   A visible aggregate of tiny water droplets and/or ice crystals in the atmosphere above the earth's surface.

Cloudburst   Any sudden and heavy rain shower.

Cloud seeding   The introduction of artificial substances (usually silver iodide or dry ice) into a cloud for the purpose of either modifying its development or increasing its precipitation.

Cloud streets   Lines or rows of cumuliform clouds.

Coalescence   The merging of cloud droplets into a single larger droplet.

Cold advection (cold air advection)   The transport of cold air by the wind from a region of lower temperatures to a region of higher temperatures.

Cold air damming   A shallow layer of cold air that is trapped between the Atlantic coast and the Appalachian Mountains.

Cold fog   See Supercooled cloud.

Cold front   A transition zone where a cold air mass advances and replaces a warm air mass.

Cold occlusion   See Occluded front.

Cold wave   A rapid fall in temperature within 24 hours that often requires increased protection for agriculture, industry, commerce, and human activities.

Collision-coalescence process   The process of producing precipitation by liquid particles (cloud droplets and raindrops) colliding and joining (coalescing).

Comma cloud   A band of organized cumuliform clouds that looks like a comma on a satellite photograph.

Computer enhancement   A process where the temperatures of radiating surfaces are assigned different shades of gray (or different colors) on an infrared picture. This allows specific features to be more clearly delineated.

Condensation   The process by which water vapor becomes a liquid.

Condensation level   The level above the surface marking the base of a cumuliform cloud.

Condensation nuclei   Also called cloud condensation nuclei. Tiny particles upon whose surfaces condensation of water vapor begins in the atmosphere. Small nuclei less than 0.2 μm in radius are called Aitken nuclei; those with radii between 0.2 and 1 μm are large nuclei, while giant nuclei have radii larger than 1 μm.

Conditionally unstable atmosphere   An atmospheric condition that exists when the environmental lapse rate is less than the dry adiabatic rate but greater than the moist adiabatic rate. Also called conditional instability.

Conduction   The transfer of heat by molecular activity from one substance to another, or through a substance. Transfer is always from warmer to colder regions.

Constant height chart (constant-level chart)   A chart showing variables, such as pressure, temperature, and wind, at a specific altitude above sea level. Variation in horizontal pressure is depicted by isobars. The most common constant-height chart is the surface chart, which is also called the sea-level chart or surface weather map.

Constant pressure chart (isobaric chart)   A chart showing variables, such as temperature and wind, on a constant pressure surface. Variations in height are usually shown by lines of equal height (contour lines).

Contact freezing The process by which contact with a nucleus such as an ice crystal causes supercooled liquid droplets to change into ice.

Continental arctic air mass An air mass characterized by extremely low temperatures and very dry air.

Continental polar air mass An air mass characterized by low temperatures and dry air. Not as cold as arctic air masses.

Continental tropical air mass An air mass characterized by high temperatures and low humidity.

Contour line A line that connects points of equal elevation above a reference level, most often sea level.

Contrail (condensation trail) A cloudlike streamer frequently seen forming behind aircraft flying in clear, cold, humid air.

Controls of temperature The main factors that cause variations in temperature from one place to another.

Convection Motions in a fluid that result in the transport and mixing of the fluid's properties. In meteorology, convection usually refers to atmospheric motions that are predominantly vertical, such as rising air currents due to surface heating. The rising of heated surface air and the sinking of cooler air aloft is often called *free convection*. (Compare with *forced convection*.)

Convective instability Instability arising in the atmosphere when a column of air exhibits warm, moist, nearly saturated air near the surface and cold, dry air aloft. When the lower part of the layer is lifted and saturation occurs, it becomes unstable.

Convergence An atmospheric condition that exists when the winds cause a horizontal net inflow of air into a specified region.

Conveyor belt model (for mid-latitude storms) A three-dimensional picture of a mid-latitude cyclone and the various air streams (called conveyor belts) that interact to produce the weather associated with the storm.

Cooling degree-day A form of degree-day used in estimating the amount of energy necessary to reduce the effective temperature of warm air. A cooling degree-day is a day on which the average temperature is one degree above a desired base temperature.

Coriolis force effects An apparent force observed on any free-moving object in a rotating system. On the earth, this deflective force results from the earth's rotation and causes moving particles (including the wind) to deflect to the right in the Northern Hemisphere and to the left in the Southern Hemisphere.

Corona (optic) A series of colored rings concentrically surrounding the disk of the sun or moon. Smaller than the halo, the corona is often caused by the diffraction of light around small water droplets of uniform size.

Country breeze A light breeze that blows into a city from the surrounding countryside. It is best observed on clear nights when the urban heat island is most pronounced.

Crepuscular rays Alternating light and dark bands of light that appear to fan out from the sun's position, usually at twilight.

Cumulonimbus An exceptionally dense and vertically developed cloud, often with a top in the shape of an anvil. The cloud is frequently accompanied by heavy showers, lightning, thunder, and sometimes hail. It is also known as a *thunderstorm cloud*.

Cumulus A cloud in the form of individual, detached domes or towers that are usually dense and well defined. It has a flat base with a bulging upper part that often resembles cauliflower. Cumulus clouds of fair weather are called *cumulus humilis*. Those that exhibit much vertical growth are called *cumulus congestus* or *towering cumulus*.

Cumulus stage The initial stage in the development of an ordinary cell thunderstorm in which rising, warm, humid air develops into a cumulus cloud.

Curvature effect In cloud physics, as cloud droplets decrease in size, they exhibit a greater surface curvature that causes a more rapid rate of evaporation.

Cut-off low A cold upper-level low that has become displaced out of the basic westerly flow and lies to the south of this flow.

Cyclogenesis The development or strengthening of middle-latitude (extratropical) cyclones.

Cyclone An area of low pressure around which the winds blow counterclockwise in the Northern Hemisphere and clockwise in the Southern Hemisphere.

Daily range of temperature The difference between the maximum and minimun temperatures for any given day.

Dart leader The discharge of electrons that proceeds intermittently toward the ground along the same ionized channel taken by the initial lightning stroke.

Dendrochronology The analysis of the annual growth rings of trees as a means of interpreting past climatic conditions.

Density The ratio of the mass of a substance to the volume occupied by it. Air density is usually expressed as g/cm^3 or kg/m^3.

Deposition A process that occurs in subfreezing air when water vapor changes directly to ice without becoming a liquid first.

Deposition nuclei Tiny particles (ice nuclei) upon which an ice crystal may grow by the process of deposition.

Derecho Strong, damaging, straight-line winds associated with a cluster of severe thunderstorms that most often form in the evening or at night.

Desertification A general increase in the desert conditions of a region.

Desert pavement An arrangement of pebbles and large stones that remains behind as finer dust and sand particles are blown away by the wind.

Dew Water that has condensed onto objects near the ground when their temperatures have fallen below the dew point of the surface air.

Dew cell An instrument used to determine the dew-point temperature.

Dew point (dew-point temperature) The temperature to which air must be cooled (at constant pressure and constant water vapor content) for saturation to occur.

Dew point hygrometer An instrument that determines the dew-point temperature of the air.

Diffraction The bending of light around objects, such as cloud and fog droplets, producing fringes of light and dark or colored bands.

Dispersion The separation of white light into its different component wavelengths.

Dissipating stage The final stage in the development of an ordinary cell thunderstorm when downdrafts exist throughout the cumulonimbus cloud.

Divergence An atmospheric condition that exists when the winds cause a horizontal net outflow of air from a specific region.

Doldrums The region near the equator that is characterized by low pressure and light, shifting winds.

Doppler lidar The use of light beams to determine the velocity of objects such as dust and falling rain by taking into account the *Doppler shift*.

Doppler radar A radar that determines the velocity of falling precipitation either toward or away from the radar unit by taking into account the *Doppler shift*.

Doppler shift (effect) The change in the frequency of waves that occurs when the emitter or the observer is moving toward or away from the other.

Downburst   A severe localized downdraft that can be experienced beneath a severe thunderstorm. (Compare *Microburst* and *Macroburst*.)

Drizzle   Small water drops between 0.2 and 0.5 mm in diameter that fall slowly and reduce visibility more than light rain.

Drought   A period of abnormally dry weather sufficiently long enough to cause serious effects on agriculture and other activities in the affected area.

Dry adiabatic rate   The rate of change of temperature in a rising or descending unsaturated air parcel. The rate of adiabatic cooling or warming is about 10°C per 1000 m (5.5°F per 1000 ft).

Dry adiabats   Lines on an adiabatic chart that show the dry adiabatic rate for rising or descending air. They represent lines of constant potential temperature.

Dry-bulb temperature   The air temperature measured by the dry-bulb thermometer of a psychrometer.

Dry climate   A climate deficient in precipitation where annual potential evaporation and transpiration exceed precipitation.

Dry haze   *See* Haze.

Dry lightning   Lightning that occurs with thunderstorms that produce little, if any, appreciable precipitation that reaches the surface.

Dryline   A boundary that separates warm, dry air from warm, moist air. It usually represents a zone of instability along which thunderstorms form.

Dry slot   On a satellite image the dry slot represents the relatively clear region (or clear wedge) that appears just to the west of the tail of a comma cloud of a mid-latitude cyclonic storm.

Dry-summer subtropical climate   A climate characterized by mild, wet winters and warm to hot, dry summers. Typically located between 30 and 45 degrees latitude on the western side of continents. Also called *Mediterranean climate*.

Dust devil (whirlwind)   A small but rapidly rotating wind made visible by the dust, sand, and debris it picks up from the surface. It develops best on clear, dry, hot afternoons.

Earth vorticity   The rotation (spin) of an object about its vertical axis brought on by the rotation of the earth on its axis. The earth's vorticity is a maximum at the poles and zero at the equator.

Easterly wave   A migratory wavelike disturbance in the tropical easterlies. Easterly waves occasionally intensify into tropical cyclones. They are also called *tropical waves*.

Eccentricity (of the earth's orbit)   The deviation of the earth's orbit from elliptical to nearly circular.

Eddy   A small volume of air (or any fluid) that behaves differently from the larger flow in which it exists.

Eddy viscosity   The internal friction produced by turbulent flow.

Ekman spiral   An idealized description of the way the wind-driven ocean currents vary with depth. In the atmosphere it represents the way the winds vary from the surface up through the friction layer or planetary boundary layer.

Ekman transport   Net surface water transport due to the Ekman spiral. In the Northern Hemisphere the transport is 90° to the right of the surface wind direction.

Electrical hygrometer   *See* Hygrometer.

Electrical thermometers   Thermometers that use elements that convert energy from one form to another (transducers). Common electrical thermometers include the electrical resistance thermometer, thermocouple, and thermistor.

Electromagnetic waves   *See* Radiant energy.

El Niño   An extensive ocean warming that begins along the coast of Peru and Ecuador and extends westward over the Tropical Pacific. Major El Niño events, or strong El Niños, occur once every 2 to 7 years as a current of nutrient-poor tropical water moves southward along the west coast of South America. (*See also* ENSO.)

Embryo   In cloud physics, a tiny ice crystal that grows in size and becomes an ice nucleus.

Energy   The property of a system that generally enables it to do work. Some forms of energy are kinetic, radiant, potential, chemical, electric, and magnetic.

Enhanced Fujita (EF) scale   A modification of the original *Fujita Scale* that describes tornado intensity by observing damage caused by the tornado.

Ensemble forecasting   A forecasting technique that entails running several forecast models (or different versions of a single model), each beginning with slightly different weather information. The forecaster's level of confidence is based on how well the models agree (or disagree) at the end of some specified time.

ENSO (El Niño/Southern Oscillation)   A condition in the tropical Pacific whereby the reversal of surface air pressure at opposite ends of the Pacific Ocean induces westerly winds, a strengthening of the equatorial countercurrent, and extensive ocean warming.

Entrainment   The mixing of environmental air into a pre-existing air current or cloud so that the environmental air becomes part of the current or cloud.

Environmental lapse rate   The rate of decrease of air temperature with elevation. It is most often measured with a radiosonde.

Equilibrium vapor pressure   The necessary vapor pressure around liquid water that allows the water to remain in equilibrium with its environment. Also called *saturation vapor pressure*.

Evaporation   The process by which a liquid changes into a gas.

Evaporation (mixing) fog   Fog produced when sufficient water vapor is added to the air by evaporation, and the moist air mixes with relatively drier air. The two common types are *steam fog*, which forms when cold air moves over warm water, and *frontal fog*, which forms as warm raindrops evaporate in a cool air mass.

Exosphere   The outermost portion of the atmosphere.

Extratropical cyclone   A cyclonic storm that most often forms along a front in middle and high latitudes. Also called a *middle-latitude cyclonic storm*, a *depression*, and a *low*. It is not a tropical storm or hurricane.

Eye   A region in the center of a hurricane (tropical storm) where the winds are light and skies are clear to partly cloudy.

Eyewall   A wall of dense thunderstorms that surrounds the eye of a hurricane.

Eyewall replacement   A situation within a hurricane (tropical cyclone) where the storm's original eyewall dissipates and a new eyewall forms outward, farther away from the center of the storm.

Fahrenheit scale   A temperature scale where 32 is assigned to the temperature where water freezes and 212 to the temperature where water boils (at sea level).

Fall streaks   Falling ice crystals that evaporate before reaching the ground.

Fall wind   A strong, cold katabatic wind that blows downslope off snow-covered plateaus.

Fata Morgana   A complex mirage that is characterized by objects being distorted in such a way as to appear as castlelike features.

Feedback mechanism   A process whereby an initial change in an atmospheric process will tend to either reinforce the process (*positive feedback*) or weaken the process (*negative feedback*).

Ferrel cell   The name given to the middle-latitude cell in the 3-cell model of the general circulation.

Fetch The distance that the wind travels over open water.

Flash flood A flood that rises and falls quite rapidly with little or no advance warning, usually as the result of intense rainfall over a relatively small area.

Flurries of snow *See* Snow flurries.

Foehn *See* Chinook wind.

Fog A cloud with its base at the earth's surface.

Forced convection On a small scale, a form of mechanical stirring taking place when twisting eddies of air are able to mix hot surface air with the cooler air above. On a larger scale, it can be induced by the lifting of warm air along a front (*frontal uplift*) or along a topographic barrier (*orographic uplift*).

Forked lightning Cloud-to-ground lightning that exhibits downward-directed crooked branches.

Formaldehyde A colorless gaseous compound (HCHO) used in the manufacture of resins, fertilizers, and dyes. Also used as an embalming fluid, a preservative, and a disinfectant.

Free convection *See* Convection.

Freeze A condition occurring over a widespread area when the surface air temperature remains below freezing for a sufficient time to damage certain agricultural crops. A freeze most often occurs as cold air is advected into a region, causing freezing conditions to exist in a deep layer of surface air. Also called *advection frost*.

Freezing nuclei Particles that promote the freezing of supercooled liquid droplets.

Freezing rain and **freezing drizzle** Rain or drizzle that falls in liquid form and then freezes upon striking a cold object or ground. Both can produce a coating of ice on objects which is called *glaze*.

Friction layer The atmospheric layer near the surface usually extending up to about 1 km (3300 ft) where the wind is influenced by friction of the earth's surface and objects on it. Also called the *atmospheric boundary layer* and *planetary boundary layer*.

Front The transition zone between two distinct air masses.

Frontal fog *See* Evaporation fog.

Frontal inversion A temperature inversion encountered upon ascending through a sloping front, usually a warm front.

Frontal thunderstorms Thunderstorms that form in response to forced convection (forced lifting) along a front. Most go through a cycle similar to those of ordinary thunderstorms.

Frontal wave A wavelike deformation along a front in the lower levels of the atmosphere. Those that develop into storms are termed *unstable waves*, while those that do not are called *stable waves*.

Frontogenesis The formation, strengthening, or regeneration of a front.

Frontolysis The weakening or dissipation of a front.

Frost (also called **hoarfrost**) A covering of ice produced by deposition on exposed surfaces when the air temperature falls below the frost point.

Frostbite The partial freezing of exposed parts of the body, causing injury to the skin and sometimes to deeper tissues.

Frost point The temperature at which the air becomes saturated with respect to ice when cooled at constant pressure and constant water vapor content.

Frozen dew The transformation of liquid dew into tiny beads of ice when the air temperature drops below freezing.

Fujita scale A scale developed by T. Theodore Fujita for classifying tornadoes according to the damage they cause and their rotational wind speed. (*See also* Enhanced Fujita Scale.)

Fulgurite A rootlike tube (or several tubes) that forms when a lightning stroke fuses sand particles together.

Funnel cloud A funnel-shaped cloud of condensed water, usually extending from the base of a cumuliform cloud. The rapidly rotating air of the funnel is not in contact with the ground; hence, it is not a tornado.

Galaxy A huge assembly of stars (between millions and hundreds of millions) held together by gravity.

Gas law The thermodynamic law applied to a perfect gas that relates the pressure of the gas to its density and absolute temperature.

General circulation of the atmosphere Large-scale atmospheric motions over the entire earth.

Geostationary satellite A satellite that orbits the earth at the same rate that the earth rotates and thus remains over a fixed place above the equator.

Geostrophic wind A theoretical horizontal wind blowing in a straight path, parallel to the isobars or contours, at a constant speed. The geostrophic wind results when the Coriolis force exactly balances the horizontal pressure gradient force.

Giant nuclei *See* Condensation nuclei.

Glaciated cloud A cloud or portion of a cloud where only ice crystals exist.

Glaze A coating of ice, often clear and smooth, that forms on exposed surfaces by the freezing of a film of supercooled water deposited by rain, drizzle, or fog. As a type of aircraft icing, glaze is called *clear ice*.

Global climate Climate of the entire globe.

Global scale The largest scale of atmospheric motion. Also called the *planetary scale*.

Global warming Increasing global surface air temperatures that show up in the climate record. The term *global warming* is usually attributed to human activities, such as increasing concentrations of greenhouse gases from automobiles and industrial processes, for example.

Glory Colored rings that appear around the shadow of an object.

Gradient wind A theoretical wind that blows parallel to curved isobars or contours.

Graupel Ice particles between 2 and 5 mm in diameter that form in a cloud often by the process of accretion. Snowflakes that become rounded pellets due to riming are called *graupel* or *snow pellets*.

Green flash A small green color that occasionally appears on the upper part of the sun as it rises or sets.

Greenhouse effect *See* Atmospheric greenhouse effect.

Ground fog *See* Radiation fog.

Growing degree-day A form of the degree-day used as a guide for crop planting and for estimating crop maturity dates.

Gulf stream A warm, swift, narrow ocean current flowing along the east coast of the United States.

Gust front A boundary that separates a cold downdraft of a thunderstorm from warm, humid surface air. On the surface its passage resembles that of a cold front.

Gustnado A relatively weak tornado associated with a thunderstorm's outflow. It most often forms along the gust front.

Gyre A large circular, surface ocean current pattern.

Haboob A dust or sandstorm that forms as cold downdrafts from a thunderstorm turbulently lift dust and sand into the air.

Hadley cell A thermal circulation proposed by George Hadley to explain the movement of the trade winds. It consists of rising air near the equator and sinking air near 30° latitude.

Hailstones Transparent or partially opaque particles of ice that range in size from that of a pea to that of golf balls.

Hailstreak   The accumulation of hail at the earth's surface along a relatively long (10 km), narrow (2 km) band.

Hair hygrometer   *See* Hygrometer.

Halons   A group of organic compounds used as fire retardants. In the stratosphere, these compounds release bromine atoms that rapidly destroy ozone. The production of halons is now banned by the Montreal Protocol.

Halos   Rings or arcs that encircle the sun or moon when seen through an ice crystal cloud or a sky filled with falling ice crystals. Halos are produced by refraction of light.

Haze   Fine dry or wet dust or salt particles dispersed through a portion of the atmosphere. Individually these are not visible but cumulatively they will diminish visibility. *Dry haze* particles are very small, on the order of 0.1 μm. *Wet haze* particles are larger.

Heat   A form of energy transferred between systems by virtue of their temperature differences.

Heat burst   A sudden increase in surface air temperature often accompanied by extreme drying. A heat burst is associated with the downdraft of a thunderstorm, or a cluster of thunderstorms.

Heat capacity   The ratio of the heat absorbed (or released) by a system to the corresponding temperature rise (or fall).

Heat Index (HI)   An index that combines air temperature and relative humidity to determine an apparent temperature—how hot it actually feels.

Heating degree-day   A form of the degree-day used as an index for fuel consumption.

Heat lightning   Distant lightning that illuminates the sky but is too far away for its thunder to be heard.

Heatstroke   A physical condition induced by a person's overexposure to high air temperatures, especially when accompanied by high humidity.

Hectopascal   Abbreviated hPa. One hectopascal is equal to 100 Newtons/m^2, or 1 millibar.

Heiligenschein   A faint white ring surrounding the shadow of an observer's head on a dew-covered lawn.

Heterosphere   The region of the atmosphere above about 85 km where the composition of the air varies with height.

High   *See* Anticyclone.

High inversion fog   A fog that lifts above the surface but does not completely dissipate because of a strong inversion (usually subsidence) that exists above the fog layer.

Homogeneous (spontaneous) freezing   The freezing of pure water. For tiny cloud droplets, homogeneous freezing does not occur until the air temperature reaches about −40°C.

Homosphere   The region of the atmosphere below about 85 km where the composition of the air remains fairly constant.

Hook echo   The shape of an echo on a Doppler radar screen that indicates the possible presence of a tornado.

Horse latitudes   The belt of latitude at about 30° to 35° where winds are predominantly light and the weather is hot and dry.

Humid continental climate   A climate characterized by severe winters and mild to warm summers with adequate annual precipitation. Typically located over large continental areas in the Northern Hemisphere between about 40° and 70° latitude.

Humidity   A general term that refers to the air's water vapor content. (*See* Relative humidity.)

Humid subtropical climate   A climate characterized by hot muggy summers, cool to cold winters, and abundant precipitation throughout the year.

Hurricane   A tropical cyclone having winds in excess of 64 knots (74 mi/hr).

Hurricane warning   A warning given when it is likely that a hurricane will strike an area within 24 hours.

Hurricane watch   A hurricane watch indicates that a hurricane poses a threat to an area (often within several days) and residents of the watch area should be prepared.

Hydrocarbons   Chemical compounds composed of only hydrogen and carbon—they are included under the general term volatile organic compounds (VOCs).

Hydrologic cycle   A model that illustrates the movement and exchange of water among the earth, atmosphere, and oceans.

Hydrophobic   The ability to resist the condensation of water vapor. Usually used to describe "water-repelling" condensation nuclei.

Hydrostatic equation   An equation that states that the rate at which the air pressure decreases with height is equal to the air density times the acceleration of gravity. The equation relates to how quickly the air pressure decreases in a column of air.

Hydrostatic equilibrium   The state of the atmosphere when there is a balance between the vertical pressure gradient force and the downward pull of gravity.

Hygrometer   An instrument designed to measure the air's water vapor content. The sensing part of the instrument can be hair (*hair hygrometer*), a plate coated with carbon (*electrical hygrometer*), or an infrared sensor (*infrared hygrometer*).

Hygroscopic   The ability to accelerate the condensation of water vapor. Usually used to describe condensation nuclei that have an affinity for water vapor.

Hypothermia   The deterioration in one's mental and physical condition brought on by a rapid lowering of human body temperature.

Hypoxia   A condition experienced by humans when the brain does not receive sufficient oxygen.

Ice Age   *See* Pleistocene epoch.

Ice-crystal (Bergeron) process   A process that produces precipitation. The process involves tiny ice crystals in a supercooled cloud growing larger at the expense of the surrounding liquid droplets. Also called the *Bergeron process*.

Ice fog   A type of fog that forms at very low temperatures, composed of tiny suspended ice particles.

Icelandic low   The subpolar low-pressure area that is centered near Iceland on charts that show mean sea-level pressure.

Ice nuclei   Particles that act as nuclei for the formation of ice crystals in the atmosphere.

Ice pellets   *See* Sleet.

Indian summer   An unseasonably warm spell with clear skies near the middle of autumn. Usually follows a substantial period of cool weather.

Inferior mirage   *See* Mirage.

Infrared hygrometer   *See* Hygrometer.

Infrared radiation   Electromagnetic radiation with wavelengths between about 0.7 and 1000 μm. This radiation is longer than visible radiation but shorter than microwave radiation.

Infrared radiometer   An instrument designed to measure the intensity of infrared radiation emitted by an object. Also called *infrared sensor*.

Insolation   The *in*coming *sol*ar radi*ation* that reaches the earth and the atmosphere.

Instrument shelter   A boxlike (often wooden) structure designed to protect weather instruments from direct sunshine and precipitation.

Interglacial period   A time interval of relatively mild climate during the Ice Age when continental ice sheets were absent or limited in extent to Greenland and the Antarctic.

Intertropical convergence zone (ITCZ) The boundary zone separating the northeast trade winds of the Northern Hemisphere from the southeast trade winds of the Southern Hemisphere.

Inversion An increase in air temperature with height.

Ion An electrically charged atom, molecule, or particle.

Ionosphere An electrified region of the upper atmosphere where fairly large concentrations of ions and free electrons exist.

Iridescence Brilliant spots or borders of colors, most often red and green, observed in clouds up to about 30° from the sun.

Isallobar A line of equal change in atmospheric pressure during a specified time interval.

Isobar A line connecting points of equal pressure.

Isobaric chart (map) *See* Constant pressure chart.

Isobaric surface A surface along which the atmospheric pressure is everywhere equal.

Isotach A line connecting points of equal wind speed.

Isotherm A line connecting points of equal temperature.

Isothermal layer A layer where the air temperature is constant with increasing altitude. In an isothermal layer, the air temperature lapse rate is zero.

Jet maximum *See* Jet streak.

Jet streak A region of high wind speed that moves through the axis of a jet stream. Also called *jet maximum*.

Jet stream Relatively strong winds concentrated within a narrow band in the atmosphere.

Katabatic (fall) wind Any wind blowing downslope. It is usually cold.

Kelvin A unit of temperature. A Kelvin is denoted by K and 1 K equals 1°C. Zero Kelvin is absolute zero, or -273.15°C.

Kelvin scale A temperature scale with zero degrees equal to the theoretical temperature at which all molecular motion ceases. Also called the *absolute scale*. The units are sometimes called "degrees Kelvin"; however, the correct SI terminology is "Kelvins," abbreviated K.

Kinetic energy The energy within a body that is a result of its motion.

Kirchhoff's law A law that states: Good absorbers of a given wavelength of radiation are also good emitters of that wavelength.

Knot A unit of speed equal to 1 nautical mile per hour. One knot equals 1.15 mi/hr.

Köppen classification system A system for classifying climates developed by W. Köppen that is based mainly on annual and monthly averages of temperature and precipitation.

Lake breeze A wind blowing onshore from the surface of a lake.

Lake-effect snows Localized snowstorms that form on the downwind side of a lake. Such storms are common in late fall and early winter near the Great Lakes as cold, dry air picks up moisture and warmth from the unfrozen bodies of water.

Laminar flow A nonturbulent flow in which the fluid moves smoothly in parallel layers or sheets.

Land breeze A coastal breeze that blows from land to sea, usually at night.

Landspout Relatively weak nonsupercell tornado that originates with a cumiliform cloud in its growth stage and with a cloud that does not contain a mid-level mesocyclone. Its spin originates near the surface. Landspouts often look like waterspouts over land.

La Niña A condition where the central and eastern tropical Pacific Ocean turns cooler than normal.

Lapse rate The rate at which an atmospheric variable (usually temperature) decreases with height. (*See* Environmental lapse rate.)

Latent heat The heat that is either released or absorbed by a unit mass of a substance when it undergoes a change of state, such as during evaporation, condensation, or sublimation.

Laterite A soil formed under tropical conditions where heavy rainfall leaches soluble minerals from the soil. This leaching leaves the soil hard and poor for growing crops.

Lee-side low Storm systems (extratropical cyclones) that form on the downwind (lee) side of a mountain chain. In the United States lee-side lows frequently form on the eastern side of the Rockies and Sierra Nevada mountains.

Lenticular cloud A cloud in the shape of a lens.

Level of free convection The level in the atmosphere at which a lifted air parcel becomes warmer than its surroundings in a conditionally unstable atmosphere.

Lidar An instrument that uses a laser to generate intense pulses that are reflected from atmospheric particles of dust and smoke. Lidars have been used to determine the amount of particles in the atmosphere as well as particle movement that has been converted into wind speed. Lidar means *li*ght *d*etection *a*nd *r*anging.

Lifting condensation level (LCL) The level at which a parcel of air, when lifted dry adiabatically, would become saturated.

Lightning A visible electrical discharge produced by thunderstorms.

Liquid-in-glass thermometer *See* Thermometer.

Little Ice Age The period from about 1550 to 1850 when average temperatures over Europe were lower, and alpine glaciers increased in size and advanced down mountain canyons.

Local winds Winds that tend to blow over a relatively small area; often due to regional effects, such as mountain barriers, large bodies of water, local pressure differences, and other influences.

Long-range forecast Generally used to describe a weather forecast that extends beyond about 8.5 days into the future.

Longwave radiation A term most often used to describe the infrared energy emitted by the earth and the atmosphere.

Longwaves in the westerlies A wave in the upper level of westerlies characterized by a long length (thousands of kilometers) and significant amplitude. Also called *Rossby waves*.

Low *See* Extratropical cyclone.

Low-level jet streams Jet streams that typically form near the earth's surface below an altitude of about 2 km and usually attain speeds of less than 60 knots.

Macroburst A strong downdraft (*downburst*) greater than 4 km wide that can occur beneath thunderstorms. A downburst less than 4 km across is called a *microburst*.

Macroclimate The general climate of a large area, such as a country.

Macroscale The normal meteorological synoptic scale for obtaining weather information. It can cover an area ranging from the size of a continent to the entire globe.

Magnetic storm A worldwide disturbance of the earth's magnetic field caused by solar disturbances.

Magnetosphere The region around the earth in which the earth's magnetic field plays a dominant part in controlling the physical processes that take place.

Mammatus clouds Clouds that look like pouches hanging from the underside of a cloud.

Marine climate A climate controlled largely by the ocean. The ocean's influence keeps winters relatively mild and summers cool.

Maritime air   Moist air whose characteristics were developed over an extensive body of water.

Maritime polar air mass   An air mass characterized by low temperatures and high humidity.

Maritime tropical air mass   An air mass characterized by high temperatures and high humidity.

Mature thunderstorm   The second stage in the three-stage cycle of an ordinary thunderstorm. This mature stage is characterized by heavy showers, lightning, thunder, and violent vertical motions inside cumulonimbus clouds.

Maunder minimum   A period from about 1645 to 1715 when few, if any, sunspots were observed.

Maximum thermometer   A thermometer with a small constriction just above the bulb. It is designed to measure the maximum air temperature.

Mean annual temperature   The average temperature at any given location for the entire year.

Mean daily temperature   The average of the highest and lowest temperature for a 24-hour period.

Mechanical turbulence   Turbulent eddy motions caused by obstructions, such as trees, buildings, mountains, and so on.

Mediterranean climate   *See* Dry-summer subtropical climate.

Medium-range forecast   Generally used to describe a weather forecast that extends from about 3 to 8.5 days into the future.

Mercury barometer   A type of barometer that uses mercury to measure atmospheric pressure. The height of the mercury column is a measure of atmospheric pressure.

Meridional flow   A type of atmospheric circulation pattern in which the north-south component of the wind is pronounced.

Mesoclimate   The climate of an area ranging in size from a few acres to several square kilometers.

Mesocyclone   A vertical column of cyclonically rotating air within a supercell thunderstorm.

Mesohigh   A relatively small area of high atmospheric pressure that forms beneath a thunderstorm.

Mesopause   The top of the mesosphere. The boundary between the mesosphere and the thermosphere, usually near 85 km.

Mesoscale   The scale of meteorological phenomena that range in size from a few km to about 100 km. It includes local winds, thunderstorms, and tornadoes.

Mesoscale Convective Complex (MCC)   A large organized convective weather system comprised of a number of individual thunderstorms. The size of an MCC can be 1000 times larger than an individual ordinary cell thunderstorm.

Mesoscale convective system (MCS)   A large cloud system that represents an ensemble of thunderstorms that form by convection, and produce precipitation over a wide area.

Mesosphere   The atmospheric layer between the stratosphere and the thermosphere. Located at an average elevation between 50 and 80 km above the earth's surface.

Meteogram   A chart that shows how one or more weather variables has changed at a station over a given period of time or how the variables are likely to change with time.

Meteorology   The study of the atmosphere and atmospheric phenomena as well as the atmosphere's interaction with the earth's surface, oceans, and life in general.

Microburst   A strong localized downdraft (downburst) less than 4 km wide that occurs beneath thunderstorms. A strong downburst greater than 4 km across is called a *macroburst*.

Microclimate   The climate structure of the air space near the surface of the earth.

Micrometer (μm)   A unit of length equal to one-millionth of a meter.

Microscale   The smallest scale of atmospheric motions.

Middle latitudes   The region of the world typically described as being between 30° and 50° latitude.

Middle-latitude cyclone   *See* Extratropical cyclone.

Mid-Holocene maximum   A warm period in geologic history about 5000 to 6000 years ago that favored the development of plants.

Milankovitch theory   A theory proposed by Milutin Milankovitch in the 1930s suggesting that changes in the earth's orbit were responsible for variations in solar energy reaching the earth's surface and climatic changes.

Millibar (mb)   A unit for expressing atmospheric pressure. Sea-level pressure is normally close to 1013 mb.

Minimum thermometer   A thermometer designed to measure the minimum air temperature during a desired time period.

Mini-swirls   Small whirling eddies perhaps 30 to 100 m in diameter that form in a region of strong wind shear of a hurricane's eyewall. They are believed to be small tornadoes.

Mirage   A refraction phenomenon that makes an object appear to be displaced from its true position. When an object appears higher than it actually is, it is called a *superior mirage*. When an object appears lower than it actually is, it is an *inferior mirage*.

Mixed cloud   A cloud containing both water drops and ice crystals.

Mixing depth   The vertical extent of the mixing layer.

Mixing layer   The unstable atmospheric layer that extends from the surface up to the base of an inversion. Within this layer, the air is well stirred.

Mixing ratio   The ratio of the mass of water vapor in a given volume of air to the mass of dry air.

Moist adiabatic rate   The rate of change of temperature in a rising or descending saturated air parcel. The rate of cooling or warming varies but a common value of 6°C per 1000 m (3.3°F per 1000 ft) is used.

Moist adiabats   Lines on an adiabatic chart that show the moist adiabatic rate for rising and descending air.

Molecular viscosity   The small-scale internal fluid friction that is due to the random motion of the molecules within a smooth-flowing fluid, such as air.

Molecule   A collection of atoms held together by chemical forces.

Monsoon depressions   Weak low-pressure areas that tend to form in response to divergence in an upper-level jet stream. The circulation around the low strengthens the monsoon wind system and enhances precipitation during the summer.

Monsoon wind system   A wind system that reverses direction between winter and summer. Usually the wind blows from land to sea in winter and from sea to land in summer.

Mountain and **valley breeze**   A local wind system of a mountain valley that blows downhill (*mountain breeze*) at night and uphill (*valley breeze*) during the day.

Multicell thunderstorm   A convective storm system composed of a cluster of convective cells, each one in a different stage of its life cycle.

Nacreous clouds   Clouds of unknown composition that have a soft, pearly luster and that form at altitudes about 25 to 30 km above the earth's surface. They are also called *mother-of-pearl clouds*.

Negative feedback mechanism   *See* Feedback mechanism.

Neutral stability (neutrally stable atmosphere)   An atmospheric condition that exists in dry air when the environmental lapse rate equals the dry adiabatic rate. In saturated air the environmental lapse rate equals the moist adiabatic rate.

NEXRAD An acronym for *Next* Generation Weather *Rad*ar. The main component of NEXRAD is the WSR 88-D, Doppler radar.

Nimbostratus A dark, gray cloud characterized by more or less continuously falling precipitation. It is rarely accompanied by lightning, thunder, or hail.

Nitric oxide (NO) A colorless gas produced by natural bacterial action in soil and by combustion processes at high temperatures. In polluted air, nitric oxide can react with ozone and hydrocarbons to form other substances. In this manner, it acts as an agent in the production of photochemical smog.

Nitrogen (N_2) A colorless and odorless gas that occupies about 78 percent of dry air in the lower atmosphere.

Nitrogen dioxide (NO_2) A reddish-brown gas, produced by natural bacterial action in soil and by combustion processes at high temperatures. In the presence of sunlight, it breaks down into nitric oxide and atomic oxygen. In polluted air, nitrogen dioxide acts as an agent in the production of photochemical smog.

Nitrogen oxides (NO_X) Gases produced by natural processes and by combustion processes at high temperatures. In polluted air, nitric oxide (NO) and nitrogen dioxide (NO_2) are the most abundant oxides of nitrogen, and both act as agents for the production of photochemical smog.

Noctilucent clouds Wavy, thin, bluish-white clouds that are best seen at twilight in polar latitudes. They form at altitudes about 80 to 90 km above the surface.

Nocturnal inversion *See* Radiation inversion.

Nonsupercell tornado A tornado that occurs with a cloud that is often in its growing stage, and one that does not contain a mid-level mesocyclone, or wall cloud. Landspouts and gustnadoes are examples of nonsupercell tornadoes.

North Atlantic Oscillation (NAO) A reversal of atmospheric pressure over the Atlantic Ocean that influences the weather over Europe and over eastern North America.

Northeaster A name given to a strong, steady wind from the northeast that is accompanied by rain and inclement weather. It often develops when a storm system moves northeastward along the coast of North America. Also called Nor'easter.

Northern lights *See* Aurora.

Nowcasting Short-term weather forecasts varying from minutes up to a few hours.

Nuclear winter The dark, cold, and gloomy conditions that presumably would be brought on by nuclear war.

Nucleation Any process in which the phase change of a substance to a more condensed state (such as condensation, deposition, and freezing) is initiated about a particle (nucleus).

Numerical weather prediction (NWP) Forecasting the weather based upon the solutions of mathematical equations by high-speed computers.

Obliquity (of the earth's axis) The tilt of the earth's axis. It represents the angle from the perpendicular to the plane of the earth's orbit.

Occluded front (occlusion) A complex frontal system that ideally forms when a cold front overtakes a warm front. When the air behind the front is colder than the air ahead of it, the front is called a *cold occlusion*. When the air behind the front is milder than the air ahead of it, it is called a *warm occlusion*.

Ocean conveyor belt The global circulation of ocean water that is driven by the sinking of cold, dense water near Greenland and Labrador in the North Atlantic.

Ocean-effect snow Localized bands of snow that occur when relatively cold air flows over a warmer ocean.

Offshore wind A breeze that blows from the land out over the water. Opposite of an onshore wind.

Omega high A ridge in the middle or upper troposphere that has the shape of a Greek letter omega (Ω).

Onshore wind A breeze that blows from the water onto the land. Opposite of an offshore wind.

Open wave The stage of development of a wave cyclone (mid-latitude cyclonic storm) where a cold front and a warm front exist, but no occluded front. The center of lowest pressure in the wave is located at the junction of the two fronts.

Orchard heaters Oil heaters placed in orchards that generate heat and promote convective circulations to protect fruit trees from damaging low temperatures. Also called *smudge pots*.

Ordinary cell thunderstorm (also called *air-mass thunderstorm*) A thunderstorm produced by local convection within a conditionally unstable air mass. It often forms in a region of low wind shear and does not reach the intensity of a severe thunderstorm.

Orographic clouds Clouds produced by lifting along rising terrain, usually mountains.

Orographic uplift The lifting of air over a topographic barrier. Clouds that form in this lifting process are called *orographic clouds*.

Outflow boundary A surface boundary formed by the horizontal spreading of cool air that originated inside a thunderstorm.

Outgassing The release of gases dissolved in hot, molten rock.

Overrunning A condition that occurs when air moves up and over another layer of air.

Overshooting top A situation in a mature thunderstorm where rising air, associated with strong convection, penetrates into a stable layer (usually the stratosphere), forcing the upper part of the cloud to rise above its relatively flat anvil top.

Oxygen (O_2) A colorless and odorless gas that occupies about 21 percent of dry air in the lower atmosphere.

Ozone (O_3) An almost colorless gaseous form of oxygen with an odor similar to weak chlorine. The highest natural concentration is found in the stratosphere where it is known as *stratospheric ozone*. It also forms in polluted air near the surface where it is the main ingredient of photochemical smog. Here, it is called *tropospheric ozone*.

Ozone hole A sharp drop in stratospheric ozone concentration observed over the Antarctic during the spring.

Pacific Decadal Oscillation (PDO) A reversal in ocean surface temperatures that occurs every 20 to 30 years over the northern Pacific Ocean.

Pacific high *See* Subtropical high.

Parcel of air An imaginary small body of air a few meters wide that is used to explain the behavior of air.

Parhelia *See* Sundog.

Particulate matter Solid particles or liquid droplets that are small enough to remain suspended in the air. Also called *aerosols*.

Pattern recognition An analogue method of forecasting where the forecaster uses prior weather events (or similar weather map conditions) to make a forecast.

P/E index (precipitation-evaporation index) An index that gives the long-range effectiveness of precipitation in promoting plant growth.

P/E ratio (precipitation-evaporation ratio) An expression devised for the purpose of classifying climates; based on monthly totals of precipitation and evaporation.

Permafrost   A layer of soil beneath the earth's surface that remains frozen throughout the year.

Persistence forecast   A forecast that the future weather condition will be the same as the present condition.

Photochemical smog   *See* Smog.

Photodissociation   The splitting of a molecule by a photon.

Photon   A discrete quantity of energy that can be thought of as a packet of electromagnetic radiation traveling at the speed of light.

Photosphere   The visible surface of the sun from which most of its energy is emitted.

Pileus cloud   A smooth cloud in the form of a cap. Occurs above, or is attached to, the top of a cumuliform cloud. Also called a *cap cloud*.

Pilot balloon   A small balloon that rises at a constant rate and is tracked by a theodolite in order to obtain wind speed and wind direction at various levels above the earth's surface.

Planetary boundary layer   *See* Atmospheric boundary layer.

Planetary scale   The largest scale of atmospheric motion. Sometimes called the *global scale*.

Plasma   *See* Solar wind.

Plate tectonics   The theory that the earth's surface down to about 100 km is divided into a number of plates that move relative to one another across the surface of the earth. Once referred to as continental drift.

Pleistocene Epoch (or **Ice Age**)   The most recent period of extensive continental glaciation that saw large portions of North America and Europe covered with ice. It began about 2 million years ago and ended about 10,000 years ago.

Polar easterlies   A shallow body of easterly winds located at high latitudes poleward of the subpolar low.

Polar front   A semipermanent, semicontinuous front that separates tropical air masses from polar air masses.

Polar front jet stream (polar jet)   The jet stream that is associated with the polar front in middle and high latitudes. It is usually located at altitudes between 9 and 12 km.

Polar front theory   A theory developed by a group of Scandinavian meteorologists that explains the formation, development, and overall life history of cyclonic storms that form along the polar front.

Polar ice cap climate   A climate characterized by extreme cold, as every month has an average temperature below freezing.

Polar low   An area of low pressure that forms over polar water behind (poleward of) the main polar front.

Polar orbiting satellite   A satellite whose orbit closely parallels the earth's meridian lines and thus crosses the polar regions on each orbit.

Polar tundra climate   A climate characterized by extremely cold winters and cool summers, as the average temperature of the warmest month climbs above freezing but remains below 10°C (50°F).

Pollutants   Any gaseous, chemical, or organic matter that contaminates the atmosphere, soil, or water.

Positive feedback mechanism   *See* Feedback mechanism.

Potential energy   The energy that a body possesses by virtue of its position with respect to other bodies in the field of gravity.

Potential evapotranspiration (PE)   The amount of moisture that, if it were available, would be removed from a given land area by evaporation and transpiration.

Potential temperature   The temperature that a parcel of dry air would have if it were brought dry adiabatically from its original position to a pressure of 1000 mb.

Precession (of the earth's axis of rotation)   The wobble of the earth's axis of rotation that traces out the path of a cone over a period of about 23,000 years.

Precipitation   Any form of water particles—liquid or solid—that falls from the atmosphere and reaches the ground.

Pressure   The force per unit area. *See also* Air pressure.

Pressure gradient   The rate of decrease of pressure per unit of horizontal distance. On the same chart, when the isobars are close together, the pressure gradient is steep. When the isobars are far apart, the pressure gradient is weak.

Pressure gradient force (PGF)   The force due to differences in pressure within the atmosphere that causes air to move and, hence, the wind to blow. It is directly proportional to the pressure gradient.

Pressure tendency   The rate of change of atmospheric pressure within a specified period of time, most often three hours. Same as *barometric tendency*.

Prevailing westerlies   The dominant westerly winds that blow in middle latitudes on the poleward side of the subtropical high-pressure areas. Also called *westerlies*.

Prevailing wind   The wind direction most frequently observed during a given period.

Primary air pollutants   Air pollutants that enter the atmosphere directly.

Probability forecast   A forecast of the probability of occurrence of one or more of a mutually exclusive set of weather conditions.

Prognostic chart (prog)   A chart showing expected or forecasted conditions, such as pressure patterns, frontal positions, contour height patterns, and so on.

Prominence   *See* Solar flare.

Psychrometer   An instrument used to measure the water vapor content of the air. It consists of two thermometers (dry bulb and wet bulb). After whirling the instrument, the dew point and relative humidity can be obtained with the aid of tables.

Radar   An electronic instrument used to detect objects (such as falling precipitation) by their ability to reflect and scatter microwaves back to a receiver. (*See also* Doppler radar.)

Radiant energy (radiation)   Energy propagated in the form of electromagnetic waves. These waves do not need molecules to propagate them, and in a vacuum they travel at nearly 300,000 km per sec (186,000 mi per sec).

Radiational cooling   The process by which the earth's surface and adjacent air cool by emitting infrared radiation.

Radiation fog   Fog produced over land when radiational cooling reduces the air temperature to or below its dew point. It is also known as *ground fog* and *valley fog*.

Radiation inversion   An increase in temperature with height due to radiational cooling of the earth's surface. Also called a *nocturnal inversion*.

Radiative equilibrium temperature   The temperature achieved when an object, behaving as a blackbody, is absorbing and emitting radiation at equal rates.

Radiative forcing   An increase (positive) or a decrease (negative) in net radiant energy observed over an area at the tropopause. An increase in radiative forcing may induce surface warming, whereas a decrease may induce surface cooling.

Radiative forcing agent   Any factor (such as increasing greenhouse gases and variations in solar output) that can change the balance between incoming energy from the sun and outgoing energy from the earth and the atmosphere.

Radiometer   *See* Infrared radiometer.

Radiosonde   A balloon-borne instrument that measures and transmits pressure, temperature, and humidity to a ground-based receiving station.

Radon A colorless, odorless, radioactive gas that forms naturally as uranium in soil and rock breaks down.

Rain Precipitation in the form of liquid water drops that have diameters greater than that of drizzle.

Rainbow An arc of concentric colored bands that spans a section of the sky when rain is present and the sun is positioned at the observer's back.

Rain gauge An instrument designed to measure the amount of rain that falls during a given time interval.

Rain shadow The region on the leeside of a mountain where the precipitation is noticeably less than on the windward side.

Rawinsonde observation A radiosonde observation that includes wind data.

Reflected light *See* reflection.

Reflection The process whereby a surface turns back a portion of the radiation that strikes it. When the radiation that is turned back (reflected) from the surface is visible light, the radiation is referred to as *reflected light*.

Refraction The bending of light as it passes from one medium to another.

Relative humidity The ratio of the amount of water vapor in the air compared to the amount required for saturation (at a particular temperature and pressure). The ratio of the air's actual vapor pressure to its saturation vapor pressure.

Relative vorticity *See* Vorticity.

Return stroke The luminous lightning stroke that propagates upward from the earth to the base of a cloud.

Ribbon lightning Lightning that appears to spread horizontally into a ribbon of parallel luminous streaks when strong winds are blowing parallel to the observer's line of sight.

Ridge An elongated area of high atmospheric pressure.

Rime A white or milky granular deposit of ice formed by the rapid freezing of supercooled water drops as they come in contact with an object in below-freezing air.

Riming *See* Accretion.

Roll cloud A dense, roll-shaped, elongated cloud that appears to slowly spin about a horizontal axis behind the leading edge of a thunderstorm's gust front.

Rossby waves *See* Longwaves in the westerlies.

Rotor cloud A turbulent cumuliform type of cloud that forms on the leeward side of large mountain ranges. The air in the cloud rotates about an axis parallel to the range.

Rotors Turbulent eddies that form downwind of a mountain chain, creating hazardous flying conditions.

Saffir-Simpson scale A scale relating a hurricane's central pressure and winds to the possible damage it is capable of inflicting.

St. Elmo's fire A bright electric discharge that is projected from objects (usually pointed) when they are in a strong electric field, such as during a thunderstorm.

Sand dune A hill or ridge of loose sand shaped by the winds.

Santa Ana wind A warm, dry wind that blows into southern California from the east off the elevated desert plateau. Its warmth is derived from compressional heating.

Saturation (of air) An atmospheric condition whereby the level of water vapor is the maximum possible at the existing temperature and pressure.

Saturation vapor pressure The maximum amount of water vapor necessary to keep moist air in equilibrium with a surface of pure water or ice. It represents the maximum amount of water vapor that the air can hold at any given temperature and pressure. (*See* Equilibrium vapor pressure.)

Savanna A tropical or subtropical region of grassland and drought-resistant vegetation. Typically found in tropical wet-and-dry climates.

Scales of motion The hierarchy of atmospheric circulations from tiny gusts to giant storms.

Scattering The process by which small particles in the atmosphere deflect radiation from its path into different directions.

Scintillation The apparent twinkling of a star due to its light passing through regions of differing air densities in the atmosphere.

Sea breeze A coastal local wind that blows from the ocean onto the land. The leading edge of the breeze is termed a *sea-breeze front*.

Sea-breeze convergence zone A region where sea breezes, having started in different regions, flow together and converge.

Sea-breeze front The horizontal boundary that marks the leading edge of cooler marine air associated with a sea-breeze.

Sea-level pressure The atmospheric pressure at mean sea level.

Secondary air pollutants Pollutants that form when a chemical reaction occurs between a primary air pollutant and some other component of air. Tropospheric ozone is a secondary air pollutant.

Secondary low A low-pressure area (often an open wave) that forms near, or in association with, a main low-pressure area.

Seiches Standing waves that oscillate back and forth over an open body of water.

Selective absorbers Substances such as water vapor, carbon dioxide, clouds, and snow that absorb radiation only at particular wavelengths.

Semi-arid climate A dry climate where potential evaporation and transpiration exceed precipitation. Not as dry as the arid climate. Typical vegetation is short grass.

Semipermanent highs and lows Areas of high pressure (anticyclones) and low pressure (extratropical cyclones) that tend to persist at a particular latitude belt throughout the year. In the Northern Hemisphere, typically they shift slightly northward in summer and slightly southward in winter.

Sensible heat The heat we can feel and measure with a thermometer.

Sensible temperature The sensation of temperature that the human body feels in contrast to the actual temperature of the environment as measured with a thermometer.

Severe thunderstorms Intense thunderstorms capable of producing heavy showers, flash floods, hail, strong and gusty surface winds, and tornadoes. The U.S. National Weather Service describes a severe thunderstorm as having at least one of the following: hail with a diameter of at least ¾ in., surface wind gusts of 50 knots or greater; or produces a tornado.

Sferics Radio waves produced by lightning. A contraction of *atmospherics*.

Shear *See* Wind shear.

Sheet lightning Occurs when the lightning flash is not seen but the flash causes the cloud (or clouds) to appear as a diffuse luminous white sheet.

Shelf cloud A dense, arch-shaped, ominous-looking cloud that often forms along the leading edge of a thunderstorm's gust front, especially when stable air rises up and over cooler air at the surface. Also called an *arcus cloud*.

Shelterbelt A belt of trees or shrubs arranged as a protection against strong winds.

Short-range forecast Generally used to describe a weather forecast that extends from about 6 hours to a few days into the future.

Shortwave (in the atmosphere) A small wave that moves around longwaves in the same direction as the air flow in the middle and upper troposphere. Shortwaves are also called *shortwave troughs*.

Shortwave radiation A term most often used to describe the radiant energy emitted from the sun, in the visible and near ultraviolet wavelengths.

Shower Intermittent precipitation from a cumuliform cloud, usually of short duration but often heavy.

Siberian high A strong, shallow area of high pressure that forms over Siberia in winter.

Sleet A type of precipitation consisting of transparent pellets of ice 5 mm or less in diameter. Same as *ice pellets*.

Smog Originally smog meant a mixture of smoke and fog. Today, smog means air that has restricted visibility due to pollution, or pollution formed in the presence of sunlight—*photochemical smog*.

Smog front (also smoke front) The leading edge of a sea breeze that is contaminated with smoke or pollutants.

Smudge pots *See* Orchard heaters.

Snow A solid form of precipitation composed of ice crystals in complex hexagonal form.

Snow-albedo feedback A positive feedback whereby increasing surface air temperatures enhance the melting of snow and ice in polar latitudes. This reduces the earth's albedo and allows more sunlight to reach the surface, which causes the air temperature to rise even more.

Snowflake An aggregate of ice crystals that falls from a cloud.

Snow flurries Light showers of snow that fall intermittently.

Snow grains Precipitation in the form of very small, opaque grains of ice. The solid equivalent of drizzle.

Snow pellets White, opaque, approximately round ice particles between 2 and 5 mm in diameter that form in a cloud either from the sticking together of ice crystals or from the process of accretion. Also called *graupel*.

Snow rollers A cylindrical spiral of snow shaped somewhat like a child's muff and produced by the wind.

Snow squall (shower) An intermittent heavy shower of snow that greatly reduces visibility.

Solar constant The rate at which solar energy is received on a surface at the outer edge of the atmosphere perpendicular to the sun's rays when the earth is at a mean distance from the sun. The value of the solar constant is about two calories per square centimeter per minute or about 1376 W/m^2 in the SI system of measurement.

Solar flare A rapid eruption from the sun's surface that emits high energy radiation and energized charged particles.

Solar wind An outflow of charged particles from the sun that escapes the sun's outer atmosphere at high speed.

Solute effect The dissolving of hygroscopic particles, such as salt, in pure water, thus reducing the relative humidity required for the onset of condensation.

Sonic boom A loud explosive-like sound caused by a shock wave emanating from an aircraft (or any object) traveling at or above the speed of sound.

Sounding An upper-air observation, such as a radiosonde observation. A vertical profile of an atmospheric variable such as temperature or winds.

Source regions Regions where air masses originate and acquire their properties of temperature and moisture.

Southern Oscillation (SO) The reversal of surface air pressure at opposite ends of the tropical Pacific Ocean that occur during major El Niño events.

Specific heat The ratio of the heat absorbed (or released) by the unit mass of the system to the corresponding temperature rise (or fall).

Specific humidity The ratio of the mass of water vapor in a given parcel to the total mass of air in the parcel.

Spin-up vortices Small whirling tornadoes perhaps 30 to 100 m in diameter that form in a region of strong wind shear in a hurricane's eyewall.

Squall line A line of thunderstorms that form along a cold front or out ahead of it.

Stable air *See* Absolutely stable atmosphere.

Standard atmosphere A hypothetical vertical distribution of atmospheric temperature, pressure, and density in which the air is assumed to obey the gas law and the hydrostatic equation. The lapse rate of temperature in the troposphere is taken as 6.5°C/1000 m or 3.6°F/1000 ft.

Standard atmospheric pressure A pressure of 1013.25 millibars (mb), 29.92 inches of mercury (Hg), 760 millimeters (mm) of mercury, 14.7 pounds per square inch (lb/in.^{2}), or 1013.25 hectopascals (hPa).

Standard rain gauge A nonrecording rain gauge with an 8-inch diameter collector funnel and a tube that amplifies rainfall by tenfold.

Stationary front A front that is nearly stationary with winds blowing almost parallel and from opposite directions on each side of the front.

Station pressure The actual air pressure computed at the observing station.

Statistical forecast A forecast based on a mathematical/statistical examination of data that represents the past observed behavior of the forecasted weather element.

Steady-state forecast A weather prediction based on the past movement of surface weather systems. It assumes that the systems will move in the same direction and at approximately the same speed as they have been moving. Also called *trend forecasting*.

Steam fog *See* Evaporation (mixing) fog.

Stefan-Boltzmann law A law of radiation which states that the amount of radiant energy emitted from a unit surface area of an object (ideally a blackbody) is proportional to the fourth power of the object's absolute temperature.

Steppe An area of grass-covered, treeless plains that has a semiarid climate.

Stepped leader An initial discharge of electrons that proceeds intermittently toward the ground in a series of steps in a cloud-to-ground lightning stroke.

Storm surge An abnormal rise of the sea along a shore; primarily due to the winds of a storm, especially a hurricane.

Stratocumulus A low cloud, predominantly stratiform, with low, lumpy, rounded masses, often with blue sky between them.

Stratosphere The layer of the atmosphere above the troposphere and below the mesosphere (between 10 km and 50 km), generally characterized by an increase in temperature with height.

Stratospheric polar night jet A jet stream that forms near the top of the stratosphere over polar latitudes during the winter months.

Stratus A low, gray cloud layer with a rather uniform base whose precipitation is most commonly drizzle.

Streamline A line that shows the wind flow pattern.

Sublimation The process whereby ice changes directly into water vapor without melting.

Subpolar climate A climate observed in the Northern Hemisphere that borders the polar climate. It is characterized by severely cold winters and short, cool summers. Also known as *taiga climate* and *boreal climate*.

Subpolar low A belt of low pressure located between 50° and 70° latitude. In the Northern Hemisphere, this "belt" consists of the *Aleutian low* in the North Pacific and the *Icelandic low* in the North Atlantic. In the Southern Hemisphere, it exists around the periphery of the Antarctic continent.

Subsidence The slow sinking of air, usually associated with high-pressure areas.

Subsidence inversion A temperature inversion produced by compressional warming—the adiabatic warming of a layer of sinking air.

Subtropical front A zone of temperature transition in the upper troposphere over subtropical latitudes, where warm air carried poleward by the Hadley cell meets the cooler air of the middle latitudes.

Subtropical high A semipermanent high in the subtropical high-pressure belt centered near 30° latitude. The *Bermuda high* is located over the Atlantic Ocean off the east coast of North America. The *Pacific high* is located off the west coast of North America.

Subtropical jet stream The jet stream typically found between 20° and 30° latitude at altitudes between 12 and 14 km.

Suction vortices Small, rapidly rotating whirls perhaps 10 m in diameter that are found within large tornadoes.

Sulfate aerosols *See* Aerosols.

Sulfur dioxide (SO_2) A colorless gas that forms primarily in the burning of sulfur-containing fossil fuels.

Summer solstice Approximately June 21 in the Northern Hemisphere when the sun is highest in the sky and directly overhead at latitude $23\frac{1}{2}°$N, the Tropic of Cancer.

Sundog A colored luminous spot produced by refraction of light through ice crystals that appears on either side of the sun. Also called *parhelia*.

Sun pillar A vertical streak of light extending above (or below) the sun. It is produced by the reflection of sunlight off ice crystals.

Sunspots Relatively cooler areas on the sun's surface. They represent regions of an extremely high magnetic field.

Supercell storm A severe thunderstorm that consists primarily of a single rotating updraft. Its organized internal structure allows the storm to maintain itself for several hours. Supercell storms can produce large hail and dangerous tornadoes.

Supercell tornadoes Tornadoes that occur within supercell thunderstorms that contain well-developed, mid-level mesocyclones.

Supercooled cloud (or **cloud droplets**) A cloud composed of liquid droplets at temperatures below 0°C (32°F). When the cloud is on the ground it is called *supercooled fog* or *cold fog*.

Superior mirage *See* Mirage.

Supersaturation A condition whereby the atmosphere contains more water vapor than is needed to produce saturation with respect to a flat surface of pure water or ice, and the relative humidity is greater than 100 percent.

Super typhoon A tropical cyclone (typhoon) in the western Pacific that has sustained winds of 130 knots or greater.

Surface inversion *See* Radiation inversion.

Surface map A map that shows the distribution of sea-level pressure with isobars and weather phenomena. Also called a *surface chart*.

Synoptic scale The typical weather map scale that shows features such as high- and low-pressure areas and fronts over a distance spanning a continent. Also called the *cyclonic scale*.

Taiga (boreal forest) The open northern part of the coniferous forest. Taiga also refers to subpolar climate.

Tangent arc An arc of light tangent to a halo. It forms by refraction of light through ice crystals.

Tcu An abbreviation sometimes used to denote a towering cumulus cloud (cumulus congestus).

Teleconnections A linkage between weather changes occurring in widely separated regions of the world.

Temperature The degree of hotness or coldness of a substance as measured by a thermometer. It is also a measure of the average speed or kinetic energy of the atoms and molecules in a substance.

Temperature inversion An increase in air temperature with height, often simply called an *inversion*.

Terminal velocity The constant speed obtained by a falling object when the upward drag on the object balances the downward force of gravity.

Texas norther A strong, cold wind from between the northeast and northwest associated with a cold outbreak of polar air that brings a sudden drop in temperature. Sometimes called a *blue norther*.

Theodolite An instrument used to track the movements of a pilot balloon.

Theory of plate tectonics *See* Plate tectonics.

Thermal A small, rising parcel of warm air produced when the earth's surface is heated unevenly.

Thermal belts Horizontal zones of vegetation found along hillsides that are primarily the result of vertical temperature variations.

Thermal circulations Air flow resulting primarily from the heating and cooling of air.

Thermal lows and **thermal highs** Areas of low and high pressure that are shallow in vertical extent and are produced primarily by surface temperatures.

Thermal tides Atmospheric pressure variations due to the uneven heating of the atmosphere by the sun.

Thermal turbulence Turbulent vertical motions that result from surface heating and the subsequent rising and sinking of air.

Thermograph An instrument that measures and records air temperature.

Thermometer An instrument for measuring temperature. The most common is liquid-in-glass, which has a sealed glass tube attached to a glass bulb filled with liquid.

Thermosphere The atmospheric layer above the mesosphere (above about 85 km) where the temperature increases rapidly with height.

Thunder The sound due to rapidly expanding gases along the channel of a lightning discharge.

Thunderstorm A convective storm (cumulonimbus cloud) with lightning and thunder. Thunderstorms can be composed of an ordinary cell, multicells, or a rapidly rotating supercell.

Tipping bucket rain gauge A rain gauge that records rainfall by collecting rain in a chamber (bucket) that tips when the chamber fills with 0.01 in. (0.025 cm) of rain.

Tornado An intense, rotating column of air that often protrudes from a cumuliform cloud in the shape of a funnel or a rope whose circulation is present on the ground. (*See* Funnel cloud.)

Tornado outbreak A series of tornadoes that forms within a particular region—a region that may include several states. Often associated with widespread damage and destruction.

Tornado vortex signature (TVS) An image of a tornado on the Doppler radar screen that shows up as a small region of rapidly changing wind directions inside a mesocyclone.

Tornado warning A warning issued when a tornado has actually been observed either visually or on a radar screen. It is also issued when the formation of tornadoes is imminent.

Tornado watch   A forecast issued to alert the public that tornadoes may develop within a specified area.

Trace (of precipitation)   An amount of precipitation less than 0.01 in. (0.025 cm).

Trade wind inversion   A temperature inversion frequently found in the subtropics over the eastern portions of the tropical oceans.

Trade winds   The winds that occupy most of the tropics and blow from the subtropical highs to the equatorial low.

Transpiration   The process by which water in plants is transferred as water vapor to the atmosphere.

Tropical cyclone   The general term for storms (cyclones) that form over warm tropical oceans.

Tropical depression   A mass of thunderstorms and clouds generally with a cyclonic wind circulation of between 20 and 34 knots.

Tropical disturbance   An organized mass of thunderstorms with a slight cyclonic wind circulation of less than 20 knots.

Tropical easterly jet   A jet stream that forms on the equatorward side of the subtropical highs near 15 km.

Tropical monsoon climate   A tropical climate with a brief dry period of perhaps one or two months.

Tropical rain forest   A type of forest consisting mainly of lofty trees and a dense undergrowth near the ground.

Tropical storm   Organized thunderstorms with a cyclonic wind circulation between 35 and 64 knots.

Tropical wave   A migratory wavelike disturbance in the tropical easterlies. Tropical waves occasionally intensify into tropical cyclones. They are also called *easterly waves*.

Tropical wet-and-dry climate   A tropical climate poleward of the tropical wet climate where a distinct dry season occurs, often lasting for two months or more.

Tropical wet climate   A tropical climate with sufficient rainfall to produce a dense tropical rain forest.

Tropopause   The boundary between the troposphere and the stratosphere.

Tropopause jets   Jet streams found near the tropopause, such as the polar front and subtropical jet streams.

Troposphere   The layer of the atmosphere extending from the earth's surface up to the tropopause (about 10 km above the ground).

Trough   An elongated area of low atmospheric pressure.

Turbulence   Any irregular or disturbed flow in the atmosphere that produces gusts and eddies.

Twilight   The time at the beginning of the day immediately before sunrise and at the end of the day after sunset when the sky remains illuminated.

Typhoon   A hurricane (tropical cyclone) that forms in the western Pacific Ocean.

Ultraviolet (UV) radiation   Electromagnetic radiation with wavelengths longer than X-rays but shorter than visible light.

Unstable air   *See* Absolutely unstable atmosphere.

Upper-air front   A front that is present aloft but usually does not extend down to the ground. Also called an *upper front* and an *upper-tropospheric front*.

Upslope fog   Fog formed as moist, stable air flows upward over a topographic barrier.

Upslope precipitation   Precipitation that forms due to moist, stable air gradually rising along an elevated plain. Upslope precipitation is common over the western Great Plains, especially east of the Rocky Mountains.

Upwelling   The rising of water (usually cold) toward the surface from the deeper regions of a body of water.

Urban heat island   The increased air temperatures in urban areas as contrasted to the cooler surrounding rural areas.

Valley breeze   *See* Mountain breeze.

Valley fog   *See* Radiation fog.

Vapor pressure   The pressure exerted by the water vapor molecules in a given volume of air. (*See also* Saturation vapor pressure.)

Veering wind   The wind that changes direction in a clockwise sense—north to northeast to east, and so on.

Vernal equinox   The equinox at which the sun approaches the Northern Hemisphere and passes directly over the equator. Occurs around March 20.

Very short range forecast   Generally used to describe a weather forecast that is made for up to a few hours (usually less than 6 hours) into the future.

Virga   Precipitation that falls from a cloud but evaporates before reaching the ground. (*See* Fall streaks.)

Viscosity   The resistance of fluid flow. (*See* Molecular viscosity *and* Eddy viscosity.)

Visible radiation (light)   Radiation with a wavelength between 0.4 and 0.7 µm. This region of the electromagnetic spectrum is called the *visible region*.

Visible region   *See* Visible radiation.

Visibility   The greatest distance at which an observer can see and identify prominent objects.

Volatile organic compounds (VOCs)   A class of organic compounds that are released into the atmosphere from sources such as motor vehicles, paints, and solvents. VOCs (which include hydrocarbons) contribute to the production of secondary pollutants, such as ozone.

Vorticity   A measure of the spin of a fluid, usually small air parcels. *Absolute vorticity* is the combined vorticity due to the earth's rotation (*earth's vorticity*) and the vorticity due to the air's circulation relative to the earth. *Relative vorticity* is due to the curving of the air flow and wind shear.

Wall cloud   An area of rotating clouds that extends beneath a supercell thunderstorm and from which a funnel cloud may appear. Also called a *collar cloud* and *pedestal cloud*.

Warm advection (or *warm air advection*)   The transport of warm air by the wind from a region of higher temperatures to a region of lower temperatures.

Warm-core low   A low-pressure area that is warmer at its center than at its periphery. Tropical cyclones exhibit this temperature pattern.

Warm front   A front that moves in such a way that warm air replaces cold air.

Warm occlusion   *See* Occluded front.

Warm sector   The region of warm air within a wave cyclone that lies between a retreating warm front and an advancing cold front.

Water equivalent   The depth of water that would result from the melting of a snow sample. Typically about 10 inches of snow will melt to 1 inch of water, producing a water equivalent of 10 to 1.

Waterspout   A column of rotating wind over water that has characteristics of a dust devil and tornado.

Water vapor   Water in a vapor (gaseous) form. Also called *moisture*.

Water vapor-greenhouse effect feedback   A positive feedback whereby increasing surface air temperatures cause an increase in the evaporation of water from the oceans. Increasing concentrations of atmospheric water vapor enhance the greenhouse effect, which causes the surface air temperature to rise even more. Also called the *water vapor-temperature rise feedback.*

Watt (W)   The unit of power in SI units where 1 watt is equivalent to 1 joule per second.

Wave cyclone   An extratropical cyclone that forms and moves along a front. The circulation of winds about the cyclone tends to produce a wavelike deformation on the front.

Wavelength   The distance between successive crests, troughs, or identical parts of a wave.

Weather   The condition of the atmosphere at any particular time and place.

Weather elements   The elements of *air temperature, air pressure, humidity, clouds, precipitation, visibility,* and *wind* that determine the present state of the atmosphere, the weather.

Weather type forecasting   A forecasting method where weather patterns are categorized into similar groups or types.

Weather types   Certain weather patterns categorized into similar groups. Used as an aid in weather prediction.

Weather warning   A forecast indicating that hazardous weather is either imminent or actually occurring within the specified forecast area.

Weather watch   A forecast indicating that atmospheric conditions are favorable for hazardous weather to occur over a particular region during a specified time period.

Weighing rain gauge   A rain gauge that records rainfall by weighing the collected water over a given time and converting the amount of water to rainfall depth.

Westerlies   The dominant westerly winds that blow in the middle latitudes on the poleward side of the subtropical high-pressure areas.

Wet-bulb depression   The difference in degrees between the air temperature (dry-bulb temperature) and the wet-bulb temperature.

Wet-bulb temperature   The lowest temperature that can be obtained by evaporating water into the air.

Wet haze   *See* Haze.

Whirlwinds   *See* Dust devils.

Wien's law   A law of radiation which states that the wavelength of maximum emitted radiation by an object (ideally a blackbody) is inversely proportional to the object's absolute temperature.

Wind   Air in motion relative to the earth's surface.

Wind chill   The cooling effect of any combination of temperature and wind, expressed as the loss of body heat. Also called *wind-chill index.*

Wind direction   The direction *from which* the wind is blowing.

Wind machines   Fans placed in orchards for the purpose of mixing cold surface air with warmer air above.

Wind profiler   A Doppler radar capable of measuring the turbulent eddies that move with the wind. Because of this, it is able to provide a vertical picture of wind speed and wind direction.

Wind rose   A diagram that shows the percent of time that the wind blows from different directions at a given location over a given time.

Wind-sculptured trees   Trees whose branches are bent, twisted, and broken off on one side by strong prevailing winds. Also called *flag trees.*

Wind shear   The rate of change of wind speed or wind direction over a given distance.

Wind sock   A tapered fabric shaped like a cone that indicates wind direction by pointing away from the wind. Also called a *wind cone.*

Wind vane   An instrument used to indicate wind direction.

Windward side   The side of an object facing into the wind.

Wind waves   Water waves that form due to the flow of air over the water's surface.

Winter chilling   The amount of time the air temperature during the winter must remain below a certain value so that fruit and nut trees will grow properly during the spring and summer.

Winter solstice   Approximately December 21 in the Northern Hemisphere when the sun is lowest in the sky and directly overhead at latitude 23½°S, the Tropic of Capricorn.

Xerophytes   Drought-resistant vegetation.

Younger-Dryas event   A cold episode that took place about 11,000 years ago, when average temperatures dropped suddenly and portions of the Northern Hemisphere reverted back to glacial conditions.

Zonal wind flow   A wind that has a predominate west-to-east component.

PERIODICALS

Selected nontechnical periodicals that contain articles on weather and climate.

Bulletin of the American Meteorological Society. Monthly. The American Meteorological Society, 45 Beacon St., Boston, MA 02108.

Meteorological Magazine. Monthly. British Meteorological Office, British Information Services, 845 Third Avenue, New York, NY.

National Weather Digest. Quarterly. National Weather Association, 4400 Stamp Road, Room 404, Marlow Heights, MD 20031. (Deals mainly with weather forecasting.)

Weather. Monthly. Royal Meteorological Society, James Glaisher House, Grenville Place, Bracknell, Berkshire, England.

Weatherwise. Bimonthly. Heldref Publications, 4000 Albermarle St., N.W., Washington, DC 20016.

Selected Technical Periodicals

EOS—Transaction of the American Geophysical Union. American Geophysical Union (AGU), Washington, DC.

Journal of Applied Meteorology. American Meteorological Society (AMS), Boston, MA.

Journal of Atmospheric and Oceanic Technology. AMS, Boston, MA.

Journal of Atmospheric Science. AMS, Boston, MA.

Journal of Climate. AMS, Boston, MA.

Journal of Geophysical Research. American Geophysical Union, Washington, DC.

Monthly Weather Review. AMS, Boston, MA.

Weather and Forecasting. AMS, Boston, MA.

Additional periodicals that frequently contain articles of meteorological interest.

American Scientist. Bimonthly. Sigma Xi, the Scientific Research Society, Inc., New Haven, CT.

Science. Weekly. American Association for the Advancement of Science, Washington, DC.

Scientific American. Monthly. Scientific American, Inc., New York, NY.

Smithsonian. Monthly. The Smithsonian Association, Washington, DC.

BOOKS

The titles listed below may be drawn upon for additional information. Many are written at the introductory level. Those that are more advanced are marked with an asterisk.

Ahrens, C. Donald. *Essentials of Meteorology* (5th ed.), Thomson/Brooks Cole, Belmont, CA, 2008.

Anthes, R. A. *Tropical Cyclones: Their Evolution, Structure, and Effect,* American Meteorological Society, Boston, MA, 1982.

Arya, Pal S. *Air Pollution Meteorology,* Oxford University Press, New York, 1998.

Bigg, Grant R. *The Oceans and Climate,* Cambridge University Press, New York, 1996.

*Bluestein, Howard B. *Synoptic-Dynamic Meteorology in Midlatitudes. Vol. 1: Principles of Kinematics and Dynamics,* Oxford University Press, New York, 1992.

———. *Synoptic-Dynamic Meteorology in Midlatitudes. Vol. II: Observations and Theory of Weather Systems,* Oxford University Press, New York, 1993.

———. *Tornado Alley: Monster Storms of the Great Plains,* Oxford University Press, New York, 1999.

Bohren, Craig F. *Clouds in a Glass of Beer: Simple Experiments in Atmospheric Physics,* Wiley, New York, 1987.

———. *What Light Through Yonder Window Breaks?,* Wiley, New York, 1991.

Boubel, Richard W., et al., *Fundamentals of Air Pollution* (3rd ed.), Academic Press, New York, 1994.

Burgess, Eric, and Douglass Torr. *Into the Thermosphere: The Atmosphere Explorers,* National Aeronautics and Space Administration, Washington, DC, 1987.

Burroughs, William J. *Watching the World's Weather,* Cambridge University Press, New York, 1991.

———. *Climate Revealed,* Cambridge University Press, Cambridge, England, 1999.

Burt, Christopher C. *Extreme Weather, A Guide and Record Book,* W.W. Norton & Company, New York, 2004.

Carlson, Toby N. *Mid-Latitude Weather Systems,* American Meteorological Society, Boston, MA, 1998.

Climate Change 2007. The Physical Science Basis. Working Group 1 contribution to the Fourth Assessment Report of the IPCC, Cambridge University Press, New York, 2007.

*Cotton, W. R., and R. A. Anthes. *Storm and Cloud Dynamics,* Academic Press, New York, 1989.

Cotton, William R. *Storms,* ASTeR Press, Fort Collins, CO, 1990.

Cotton, William R., and Roger A. Pielke. *Human Impacts on Weather and Climate,* Cambridge University Press, New York, 1995.

Crowley, Thomas J. and Gerald R. North. *Paleoclimatology,* Oxford University Press, New York, 1991.

De Blij, H. J. *Nature on the Rampage,* Smithsonian Books, Washington, DC, 1994.

Doswell, Charles A. III, editor. *Severe Convection Storms,* American Meteorological Society, Boston, MA, 2001.

Elsner, James B., and A. Biral Kara. *Hurricanes of the North Atlantic,* Oxford University Press, New York, 1999.

Elsom, Derek M. *Atmospheric Pollution: A Global Problem* (2nd ed.), Blackwell Publishers, Oxford, England, 1992.

Emanuel, Kerry. *Divine Wind-The History and Science of Hurricanes,* Oxford University Press, Oxford, New York, 2005.

Encyclopedia of Climate and Weather, Vol. 1 and Vol. 2, Stephen H. Schneider, Ed., Oxford University Press, New York, 1996.

Energy and Climate Change. Report of the DOE Multi-Laboratory Climate Change Committee, Lewis Publishers, Chelsea, MI, 1991.

England, Gary A. *Weathering the Storm,* University of Oklahoma Press, Norman, OK, 1996.

Firor, John. *The Changing Atmosphere: A Global Challenge,* Yale University Press, New Haven, CT, 1990.

Fujita, T. T. *The Downburst–Microburst and Macroburst,* University of Chicago Press, Chicago, 1985.

Glossary of Meteorology. Todd S. Glickman, Managing Ed., American Meteorological Society, Boston, MA, 2000.

Glossary of Weather and Climate. Ira W. Geer, Ed., American Meteorological Society, Boston, MA, 1996.

*Graedel, T. E. and Paul J. Crutzen. *Atmospheric Change: An Earth System Perspective,* W. H. Freeman, New York, 1993.

Graedel, Thomas E., and Paul J. Crutzen. *Atmosphere, Climate, and Change,* W. H. Freeman, New York, 1995.

Greenler, Robert. *Rainbows, Halos and Glories,* Cambridge University Press, New York, 1980.

*Grotjahn, Richard. *Global Atmospheric Circulations: Observations and Theories,* Oxford University Press, Oxford, England, 1993.

Henson, Robert, *The Rough Guide to Climate Change,* (2nd ed.), Rough Guide, New York, 2008.

———. *The Rough Guide to Weather,* Rough Guide, New York, 2007.

*Hobbs, Peter V. *Basic Physical Chemistry for Atmospheric Sciences,* Cambridge University Press, New York, 1995.

Hoyt, Douglas V. and Kenneth H. Schatten. *The Role of the Sun in Climate Change,* Oxford University Press, New York, 1997.

International Cloud Atlas. World Meteorological Organization, Geneva, Switzerland, 1987.

James, Bruce P., et al., Eds. *Climate Change 1995. Economic and Social Dimensions of Climate Change,* Cambridge University Press, Cambridge, England, 1996.

*Karoly, David J., and Dayton G. Vincent, Eds. *Meteorology of the Southern Hemisphere,* American Meteorological Society, Boston, MA, 1998.

Keen, Richard A. *Skywatch: The Western Weather Guide,* Fulcrum Incorporated, Golden, CO, 1987.

———. *Skywatch East: A Weather Guide,* Fulcrum Incorporated, Golden, CO, 1992.

Kessler, Edwin. *Thunderstorm Morphology and Dynamics* (2nd ed.), University of Oklahoma Press, Norman, OK, 1986.

Kocin, Paul J., and L. W. Uccellini. *Snowstorms along the Northeastern Coast of the United States: 1955 to 1985,* American Meteorological Society, Boston, MA, 1990.

Laskin, David. *Braving the Elements: The Stormy History of American Weather,* Doubleday, New York, 1996.

Ludlum, D. M. *The Audubon Society Field Guide to North American Weather,* Alfred A. Knopf, New York, 1991.

Lynch, David K., and William Livingston. *Color and Light in Nature,* Cambridge University Press, New York, 1995.

Mason, B. J. *Acid Rain: Its Causes and Its Effects on Inland Waters,* Oxford University Press, New York, 1992.

Meinel, Aden, and Marjorie Meinel. *Sunsets, Twilights and Evening Skies.* Cambridge University Press, New York, 1983.

Nelson, Mike. *The Colorado Weather Book,* Westcliff Publishers, Englewood, CO, 1999.

Pretor-Pinney, Gavin. *The Cloud Spotters Guide,* Penguin Group, New York 2006.

Prospects for Future Climate. Special US/USSR Report on Climate and Climate Change, Lewis Publishers, Chelsea, MI, 1990.

Righter, Robert W. *Wind Energy in America,* University of Oklahoma Press, Norman, OK, 1996.

*Rogers, R. R. *A Short Course in Cloud Physics* (3rd ed.), Pergamon Press, Oxford, England, 1989.

Schaefer, Vincent J. *Peterson First Guide to Clouds and Weather,* Houghton Mifflin, Boston, MA, 1991.

Scorer, Richard S., and Arjen Verkaik. *Spacious Skies,* David and Charles Publishers, London, 1989.

*Stull, Roland B. *Meteorology Today for Scientists and Engineers* (2nd Ed.), Brooks/Cole Publishing Co., Pacific Grove, CA, 2000.

Van Andel, Tjeerd H. *New Views on an Old Planet—A History of Global Change,* Cambridge University Press, Cambridge, England, 1994.

Vasquez, Tim. *Storm Chasing Handbook,* Weather Graphics Technologies, Austin, TX, 2002.

Vital Signs 2007-2008. The World Watch Institute, New York, 2007.

Wallace, John M. and Peter V. Hobbs. *Atmospheric Science: An Introductory Survey* (2nd ed.), Academic Press, Burlington, MA, 2006.

*Watson, Robert T., et al., Eds. *Climate Change 1995. Impacts, Adaptations and Mitigation of Climate Change: Scientific-Technical Analyses,* Cambridge University Press, Cambridge, England, 1996.

Weather. Smithsonian Field Guide, Harper Collins Publishers, New York, 2006.

Williams, Jack. *The USA Today Weather Book,* Vintage Books, Random House, New York, 1992.

INDEX